Metal Fabrication Technology for Agriculture

Metal Fabrication Technology for Agriculture

Larry Jeffus

THOMSON
DELMAR LEARNING

Australia Canada Mexico Singapore Spain United Kingdom United States

THOMSON
DELMAR LEARNING

Metal Fabrication Technology for Agriculture
Larry Jeffus

Executive Director:
Alar Elken

Executive Editor:
Sandy Clark

Acquisitions Editor:
Sanjeev Rao

Developmental Editor:
Alison S. Weintraub

Executive Marketing Manager:
Maura Theriault

Channel Manager:
Fair Huntoon

Marketing Coordinator:
Sarena Douglass

Executive Production Manager:
Mary Ellen Black

Production Coordinator:
Sharon Popson

Project Editor:
Barbara L. Diaz

Senior Art/Design Coordinator:
Mary Beth Vought

Technology Project Manager:
David Porush

Technology Project Specialist:
Kevin Smith

Editorial Assistant:
Jill Carnahan

Printed in the United States of America
1 2 3 4 5 XX 06 05 04 03

For more information contact
Delmar Learning
Executive Woods
5 Maxwell Drive, PO Box 8007,
Clifton Park, NY 12065-8007
Or find us on the World Wide Web at
http://www.delmarlearning.com

Library of Congress Cataloging-in-Publication Data:

Jeffus, Larry F.
Metal fabrication technology for agriculture / Larry Jeffus.—1st ed.
p. cm.
ISBN 1-4018-1563-4
1. Welding. 2. Farm equipment. I. Title.
TS227.J4175 2002
671.5'2—dc21
2002032782

NOTICE TO THE READER

Publisher does not warrant or guarantee any of the products described herein or perform any independent analysis in connection with any of the product information contained herein. Publisher does not assume, and expressly disclaims, any obligation to obtain and include information other than that provided to it by the manufacturer.

The reader is expressly warned to consider and adopt all safety precautions that might be indicated by the activities herein and to avoid all potential hazards. By following the instructions contained herein, the reader willingly assumes all risks in connection with such instructions.

The publisher makes no representation or warranties of any kind, including but not limited to, the warranties of fitness for particular purpose or merchantability, nor are any such representations implied with respect to the material set forth herein, and the publisher takes no responsibility with respect to such material. The publisher shall not be liable for any special, consequential, or exemplary damages resulting, in whole or part, from the readers' use of, or reliance upon, this material.

Contents

Preface

Introduction

In much of our society, many items are considered to be disposable. On the farm or ranch, however, that is not true. In addition to the replacement cost, it is not always possible to simply run into town and pick up a new shovel because the blade on the one you have has split. Replacement time, more than money, is often the controlling factor for the farmer or rancher. The farmers and ranchers of today need to be proficient in welding and fabrication for a number of reasons:

- Time—The time it takes to go into town and pick up a replacement part during a busy time can be more costly than the replacement part.
- Availability—Often the parts needed are not stocked and have to be ordered.
- Cost—The replacement part will always cost more than the welding materials needed to make the repair.
- Convenience—Often the broken part on a tractor or trailer can be repaired in place, so removal and replacement time and effort are not needed.
- Size—Sometimes the part needing repair or fabrication is too large to be transported to and from the farm or ranch.

Studying *Metal Fabrication Technology for Agriculture* in the classroom or shop setting will help students prepare for the challenges facing today's farmers and ranchers. The comprehensive technical content combines the basics of agricultural fabrication with the must-know welding and fabrication techniques. The extensive descriptions of equipment and supplies with in-depth explanations focused squarely on their agricultural applications will make the student familiar with their uses.

The book's complete instructions for setup in preparation for welding make it easier for students to be successful. These comprehensive instructions can later be used as a reference book by graduates as they work on the farm or ranch. Up-close shots of actual welding included throughout the book provide a realistic look at each of the processes. Students can see exactly what their welding should look like, which will enable them to make better welds.

Extensive coverage of brazing and specialized nonmetallic fabrication is designed to lead readers step by step in developing the skills necessary for welding all types of agricultural machinery. This book is an effective learning aid; it is a how-to and reference manual and a key resource for today's farmers, ranchers, and students participating in agriculture education.

Organization

Each chapter begins with a list of *learning objectives* that tell the student and instructor what is to be learned while studying the chapter. A survey of the objectives will show that the student will have the opportunity to develop a full range of welding skills, depending upon the topics selected for the program. Each major process is presented in such a way that the instructor can eliminate processes having little economic value in the market served by the program. However, the student will still learn all essential information needed for a thorough understanding of all processes studied.

In each chapter, *Key Terms* are highlighted in color and defined. In addition, the new terms are listed at the beginning of the chapter to enable students to recognize the terms when they appear. Terms and definitions used throughout the text are based on the American Welding Society's standards. Industry jargon has also been included where appropriate. The *Bilingual Glossary* includes a Spanish equivalent for each term, and many definitions feature additional drawings to assist all learners in gaining a complete understanding of the new term.

Cautions for the student are given throughout the text. *Metric equivalents* are listed in parentheses for dimensions. The metric equivalent in most cases has been rounded to the nearest whole number. Numerous full-color photographs, line drawings, and plans illustrate concepts and clarify the discussions.

Most of the chapters contain learning activities in the form of *Experiments* and *Practices*. The end of the experiments are identified by the (♦) and the end of the practices are identified by the (♦) symbol.

By completing the *Experiments,* the student learns the parameters of each welding process. Often, because it is hard both to perform the experiment and to observe the results closely, students will do most of the experiments in a small group. This will allow students both to perform the activity and to observe the reactions. In the experiments, the student changes the parameters to observe the effect on the process. In this way, the student learns to manipulate the variables to obtain the desired welding outcome for given conditions. The experiments provided in the chapters do not have right or wrong answers. They are designed to allow the student to learn the operating limitations or the effects of changes that may occur during the welding process.

A large selection of *Practices* are included to enable the student to develop the required manipulative skills, using different materials and material thicknesses in different positions. A sufficient number of practices is provided so that, after the basics are learned, the student may choose an area of specialization. Materials specified in the practices may be varied in both thickness and length to accommodate those supplies that students have in their lab. Changes within a limited range of both thickness and length will not affect the learning process designed for the practice. A chapter-end summary recaps the significant material covered in the chapter. This summary will help the student more completely understand the chapter material and will serve as a handy study tool.

The multiple-choice *Review* questions at the end of each chapter can be used as indicators of how well the student has learned the material in each chapter.

Computers in Welding

As in every skilled trade in today's ever changing world, computers are becoming more commonly used in welding. Some of the basic programs provide a cross-reference to welding filler metals, whereas others aid in weld symbol selection. More complex programs allow welding engineers to design structures and test them for strength without ever building them. These programs aid in proper design and make more effective use of materials, resulting in better, more cost–effective construction. The most commonly used programs are ones such as Arc Works™, published by Lincoln Electric Company, which are used to help write Welding Procedure Specifications (WPS), Procedure Qualification Records (PQR), and Welder Qualification Test Records (WQTR). These documents are extensively used throughout the welding industry.

Most of the welding programs operate on a variety of platforms, but the most popular ones use a version of Microsoft Windows. Having a good basic understanding of the Windows operating platform will give you a great start with these programs. In addition you should become familiar with one of the commonly used word processing programs, such as Microsoft Word. This will aid you in producing high-quality reports both in school and later on the job.

FEATURES OF THE TEXT

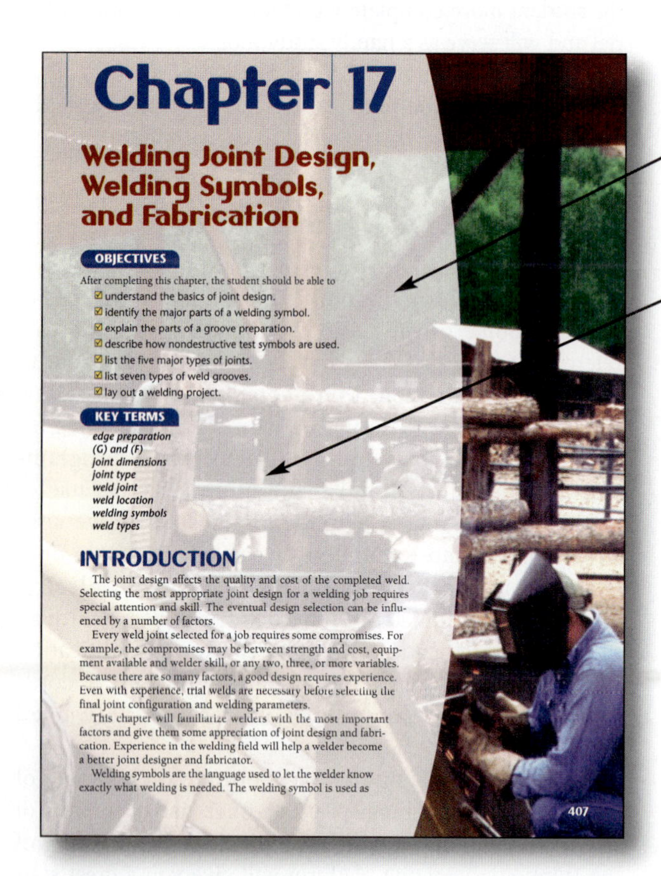

Chapter 17

Welding Joint Design, Welding Symbols, and Fabrication

OBJECTIVES

After completing this chapter, the student should be able to

- understand the basics of joint design.
- identify the major parts of a welding symbol.
- explain the parts of a groove preparation.
- describe how nondestructive test symbols are used.
- list the five major types of joints.
- list seven types of weld grooves.
- lay out a welding project.

KEY TERMS

edge preparation
(G) and (F)
joint dimensions
joint type
weld joint
weld location
welding symbols
weld types

INTRODUCTION

The joint design affects the quality and cost of the completed weld. Selecting the most appropriate joint design for a welding job requires special attention and skill. The eventual design selection can be influenced by a number of factors.

Every weld joint selected for a job requires some compromises. For example, the compromises may be between strength and cost, equipment available and welder skill, or any two, three, or more variables. Because there are so many factors, a good design requires experience. Even with experience, trial welds are necessary before selecting the final joint configuration and welding parameters.

This chapter will familiarize welders with the most important factors and give them some appreciation of joint design and fabrication. Experience in the welding field will help a welder become a better joint designer and fabricator.

Welding symbols are the language used to let the welder know exactly what welding is needed. The welding symbol is used as

407

Objectives, found at the beginning of each chapter, are a brief list of the most important topics to study in the chapter.

Key terms are the most important technical words you will learn in the chapter. These are listed at the beginning of each chapter following the objectives and appear in color print where they are first defined. These terms are also defined in the glossary at the end of the book.

Cautions summarize critical safety rules. They alert you to operations that could hurt you or someone else. Not only are they covered in the safety chapter, but you will find them throughout the text when they apply to the discussion, practice, or experiment.

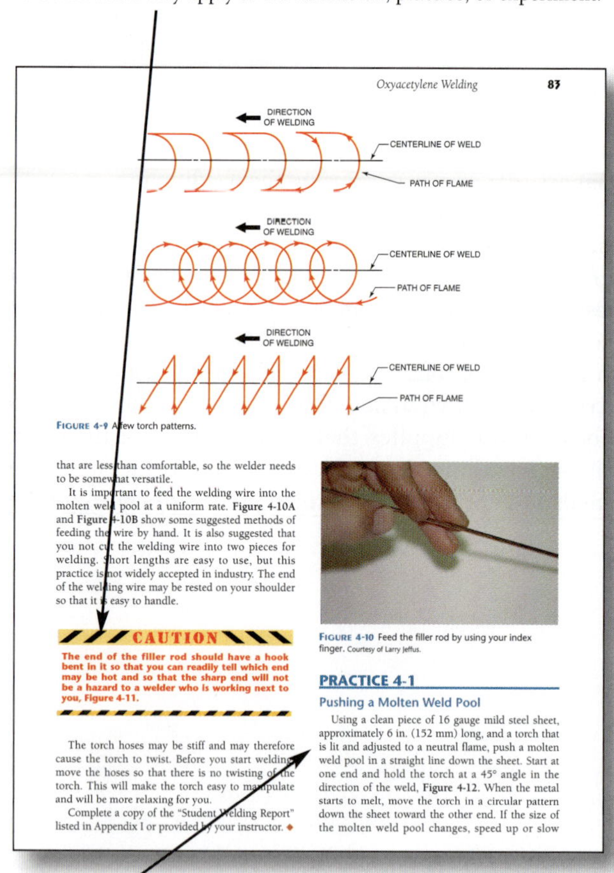

Oxyacetylene Welding 83

FIGURE 4-9 A few torch patterns.

that are less than comfortable, so the welder needs to be somewhat versatile.

It is important to feed the welding wire into the molten weld pool at a uniform rate. Figure 4-10A and Figure 4-10B show some suggested methods of feeding the wire by hand. It is also suggested that you not cut the welding wire into two pieces for welding. Short lengths are easy to use, but this practice is not widely accepted in industry. The end of the welding wire may be rested on your shoulder so that it is easy to handle.

CAUTION

The end of the filler rod should have a hook bent in it so that you can readily tell which end may be hot and so that the sharp end will not be a hazard to a welder who is working next to you, Figure 4-11.

The torch hoses may be stiff and may therefore cause the torch to twist. Before you start welding, move the hoses so that there is no twisting of the torch. This will make the torch easy to manipulate and will be more relaxing for you.

Complete a copy of the "Student Welding Report" listed in Appendix I or provided by your instructor. ◆

FIGURE 4-10 Feed the filler rod by using your index finger. Courtesy of Larry Jeffus.

PRACTICE 4-1

Pushing a Molten Weld Pool

Using a clean piece of 16 gauge mild steel sheet, approximately 6 in. (152 mm) long, and a torch that is lit and adjusted to a neutral flame, push a molten weld pool in a straight line down the sheet. Start at one end and hold the torch at a 45° angle in the direction of the weld, Figure 4-12. When the metal starts to melt, move the torch in a circular pattern down the sheet toward the other end. If the size of the molten weld pool changes, speed up or slow

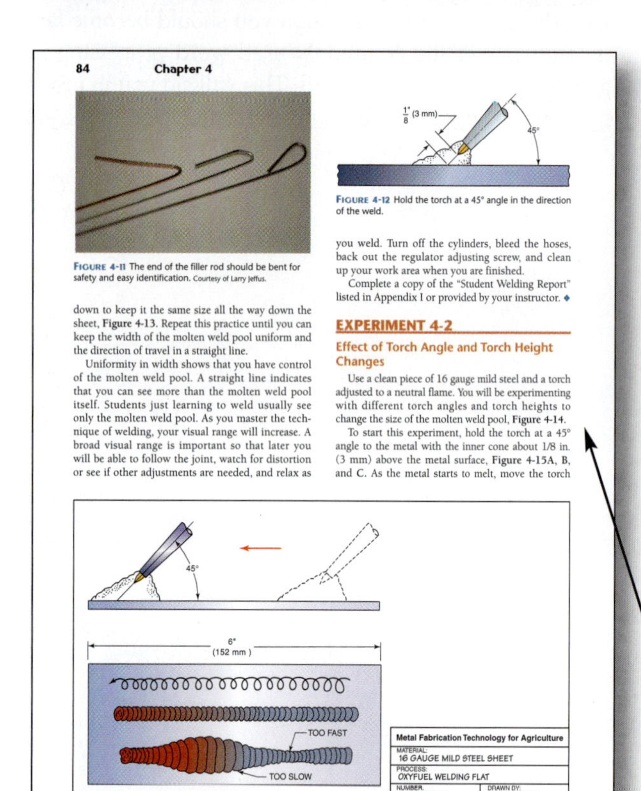

84 Chapter 4

FIGURE 4-11 The end of the filler rod should be bent for safety and easy identification. Courtesy of Larry Jeffus.

FIGURE 4-12 Hold the torch at a 45° angle in the direction of the weld.

down to keep it the same size all the way down the sheet, Figure 4-13. Repeat this practice until you can keep the width of the molten weld pool uniform and the direction of travel in a straight line.

Uniformity in width shows that you have control of the molten weld pool. A straight line indicates that you can see more than the molten weld pool itself. Students just learning to weld usually see only the molten weld pool. As you master the technique of welding, your visual range will increase. A broad visual range is important so that later you will be able to follow the joint, watch for distortion or see if other adjustments are needed, and relax as

you weld. Turn off the cylinders, bleed the hoses, back out the regulator adjusting screw, and clean up your work area when you are finished.

Complete a copy of the "Student Welding Report" listed in Appendix I or provided by your instructor. ◆

EXPERIMENT 4-2

Effect of Torch Angle and Torch Height Changes

Use a clean piece of 16 gauge mild steel and a torch adjusted to a neutral flame. You will be experimenting with different torch angles and torch heights to change the size of the molten weld pool, Figure 4-14.

To start this experiment, hold the torch at a 45° angle to the metal with the inner cone about 1/8 in. (3 mm) above the metal surface, Figure 4-15A, B, and C. As the metal starts to melt, move the torch

FIGURE 4-13 Effect of changing the rate of travel.

Practices are hands-on exercises designed to build your welding skills. Each practice describes in detail what skill you will learn and what equipment, supplies, and tools you will need to complete the exercise.

Experiments are designed to allow you to see what effect changes in the process settings, operation, or techniques have on the type of weld produced. Many are group activities and will help you learn as a team.

Summaries review the important points in the chapter and serve as a useful study tool.

Use some flux and solder and tin the penny. Place the penny on the aluminum plate so the areas of solder on both the penny and plate are touching each other. Heat the two until the solder melts and flows out from between the pennies and plate. When the parts cool, to check the bond, try to break the joint apart.

This process will work on other types of metals that have a strong oxide layer that prevents the solder from bonding. By breaking the oxide layer free with the mechanical action of the steel wool the metals can join. This process can allow a copper patch over an aluminum tube such as those used in air conditioning or refrigeration, **Figure 5-60**. Turn off the cylinders, bleed the hoses, back out the regulator adjusting screw, and clean up your work area when you are finished.

Complete a copy of the "Student Welding Report" listed in Appendix I or provided by your instructor. ◆

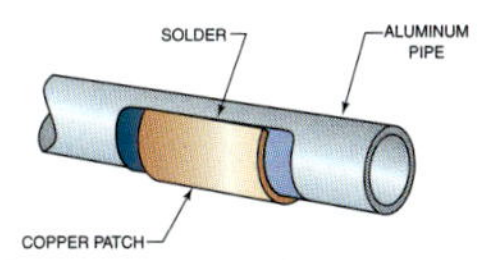

Figure 5-60 Aluminum pipe patch.

Summary

Soldering and brazing are excellent processes for repair work on thin sheet metal and small parts when overheating caused by welding might damage surrounding material. The lower heat input required for soldering, for example, means that some parts can be repaired even if there are electrical wires near the repair. Some metals such as copper and brass are too easily melted to be welded successfully, but they can be soldered or brazed. Gas welding of stainless steel and aluminum is not practical, but they are easily soldered or brazed.

A good example of a job that can be done with brazing is the work of a cattle rancher in the high desert of New Mexico who says he works in his small welding shop when it is "snowing sideways." He uses gas welding and brazing to make silver saddle buckles as a side business.

Review

1. Soldering and brazing are much weaker joining processes than welding.
   A. true
   B. false
2. What type of joint design gives a solder joint the greatest strength?
   A. butt joint
   B. lap joint
   C. tee joint
   D. outside corner joint
3. What force draws the liquid braze metal into the narrow joint gap?
   A. heat
   B. gravity
   C. chemical attraction
   D. capillary attraction

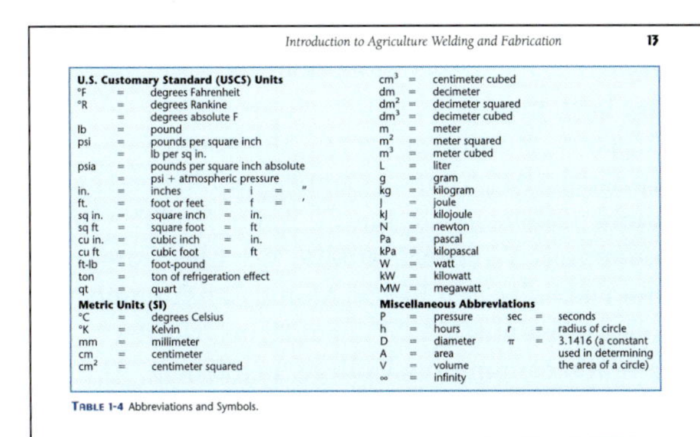

U.S. Customary Standard (USCS) Units											
°F	=	degrees Fahrenheit					cm³	=	centimeter cubed		
°R	=	degrees Rankine					dm	=	decimeter		
	=	degrees absolute F					dm²	=	decimeter squared		
lb	=	pound					dm³	=	decimeter cubed		
psi	=	pounds per square inch					m	=	meter		
	=	lb per sq in.					m²	=	meter squared		
psia	=	pounds per square inch absolute					m³	=	meter cubed		
	=	psi + atmospheric pressure					L	=	liter		
in.	=	inches	=	i	=	"	g	=	gram		
ft.	=	foot or feet	=	f	=	'	kg	=	kilogram		
sq in.	=	square inch	=	in.			J	=	joule		
sq ft	=	square foot	=	ft			kJ	=	kilojoule		
cu in.	=	cubic inch	=	in.			N	=	newton		
cu ft	=	cubic foot	=	ft			Pa	=	pascal		
ft-lb	=	foot-pound					kPa	=	kilopascal		
ton	=	ton of refrigeration effect					W	=	watt		
qt	=	quart					kW	=	kilowatt		
							MW	=	megawatt		
Metric Units (SI)							**Miscellaneous Abbreviations**				
°C	=	degrees Celsius					P	=	pressure	sec	= seconds
°K	=	Kelvin					h	=	hours	r	= radius of circle
mm	=	millimeter					D	=	diameter	π	= 3.1416 (a constant used in determining the area of a circle)
cm	=	centimeter					A	=	area		
cm²	=	centimeter squared					V	=	volume		
							∞	=	infinity		

Table 1-4 Abbreviations and Symbols.

Summary

Agriculture welding is a very diverse job. Almost every farm and ranch utilizes welding. Products that are produced by welding range from small objects to larger structures. Your knowledge and understanding of the various processes and their applications will provide you with employable skills that can result in a rich and rewarding career. The art and science of joining metals has been around for centuries, and with minor changes and improvements in materials, equipment, and supplies, it will be with us through the remainder of the twenty-first century.

Review

1. One of the earliest forms of welding used a mold made from _____ to hold the molten metal as it was being cast on the base plate.
   A. sand
   B. steel
   C. clay
   D. brass
2. How does fusion take place in forge welding?
   A. Liquid metal is poured between the metal ends until they melt and flow together.
   B. Electricity causes heat through resistance and the parts fuse from the heat.
   C. A forge is used to melt the metal edges so they flow together.
   D. The red hot parts are hammered until fusion takes place.
3. Pressure may or may not be used during welding.
   A. true
   B. false

Review questions in multiple-choice format help measure the skills and knowledge you learned in the chapter. Each question is designed to help you apply and understand the information in the chapter.

Bilingual Glossary

The terms and definitions in this glossary are extracted from the American Welding Society publication AWS A3.0-80 Welding Terms and Definitions. The terms with an asterisk (*) are from a source other than the American Welding Society. Note: The English term and definition are given first, followed by the same term and definition in Spanish.

A

acceptable criteria. Agreed-upon standards that must be satisfactorily met.
criterios aceptables. Las normas sobre las que se ha llegado a un acuerdo y que deben cumplirse en forma satisfactoria.

acceptable weld. A weld that meets all the requirements and the acceptance criteria prescribed by welding specifications.
soldadura aceptable. Una soldadura que satisface los requisitos y el criterio aceptable prescribida por las especificaciónes de la soldadura.

***acetone.** A fragrant liquid chemical used in acetylene cylinders. The cylinder is filled with a porous material and acetone is then added to fill. Acetylene is then added and absorbed by the acetone, which can absorb up to 28 times its own volume of the gas.
acetona. Un liquido fragante químico que se usa en los cilindros del acetileno. El cilindro se llena de un material poroso y luego se le agrega la acetona hasta que se llene. El acetileno es absorbido por la acetona, la cual puede absorber 28 veces el propio volumen del gas.

***acetylene.** A fuel gas used for welding and cutting. It is produced as a result of the chemical reaction between calcium carbide and water. The chemical formula for acetylene is C_2H_2. It is colorless, is lighter than air, and has a strong garlic-like smell. Acetylene is unstable above pressures of 15 psig (1.05 kg/cm^2 g). When burned in the presence of oxygen, acetylene produces one of the highest flame temperatures available.
acetileno. Un gas combustible que se usa para soldar y cortar. Es producido a consecuencia de una reacción química de agua y calcio y carburo. La fórmula química para el acetileno es C_2H_2. No tiene color, es más ligero que el aire, y tiene un olor fuerte como a ajo. El acetileno es inestable en presiones más altas de 15 psig (1.05 kg/cm^2 g). Cuando se quema en presencia del oxígeno, el acetileno produce una de las llamás con una temperatura más alta que la que se utiliza.

acid copper chromate. A wood preservative.
cromato de cobre acido. Preservativo de la madera.

actual throat. See throat of a fillet weld.
garganta actual. Vea garganta de soldadura filete.

***adaptable.** Capable of making self-directed corrections; in a robot, this is often accomplished with visual, force, or tactile sensors.
adaptable. Capaz de hacer correcciones por instrucción propia de un robot, esto se lleva a cabo con sensores tangibles visuales, o de fuerza.

agriculture. Enterprises involving the production of plants and animals, along with supplies, services, mechanics, products, processing, and marketing related to those enterprises.
agricultura. Empresas que compreden la producción de plantas y animales, juntos con los articulos, servicios, mecanismos, productos, el elaborar, y la venta relativos a esas empresas.

air acetylene welding (AAW). An oxyfuel gas welding process that uses an air-acetylene flame. The process is used without the application of pressure. This is an obsolete or seldom-used process.
soldadura de aire acetileno. Un proceso de soldar con gas (oxi/combustible) que usa aire-acetileno sin aplicarse presión. Un proceso anticuado que es una rareza.

air carbon arc cutting (CAC-A). A carbon arc cutting process variation that removes molten metal with a jet of air.
arco de carbón con aire. Un proceso de cortar con arco de carbón variante que quita el metal derretido con un chorro de aire.

air compressor. A pump that increases pressure on air.
compresor de aire. Bomba que aumenta la presion sobre el aire.

air-dried lumber. Sawed lumber separated with wooden strips and protected from rain and snow for six months or more.
madera de construccion secado al aire libre. Madera aserrada separada con listones y protegida de la lluvia y la nieve por seis meses or mas.

Allen screw. A screw with a six-sided hole in the head.
tornillo de cabesa allen. Un tornillo cuyo cabeza consiste en agujero hexagonal.

***alloy.** A metal with one or more elements added to it, resulting in a significant change in the metal's properties.
aleación. Un metal en que se le agrega uno o más elementos resultando en un cambio significante en las propiedades del metal.

Bilingual glossary definitions provide a Spanish equivalent for each new term. Additional line art in the glossary will also help you gain a greater understanding of challenging terms.

An **Instructor's Answer Key** combining answers to all the text's review questions is also available.

For more information about Skills USA-VICA visit their web site at www.skillsusa.org.

Experiments and Practices

A number of the chapters in this book contain both experiments and practices. These are intended to help you develop your welding knowledge and skills.

The experiments are designed to allow you to see what effect changes in the process settings, operation, or techniques have on the type of weld produced. When you do an experiment, you should observe and possibly take notes on how the change affected the weld. Often as you make a weld it will be necessary for you to make changes in your equipment settings or your technique in order to ensure that you are making an acceptable weld. By watching what happens when you make the changes in the welding shop, you will be better prepared to decide on changes required to make good welds on the job.

It is recommended that you work in a small group as you try the experiments. When doing the experiments in a small group one person can be welding, one adjusting the equipment, and the others recording the machine settings and weld effects. It also allows you to watch the weld change more closely if someone is welding as you look on. Then, as a group member, changing places will reinforce your learning.

The practices are designed to build your welding skills. Each practice tells you in detail what equipment, supplies, and tools you will need as you develop the specific skill. In most chapters the practices start off easy and become progressively harder. Welding is a skill that requires that you develop in stages from the basic to the more complex.

Each practice gives the evaluation or acceptable limits for the weld. All welds have some discontinuities, but if they are within the acceptable limits, they are not defects but are called flaws. As you practice your welding, keep in mind the acceptable limits so that you can progress to the next level when you have mastered the process and weld you are working on.

Metric Units

Both standard and metric (SI) units are given in this text. The SI units are in parentheses ( ) following the standard unit. When nonspecific values are used—for example, "set the gauge at 2 psig" where 2 is an approximate value—the SI units have been rounded off to the nearest whole number. Round-off occurs in these cases to agree with the standard value and because whole numbers are easier to work with. SI units are not rounded off only when the standard unit is an exact measurement.

Often students have difficulty understanding metric units because exact conversions are used even when the standard measurement was an approximation. Rounding off the metric units makes understanding the metric system much easier, **Table 1-2**. By using this approximation method, you can make most standard-to-metric conversions in your head without needing to use a calculator.

Once you have learned to use approximations for metric, you will find it easier to make exact conversions whenever necessary. Conversions must be exact in the shop when a part is dimensioned with one system's units and the other system must be used to fabricate the part. For that reason you must be able to make those conversions. **Table 1-3** and **Table 1-4** are set up to be used with or without the aid of a calculator. Many calculators today have built-in standard–metric conversions. It is a good idea to know how to make these conversions with and without these aids, of course. Practice making such conversions whenever the opportunity arises.

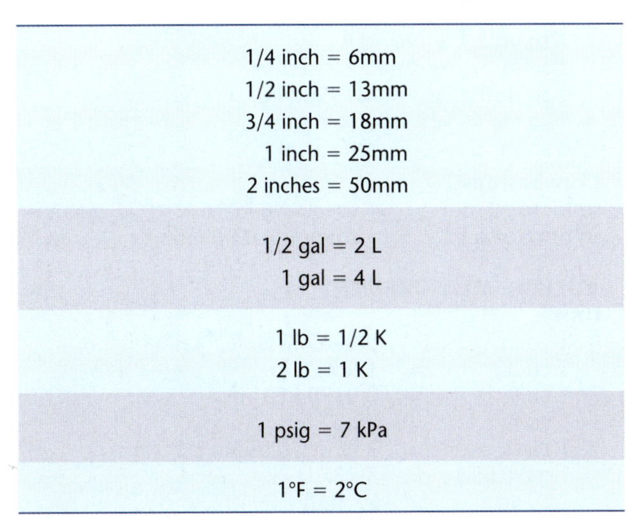

1/4 inch = 6mm
1/2 inch = 13mm
3/4 inch = 18mm
1 inch = 25mm
2 inches = 50mm
1/2 gal = 2 L
1 gal = 4 L
1 lb = 1/2 K
2 lb = 1 K
1 psig = 7 kPa
1°F = 2°C

Table 1-2 Metric Conversions Approximations. By using an approximation for converting standard units to metric it is possible to quickly have an idea of how large or heavy an object is in the other units. For estimating it is not necessary to be concerned with the exact conversions.

TEMPERATURE

Units

°F (each 1° change)	=	0.555°C (change)
°C (each 1° change)	=	1.8°F (change)
32°F (ice freezing)	=	0°Celsius
212°F (boiling water)	=	100°Celsius
–460°F (absolute zero)	=	0°Rankine
–273°C (absolute zero)	=	0°Kelvin

Conversions

°F to °C	______	°F – 32 = ______	× .555 = ______	°C
°C to °F	______	°C × 1.8= ______	+ 32 = ______	°F

LINEAR MEASUREMENT

Units

1 inch	=	25.4 millimeters
1 inch	=	2.54 centimeters
1 millimeter	=	0.0394 inch
1 centimeter	=	0.3937 inch
12 inches	=	1 foot
3 feet	=	1 yard
5280 feet	=	1 mile
10 millimeters	=	1 centimeter
10 centimeters	=	1 decimeter
10 decimeters	=	1 meter
1,000 meters	=	1 kilometer

Conversions

in. to mm	______	in.	×	25.4	=	______	mm
in. to cm	______	in.	×	2.54	=	______	cm
ft to mm	______	ft	×	304.8	=	______	mm
ft to m	______	ft	×	0.3048	=	______	m
mm to in.	______	mm	×	0.0394	=	______	in.
cm to in.	______	cm	×	0.3937	=	______	in.
mm to ft	______	mm	×	0.00328	=	______	ft
m to ft	______	m	×	3.28	=	______	ft

AREA MEASUREMENT

Units

1 sq in.	=	0.0069 sq ft
1 sq ft	=	144 sq in.
1 sq ft	=	0.111 sq yd
1 sq yd	=	9 sq ft
1 sq in.	=	645.16 sq mm
1 sq mm	=	0.00155 sq in.
1 sq cm	=	100 sq mm
1 sq m	=	1,000 sq cm

Conversions

sq in. to sq mm	______	sq in. × 645.16 =	______	sq mm
sq mm to sq in.	______	sq mm × 0.00155 =	______	sq in.

VOLUME MEASUREMENT

Units

1 cu in.	=	0.000578 cu ft
1 cu ft	=	1728 cu in.
1 cu ft	=	0.03704 cu yd
1 cu ft	=	28.32 L
1 cu ft	=	7.48 gal (U.S.)
1 gal (U.S.)	=	3.737 L
1 cu yd	=	27 cu ft
1 gal	=	0.1336 cu ft
1 cu in.	=	16.39 cu cm
1 L	=	1,000 cu cm
1 L	=	61.02 cu in.
1 L	=	0.03531 cu ft
1 L	=	0.2642 gal (U.S.)
1 cu yd	=	0.769 cu m
1 cu m	=	1.3 cu yd

Conversions

cu in. to L	______	cu in.	×	0.01638	=	______	L
L to cu in.	______	L	×	61.02	=	______	cu in.
cu ft to L	______	cu ft	×	28.32	=	______	L
L to cu ft	______	L	×	0.03531	=	______	cu ft
L to gal	______	L	×	0.2642	=	______	gal
gal to L	______	gal	×	3.737	=	______	L

WEIGHT (MASS) MEASUREMENT

Units

1 oz	=	0.0625 lb
1 lb	=	16 oz
1 oz	=	28.35 g
1 g	=	0.03527 oz
1 lb	=	0.0005 ton
1 ton	=	2,000 lb
1 oz	=	0.283 kg
1 lb	=	0.4535 kg
1 kg	=	35.27 oz
1 kg	=	2.205 lb
1 kg	=	1,000 g

Conversions

lb to kg	______	lb	×	0.4535	=	______	kg
kg to lb	______	kg	×	2.205	=	______	lb
oz to g	______	oz	×	0.03527	=	______	g
g to oz	______	g	×	28.35	=	______	oz

PRESSURE AND FORCE MEASUREMENTS

Units

1 psig	=	6.8948 kPa
1 kPa	=	0.145 psig
1 psig	=	0.000703 kg/sq mm
1 kg/sq mm	=	6894 psig
1 lb (force)	=	4.448 N
1 N (force)	=	0.2248 lb

Conversions

psig to kPa	______	psig	×	6.8948	=	______	kPa
kPa to psig	______	kPa	×	0.145	=	______	psig
lb to N	______	lb	×	4.448	=	______	N
N to lb	______	N	×	0.2248	=	______	psig

VELOCITY MEASUREMENTS

Units

1 in./sec	=	0.0833 ft/sec
1 ft/sec	=	12 in./sec
1 ft/min	=	720 in./sec
1 in./sec	=	0.4233 mm/sec
1 mm/sec	=	2.362 in./sec
1 cfm	=	0.4719 L/min
1 L/min	=	2.119 cfm

Conversions

ft/min to in./sec	______	ft/min	× 720	=	______ in./sec
in./min to mm/sec	______	in./min	× .4233	=	______ mm/sec
mm/sec. to in./min	______	mm/sec	× 2.362	=	______ in./min
cfm to L/min	______	cfm	× 0.4719	=	______ L/min
L/min to cfm	______	L/min	× 2.119	=	______ cfm

TABLE 1-3 Table of Conversions: U.S. Customary Standard (USCS) Units and Metric Units (SI).

U.S. Customary Standard (USCS) Units

°F	=	degrees Fahrenheit				
°R	=	degrees Rankine				
	=	degrees absolute F				
lb	=	pound				
psi	=	pounds per square inch				
	=	lb per sq in.				
psia	=	pounds per square inch absolute				
	=	psi + atmospheric pressure				
in.	=	inches	=	i	=	″
ft.	=	foot or feet	=	f	=	′
sq in.	=	square inch	=	in.		
sq ft	=	square foot	=	ft		
cu in.	=	cubic inch	=	in.		
cu ft	=	cubic foot	=	ft		
ft-lb	=	foot-pound				
ton	=	ton of refrigeration effect				
qt	=	quart				

Metric Units (SI)

°C	=	degrees Celsius
°K	=	Kelvin
mm	=	millimeter
cm	=	centimeter
cm^2	=	centimeter squared
cm^3	=	centimeter cubed
dm	=	decimeter
dm^2	=	decimeter squared
dm^3	=	decimeter cubed
m	=	meter
m^2	=	meter squared
m^3	=	meter cubed
L	=	liter
g	=	gram
kg	=	kilogram
J	=	joule
kJ	=	kilojoule
N	=	newton
Pa	=	pascal
kPa	=	kilopascal
W	=	watt
kW	=	kilowatt
MW	=	megawatt

Miscellaneous Abbreviations

P	=	pressure	sec	=	seconds
h	=	hours	r	=	radius of circle
D	=	diameter	π	=	3.1416 (a constant used in determining the area of a circle)
A	=	area			
V	=	volume			
∞	=	infinity			

TABLE 1-4 Abbreviations and Symbols.

Summary

Agriculture welding is a very diverse job. Almost every farm and ranch utilizes welding. Products that are produced by welding range from small objects to larger structures. Your knowledge and understanding of the various processes and their applications will provide you with employable skills that can result in a rich and rewarding career. The art and science of joining metals has been around for centuries, and with minor changes and improvements in materials, equipment, and supplies, it will be with us through the remainder of the twenty-first century.

Review

1. One of the earliest forms of welding used a mold made from ______ to hold the molten metal as it was being cast on the base plate.
   A. sand
   B. steel
   C. clay
   D. brass
2. How does fusion take place in forge welding?
   A. Liquid metal is poured between the metal ends until they melt and flow together.
   B. Electricity causes heat through resistance and the parts fuse from the heat.
   C. A forge is used to melt the metal edges so they flow together.
   D. The red hot parts are hammered until fusion takes place.
3. Pressure may or may not be used during welding.
   A. true
   B. false

Review continued

4. The most popular agricultural welding processes are
   A. AAW, AB, FW, and LBC
   B. OAW, SMAW, GMAW, and FCAW
   C. FS, EASP, CAC, and IW
   D. GTAW, SW, HFRW, and GMAC
5. Which process is easily used on thin-gauge metal as well as heavy plate?
   A. GMAW
   B. GTAW
   C. SMAW
   D. BT
6. FCAW uses the same type of equipment that is used by GMAW.
   A. true
   B. false
7. What would the type, capacity, and condition of welding equipment affect?
   A. when the part was welded
   B. the selection of the process
   C. how hot you have to get the metal to make a weld
   D. who makes the weld
8. Which of these is an example of semiautomatic welding?
   A. FCAW
   B. SMAW
   C. TB
   D. OFW
9. Generally who performs most welding?
   A. fitters
   B. helpers
   C. welders
   D. tack welders
10. A cutting torch is used to weld parts together.
    A. true
    B. false
11. Without which device could most agricultural welding repairs not be made?
    A. shop hoist
    B. hammer and chisel
    C. cutting torch
    D. grinder
12. Most people can learn to weld very well in a week.
    A. true
    B. false
13. A person planning a career in welding needs good ______.
    A. eyesight
    B. manual dexterity
    C. hand-eye coordination
    D. all of the above
14. The letters FFA stand for ________.
    A. For Fine Art
    B. Forge, Fire and Arc welding
    C. Future Farmers of America
    D. Future Firefighters Association
15. FFA has chapters in ______ states.
    A. 50
    B. 48
    C. 52
    D. most
16. How does 4-H meet the needs of individual communities?
    A. by having one very good program
    B. encouraging everyone to work on the same type of project
    C. by tailoring its programs to meet the needs of a community
    D. by having very low admission fee
17. Dignity of work, fair play, and high moral standards are all part of Skills USA-VICA's creed.
    A. true
    B. false
18. When you are doing an experiment, it is recommended that you work alone.
    A. true
    B. false
19. Two inches is about ______ in metric.
    A. 2 cm
    B. 4 km
    C. 25 mm
    D. 50 mm
20. One-half gallon is about ______ in metric.
    A. 2 L
    B. 4 k
    C. 7 kPa
    D. 300 mL

Chapter 2

Safety in Welding and Fabrication

OBJECTIVES

After completing this chapter, the student should be able to

- ☑ describe the type of protection that should be worn for welding.
- ☑ describe the proper method of handling, storing, and setting up cylinders.
- ☑ discuss the proper way to ventilate a welding area.
- ☑ explain how to avoid electric shock.
- ☑ describe how to avoid possible health hazards for welding.
- ☑ explain how to prevent fires in the welding shop.
- ☑ describe how to safely use a circular saw to make a cut.
- ☑ demonstrate the proper care and use of agricultural hand tools.

KEY TERMS

acetone
acetylene
earmuffs
earplugs
electrical ground
electrical resistance
electric shock
exhaust pickups
flash burn
flash glasses
forced ventilation
full face shield
goggles
infrared light
material specification data sheet (MSDS)
natural ventilation
safety glasses
type A fire extinguisher
type B fire extinguisher
type C fire extinguisher
type D fire extinguisher
ultraviolet light
valve protection cap
ventilation
visible light
warning label
welding helmet

INTRODUCTION

Accident prevention is the main intent of this chapter. The safety information included in this text is intended as a guide. There is no substitute for caution and common sense. A safe job is no accident; it takes work to make the job safe. Each person working must do what it takes to keep the job safe.

Agriculture is a very large and diverse industry. This chapter will concentrate on only that portion of agriculture safety related to the areas of light metal and wood fabrication. Before beginning work in other areas of agriculture, it is your responsibility to obtain appropriate safety information for those areas. You must read, learn, and follow all safety rules, regulations, and procedures for those areas.

Light agricultural fabrication, like all other areas of agriculture work, has a number of potential safety hazards. These hazards need not result in anyone's being injured. Learning to work safely is as important as learning to be a skilled agricultural fabrication worker.

You must approach new jobs with your safety in mind. Your safety is your own responsibility, and you must take that responsibility. It is not possible to anticipate all of the possible dangers in every job. There may be some dangers not covered in this text. You can get specific safety information from your local agricultural extension office; equipment manufacturers and their local suppliers; your local, state, and national departments of agriculture; college and university extension and regular courses; and the World Wide Web.

Burns

Burns are one of the most common and painful injuries that occur in welding fabrication. Burns can be caused by ultraviolet light rays as well as by contact with hot welding material. The chance of infection is high with burns because of the dead tissue. It is important that all burns receive proper medical treatment to reduce the chance of infection. Burns are divided into three classifications, depending upon the degree of severity. The three classifications include first-degree, second-degree, and third-degree burns.

First-degree Burns First-degree burns have occurred when the surface of the skin is reddish in color, tender, and painful and does not involve any broken skin, **Figure 2-1**. The first step in treating a first-degree burn is to immediately put the burned area under cold water (not iced) or apply cold water compresses (clean towel, washcloth, or handkerchief soaked in cold water) until the pain decreases. Then cover the area with sterile bandages or a clean cloth. Do not apply butter or grease, or any other home remedies or medications without a doctor's recommendation.

Second-degree Burns Second-degree burns have occurred when the surface of the skin is severely damaged, resulting in the formation of blisters and possible breaks in the skin, **Figure 2-2**. Again, the most important first step in treating a second-degree burn is to put the area under cold water (not iced) or apply cold water compresses until the pain decreases. Gently pat the area dry with a clean towel, and cover the area with a sterile bandage or clean cloth to prevent infection. Seek medical attention. If the burns are around the mouth or nose, or involve singed nasal hair, breathing problems may develop. Do not apply ointments, sprays, antiseptics, or home remedies. Note: in an emergency any cold liquid you drink—for example, water, cold tea, soft drinks, or milk shake—can be poured on a burn. The purpose is to reduce the skin temperature as quickly as possible to reduce tissue damage.

Third-degree Burns Third-degree burns have occurred when the surface of the skin and possibly the tissue below the skin appear white or charred, **Figure 2-3**. Initially, little pain is present because nerve endings have been destroyed. Do not remove any clothes that are stuck to the burn. Do not put ice water or ice on the burns; this could intensify the shock reaction. Do not

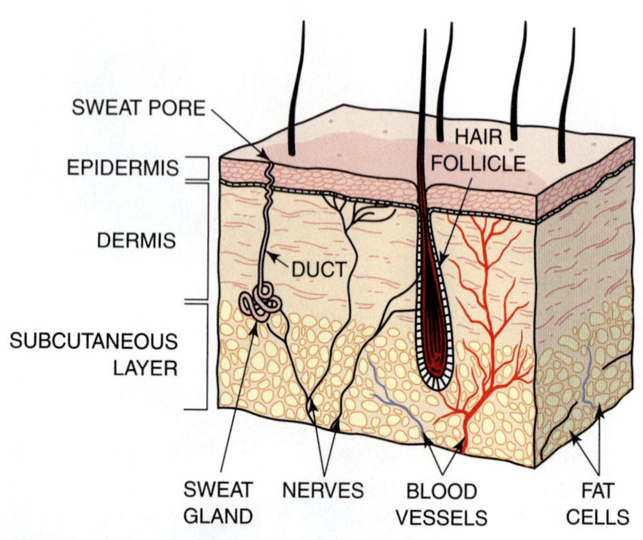

FIGURE 2-1 First-degree burn — only the skin surface (epidermis) is affected.

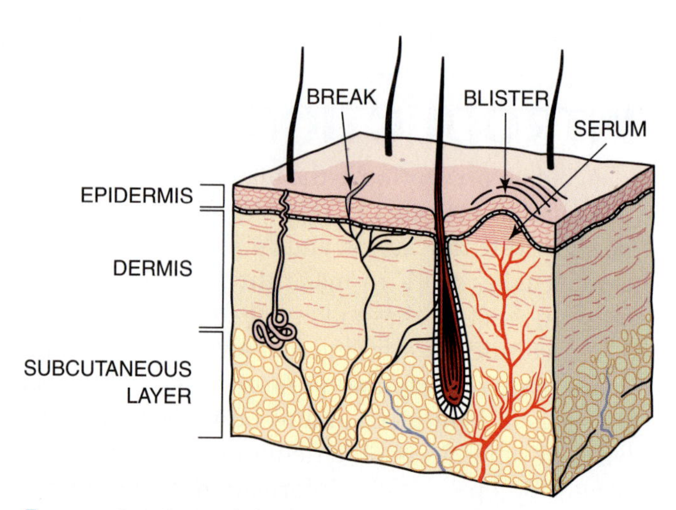

FIGURE 2-2 Second-degree burn — the epidermal layer is damaged, forming blisters or shallow breaks.

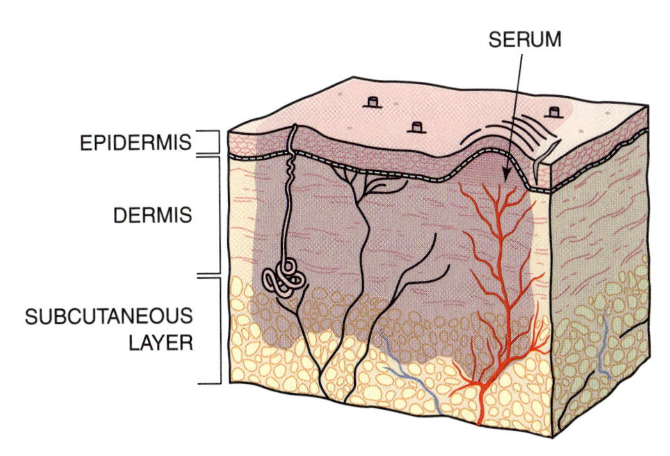

FIGURE 2-3 Third-degree burn — the epidermis, dermis, and the subcutaneous layers of tissue are destroyed.

FIGURE 2-4 Truck was parked so horses could not look at arc. Courtesy of Larry Jeffus.

apply ointments, sprays, antiseptics, or home remedies to burns. If the victim is on fire, smother the flames with a blanket, rug, or jacket. Breathing difficulties are common with burns around the face, neck, and mouth; be sure that the victim is breathing. Place a cold cloth or cool (not iced) water on burns of the face, hands, or feet to cool the burned areas. Cover the burned area with thick, sterile, nonfluffy dressings. Call for an ambulance immediately; people with even small third-degree burns need to consult a doctor.

Burns Caused by Light Some types of light can cause burns. There are three types of light—ultraviolet, infrared, and visible. Ultraviolet and infrared are not visible to the unaided human eye and can cause burns. During welding, one or more of the three types of light may be present. Arc welding and cutting produces all three kinds of light, but gas welding produces visible and infrared light only.

The light from the welding process can be reflected from walls, ceilings, floors, or any other large surface. This reflected light is as dangerous as the direct welding light. To reduce the danger from reflected light, welding shops, if possible, should be painted flat black. Flat black will reduce the reflected light by absorbing more of it than any other color. When the welding is to be done around livestock, especially horses, precautions must be taken to protect them from their natural curiosity to want to see what you are doing, **Figure 2-4**. If the welding cannot be moved away from other workers and animals, or screened off with equipment, then welding curtains can be placed to absorb the welding light, **Figure 2-5**. These special portable welding curtains may be either transparent or opaque. Transparent welding curtains are made of a special high-temperature, flame-resistant plastic that will prevent the harmful light from passing through.

FIGURE 2-5 Portable welding curtains. Courtesy of Frommelt.

CAUTION

Welding curtains must always be used to protect other workers and livestock in an area that might be exposed to the welding light.

Ultraviolet Light

Ultraviolet light waves are the most dangerous. They can cause first-degree and second-degree burns to eyes or to any exposed skin. Because you and your animals cannot see or feel ultraviolet light while being exposed to it, you must stay protected and keep the animals protected when they are in the area of any of the arc welding processes. The closer you or they are to the arc and the higher the current, the quicker a burn may occur. The ultraviolet light is so intense during some welding processes that eyes can receive a **flash burn** within seconds, and the skin can be burned within minutes. Ultraviolet light can pass through loosely woven clothing, thin clothing, light-colored clothing, and damaged or poorly maintained arc welding helmets.

Infrared Light

Infrared light is the light wave that is felt as heat. Although infrared light can cause burns, a person will immediately feel this type of light. Therefore, burns can easily be avoided.

Visible Light

Visible light is the light that we see. It is produced in varying quantities and colors during welding. Too much visible light may cause temporary night blindness (poor eyesight under low light levels). Too little visible light may cause eye strain, but visible light is not hazardous.

Whether burns are caused by ultraviolet light or hot material, they can be avoided if proper clothing and other protection are worn.

Eye and Ear Protection

Face and Eye Protection Eye protection must be worn in the shop at all times. Eye protection can be **safety glasses,** with side shields, **Figure 2-6**, **goggles,** or a **full face shield.** To give better protection when working in brightly lit areas or outdoors, some welders wear **flash glasses,** which are special, lightly tinted, safety glasses. These safety glasses provide protection from both flying debris and reflected light.

Suitable eye protection is important because eye damage caused by excessive exposure to arc light is not noticed. Welding light damage occurs often without warning, like a sunburn's effect that is felt the following day. Therefore, welders must take appropriate precautions in selecting filters or goggles that are suitable for the process being used, **Table 2-1**. Selecting the correct shade lens is also important, because both extremes of too light or too dark can cause eye strain. New welders often select too dark a lens, assuming it will give them better protection, but this results in eye strain in the same manner as if they were trying to read in a poorly lit room. In reality, any approved arc welding lenses will filter out the harmful ultraviolet light. Select a lens that lets you see comfortably. At the very least, the welder's eyes must not be strained by excessive glare from the arc.

Ultraviolet light can burn the eye in two ways. This light can injure either the white of the eye or the retina, which is the back of the eye. Burns on the retina are not painful but may cause some loss of eyesight. The whites of the eyes can also be burned by ultraviolet light, **Figure 2-7**. The whites of the eyes are very sensitive, and burns are very painful. The eyes are easily infected because, as with any burn, many cells are killed. These dead cells in the moist environment of the eyes will promote the growth of bacteria that cause infection. When the eye is burned, it feels as though there is something in the eye, but without a professional examination, it is impossible to know. Because there may be something in the eye and because of the high risk of infection, home remedies or other medicines should never be used for eye burns. Anytime you receive an eye injury you should see a doctor.

Even with quality **welding helmets,** like those shown in **Figure 2-8**, the welder must check for potential problems that may occur from accidents or daily use. Small, undetectable leaks of ultraviolet light in an arc welding helmet can cause a welder's eyes to itch or feel sore after a day of welding. To prevent these leaks, make sure the lens gasket is installed

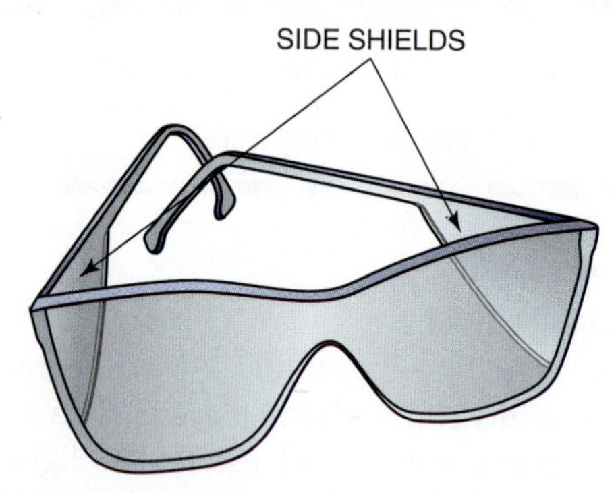

FIGURE 2-6 Safety glasses with side shields.

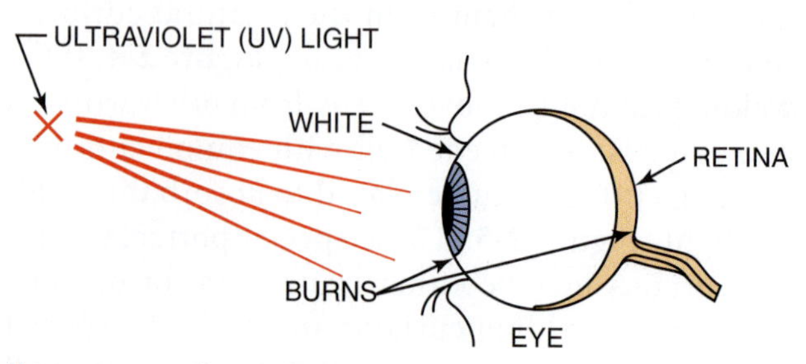

FIGURE 2-7 The eye can be burned on the white or on the retina by ultraviolet light.

1
Goggles, flexible fitting, regular ventilation

2
Goggles, flexible fitting, hooded ventilation

3
Goggles, cushioned fitting, rigid body

4
Spectacles

5
Spectacles, eyecup type eyeshields

6
Spectacles, semi-flat-fold sideshields

7
Welding goggles, eyecup type, tinted lenses

7A
Chipping goggles, eyecup type, tinted lenses

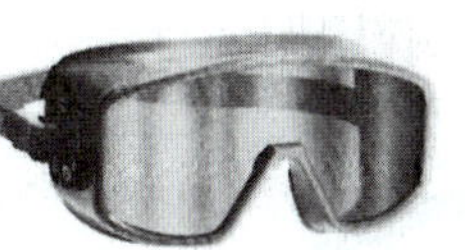

8
Welding goggles, coverspec type, tinted lenses

8A
Chipping goggles, coverspec type, clear safety lenses

9
Welding goggles, coverspec type, tinted plate lens

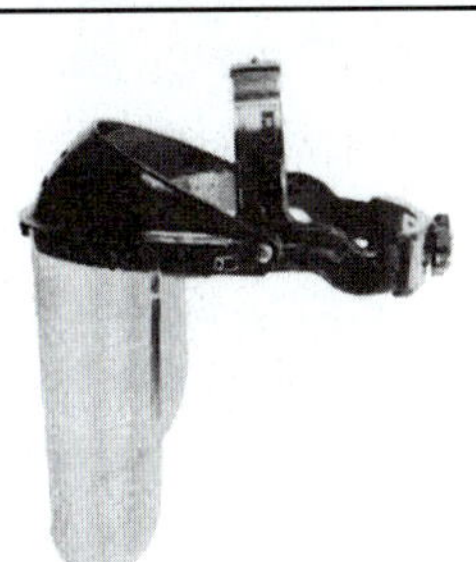

10
Face shield, plastic or mesh window (see caution note)

11
Welding helmet

**Non-sideshield spectacles are available for limited hazard use requiring only frontal protection.*

Applications

Operation	Hazards	Protectors
Acetylene-Burning Acetylene-Cutting Acetylene-Welding	Sparks, Harmful Rays, Molten Metal, Flying Particles	7,8,9
Chemical Handling	Splash, Acid Burns, Fumes	2 (for severe exposure add 10)
Chipping	Flying Particles	1,2,4,5,6,7A,8A
Electric (Arc) Welding	Sparks, Intense Rays, Molten Metal	11 (in combination with 4,5,6 in tinted lenses advisable)
Furnace Operations	Glare, Heat, Molten Metal	7,8,9 (for severe exposure add 10)
Grinding-Light	Flying Particles	1,3,5,6 (for severe exposure add 10)
Grinding-Heavy	Flying Particles	1,3,7A,8A (for severe exposure add 10)
Laboratory	Chemical Splash, Glass Breakage	2 (10 when in combination with 5,6)
Machining	Flying Particles	1,3,5,6 (for severe exposure add 10)
Molten Metals	Heat, Glare, Sparks, Splash	7,8 (10 in combination with 5,6 in tinted lenses)
Spot Welding	Flying Particles, Sparks	1,3,4,5,6 (tinted lenses advisable, for severe exposure add 10)

CAUTION:
Face shields alone do not provide adequate protection. Plastic lenses are advised for protection against molten metal splash. Contact lenses, of themselves, do not provide eye protection in the industrial sense and shall not be worn in a hazardous environment without appropriate covering safety eyewear.

TABLE 2-1 Huntsman® Selector Chart. Courtesy of Kedman Co., Huntsman Product Division.

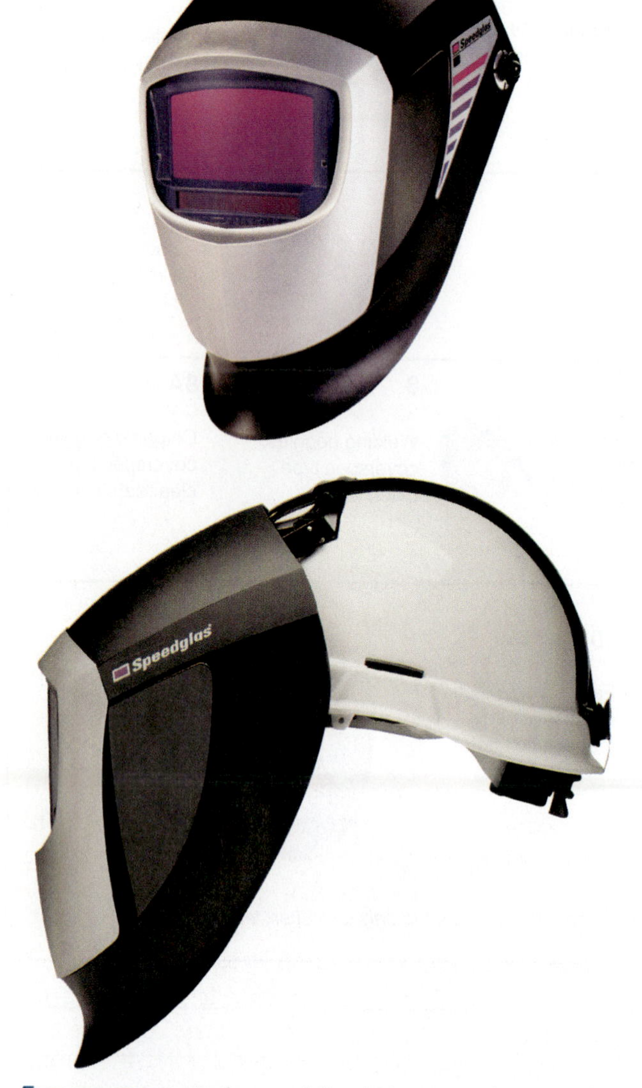

FIGURE 2-8 Typical arc welding helmets used to provide eye and face protection during welding. Courtesy of Hornell, Inc.

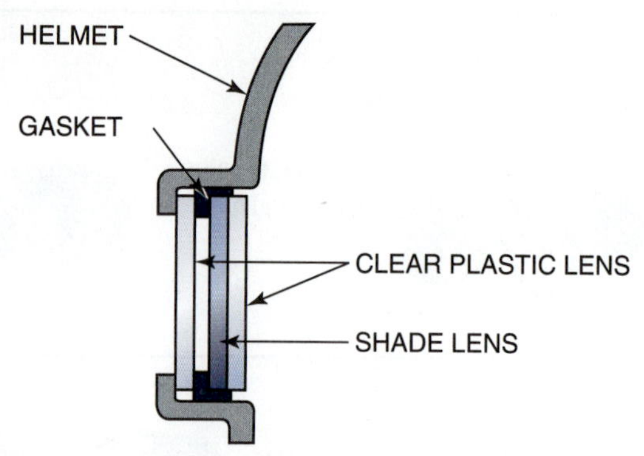

FIGURE 2-9 The correct placement of the gasket around the shade lens is important because it can stop ultraviolet light from bouncing around the lens assembly.

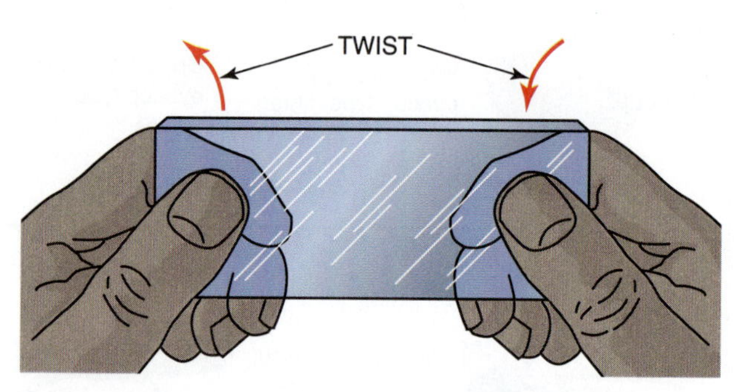

FIGURE 2-10 To check the shade lens for possible cracks, gently twist it.

FIGURE 2-11 Full face shield.

correctly, **Figure 2-9**. The outer and inner clear lens must be plastic. As shown in **Figure 2-10**, the lens can be checked for cracks by twisting it between your fingers. Worn or cracked spots on a helmet must be repaired. Tape can be used as a temporary repair until the helmet can be replaced or permanently repaired.

Safety glasses with side shields are adequate for general use, but if heavy grinding, chipping, or overhead work is being done, goggles or a full face shield should be worn in addition to safety glasses, **Figure 2-11**. Safety glasses are best for general protection because they must be worn under an arc welding helmet at all times.

Ear Protection The welding environment can be very noisy. The sound level is at times high enough to cause pain and some loss of hearing if the welder's ears are unprotected. Hot sparks can also drop into an open ear, causing severe burns.

Ear protection is available in several forms. One form of protection is earmuffs that cover the outer ear completely, **Figure 2-12**. Another form of protec-

FIGURE 2-12 Earmuffs provide complete ear protection and can be worn under a welding helmet. Courtesy of Mine Safety Appliances Company.

FIGURE 2-13 Earplugs used as protection from noise only. Courtesy of Mine Safety Appliances Company.

tion is earplugs that fit into the ear canal, **Figure 2-13**. Both of these protect a person's hearing, but only the earmuffs protect the outer ear from burns.

CAUTION

Damage to your hearing caused by high sound levels may not be detected until later in life, and the resulting loss in hearing is nonrecoverable. Your hearing will not improve with time, and each exposure to high levels of sound will further damage your hearing.

Respiratory Protection

All welding and cutting processes produce undesirable by-products, such as harmful dusts, fumes, mists, gases, smokes, sprays, or vapors. For your safety and the safety of others your primary objective will be to prevent these contaminants from forming and collecting in the area's atmosphere. This will be accomplished as much as possible by thorough cleaning of surface contaminants before starting work, and confinement of the operation to outdoor or open spaces.

Production of welding by-products cannot be avoided. They are created when the temperature of metals and fluxes is raised above the temperatures at which they boil or decompose. Most of the by-products are recondensed in the weld. However, some do escape into the atmosphere, producing the haze that occurs in improperly ventilated welding shops. Some fluxes used in welding electrodes produce fumes that may irritate the welder's nose, throat, and lungs.

When welders must work in an area where effective controls to remove air-borne welding by-products are not feasible, respirators shall be provided by their employers when this equipment is necessary to protect their health. The respirators must be applicable and suitable for the purpose intended. Where respirators are necessary to protect welders' health or whenever respirators are required by the welding shop, the shop will establish and implement a written respiratory protection program with worksite-specific procedures. Welders are responsible for following the welding shop's established written respiratory protection program. Guidelines for the respiratory protection program are available from the Occupational Safety and Health Administration (OSHA) office in Washington, DC.

Training must be a part of the welding shop's respiratory protection program. This training should include instruction on any and/or all of the following procedures for

- proper use of respirators, including techniques for putting them on and removing them,
- schedules for cleaning, disinfecting, storing, inspecting, repairing, discarding, and performing other aspects of maintenance of the respiratory protection equipment,
- selection of the proper respirators for use in the workplace, and any respiratory equipment limitations,
- procedures for testing for tight-fitting respirators,

- proper use of respirators in both routine and reasonably foreseeable emergency situations, and
- regular evaluation of the effectiveness of the program.

All respiratory protection equipment used in a welding shop should be certified by the National Institution for Occupational Safety and Health (NIOSH). Some of the types of respiratory protection equipment that may be used are the following:

- Air-purifying respirators have an air-purifying filter, cartridge, or canister that removes specific air contaminants by passing ambient air through the air-purifying element.
- Atmosphere-supplying respirators supply breathing air from a source independent of the ambient atmosphere; this includes both the supplied-air respirators (SARs) and self-contained breathing apparatus (SCBA) type units.
- Demand respirators are atmosphere-supplying respirators that admit breathing air to the facepiece only when a negative pressure is created inside the facepiece by inhalation.
- Positive pressure respirators are respirators in which the pressure inside the respiratory inlet covering exceeds the ambient air pressure outside the respirator.
- Powered air-purifying respirators (PAPR) are air-purifying respirators that use a blower to force the ambient air through air-purifying elements to the inlet covering, **Figure 2-14.**
- Self-contained breathing apparatus (SCBA) are atmosphere-supplying respirators for which the breathing air source is designed to be carried by the user.
- Supplied-air respirators (SAR), or airline respirators, are atmosphere-supplying respirators for which the source of breathing air is not designed to be carried by the user.

FIGURE 2-14 Filtered fresh air is forced into the welder's breathing area. Courtesy of Hornell, Inc.

Respiratory protection equipment used in many welding applications is of the filtering facepiece (dust mask) type, **Figure 2-15.** These mask types use the negative pressure as you inhale to draw air through a filter, which is an integral part of the facepiece. In areas of severe contamination you may use a hood-type respirator, which covers your head and neck and may cover portions of your shoulders and torso.

CAUTION

Welding or cutting must never be performed on drums, barrels, tanks, vessels, or other containers until they have been emptied and cleaned thoroughly, eliminating all flammable materials and all substances (such as detergents, solvents, greases, tars, or acids) that might produce flammable, toxic, or explosive vapors when heated.

Some materials that can cause respiratory problems are used as paints, coating, or plating on metals to prevent rust or corrosion. Other potentially hazardous materials might be used as alloys in metals to give them special properties.

Before welding or cutting, any metal that has been painted or has any grease, oil, or chemicals on its surface must be thoroughly cleaned. This cleaning

FIGURE 2-15 Typical respirator for contaminated environments. The filters can be selected for specific types of contaminants. Courtesy of Mine Safety Appliances Company.

may be done by grinding, sand blasting, or applying an approved solvent.

CAUTION

Never weld, cut, or braze on any metal that has pesticide, herbicide, or fertilizer residue. Even small amounts trapped in crevasses, such as those in the hopper or between parts must be thoroughly removed before any welding, cutting, or brazing can safely be performed. All of these chemicals can release deadly gases when heated. Some fertilizers are explosive.

Most paints containing lead have been removed from the market. But some industries, such as marine or ship applications, still use these lead-based paints. Often old machinery and farm equipment surfaces still have lead-based paint coatings. Solder often contains lead alloys. The welding and cutting of lead-bearing alloys or metals whose surfaces have been painted with lead-based paint can generate lead oxide fumes. Inhalation and ingestion of lead oxide fumes and other lead compounds will cause lead poisoning. Symptoms include metallic taste in the mouth, loss of appetite, nausea, abdominal cramps, and insomnia. In time, anemia and a general weakness, chiefly in the muscles of the wrists, develop.

Both cadmium and zinc are plating materials used to prevent iron or steel from rusting. Cadmium is often used on bolts, nuts, hinges, and other hardware items, and it gives the surface a yellowish-gold appearance. Acute exposure to high concentrations of cadmium fumes can produce severe lung irritation. Long-term exposure to low levels of cadmium in air can result in emphysema (a disease affecting the lung's ability to absorb oxygen) and can damage the kidneys.

Zinc, often in the form of galvanizing, may be found on pipes, sheet metal, water troughs, bolts, nuts, and many other types of hardware. Zinc plating that is thin may appear as a shiny, metallic patchwork or crystal pattern; thicker, hot-dipped zinc appears rough and may look dull. Zinc is used in large quantities in the manufacture of brass and is found in brazing rods. Inhalation of zinc oxide fumes can occur when welding or cutting on these materials. Exposure to these fumes is known to cause metal fume fever, whose symptoms are very similar to those of common influenza.

Some concern has been expressed about the possibility of lung cancer being caused by some of the chromium compounds that are produced when welding stainless steels.

CAUTION

Extreme caution must be taken to avoid the fumes produced when welding is done on dirty or used metal. Any chemicals that are on the metal will become mixed with the welding fumes, a combination that can be extremely hazardous. All metal must be cleaned before welding to avoid this potential problem.

Rather than take chances, welders should recognize that fumes of any type, regardless of their source, should not be inhaled. The best way to avoid problems is to provide adequate ventilation. If this is not possible, breathing protection should be used. Protective devices for use in poorly ventilated or confined areas are shown in **Figure 2-14** and **Figure 2-15**.

Potentially dangerous gases also can be present in a welding shop. Proper ventilation or respirators are necessary when welding in confined spaces, regardless of the welding process being used. Ozone is a gas that is produced by the ultraviolet radiation in the air in the vicinity of arc welding and cutting operations. Ozone is very irritating to all mucous membranes, with excessive exposure producing pulmonary edema. Other effects of exposure to ozone include headache, chest pain, and dryness in the respiratory tract.

Phosgene is formed when ultraviolet radiation decomposes chlorinated hydrocarbon. Fumes from chlorinated hydrocarbons can come from solvents such as those used for degreasing metals and from refrigerants from air-conditioning systems. They decompose in the arc to produce a potentially dangerous chlorine acid compound. This compound reacts with the moisture in the lungs to produce hydrogen chloride, which in turn destroys lung tissue. For this reason, any use of chlorinated solvents should be well away from welding operations in which ultraviolet radiation or intense heat is generated. Any welding or cutting on refrigeration or air-conditioning piping must be done only after the refrigerant has been completely removed in accordance with EPA regulations.

Care also must be taken to avoid the infiltration of any fumes or gases, including argon or carbon dioxide, into a confined working space, such as when welding in tanks. The collection of some fumes and gases in a work area can go unnoticed by the welders. Concentrated fumes or gases can cause a

FIGURE 3-10 Left-hand threaded fittings are identified with a notch. Courtesy of Larry Jeffus.

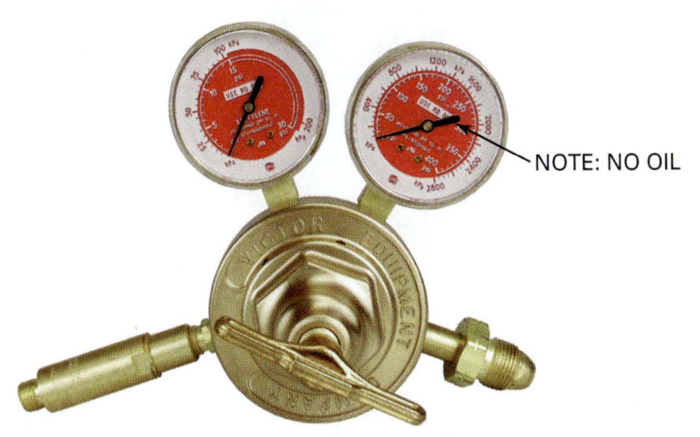

FIGURE 3-11 Never oil a regulator. Courtesy of Victor Equipment, a Thermadyne Company.

If the adjusting screw becomes tight and difficult to turn, it can be removed and cleaned with a dry, oil-free rag. When replacing the adjusting screw, be sure it does not become cross-threaded. Many regulators use a nylon nut in the regulator body, and the nylon is easily cross-threaded.

When welding is finished and the cylinders are turned off, the gas pressure must be released and the adjusting screw backed out. This is required both by federal regulation and to prevent damage to the diaphragm, gauges, and adjusting spring if they are left under a load. A regulator that is left pressurized causes the diaphragm to stretch, the Bourdon tube to straighten, and the adjusting spring to compress. These changes result in a less accurate regulator with a shorter life expectancy.

Welding and Cutting Torches Design and Service

The oxyacetylene hand torch is the most common type of oxyfuel gas torch used in industry. The hand torch may be either a combination welding and cutting torch or a cutting torch only, **Figure 3-12** and **Figure 3-13.**

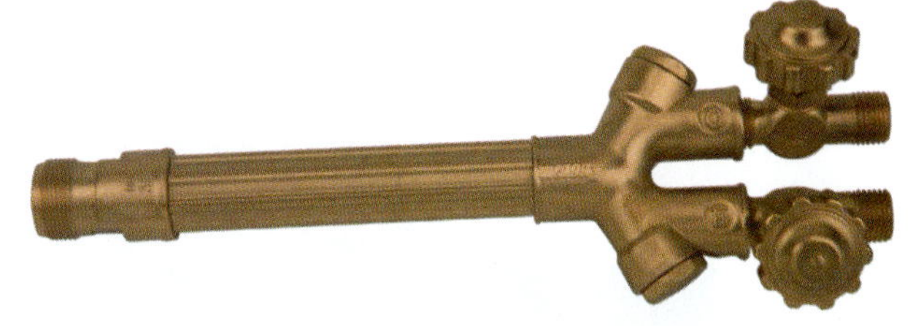

FIGURE 3-12 A torch body or handle used for welding or cutting. Courtesy of Victor Equipment, a Thermadyne Company.

FIGURE 3-13 A torch used for cutting only. Courtesy of Victor Equipment, a Thermadyne Company.

The combination welding and cutting torch offers more flexibility because a cutting head, welding tip, or heating tip can be attached quickly to the same torch body, **Figure 3-14.** Combination torch sets are often used in schools, automotive repair shops, auto body shops, small welding shops, or any other situation where flexibility is needed. The combination torch sets usually are more practical for portable welding since the one unit can be used for both cutting and welding.

Straight cutting torches are usually longer than combination torches. The longer length helps keep the operator farther away from heat and sparks. In addition, thicker material can be cut with greater comfort.

Most manufacturers make torches in a variety of sizes for different types of work. There are small torches for jewelry work, **Figure 3-15**, and large torches for heavy plates. Specialty torches for heating, brazing, or soldering are also available. Some of these use a fuel-air mixture, **Figure 3-16.** Fuel-air torches are often used by plumbers and air-conditioning technicians for brazing and soldering copper pipe and tubing. There are no industrial standards for tip size identification, tip threads, or seats. Therefore, each style, size, and type of torch can be used only with the tips made by the same manufacturer to fit the specific torch.

Torch Care and Use The torch body contains threaded connections for the hoses and tips. These connections must be protected from any damage. Most torch connections are external and made of soft brass that is easily damaged. Some connections, however, are more protected because they have either internal threads or stainless steel threads for the tips.

FIGURE 3-14 A combination welding and cutting torch kit. Courtesy of Victor Equipment, a Thermadyne Company.

FIGURE 3-15 Lightweight torch for small, delicate jobs. Courtesy of Victor Equipment, a Thermadyne Company.

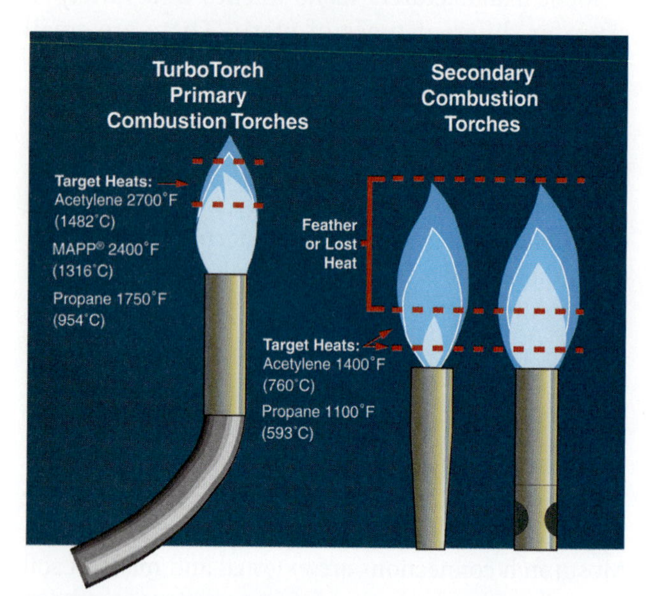

FIGURE 3-16 Some air/gas torches use a special tip that improves the combustion for a hotter, more effective flame. Courtesy of TurboTorch ®/Victor® Equipment, A Thermadyne Company.

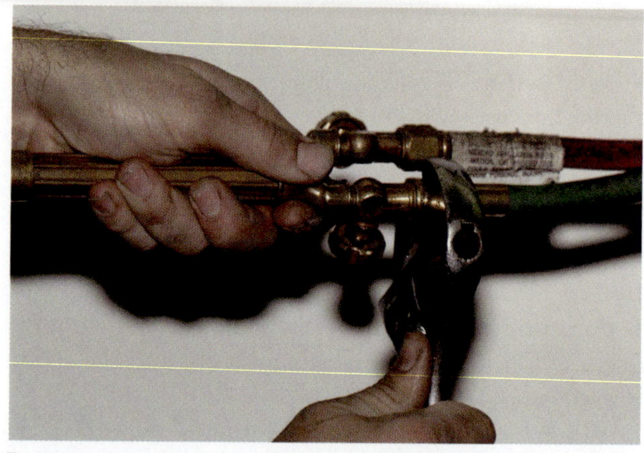

FIGURE 3-17 One hose-fitting nut will protect the threads when the other nut is loosened or tightened. Courtesy of Larry Jeffus.

The best protection against damage and dirt is to leave the tip and hoses connected when the torch is not in use.

Because the hose connections are close to each other, a wrench should never be used on one nut unless the other connection is protected with a hose-fitting nut, **Figure 3-17**.

The hose connections should not leak after they are tightened with a wrench. If leaks are present, the seat should be repaired or replaced. Some torches have removable hose connection fittings so that replacement is possible.

FIGURE 3-18 Check all connections for possible leaks and tighten if necessary. Courtesy of Larry Jeffus.

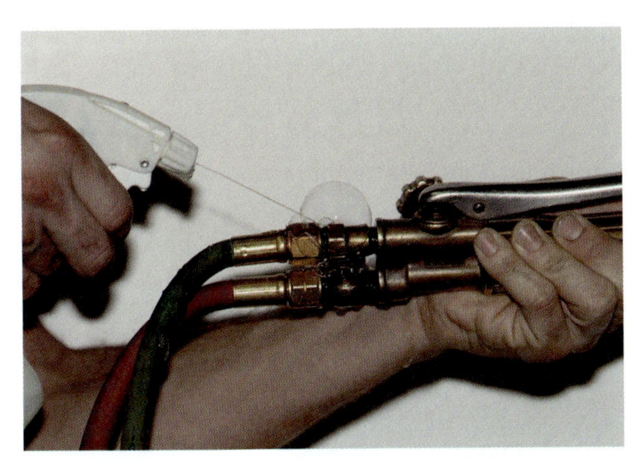

FIGURE 3-19 The torch valves should be checked for leaks, and the valve packing nut should be tightened if necessary. Courtesy of Larry Jeffus.

The valves should be easily turned on and off and should stop all gas flowing with minimum finger pressure. To find leaking valve seats, set the regulators to a working pressure. With the torch valves off, spray the tip with a leak-detecting solution. The presence of bubbles indicates a leaking valve seat, **Figure 3-18**. The gas should not leak past the valve stem packing when the valve is open or when it is closed. To test leaks around the valve stem, set the regulator to a working pressure. With the valves off, spray the valve stem with a leak-detecting solution and watch for bubbles, indicating a leaking valve packing. The valve stem packing can now be tested with the valve open. Place a finger over the hole in the tip and open the valve. Spray the stem and watch for bubbles, which would indicate a leaking valve packing, **Figure 3-19**. If either test indicates a leak, the valve stem packing nut can be tightened until the leak stops. After the leak stops, turn the valve knob. It should still turn freely. If it does not, or if the leak cannot be stopped, replace the valve packing.

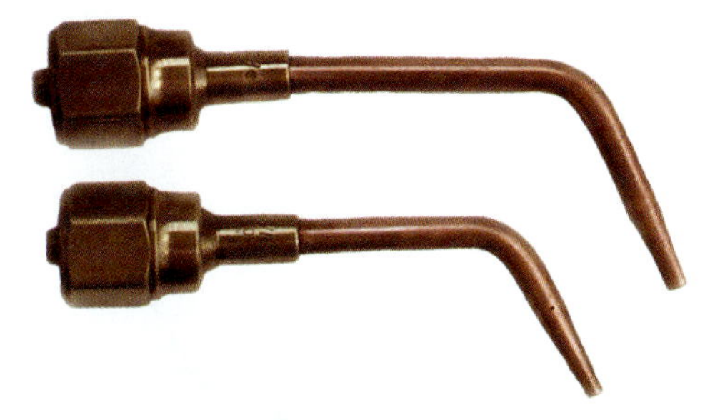

FIGURE 3-20 A variety of tip styles and sizes for one torch body. Courtesy of Larry Jeffus.

Tip Cleaner Standard Set			
	Use Cleaner	For Drill	
			77 = .0160"
Smallest	1	77-76	(0.4064 mm)
↓	2	75-74	
	3	73-72-71	
	4	70-69-68	
	5	67-66-65	
	6	64-63-62	
	7	61-60	
	8	59-58	
	9	57	
	10	56	
	11	55-54	
	12	53-52	49 = .0730"
Largest	13	51-50-49	(1.8542 mm)

TABLE 3-2 Tip Cleaner Size Compared to Drill Size Found on Most Standard Tip-Cleaning Sets.

The valve packing and valve seat can be easily repaired on most torches by following the instructions given in the repair kit. On some torches, the entire valve assembly can be replaced, if necessary.

Welding and Heating Tips

Because no industrial standard tip size identification system exists, the student must become familiar with the size of the orifice (hole) in the tip and the thickness range for which it can be used. Comparing the overall size of the tip can be done only for tips made by the same manufacturer for the same type and style of torch, **Figure 3-20**. Learning a specific manufacturer's system is not always the answer because on older, worn tips the orifice may have been enlarged by repeated cleaning.

Tip sizes can be compared to the numbered drill size used to make the hole, **Table 3-2**. The sizes of tip cleaners are given according to the drill size of the hole they fit. By knowing the tip cleaner size commonly used to clean a tip, the welder can find

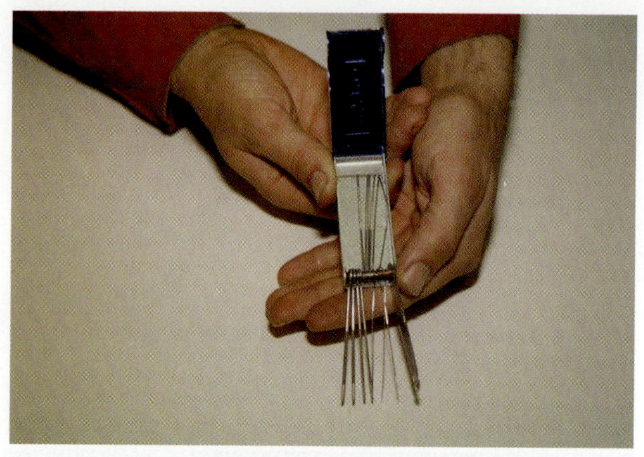

FIGURE 3-21 Standard set of tip cleaners. Courtesy of Larry Jeffus.

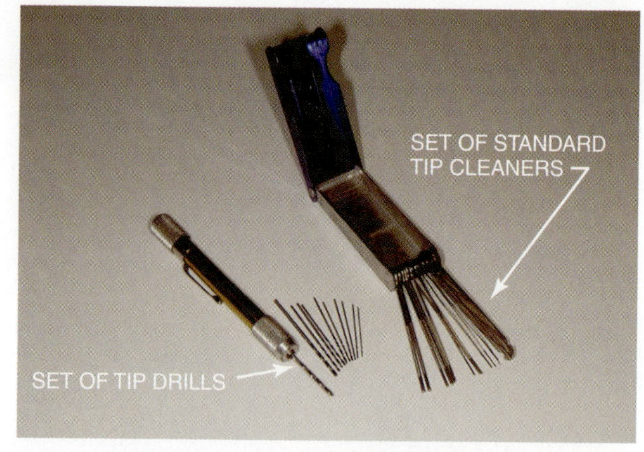

FIGURE 3-23 Tools used to repair tips. Courtesy of Larry Jeffus.

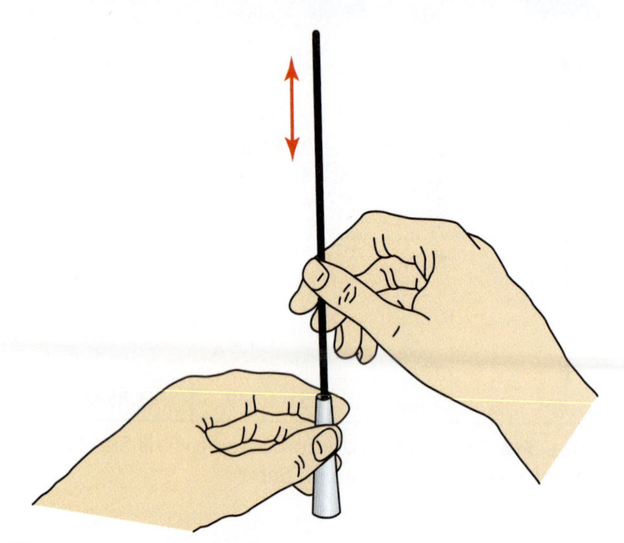

FIGURE 3-22 Cleaning a tip with a standard tip cleaner.

the same-size tip made by a different manufacturer. The tip size can also be determined by trial and error.

On some torch sets, each tip has its own mixing chamber. On other torch sets, however, one mixing chamber may be used with a variety of tip sizes.

Tip Care and Use Torch tips may have metal-to-metal seals, or they may have an O ring or a gasket between the tip and the torch seat. Metal-to-metal seal tips must be tightened with a wrench. Tips with an O ring or a gasket are tightened by hand. Using the wrong method of tightening the tip fitting may result in damage to the torch body or the tip.

Dirty tips can be cleaned using a set of tip cleaners. Using the file provided in the tip-cleaning set, **Figure 3-21**, file the end of the tip smooth and square. Next, select the size of tip cleaner that fits easily into the orifice. The tip cleaner is a small, round file and should only be moved in and out of the orifice a few times, **Figure 3-22**. Be sure the tip cleaner is straight and that it is held in a steady position to prevent it from bending or breaking off in the tip. Excessive use of the tip cleaner tends to ream the orifice, making it too large. Therefore, use the tip cleaner only as required. Once the tip is cleaned, turn on the oxygen for a moment to blow out any material loosened during the cleaning.

Damaged tips or tips with cleaners broken in them can be reconditioned, but they require a good deal of work and some specialized tools, **Figure 3-23**.

Reverse Flow and Flashback Valves

The purpose of the reverse flow valve is to prevent gases from accidentally flowing through the torch and into the wrong hose. If the gases being used are allowed to mix in the hose or regulator, they might explode. The reverse flow valve is a spring-loaded check valve that closes when gas pressure from a back flow tries to occur through the torch valves, **Figure 3-24**. Some torches have reverse flow valves built into the torch body, but most torches must have these safety devices added. If the torch does not come with a reverse flow valve, it must be added to either the torch end or regulator end of the hose.

A reverse flow of gas will occur if the torch is not turned off or bled properly. The torch valves must be opened one at a time so that the gas pressure in that hose will be vented into the atmosphere and not into the other hose, **Figure 3-25**.

CAUTION

If both valves are opened at the same time, one gas may be pushed back up the hose of the other gas.

A reverse flow valve will not stop the flame from a flashback from continuing through the hoses. A flashback arrestor will do the job of a reverse flow valve, and it will stop the flame of a flashback, Figure 3-26. The flashback arrestor is designed to quickly stop the flow of gas during a flashback. These valves work on a similar principle to the gas valve at a service station. They are very sensitive to any back pressure in the hose and stop the flow if any back pressure is detected.

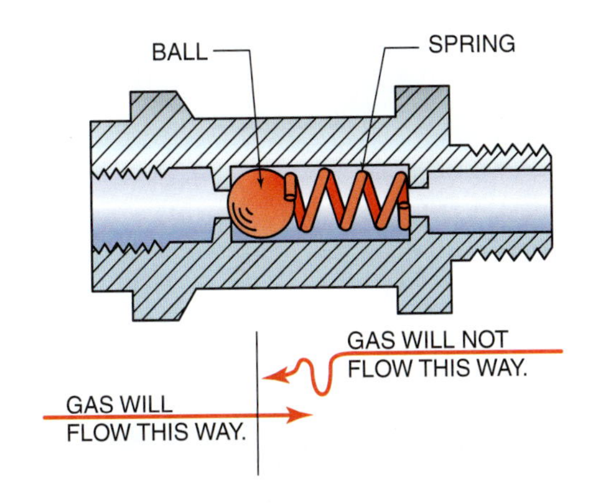

FIGURE 3-24 Reverse flow valve only. Photo courtesy of ESAB Welding & Cutting Products.

Care of the Reverse Flow Valve and Flashback Arrestor Both devices must be checked on a regular basis to see that they are working correctly. The internal valves may become plugged with dirt or they may become sticky and not operate correctly. To test the reverse flow valve, you can try to blow air backwards through the valve. To test the flashback arrestor, follow the manufacturer's recommended procedure. If the safety device does not function correctly, it should be replaced.

Hoses and Fittings

Most welding hoses used today are molded together as one piece and are referred to as Siamese hose. Hoses that are not of the Siamese type, or hose ends that have separated, may be taped together. When taping the hoses, they must not be wrapped solidly. The hoses should be wrapped for about 2 in. (51 mm) every 12 in. (305 mm), allowing the colors of the hose to be seen.

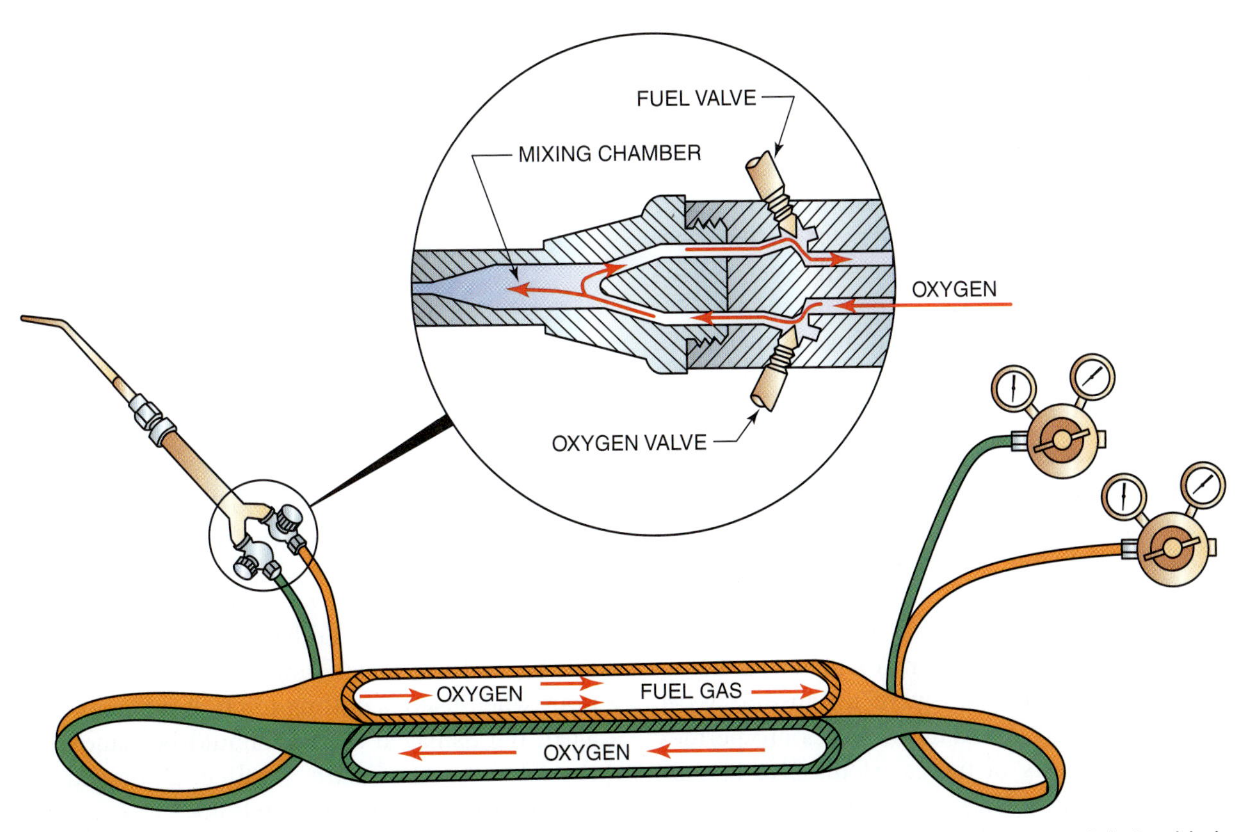

FIGURE 3-25 Gas may flow back up the hose if both valves are opened at the same time when the system is being bled down after use. Installing reverse flow valves on the torch can prevent this from occurring.

FIGURE 3-26 (A) Acetylene. (B) Oxygen combination flashback arrestors and check valves. (C) Replacement cartridge for flashback arrestor. (D) Torch designed with flashback arrestors and check valves built into the torch body. (A & B) Courtesy of ESAB Welding & Cutting Products. (C & D) Courtesy of Victor Equipment, a Thermadyne Company.

Acetylene and other fuel gas hoses must be red and have left-hand threaded fittings. Oxygen hoses must be green and have right-hand threaded fittings.

Hoses are available in four sizes: 3/16 in. (4.8 mm), 1/4 in. (6 mm), 5/16 in. (8 mm), and 3/8 in. (10 mm). The size given is the inside diameter of the hose. Larger sizes offer less resistance to gas flow and should be used where long hose lengths are required. The smaller sizes are more flexible and easier to handle for detailed work.

The three sizes of hose end fittings available are A (small), B (standard), and C (large). The three sizes are made to fit all hose sizes.

Hose Care and Use When hoses are not in use, the gas must be turned off and the pressure bled off. Turning off the equipment and releasing the pressure prevents any undetected leaks from causing a fire or an explosion. This action also eliminates a dangerous situation that would be created if a hose were cut by equipment or materials being handled by workers who were unfamiliar with welding equipment. In addition, hoses are permeable to gases (ability of the gas to dissolve into or through the hose walls). Thus, gases left under pressure for long periods of time can migrate through the hose walls and mix with each other,

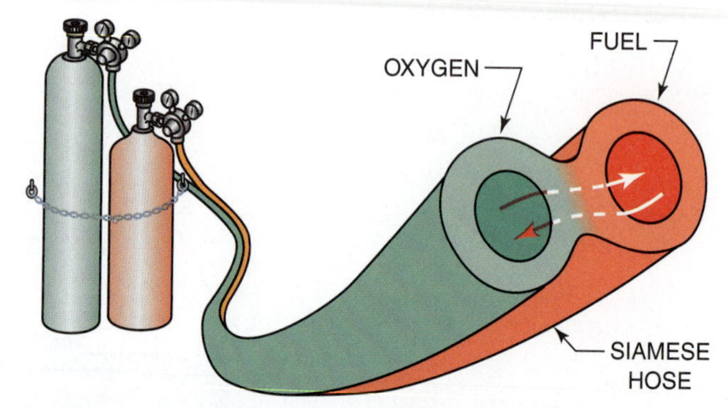

FIGURE 3-27 Gas left under pressure may migrate through the hose walls.

Figure 3-27. If the gases mix and the torch is lit without first purging the lines, the hoses can explode. For this reason, if the welder is not certain that the hoses were bled, it is recommended that they be purged before the torch is lit.

Hoses are resistant to burns, but they are not burn-proof. They should be kept out of direct flame, sparks, and hot metal. You must be especially cautious when using a cutting torch. If it becomes damaged, the damaged section should be removed and the hose repaired with a splice. Damaged hoses should never be taped to stop leaks.

Hoses should be checked periodically for leaks. To test a hose for leaks, adjust the regulator to a

FIGURE 4-18 The hot end of the filler rod is protected from the atmosphere by the outer flame envelope. Courtesy of Larry Jeffus.

FIGURE 4-20 Filler metal being incorrectly added by allowing the rod to melt and drip into the molten weld pool. Courtesy of Larry Jeffus.

FIGURE 4-19 Filler metal being correctly added by dipping the rod into the leading edge of the molten weld pool. Courtesy of Larry Jeffus.

The end of the welding rod should always be kept inside the protective envelope of the flame, **Figure 4-18**. The hot end of the welding rod oxidizes each time it is removed from the protection of the flame. This oxide is deposited in the molten weld pool, causing sparks and a weak or brittle weld.

When the rod is added to the molten weld pool, the flame can be moved back as the end of the rod is dipped into the leading edge of the molten weld pool, **Figure 4-19**. If the torch is not moved back, the rod may melt and drip into the molten weld pool; this is an incorrect way of adding filler metal, **Figure 4-20**. The major problems with adding rod in this manner are (1) the drop of metal tends to overheat, resulting in important alloys being burned out; (2) the metal cannot always be added where it is needed; and (3) the method works only in the flat position. When dipping the rod into the molten weld pool, if the rod touches the hot metal around the molten weld pool, it will stick. When this happens, move the flame directly to the end of the rod to melt and free it. Turn off the cylinders, bleed the hoses, back out the regulator adjusting screw, and clean up your work area when you are finished.

Complete a copy of the "Student Welding Report" listed in Appendix I or provided by your instructor. ◆

EXPERIMENT 4-3

Effect of Rod Size on the Molten Weld Pool

Use a properly lit and adjusted torch, 6 in. (152 mm) of 16 gauge mild steel, and three different diameters of RG45 gas welding rods, 1/8 in. (3 mm), 3/32 in. (2.4 mm), and 1/16 in. (2 mm), by 36 in. (914 mm) long. In this experiment, you will observe the effect on the molten weld pool of changing the size of filler metal. You also will practice adding the filler metal to the molten weld pool.

Starting with the 1/8-in. (3-mm) filler metal, make a weld 6 in. (152 mm) long. Next to this weld, make another one with the 3/32-in. (2.4-mm) rod, then one with the 1/16-in. (2-mm) rod. Try to keep the angle, speed, height, and pattern of the torch the same during each of the welds. Observe the differences in each of the weld sizes.

Complete a copy of the "Student Welding Report" listed in Appendix I or provided by your instructor. ◆

PRACTICE 4-3

Stringer Bead, Flat Position

Repeat Experiment 4-3 until you have mastered a straight and uniform weld with any of the three sizes of filler rod.

Joining two or more clean pieces of metal to form a welded joint is the next step in learning to

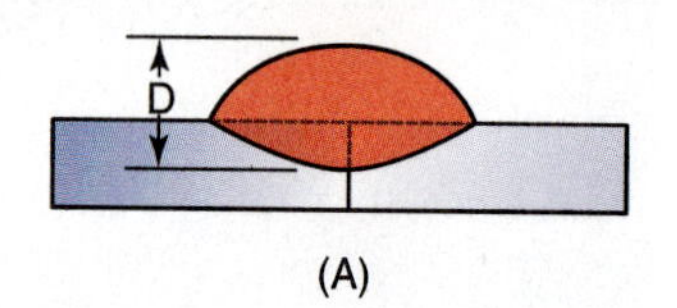

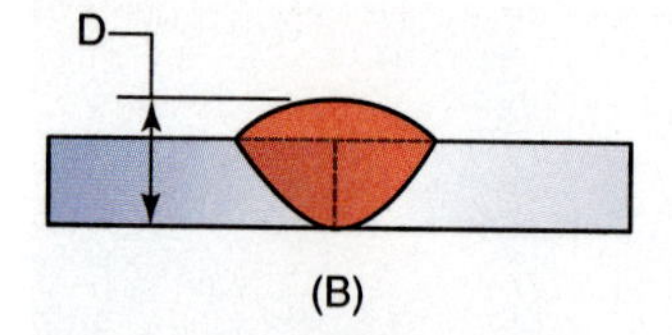

FIGURE 4-21 Welds (A) and (B) both have approximately the same strength — but only if the reinforcement does not have to be removed.

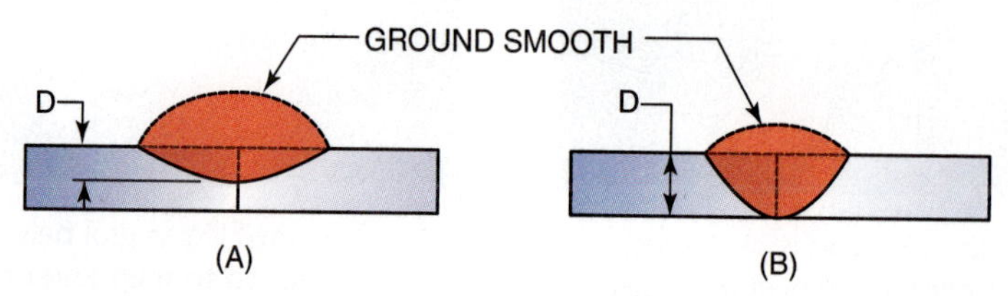

FIGURE 4-22 If the reinforcement on both welds is removed, weld (B) would be the stronger weld.

weld. The joints must be uniform in width and reinforcement so that they will have maximum strength. For each type of gas welded joint, the amount of penetration required to give maximum strength will vary and may not be 100%, **Figure 4-21**. For example, in thin sheet metal there is usually enough reinforcement on the weld to give the weld adequate strength. But if that reinforcement has to be removed, then 100% penetration is important, **Figure 4-22**. Some joints, such as the lap joint, may never need 100% penetration. However, they do need 100% fusion. Turn off the cylinders, bleed the hoses, back out the regulator adjusting screw, and clean up your work area when you are finished.

Complete a copy of the "Student Welding Report" listed in Appendix I or provided by your instructor. ◆

Outside Corner Joint

The flat outside corner joint can be made with or without the addition of filler metal. This joint is one of the easiest welded joints to make. If the sheets are tacked together properly, the addition of filler metal is not needed. However, if filler metal is added, it should be added uniformly, as in the stringer beads in Practice 4-3.

PRACTICE 4-4

Outside Corner Joint, Flat Position

Using a properly lit and adjusted torch, two clean pieces of 16 gauge mild steel 6 in. (152 mm) long, and filler metal, you will make a flat outside corner welded joint, **Figure 4-23**.

Place one of the pieces of metal in a jig or on a firebrick and hold or brace the other piece of metal vertically on it, as shown in **Figure 4-24** and **Figure 4-25**. Tack the ends of the two sheets together. Then set it upright and put two or three more tacks on the joint, **Figure 4-26**. Holding the torch as shown in **Figure 4-27**, make a uniform weld along the joint. Repeat this until the weld can be made without defects. Turn off the cylinders, bleed the hoses, back out the regulator adjusting screw, and clean up your work area when you are finished.

Complete a copy of the "Student Welding Report" listed in Appendix I or provided by your instructor. ◆

Butt Joint

The flat butt joint is a welded joint and one of the easiest to make. To make the butt joint, place two clean pieces of metal flat on the table and tack weld both ends together, as illustrated in **Figure 4-28**. Tack welds may also be placed along the joint before welding begins. Point the torch so that the flame is distributed equally on both sheets. The flame is to be in the direction that the weld is to progress. If the sheets to be welded are of different sizes or thicknesses, the torch should be pointed so that both pieces melt at the same time, **Figure 4-29**.

When both sheet edges have melted, add the filler rod in the same manner as in Practice 4-3.

PRACTICE 4-5

Butt Joint, Flat Position

Using a properly lit and adjusted torch, two clean pieces of 16 gauge mild steel 6 in. (152 mm) long, and filler metal, you will make a welded butt joint, **Figure 4-30**.

Place the two pieces of metal in a jig or on a fire brick and tack weld both ends together. The tack

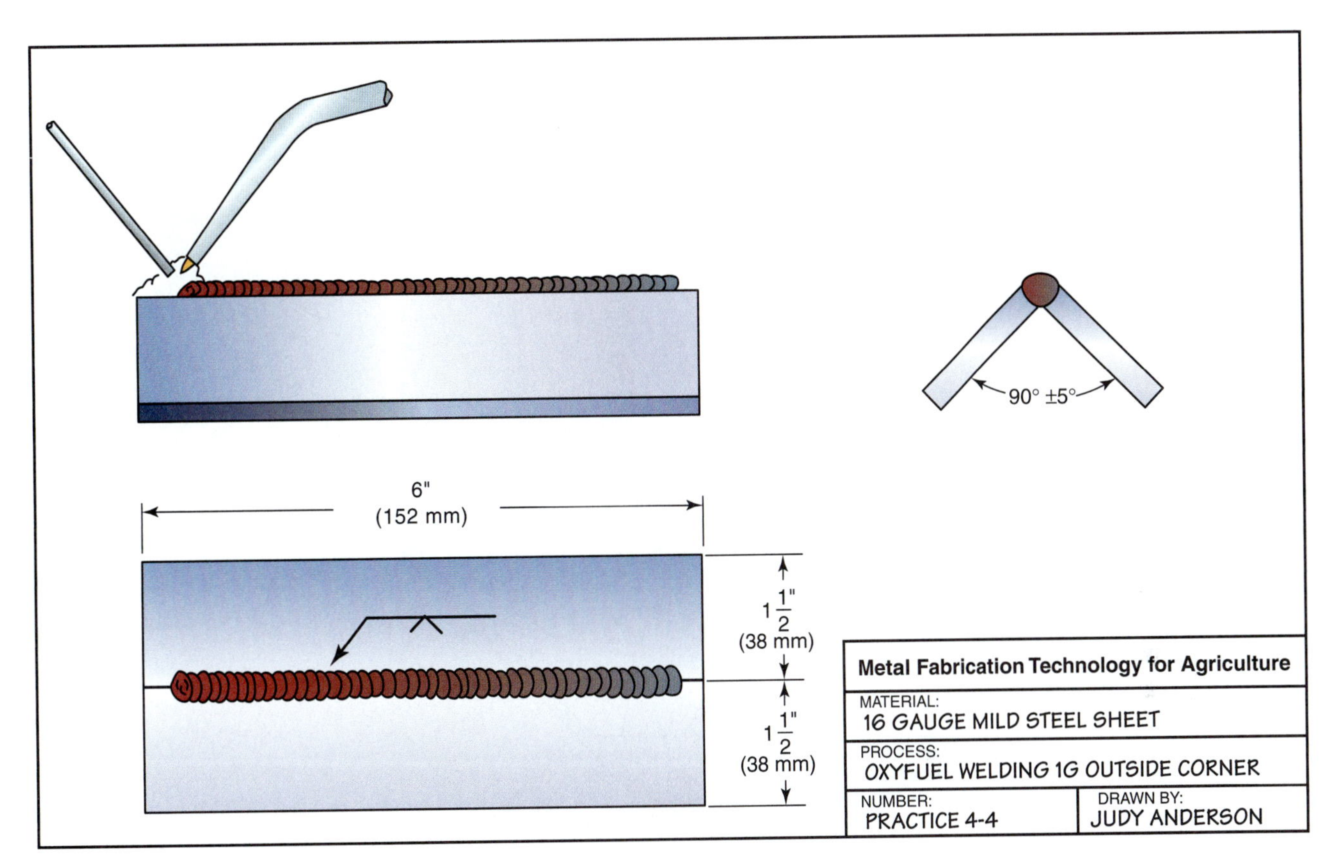

FIGURE 4-23 Outside corner.

FIGURE 4-24 Angle iron jig for holding metal so it can be tack welded.

on the ends can be made by simply heating the ends and allowing them to fuse together or by placing a small drop of filler metal on the sheet and heating the filler metal until it fuses to the sheet. The latter method is especially convenient if you have to use one hand to hold the sheets together and the other to hold the torch. After both ends are tacked together, place one or two small tacks along the joint to prevent warping during welding.

With the sheets tacked together, start welding from one end to the other using the technique learned in Practice 4-3. Repeat this weld until you can make a welded butt joint that is uniform in width and reinforcement and has no visual defects.

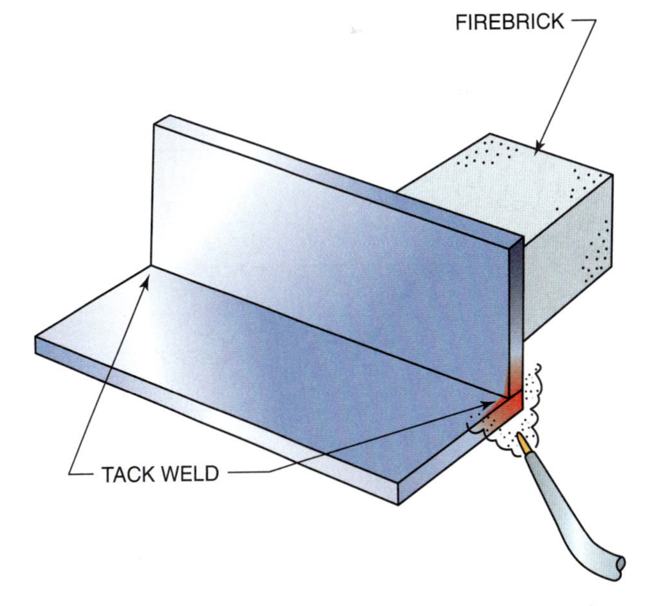

FIGURE 4-25 Tack welding the outside corner joint using a firebrick to support the metal.

The penetration of this practice weld may vary. Turn off the cylinders, bleed the hoses, back out the regulator adjusting screw, and clean up your work area when you are finished.

Complete a copy of the "Student Welding Report" listed in Appendix I or provided by your instructor. ◆

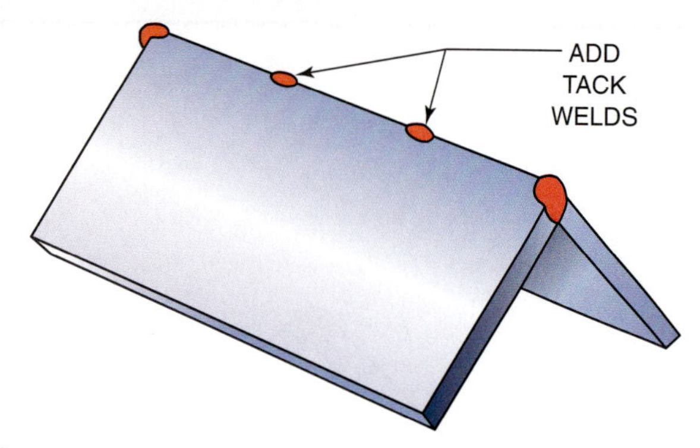

FIGURE 4-26 Tack welding.

FIGURE 4-27 Outside corner joint. Courtesy of Larry Jeffus.

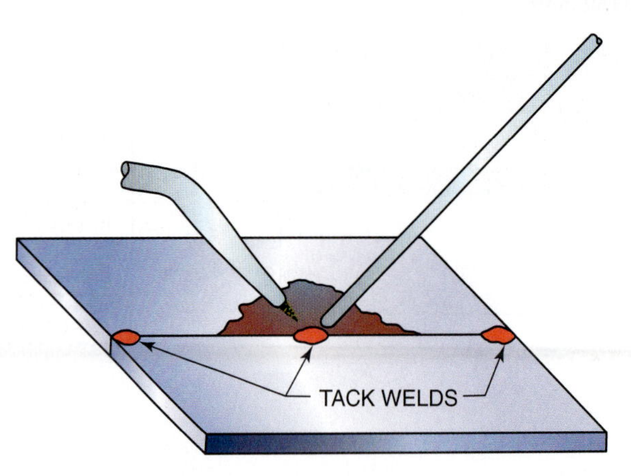

FIGURE 4-28 Making a tack weld.

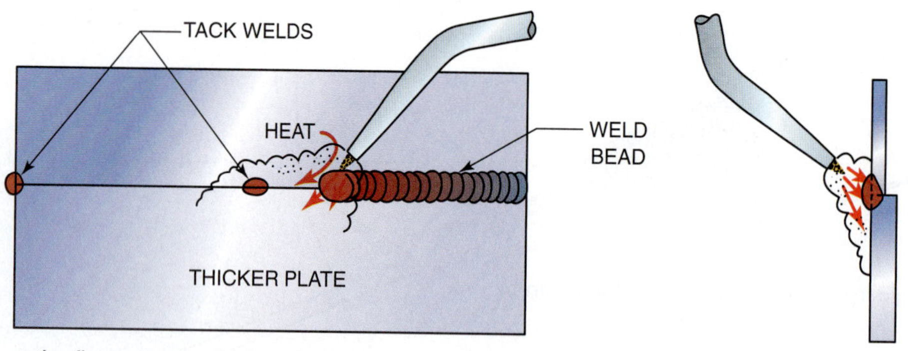

FIGURE 4-29 Direct the flame on the thicker plate.

Lap Joint

The flat lap joint can be easily welded if some basic manipulations are used. When heating the two clean sheets, caution must be exercised to ensure that both sheets start melting at the same time. Heat is not distributed uniformly in the lap joint, **Figure 4-31.** Because of this difference in heating rate, the flame must be directed on the bottom sheet and away from the metal top sheet, **Figure 4-32.** The filler rod should be added to the top sheet. Gravity will pull the molten weld pool down to the bottom sheet, so it is therefore not necessary to put metal on the bottom sheet. If the filler metal is not added to the top sheet or if it is not added fast enough, surface tension will pull the molten weld pool back from the joint, **Figure 4-33.** When this happens, the rod should be added directly into this notch, and it will close. The weld appearance and strength will not be affected.

PRACTICE 4-6

Lap Joint, Flat Position

Using a properly lit and adjusted torch, two clean pieces of 16 gauge mild steel 6 in. (152 mm) long, and filler metal, you will make a welded lap joint,

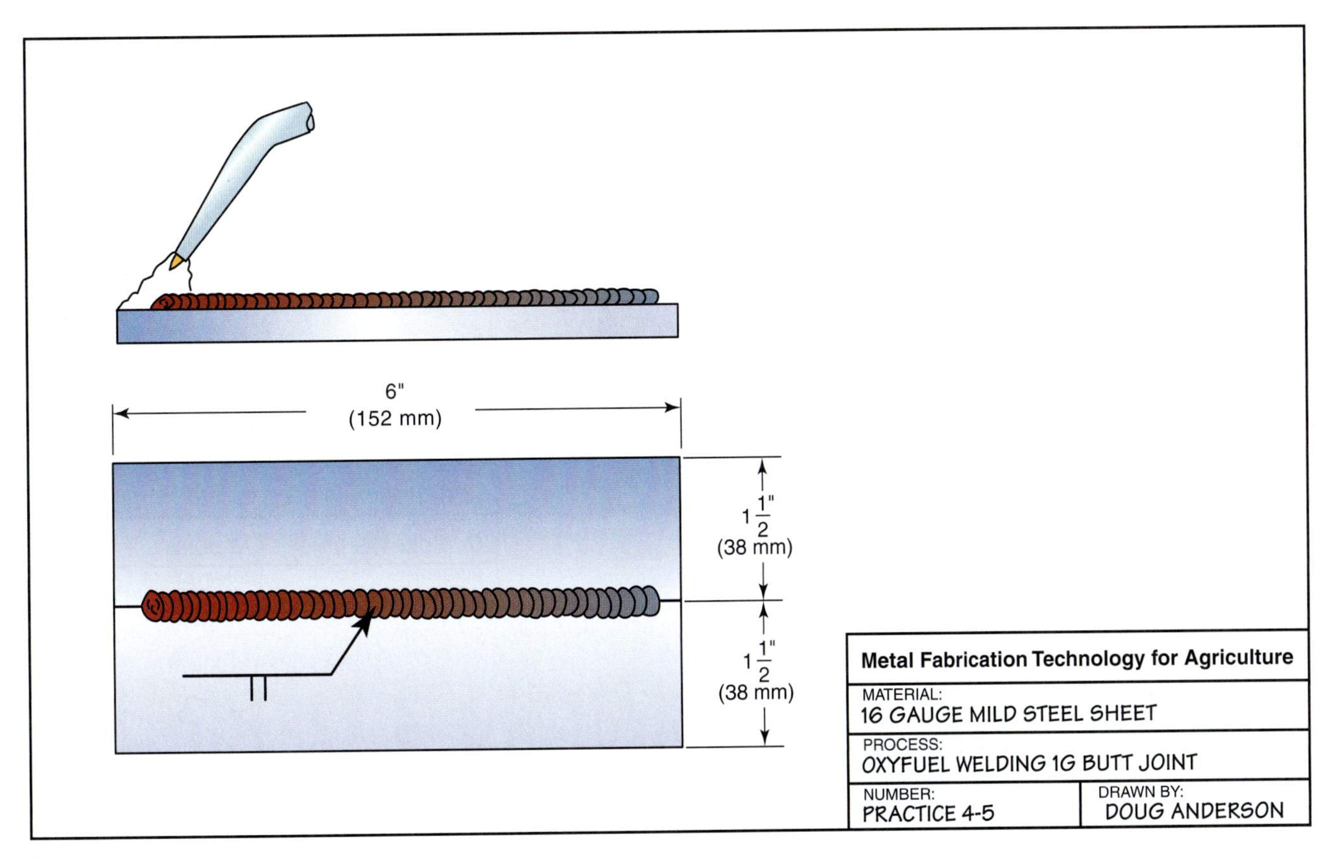

FIGURE 4-30 Butt joint.

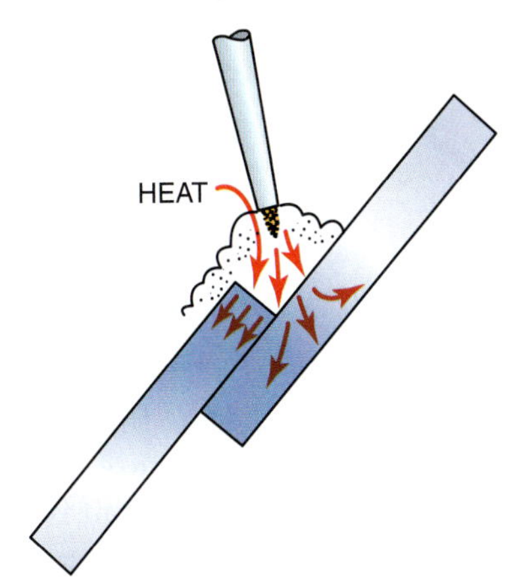

FIGURE 4-31 Heat is conducted away faster in the bottom plate, resulting in the top plate melting more quickly.

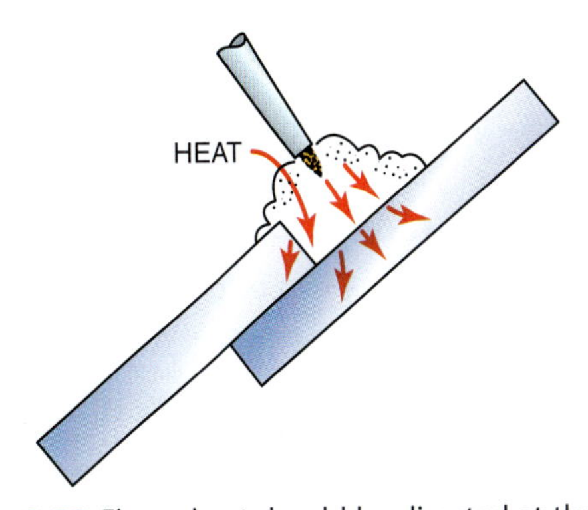

FIGURE 4-32 Flame heat should be directed at the bottom plate to compensate for thermal conductivity.

Figure 4-34. Place the two pieces of metal on a fire brick and tack both ends, as shown in **Figure 4-35** and **Figure 4-36**. Make two or three more tack welds. Starting at one end, make a uniform weld along the joint. Both sides of the joint can be welded. Repeat this practice until the weld can be made without defects. Turn off the cylinders, bleed the hoses, back out the regulator adjusting screw, and clean up your work area when you are finished.

Complete a copy of the "Student Welding Report" listed in Appendix I or provided by your instructor. ◆

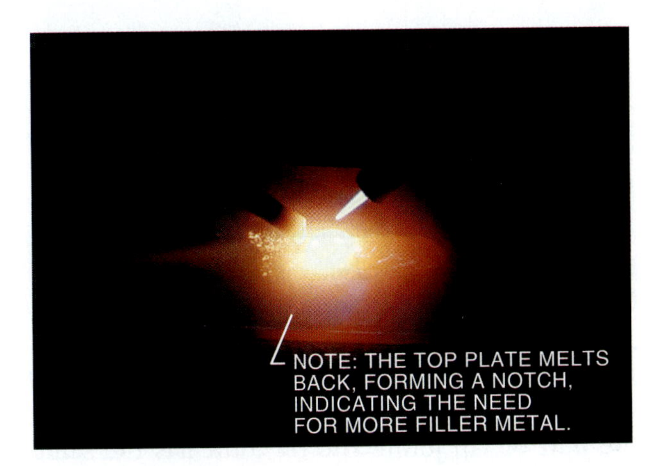

FIGURE 4-33 Add filler metal. Courtesy of Larry Jeffus.

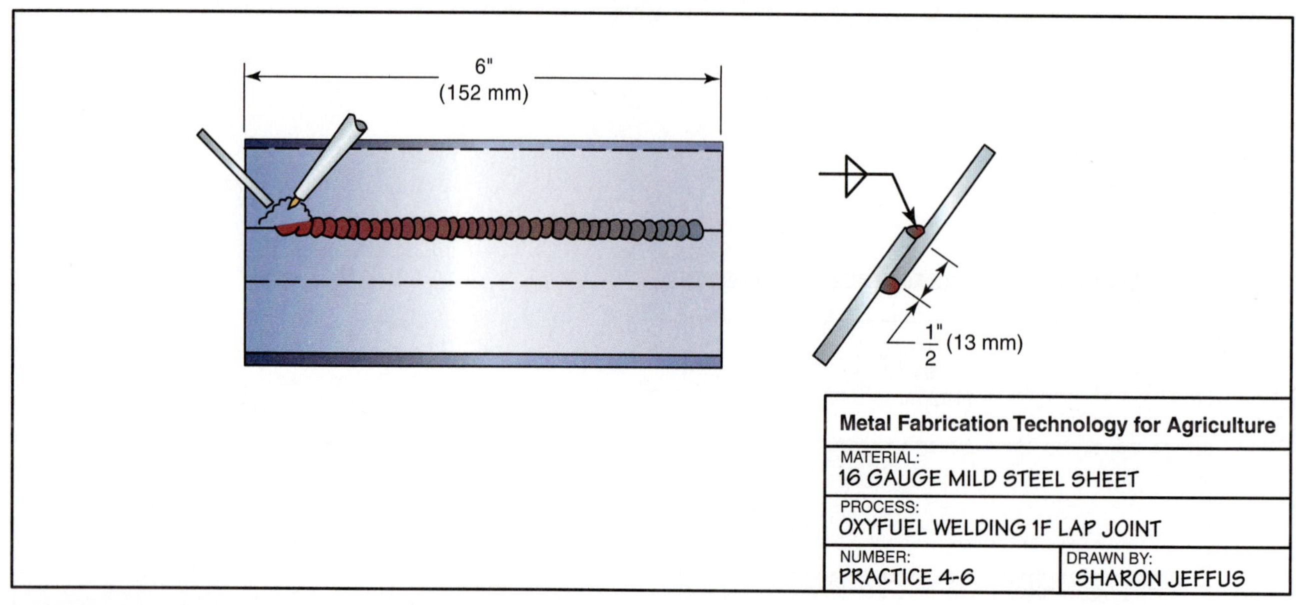

FIGURE 4-34 Lap joint.

FIGURE 4-35 Heating the joint before tacking. Courtesy of Larry Jeffus.

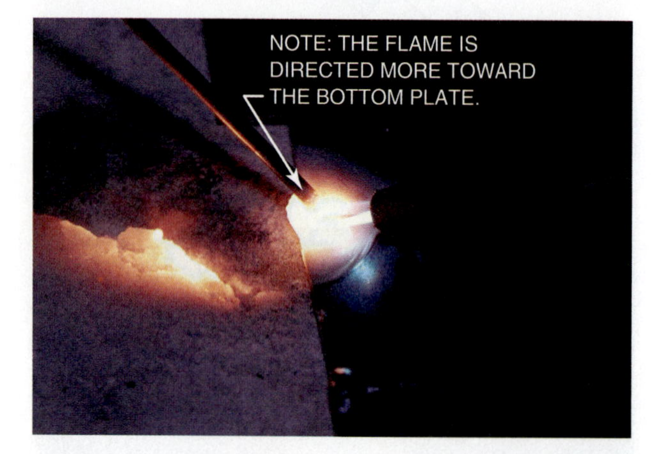

FIGURE 4-36 Filler rod is added after both pieces are heated to a melt. Courtesy of Larry Jeffus.

Tee Joint

The flat tee joint is more difficult to make than the butt or lap joint. The tee joint has the same problem with uneven heating that the lap joint does. It is important to hold the flame so that both sheets melt at the same time, **Figure 4-37**. Another problem that is unique to the tee joint is that a large percentage of the welding heat is reflected back on the torch. This reflected heat can cause even a properly cleaned and adjusted torch to backfire or pop. To help prevent this from happening, angle the torch more in the direction of the weld travel. Because of the slightly restricted atmosphere of the tee joint, it may be necessary to adjust the flame so that it is somewhat oxidizing. The beginning student should not be overly concerned with this.

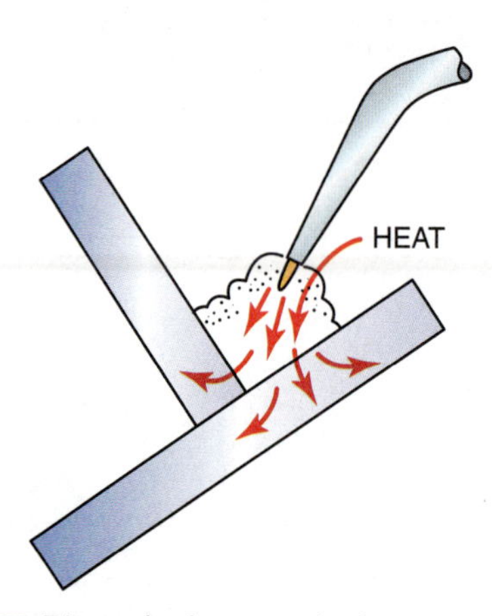

FIGURE 4-37 Direct the heat on the bottom plate to equalize the heating rates.

PRACTICE 4-7

Tee Joint, Flat Position

Using a properly lit and adjusted torch, two clean pieces of 16 gauge mild steel 6 in. (152 mm) long, and filler metal, you will weld a flat tee joint, **Figure 4-38**.

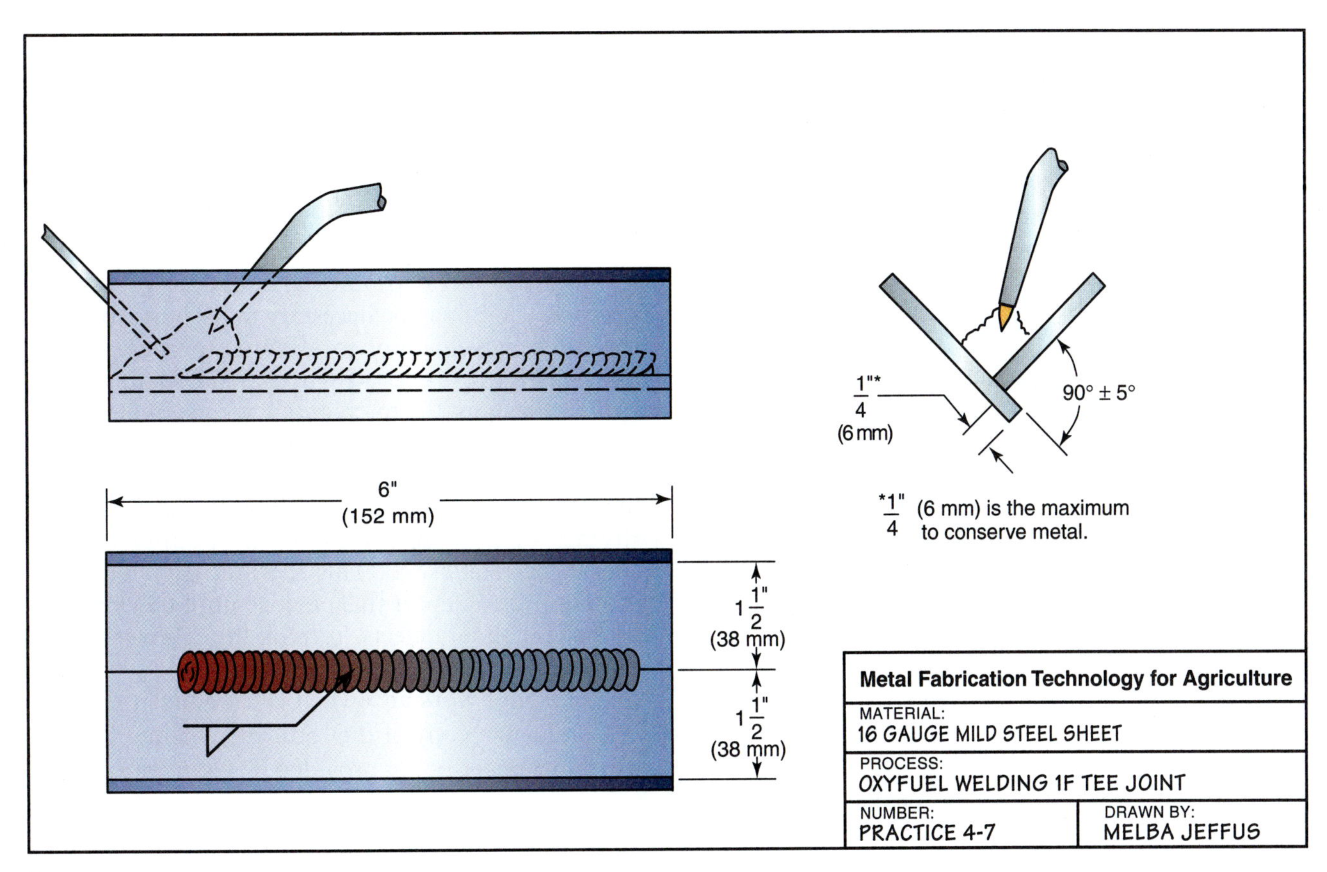

FIGURE 4-38 Tee joint.

Place the first piece of metal flat on a firebrick and hold or brace the second piece vertically on the first piece. The vertical piece should be within 5° of square to the bottom sheet. Tack the two sheets at the ends. Then put two or three more tacks along the joint and brace the tee joint in position. Turn off the cylinders, bleed the hoses, back out the regulator adjusting screw, and clean up your work area when you are finished.

Complete a copy of the "Student Welding Report" listed in Appendix I or provided by your instructor. ◆

Out-of-position Welding

A part to be welded cannot always be positioned so that it can be welded in the flat position. Whenever a weld is performed in a position other than flat, it is said to be out-of-position welding. Welds made in the vertical, horizontal, or overhead position are out of position and somewhat more difficult than flat welds.

Vertical Welds

A vertical weld is the most common out-of-position weld that a welder is required to perform. When making a vertical weld, it is important to control the size of the molten weld pool. If the molten weld pool size increases beyond that which the shelf will support, **Figure 4-39**, the molten weld pool will overflow and drip down the weld. These drops, when cooled, look like the drips of wax on a candle. To prevent the molten weld pool from dripping, the trailing edge of the molten weld pool must be watched. The trailing edge will constantly be solidifying, forming a new shelf to support the molten weld pool as the weld progresses upward, **Figure 4-40**. Small molten weld pools are less likely than large ones to drip.

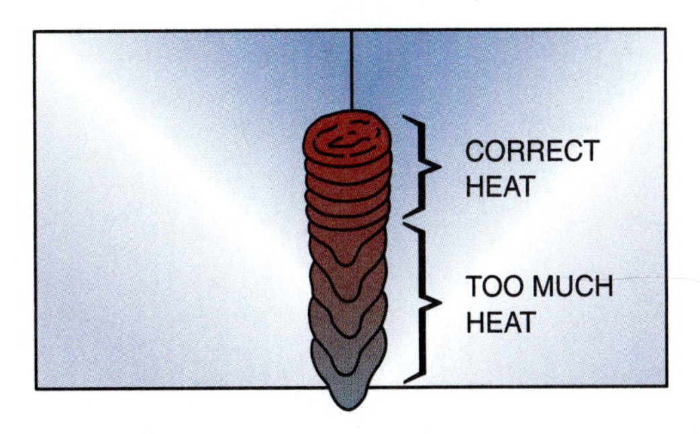

FIGURE 4-39 Vertical weld showing effect of too much heat.

The less vertical the sheet, the easier the weld is to make, but the type of manipulation required is the same. Welding on a sheet at a 45° angle requires the same manipulation and skill as welding on a vertical sheet. However, the speed of manipulation is slower, and the skill is less critical than at 90° vertical. This welding technique should be mastered at a 45° angle. Then the angle of the sheet is increased until it is possible to make totally vertical welds. Each practice weld in this section should be started on an incline, and, as skill is gained, the angle should be increased until the sheets are vertical.

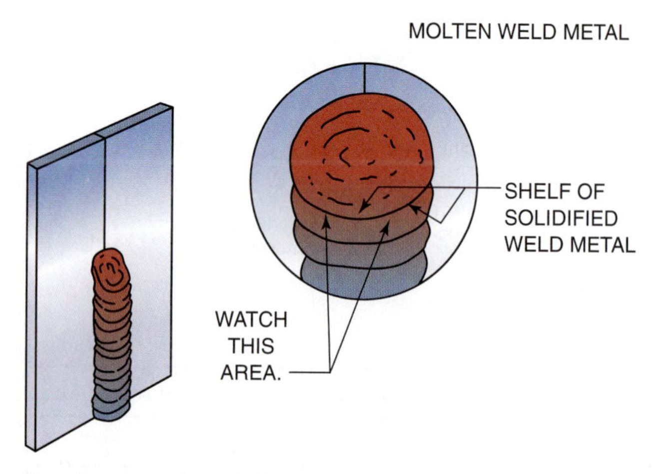

FIGURE 4-40 Watch the trailing edge to see that the molten pool stays properly supported on the shelf.

PRACTICE 4-8

Stringer Bead at a 45° Angle

Using a properly lit and adjusted torch, two clean pieces of 16 gauge mild steel 6 in. (152 mm) long, and filler metal, you will make a bead at a 45° angle.

The filler metal should be added as you did in Practice 4-3. It may be necessary to flash the torch off the molten weld pool to allow it to cool, **Figure 4-41A**, **Figure 4-41B**, **Figure 4-41C**, and **Figure 4-41D**. Flashing the torch off allows the molten weld pool to cool by moving the hotter inner cone away from the molten weld pool itself. While still protecting the molten metal with the outer flame envelope, a rhythm of moving the torch and adding the rod should be established. This rhythm helps make the bead uniform. Repeat this practice until the weld can be made without defects. Turn off the cylinders, bleed the hoses, back out the regulator adjusting screw, and clean up your work area when you are finished.

Complete a copy of the "Student Welding Report" listed in Appendix I or provided by your instructor. ◆

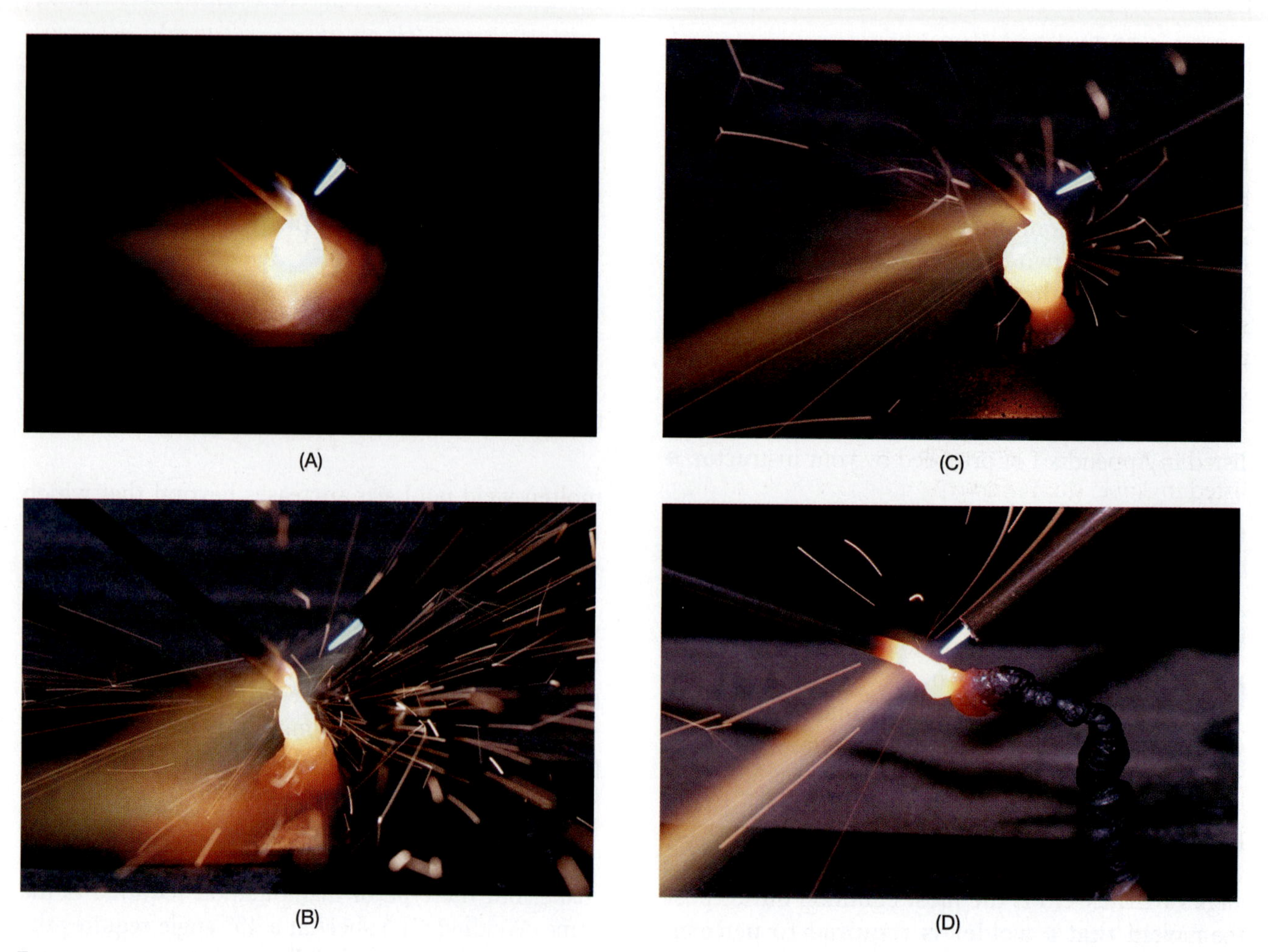

FIGURE 4-41 By flashing the flame off and controlling the pool size, a weld can be built up (A), and up (B), and over (C), and over (D). Courtesy of Larry Jeffus.

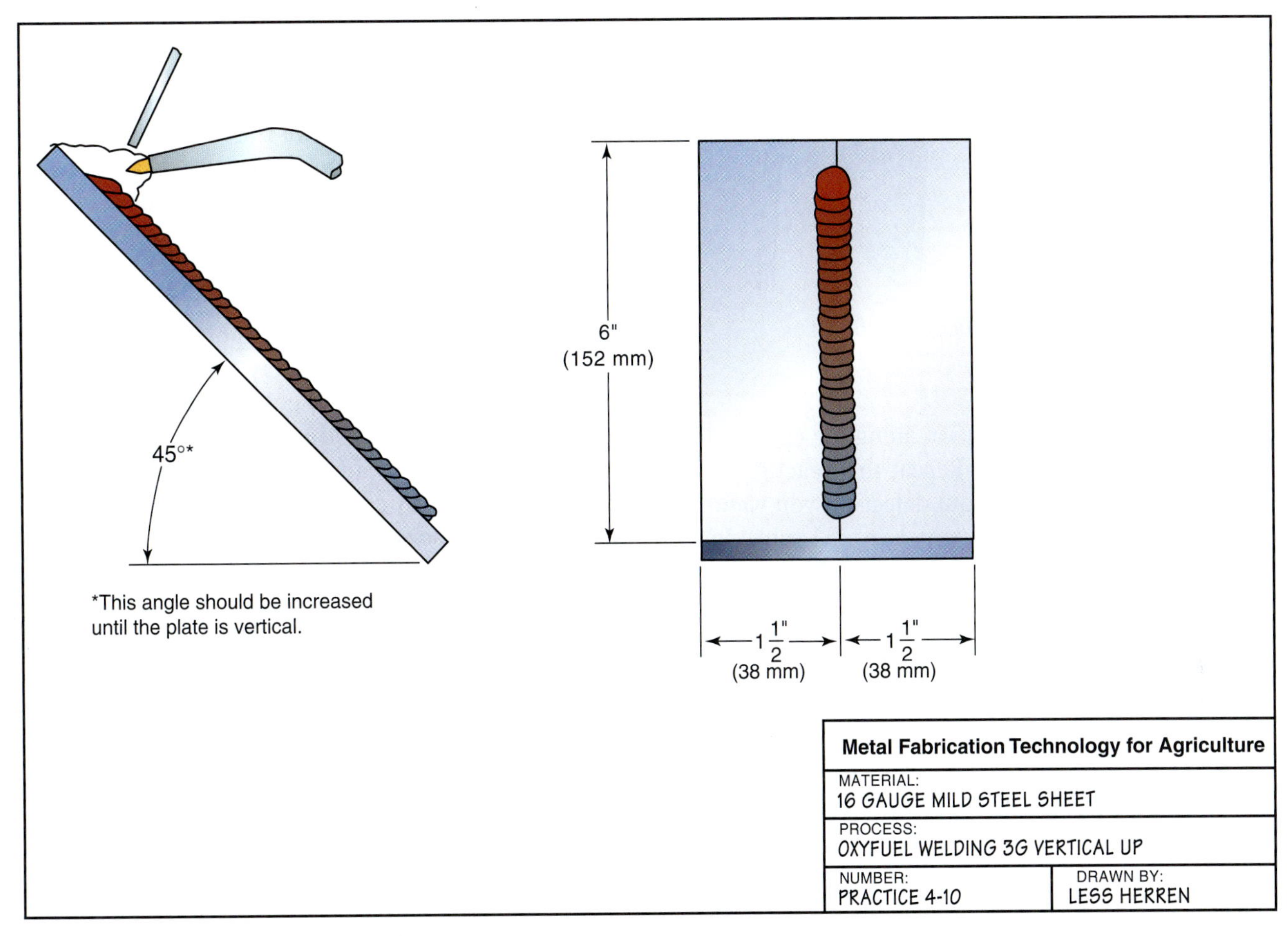

FIGURE 4-42 Butt joint at a 45° angle.

PRACTICE 4-9

Stringer Bead, Vertical Position

Repeat Practice 4-8 until you have mastered a straight and uniform weld bead in a vertical position. Turn off the cylinders, bleed the hoses, back out the regulator adjusting screw, and clean up your work area when you are finished.

Complete a copy of the "Student Welding Report" listed in Appendix I or provided by your instructor. ◆

Butt Joint

PRACTICE 4-10

Butt Joint at a 45° Angle

Using a properly lit and adjusted torch, two clean pieces of 16 gauge mild steel 6 in. (152 mm) long, and filler metal, you will make a welded butt joint at a 45° angle, **Figure 4-42.**

Tack the sheets together and support them at a 45° angle. Weld using the method of flashing the torch off the molten weld pool to control penetration and weld contour. Make a weld that has uniform width and reinforcement. Repeat this practice until the weld can be made without defects. Turn off the cylinders, bleed the hoses, back out the regulator adjusting screw, and clean up your work area when you are finished.

Complete a copy of the "Student Welding Report" listed in Appendix I or provided by your instructor. ◆

PRACTICE 4-11

Butt Joint, Vertical Position

Using the same equipment, materials, and setup as described in Practice 4-10 make a welded butt joint in the vertical position. Make a weld that is uniform in width and reinforcement and has no visual detects. The penetration of this practice weld may vary. Turn off the cylinders, bleed the hoses, back out the regulator adjusting screw, and clean up your work area when you are finished.

Complete a copy of the "Student Welding Report" listed in Appendix I or provided by your instructor. ◆

PRACTICE 4-12

Butt Joint, Vertical Position, with 100% Penetration

Using the same equipment, materials, and setup as listed in Practice 4-10 weld a butt joint in the vertical

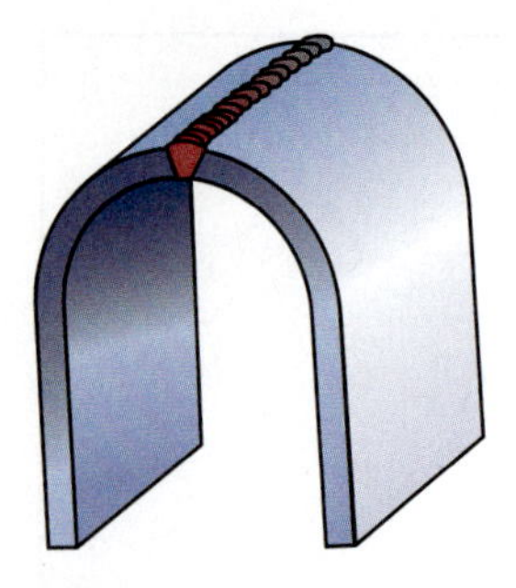
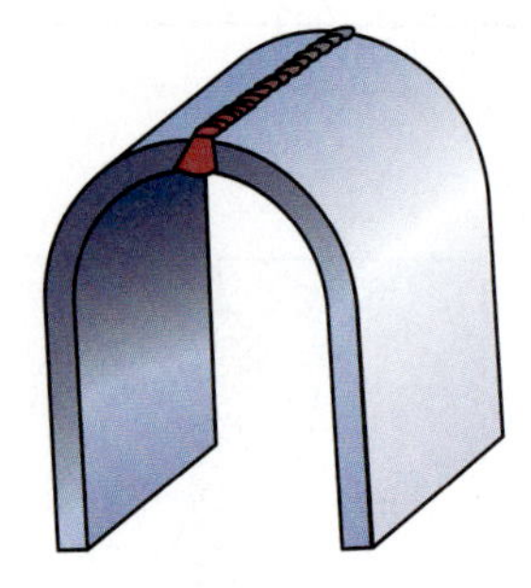

FIGURE 4-43 Bend strips to check a weld.

position with 100% penetration along the entire 6 in. (152 mm) of welded joint. Repeat this practice until this weld can be made without defects. If you want to test your skill, shear out a strip 1 in. (25 mm) wide and test it for 100% root penetration, **Figure 4-43.** Turn off the cylinders, bleed the hoses, back out the regulator adjusting screw, and clean up your work area when you are finished.

Complete a copy of the "Student Welding Report" listed in Appendix I or provided by your instructor. ◆

Lap Joint

PRACTICE 4-13

Lap Joint at a 45° Angle

Using a properly lit and adjusted torch, two clean pieces of 16 gauge mild steel 6 in. (152 mm) long, and filler metal, you will weld a lap joint at a 45° angle.

After tacking the sheets together and supporting them at a 45° angle, use the same method of adding rod as you did for the flat lap joint. Again, flash off the torch as needed to control the molten weld pool.

Repeat this weld until you can make a weld that is uniform in width and reinforcement and has no visual defects. Both sides of the joint can be welded. Turn off the cylinders, bleed the hoses, back out the regulator adjusting screw, and clean up your work area when you are finished.

Complete a copy of the "Student Welding Report" listed in Appendix I or provided by your instructor. ◆

PRACTICE 4-14

Lap Joint, Vertical Position

Using the same equipment, materials, and setup as listed in Practice 4-13, weld a lap joint in the vertical position. Make a weld that is uniform in width and reinforcement and has no visual defects. Both sides of the joint can be welded. Repeat this practice until the weld can be made without defects. Turn off the cylinders, bleed the hoses, back out the regulator adjusting screw, and clean up your work area when you are finished.

Complete a copy of the "Student Welding Report" listed in Appendix I or provided by your instructor. ◆

Tee Joint

The vertical tee joint has a right and a left side. **Figure 4-44** shows the best way to place the sheets depending upon whether the welder is right-handed or left-handed. It is a good idea to try both the right-hand and left-hand joints because in the field you may not be able to change the joint direction. Use the same method of adjusting the torch and torch angle that was practiced for the flat tee joint. In addition, use the method of flashing the torch off for molten weld pool control. Surface tension in the molten weld pool of a tee joint enables a larger weld to be made than either the butt or lap joint without as severe a problem with dripping.

PRACTICE 4-15

Tee Joint at a 45° Angle

Using a properly lit and adjusted torch, two clean pieces of 16 gauge mild steel 6 in. (152 mm) long, and filler metal, you will weld a tee joint at a 45° angle.

After tacking the sheets together and supporting them at a 45° angle, make a fillet weld that has uniform width and reinforcement and no visual defects. It is often best to weld only one side of the practice tee joint unless the oxides can be easily removed from the back side of the previous weld. Repeat this practice until the weld can be made without defects. Turn off the cylinders, bleed the hoses, back out the regulator adjusting screw, and clean up your work area when you are finished.

Complete a copy of the "Student Welding Report" listed in Appendix I or provided by your instructor. ◆

PRACTICE 4-16

Tee Joint, Vertical Position

Using the same equipment, materials, and setup as listed in Practice 4-15, make a fillet weld. Repeat this practice until the weld passes the test. Turn off the cylinders, bleed the hoses, back out the regulator adjusting screw, and clean up your work area when you are finished.

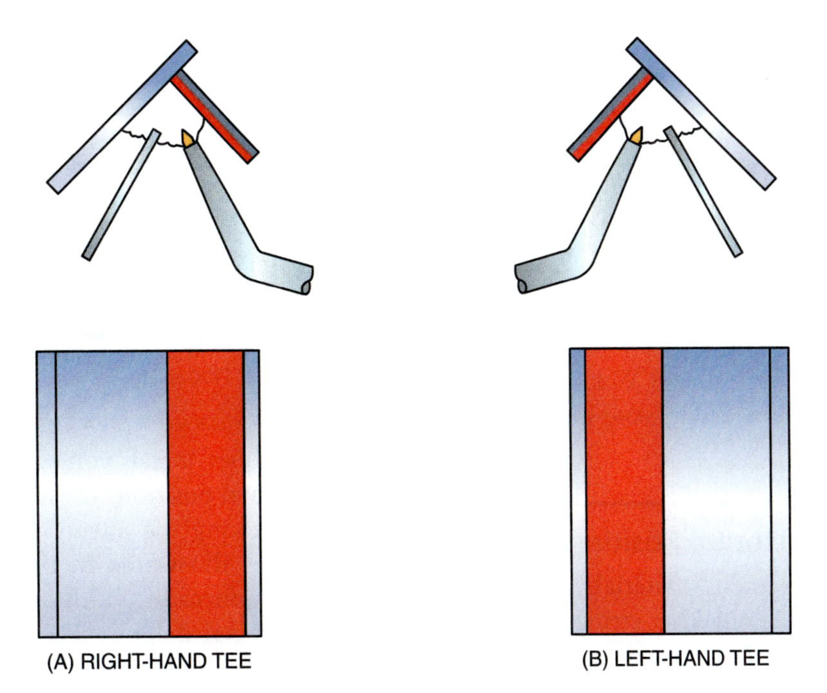

FIGURE 4-44 Some vertical tee joints are easier for right-handed or left-handed welders.

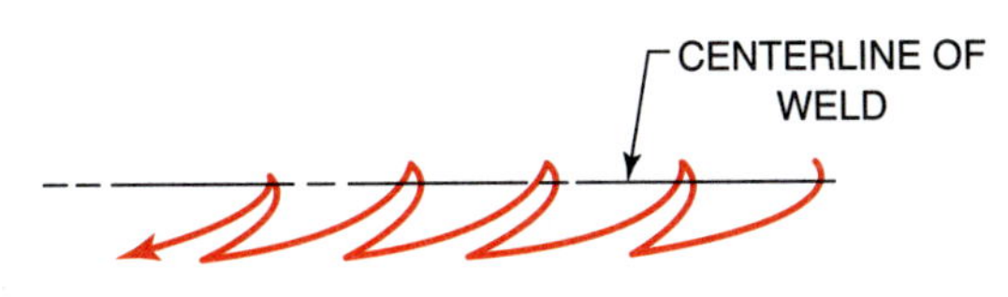

FIGURE 4-45 A "J" weave pattern for horizontal welds.

Complete a copy of the "Student Welding Report" listed in Appendix I or provided by your instructor. ◆

Horizontal Welds

Horizontal welds, like vertical welds, must rely on some part of the weld bead to support the molten weld pool as the weld is made. The shelf that supports a horizontal weld must be built up under the molten weld pool and at the same time must keep the weld bead uniform. The weave pattern required for a horizontal weld is completely different from that of any of the other positions. The pattern, **Figure 4-45**, builds a shelf on the bottom side of the bead to support the molten weld pool, which is elongated across the top. The sheet may be tipped back at a 45° angle for the stringer bead. Doing this allows the student to acquire the needed skills before proceeding to the more difficult, fully horizontal position. As with the vertically inclined sheet, the skills required are the same.

Horizontal Stringer Bead

When starting a horizontal bead, it is important to start with a small bead and build it to the desired size. If too large a molten weld pool is started, the shelf does not have time to form properly. The weld bead will tend to sag downward and not be uniform. As a result, there may be an undercut of the top edge and an overlap on the bottom edge, **Figure 4-46**.

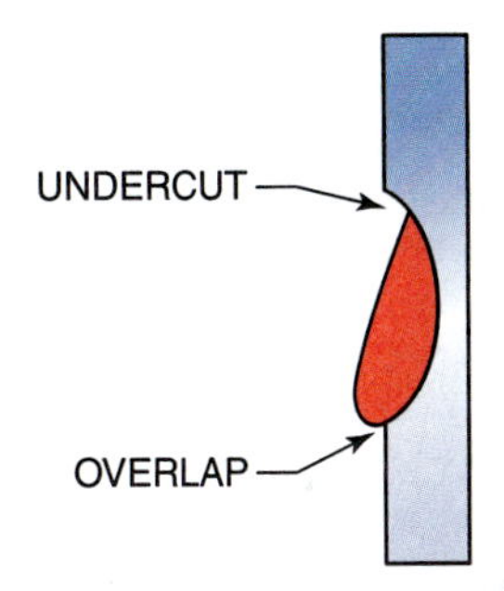

FIGURE 4-46 Too large a molten weld pool.

PRACTICE 4-17

Horizontal Stringer Bead at a 45° Sheet Angle

Using a properly lit and adjusted torch, one clean piece of 16 gauge mild steel 6 in. (152 mm) long, and filler metal, you will make a horizontal bead at a 45° reclining angle, **Figure 4-47**.

Add the filler metal along the top leading edge of the molten weld pool. Surface tension will help hold it on the top. The weld should be uniform in width and reinforcement and have no visual defects. Repeat this practice until the weld can be made without defects. Turn off the cylinders, bleed the hoses, back out the regulator adjusting screw, and clean up your work area when you are finished.

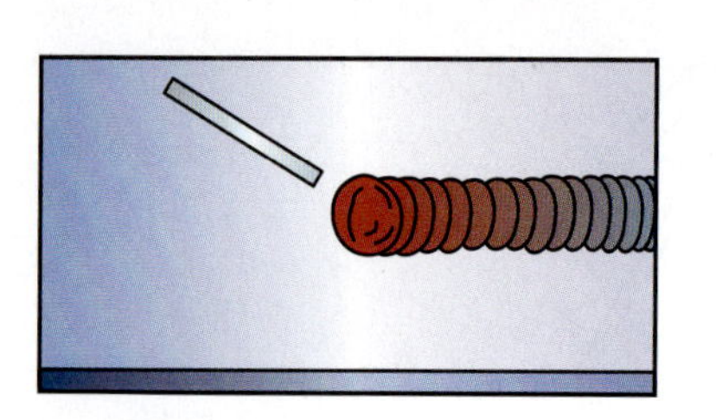

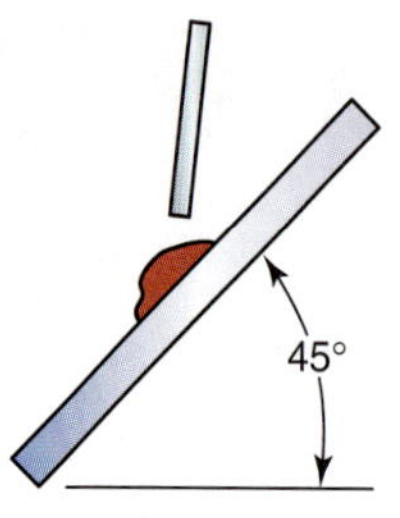

FIGURE 4-47 Horizontal stringer bead at a reclining angle.

Complete a copy of the "Student Welding Report" listed in Appendix I or provided by your instructor. ◆

PRACTICE 4-18

Stringer Bead, Horizontal Position

Using the same equipment, materials, and setup as listed in Practice 4-17, make a stringer bead in the horizontal position. The stringer bead should be uniform in width and reinforcement and have no visual defects. Turn off the cylinders, bleed the hoses, back out the regulator adjusting screw, and clean up your work area when you are finished.

Complete a copy of the "Student Welding Report" listed in Appendix I or provided by your instructor. ◆

Butt Joint

PRACTICE 4-19

Butt Joint, Horizontal Position

Using a properly lit and adjusted torch, two clean pieces of 16 gauge mild steel 6 in. (152 mm) long, and filler metal, you will weld a butt joint in the horizontal position.

After tacking the sheets together and supporting them in the horizontal position, make a weld using the same technique as practiced in the horizontal beading, Practice 4-18. The weld must be uniform in width and reinforcement and have no visual defects. Repeat this practice until the weld can be made without defects. Turn off the cylinders, bleed the hoses, back out the regulator adjusting screw, and clean up your work area when you are finished.

Complete a copy of the "Student Welding Report" listed in Appendix I or provided by your instructor. ◆

Lap Joint

PRACTICE 4-20

Lap Joint, Horizontal Position

Using a properly lit and adjusted torch, two clean pieces of 16 gauge mild steel 6 in. (152 mm) long, and filler metal, you will weld a lap joint in the horizontal position.

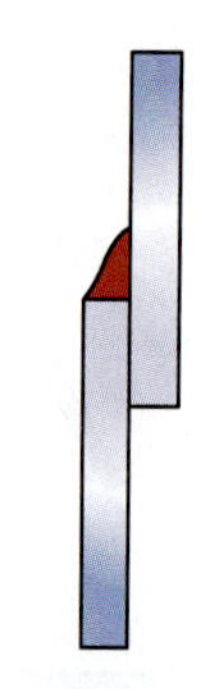

FIGURE 4-48 Horizontal lap joint.

After tacking the sheets together, support the assembly as illustrated in **Figure 4-48**. The weld must be uniform in width and reinforcement and have no visual defects. The sheet can be turned over, and the other side can be welded. Repeat this practice until the weld can be made without defects. Turn off the cylinders, bleed the hoses, back out the regulator adjusting screw, and clean up your work area when you are finished.

Complete a copy of the "Student Welding Report" listed in Appendix I or provided by your instructor. ◆

Tee Joint

The horizontal and flat tee joints are very similar in relation to the types of skills required to perform metal welds. The horizontal fillet weld tends to flow down toward the horizontal sheet from the vertical sheet. To correct this problem, the filler metal should be added to the top edge of the molten weld pool.

PRACTICE 4-21

Tee Joint, Horizontal Position

Using a properly lit and adjusted torch, two clean pieces of 16 gauge mild steel 6 in. (162 mm) long, and filler metal, you will weld one side of a tee joint in the horizontal position. After tacking the sheets together, make a fillet weld that is uniform in width and reinforcement and has no visual defects. Repeat this practice until the weld can be made without defects. Turn off the cylinders, bleed the hoses, back out the regulator adjusting screw, and clean up your work area when you are finished.

Complete a copy of the "Student Welding Report" listed in Appendix I or provided by your instructor. ◆

Overhead Welds

When welding in the overhead position, it is important to wear the proper personal protection, including leather gloves, leather sleeves, a leather apron, and a cap. The possibility of being burned increases greatly when welding in the overhead position. However, with the proper protective clothing you should avoid being burned.

With the overhead weld, the molten weld pool is held to the sheet by surface tension in the same manner that a drop of water is held to the bottom of a glass sheet. If the molten weld pool gets too large, big drops of metal may fall. If the welding rod is not dipped into the molten weld pool, but is allowed to melt in the flame, it also may drip. As long as the molten weld pool is controlled and the rod is added properly, overhead welding is safe.

The direction that you choose to weld in the overhead position is one of personal preference. It is a good idea to try several directions before deciding on one. The height of the sheet also affects your skill and progress. Welders often prefer to stand while overhead welding so that sparks do not land in their laps. If you decide to stand, you need to somehow brace yourself to help your stability.

Stringer Bead

Place the metal at a height recommended by your instructor. With the torch off, your goggles down, and a rod in your hand, try to progress across the sheet in a straight line. Use several directions until you find the direction that best suits you. Change the height of the sheet up and down to determine the height at which welding is most comfortable.

PRACTICE 4-22

Stringer Bead, Overhead Position

Using a properly lit and adjusted torch and one clean piece of 16 gauge mild steel 6 in. (152 mm) long, you will make a bead in the overhead position.

Heat the sheet until it melts and forms a molten weld pool. Put the welding rod into the molten weld pool as the torch tip is moved away from the molten weld pool. Return the flame to the molten weld pool as you remove the rod, **Figure 4-49**. Continue repeating this sequence as you move along the sheet. The weld bead should be uniform in width and reinforcement and have no visual defects. Repeat this practice until the weld can be made without defects. Turn off the cylinders, bleed the hoses, back out the regulator adjusting screw, and clean up your work area when you are finished.

Complete a copy of the "Student Welding Report" listed in Appendix I or provided by your instructor. ◆

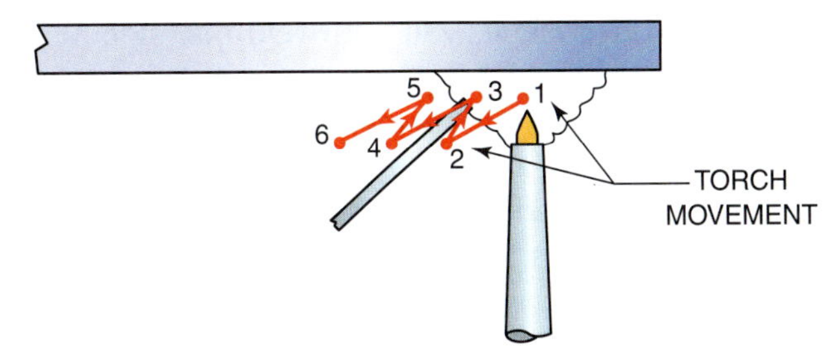

FIGURE 4-49 Overhead.

PRACTICE 4-23

Butt Joint, Overhead Position

Using a properly lit and adjusted torch, two clean pieces of 16 gauge mild steel 6 in. (152 mm) long, and filler metal, you will weld a butt joint in the overhead position.

After tacking the sheets together, put them in the overhead position. Following the sequence used in Practice 4-22 for the overhead stringer bead, make a weld along the joint. The weld should be uniform in width and reinforcement and have no visual defects. Repeat this practice until the weld can be made without defects. Turn off the cylinders, bleed the hoses, back out the regulator adjusting screw, and clean up your work area when you are finished.

Complete a copy of the "Student Welding Report" listed in Appendix I or provided by your instructor. ◆

PRACTICE 4-24

Lap Joint, Overhead Position

Using a properly lit and adjusted torch, two clean pieces of 16 gauge mild steel 6 in. (152 mm) long, and filler metal, you will weld a lap joint in the overhead position.

After tacking the sheets together, put them in the overhead position. Using the sequence for the overhead stringer bead, make a weld down the joint. The filler metal should be added to the leading edge of the molten weld pool on the top sheet. The weld should be uniform in width and reinforcement and have no visual defects. Repeat this practice until the weld can be made without defects. Turn off the cylinders, bleed the hoses, back out the regulator adjusting screw, and clean up your work area when you are finished.

Complete a copy of the "Student Welding Report" listed in Appendix I or provided by your instructor. ◆

PRACTICE 4-25

Tee Joint, Overhead Position

Using a properly lit and adjusted torch, two clean pieces of 16 gauge mild steel 6 in. (152 mm) long, and filler metal, you will weld one side of a tee joint in the overhead position.

After tacking the sheets together, put them in the overhead position and make a fillet weld. The filler metal should be added to the top sheet. The weld should be uniform in width and reinforcement and have no visual defects. Repeat this practice until the weld can be made without defects. Turn off the cylinders, bleed the hoses, back out the regulator adjusting screw, and clean up your work area when you are finished.

Complete a copy of the "Student Welding Report" listed in Appendix I or provided by your instructor. ◆

Mild Steel Pipe and Tubing

Mild steel pipe and tubing, both small diameter and thin wall, can be gas welded. The welding process for both pipe and tubing are usually the same. Thin-wall material does not require a grooved preparation. Gas welding is used to fabricate piping systems. It is also used on both pipe and tubing to make structures, such as frames, gates, hand rails, and works of art, **Figure 4-50**.

Horizontal Rolled Position 1G

The experiments and practices that follow (through Practice 4-27) will give the student the opportunity to gain skill in making welds in the 1G position, **Figure 4-51**.

EXPERIMENT 4-4

Effect of Changing Angle on Molten Weld Pool

This experiment will show how the molten weld pool is affected by changing the surface angle. With a piece of pipe having a diameter of approximately 1 in. to 2 in. (25 mm to 50 mm), you will push a molten weld pool across the top of the pipe. The pipe is in the 1G position, **Figure 4-52**. Starting at the 2 o'clock position, weld upward and across to the 11 o'clock position. Use the same torch angle and distances that you learned in Practice 4-2 and Experiment 4-3. Repeat this experiment until you

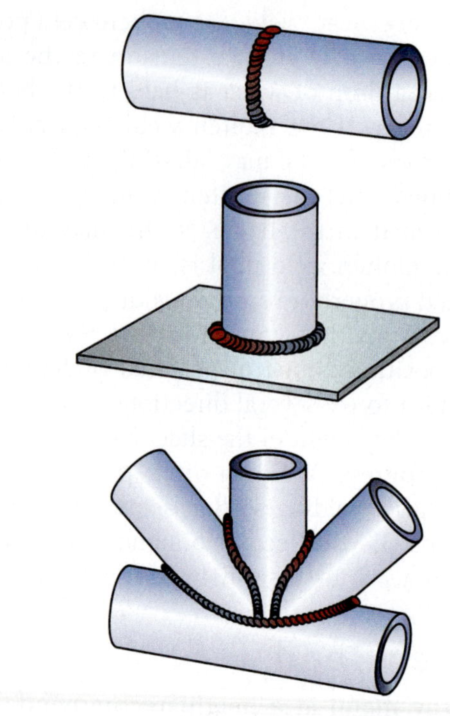

FIGURE 4-50 Examples of tube joints commonly used in industry.

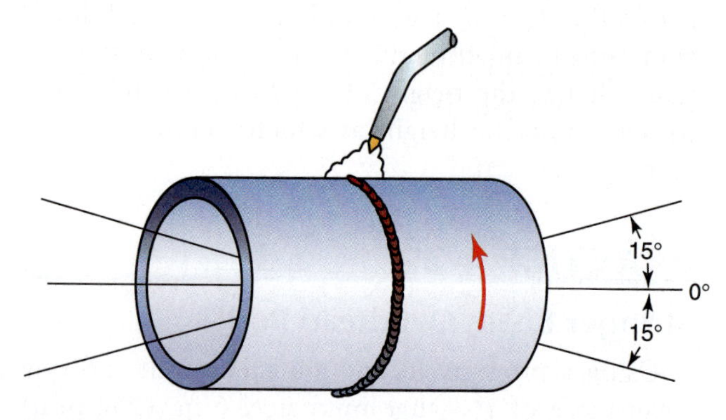

FIGURE 4-51 1G position. The pipe is rolled horizontally. The weld is made in the flat position (approximately 12 o'clock as the pipe is rolled).

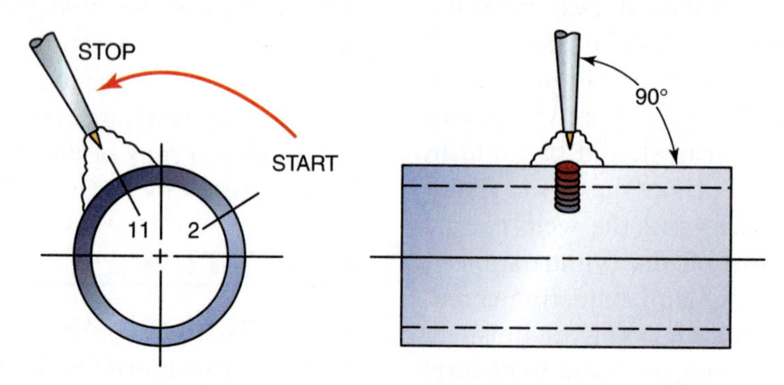

FIGURE 4-52 When the weld is finished, stop and roll the pipe so that the end of the weld is at the 2 o'clock position.

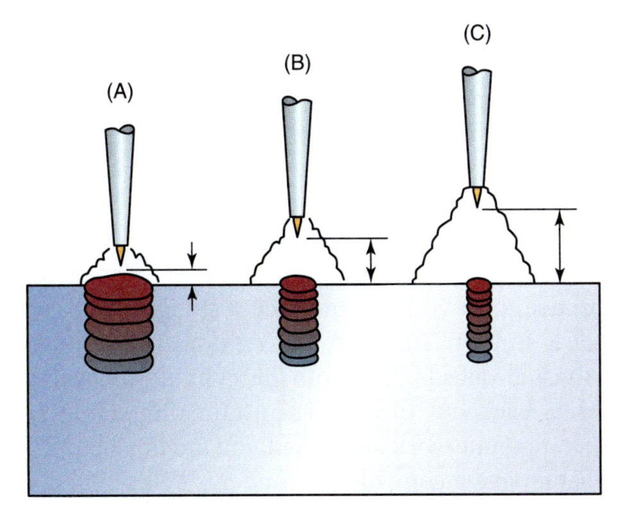

Figure 4-53 As the distance between the inner core and pipe changes, the size of the molten weld pool changes.

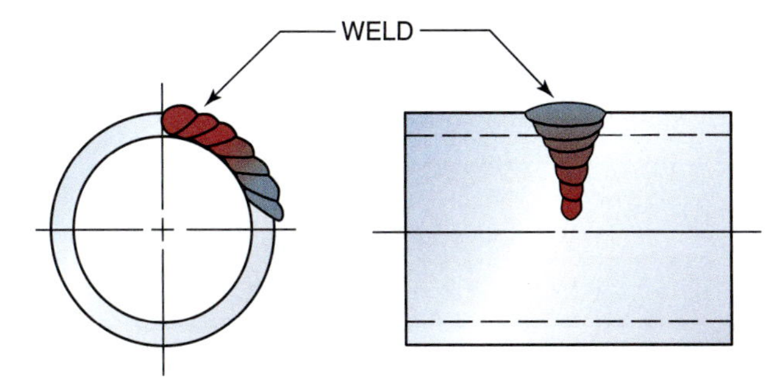

Figure 4-54 The weld bead shape is affected by its relative position on the pipe.

can make a straight weld bead that is uniform in width.

Keeping the torch at the correct angle to the pipe surface requires some practice. Changes in the relative torch angle will greatly affect the size of the molten weld pool. The distance of the inner core from the pipe surface is also important, **Figure 4-53.** You must be comfortable and able to freely move around. Your free hand should not be used for supporting or steadying the torch or yourself. Later, when you need this hand to add filler metal, you will not have to learn how to be steady without the use of your hand.

Complete a copy of the "Student Welding Report" listed in Appendix I or provided by your instructor. ◆

EXPERIMENT 4-5

Stringer Bead, 1G Position

Using a properly lit and adjusted torch, one clean piece of pipe approximately 1 in. to 2 in. (25 mm to 50 mm) in diameter, and three different sizes of filler metal, you will make a stringer bead on the pipe.

With the pipe in the 1G position, start a molten weld pool at the 2 o'clock position and weld to the 11 o'clock position. When you are at the 2 o'clock position, gravity will pull the molten metal outward, making the bead high and narrow, **Figure 4-54.** As the weld progresses toward the 12 o'clock position, if you keep using the same technique, the weld metal is pulled down, making the bead flat and wide. Repeat this experiment until you can make a straight weld bead that is uniform in width and reinforcement.

To keep the contour and width of the weld bead uniform, you must adjust your technique as the weld progresses. The following are some methods of keeping the buildup uniform:

- Move the flame farther from the surface of the pipe.
- Decrease the torch angle relative to the pipe surface.
- Travel at an increasing rate of speed.

A combination of these methods can be used to control the spreading weld bead. The spreading weld bead should be treated as if it is becoming too hot and you must cool it down. However, gravity rather than temperature is the problem, but decreasing the temperature can solve the problem.

Complete a copy of the "Student Welding Report" listed in Appendix I or provided by your instructor. ◆

EXPERIMENT 4-6

Stops and Starts

Using a properly lit and adjusted torch, one clean piece of pipe having a diameter of about 1 in. to 2 in. (25 mm to 50 mm), and filler metal, you will learn how to make good starts and stops. When welding pipe, you will frequently need to stop and restart. With practice and the proper technique, it is possible to make uniform stops and starts that are as strong as the surrounding weld bead.

To make a proper stop, you should slightly taper down the molten weld pool by flashing the torch off. This allows the molten weld pool to solidify before the flame is totally removed from the weld pool. If the flame is removed too soon from the molten weld pool, it will rapidly oxidize, throwing out a burst of sparks. The pocket of oxides will greatly weaken the weld at this point.

Restarting the weld requires that a molten weld pool equal in size to the original one be reestablished. To restart the molten weld pool, point the flame slightly ahead of the crater at the end of the

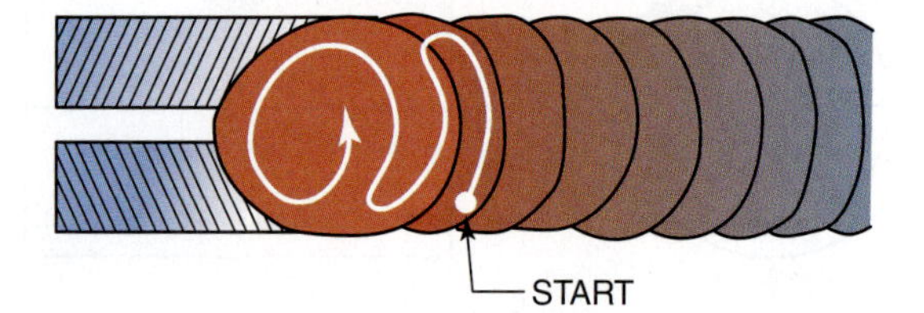

FIGURE 4-55 When restarting a weld pool, be sure the entire pool area is remelted before starting to add weld metal.

weld bead, **Figure 4-55.** When this metal starts to melt, move the flame back to the weld crater and melt the entire crater. Once the crater melts, start adding filler metal and continue with the weld. If the metal ahead of the crater is not heated up first when the weld metal is added to the crater, it may form a cold lap over the base metal.

Practice stops and starts until you can make them so that they are uniform and unnoticeable on the weld bead.

Complete a copy of the "Student Welding Report" listed in Appendix I or provided by your instructor. ◆

PRACTICE 4-26

Stringer Bead, 1G Position

Using a properly lit and adjusted torch and one clean piece of pipe approximately 1 in. to 2 in. (25 mm to 50 mm) in diameter, you will make a stringer bead around the pipe.

With the pipe in the 1G position, start a welding bead at the 2 o'clock position and weld toward the 12 o'clock position. When you must stop to change positions, roll the pipe so the ending crater is at the 2 o'clock position. Start the weld bead as practiced in Experiment 4-6 and proceed with the weld as before, **Figure 4-56.** Repeat this process until the weld bead extends all the way around the pipe. Repeat this practice until you can produce a straight weld bead that is uniform in width and reinforcement and has no visual defects. Turn off the cylinders, bleed the hoses, back out the regulator adjusting screw, and clean up your work area when you are finished.

Complete a copy of the "Student Welding Report" listed in Appendix I or provided by your instructor. ◆

PRACTICE 4-27

Butt Joint, 1G Position

Using a properly lit and adjusted torch, two clean pieces of schedule 40 pipe approximately 1 in. to 2 in. (25 mm to 50 mm) in diameter, and filler metal, you will weld a butt joint in the 1G position.

The pipe ends should be prepared as shown in **Figure 4-57.** The weld will be made with one pass. Tack weld the pipe together and place it on a firebrick. Using the principles and applications you learned in Practice 4-27, make this weld. Repeat this weld until it can be made without defects. Turn off the cylinders, bleed the hoses, back out the regulator adjusting screw, and clean up your work area when you are finished.

Complete a copy of the "Student Welding Report" listed in Appendix I or provided by your instructor. ◆

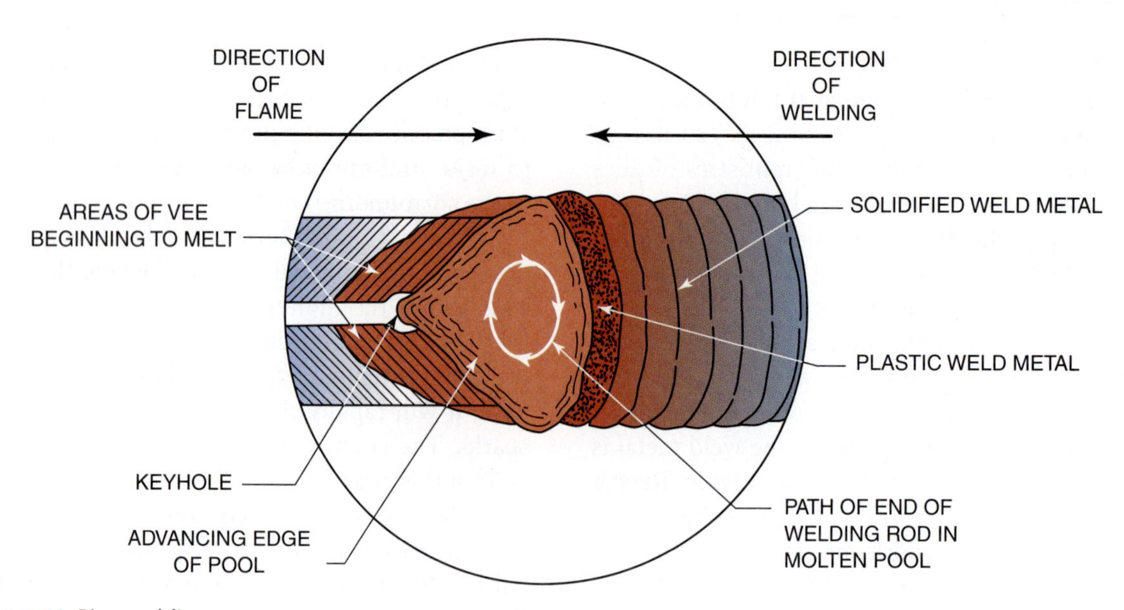

FIGURE 4-56 Pipe welding.

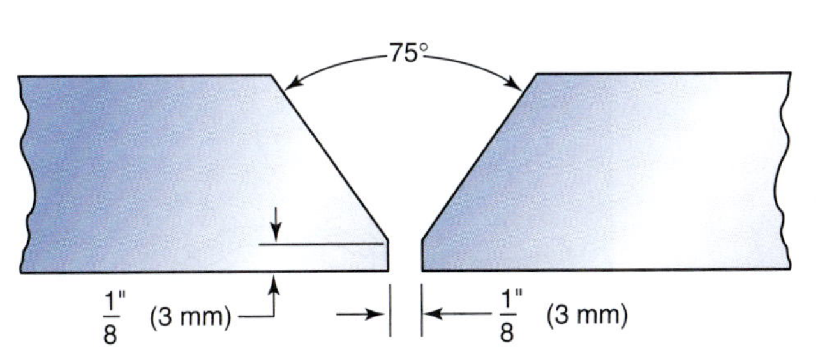

FIGURE 4-57 The bevel on the pipe may be oxyfuel flame-cut, ground, or machined.

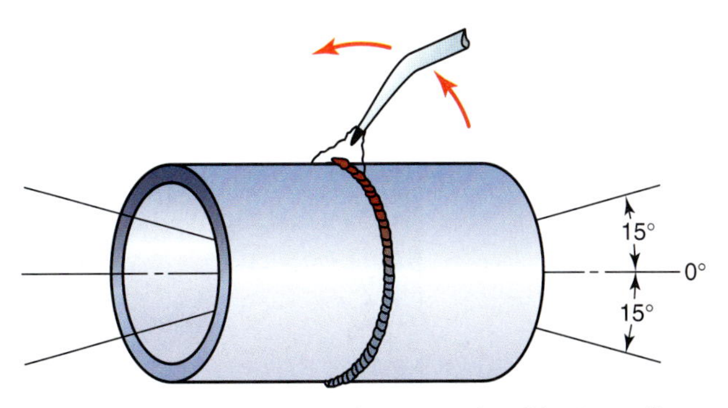

FIGURE 4-58 5G position. The pipe is fixed horizontally.

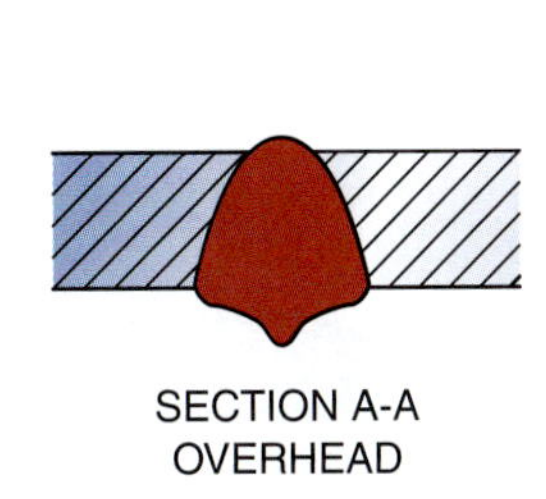

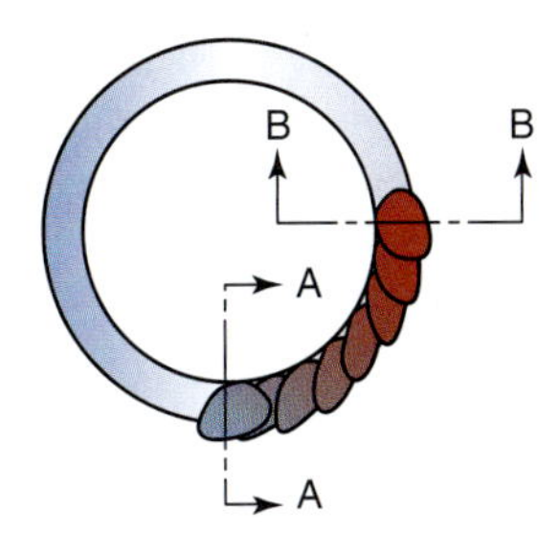

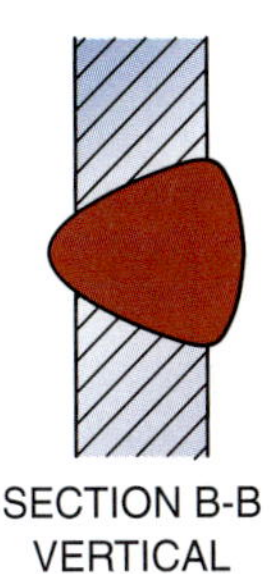

FIGURE 4-59 Changes in weld bead shape at different locations.

Horizontal Fixed Position 5G

The horizontal fixed position 5G weld, **Figure 4-58**, requires little skill development after completing the horizontal rolled position 1G welds. The torch height and angle skills you have developed will help you master the 5G position.

As the weld changes from overhead (6 o'clock) to vertical (3 o'clock), the weld contour changes. At the overhead position, the weld bead tends to be shaped as shown in **Figure 4-59, Section A-A**. Note that the bead is high in the center and recessed and possibly undercut on the sides. The vertical position has a high and narrow bead shape, **Figure 4-59, Section B-B**.

The overhead bead shape can be controlled by stepping the molten weld pool and moving the flame and rod back and forth at the same time. This process will deposit the metal and allow it to cool before it can sag. It allows the surface tension to hold the metal in place.

As the weld progresses toward the vertical section, the need to step the weld decreases. When the weld reaches the vertical section, the bead shape should be controlled by torch angle and flame distance.

EXPERIMENT 4-7

5G Position

In this experiment, you are going to push a molten weld pool from the bottom of a fixed pipe up to the side of the pipe to see how torch manipulation affects the bead shape.

Using a properly lit and adjusted torch and one clean piece of pipe approximately 1 in. to 2 in. (25 mm to 50 mm) in diameter, start by establishing a molten weld pool at the 6 o'clock position. Then move the molten weld pool forward toward the 3 o'clock position. Observe what effect torch angle, flame distance, and stepping have on the molten weld pool. Turn the pipe and repeat this experiment until you can control the bead width.

Complete a copy of the "Student Welding Report" listed in Appendix I or provided by your instructor. ◆

PRACTICE 4-28

Stringer Bead, 5G Position

Using a properly lit and adjusted torch, one clean piece of pipe having a diameter of about 1 in. to 2 in. (25 mm to 50 mm), and filler metal, you will make a stringer bead upward around one side of the pipe.

With the pipe in the 5G position, start a welding bead at the 6 o'clock position and weld toward the 12 o'clock position. Use all procedures necessary to keep the weld bead uniform. Repeat this practice until you can produce a straight weld bead that is uniform in width and reinforcement and has no visual defects. Turn off the cylinders, bleed the hoses, back out the regulator adjusting screw, and clean up your work area when you are finished.

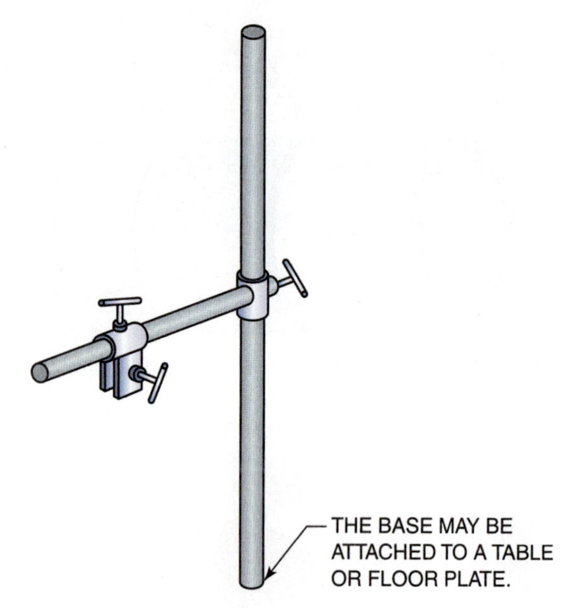

FIGURE 4-60 Pipe stand.

Complete a copy of the "Student Welding Report" listed in Appendix I or provided by your instructor. ◆

PRACTICE 4-29

Butt Joint, 5G Position

Using a properly lit and adjusted torch, two clean pieces of pipe having a diameter of approximately 1 in. to 2 in. (25 mm to 50 mm), and filler metal, you will weld a butt joint in the 5G position.

The pipe ends should be beveled as shown in **Figure 4-57** and tack welded together. Secure the pipe on a stand, such as the one shown in **Figure 4-60**, and start welding at the 6 o'clock position, moving toward the 12 o'clock position. When you reach the top, stop, restart back at the 6 o'clock position, and continue up the other side. Repeat this weld until it can be made without defects. Turn off the cylinders, bleed the hoses, back out the regulator adjusting screw, and clean up your work area when you are finished.

Complete a copy of the "Student Welding Report" listed in Appendix I or provided by your instructor. ◆

Vertical Fixed Position 2G

The vertically fixed pipe requires a horizontal weld, **Figure 4-61**. The welding manipulative skill required for the 2G position is similar to that for a horizontal butt joint. The pipe may be rotated around its vertical axis but may not be turned end for end after you have started welding, **Figure 4-62**.

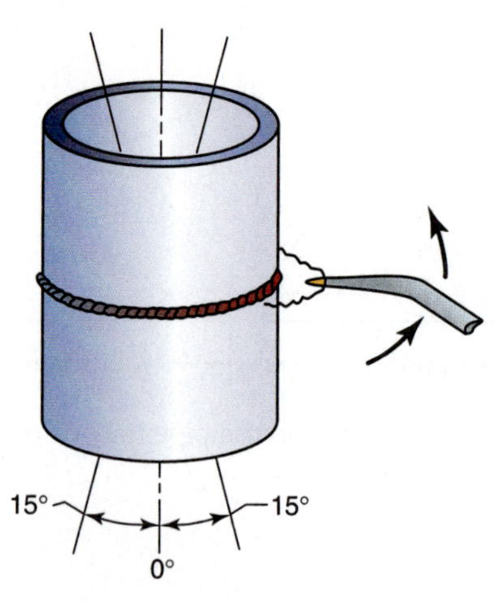

FIGURE 4-61 2G position. The pipe is fixed vertically and welded horizontally.

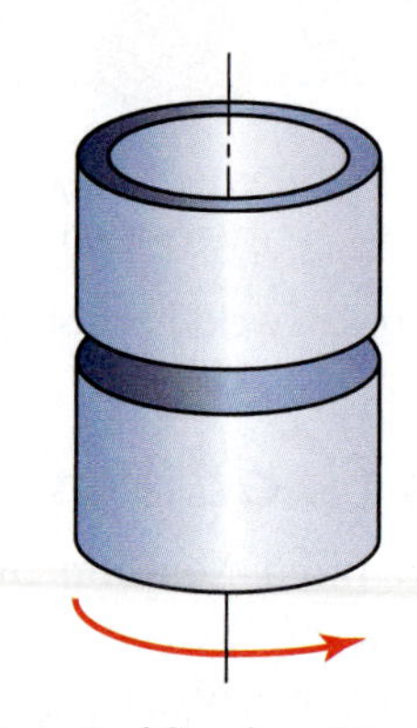

FIGURE 4-62 2G vertical fixed position.

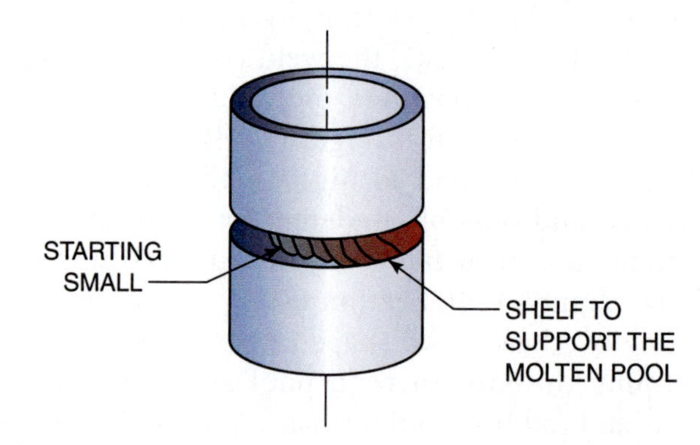

FIGURE 4-63 The proper starting technique will aid with tying in the weld when it is completed around the pipe.

PRACTICE 4-30

Stringer Bead, 2G Position

Using a properly lit and adjusted torch and one clean piece of pipe having a diameter of approximately 1 in. to 2 in. (25 mm to 50 mm), you will make a stringer bead around the pipe.

With the pipe in the 2G position, start with a small bead, as illustrated in **Figure 4-63**, and then increase

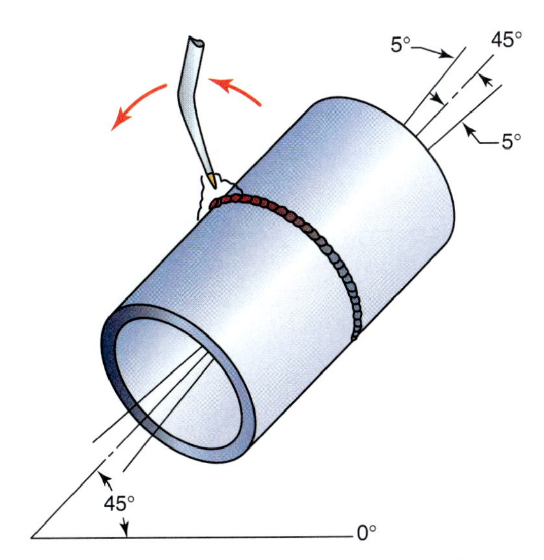

FIGURE 4-64 6G position. The pipe is inclined at a 45° angle.

the bead to the desired size. Starting small will allow you to build a shelf to support the molten weld pool. It also will let you tie the end of the weld into the start of the weld so that a stronger weld is obtained. Repeat this weld until it can be made without defects. Turn off the cylinders, bleed the hoses, back out the regulator adjusting screw, and clean up your work area when you are finished.

Complete a copy of the "Student Welding Report" listed in Appendix I or provided by your instructor. ◆

PRACTICE 4-31

Butt Joint, 2G Position

Using a properly lit and adjusted torch, two clean pieces of pipe having a diameter of approximately 1 in. to 2 in. (25 mm to 50 mm), and filler metal, you will weld a butt joint in the 2G position. The pipe ends should be beveled and tack welded together. Secure the pipe in the vertical position and start welding. Repeat this weld until it can be made without defects. Turn off the cylinders, bleed the hoses, back out the regulator adjusting screw, and clean up your work area when you are finished.

Complete a copy of the "Student Welding Report" listed in Appendix I or provided by your instructor. ◆

45° Fixed Position 6G

The 45° fixed pipe position, **Figure 4-64**, requires careful manipulation of the molten weld pool to ensure a uniform and satisfactory weld. The weld progresses around the pipe, changing from vertical to horizontal to overhead to flat and not completely in any one position. It is the combination of compound angles that makes the 6G position particularly difficult.

PRACTICE 4-32

Stringer Bead, 6G Position

Using a properly lit and adjusted torch, one clean piece of pipe having a diameter of approximately 1 in. to 2 in. (25 mm to 50 mm), and filler metal, you will make a stringer bead around the pipe with the pipe in the 6G position.

Start at the bottom and weld upward toward the top of the pipe, **Figure 4-65**. The weld bead shape will change as you move around the pipe. To prevent the bottom from pulling to one side and slight movement to the high side, the side movement will create a shelf to hold the metal in place.

The side of the bead will pull to one side, but not as severely as at the bottom. To keep the side from being pulled out of shape, simply add the filler metal on the high side of the joint. Some side movement may be required, but not as much as for the bottom.

The top will pull down to one side and tend to be flatter than the other parts of the bead, especially

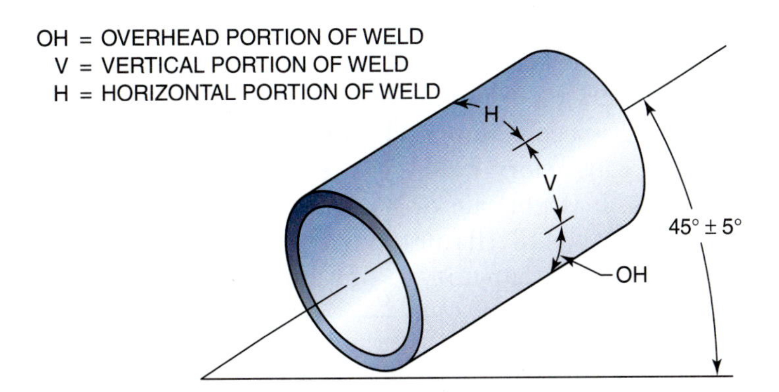

FIGURE 4-65 6G position.

along the side of the pipe. To prevent one-sidedness, the filler metal is added to the top side. To add to the buildup of the bead, change the torch angle or the flame height.

Repeat this weld until it can be made straight and uniform in width and reinforcement. Turn off the cylinders, bleed the hoses, back out the regulator adjusting screw, and clean up your work area when you are finished.

Complete a copy of the "Student Welding Report" listed in Appendix I or provided by your instructor. ◆

PRACTICE 4-33

Butt Joint, 6G Position

Using a properly lit and adjusted torch, one clean piece of pipe approximately 1 in. to 2 in. (25 mm to 50 mm) in diameter, and filler metal, you will weld a butt joint in the 6G position. The pipe ends should be beveled and tack welded together. Secure the pipe at a 45° angle and start welding. Repeat this weld until it can be made without defects. Turn off the cylinders, bleed the hoses, back out the regulator adjusting screw, and clean up your work area when you are finished.

Complete a copy of the "Student Welding Report" listed in Appendix I or provided by your instructor. ◆

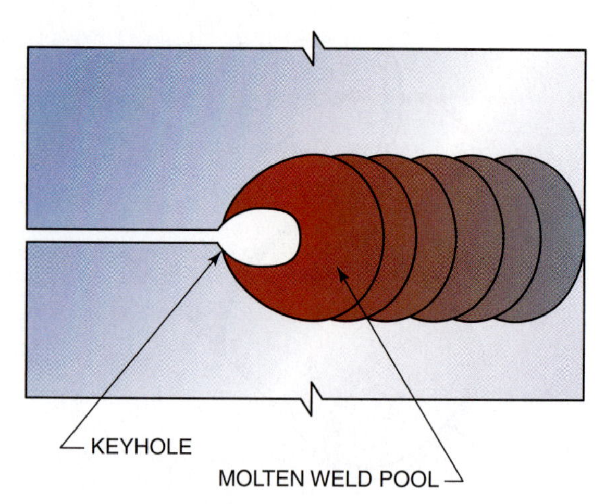

FIGURE 4-66 Keyhole.

Thin-wall Tubing

Welding thin-wall tubing requires a technique similar to welding a stringer bead around pipe. If penetration is not a concern, the welding proceeds as if you were making a stringer bead on pipe. However, if penetration is required, then the weld will probably have a **keyhole** to ensure 100% penetration, **Figure 4-66.** This will be similar to welding the root pass.

If you want to practice on tubing, but it is not readily available, then 16 gauge sheet metal can be rolled up and tack welded. Since 16 gauge is a standard wall thickness for tubing, fabricating your own will give you a realistic experience.

Summary

In an ideal world we would all have the correct tip size for every job we attempted; however, in reality often the oxyfuel welding process is used for repair work when it is not possible to have an infinite selection of tip sizes. Learning to control the heat input to the weld by changing the torch angle, height, or travel speed is important, so that you can produce satisfactory welds under these service conditions. It is not safe or recommended that you turn down the flame below the tip's optimal operating capacity. Therefore, you must learn to control the heat by changing the angle, tip height, or travel speed.

Gas welding has widespread use in agriculture because the equipment is so readily available. Almost every farm or ranch has an oxyacetylene cutting rig. Just by changing the tip from a cutting tip to a welding tip, this same setup can be used to make gas welds.

By learning to gas weld, you are able to build many small items with the torch that could not be built with most other welding processes,

Summary continued

Figure 4-67. The control you have allows very fine work; in fact some of the finish welding for jewelry making and watch repair is done with a torch's flame.

TIG or Heliarc welding can do as fine a weld as gas welding, but a higher skill level and longer training time is required to do the same work. In addition, the equipment for TIG or Heliarc welding is much more expensive and far less readily available than gas welding equipment. However, the eye-hand coordination and control you have learned in gas welding will be useful if you learn to TIG or Heliarc weld.

FIGURE 4-67 This elephant was made from filler rods.
Courtesy of Larry Jeffus.

Review

1. Which of the following should not be used to control the weld pool size?
   A. changing tip size
   B. changing torch angle
   C. changing travel speed
   D. changing gas pressures
2. At which angle is the most flame heat transferred into the weld?
   A. 90°
   B. 60°
   C. 45°
   D. 180°
3. What is meant by the term flashing off the flame?
   A. quickly turning off the torch and relighting the flame
   B. manipulating the flame so it is momentarily pointed away from the weld
   C. moving the flame in a small counterclockwise circle
   D. looking away from the flame so it does not burn your eyes
4. Changing filler rod sizes will not affect the size of the molten weld pool.
   A. true
   B. false
5. What protects the molten weld pool from atmospheric contamination?
   A. the heat from the flame
   B. the acetylene gas
   C. the primary flame
   D. the secondary flame
6. What does kindling temperature mean?
   A. the temperature at which something can begin to burn
   B. how hot kindling gets when it burns to start a fire
   C. the maximum temperature at which kindling can burn
   D. the temperature of sparks leaving the weld pool
7. What is a flame called that has too much acetylene?
   A. neutral
   B. carbonizing
   C. oxidizing
   D. balanced

Review continued

8. Which flame produces foam around the molten weld pool?
   A. neutral
   B. carbonizing
   C. oxidizing
   D. balanced
9. Which flame produces a weld bead with a smooth appearance?
   A. neutral
   B. carbonizing
   C. oxidizing
   D. balanced
10. As the flame height above the plate increases, the weld bead width will decrease.
    A. true
    B. false
11. A larger diameter filler rod can be used as a __________ to limit weld penetration.
    A. heat shield
    B. flame spreader
    C. heat sink
    D. flame reducer
12. Why should the end of the filler rod be kept inside the flame?
    A. to keep it warm
    B. so it will be close to the weld when needed
    C. to keep it in view
    D. so it will not oxidize
13. How can a notch that forms on a lap joint be closed?
    A. by adding more filler metal
    B. by moving slower
    C. by welding faster
    D. by angling the flame into the notch
14. Which welded joint can be easily made without adding filler metal?
    A. butt joint
    B. tee joint
    C. lap joint
    D. outside corner joint
15. What force moves the molten weld down to the bottom plate in a lap joint?
    A. gravity
    B. surface tension
    C. magnetism
    D. electrolysis
16. The plates of a lap joint are at a 45° angle.
    A. true
    B. false
17. Why might the welding tip overheat when doing a tee joint?
    A. The weld is being done too fast.
    B. More heat is reflected back on the tip.
    C. The molten end of the welding rod keeps hitting the tip.
    D. The tip has to be held closer to the weld to get it hot.
18. What out-of-position welding position is the most common?
    A. flat
    B. vertical
    C. horizontal
    D. overhead
19. One of the most difficult parts of making a pipe weld is that the weld position is constantly changing.
    A. true
    B. false
20. A pipe in the 5G position is __________ .
    A. horizontal and rolled
    B. horizontal and fixed
    C. vertical
    D. mounted at a 45° angle

Chapter 5

Soldering and Brazing

OBJECTIVES

After completing this chapter, the student should be able to

- ☑ define the terms *soldering, brazing,* and *braze welding.*
- ☑ explain the advantages and disadvantages of liquid-solid phase bonding.
- ☑ demonstrate an ability to properly clean, assemble, and perform required practice joints.
- ☑ describe the functions of fluxes in making proper liquid-solid phase bonded joints.

KEY TERMS

braze buildup
braze welding
brazing
brazing alloys
capillary action
eutectic composition
fluxes
low-fuming alloys
paste range
silver braze
soldering
soldering alloys

INTRODUCTION

Brazing and soldering both fall under the same American Welding Society classification. Only the temperature required to melt the filler metal separates soldering from brazing. Soldering takes place at temperatures below 840°F (450°C), and brazing takes place at temperatures above 840°F (450°C).

One of the advantages of soldering and brazing is that it can make either a permanent or a temporary joint. For example, a soldered copper water pipe fitting can last for years or it can be easily heated and separated within a few moments. The reusability of parts can cut time and expense. For instance, suppose a bolt or nut keeps vibrating loose on a tractor. The bolt can be tinned with a thin layer of solder before assembly and a thin layer of flux can be applied inside the nut. Once the two are tight, warming them with a torch will solder them in place. Warming them again later allows them to be removed, **Figure 5-1**. This works best on small diameter screws and nuts; it is an easy way to make them self-locking.

FIGURE 5-1(A) This worn bolt's threads are too worn to hold without vibrating loose. Courtesy of Larry Jeffus.

FIGURE 5-1(B) Solder has been applied to the threads to act as a lock tight. Courtesy of Larry Jeffus.

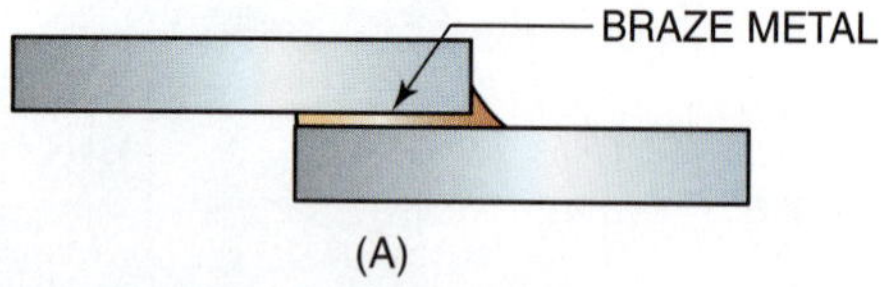

FIGURE 5-2 A brazed lap joint (A) and a braze welded lap joint (B).

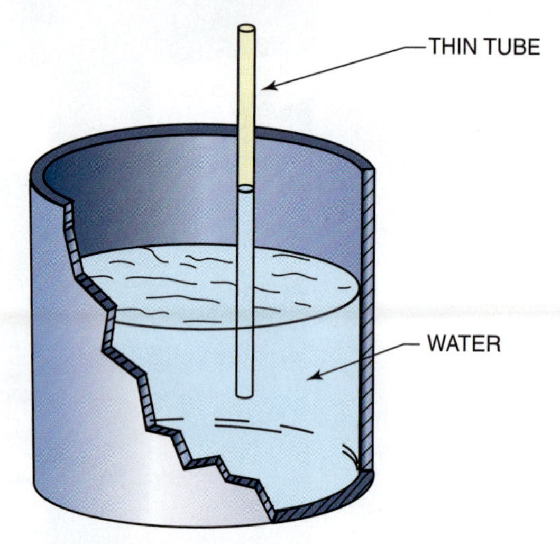

FIGURE 5-3 Capillary action pulls water into a thin tube.

Sometimes soldering and brazing are assumed to be a weaker way of joining or repairing parts. That is not necessarily true. When done correctly, both soldering and brazing can produce joints that are several times stronger than the filler metal itself. The higher strength comes from using a lap joint so that the area of the joint is greater than the thickness of the parts being joined, **Figure 5-2**.

To obtain this higher strength, the parts being joined must be fitted so that the joint spacing is very small. This small spacing allows capillary action to draw the filler metal into the joint when the parts reach the proper phase temperature.

Capillary action is the force that pulls water up into a paper towel or pulls a liquid into a very fine straw, **Figure 5-3**. Braze welding does not need capillary action to pull filler metal into the joint. Examples of brazing and braze welding joint designs are shown in **Figure 5-4**.

Advantages of Soldering and Brazing

Some advantages of soldering and brazing as compared to other methods of joining include

- Low temperature—Since the base metal does not have to melt, a low-temperature heat source can be used.
- May be permanently or temporarily joined—Since the base metal is not damaged, parts may be disassembled at a later time by simply reapplying heat. The parts then can be reused. However, the joint is solid enough to be permanent, **Figure 5-5**.
- Dissimilar materials can be joined—It is easy to join dissimilar metals, such as copper to steel, aluminum to brass, and cast iron to stainless steel, **Figure 5-6**. It is also possible to join nonmetals to each other or nonmetals to metals. Ceramics are easily brazed to each other or to metals.
- Speed of joining
  a. Parts can be preassembled and dipped or furnace soldered or brazed in large quantities, **Figure 5-7**.
  b. A lower temperature means less time in heating.
- Less chance of damaging parts—A heat source can be used that has a maximum temperature below the temperature that may cause damage to the parts. With the controlled temperature sufficiently low, even damage from unskilled or semiskilled workers can be eliminated, **Figure 5-8**.

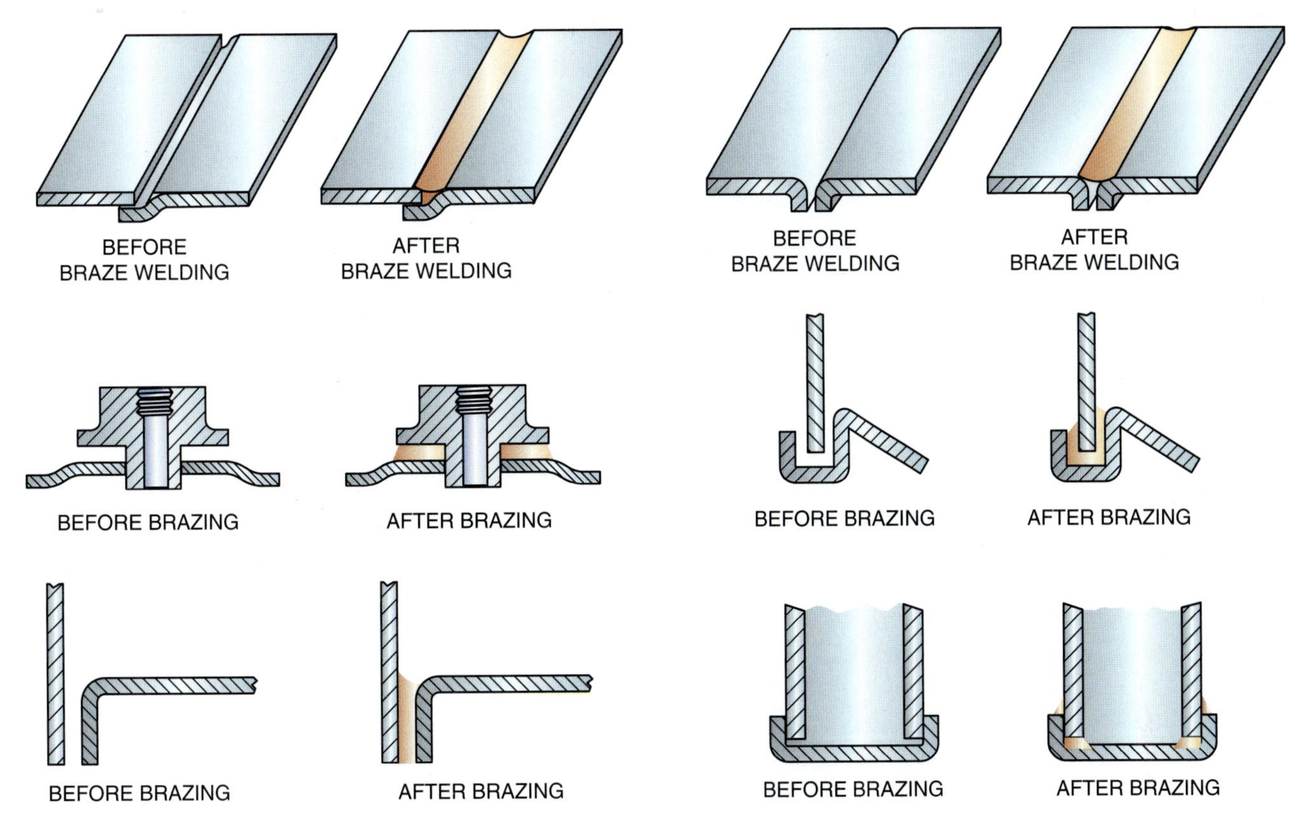

FIGURE 5-4 Examples of brazing and braze welded joints.

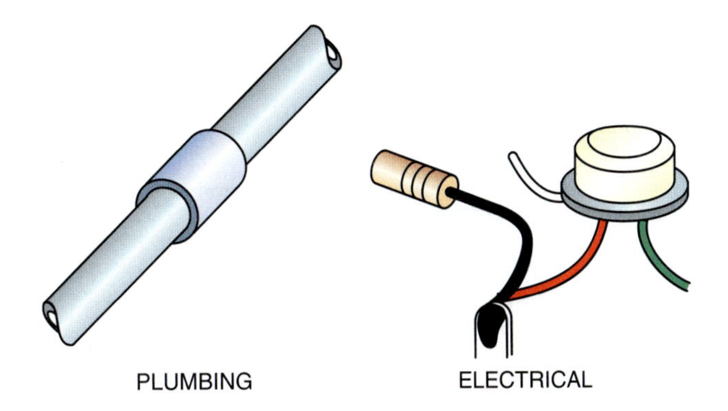

FIGURE 5-5 Examples of permanent joints that can easily be disassembled and the parts reused.

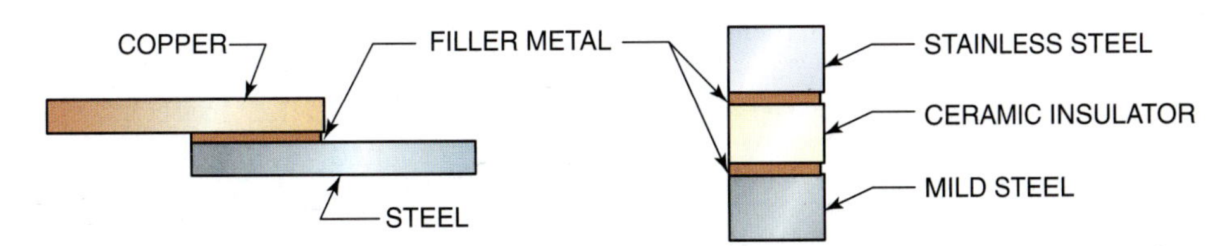

FIGURE 5-6 Dissimilar materials joined.

- Preassembled for better alignment and fit.
- Slow rate of heating and cooling—Because it is not necessary to heat a small area to its melting temperature and then allow it to cool quickly to a solid, the internal stresses caused by rapid temperature changes can be reduced.
- Parts of varying thicknesses can be joined—Very thin parts or a thin part and a thick part can be joined without burning or overheating them.
- Preassembly—All of the parts can be cut and assembled for a better fit and alignment. They can also be reheated and repositioned as needed.

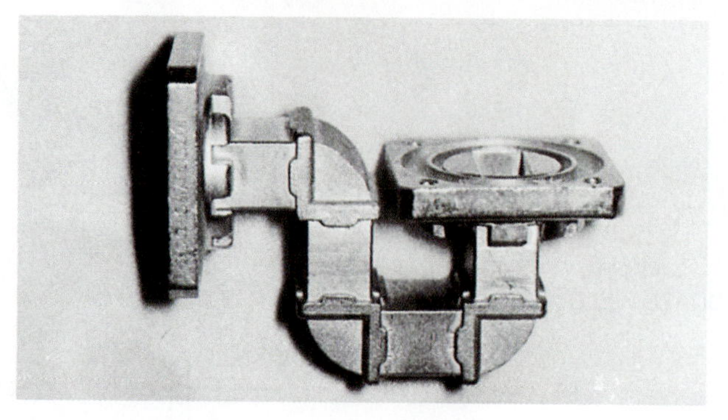

FIGURE 5-7 Furnace brazed part. Courtesy of Larry Jeffus.

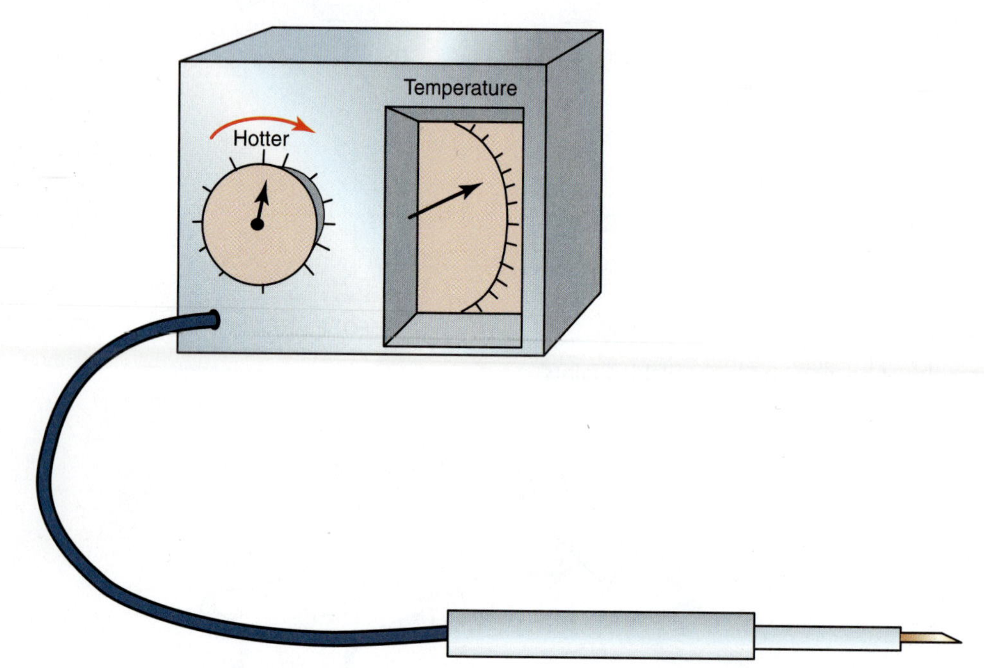

FIGURE 5-8 Control console for resistance soldering.

Fluxes

General Fluxes used in soldering and brazing have three major functions:

- They must remove any oxides that form as a result of heating the parts.
- They must promote wetting.
- They should aid in capillary action.

The flux, when heated to its reacting temperature, must be thin and flow through the gap provided at the joint. As it flows through the joint, the flux absorbs and dissolves oxides, allowing the molten filler metal to be pulled in behind it, **Figure 5-9**. After the joint is complete, the flux residue should be easily removable.

Fluxes are available in many forms, such as solids, powders, pastes, liquids, sheets, rings, and washers, **Figure 5-10**. They are also available mixed with the filler metal, inside the filler metal, or on the outside of the filler metal, **Figure 5-11**. Sheets, rings, and washers may be placed within the joints of an assembly before heating so that a good bond inside the joint can be assured. Paste and liquids can be injected into a joint from tubes using a special gun, **Figure 5-12**. Paste, powders, and liquids may be brushed on the joint before or after the material is heated. Paste and powders may also be applied to the end of the rod by heating the rod and dipping it in the flux. Most powders can be made into a paste, or a paste can be thinned by adding distilled water or alcohol; see manufacturers' specifications for details. If water is used, it should be distilled because tap water may contain minerals that will weaken the flux.

Some liquid fluxes may also be added to the gas when using an oxyfuel gas torch for soldering or brazing. The flux is picked up by the fuel gas as it is bubbled through the flux container and is then carried to the torch where it becomes part of the flame.

Flux and filler metal combinations are most convenient and easy to use, **Figure 5-13**. It may be necessary to stock more than one type of flux-filler metal combination for different jobs. These combinations are more expensive than buying the filler and flux separately. In cases where the flux covers the outside of the filler metal, it may be damaged by humidity or chipped off during storage.

Using excessive flux in a joint may result in flux being trapped in the joint, weakening the joint, or causing the joint to leak or fail.

Note: Most of the fluxes used for brazing and soldering are not harmful to the environment. However, large quantities of even the most benign materials introduced accidentally or intentionally into the environment can cause damage. Even lemon juice, a common electronic soldering flux, in large enough quantities can cause harm to the environment. Before

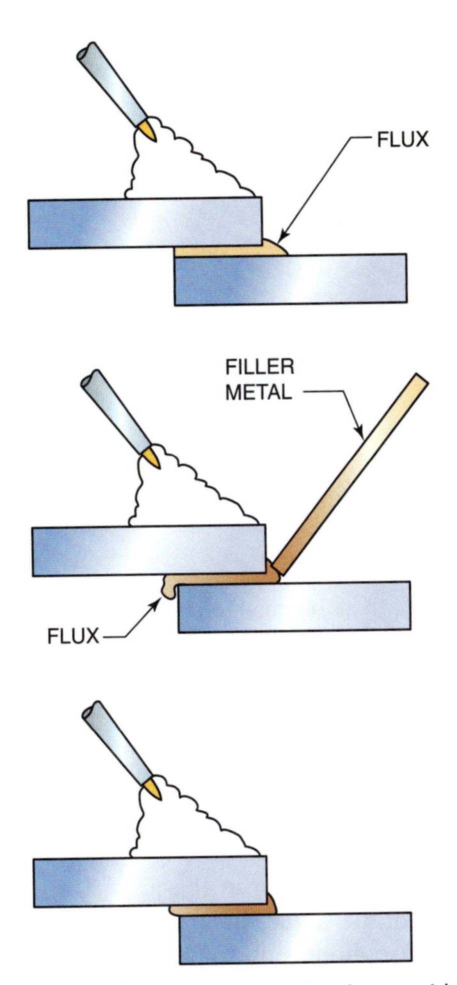

FIGURE 5-9 Flux flowing into a joint reduces oxides to clean the surfaces and gives rise to a capillary action that causes the filler metal to flow behind it.

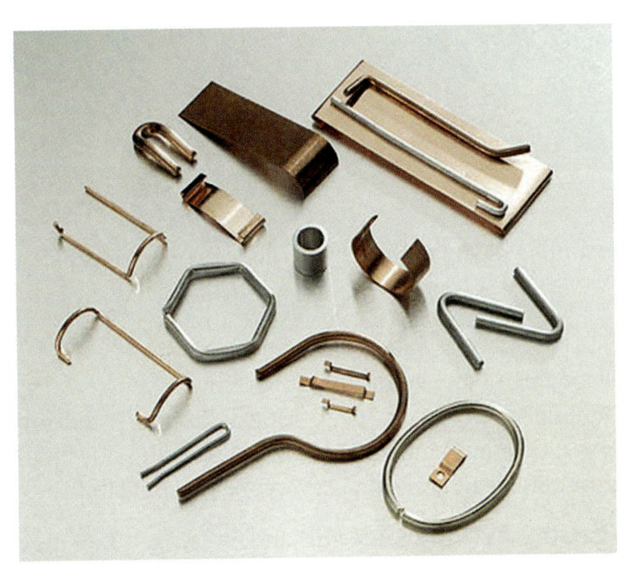

FIGURE 5-10 Braze/solder forms that can be preplaced in a braze/solder joint. Courtesy of Prince & Izant Co.

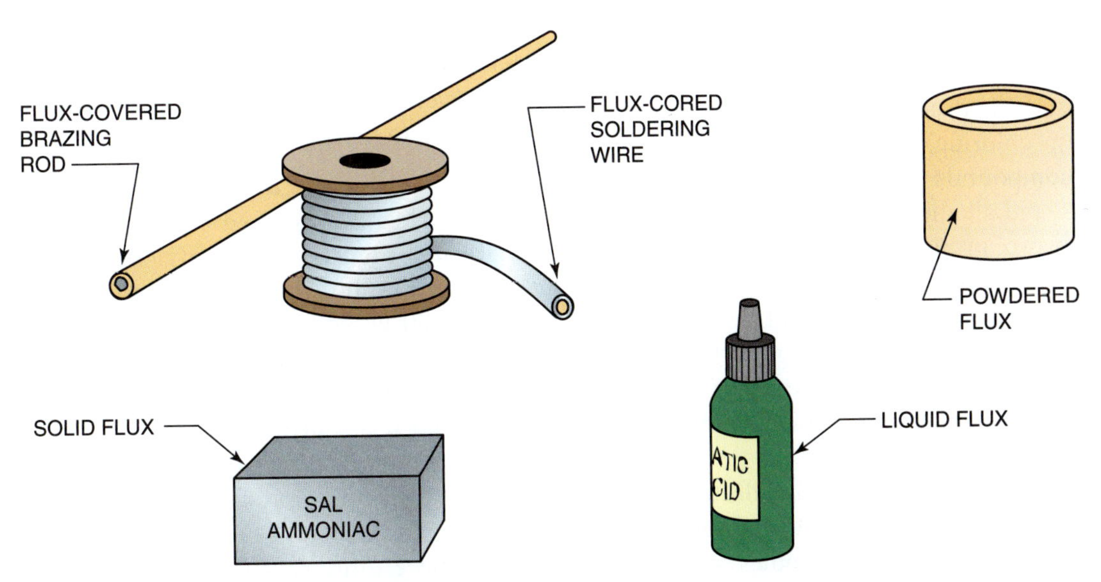

FIGURE 5-11 Flux can be purchased with the filler metal or separately.

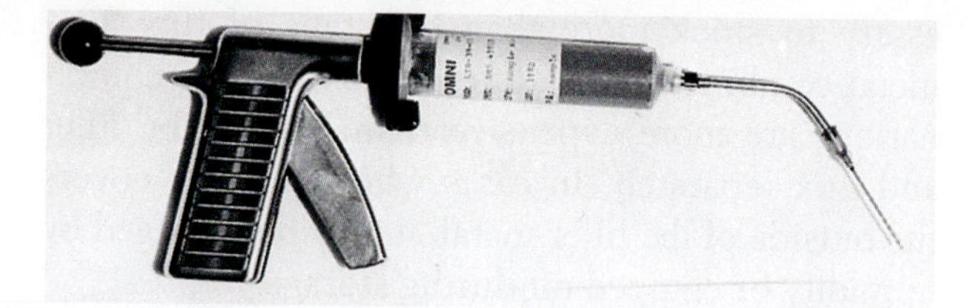

FIGURE 5-12 Gun for injecting flux into joint. Courtesy of Larry Jeffus.

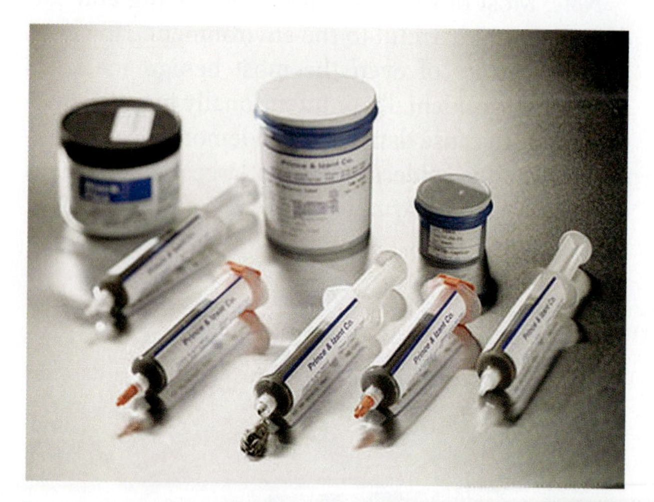

FIGURE 5-13 Tubes that contain flux filler metal mixtures. Courtesy of Prince & Izant Co.

FIGURE 5-14 An air propane torch can be used in soldering joints. Courtesy of ESAB Welding & Cutting Products.

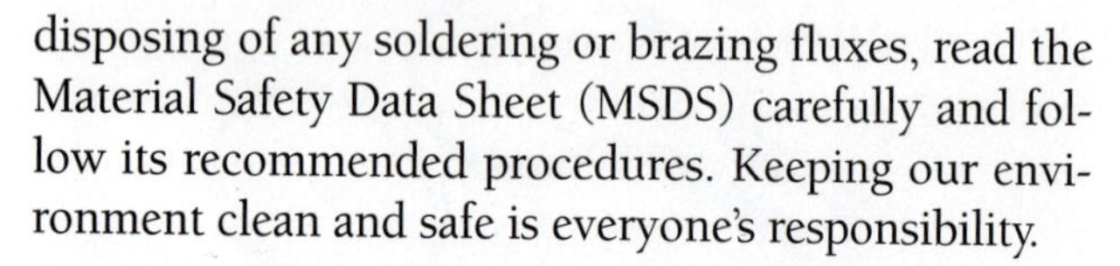

disposing of any soldering or brazing fluxes, read the Material Safety Data Sheet (MSDS) carefully and follow its recommended procedures. Keeping our environment clean and safe is everyone's responsibility.

Fluxing Action Soldering and brazing fluxes will remove light surface oxides, promote wetting, and aid in capillary action. The use of fluxes does not eliminate the need for good joint cleaning. Fluxes will not remove oil, dirt, paint, glues, heavy oxide layers, or other surface contaminants.

Soldering fluxes are chemical compounds such as muriatic acid (hydrochloric acid), sal ammoniac (ammonium chloride), or rosin. Brazing fluxes are chemical compounds such as fluorides, chlorides, boric acids, and alkalies. These compounds react to dissolve, absorb, or mechanically break up thin surface oxides that are formed as the parts are being heated. They must be stable and remain active through the entire temperature range of the solder or braze filler metal. The chemicals in the flux react with the oxides as either acids or alkaline (bases). Some dip fluxes are salts.

The reactivity of a flux is greatly affected by temperature. As the parts are heated to the soldering or brazing temperature, the flux becomes more active. Some fluxes are completely inactive at room temperature. Most fluxes have a temperature range within which they are most effective. Care should be taken to avoid overheating fluxes. If they become overheated or burned, they will stop working as fluxes, and they become a contamination in the joint. If overheating has occurred, the welder must stop and clean off the damaged flux before continuing.

Fluxes that are active at room temperature must be neutralized (made inactive) or washed off after the job is complete. If these fluxes are left on the joint, premature failure may result due to flux-induced corrosion. Fluxes that are inactive at room temperature do not have to be cleaned off the part. However, if the part is to be painted or auto body plastic is to be applied, fluxes must be removed.

Torch Soldering and Brazing Oxyfuel or air fuel torches can be used either manually or automatically, **Figure 5-14**. Acetylene is often used as the fuel gas, but it is preferable to use one of the other fuel gases having a higher heat level in the secondary flame, **Figure 5-15**. The oxyacetylene flame has a very high temperature near the inner cone, but it has little heat in the outer flame. This often results in the parts being overheated in a localized area. Such fuel gases as MAPP®, propane, butane, and natural gas have a flame that will heat parts more uniformly. Often torches are used that mix air with the fuel gas in a swirling or turbulent

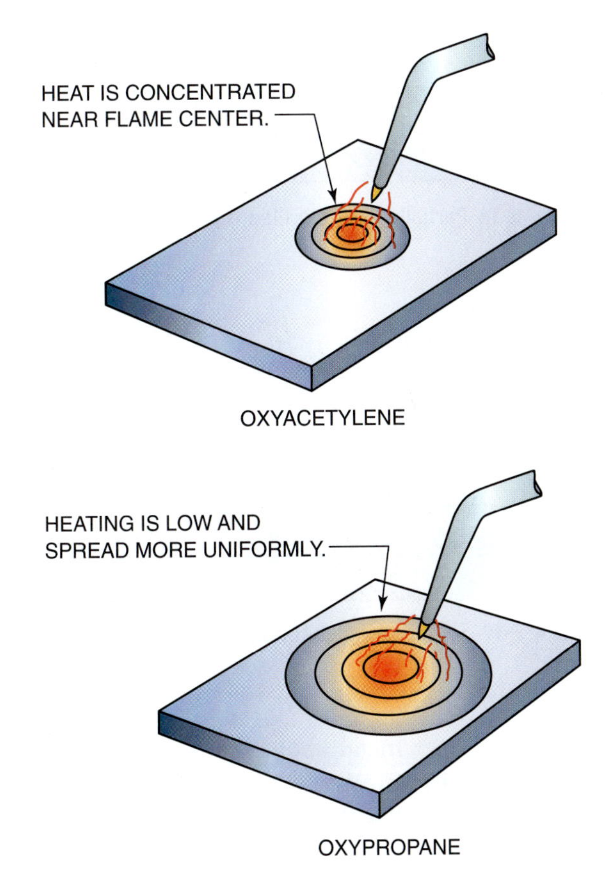

FIGURE 5-15 The high temperature of an oxyacetylene flame may cause localized overheating.

manner to increase the flame's temperature, **Figure 5-16**. The flame may even completely surround a small-diameter pipe, heating it from all sides at once, **Figure 5-17**.

Some advantages of using a torch include the following:

- Versatility—Using a torch is the most versatile method. Both small and large parts in a wide variety of materials can be joined with the same torch.
- Portability—A torch is very portable. Anyplace a set of cylinders can be taken or anywhere the hoses can be pulled into can be soldered or brazed with a torch.
- Speed—The flame of the torch is one of the quickest ways of heating the material to be joined, especially on thicker sections.

Some of the disadvantages of using a torch include the following:

- Overheating—When using a torch, it is easy to overheat or burn the parts, flux, or filler metal.
- Skill—A high level of skill with a torch is required to produce consistently good joints.

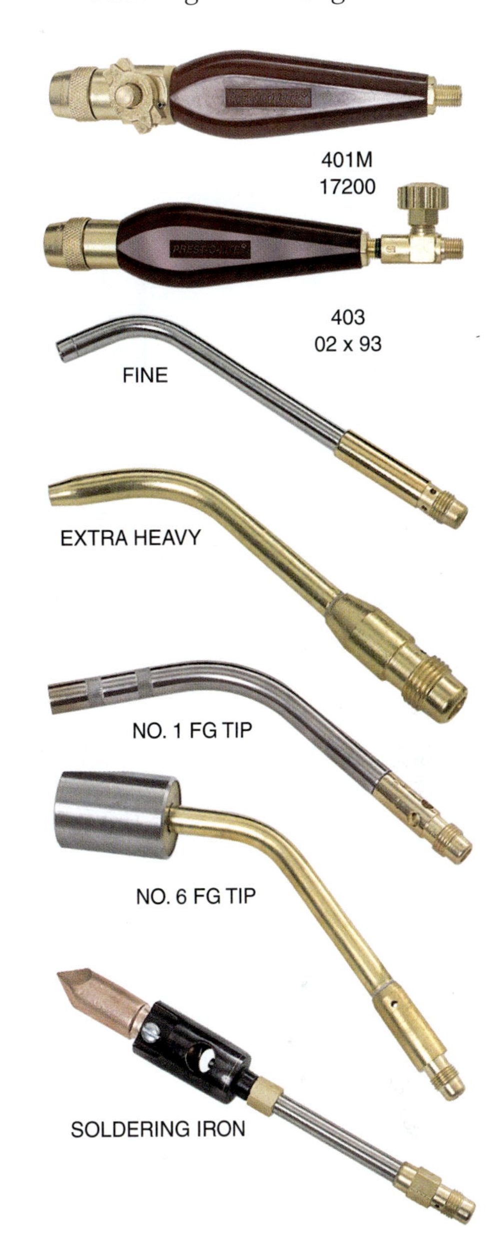

FIGURE 5-16 Examples of torch tips and handles that use air-fuel mixtures for brazing. Courtesy of ESAB Welding & Cutting Products.

- Fires—It is easy to start a fire if a torch is used around combustible (flammable) materials.

Filler Metals

General The type of filler metal used for any specific joint should be selected by considering as many as possible of the criteria listed in **Table 5-1**. It would be impossible to consider each of these items with the same importance. Welders must decide which things they feel are the most important and then base their selection on that decision.

Soldering and brazing metals are alloys—that is, a mixture of two or more metals. Each alloy is available in a variety of percentage mixtures. Some

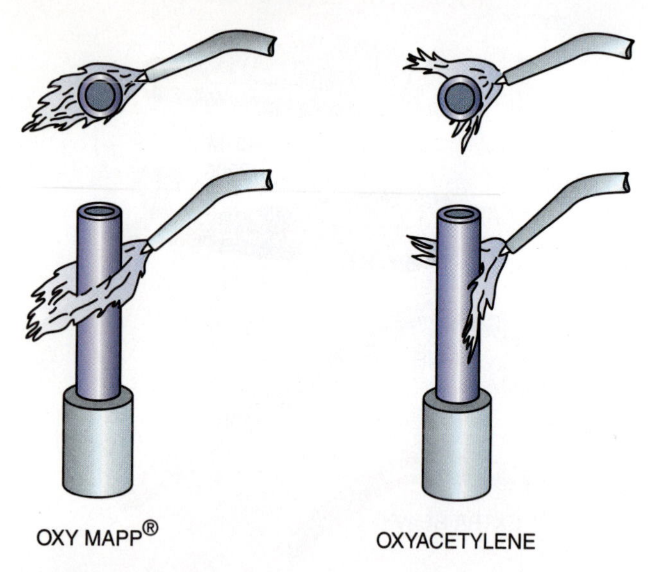

FIGURE 5-17 Heating characteristics of oxy MAPP® compared with oxyacetylene on round materials.

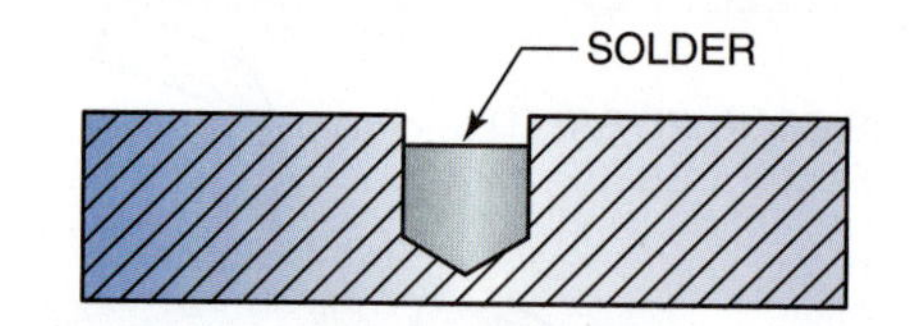

FIGURE 5-18 Partially fill the drilled hole with solder.

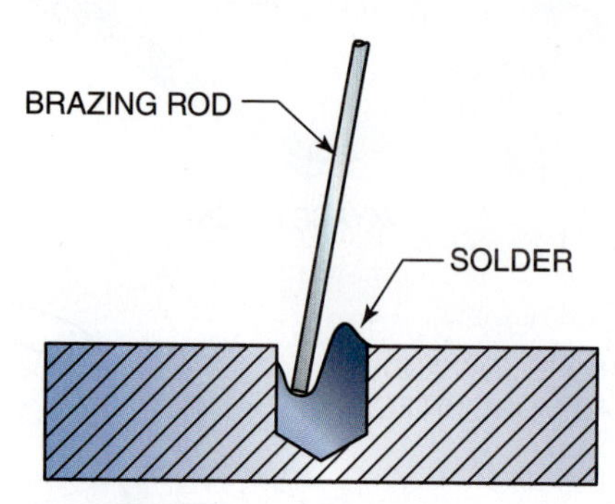

FIGURE 5-19 Solder being shaped as it cools to its paste range.

- Material being joined
- Strength required
- Joint design
- Availability and cost
- Appearance
- Service (corrosion)
- Heating process to be used
- Cost

TABLE 5-1 Criteria for Selecting Filler Metal.

mixtures are stronger, and some melt at lower temperatures than other mixtures. Each one has specific properties. Almost all of the alloys used for soldering or brazing have a paste range. A paste range is the temperature range in which a metal is partly solid and partly liquid as it is heated or cooled. As the joined part cools through the paste range, it is important that the part not be moved. If the part is moved, the solder or braze metal may crumble like dry clay, destroying the bond.

EXPERIMENT 5-1

Paste Range

This experiment shows the effect on bonding of moving a part as the filler metal cools through its paste range. The experiment also shows how metal can be "worked" using its paste range. You will need tin-lead solder composed of 20% to 50% tin, with the remaining percentage being lead. You also will need a properly lit and adjusted torch, a short piece of brazing rod, and a piece of sheet metal. Using a hammer make a dent in the sheet metal about the size of a quarter (25¢), **Figure 5-18.**

In a small group, watch the effects of heating and cooling solder as it passes through the paste range. Using the torch, melt a small amount of the solder into the dent and allow it to harden. Remelt the solder slowly, frequently flashing the torch off and touching the solder with the brazing rod until it is evident the solder has all melted. Once the solder has melted, stick the brazing rod in the solder and remove the torch. As the solder cools, move the brazing rod in the metal and observe what happens, **Figure 5-19.**

As the solder cools to the uppermost temperature of its paste range, it will have a rough surface appearance as the rod is moved. When the solder cools more, it will start to break up around the rod. Finally, as it becomes a solid, it will be completely broken away from the rod.

Now slowly reheat the solder and work the surface with the rod until it can be shaped like clay. If the surface is slightly rough, a quick touch of the flame will smooth it. This is the same way in which "lead" is applied to some body panel joints on a new car so that the joints are not seen on the car when it is finished. The lead used is actually a tin-lead alloy or solder. A large area can be made as smooth as glass without sanding by simply flashing the area with the flame.

Complete a copy of the "Student Welding Report" listed in Appendix I or provided by your instructor. ◆

Soldering Alloys Soldering alloys are usually identified by their major alloying elements. **Table 5-2** lists the major types of solder and the materials they will

Tin-lead	Copper and copper alloys Mild steel Galvanized metal
Tin-antimony	Copper and copper alloys Mild steel
Cadmium-silver	High strength for copper and copper alloys Mild steel Stainless steel
Cadmium-zinc	Aluminum and aluminum alloys

TABLE 5-2 Soldering Alloys.

join. In many cases, a base material can be joined by more than one solder alloy. In addition to the considerations for selecting filler metal, specific factors are listed in the following sections for the major soldering alloys.

Tin-lead

This is the most popular solder and is the least expensive one. An alloy of 61.9% tin and 38.1% lead melts at 362°F (183°C) and has no paste range. This is the eutectic composition (lowest possible melting point of an alloy) of the tin-lead solder. An alloy of 60% tin and 40% lead is commercially available and is close enough to the eutectic alloy to have the same low melting point with only a 12°F (7.8°C) paste range. The widest paste range is 173°F (78°C) for a mixture of 19.5% tin and 80.5% lead. This mixture begins to solidify at 535°F (289°C) and is totally solid at 362°F (193°C). The closest mixture that is commercially available is a 20% tin and 80% lead alloy. **Table 5-3** lists the percentages, temperatures, and paste ranges for tin-lead solders. Tin-lead solders are most commonly used on electrical connections but must never be used for water piping. Most health and construction codes will not allow tin-lead solders for use on water or food-handling equipment.

CAUTION

Tin-lead solders must not be used where lead could become a health hazard in things such as food and water.

Tin-antimony

This family of solder alloys has a higher tensile strength and lower creep than the tin-lead solders. The most common alloy is 95/5, 95% tin and 5% antimony. This is often referred to as "hard solder."

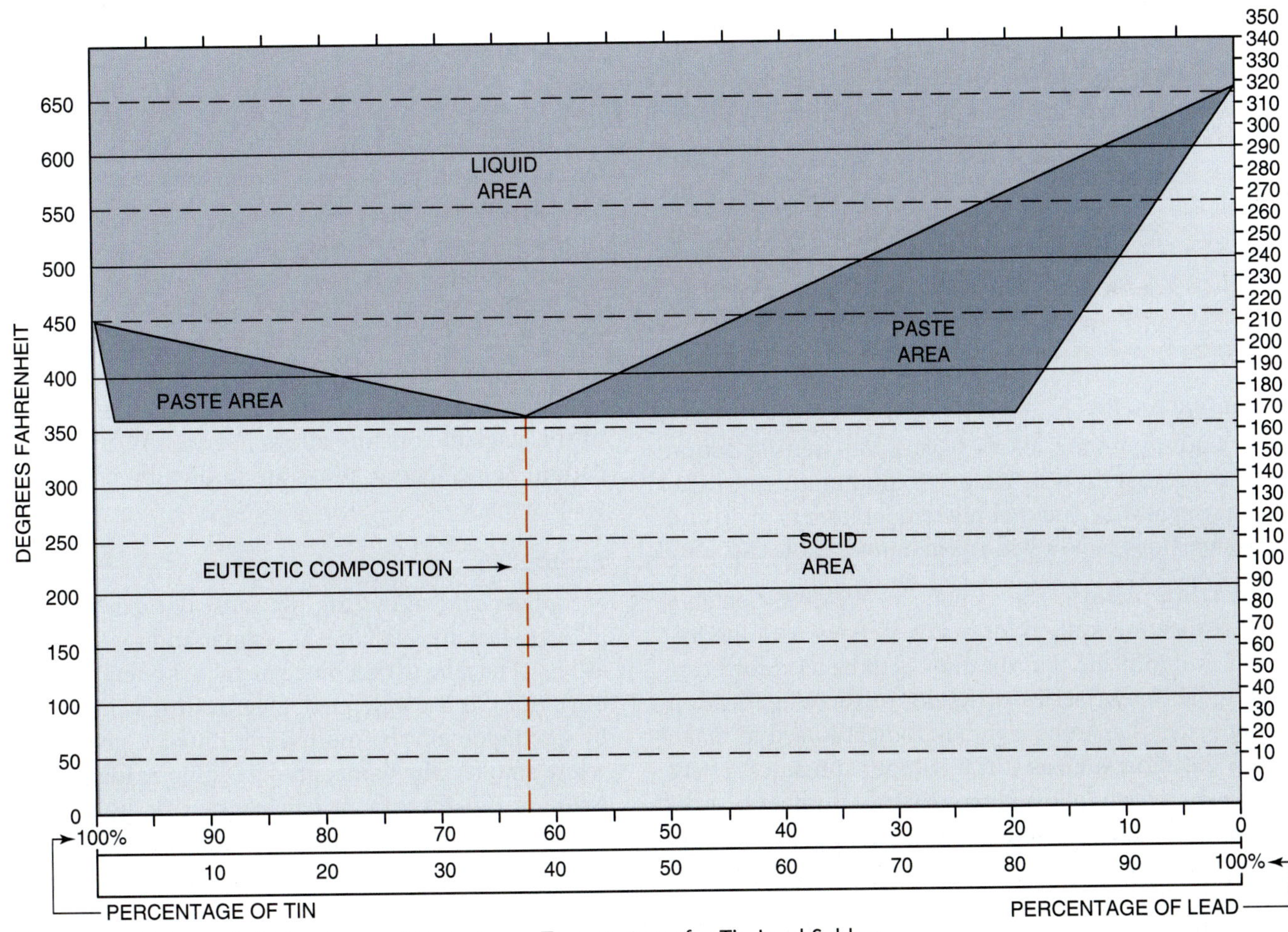

TABLE 5-3 Melting, Solidification, and Paste Range Temperatures for Tin-Lead Solders.

Cadmium	Zinc	Completely Liquid	Completely Solid	Paste Range
82.5%	17.5%	509°F (265°C)	509°F (265°C)	No paste range
40.0%	60.0%	635°F (335°C)	509°F (265°C)	126°F (52°C)
10.0%	90.0%	750°F (399°C)	509°F (265°C)	241°F (116°C)

TABLE 5-4 Cadmium-Zinc Alloys.

This is the most common solder used in plumbing because it is lead free. The use of "C" flux, which is a mixture of flux and small flakes of solder, makes it easier to fabricate quality joints. This mixture of flux and solder will draw additional solder into the joint as it is added.

Cadmium-silver

These solder alloys have excellent wetting, flow, and strength characteristics, but they are expensive. The silver in this solder helps improve wetting and strength. Cadmium-silver alloys melt at a temperature of around 740°F (393°C); they are called high-temperature solders because they retain their strength at temperatures above most other solders. These solder alloys can be used to join aluminum to itself or other metals—for example, to piping that is used in air-conditioning equipment.

Cadmium-zinc

Cadmium-zinc alloys have good wetting action and corrosion resistance on aluminum and aluminum alloys. The melting temperature is high, and some alloys have a wide paste range, **Table 5-4**.

Brazing Alloys The American Welding Society's classification system for brazing alloys uses the letter *B* to indicate that the alloy is to be used for brazing. The next series of letters in the classification indicates the atomic symbol of metals used to make up the alloy, such as CuZn (copper and zinc). There may be a dash followed by a letter or number to indicate a specific alloyed percentage. The letter *R* may be added to indicate that the braze metal is in rod form. An example of a filler metal designation is

Base Metal	Brazing Filler Metal
Aluminum	BAlSi, aluminum silicon
Carbon steel	BCuZn, brass (copper-zinc)
	BCu, copper alloy
	BAg, silver alloy
Alloy steel	BAg, silver alloy
	BNi, nickel alloy
Stainless steel	BAg, silver alloy
	BAu, gold base alloy
	BNi, nickel alloy
Cast iron	BCuZn, brass (copper-zinc)
Galvanized iron	BCuZn, brass (copper-zinc)
Nickel	BAu, gold base alloy
	BAg, silver alloy
	BNi, nickel alloy
Nickel-copper alloy	BNi, nickel alloy
	BAg, silver alloy
	BCuZn, brass (copper-zinc)
Copper	BCuZn, brass (copper-zinc)
	BAg, silver alloy
	BCuP, copper-phosphorus
Silicon bronze	BCuZn, brass (copper-zinc)
	BAg, silver alloy
	BCuP, copper-phosphorus
Tungsten	BCuP, copper-phosphorus

TABLE 5-5 Base Metals and Common Brazing Filler Metals Used to Join the Base Metals.

BRCuZn-A, which indicates a copper-zinc brazing rod with 59.25% copper, 40% zinc, and 0.75% tin; **Table 5-5** is a list of the base metals and the most common alloys used to join the base metals. Not all of the available brazing alloys have an AWS classification. Some special alloys are known by registered trade names.

Copper-zinc

Copper-zinc alloys are the most popular brazing alloys. They are available as regular and low-fuming alloys. The zinc in this braze metal has a tendency to burn out if it is overheated. Overheating is indicated by a red glow on the molten pool, which gives off a white smoke. The white smoke is zinc oxide. If zinc oxide is breathed in, it can cause zinc poisoning. Using a low-fuming alloy will help eliminate this problem. Examples of low-fuming alloys are RCuZn-B and RCuZn-C.

Breathing zinc oxide can cause zinc poisoning. If you think you have zinc poisoning, get medical treatment immediately.

Copper-zinc and Copper-phosphorus A5.8

The copper-zinc filler rods are often grouped together and known as brazing rods. The copper-phosphorus rods are referred to as phos-copper. Both terms do not adequately describe the metals in this group. There are vast differences among the five major classifications of the copper-zinc filler metals, as well as among the five major classifications of the copper-phosphorus filler metals. The following material describes the five major classifications of copper-zinc filler rods.

Class BRCuZn is used for the same application as BCu fillers. The addition of 40% zinc (Zn) and 60% copper (Cu) improves the corrosion resistance and aids in this rod's use with silicon-bronze, copper-nickel, and stainless steel.

CAUTION

Care must be exercised in order to prevent overheating this alloy, as the zinc will vaporize, causing porosity and poisonous zinc fumes.

Class BRCuZn-A is commonly referred to as naval brass and can be used to fuse weld naval brass. The addition of 17% tin (Sn) to the alloy adds strength and corrosion resistance. The same types of metal can be joined with this rod as could be joined with BRCuZn.

Class BRCuZn-B is a manganese-bronze filler metal. It has a relatively low melting point and is free flowing. This rod can be used to braze weld steel, cast iron, brass, and bronze. The deposited metal is higher than BRCuZn or BRCuZn-A in strength, hardness, and corrosion resistance.

Class BRCuZn-C is a low-fuming, high silicon (Si) bronze rod. It is especially good for general-purpose work due to the low-fuming characteristic of the silicon on the zinc.

Class BRCuZn-D is a nickel-bronze rod with enough silicon to be low-fuming. The nickel gives the deposit a silver-white appearance and is referred to as white brass. This rod is used to braze and braze weld steel, malleable iron, and cast iron and for building up wear surfaces on bearings.

Copper-phosphorus

This alloy is sometimes referred to as phos-copper. It is a good alloy to consider for joints where silver braze alloys may have been used in the past. Phos-copper has good fluidity and wettability on copper and copper alloys. The joint spacing should be from .001 in. (0.03 mm) to .005 in. (0.12 mm) for the strongest joints. Heavy buildup of this alloy may cause brittleness in the joint. Phosphorus forms brittle iron phosphide at brazing temperatures on steel. Copper-phos or copper-phos-silver should not be used on copper-clad fittings with ferrous substrates because the copper can easily be burned off, exposing the underlying metal to phosphorus embrittlement.

The copper-phosphorus (BCuP group) rods are used in air-conditioning applications and in plumbing to join copper piping. The phosphorus makes the rod self-fluxing on copper. This feature is one of the major advantages of copper-phosphorus rods. The addition of a small amount of silver, approximately 2%, helps with wetting and flow into joints.

Class BCuP-1 has a low wetting characteristic and a lower flow rate than the other phos-copper alloys. This type of filler metal should be preplaced in the joint. The major advantage of this type of filler metal is its increased ductility.

Classes BCuP-2 and BCuP-4 both have good flow into the joint. The high phosphorus content of the rods makes them self-fluxing on copper. Both of these classes are used often for plumbing installations.

Classes BCuP-3 and BCuP-5 both have high surface tension and low flow so that they are used when close fit-ups are not available.

Copper-phosphorus-silver

This alloy is sometimes referred to as sil-phos. Its characteristics are similar to copper-phosphorus except the silver gives this alloy a little better wetting and flow characteristic. Often it is not necessary to use flux with alloys containing 5% or more of silver when joining copper pipe. This is the most common brazing alloy used in air-conditioning and refrigeration work. When sil-phos is used on air-conditioning compressor fittings that are copper-clad steel, care must be taken to make the braze quickly. If the fitting is heated too much or for too long, the copper cladding can be burned off. With this burn-off, the phosphorus can make the steel fitting very brittle, and embrittlement can cause the fitting to crack and leak sometime later.

Silver-copper

Silver-copper alloys can be used to join almost any metal, ferrous or nonferrous, except aluminum, magnesium, zinc, and a few other low-melting metals. This alloy is often referred to as silver braze and is the most versatile. It is among the most expensive alloys, except for the gold alloys.

Nickel

Nickel alloys are used for joining materials that need high strength and corrosion resistance at an elevated temperature. Some applications of these alloys include joining turbine blades in jet engines, torch parts, furnace parts, and nuclear reactor tubing. Nickel will wet and flow acceptably on most metals. When used on copper-based alloys, nickel may diffuse into the copper, stopping its capillary flow.

Nickel and Nickel Alloys A5.14

Nickel and nickel alloys are increasingly used as a substitute for silver-based alloys. Nickel is generally more difficult than silver to use because it has lower wetting and flow characteristics. However, nickel has much higher strength than silver.

Class BNi-1 is a high-strength, heat-resistant alloy that is ideal for brazing jet engine parts and for other similar applications.

Class BNi-2 is similar to BNi-1 but has a lower melting point and a better flow characteristic.

Class BNi-3 has a high flow rate that is excellent for large areas and close-fitted joints.

Class BNi-4 has a higher surface tension than the other nickel filler rods, which allows larger fillets and poor-fitted joints to be filled.

Class BNi-5 has a high oxidation resistance and high strength at elevated temperatures and can be used for nuclear applications.

Class BNi-6 is extremely free flowing and has good wetting characteristics. The high corrosion resistance gives this class an advantage when joining low chromium steels in corrosive applications.

Class BNi-7 has a high resistance to erosion and can be used for thin or honeycomb structures.

Aluminum-silicon

(BAlSi) brazing filler metals can be used to join most aluminum sheet and cast alloys. The AWS type number 1 flux must be used when brazing aluminum. It is very easy to overheat the joint. If the flux is burned by overheating, it will obstruct wetting. Use standard torch brazing practices but guard against overheating.

Copper and Copper Alloys A5.7

Although pure copper (Cu) can be gas fusion welded successfully using a neutral oxyfuel flame without a flux, most copper filler metals are used to join other metals in a brazing process.

Class BCu-1 can be used to join ferrous, nickel, and copper-nickel metals with or without a flux. BCu-1 is also available as a powder that is classified as BCu-1a. This material has the same applications as Bcu-1. The AWS type number 3B flux must be used with metals that are prone to rapid oxidation or with heavy oxides such as chromium, titanium, manganese, and others.

Class BCu-2 has applications similar to those for BCu-1. However, BCu-2 contains copper oxide suspended in an organic compound. Since copper oxides can cause porosity, tying up the oxides with the organic compounds reduces the porosity.

Silver and Gold

Silver and gold are both used in small quantities when joining metals that will be used under corrosive conditions, when high joint ductility is needed, or when low electrical resistance is important. Because of the ever increasing price and reduced availability of these precious metals, other filler metals should first be considered. In many cases, other alloys can be used with great success. When substituting a different filler metal for one that has been used successfully, the new metal and joint should first be extensively tested.

Joint Design

General The spacing between the parts being joined greatly affects the tensile strength of the finished part. **Table 5-6** lists the spacing requirements at the joining temperature for the most common alloys. As the parts are heated, the initial space may increase or decrease, depending upon the joint design and fixturing. The changes due to expansion can be calculated, but trial and error also works.

The strongest joints are obtained when the parts use lap or scarf joints where the joining area is equal to three times the thickness of the thinnest joint member, **Figure 5-20**. The strength of a butt joint can be increased if the area being joined can be increased. Parts that are 1/4 in. (6 mm) thick should

Filler Metal	Joint Spacing	
	in.	mm
BAlSi	.006–.025	(0.15–0.61)
BAg	.002–.005	(0.05–0.12)
BAu	.002–.005	(0.05–0.12)
BCuP	.001–.005	(0.03–0.12)
BCuZn	.002–.005	(0.05–0.12)
BNi	.002–.005	(0.05–0.12)

TABLE 5-6 Brazing Alloy Joint Tolerance.

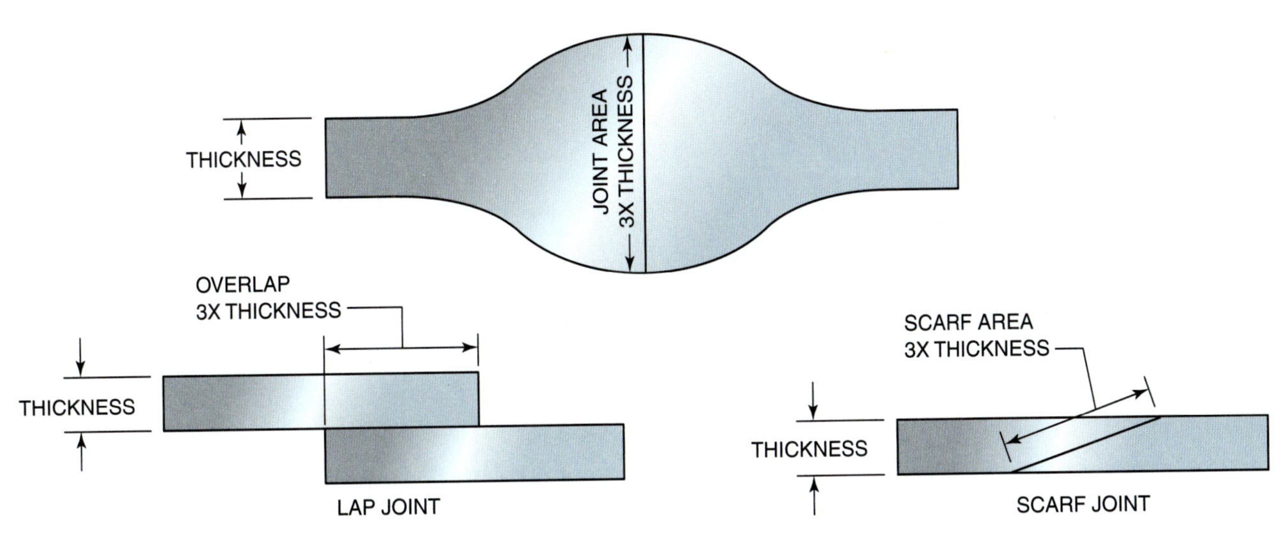

FIGURE 5-20 The joining area should be three times the thickness of the thinnest joint member.

not be considered for brazing or soldering if another process will work successfully.

Some joints can be designed so that the flux and filler metal may be preplaced. When this is possible, visual checking for filler metal around the outside of the joint is easy. Evidence of filler metal around the outside is a good indication of an acceptable joint.

Joint preparation is also very important to a successful soldered or brazed part. The surface must be cleaned of all oil, dirt, paint, oxides, or any other contaminants. The surface can be mechanically cleaned by using a wire brush or by sanding, sandblasting, grounding, scraping, or filing, or it can be cleaned chemically with an acid, alkaline, or salt bath. Soldering or brazing should start as soon as possible after the parts are cleaned to prevent any additional contamination of the joint.

EXPERIMENT 5-2

Fluxing Action

In this experiment, as part of a small group, you will observe oxide removal by a flux as the flux reaches its effective temperature. For this experiment, you will need a piece of copper, either tubing or sheet, rosin or C flux, and a properly lit and adjusted torch.

Any paint, oil, or dirt must first be removed from the copper. Do not remove the oxide layer unless it is blue-black in color. Put some flux on the copper and start heating it with the torch. When the flux becomes active, the copper that is covered by the flux will suddenly change to a bright coppery color. The copper that is not covered by the flux will become darker and possibly turn blue-black, **Figure 5-21**. Continue heating the copper until the flux is burned off and the once clean spot quickly builds an oxide layer.

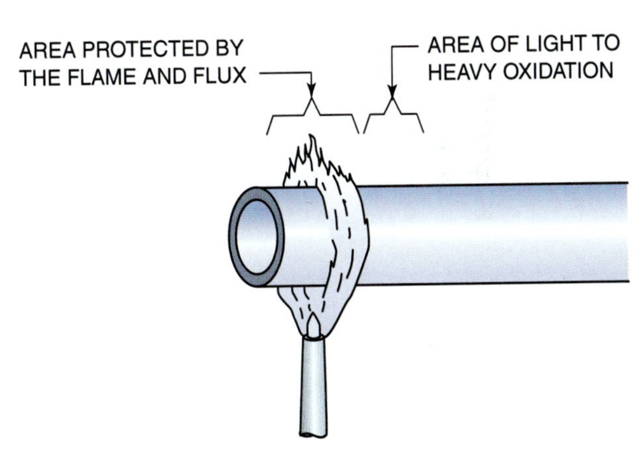

FIGURE 5-21 Copper pipe fluxed and exposed to heat.

Repeat this experiment, but this time hold the torch farther from the metal's surface. When the flux begins to clean the copper, flash the torch off the metal (quickly move the flame off and back onto the same spot). Try to control the heat so that the flux does not burn off.

Complete a copy of the "Student Welding Report" listed in Appendix I or provided by your instructor. ◆

EXPERIMENT 5-3

Uniform Heating

In this experiment, as part of a small group, you will learn how to control the flame direction so that two pieces of metal of unequal size are heated at the same rate to the same temperature. You will need, for this experiment, two pieces of mild steel, one 16 gauge and the other 1/8 in. (3 mm) thick, and a properly lit and adjusted torch.

Place the two pieces of metal on a firebrick to form a butt joint. Then take the torch and point the flame toward the thicker piece of metal, moving it as

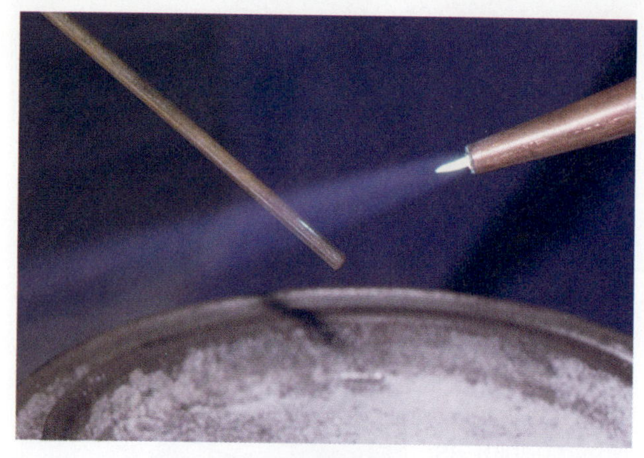

FIGURE 5-22(A) Heating a brazing rod. Courtesy of Larry Jeffus.

FIGURE 5-22(B) Dipping the heated rod into the flux. Courtesy of Larry Jeffus.

FIGURE 5-22(C) Flux stuck to rod ready for brazing. Courtesy of Larry Jeffus.

FIGURE 5-22(D) Prefluxed brazing rods. Courtesy of Larry Jeffus.

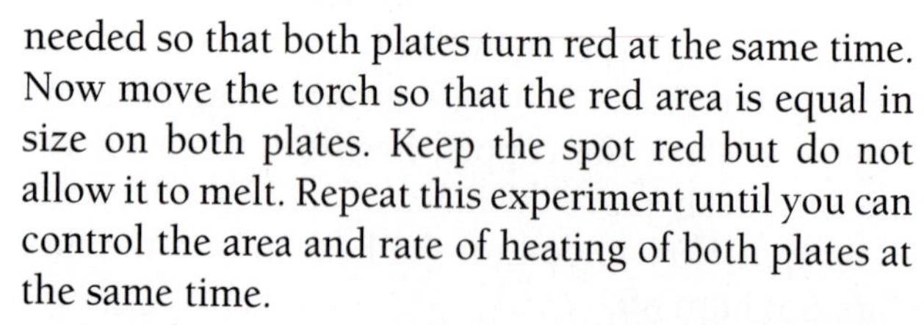

needed so that both plates turn red at the same time. Now move the torch so that the red area is equal in size on both plates. Keep the spot red but do not allow it to melt. Repeat this experiment until you can control the area and rate of heating of both plates at the same time.

Complete a copy of the "Student Welding Report" listed in Appendix I or provided by your instructor. ◆

EXPERIMENT 5-4

Tinning or Phase Temperature

In this experiment, as part of a small group, you will observe the wetting of a piece of metal by a filler metal. For this experiment, you will need one piece of 16 gauge mild steel, BRCuZn filler metal rod, powdered flux, and a properly lit and adjusted torch.

Place the sheet flat on a firebrick. Heat the end of the rod and dip it in the flux so that some flux sticks on the rod, **Figure 5-22A**, **Figure 5-22B**, and **Figure 5-22C**. BRCuZn brazing rods are available as both

FIGURE 5-23 Deposit a spot of braze on the plate and continue heating the plate until the braze flows onto the surface. Courtesy of Larry Jeffus.

bare rods and prefluxed, **Figure 5-22D**. Direct the flame onto the plate. When the sheet gets hot, hold the brazing rod in contact with the sheet, directing the flame so that a large area of the sheet is dull red and the rod starts to melt, **Figure 5-23**. After a

FIGURE 5-24 Checking the surface temperature with a spot of braze metal. Courtesy of Larry Jeffus.

FIGURE 5-25(A) Once the plate is up to temperature, start adding more filler. Courtesy of Larry Jeffus.

molten pool of braze metal is deposited on the sheet, remove the rod and continue heating the sheet and molten pool until the braze metal flows out. Repeat this experiment until you can get the braze metal to flow out in all directions equally at the same time.

Complete a copy of the "Student Welding Report" listed in Appendix I or provided by your instructor. ◆

Brazing The brazing practices that follow use copper-zinc alloys (BRCuZn) for brazing joints on mild steel. Using prefluxed rods is easier for students than using powdered flux, which has to have the hot tip of the brazing rod dipped to apply. Because you may be asked to braze with bare brazing rods, the practices explain how to use them. If you are using prefluxed rods, just ignore this step.

FIGURE 5-25(B) Dip the brazing rod into the leading edge of the molten weld pool. Courtesy of Larry Jeffus.

PRACTICE 5-1

Brazed Stringer Bead

Using a properly lit and adjusted torch, 6 in. (152 mm) of clean 16 gauge mild steel, brazing flux, and BRCuZn brazing rod, you will make a straight bead the length of the sheet.

Place the sheet flat on a firebrick and hold the flame at one end until the metal reaches the proper temperature. Then touch the flux covered rod to the sheet and allow a small amount of brazing rod to melt onto the hot sheet, **Figure 5-24**. Once the molten brazing metal wets the sheet, start moving the torch in a circular pattern while dipping the rod into the molten braze pool as you move along the sheet. If the size of the molten pool increases, you can control it by reducing the torch angle, raising the torch, traveling at a faster rate, or flashing the flame off the molten braze pool, **Figure 5-25A**, **Figure 5-25B**, and **Figure 5-25C**. Flashing the torch off a braze joint will not cause oxidation problems as it does when welding, because the molten metal is protected by a layer of flux.

FIGURE 5-25(C) Remove the rod from the flame area when it is not being added to the molten weld pool. Courtesy of Larry Jeffus.

FIGURE 5-26 Observe the molten flux flowing ahead of the molten weld pool. Courtesy of Larry Jeffus.

As the braze bead progresses across the sheet, dip the end of the rod back in the flux, if a powdered flux is used, as often as needed to keep a small molten pool of flux ahead of the bead, **Figure 5-26.**

The object of this practice is to learn how to control the size and direction of the braze bead. Controlling the width, buildup, and shape shows that you have a good understanding and control of the process. Keeping the braze bead in a straight line indicates that you have mastered the bead well enough to watch the bead and the direction at the same time. Turn off the cylinders, bleed the hoses, back out the regulator adjusting screw, and clean up your work area when you are finished.

Complete a copy of the "Student Welding Report" listed in Appendix I or provided by your instructor. ◆

PRACTICE 5-2

Brazed Butt Joint

Using the same equipment and setup as listed in Practice 5-1, make a braze butt joint on two pieces of 16 gauge mild steel sheet, 6 in. (152 mm) long.

Place the metal flat on a firebrick, hold the plates tightly together, and make a tack braze at both ends of the joint. If the plates become distorted, they can be bent back into shape with a hammer before making another tack weld in the center. Align the sheets so that you can comfortably make a braze bead along the joint. Starting as you did in Practice 5-1, make a uniform braze along the joint. Repeat this practice until a uniform braze can be made without defects. Turn off the cylinders, bleed the hoses, back out the regulator adjusting screw, and clean up your work area when you are finished.

Complete a copy of the "Student Welding Report" listed in Appendix I or provided by your instructor. ◆

PRACTICE 5-3

Brazed Tee Joint

Using the same equipment, material, and setup as listed in Practice 5-2, you will make a brazed tee joint with 100% root penetration, **Figure 5-27.**

Tack the pieces of metal into a tee joint as you did in Practice 5-2. To obtain 100% penetration, direct the flame on the sheets just ahead of the braze bead, being careful not to overheat the braze metal. If the bead has a notch, the root of the joint is still not hot enough to allow the braze metal to flow properly. If the braze metal appears to have flowed properly after you have completed the joint, look at the back of the joint for a line of braze metal that flowed through. Repeat this practice until the joint can be made without defects. Turn off the cylinders, bleed the hoses, back out the regulator adjusting screw, and clean up your work area when you are finished.

Complete a copy of the "Student Welding Report" listed in Appendix I or provided by your instructor. ◆

PRACTICE 5-4

Brazed Lap Joint

Using the same equipment, material, and setup as listed in Practice 5-2, you will make a brazed lap joint in the flat position.

Place the pieces of sheet metal on a firebrick so that they overlap each other by approximately 1/2 in. (13 mm). It is important that the pieces be held flat relative to each other, **Figure 5-28.** Make a small tack braze on both ends and then one or two tack brazes along the joint. Hold the torch so the flame moves along the joint and heats up both pieces at the same time. When the sheets are hot, touch the rod to the sheets and make a bead similar to the butt joint. After completing the brazed joint, it should be uniform in width and appearance. Repeat this practice until the joint can be made without defects. Turn off the cylinders, bleed the hoses, back out the regulator adjusting screw, and clean up your work area when you are finished.

Complete a copy of the "Student Welding Report" listed in Appendix I or provided by your instructor. ◆

PRACTICE 5-5

Brazed Tee Joint, Thin to Thick Metal

Using a properly lit and adjusted torch, two pieces of mild steel 6 in. (152 mm) long, one 16 gauge and the other 1/4 in. (6 mm) thick, brazing flux, and BRCuZn brazing rod, you will make a tee joint in the flat position.

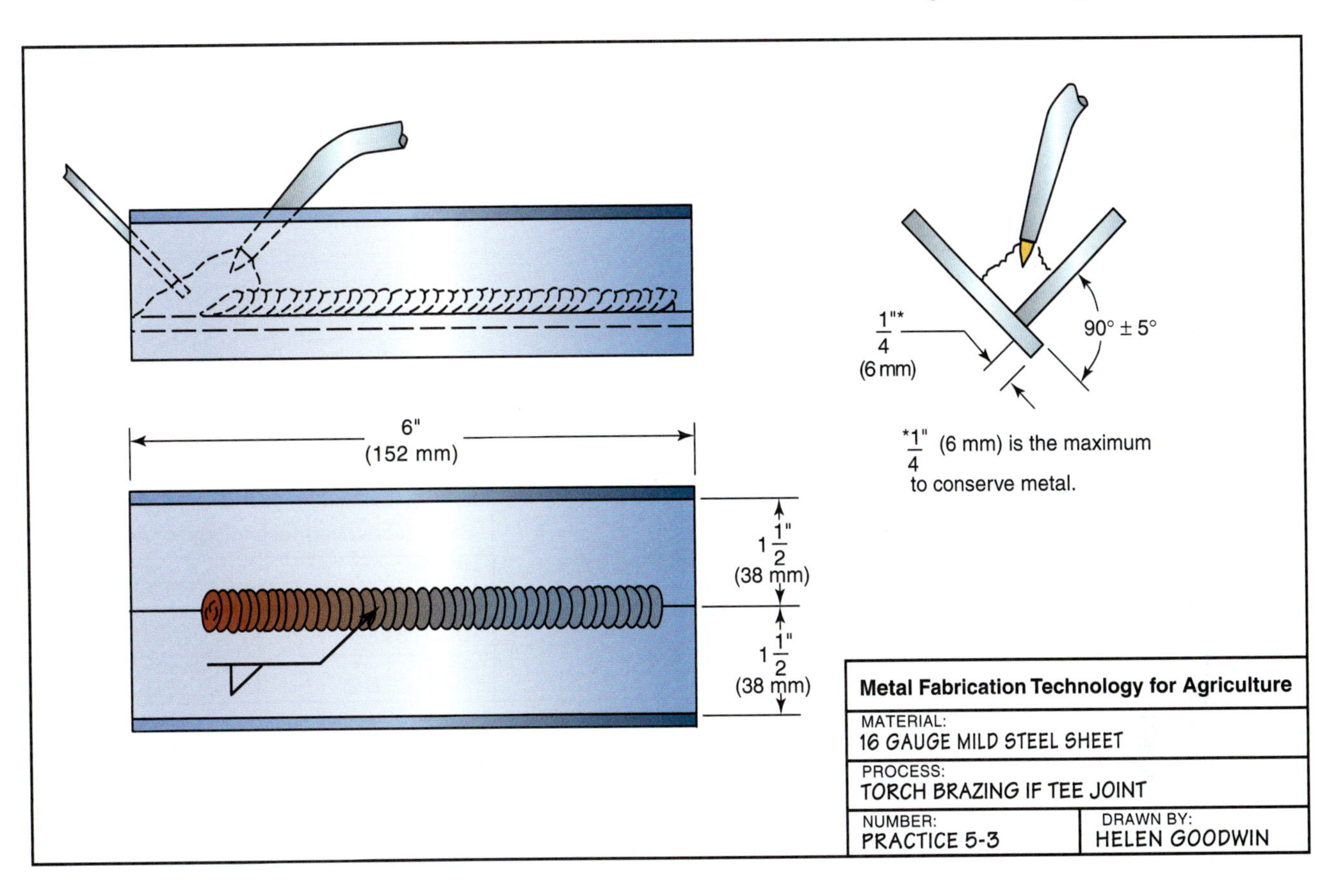

FIGURE 5-27 Brazed tee joint.

FIGURE 5-28 Tack braze lap joint. Courtesy of Larry Jeffus.

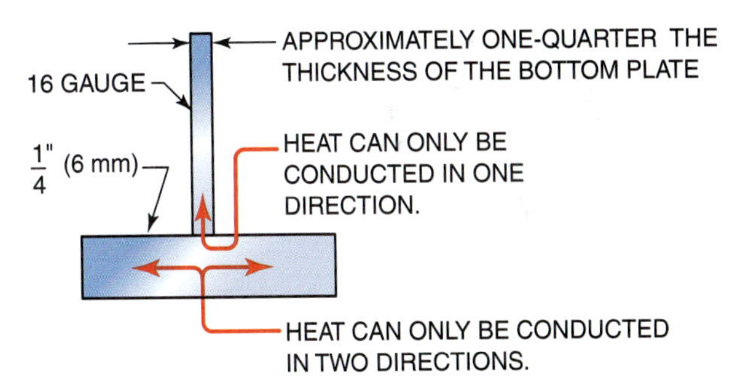

FIGURE 5-29 Unequal rate of heating due to a difference in mass.

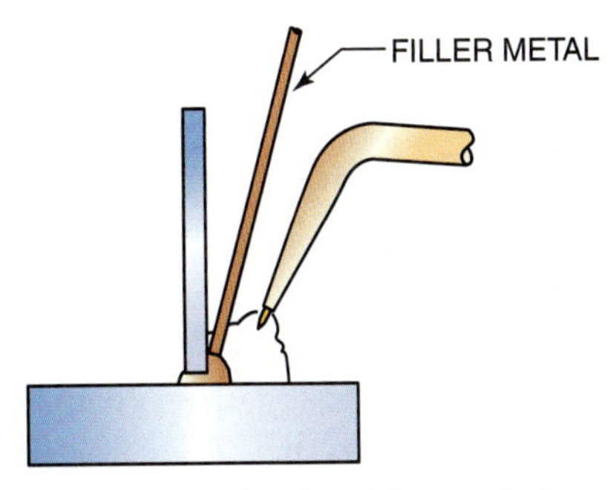

FIGURE 5-30 Torch and rod positions to balance heating between parts of unequal thickness.

Hold the 16 gauge metal vertically on the 1/4 in. (6 mm) plate and tack braze both ends. The vertical member of a tee joint heats up faster than the flat member because the heat on the vertical member can only be conducted, **Figure 5-29**. The thin plate will heat up faster than the thick plate because there is less mass (metal). For the braze bead to be equal, both plates must be heated equally. Direct the flame on the thicker plate, as shown in **Figure 5-30**, and add the brazing rod on the thinner plate. This action will keep the thin plate from overheating. Make a braze along the joint that is uniform in appearance. Repeat this practice until the joint can be made without defects. Turn off the cylinders, bleed the hoses, back out the regulator adjusting

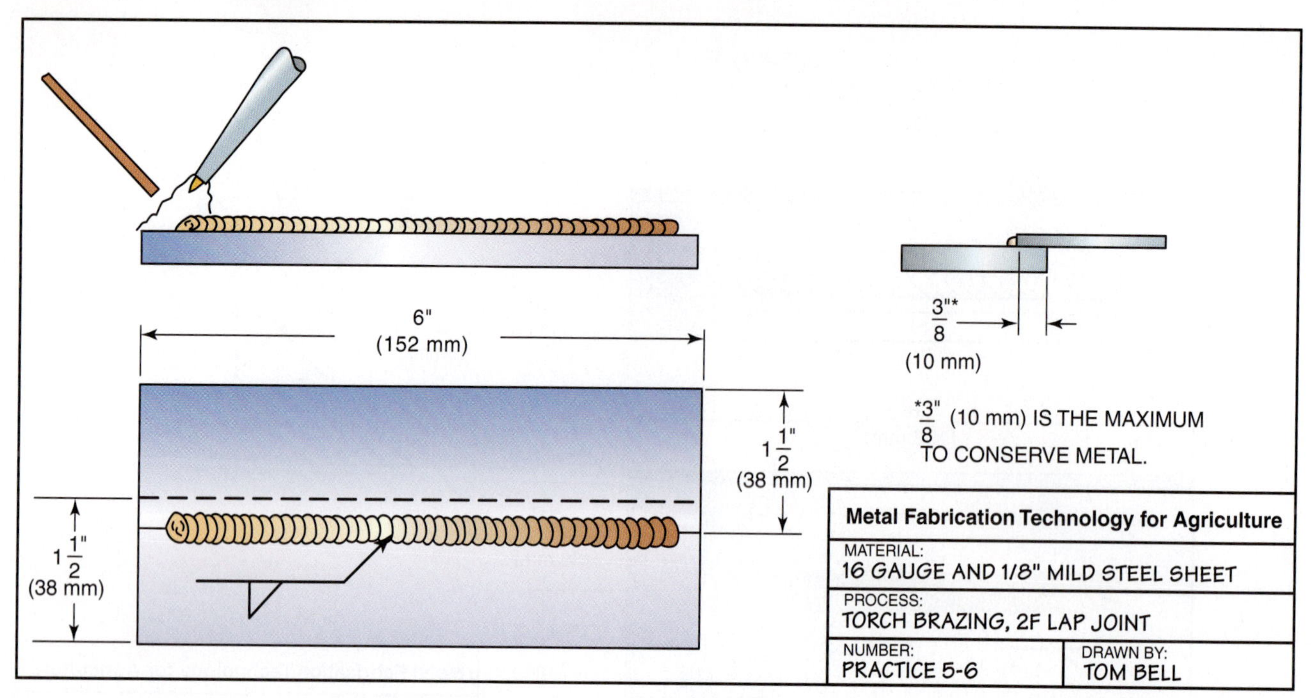

FIGURE 5-31 Brazed lap joint, thin to thick metal.

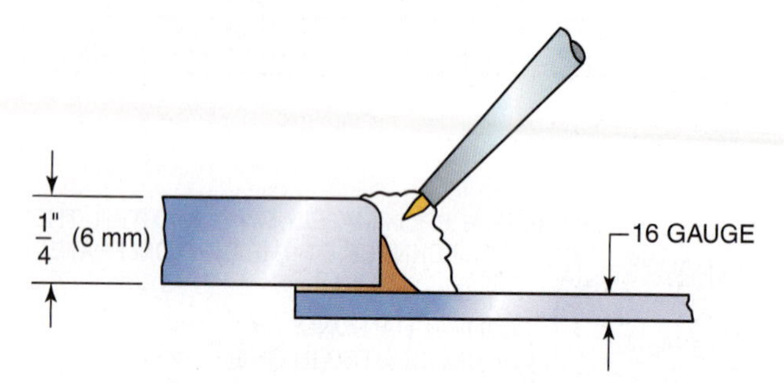

FIGURE 5-32 Direct the heat on the thicker section to ensure proper bonding.

screw, and clean up your work area when you are finished.

Complete a copy of the "Student Welding Report" listed in Appendix I or provided by your instructor. ◆

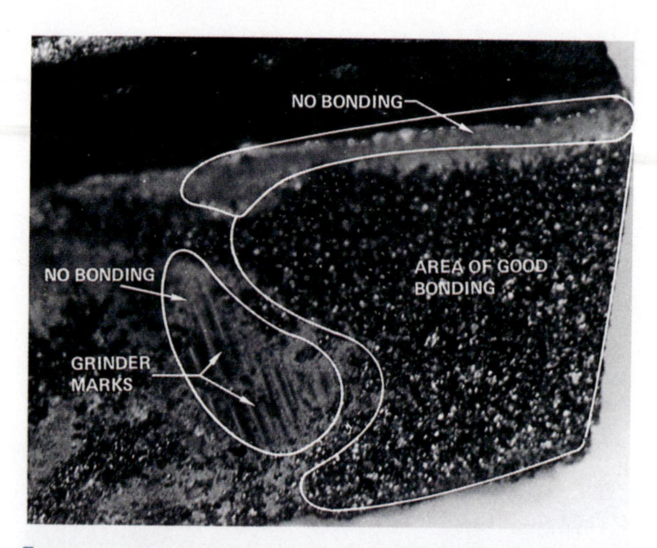

FIGURE 5-33 Only part of this braze joint is bonded properly. Grinder marks can be seen casted into the braze. The base metal was too cool for proper bonding to occur.
Courtesy of Larry Jeffus.

PRACTICE 5-6

Brazed Lap Joint, Thin to Thick Metal

Using the same equipment, materials, and setup as described in Practice 5-5, make a lap joint in the horizontal position, **Figure 5-31**.

Tack braze the pieces together, being sure that they are held tightly together. Place the metal on a firebrick with the thin metal up. Apply heat to the exposed thick metal and more slowly to the overlapping thin metal so that conduction from the thin metal will heat the thick metal at the lap, **Figure 5-32**. If the braze is started before the thick metal is sufficiently heated, the filler metal will be chilled, and a bond will not occur. After the joint is completed and cooled, tap the joint with a hammer to see if there is a good bonded joint. **Figure 5-33** shows a broken braze joint that bonded properly only in certain areas. Repeat this practice until the joint can be made without defects. Turn off the cylinders, bleed the hoses, back out the regulator adjusting screw, and clean up your work area when you are finished.

Complete a copy of the "Student Welding Report" listed in Appendix I or provided by your instructor. ◆

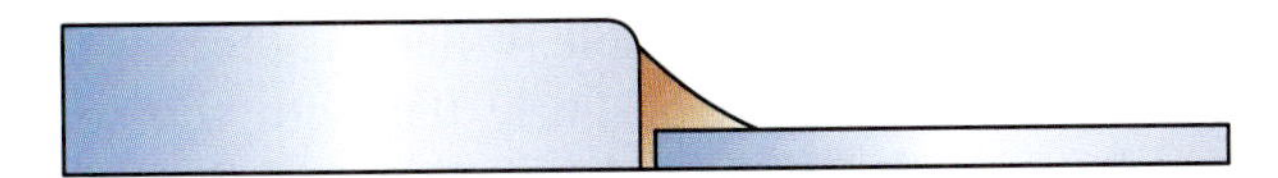

FIGURE 5-34 Joint preparation.

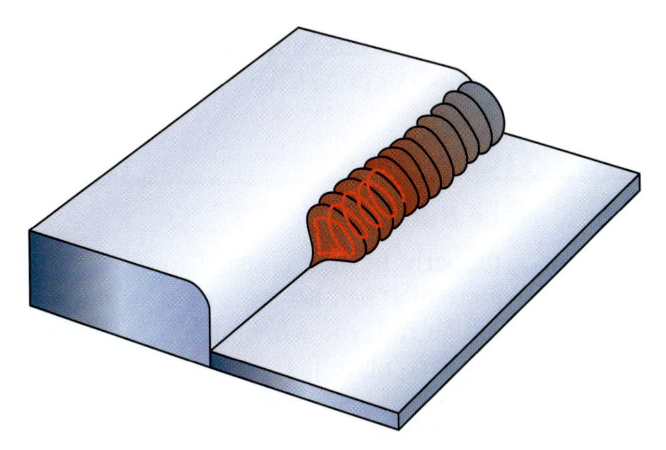

FIGURE 5-35 Flame movement to ensure a 100% root penetration.

PRACTICE 5-7

Braze Welded Butt Joint, Thick Metal

Using a properly lit and adjusted torch, 6 in. (152 mm) of mild steel plate 1/4 in. (6 mm) thick, flux, and BRCuZn filler metal, you will make a flat braze welded butt joint.

Grind the edges of the plate so that they are slightly rounded, **Figure 5-34**. The rounded edges are better than the shape used for brazing because they distribute the strain on the braze more uniformly. After the plates have been prepared, tack braze both ends together. Because of the mass of the plates, it may be necessary to preheat the plates. This helps the penetration and eliminates cold lap at the root. The flame should be moved in a triangular motion so that the root is heated, as well as the top of the bead, **Figure 5-35**. When the joint is complete and cool, bend the joint to check for complete root bonding, **Figure 5-36**. Repeat this practice until the joint can be made without defects. Turn off the cylinders, bleed the hoses, back out the regulator adjusting screw, and clean up your work area when you are finished.

Complete a copy of the "Student Welding Report" listed in Appendix I or provided by your instructor. ◆

PRACTICE 5-8

Braze Welded Tee Joint, Thick Metal

Using the same equipment, materials, and setup as listed in Practice 5-7, you will make a braze welded tee joint in the flat position.

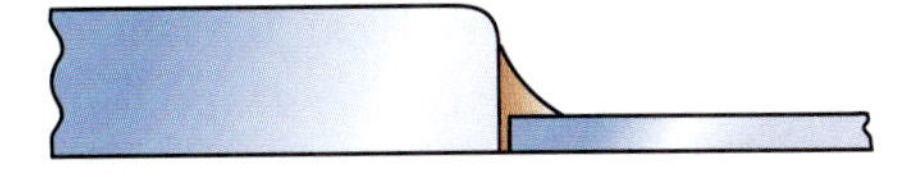

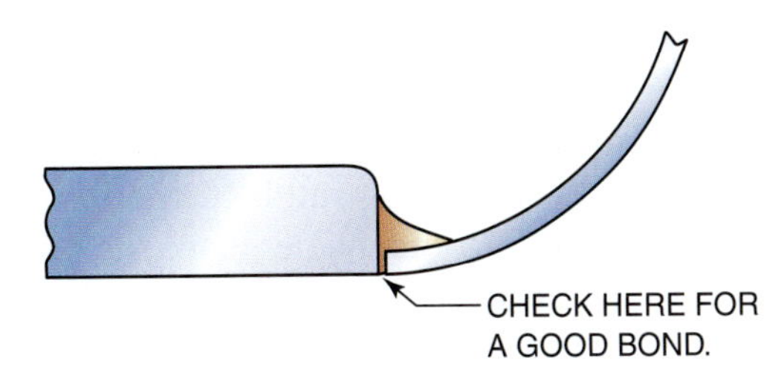

FIGURE 5-36 Bend the braze joint to check for a good bond.

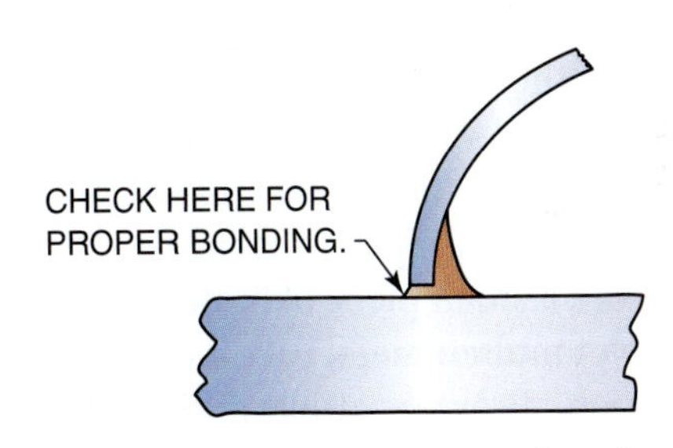

FIGURE 5-37 Bend test to check for root bonding.

As with the butt joint, the edge of the plate is to be slightly rounded, and the metal may have to be preheated to get a good bond at the root. Direct the flame toward the bottom plate and into the root and add the filler metal back into the root. Watch for a notch, which indicates lack of bonding at the root. When the braze is complete and cool, bend it with a hammer to check for a bond at the root, **Figure 5-37**. Repeat this practice until the joint can be made without defects. Turn off the cylinders, bleed the hoses, back out the regulator adjusting screw, and clean up your work area when you are finished.

Complete a copy of the "Student Welding Report" listed in Appendix I or provided by your instructor. ◆

Building up Surfaces and Filling Holes

Surfaces on worn parts are often built up again with braze metal. Braze metal is ideal for parts that receive limited abrasive wear because the buildup is easily machinable. Unlike welding or hardsurfacing, **braze buildup** has no hard spots that make remachining difficult. Braze buildups are good both for flat and round stock. The lower temperature used in brazing does not tend to harden the base metal as much as in welding.

Holes in light-gauge metal can be filled using braze metal. The filled hole can be ground flush if it is required for clearance, leaving a strong patch with minimum distortion.

PRACTICE 5-9

Braze Welding to Fill a Hole

Using a properly lit and adjusted torch, one piece of 16 gauge mild steel, flux, and BRCuZn filler rod, you will fill a 1-in. (25-mm) hole. Place the piece of metal on two firebricks so that the hole is between them. Start by running a stringer bead around the hole, **Figure 5-38.** Once the bead is complete, turn the torch at a very sharp angle and point it at the edge of the hole nearest the torch. Hold the end of the filler rod in the flame so that both the bead around the hole and the rod meet at the same time, **Figure 5-39.** Put the rod in the molten bead and flash the torch off to allow the molten braze pool to cool. When it has cooled, repeat this process. Surface tension will hold a small piece of molten metal in place. If the piece of molten metal becomes too large, it will drop through. Progress around the hole as many times as needed to fill the hole. When the braze weld is complete, it should be fairly flat with the surrounding metal. Repeat this practice until it can be made without defects. Turn off the cylinders, bleed the hoses, back out the regulator adjusting screw, and clean up your work area when you are finished.

Complete a copy of the "Student Welding Report" listed in Appendix I or provided by your instructor. ◆

PRACTICE 5-10

Flat Surface Buildup

Using a properly lit and adjusted torch, a 3-in. (76-mm) square of 1/4-in. (6-mm) mild steel, flux, and BRCuZn filler rod, you will build up a surface.

Place the square plate flat on a firebrick. Start along one side of the plate and make a braze weld down that side. When you get to the end, turn the plate 180° and braze back alongside the first braze, **Figure 5-40**, covering about one-half of the first braze bead. Repeat this procedure until the side is covered with braze metal.

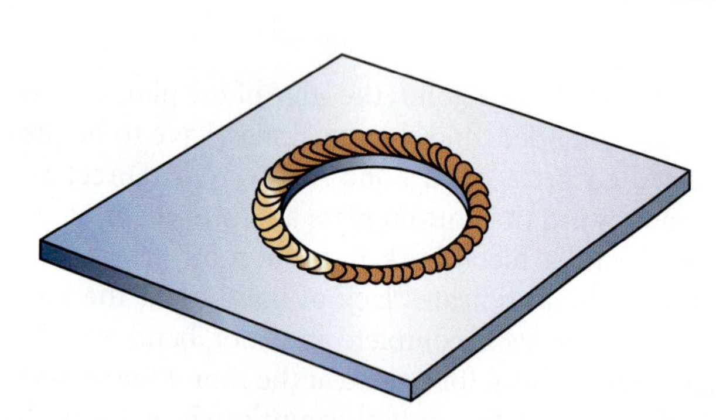

FIGURE 5-38 Filling a hole with braze. First run a bead around the outside of the hole.

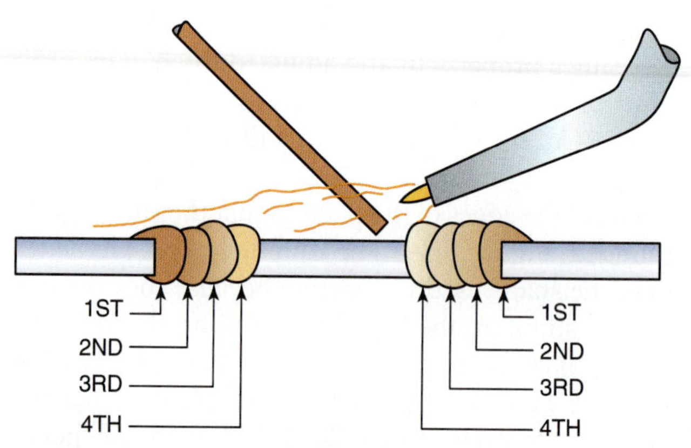

FIGURE 5-39 Keep running beads around the hole until it is closed.

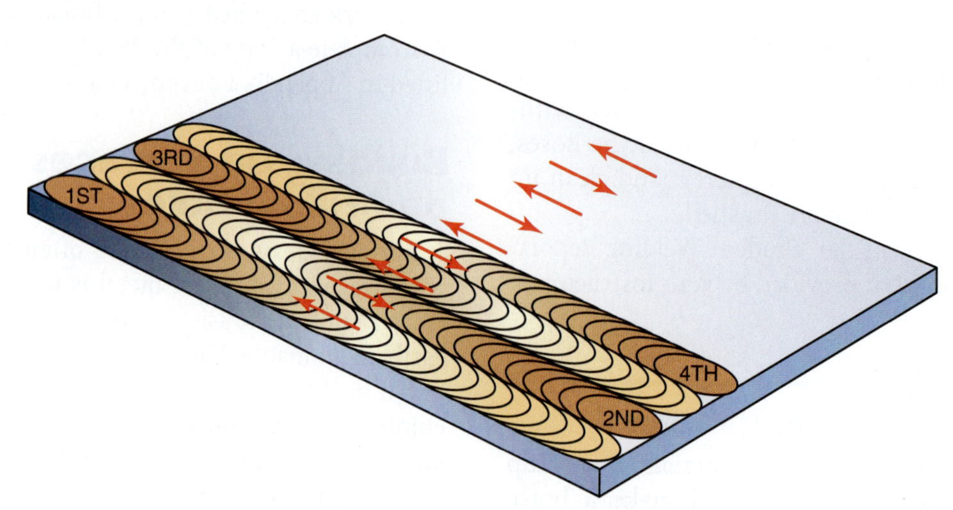

FIGURE 5-40 Braze buildup, first layer.

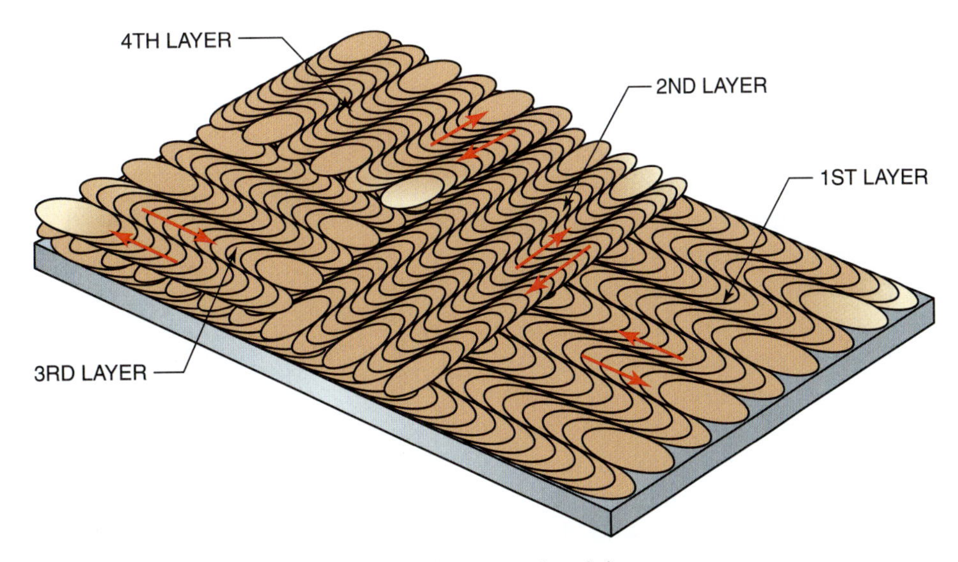

FIGURE 5-41 When building up a surface, alternate the direction of each layer.

Turn the plate 90° and repeat the process, **Figure 5-41.** Be sure that you are getting good fusion with the first layer and that there are no slag deposits trapped under the braze. Be sure to build up the edges so that they could be cut back square. This process should be repeated until there is at least 1/4 in. (6 mm) of buildup on the surface. The surfacing can be checked visually or by machining the square to a 3-in. (76-mm) × 1/2-in. (13-mm) block and checking for slag inclusions. The plate will warp as a result of this buildup. If the plate is going to be used, it should be clamped down to prevent distortion. Turn off the cylinders, bleed the hoses, back out the regulator adjusting screw, and clean up your work area when you are finished.

Complete a copy of the "Student Welding Report" listed in Appendix I or provided by your instructor. ◆

FIGURE 5-42 Round shaft built up with braze. Courtesy of Larry Jeffus.

PRACTICE 5-11

Round Surface Buildup

Using a properly lit and adjusted torch, one piece of mild steel rod 1/2 in. (13 mm) in diameter × 3 in. (76 mm) long, flux, and BRCuZn brazing rod, you will build up a round surface.

In the flat position, start at one end and make a braze weld bead, 1 1/2 in. (38 mm) long, along the side of the steel rod. Turn the rod and make another bead next to the first bead, covering about one-half of the first, **Figure 5-42** and **Figure 5-43.** Repeat this procedure until the rod is 1 in. (25 mm) in diameter. It may be necessary to make a braze bead around both ends of the buildup to keep it square. The buildup can be visually inspected, or it can be turned down in a lathe. Repeat this practice until it can be made without defects. Turn off the cylinders, bleed the hoses, back out the regulator adjusting screw, and clean up your work area when you are finished.

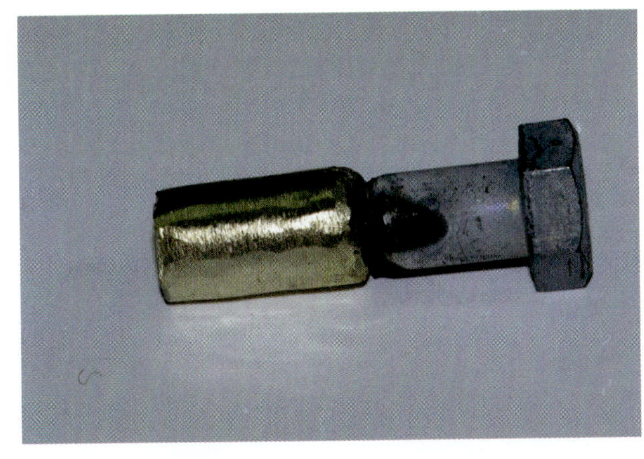

FIGURE 5-43 Shaft turned down to check for slag inclusions or poor bonding. Courtesy of Larry Jeffus.

Complete a copy of the "Student Welding Report" listed in Appendix I or provided by your instructor. ◆

Silver Brazing

The silver brazing practices that follow will use BCuP-2 to BCuP-5 alloys to make brazed copper pipe joints. The melting temperature for the alloys is around 1400°F (760°C). At these temperatures the copper pipe will be glowing a dull red. The best types of flame to use for this type of brazing are air acetylene, air MAPP®, air propane, or any air fuel-gas mixture. The most popular types are air acetylene and air MAPP®. If an oxyacetylene torch is used, it is easy to overheat the alloy. To prevent overheating with an oxyacetylene flame, keep the torch moving and hold the flame so that the inner cone is about 1″ (25 mm) from the surface.

When using BCuP silver brazing alloys on clean copper it is not necessary to use a flux. The phosphorus in these alloys makes them self-fluxing. It is the phosphorus that promotes the wetting and enhances the flow of the alloy into the joint space.

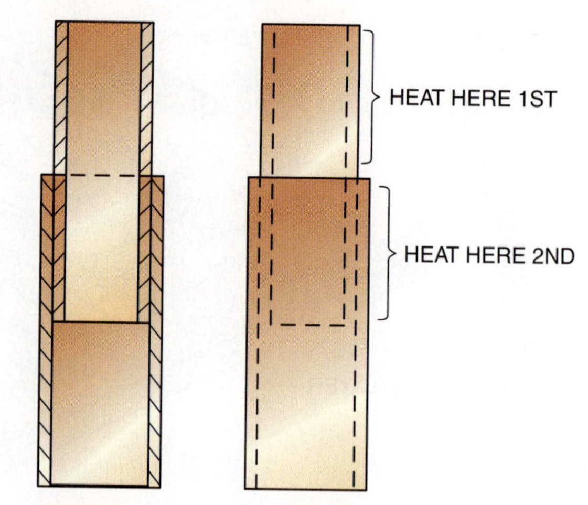

FIGURE 5-44 2G vertical down.

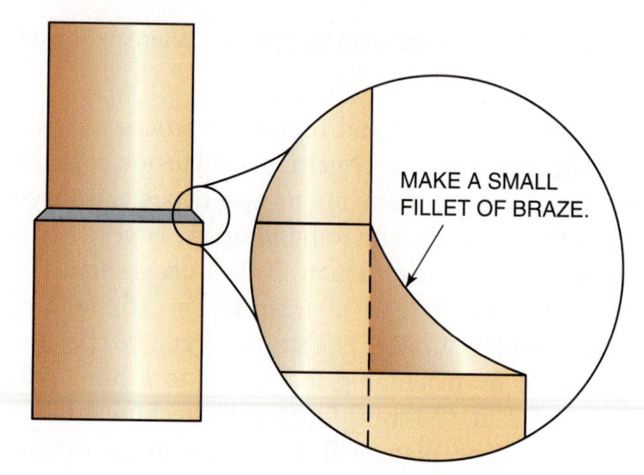

FIGURE 5-45 Making a smooth brazed fillet is a good way to prevent pin hole leaks.

PRACTICE 5-12

Silver Brazing Copper Pipe, 2G Vertical Down Position

Using a properly lit and adjusted torch with air acetylene, air MAPP®, air propane, or any air fuel-gas mixture; two or more pieces of 1/2 in. to 1 in. (13 mm to 25 mm) copper pipe with matching copper pipe fittings; BCuP-2 to BCuP-5 brazing metal; steel wool; sand cloth and/or wire brush; safety glasses with side shields; gloves; proper protective clothing; pliers; and any other required personal safety equipment, you will make a silver brazed joint in copper pipe in the 2G vertical down position, **Figure 5-44.**

1. Set the regulator pressures according to the manufacturer's specifications for your fuel type and tip size.
2. Clean the pipe and fitting using steel wool, sand cloth, or a wire brush. *Note:* Do not touch the cleaned surfaces with your hands. Oils from your skin can prevent the braze metal from wetting. This can result in leaks.
3. Slide the fitting onto the pipe. Be sure that the pipe is completely seated at the bottom of the fitting.
4. Heat the brazing rod and make a bend in the wire about 3/4 in. (19 mm) from the end. This will give you a gauge so that you do not put too much brazing metal in the joint.
5. Heat the pipe first, but not too much. *Note:* As the pipe is heated it will expand, forming a tighter fit in the fitting. This will aid in conducting heat deeper into the fitting socket, and the tighter fit will aid in capillary attraction of the braze metal.
6. Once the pipe is hot but not glowing red, start heating the fitting. Keep the torch moving to uniformly heat the entire joint.
7. The joint is at the correct temperature for brazing when the braze metal starts to wet the surface. Touch the tip of the brazing rod to check the parts, so that you know when they reach the correct temperature.
8. Move the flame to the back side of the joint and feed the brazing rod into the joint space. *Note:* A slight pressure directly into the joint with the brazing rod will help it flow deeper into the fitting. This is especially helpful on larger diameter pipes.
9. Next, move the torch and brazing rod around the pipe and fitting, so that the entire joint will be filled. There should be a slight fillet showing all the way around the joint, **Figure 5-45.** *Note:* The braze metal fillet adds very lit-

FIGURE 5-46 Saw the joint in two. Courtesy of Larry Jeffus.

FIGURE 5-47 Quarter saw the joint. Courtesy of Larry Jeffus.

FIGURE 5-48 Flatten the sawed quarters. Courtesy of Larry Jeffus.

tle strength to the joint, but it is an easy way to make sure that the joint is leak free.

After the pipe has cooled, test the joint, by sawing the fitting in to just past the end of the inside pipe, **Figure 5-46.** Place the pipe vertically in a vise and saw it into quarters, **Figure 5-47.** Use a hammer and flat surface to flatten out each quarter section, **Figure 5-48.**

FIGURE 5-49 Complete joint fill. Courtesy of Larry Jeffus.

FIGURE 5-50 Incomplete joint fill. Courtesy of Larry Jeffus.

With the joint prepared, check for complete penetration of the braze alloy, **Figure 5-49.** Check the sides of the flattened quarters for braze alloy, **Figure 5-50**, or other defects.

Repeat this practice until you can make defect-free joints. Turn off the cylinder, bleed the hoses, back out the regulator adjusting screw, and clean up your work area when you are finished.

Complete a copy of the "Student Welding Report" listed in Appendix I or provided by your instructor. ◆

PRACTICE 5-13

Silver Brazing Copper Pipe, 5G Horizontal Fixed Position

Using the same equipment, materials, setup, and procedures as described in Practice 5-12, make a silver brazing copper pipe joint in the 5G horizontal fixed position.

1. Mount the pipe in a horizontal position, **Figure 5-51.**
2. Start by heating entirely around the pipe. Move the flame back and forth around the entire

FIGURE 5-51 5G horizontal fixed position.

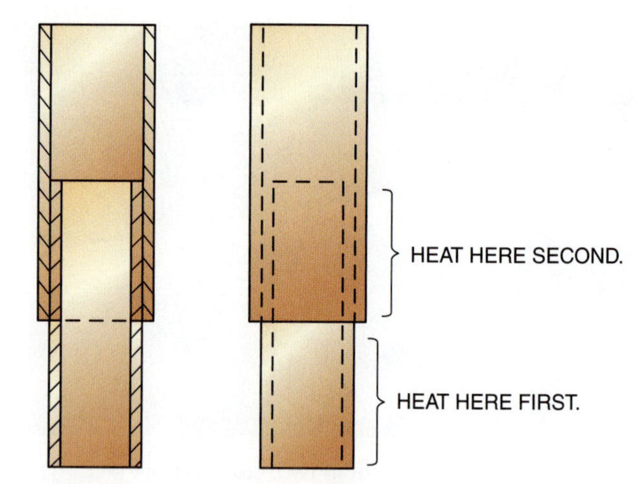

FIGURE 5-52 2G vertical up position.

diameter of the pipe. *Note:* On pipe that is larger than 1 in. (25 mm) concentrate most of your heating on the top of the pipe because the flame may not have enough heat to heat up the entire pipe to the brazing temperature.

3. Once the pipe is hot but not glowing red start heating the fitting. Keep the torch moving to heat the entire joint uniformly.
4. The joint is at the correct temperature for brazing when the braze metal starts to wet the surface. Touch the tip of the brazing rod to check the parts, so that you know when they reach the correct temperature.
5. Move the flame to the back side of the joint and feed the brazing rod into the joint space with a slight pressure into the joint space.
6. Gravity and capillary action will pull the braze metal down to the bottom of the joint if the entire diameter of the pipe is at the correct temperature. If necessary, move the torch slowly down the sides of the pipe to bring this part of the joint up to the brazing temperature. Add braze metal as needed, but do not overfill the joint. An indication that the joint has been overfilled is a bump of braze metal hanging from the bottom of the finished joint.

Repeat this practice until you can make defect-free joints. Turn off the cylinder, bleed the hoses, back out the regulator adjusting screw, and clean up your work area when you are finished.

Complete a copy of the "Student Welding Report" listed in Appendix I or provided by your instructor. ◆

PRACTICE 5-14

Silver Brazing Copper Pipe, 2G Vertical Up Position

Using the same equipment, materials, setup, and procedures as described in Practice 5-12, make a silver brazing copper pipe joint in the 2G vertical up position.

1. Mount the pipe in a vertical up position, **Figure 5-52.**
2. Start by heating entirely around the pipe. Move the flame back and forth around the entire diameter of the pipe.
3. Once the pipe is hot but not glowing red move the flame up onto the fitting to start it heating. Keep the torch moving to uniformly heat the entire joint.
4. The joint is at the correct temperature for brazing when the braze metal starts to wet the surface. Touch the tip of the brazing rod to check the parts, so that you know when they reach the correct temperature.
5. Move the flame to the top of the joint and feed the brazing rod into the joint space with a slight pressure into the joint space.
6. The heat and capillary action will pull the braze metal up into the top of the joint space if the pipe is at the correct temperature. Add braze metal as needed, but do not overfill the joint. An indication that the joint has been overfilled are drips of braze metal running down the vertical pipe below the joint.

Repeat this practice until you can make defect-free joints. Turn off the cylinder, bleed the hoses, back out the regulator adjusting screw, and clean up your work area when you are finished.

Complete a copy of the "Student Welding Report" listed in Appendix I or provided by your instructor. ◆

Soldering

The soldering practices that follow use tin-lead or tin-antimony solders. Both solders have low melting temperature. If an oxyacetylene torch is used, its very easy to overheat the solder. Caution is necessary because most of the fluxes used with this type of solder are easily overheated. The best type of flame to use for this type of soldering is air acetylene, air MAPP®, air propane, or any air fuel-gas mixture. The most popular types are air acetylene and air propane. If galvanized metal is used, additional ventilation should be used to prevent zinc oxide poisoning.

PRACTICE 5-15

Soldered Tee Joint

Using a properly lit and adjusted torch, two pieces of 18 gauge to 24 gauge mild steel sheet 6 in. (152 mm) long, flux, and tin-lead or tin-antimony solder wire, you will solder a flat tee joint.

Hold one piece of metal vertical on the other piece and spot solder both ends. If flux cored wire is not being used, paint the flux on the joint at this time. Hold the torch flame so it moves down the joint in the same direction you will be soldering. Continue flashing the torch off and touching the solder wire to the joint until the solder begins to melt. Keeping the molten pool small enough to work with is a major problem with soldering. The flame must be flashed off frequently to prevent overheating. When the joint is completed, the solder should be uniform. Repeat this practice until it can be made without defects. Turn off the cylinders, bleed the hoses, back out the regulator adjusting screw, and clean up your work area when you are finished.

Complete a copy of the "Student Welding Report" listed in Appendix I or provided by your instructor. ◆

PRACTICE 5-16

Soldered Lap Joint

Using the same equipment, materials, and setup as listed in Practice 5-15, you will solder a lap joint in the horizontal position.

Tack the pieces of metal together as shown in **Figure 5-53.** Apply the flux and heat the metal slowly, checking the temperature by touching the solder wire to the metal often. When the work gets hot enough, flash the flame off frequently to prevent overheating and proceed along the joint. When the joint is completed, the solder should be uniform. Repeat this practice until it can be made without defects. Turn off the cylinders, bleed the hoses, back out the regulator adjusting screw, and clean up your work area when you are finished.

Complete a copy of the "Student Welding Report" listed in Appendix I or provided by your instructor. ◆

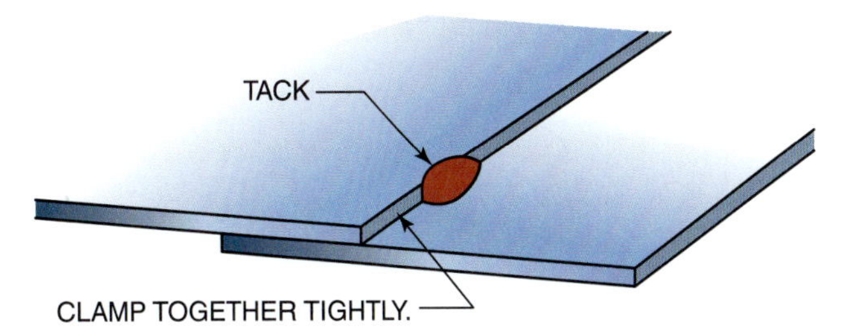

FIGURE 5-53 Tacking metal together for soldering.

PRACTICE 5-17

Soldering Copper Pipe, 2G Vertical Down Position

Using a properly lit and adjusted torch, a piece of 1/2-in. to 1-in. (13-mm to 25-mm) copper pipe, a copper pipe fitting, steel wool, flux, and tin-lead or tin-antimony solder wire, you will solder a pipe joint in the vertical down position, **Figure 5-54.**

Clean the pipe and fitting using steel wool and apply the flux to both parts. Slide the fitting onto the pipe and twist the fitting to insure that the flux is applied completely around the inside of the joint.

Make a bend in the solder wire about 3/4 in. (19 mm) from the end. This will give you a gauge so that you do not put too much solder in the joint. Excessive solder will flow inside the pipe, and it may cause problems to the system later.

Heat the pipe and the fitting with the torch. As the parts become hot, keep checking the parts with the solder wire so that you know when they reach the correct temperature. When the solder starts to wet, remove the flame and wipe the joint with the end of the solder wire, **Figure 5-55.** Next, heat the fitting more than the pipe so that the solder will be drawn into the joint. Rewipe the joint with the solder as needed. There should be a small fillet of solder around the joint. This fillet adds very little strength to the joint, but it is an easy way to make sure that the joint is leak free.

After the pipe has cooled, notch it diagonally with a hacksaw, put a screw driver in the cut, and twist the joint apart, **Figure 5-56.** With the joint separated, check for (1) complete penetration, (2) small porosity caused by overheating the solder, (3) drops of solder inside of the pipe, or (4) other defects. Repeat this practice until it can be done without defects. Turn off the cylinders, bleed the hoses, back out the regulator adjusting screw, and clean up your work area when you are finished.

Complete a copy of the "Student Welding Report" listed in Appendix I or provided by your instructor. ◆

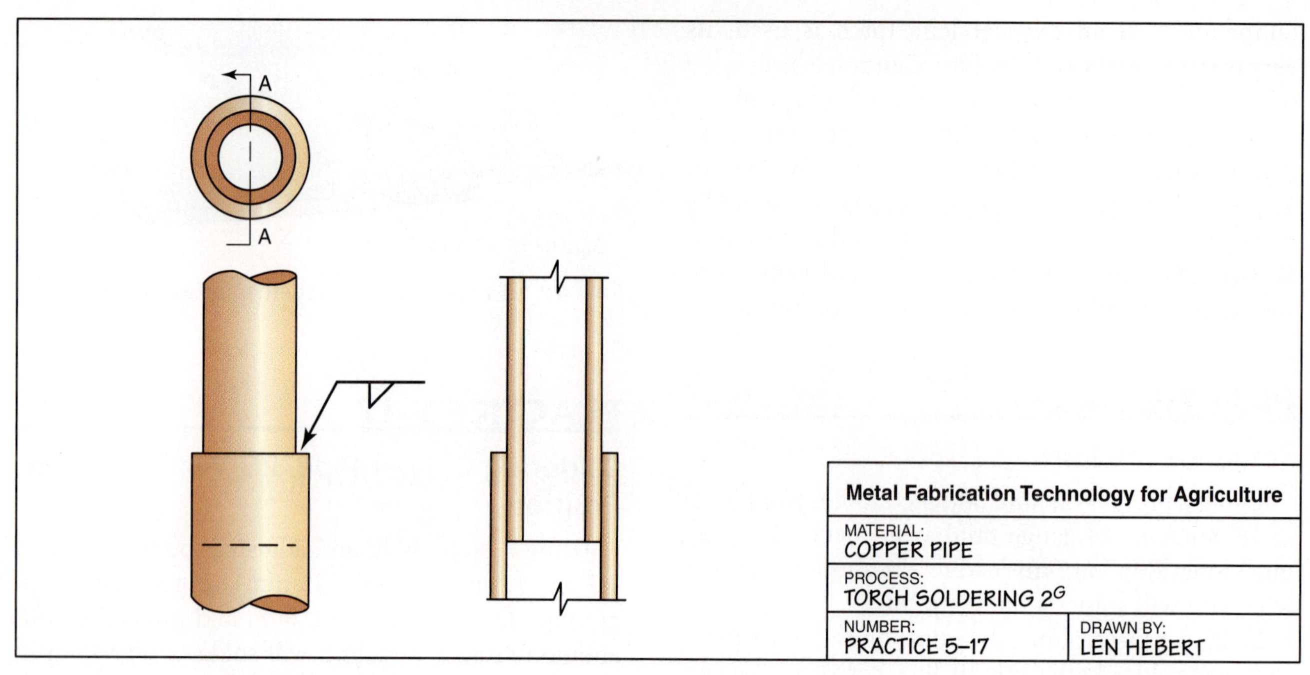

FIGURE 5-54 2G vertical down position.

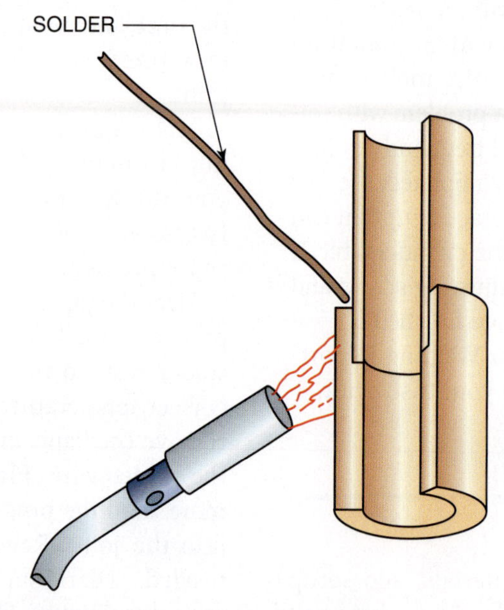

FIGURE 5-55 Soldering copper fitting to copper pipe. Courtesy of Praxair, Inc.

PRACTICE 5-18

Soldering Copper Pipe, 1G Position

Using the same equipment and setup as listed in Practice 5-17, make a soldered joint with the pipe held horizontal, **Figure 5-57**. Test the joint as before and repeat as necessary until it passes the test. Turn off the cylinders, bleed the hoses, back out the regulator adjusting screw, and clean up your work area when you are finished.

Complete a copy of the "Student Welding Report" listed in Appendix I or provided by your instructor. ◆

PRACTICE 5-19

Soldering Copper Pipe, 4G Vertical Up Position

Using the same equipment and setup as listed in Practice 5-17, make a soldered joint with the pipe held in the vertical position with the solder flowing uphill. Test the joint as before and repeat as necessary until it can be made without defects. Turn off the cylinders, bleed the hoses, back out the regulator adjusting screw, and clean up your work area when you are finished.

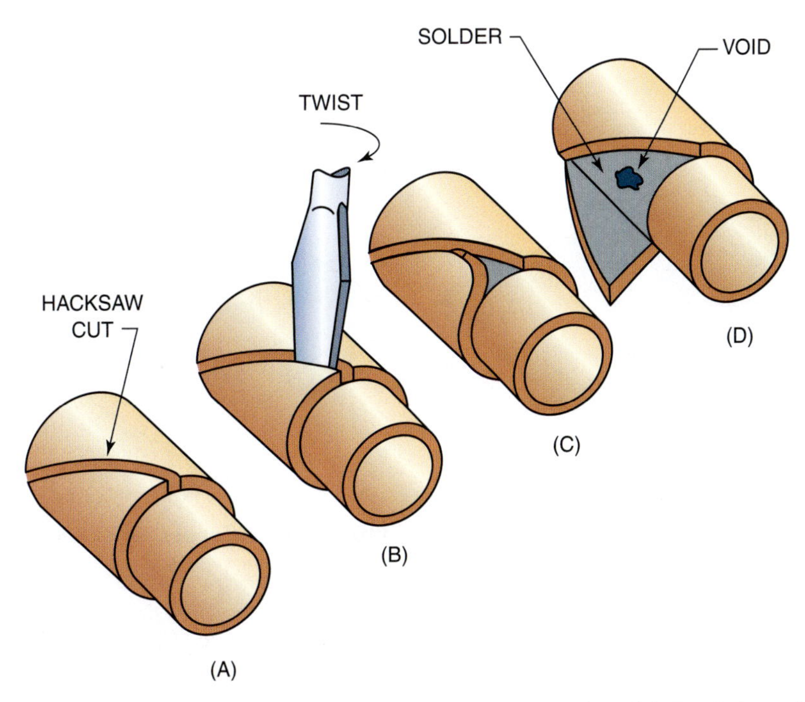

FIGURE 5-56 (A) Use a hacksaw to cut a groove through the outside copper pipe. (B) Carefully push a flat blade screwdriver into the saw cut groove. (C) Twist the blade to open up the groove. (D) Unwrap the outer copper pipe to reveal the solder in the joint.

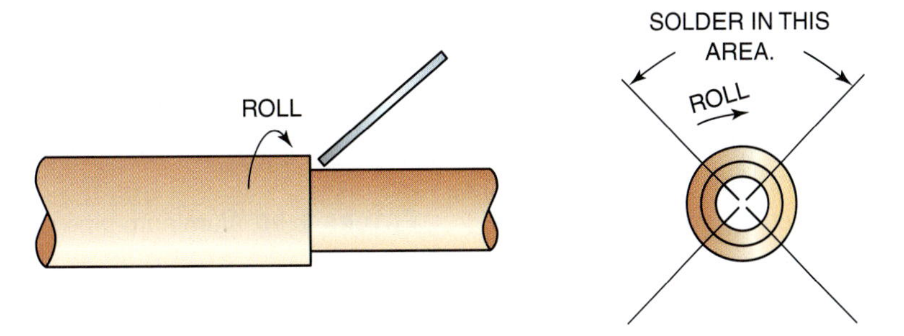

FIGURE 5-57 1G soldering.

Complete a copy of the "Student Welding Report" listed in Appendix I or provided by your instructor. ◆

PRACTICE 5-20

Soldering Aluminum to Copper

Using a properly lit and adjusted torch, a piece of aluminum plate, a copper penny, steel wool, flux, and tin-lead or tin-antimony solder wire, you will tin both the aluminum and the copper with solder and then join both together.

The surface of the aluminum must be clean and free of paint, oils, dirt, and coatings such as anodizes. Hold the flame on the aluminum until it warms up slightly. Hold the solder in the flame and allow a small amount to melt and drop on the aluminum plate; do not add flux, **Figure 5-58.** Move the flame off the plate and rub the liquid solder with the steel wool. Be careful not to burn your fingers or allow the flame to touch the steel wool. The solder should be stuck in the steel wool when it is lifted off the plate. Alternately heat the plate and rub it with the steel wool–solder. When the plate becomes hot enough, it will melt the solder, and the solder will tin the aluminum surface, **Figure 5-59.**

Use some flux and solder and tin the penny. Place the penny on the aluminum plate so the areas of solder on both the penny and plate are touching each other. Heat the two until the solder melts and flows out from between the pennies and plate. When the parts cool, to check the bond, try to break the joint apart.

This process will work on other types of metals that have a strong oxide layer that prevents the solder from bonding. By breaking the oxide layer free with the mechanical action of the steel wool the metals can join. This process can allow a copper patch over an aluminum tube such as those used in air conditioning or

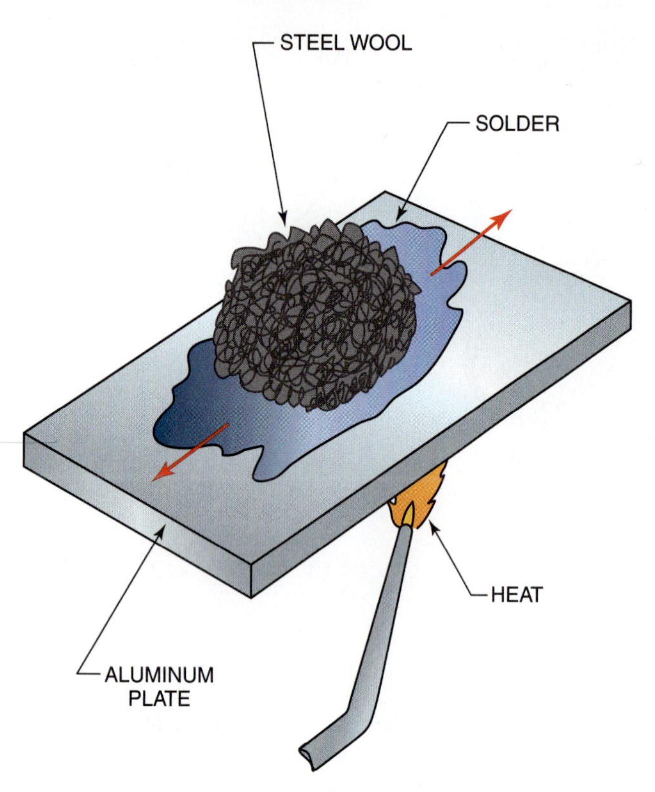

FIGURE 5-58 Tinning aluminum with solder.

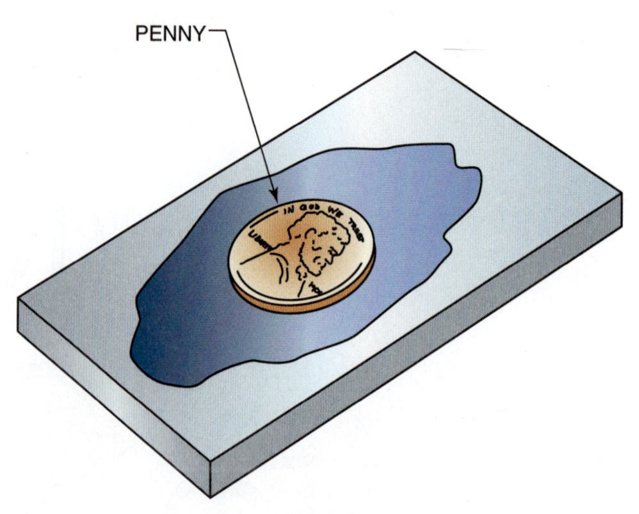

FIGURE 5-59 Copper penny soldered to aluminum using solder.

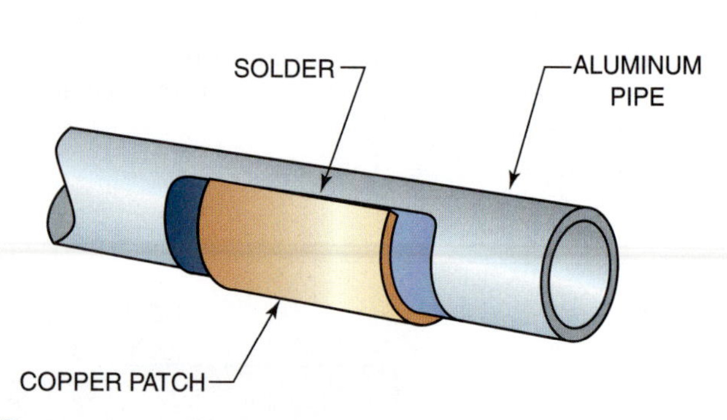

FIGURE 5-60 Aluminum pipe patch.

refrigeration, **Figure 5-60**. Turn off the cylinders, bleed the hoses, back out the regulator adjusting screw, and clean up your work area when you are finished.

Complete a copy of the "Student Welding Report" listed in Appendix I or provided by your instructor. ◆

Summary

Soldering and brazing are excellent processes for repair work on thin sheet metal and small parts when overheating caused by welding might damage surrounding material. The lower heat input required for soldering, for example, means that some parts can be repaired even if there are electrical wires near the repair. Some metals such as copper and brass are too easily melted to be welded successfully, but they can be soldered or brazed. Gas welding of stainless steel and aluminum is not practical, but they are easily soldered or brazed.

A good example of a job that can be done with brazing is the work of a cattle rancher in the high desert of New Mexico who says he works in his small welding shop when it is "snowing sideways." He uses gas welding and brazing to make silver saddle buckles as a side business.

Review

1. Soldering and brazing are much weaker joining processes than welding.
   A. true
   B. false
2. What type of joint design gives a solder joint the greatest strength?
   A. butt joint
   B. lap joint
   C. tee joint
   D. outside corner joint
3. What force draws the liquid braze metal into the narrow joint gap?
   A. heat
   B. gravity
   C. chemical attraction
   D. capillary attraction
4. An advantage of soldering is that the joint can be permanent or temporary.
   A. True
   B. False
5. Which of the following is not an advantage of brazing?
   A. It can join dissimilar metals.
   B. Parts can be preassembled for better alignment.
   C. Metal being joined must have the same thickness.
   D. There is less chance of damaging parts because they do not melt.
6. How do fluxes remove oxides from metal?
   A. They absorb and dissolve them.
   B. They push them out of the way.
   C. Fluxes cover up oxides.
   D. They cannot do anything with oxides, which is why you must sand joints before brazing.
7. Which of the following is not a form of flux?
   A. solid
   B. liquid
   C. paste
   D. gas
8. What can happen if a flux is overheated?
   A. It can evaporate from the joint.
   B. It can eat away the base metal.
   C. It can become burned and stop working.
   D. Nothing—fluxes are designed to take a lot of heat.
9. All fluxes are inactive at room temperature.
   A. true
   B. false
10. Why would MAPP® or propane be a better fuel gas to use with oxygen for some brazing?
    A. The acetylene flame heat is too concentrated.
    B. Acetylene is too dangerous to use for brazing
    C. The oxyacetylene torch tips are too large for brazing.
    D. They are much cheaper gases.
11. What does the term 'past range' refer to?
    A. the way a flux reacts with heat
    B. the temperature range where solder is both a liquid and solid
    C. the temperature at which solder flows best
    D. a type of flux that is a mixture of past flux and powdered metal solder
12. Which solder is recommended for soldering stainless steel?
    A. tin-lead
    B. tin-antimony
    C. cadmium-silver
    D. cadmium-zinc
13. All mixtures of tin-lead solder melt and solidify at the same temperature.
    A. true
    B. false

Review continued

14. Which of the following is not a brazing alloy?
    A. PbSn
    B. BRCuZn
    C. Bag
    D. BNi
15. According to Table 5-5 which brazing alloy could be used to repair cast iron?
    A. BAlSi
    B. BCu
    C. BAu
    D. BCuZn
16. Which brazing alloy is considered to be low-fuming?
    A. BRCuZn-A
    B. BRCuZn-B
    C. BRCuZn-C
    D. BRCuZn-D
17. What is the only thing that can be left on the surface of metal before brazing begins?
    A. light surface oxides
    B. oil
    C. paint
    D. dirt
18. For greatest joint strength, a part should be overlapped.
    A. 1 times its thickness
    B. 2 times its thickness
    C. 3 times its thickness
    D. 5 times its thickness
19. Most of a brazed joint's strength comes from the fillet around the joint edge.
    A. true
    B. false
20. To make a brazed joint when there are two different thicknesses of metal you must __________.
    A. direct the flame on the thinner metal
    B. direct the heat on the thicker metal
    C. use more flux on the thicker metal
    D. place the filler rod so it protects the thinner metal from the heat

Chapter 6

Oxyacetylene Cutting

OBJECTIVES

After completing this chapter, the student should be able to

- ☑ explain how the flame-cutting process works.
- ☑ demonstrate how to properly set up and use an oxyfuel gas cutting torch.
- ☑ safely use an oxyfuel gas cutting torch to make a variety of cuts.

KEY TERMS

coupling distance
cutting lever
cutting tips
drag
drag lines
equal-pressure torches
hard slag
kindling point
MPS gases
orifice
oxyacetylene hand torch
oxyfuel gas cutting (OFC)
preheat flame
preheat holes
slag
soapstone
soft slag
tip cleaners
venturi

INTRODUCTION

Oxyacetylene cutting (OFC-A) is the primary cutting process in a larger group called oxyfuel gas cutting (OFC). OFC is a group of oxygen cutting processes that uses heat from an oxygen fuel gas flame to raise the temperature of the metal to its kindling temperature. When the metal is hot enough, a high-pressure stream of oxygen is directed onto the metal, causing it to be cut. The kindling temperature of a material is the temperature at which combustion (rapid oxidation) can begin. The kindling temperature of steel in pure oxygen is a dull red temperature of about 1600°F to 1800°F (870°C to 900°C). The processes in this group are identified by the type of fuel gas mixed with oxygen to produce the preheat flame. Oxyfuel gas cutting is most commonly performed with OFC-A. **Table 6-1** lists a number of other fuel gases used for OFC. Although other fuel gases are being used, only acetylene and oxygen

Fuel Gas	Flame (Fahrenheit)	Temperature* (Celsius)
Acetylene	5589°	3087°
MAPP®	5301°	2927°
Natural gas	4600°	2538°
Propane	4579°	2526°
Propylene	5193°	2867°
Hydrogen	4820°	2660°

*Approximate neutral oxyfuel flame temperature.

TABLE 6-1 Fuel Gases Used for Flame Cutting.

can be used to weld. Because it can be used for both cutting and welding, acetylene will remain the primary agricultural fuel gas.

More farms and ranches use the oxyacetylene cutting torch than any other welding equipment, **Figure 6-1**. Unfortunately it is one of the most commonly misused processes. Many agricultural workers know how to light the torch and make a cut, but their cuts are very poor quality. Often, in addition to making bad cuts, they use unsafe torch techniques. A good oxyacetylene cut should not only be straight and square, but it also should require little or no postcut cleanup.

Metals Cut by the Oxyfuel Process

Oxyfuel gas cutting is used to cut iron base alloys. Low carbon steels (up to 0.3% carbon) are easy to cut. High nickel steels, cast iron, and stainless steel are considered uncuttable with OCF-A. Most nonferrous metals—such as brass, copper, and aluminum—cannot be cut by oxyacetylene cutting.

Eye Protection for Flame Cutting

Cutting goggles or cutting glasses are required anytime you are using a cutting torch. You must protect your eyes from both the bright light and flying sparks. The National Bureau of Standards has identified proper filter plates and uses. The recommended filter plates are identified by shade number and are related to the type of cutting operation being performed.

(A)

(B)

FIGURE 6-1 (A) Oxyacetylene portable cutting rig mounted on a hand cart for use in the shop and barn area. (B) Oxyacetylene setup mounted on a farm tractor for use anywhere around the place. Courtesy of Larry Jeffus.

Type of Cutting Operation	Hazard	Suggested Shade Number
Light cutting, up to 1 in.	Sparks, harmful	3 or 4
Medium cutting, 1–6 in.	rays, molten metal,	4 or 5
Heavy cutting, over 6 in.	flying particles	5 or 6

Table 6-2 A General Guide for the Selection of Eye and Face Protection Equipment.

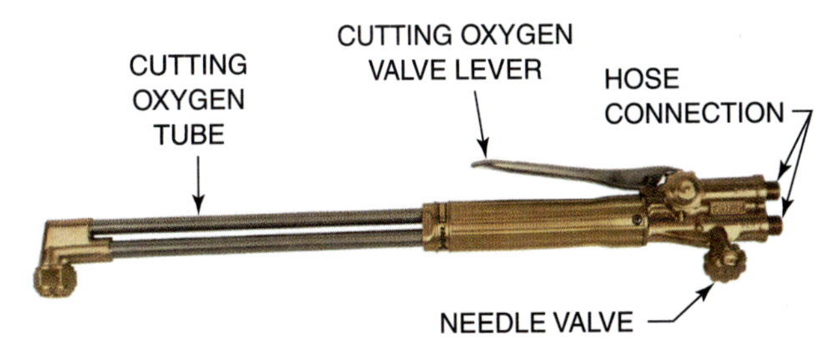

Figure 6-2 Oxyfuel cutting torch. Courtesy of Victor Equipment, a Thermadyne Company.

Goggles, glasses, or other suitable eye protection must be used for flame cutting. Goggles should have vents near the lenses to prevent fogging. Cover lenses or plates should be provided to protect the filter lens. Filter lenses must be marked so that the shade number can be readily identified, **Table 6-2**.

Sunglasses or other dark shaded glasses are not safe to use for oxyacetylene cutting.

Cutting Torches

The oxyacetylene hand torch is the most common type of oxyfuel gas cutting torch used. The hand torch, as it is often called, may be either a part of a combination welding and cutting torch set or a cutting torch only, **Figure 6-2**. The combination welding-cutting torch offers more flexibility because a cutting head, welding tip, or heating tip can be attached quickly to the same torch body, **Figure 6-3**. Combination torch sets are often used. A cut made with either type of torch has the same quality; however, the dedicated cutting torches are usually longer and have larger gas flow passages than the combination torches. The added length of the dedicated cutting torch helps keep the operator farther away from the heat and sparks and allows thicker material to be cut.

Oxygen is mixed with the fuel gas to form a high-temperature preheating flame. The two gases must be completely mixed before they leave the tip and create the flame. Two methods are used to mix the gases. One method uses a mixing chamber, and the other method uses an injector chamber.

The mixing chamber may be located in the torch body or in the tip, **Figure 6-4**. Torches that use a mixing chamber are known as equal-pressure torches because the gases must enter the mixing chamber under the same pressure. The mixing chamber is larger than both the gas inlet and the

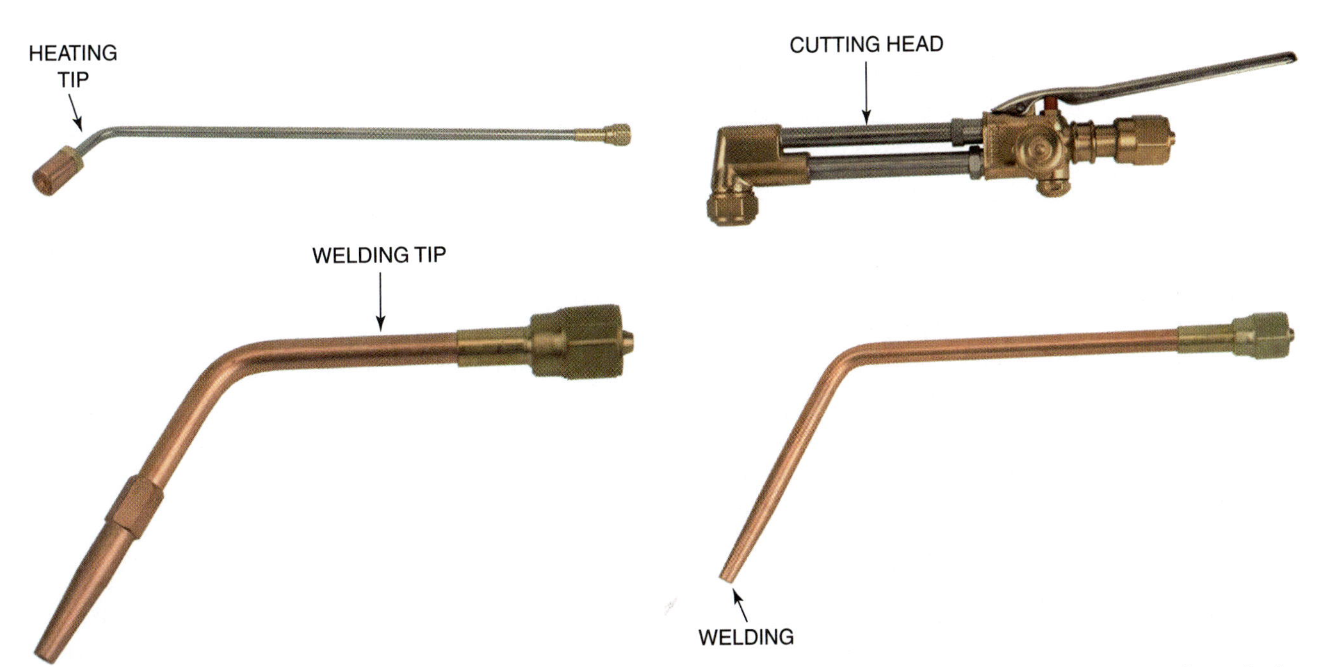

Figure 6-3 The attachments that are used for heating, cutting, welding, or brazing make the combination torch set flexible. Courtesy of Victor Equipment, a Thermadyne Company.

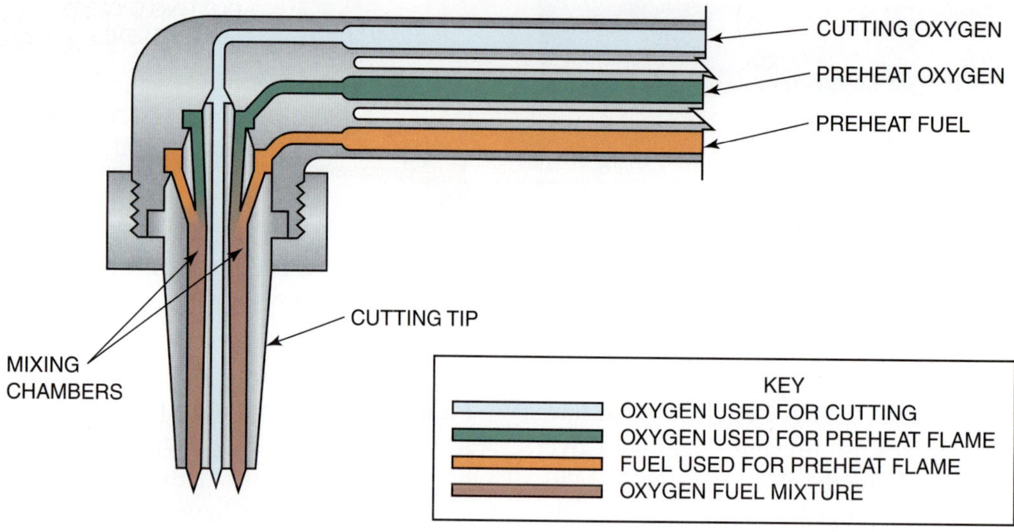

FIGURE 6-4 A mixing chamber located in the tip.

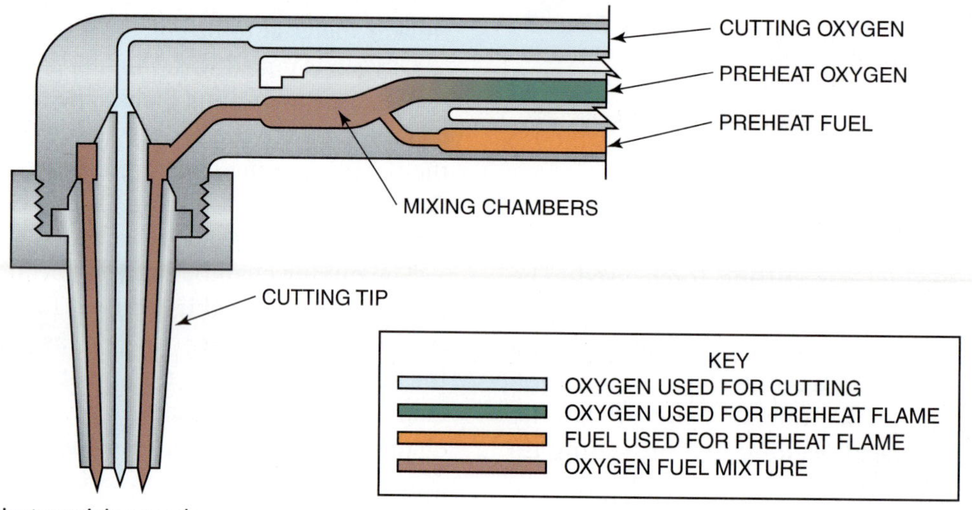

FIGURE 6-5 Injector mixing torch.

gas outlet. This larger size causes turbulence in the gases, resulting in the gases mixing thoroughly.

Injector torches will work with both equal gas pressures and low fuel-gas pressures, **Figure 6-5.** The injector allows the oxygen to draw the fuel gas into the chamber even if the fuel gas pressure is as low as 6 oz/in.^{2} (26 g/cm^2). The injector works by passing the oxygen through a venturi, which creates a low-pressure area that pulls the fuel gases in and mixes them together. An injector-type torch must be used if a low-pressure acetylene generator or low-pressure residential natural gas is used as the fuel gas supply.

The cutting head may hold the cutting tip at a right angle to the torch body or it may be held at a slight angle. Torches with the tip slightly angled are easier for the welder to use when cutting flat plate. Torches with a right-angle tip are easier to use when cutting pipe, angle iron, I-beams, or other uneven material shapes. Both types of torches can be used for any type of material being cut, but practice is needed to keep the cut square and accurate.

The location of the cutting lever may vary from one torch to another, **Figure 6-6.** Most cutting levers pivot from the front or back end of the torch body. Personal preference will determine which one the welder uses.

Cutting Tips

Most cutting tips are made of copper alloy, but some tips are chrome. Chrome plating prevents spatter from sticking to the tip, thus prolonging its usefulness. Tip designs change for the different types of

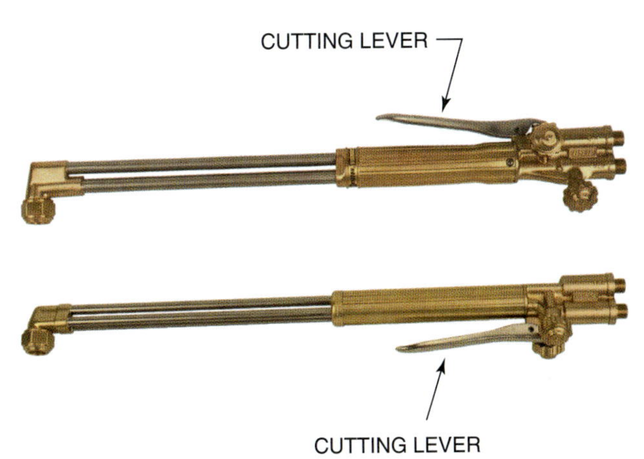

FIGURE 6-6 The cutting lever may be located on the front or back of the torch body. Courtesy of Victor Equipment, a Thermadyne Company.

uses and gases, and from one torch manufacturer to another, **Figure 6-7**.

The diameter, or size of the center cutting orifice, determines the thickness of the metal that can be cut. A larger diameter oxygen orifice is required for cutting thick metal. There is no standard numbering system for sizing cutting tips. Each manufacturer uses its own system. Some systems are similar, some are not. **Table 6-3** lists several manufacturers' tip numbering systems. As a way of comparing the size of one manufacturer's tip size to another, the center hole diameter in inches is given below the tip number. For example, on Table 6-3 you can see that Airco's tip number 00 has a center orifice size equal to a number 70 drill size. You can also see that this cutting tip is designed for cutting metal approximately 1/8″ (3 mm) thick. Other manufacturer tip numbers designed for this thickness have the following numbers: 000, 00, 1/4, 2, and 3.

Finding the correctly sized tip for a job can be confusing, especially if you are using the cutting unit for the first time. To make it easier to select a tip, you can use a standard set of tip cleaners to find the size of the center cutting orifice. **Table 6-4** lists the material thickness being cut with the tip cleaner size.

FIGURE 6-7 Different cutting torch seals for different manufacturers' torches. Courtesy of American Torch Tip.

If the manufacturers' recommendations for gas pressure are not available, you can use Table 6-4 to find the approximate pressures to be used with the tip. Actual gas pressures vary, depending on a number of factors, such as the equipment manufacturer, the condition of the equipment, hose length, hose diameter, regulator size, and operator skill. In all cases start out with the pressure recommended by the particular manufacturer of the equipment being used. Adjust the pressure to fit the job being cut.

A wide variety of tip shapes are also available for specialized cutting jobs. Each tip, of course, also comes in several sizes. Some tips are specialized for the kind of fuel gas being used. Different means are

Manufacturer	Metal Thickness, Inches (mm)										
	1/8 (3)	1/4 (6)	1/2 (13)	3/4 (19)	1 (24)	1 1/2 (37)	2 (49)	2 1/2 (61)	3 (74)	4 (98)	5 (123)
Cutting orifice drill number	70	68	60	56	54	53	50	47	45	39	31
Airco	00	0	1	1	2	2	3	4	4	5	6
ESAB	1/4	1/4	1/2	1 1/2	1 1/2	1 1/2	4	4	4	4	8
Harris	000	00	0	1	1	2	2	3	3	3	4
Oxweld	2	3	4	6	6	6	8	8	8	8	8
Purox	3	3	4	4	5	5	7	7	7	7	9
Smith	00	0	1	2	2	3	3	4	4	4	5
Victor	000	00	0	1	2	2	3	4	4	5	6

TABLE 6-3 Comparison of Some Manufacturers' Oxyacetylene Cutting Tip Identifications.

Tip Size	Metal Thickness, Inches (mm)										
	1/8 (3)	1/4 (6)	1/2 (13)	3/4 (19)	1 (24)	1 1/2 (37)	2 (49)	2 1/2 (61)	3 (74)	4 (98)	5 (123)
Cutting orifice drill number	70	68	60	56	54	53	50	47	45	39	31
WYPO tip cleaner number*	10	10	15	18	22	24	26				
Campbell Hausfeld tip cleaner number*	3	3	6	9	10	11	12				
Oxygen pressure, psi**	20 25	20 25	25 30	30 35	35 40	35 40	40 45	40 45	40 45	45 55	45 55
Oxygen pressure, kPa**	140 170	140 170	170 200	200 240	240 275	240 275	275 310	275 310	275 310	310 380	310 380
Acetylene pressure, psi**	3 5	3 5	3 5	3 5	3 5	3 5	4 8	4 8	5 11	6 13	8 14
Acetylene pressure, kPa**	20 35	20 35	20 35	20 35	20 35	20 35	30 55	30 55	35 75	40 90	55 95

*There is no standard numbering system for tip cleaners, so numbers can differ from one manufacturer to another.
** Tip size and pressures are approximate. Use the manufacturer's specification for equipment being used when available.

TABLE 6-4 Center Cutting Orifice Size, Metal Thickness, and Gas Pressures for Oxyacetylene Cutting.

used to attach the cutting tip to the torch head. Some tips screw in; others have a push fitting.

Always choose the correct type and size of tip for the specific cutting job. Check the manufacturer's literature for tip size and type recommendations. Make sure the tip is designed for the type of fuel gas being used. Inspect the tip before using it. If the tip is clogged or dirty, clean the tip and clean out the orifices with the proper size drill. Check to make sure there is no damage to the threads. If the threads or the tapered seat is damaged, do not use the tip.

The amount of preheat flame required to make a perfect cut is determined by the type of fuel gas used and by the material thickness, shape, and surface condition. Materials that are thick, are round, or have surfaces covered with rust, paint, oil, and so on require more preheat flame, **Figure 6-8**.

Different cutting tips are available for each of the major types of fuel gases. The differences in the type or number of preheat holes determine the type of fuel gas to be used in the tip. **Table 6-5** lists the fuel gas and range of preheat holes or tip designs used with each gas. Acetylene is used in tips having from one to six preheat holes. Some large acetylene cutting tips may have eight or more preheat holes.

Fuel Gas	Number of Preheat Holes
Acetylene	One to six
MPS (MAPP ®)	Eight to two-piece tip
Propane and natural gas	Two-piece tip

TABLE 6-5 Fuel Gas and Number of Preheat Holes Needed in the Cutting Tip.

If acetylene is used in a tip that was designed to be used with one of the other fuel gases, the tip may overheat, causing a backfire or the tip to explode.

MPS gases are used in tips having eight preheat holes or in a two-piece tip that is not recessed, **Figure 6-9**. These gases have a slower flame combustion rate than acetylene. For tips with less than eight preheat holes, there may not be enough heat to start a cut, or the flame may pop out when the cutting lever is pressed.

CAUTION

If MPS gases are used in a deeply recessed, two-piece tip, the tip will overheat, causing a backfire or the tip to explode.

Propane and natural gas may be used in a two-piece tip that is deeply recessed, **Figure 6-9**. The flame burns at such a slow rate that it may not stay lit on any other tip.

Some cutting tips have metal-to-metal seals. When they are installed in the torch head, a wrench must be used to tighten the nut. Other cutting tips have fiber packing seats to seal the tip to the torch. If

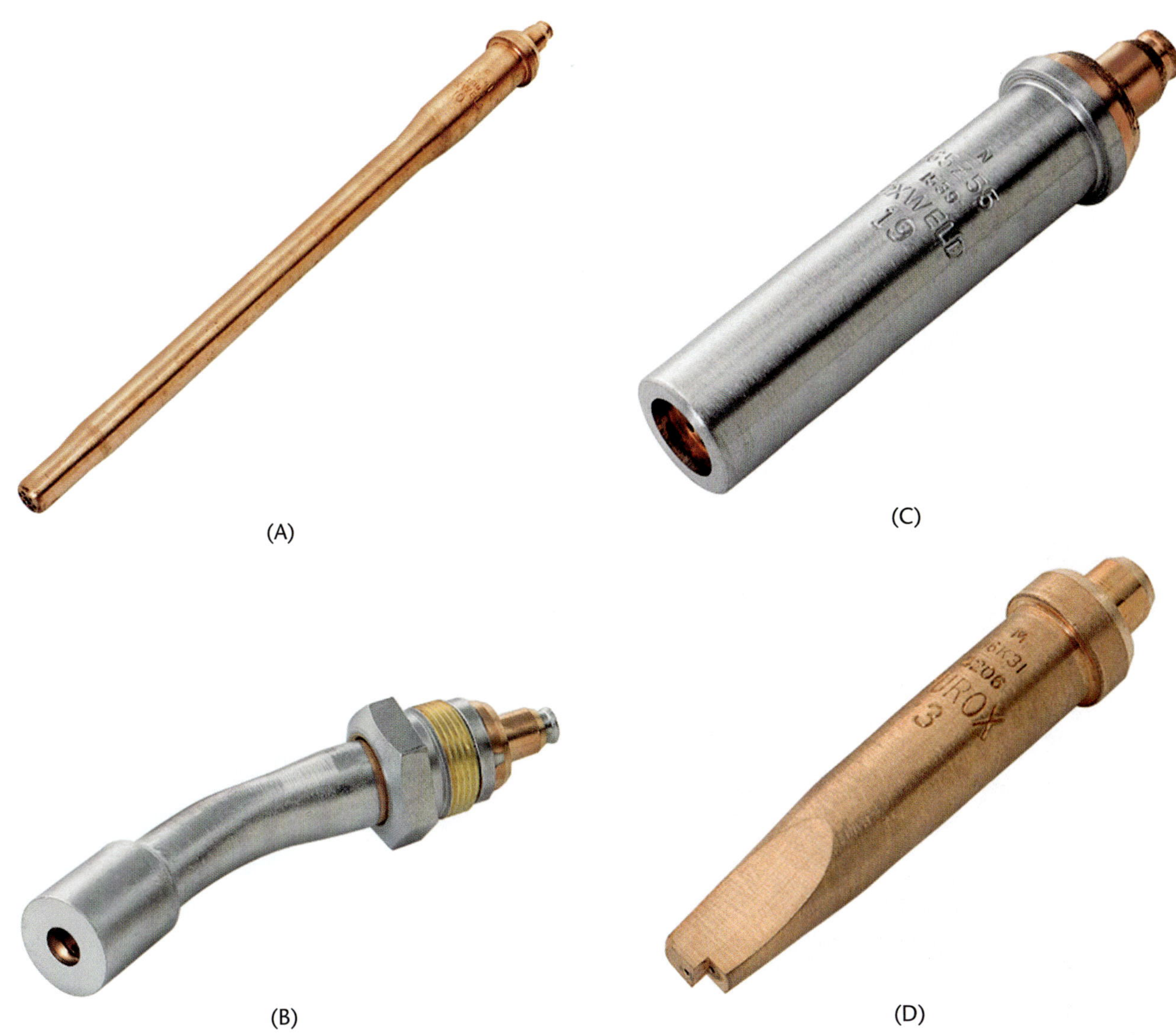

FIGURE 6-8 Special cutting tips: (A) 10-in-long cutting tip, (B) water cooled cutting tip, (C) two piece cutting tip, and (D) sheet metal cutting tip. Courtesy of ESAB Welding & Cutting Products.

a wrench is used to tighten the nut for this type of tip, the tip seat may be damaged, **Figure 6-10**. A torch owner's manual should be checked or a welding supplier should be asked about the best way to tighten various torch tips.

When removing a cutting tip, if the tip is stuck in the torch head, tap the back of the head with a plastic hammer, **Figure 6-11**. Any tapping on the side of the tip may damage the seat.

To check the assembled torch tip for a good seal, turn on the oxygen valve and spray the tip with a leak-detecting solution, **Figure 6-12**.

CAUTION

Carefully handle and store the tips to prevent damage to the tip seats and to keep dirt from becoming stuck in the small holes.

If the cutting tip seat or the torch head seat is damaged, it can be repaired by using a reamer designed for the specific torch tip and head, **Figure 6-13**, or it can be sent out for repair. New fiber packings are available for tips with packings. The original leak-checking test should be repeated to be sure the new seal is good.

Oxyfuel Cutting, Setup, and Operation

The setting up of a cutting torch system is exactly like setting up oxyfuel welding equipment except for the adjustment of gas pressures. This chapter covers gas pressure adjustments and cutting equipment operations. Chapter 3, Oxyfuel Welding and Cutting, Equipment, Setup, and Operation, gives detailed technical information and instructions for

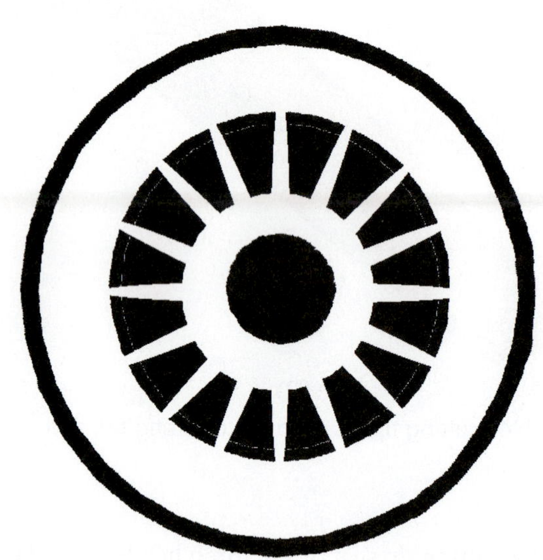

FIGURE 6-9 Parts of a two-piece cutting tip. Courtesy of ESAB Welding & Cutting Products.

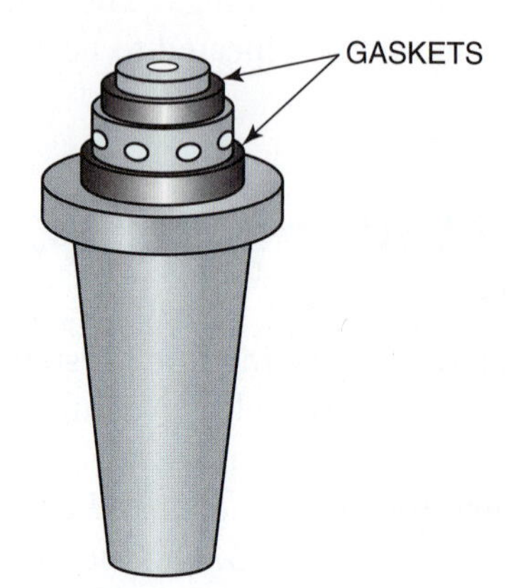

FIGURE 6-10 Some cutting tips use gaskets to make a tight seal.

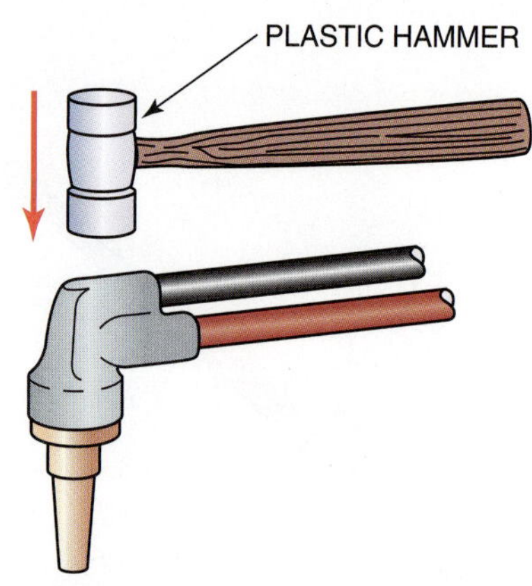

FIGURE 6-11 Tap the back of the torch head to remove a tip that is stuck. The tip itself should never be tapped.

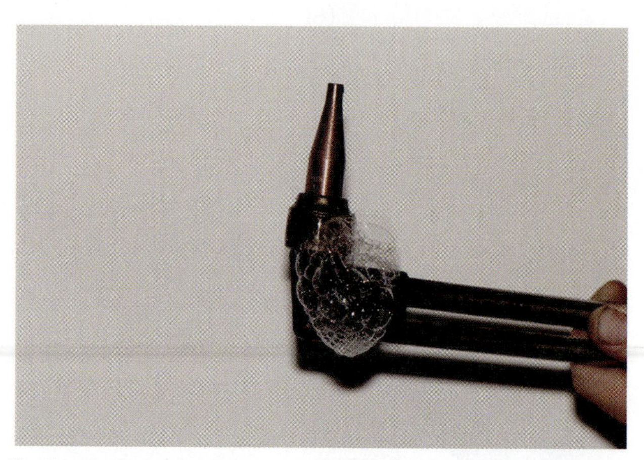

FIGURE 6-12 Checking a cutting tip for leaks. Courtesy of Larry Jeffus.

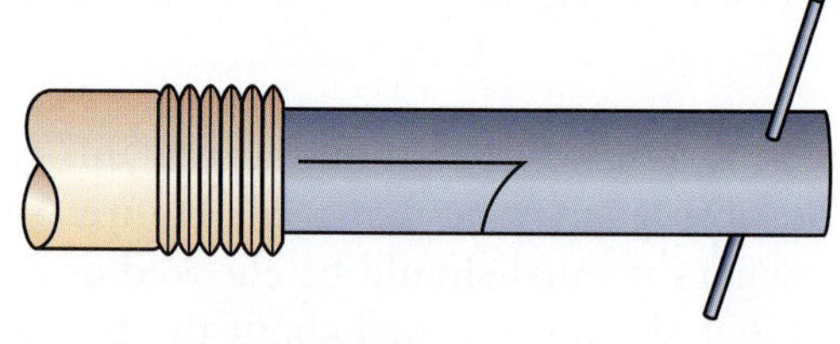

FIGURE 6-13 Damaged torch seats can be repaired by using a reamer.

oxyfuel systems. The chapter covers the following topics:

- Safety
- Pressure regulator setup and operation
- Welding and cutting torch design and service
- Reverse flow and flashback valves
- Hoses and fittings
- Types of flames
- Leak detection

PRACTICE 6-1

Setting Up a Cutting Torch

Demonstrate to other students and your instructor the proper method of setting up cylinders, regulators, hoses, and the cutting torch.

1. The oxygen and acetylene cylinders must be securely chained to a cart or wall before the safety caps are removed.
2. After removing the safety caps, stand to one side and crack (open and quickly close) the cylinder valves, being sure there are no sources of possible ignition that may start a fire. Cracking the cylinder valves is done to blow out any dirt that may be in the valves.
3. Visually inspect all of the parts for any damage, needed repair, or cleaning.
4. Attach the regulators to the cylinder valves and tighten them securely with a wrench.
5. Attach a reverse flow valve or flashback arrestor, if the torch does not have them built in, to the hose connection on the regulator or to the hose connection on the torch body, depending on the type of reverse flow valve in the set. Occasionally, test each reverse flow valve by blowing through it to make sure it works properly.
6. If the torch you will be using is a combination-type torch, attach the cutting head at this time.
7. Last, install a cutting tip on the torch.
8. Before the cylinder valves are opened, back out the pressure regulating screws so that when the valves are opened the gauges will show zero pounds working pressure.
9. Stand to one side of the regulators' face as the cylinder valves are opened slowly.
10. The oxygen valve is opened all the way until it becomes tight, but not overtight, and the acetylene valve is opened no more than one-half turn.
11. Open one torch valve and then turn the regulating screw in slowly until 2 psig to 4 psig (14 kPag to 30 kPag) shows on the working pressure gauge. Allow the gas to escape so that the line is completely purged.
12. If you are using a combination welding and cutting torch, the oxygen valve nearest the hose connection must be opened before the flame adjusting valve or cutting lever will work.

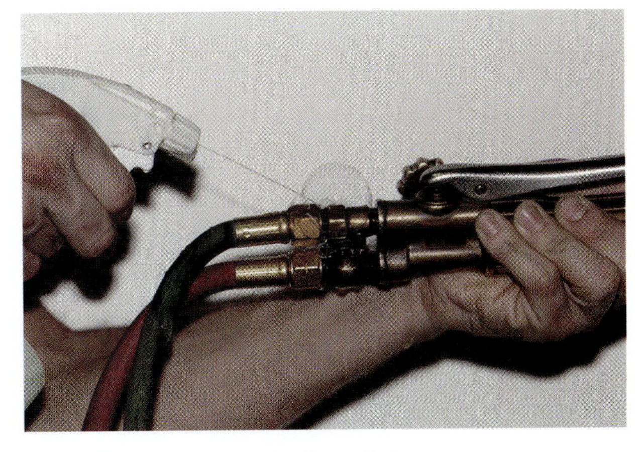

FIGURE 6-14 Leak-check all gas fittings. Courtesy of Larry Jeffus.

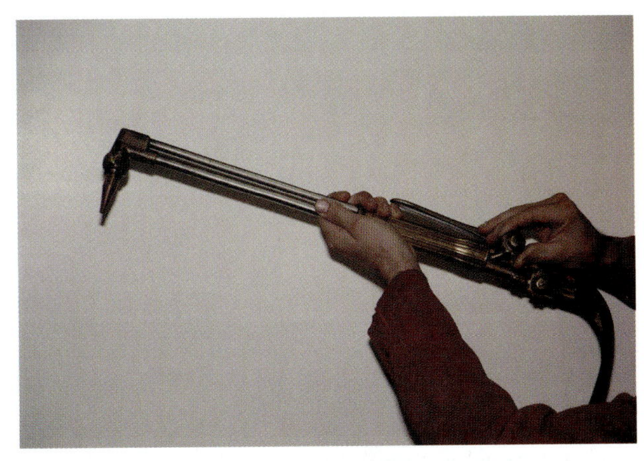

FIGURE 6-15 Turn on the oxygen valve. Courtesy of Larry Jeffus.

13. Close the torch valve and repeat the purging process with the other gas.
14. Be sure there are no sources of possible ignition that may result in a fire.
15. With both torch valves closed, spray a leak-detecting solution on all connections, including the cylinder valves. Tighten any connection that shows bubbles, **Figure 6-14**.

Complete a copy of the "Student Welding Report" listed in Appendix I or provided by your instructor. ◆

PRACTICE 6-2

Cleaning a Cutting Tip

Using a cutting torch set that is assembled and adjusted as described in Practice 6-1, and a set of tip cleaners, you will clean the cutting tip.

1. Turn on a small amount of oxygen, **Figure 6-15**. This procedure is done to blow out any dirt loosened during the cleaning.
2. The end of the tip is first filed flat, using the file provided in the tip cleaning set, **Figure 6-16**.

FIGURE 6-16 File the end of the tip flat. Courtesy of Larry Jeffus.

FIGURE 6-17 A tip cleaner should be used to clean the flame and center cutting holes. Courtesy of Larry Jeffus.

3. Try several sizes of tip cleaners in a preheat hole until the correct size cleaner is determined. It should easily go all the way into the tip, **Figure 6-17**
4. Push the cleaner in and out of each preheat hole several times. Tip cleaners are small, round files. Excessive use of them will greatly increase the orifice (hole) size.
5. Next, depress the cutting lever and, by trial and error, select the correct size tip cleaner for the center cutting orifice.

A tip cleaner should never be forced. If the tip needs additional care, refer to the section on tip care in Chapter 15.

Complete a copy of the "Student Welding Report" listed in Appendix I or provided by your instructor. ◆

PRACTICE 6-3

Lighting the Torch

Wearing welding goggles, gloves, and any other required personal protective clothing, and with a cutting torch set that is safely assembled, you will light the torch.

1. Set the regulator working pressure for the tip size. If you do not know the correct pressure for the tip, start with the fuel set at 5 psig (35 kPag) and the oxygen set at 25 psig (170 kPag).
2. Point the torch tip upward and away from any equipment or other students.
3. Turn on just the acetylene valve and use only a sparklighter to ignite the acetylene. The torch may not stay lit. If this happens, close the valve slightly and try to relight the torch.
4. If the flame is small, it will produce heavy black soot and smoke. In this case, turn the flame up to stop the soot and smoke. The welder need not be concerned if the flame jumps slightly away from the torch tip.
5. With the acetylene flame burning smoke free, slowly open the oxygen valve and by using only the oxygen valve adjust the flame to a neutral setting, **Figure 6-18.**
6. When the cutting oxygen lever is depressed, the flame may become slightly carbonizing. This may occur because of a drop in line pressure due to the high flow of oxygen through the cutting orifice.
7. With the cutting lever depressed, readjust the preheat flame to a neutral setting.

The flame will become slightly oxidizing when the cutting lever is released. Since an oxidizing flame is hotter than a neutral flame, the metal being cut will be preheated faster. When the cut is started by depressing the lever, the flame automatically returns to the neutral setting and does not oxidize the top of the plate. Extinguish the flame by first turning off the oxygen and then the acetylene.

Complete a copy of the "Student Welding Report" listed in Appendix I or provided by your instructor. ◆

CAUTION

Sometimes with large cutting tips the tip will pop when the acetylene is turned off first. If that happens, turn the oxygen off first.

Hand Cutting

When making a cut with a hand torch, it is important for the welder to be steady in order to make the cut as smooth as possible. A welder must also be comfortable and free to move the torch along the line to be

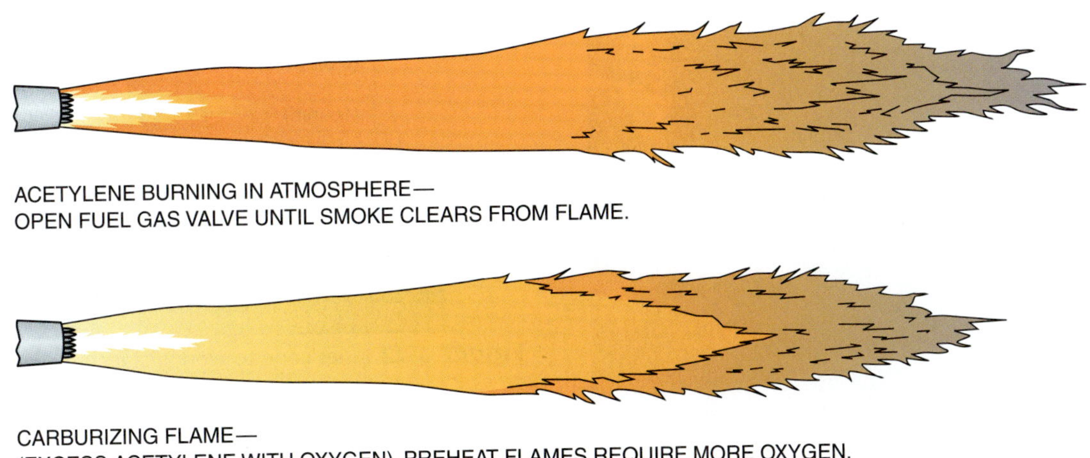

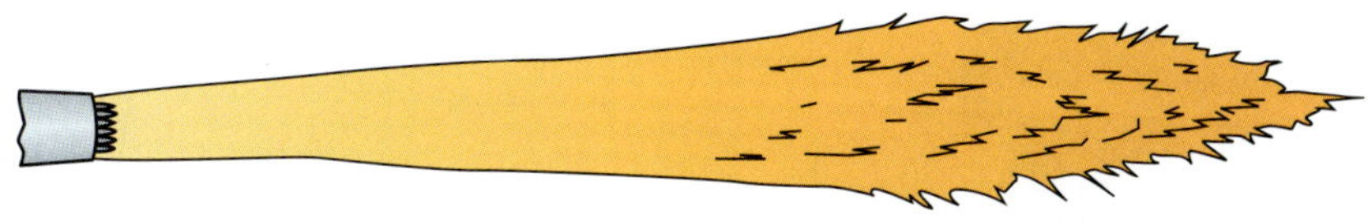

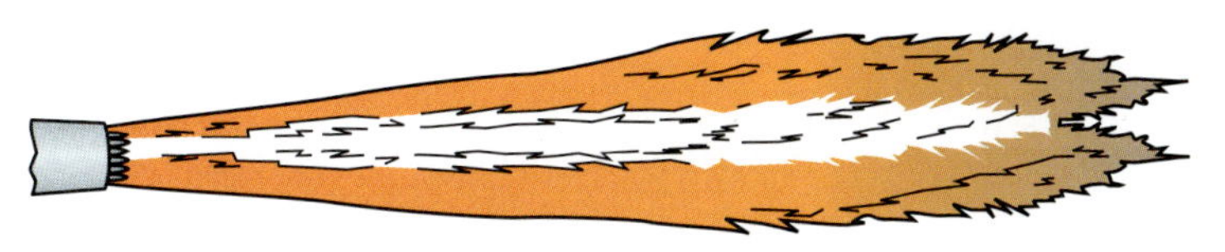

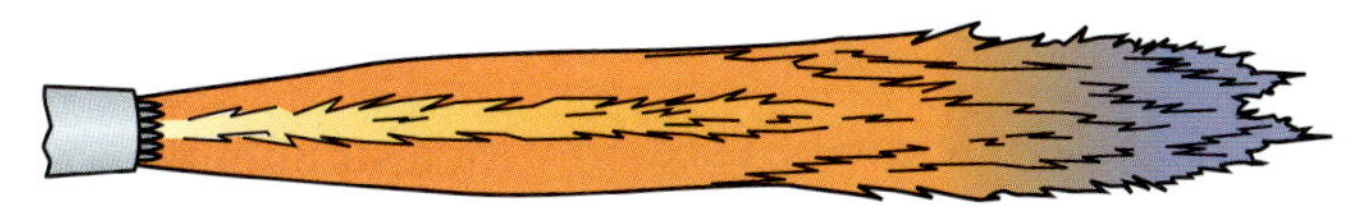

FIGURE 6-18 Oxyacetylene flame adjustments for the cutting torch.

cut. It is a good idea for a welder to get into position and practice the cutting movement a few times before lighting the torch. Even when the welder and the torch are braced properly, a tiny movement such as a heartbeat will cause a slight ripple in the cut. Attempting a cut without leaning on the work, to brace oneself, is tiring and causes inaccuracies.

The torch should be braced with the left hand if the welder is right-handed or with the right hand if the welder is left-handed. The torch may be moved by sliding it toward you over your supporting hand, **Figures 6-19** and **Figure 6-20**. The torch can also be pivoted on the supporting hand. If the pivoting method is used, care must be taken to prevent the cut from becoming a series of arcs.

FIGURE 6-19 For short cuts, the torch can be drawn over the gloved hand. Courtesy of Larry Jeffus.

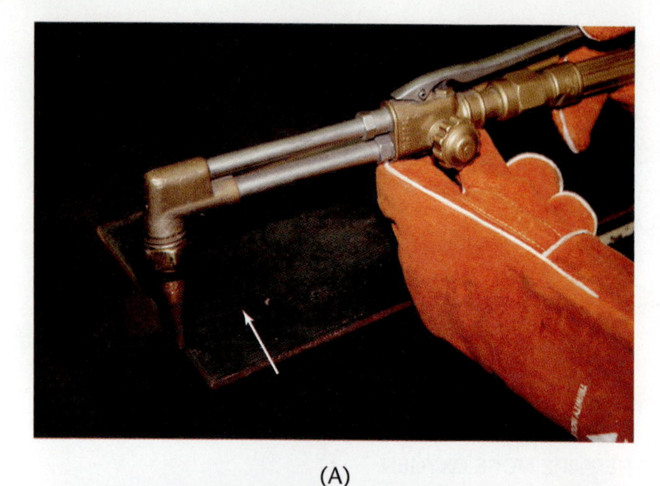

(A)

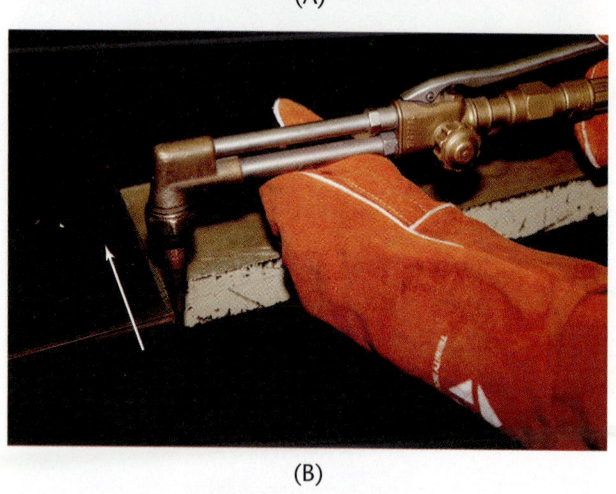

(B)

FIGURE 6-20 For longer cuts, the torch can be moved by sliding your gloved hand along the plate parallel to the cut: (A) start and (B) finish. Always check for free and easy movement before lighting the torch. Courtesy of Larry Jeffus.

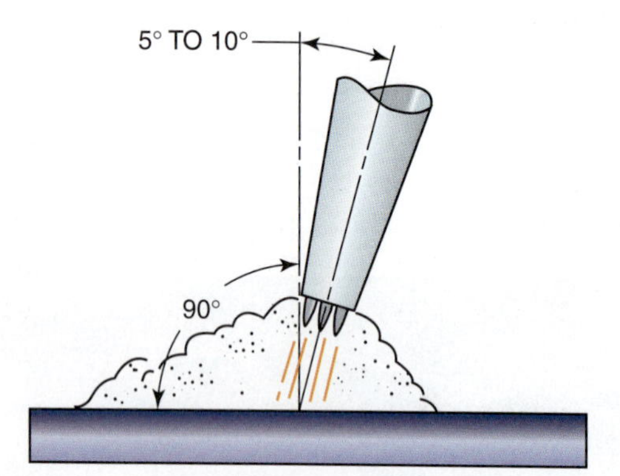

FIGURE 6-21 A slight forward angle helps when cutting thin material.

A slight forward torch angle helps the flame preheat the metal, keeps some of the reflected flame heat off the tip, aids in blowing dirt and oxides away from the cut, and keeps the tip clean for a longer period of time because slag is less likely to be blown back on it, **Figure 6-21**. The forward

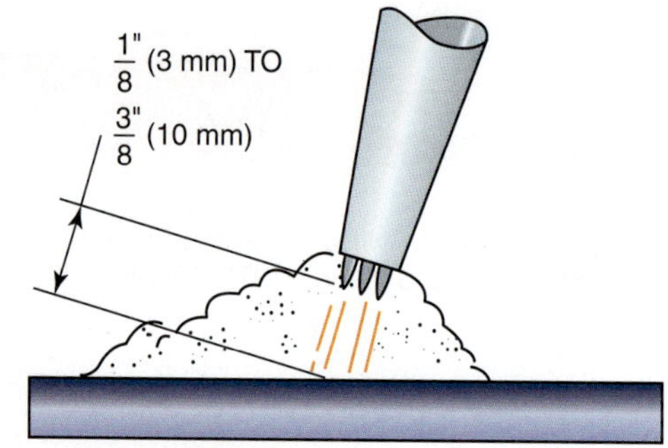

FIGURE 6-22 Inner cone to work distance.

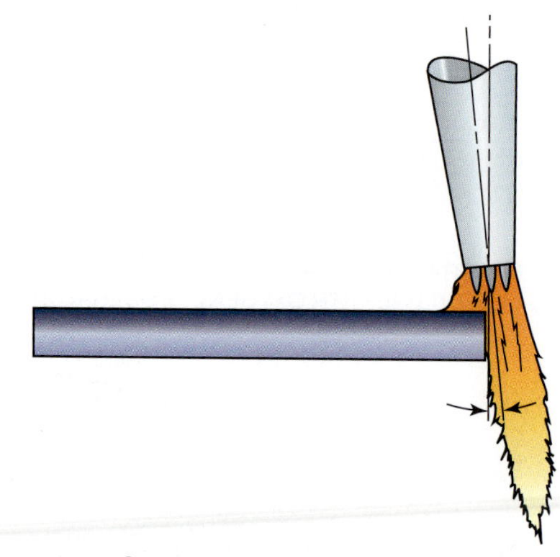

FIGURE 6-23 Starting a cut on the edge of a plate. Notice how the torch is pointed at a slight angle away from the edge.

angle can be used only for a straight line square cut. If shapes are cut using a slight angle, the part will have beveled sides.

When making a cut, the inner cones of the flame should be kept 1/8 in. (3 mm) to 3/8 in. (10 mm) from the surface of the plate, **Figure 6-22**. This distance is known as the coupling distance.

To start a cut on the edge of a plate, hold the torch at a right angle to the surface or pointed slightly away from the edge, **Figure 6-23**. The torch must also be

CAUTION

NEVER USE A CUTTING TORCH TO CUT OPEN A USED CAN, DRUM, TANK, OR OTHER SEALED CONTAINER. The sparks and oxygen cutting stream may cause even nonflammable residue inside to burn or explode. If a used container must be cut it must first have one end removed and all residue cleaned out. In addition to the possibility of a fire or an explosion you might be exposing yourself to hazardous fumes. Before making a cut check the material specification data sheet (MSDS) for safety concerns.

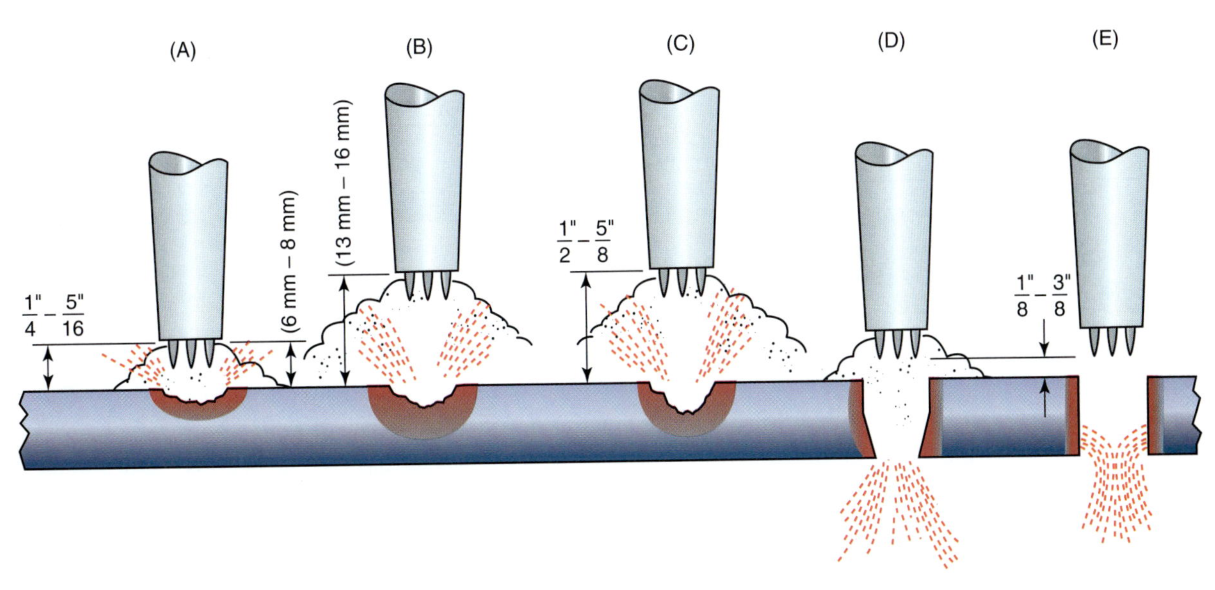

FIGURE 6-24 Sequence for piercing plate.

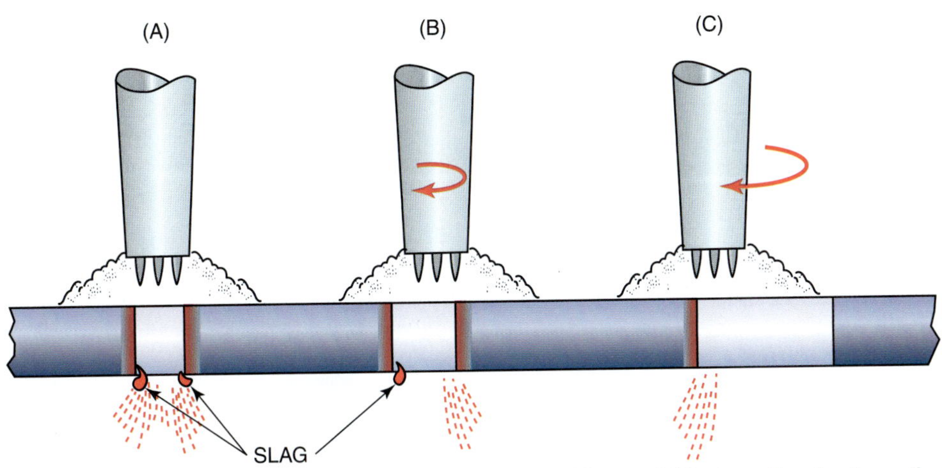

FIGURE 6-25 A short, backward movement (A) before the cut is carried forward (B) clears the slag from the kerf (C) slag left in the kerf may cause the cutting stream to gouge into the base metal, resulting in a poor cut.

pointed so that the cut is started at the very edge. The edge of the plate heats up more quickly and allows the cut to be started sooner. Also, fewer sparks will be blown around the shop. Once the cut is started, the torch should be rotated back to a right angle to the surface or to a slight leading angle.

If a cut is to be started in a place other than the edge of the plate, the inner cones should be held as close as possible to the metal. Having the inner cones touch the metal will speed up the preheat time. When the metal is hot enough to allow the cut to start, the torch should be raised as the cutting lever is slowly depressed. When the metal is pierced, the torch should be lowered again, **Figure 6-24.** By raising the torch tip away from the metal, the amount of sparks blown into the air is reduced, and the tip is kept cleaner. If the metal being cut is thick, it may be necessary to move the torch tip in a small circle as the hole goes through the metal. If the metal is to be cut in both directions from the spot where it was pierced, the torch should be moved backward a short distance and then forward, **Figure 6-25.** This prevents slag from refilling the kerf at the starting point, thus making it difficult to cut in the other direction. The kerf is the space produced during any cutting process.

Starts and stops can be made more easily and better if one side of the metal being cut is scrap. When it is necessary to stop and reposition oneself before continuing the cut, the cut should be turned out, a short distance, into the scrap side of the metal, **Figure 6-26.** The extra space that this procedure provides will allow a smoother and more even start with less chance that slag will block the cut. If neither side of the cut is to be scrap, the forward movement should be stopped for a moment before releasing the cutting lever. This action will allow the drag,

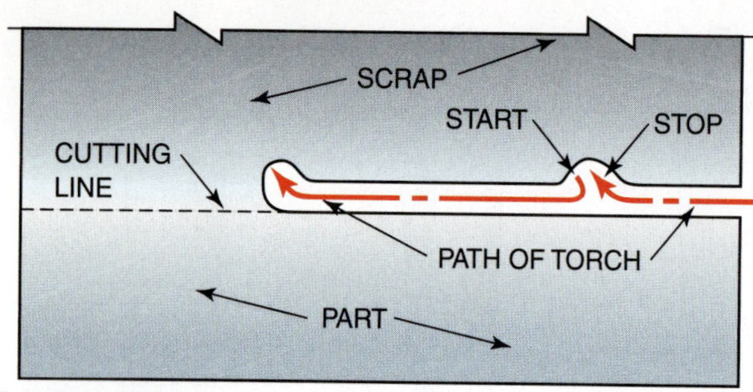

FIGURE 6-26 Turning out into scrap to make stopping and starting points smoother.

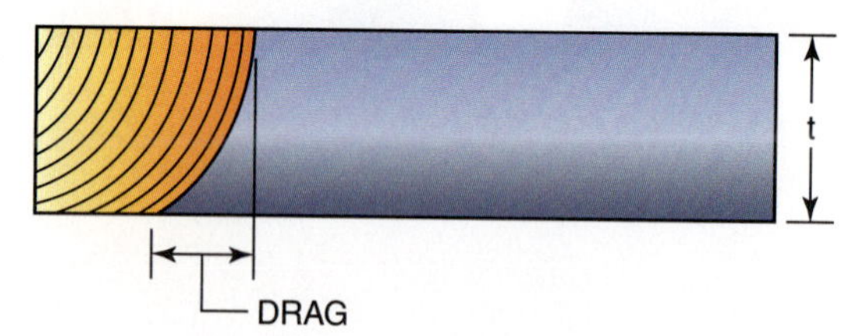

FIGURE 6-27 Drag is the distance by which the bottom of a cut lags behind the top.

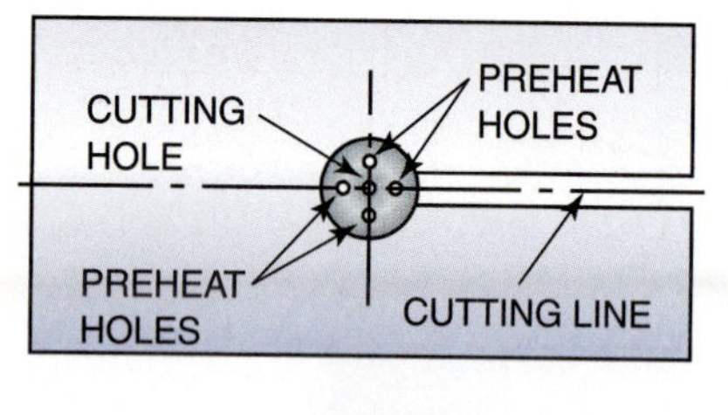

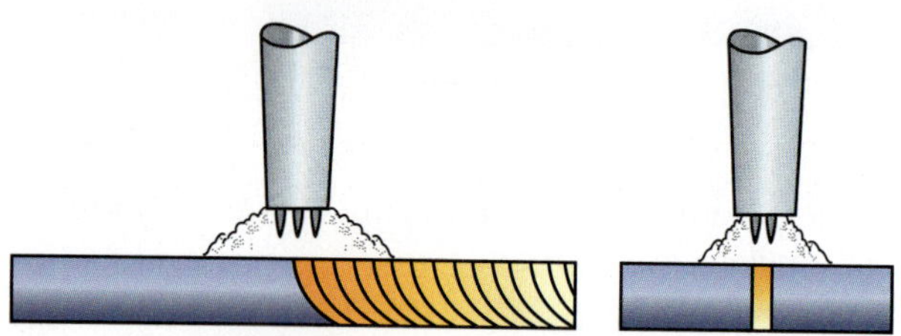

FIGURE 6-28 Tip alignment for a square cut.

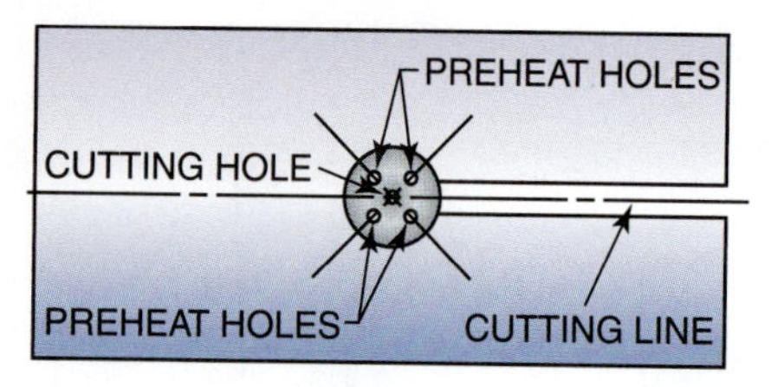

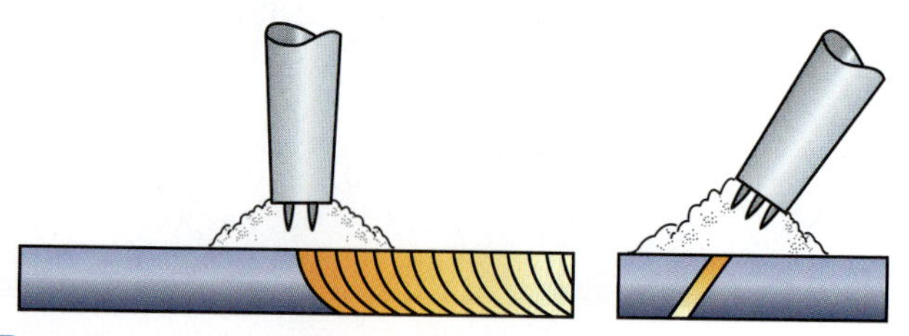

FIGURE 6-29 Tip alignment for a bevel cut.

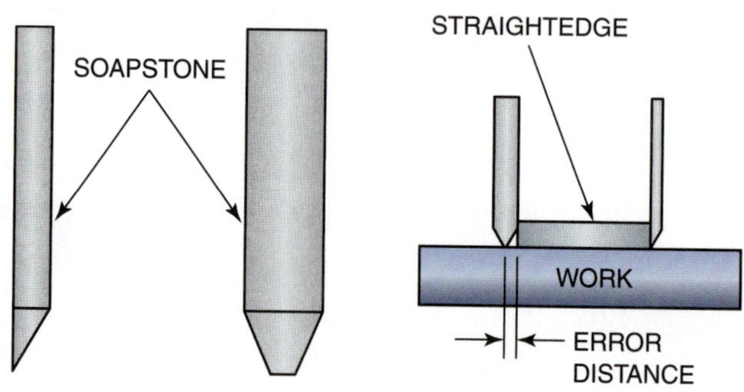

FIGURE 6-30 Proper method of sharpening a soapstone.

FIGURE 6-31 Holding the punch slightly above the surface allows the punch to be struck rapidly and moved along a line to mark it for cutting. Courtesy of Larry Jeffus.

or the distance that the bottom of the cut is behind the top, to catch up before stopping, **Figure 6-27**. To restart, use the same procedure that was given for starting a cut at the edge of the plate.

The proper alignment of the preheat holes will speed up and improve the cut. The holes should be aligned so that one is directly on the line ahead of the cut and another is aimed down into the cut when making a straight line square cut, **Figure 6-28**. The flame is directed toward the smaller piece and the sharpest edge when cutting a bevel. For this reason, the tip should be changed so that at least two of the flames are on the larger plate and none of the flames are directed on the sharp edge, **Figure 6-29**. If the preheat flame is directed at the edge, it will be rounded off as it is melted off.

Layout

Laying out a line to be cut can be done with a piece of soapstone or a chalk line. To obtain an accurate line, a scribe or a punch can be used. If a piece of soapstone is used, it should be sharpened properly to increase accuracy, **Figure 6-30**. A chalk line will make a long, straight line on metal and is best used on large jobs. The scribe and punch can both be used to lay out an accurate line, but the punched line is easier to see when cutting. A punch can be held as shown in **Figure 6-31**,

Metal Thickness in (mm)	Center Orifice Size No. Drill Size	Tip Cleaner No.*	Oxygen Pressure lb/in² (kPa)	Acetylene lb/in² (kPa)
1/8 (3)	60	7	10 (70)	3 (20)
1/4 (6)	60	7	15 (100)	3 (20)
3/8 (10)	55	11	20 (140)	3 (20)
1/2 (13)	55	11	25 (170)	4 (30)
3/4 (19)	55	11	30 (200)	4 (30)
1 (25)	53	12	35 (240)	4 (30)
2 (51)	49	13	45 (310)	5 (35)
3 (76)	49	13	50 (340)	5 (35)
4 (102)	49	13	55 (380)	5 (35)
5 (127)	45	**	60 (410)	5 (35)

*The tip cleaner number when counted from the small end toward the large end in a standard tip cleaner set
**Larger than normally included in a standard tip cleaner set

TABLE 6-6 Cutting Pressure and Tip Size.

with the tip just above the surface of the metal. When the punch is struck with a lightweight hammer, it will make a mark. If you move your hand along the line and rapidly strike the punch, it will leave a series of punch marks for the cut to follow.

Selecting the Correct Tip and Setting the Pressure

Each welding equipment manufacturer uses its own numbering system to designate the tip size. It would be impossible to remember each of the systems. Each manufacturer, however, does relate the tip number to the numbered drill size used to make the holes. On the back of most tip cleaning sets, the manufacturer lists the equivalent drill size of each tip cleaner. By remembering approximately which tip cleaner was used on a particular tip for a metal thickness range, a welder can easily select the correct tip when using a new torch set. Using the tip cleaner that you are familiar with, try it in the various torch tips until you find the correct tip that the tip cleaner fits. **Table 6-6** lists the tip drill size, pressure range, and metal thickness range for which the tip can be used.

PRACTICE 6-4

Setting the Gas Pressures

Setting the working pressure of the regulators can be done by following a table, or it can be set by watching the flame.

1. To set the regulator by watching the flame, first set the acetylene pressure at 2 psig to 4 psig (14 kPag to 30 kPag) and then light the acetylene flame.
2. Open the acetylene torch valve one to two turns and reduce the regulator pressure by backing out the setscrew until the flame starts to smoke.
3. Increase the pressure until the smoke stops and then increase it just a little more.

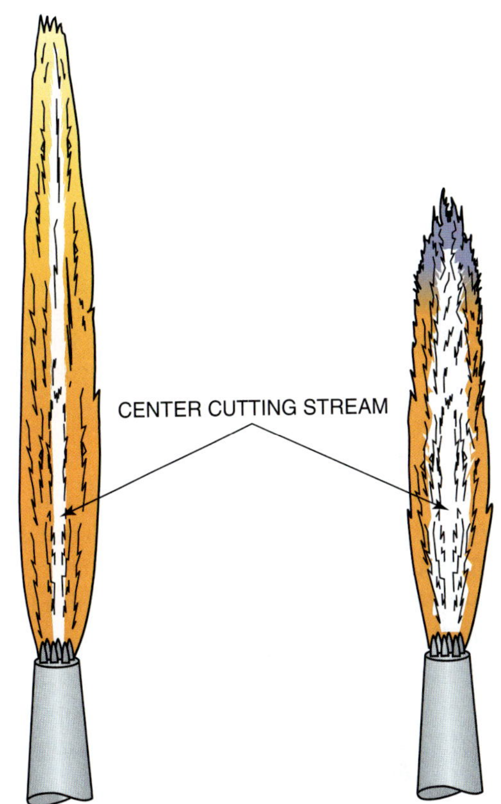

FIGURE 6-32 A clean cutting tip will have a long, well-defined oxygen stream.

This is the maximum fuel gas pressure the tip needs. With a larger tip and a longer hose, the pressure must be set higher. This is the best setting, and it is the safest one to use. With this lowest possible setting, there is less chance of a leak. If the hoses are damaged, the resulting fire will be much smaller than a fire burning from a hose with a higher pressure. There is also less chance of a leak with the lower pressure.

4. With the acetylene adjusted so that the flame just stops smoking, slowly open the torch oxygen valve.
5. Adjust the torch to a neutral flame. When the cutting lever is depressed, the flame will become carbonizing, not having enough oxygen pressure.
6. While holding the cutting lever down, increase the oxygen regulator pressure slightly. Readjust the flame, as needed, to a neutral setting by using the oxygen valve on the torch.
7. Increase the pressure slowly and readjust the flame as you watch the length of the clear cutting stream in the center of the flame, **Figure 6-32**.

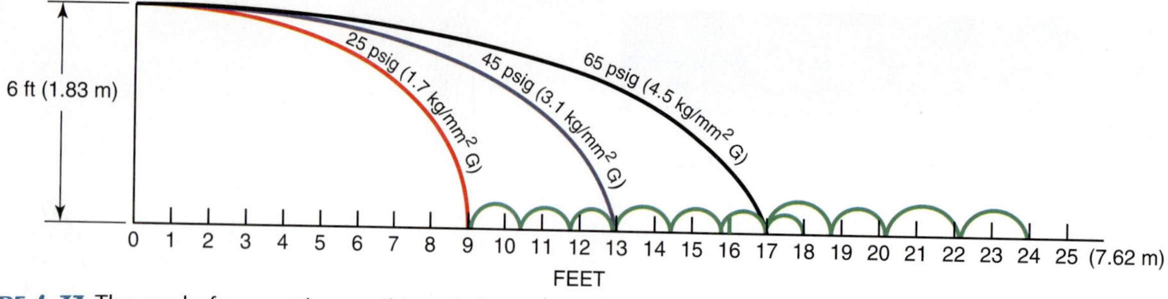

FIGURE 6-33 The sparks from cutting a mild steel plate, 3/8 in. (10 mm) thick, 6 ft (1.8 m) from the floor, will be thrown much farther if the cutting pressure is too high for the plate thickness. These cuts were made with a Victor cutting tip no. 0–1–101 using 25 psig (1.7 kg/mm^2) as recommended by the manufacturer and by excessive pressures of 45 psig (3.1 kg/mm^2) and 65 psig (4.5 kg/mm^2).

The center stream will stay fairly long until a pressure is reached that causes turbulence disrupting the cutting stream. This turbulence will cause the flame to shorten in length considerably, **Figure 6-32.**

8. With the cutting lever still depressed, reduce the oxygen pressure until the flame lengthens once again. This is the maximum oxygen pressure that this tip can use without disrupting turbulence in the cutting stream. This turbulence will cause a very poor cut. The lower pressure also will keep the sparks from being blown a longer distance from the work, **Figure 6-33.**

Complete a copy of the "Student Welding Report" listed in Appendix I or provided by your instructor. ◆

The Chemistry of a Cut

The oxyfuel gas cutting torch works when the metal being cut rapidly oxidizes or burns. This rapid oxidization or burning occurs when a high-pressure stream of pure oxygen is directed on the metal after it has been preheated to a temperature above its kindling point. **Kindling point** is the lowest temperature at which a material will burn. The kindling temperature of iron is 1600°F (870°C), which is a dull red color. Note that iron is the pure element and cast iron is an alloy primarily of iron and carbon. The process will work easily on any metal that will rapidly oxidize, such as iron, low carbon steel, magnesium, titanium, and zinc.

CAUTION

Some metals release harmful oxides when they are cut. Extreme caution must be taken when cutting used, oily, dirty, or painted metals. They often produce very dangerous fumes when they are cut. You may need extra ventilation and a respirator to be safe.

The process is most often used to cut iron and low carbon steels, because unlike most of the metals, little or no oxides are left on the metal, and it can easily be welded.

The burning away of the metal is a chemical reaction with iron (Fe) and oxygen (O). The oxygen forms an iron oxide, primarily Fe_3O_4, that is light gray in color. Heat is produced by the metal as it burns. This heat helps carry the cut along. On thick pieces of metal, once a small spot starts burning (being cut), the heat generated helps the cut continue quickly through the metal. With some cuts the heat produced may overheat small strips of metal being cut from a larger piece. As an example, the center piece of a hole being cut will quickly become red hot and will start to oxidize with the surrounding air, **Figure 6-34.** This heat produced by the cut makes it difficult to cut out small or internal parts.

EXPERIMENT 6-1

Observing Heat Produced during a Cut

This experiment may require more skill than you have developed by this time. You may wish to observe your instructor performing the experiment or try it at a later time.

Using a properly lit and adjusted cutting torch, welding gloves, appropriate eye protection and clothing, and one piece of clean mild steel plate 6 in. (152 mm) long × 1/4 in. (6 mm) to 1/2 in. (13 mm) thick, you will make an oxyfuel gas cut without the preheat flame.

Place the piece of metal so that the cutting sparks fall safely away from you. With the torch lit, pass the flame over the length of the plate until it is warm, but not hot. Brace yourself and start a cut near the edge of the plate. When the cut has been established, have another student turn off the acetylene regulator. The

FIGURE 6-34 As a hole is cut, the center may be overheated. Courtesy of Larry Jeffus.

cut should continue if you remain steady and the plate is warm enough. *Hint: Using a slightly larger tip size will make this easier.*

Complete a copy of the "Student Welding Report" listed in Appendix I or provided by your instructor. ◆

The Physics of a Cut

As a cut progresses along a plate, a record of what happened during the cut is preserved along both

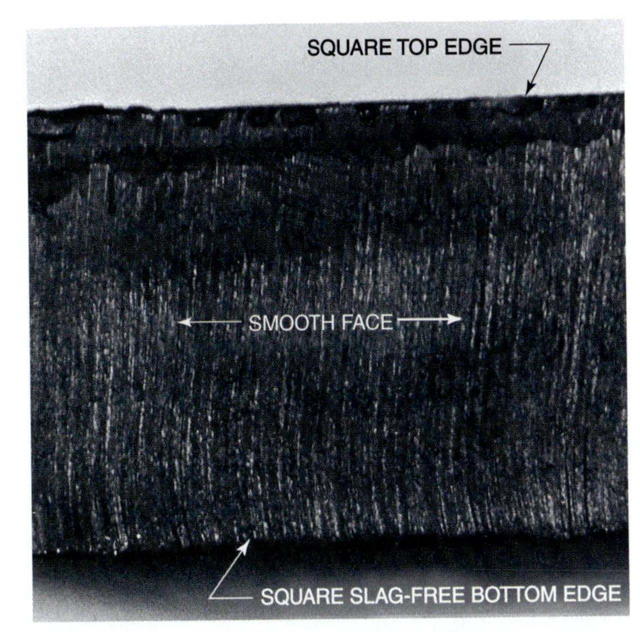

FIGURE 6-35 Correct cut. Courtesy of Larry Jeffus.

sides of the kerf. This record indicates to the welder what was correct or incorrect with the preheat flame, cutting speed, and oxygen pressure.

Preheat The size and number of preheat holes in a tip has an effect on both the top and bottom edges of the metal. An excessive amount of preheat flame results in the top edge of the plate being melted or rounded off. In addition, an excessive amount of hard-to-remove slag is deposited along the bottom edge. If the flame is too small, the travel speed must be slower. A reduction in speed may result in the cutting stream wandering from side to side. The torch tip can be raised slightly to eliminate some of the damage caused by too much preheat. However, raising the torch tip causes the cutting stream of oxygen to be less forceful and less accurate.

Speed The cutting speed should be fast enough so that the drag lines have a slight slant backward if the tip is held at a 90° angle to the plate, **Figure 6-35**. If the cutting speed is too fast, the oxygen stream may not have time to go completely through the metal, resulting in an incomplete cut, **Figure 6-36**. Too slow a cutting speed results in the cutting stream wandering, thus causing gouges in the side of the cut, **Figure 6-37** and **Figure 6-38**.

Pressure A correct pressure setting results in the sides of the cut being flat and smooth. A pressure setting that is too high causes the cutting stream to expand as it leaves the tip, resulting in the sides of

FIGURE 6-36 Too fast a travel speed resulting in an incomplete cut; too much preheat and the tip is too close, causing the top edge to be melted and removed. Courtesy of Larry Jeffus.

FIGURE 6-37 Too slow a travel speed results in the cutting stream wandering, thus causing gouges in the surface; preheat flame is too close, melting the top edge. Courtesy of Larry Jeffus.

FIGURE 6-38 Too slow a travel speed at the start; too much preheat. Courtesy of Larry Jeffus.

the kerf being slightly dished, **Figure 6-39**. When the pressure setting is too low, the cut may not go completely through the metal.

EXPERIMENT 6-2

Effect of Flame, Speed, and Pressure on a Machine Cut

Using a properly lit and adjusted automatic cutting machine, welding gloves, appropriate eye protection and clothing, a variety of tip sizes, and one piece of mild steel plate 6 in. (152 mm) long × 1/2 in. (13 mm) to 1 in. (25 mm) thick, you will observe the effect of the preheat flame, travel speed, and pressure on the metal being cut.

Using the variety of tips, speeds, and oxygen pressures, make a series of cuts on the plate. As the cut is being made, listen to the sound it makes. Also look at the stream of sparks coming off the bottom. A good cut should have a smooth, even sound, and the sparks should come off the bottom of the metal more like a stream than a spray, **Figure 6-40**. When the cut is complete, look at the drag lines to determine what was correct or incorrect with the cut, **Figure 6-41**.

Repeat this experiment until you know a good cut by the sound it makes and the stream of sparks. A good cut has little or no slag left on the bottom of the plate.

Complete a copy of the "Student Welding Report" listed in Appendix I or provided by your instructor. ◆

EXPERIMENT 6-3

Effect of Flame, Speed, and Pressure on a Hand Cut

Using a properly lit and adjusted hand torch, welding gloves, appropriate eye protection and clothing, and the same tip sizes and mild steel plate, repeat Experiment 6-2 to note the effects of the preheat flame, travel speed, and pressure on hand cutting. Complete a copy of the "Student Welding Report" listed in Appendix I or provided by your instructor.

Slag The two types of slag produced during a cut are soft slag and hard slag. **Soft slag** is very porous, brittle, and easily removed from a cut. There is little or no unoxidized iron in it. It may be found on some good cuts. Hard slag may be mixed with soft slag. **Hard slag** is attached solidly to the bottom edge of a cut, and it requires a lot of chipping and grinding to be removed. There is 30% to 40% or

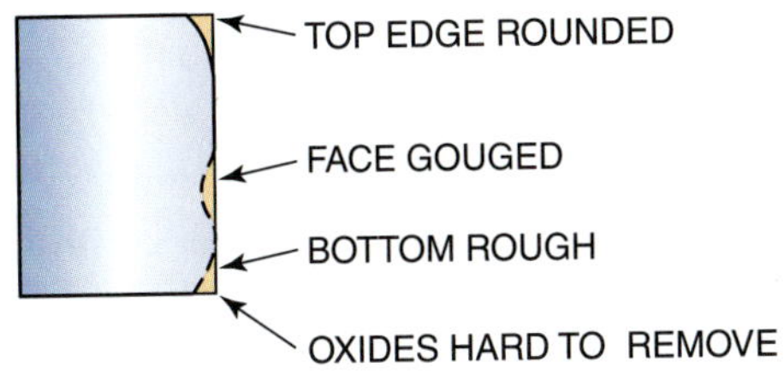

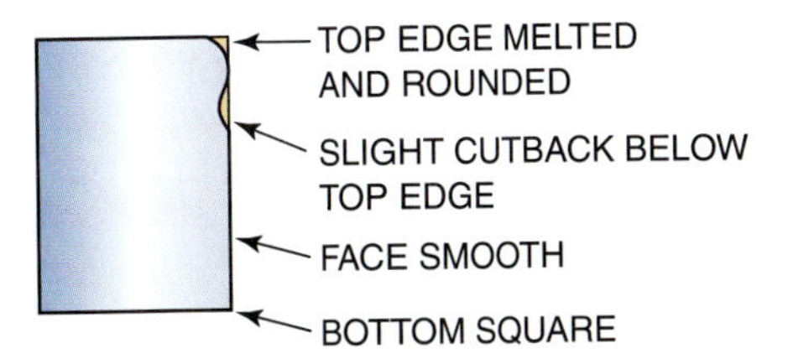

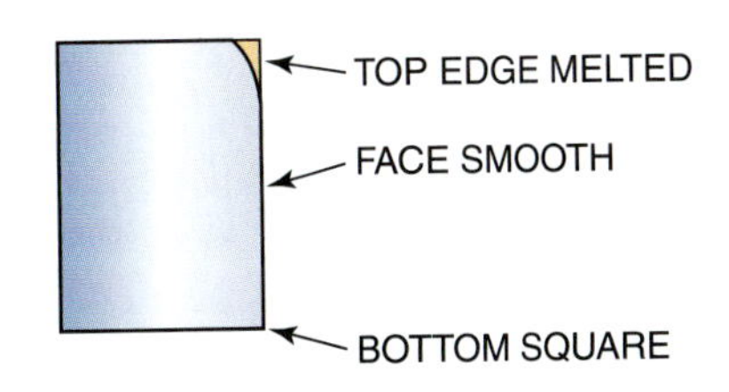

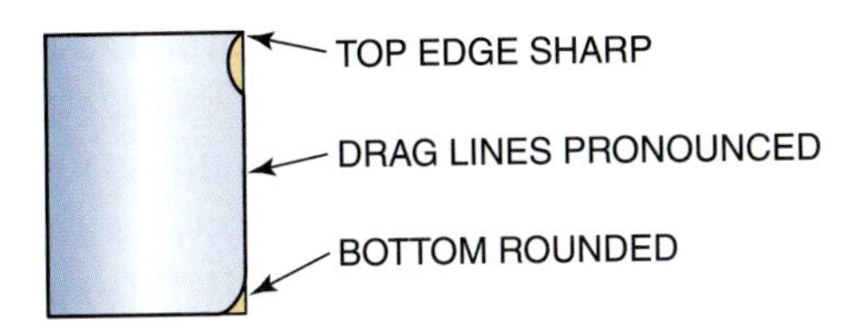

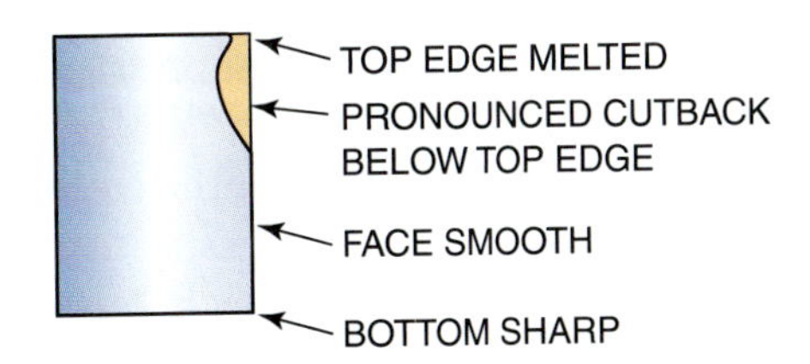

FIGURE 6-39 Profile of flame-cut plates.

FIGURE 6-40 A good cut showing a steady stream of sparks flying out from the bottom of the cut. Courtesy of Larry Jeffus.

FIGURE 6-41 Poor cut. The slag is backing up because the cut is not going through the plate. Courtesy of Larry Jeffus.

more unoxidized iron in hard slag. The higher the unoxidized iron content, the more difficult the slag is to remove. Slag is found on bad cuts, due to dirty tips, too much preheat, too slow a travel speed, too short a coupling distance, or incorrect oxygen pressure.

The slag from a cut may be kept off one side of the plate being cut by slightly angling the cut toward the scrap side of the cut, **Figure 6-42**. The angle needed to force the slag away from the good side of the plate may be as small as 2° or 3°. This technique works best on thin sections; on thicker sections the bevel may show. ◆

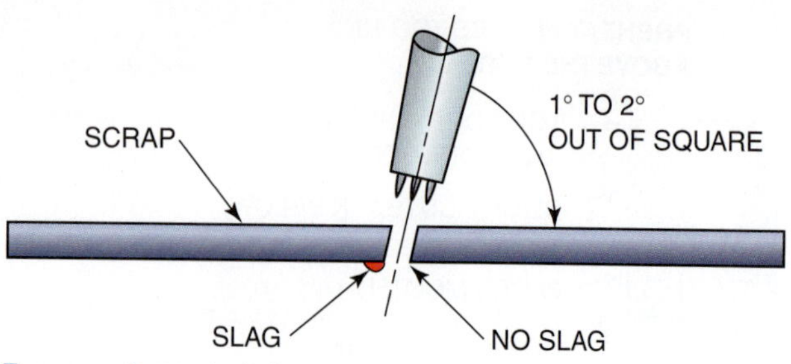

FIGURE 6-42 A slight angle on the torch will put the slag on the scrap side of the cut.

Plate Cutting

Low carbon steel plate can be cut quickly and accurately, whether thin-gauge sheet metal or sections more than 4 feet (1.2 m) thick are used. It is possible to achieve cutting speeds as fast as 32 inches per minute (13.5 mm/s), in 1/8-in. (3-mm) plate, and accuracy on machine cuts of ±3/64 in. Some very large hand-cutting torches with an oxygen cutting volume of 600 cfh (2830 L/min) can cut metal that is 4 ft (1.2 m) thick, **Figure 6-43.** Most hand torches will not easily cut metal that is more than 7 in. (178 mm) to 10 in. (254 mm) thick.

The thicker the plate, the more difficult the cut is to make. Thin plate, 1/4 in. (6 mm) or less, can be cut and the pieces separated even if poor techniques and incorrect pressure settings are used. Thick plate, 1/2 in. (13 mm) or thicker, often cannot be separated if the cut is not correct. For very heavy cuts, on plate 12 in. (305 mm) or thicker, the equipment and operator technique must be near perfection or the cut will be faulty.

Plate that is properly cut can be assembled and welded with little or no postcut cleanup. Poor-quality cuts require more time to clean up than is needed to make the required adjustments to make a good weld.

Cutting Table

Because of the nature of the torch cutting process, special consideration is given to the flame cutting support. Any piece being cut should be supported so the torch flame will not cut through the piece and into the table. Special cutting tables are used that expose only a small metal area to the torch flame. Some tables use parallel steel bars of metal and others use cast iron pyramids. All cutting should be set up so the flame and oxygen stream runs between the support bars or over the edge of the table.

If an ordinary welding table or another steel table is used, special care must be taken to avoid cutting through the table top. The piece being cut may be supported above the support table by firebrick. Another method is to cut the metal over the edge of the table.

Torch Guides

In manual torch cutting a guide or support is frequently used to allow for better control and more even cutting. It takes a very skilled welder to make a straight, clean cut even when following a marked line. It is even more difficult to make a radius cut to any accuracy. Guides and supports allow the height and angle of the torch head to remain constant. The speed of the cut, which is very important to making a clean, even kerf, must be controlled by the welder.

Since the torch must be held in an exact position while making any accurate cut, the welder normally supports the torch weight with the hand. Supporting the torch weight this way not only allows for more accurate work but also cuts down on fatigue. A rest, such as a firebrick, is also used to support the torch.

Various types of guides can be used to guide the torch in a straight line. **Figure 6-44** shows one type of guide using angle iron. The edge of the angle is followed to give the straight cut. Bevel cuts can be made freehand with the torch, but it is very difficult to keep them uniform. More accurate bevel cuts are made by resting the torch against the angle side of an angle iron.

Special roller guides, **Figure 6-45A**, can also be attached to the torch head. The attachment holds the torch cutting tip at an exact height.

When cutting circles, a circle cutting attachment is used. **Figure 6-45B** shows how the attachment fits on the torch head. The radius can be preset to any required distance. The cutter revolves around the

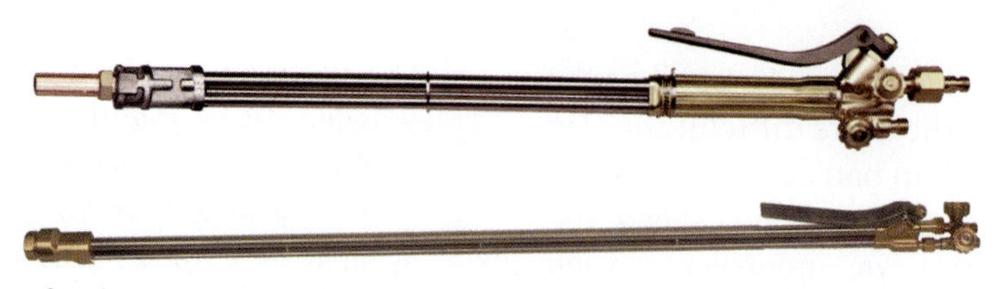

FIGURE 6-43 Hand torches for thick sections. Courtesy of Victor Equipment, a Thermadyne Company.

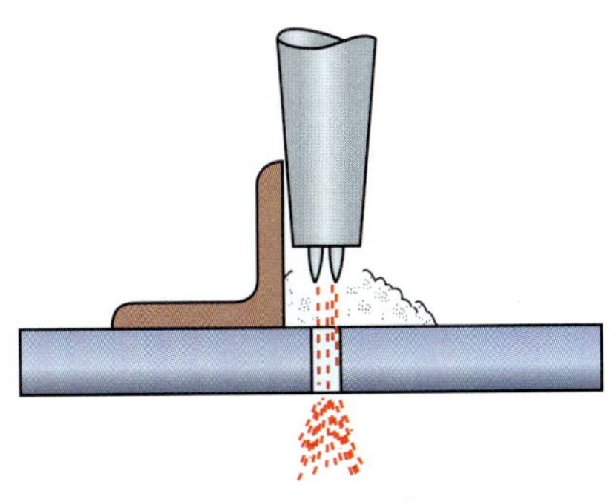

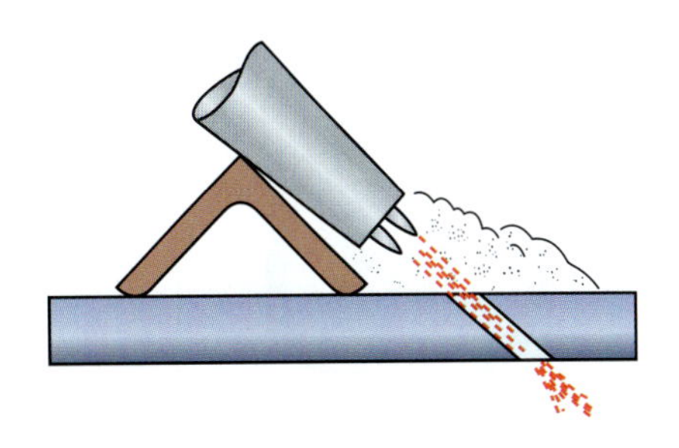

FIGURE 6-44 Using angle irons to aid in making cuts.

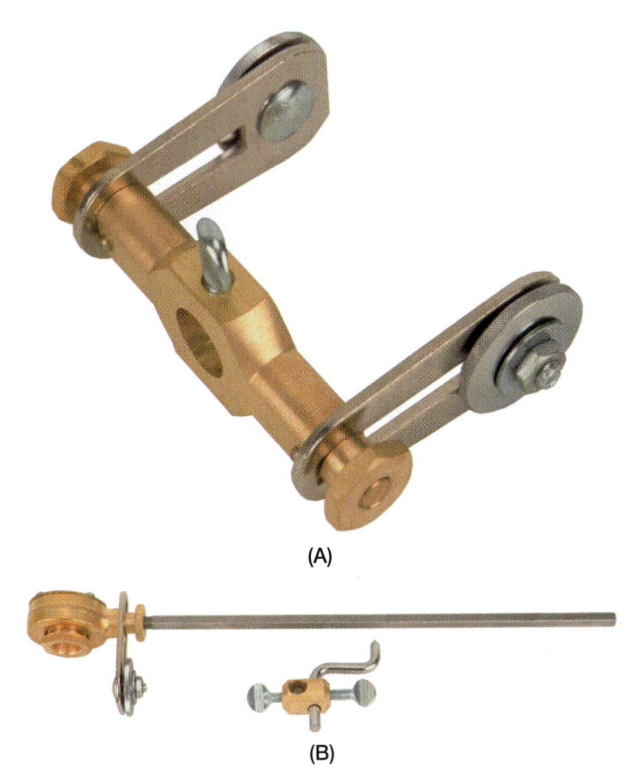

FIGURE 6-45 Devices that are used to improve hand cutting. Courtesy of Victor Equipment, a Thermadyne Company.

center point when making the cut. The roller controls the torch tip height above the plate surface.

PRACTICE 6-5

Flat, Straight Cut in Thin Plate

Using a properly lit and adjusted cutting torch and one piece of mild steel plate 6 in. (152 mm) long × 1/4 in. (6 mm) thick, you will cut off 1/2-in. (13-mm) strips.

Using a straightedge and soapstone, make several straight lines 1/2 in. (13 mm) apart. Starting at one end, make a cut along the entire length of plate. The strip must fall free, be slag free, and be within ±3/32 in. (2 mm) of a straight line and ±5° of being square. Repeat this procedure until the cut can be made straight and slag free. Turn off the cylinder valves, bleed the hoses, back out the pressure regulators, and clean up your work area when you are finished cutting.

Complete a copy of the "Student Welding Report" listed in Appendix I or provided by your instructor. ◆

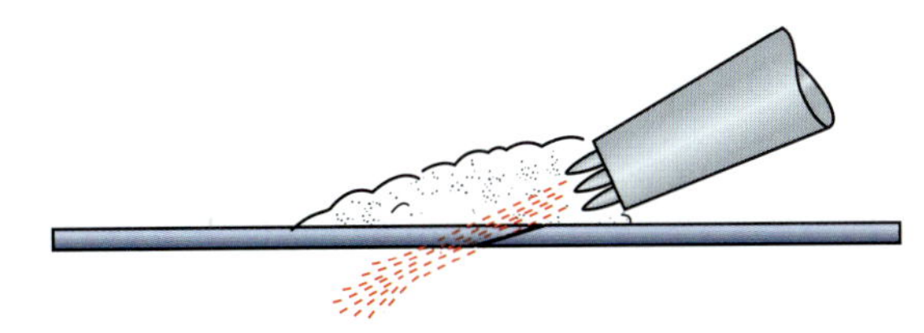

FIGURE 6-46 Cut the sheet metal at a very sharp angle.

PRACTICE 6-6

Flat, Straight Cut in Thick Plate

Using a properly lit and adjusted cutting torch and one piece of mild steel plate 6 in. (152 mm) long × 1/2 in. (13 mm) thick or thicker, you will cut off 1/2-in. (13-mm) strips. *Note: Remember that starting a cut in thick plate will take longer, and the cutting speed will be slower.* Lay out, cut, and evaluate the cut as was done in Practice 6-5. Repeat this procedure until the cut can be made straight and slag free. Turn off the cylinder valves, bleed the hoses, back out the pressure regulators, and clean up your work area when you are finished cutting.

Complete a copy of the "Student Welding Report" listed in Appendix I or provided by your instructor. ◆

PRACTICE 6-7

Flat, Straight Cut in Sheet Metal

Use a properly lit and adjusted cutting torch and a piece of mild steel sheet that is 10 in. (254 mm) long and 18 gauge to 11 gauge thick. Holding the torch at a very sharp leading angle, **Figure 6-46**, cut the sheet along the line. The cut must be smooth and straight with as little slag as possible. Repeat this procedure until the cut can be made flat, straight, and slag free. Turn off the cylinder valves, bleed the hoses, back out

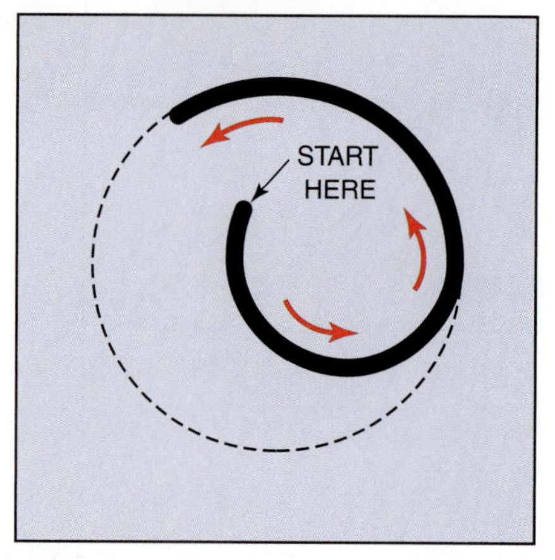

FIGURE 6-47 Start a cut for a hole near the middle.

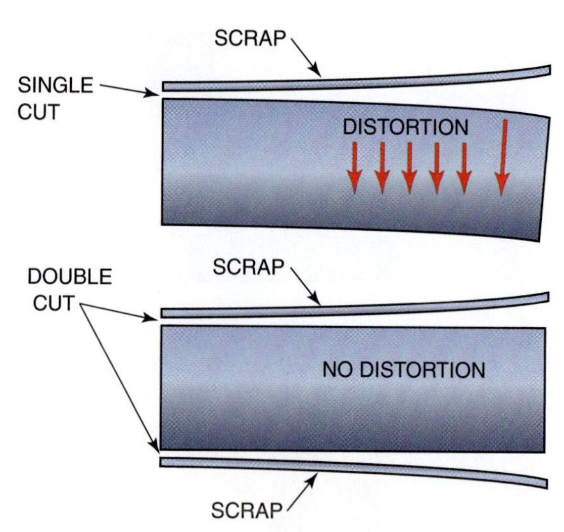

FIGURE 6-48 Making two parallel cuts at the same time will control distortion.

the pressure regulators, and clean up your work area when you are finished cutting.

Complete a copy of the "Student Welding Report" listed in Appendix I or provided by your instructor. ◆

PRACTICE 6-8

Flame Cutting Holes

Using a properly lit and adjusted cutting torch, welding gloves, appropriate eye protection and clothing, and one piece of mild steel plate 1/4 in. (6 mm) thick, you will cut holes with diameters of 1/2 in. (13 mm) and 1 in. (25 mm). Using the technique described for piercing a hole, start in the center and make an outward spiral until the hole is the desired size, **Figure 6-47**. The hole must be within ±3/32 in. (2 mm) of being round and ±5° of being square. The hole may have slag on the bottom. Repeat this procedure until both small and large sizes of holes can be made within tolerance. Turn off the cylinder valves, bleed the hoses, back out the pressure regulators, and clean up your work area when you are finished cutting.

Complete a copy of the "Student Welding Report" listed in Appendix I or provided by your instructor. ◆

Distortion

Distortion is when the metal bends or twists out of shape as a result of being heated during the cutting process. This is a major problem when cutting a plate. If the distortion is not controlled, the end product might be worthless. There are two major methods of controlling distortion. One method involves making two parallel cuts on the same plate at the same speed and time, **Figure 6-48**. Because the plate is heated evenly, distortion is kept to a minimum, **Figure 6-49**.

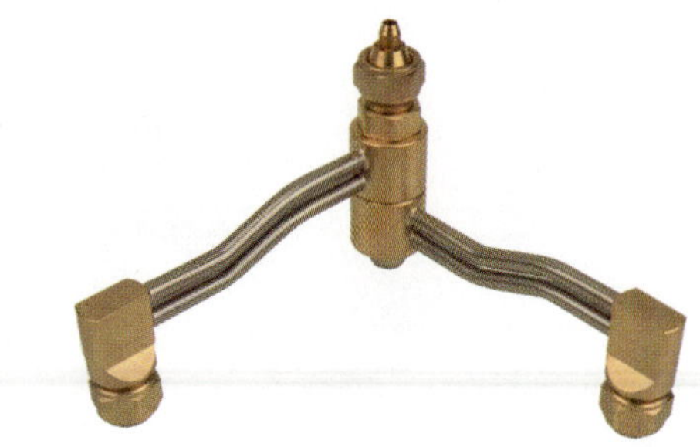

FIGURE 6-49 Slitting adaptor for cutting machine. It can be used for parallel cuts from 1" (38 mm) to 12" (500 mm). Ideal for cutting test coupons. Courtesy of Victor Equipment, a Thermadyne Company.

The second method involves starting the cut a short distance from the edge of the plate, skipping other short tabs every 2 ft (0.6 m) to 3 ft (0.9 m) to keep the cut from separating. Once the plate cools, the remaining tabs are cut, **Figure 6-50**.

EXPERIMENT 6-4

Minimizing Distortion

Using a properly lit and adjusted cutting torch, welding gloves, appropriate eye protection and clothing, and two pieces of mild steel 10 in. (254 mm) long × 1/4 in. (6 mm) thick, you will make two cuts and then compare the distortion. Lay out and cut both pieces of metal as shown in **Figure 6-51**. Allow the metal to cool, and then cut the remaining tabs. Compare the four pieces of metal for distortion.

Complete a copy of the "Student Welding Report" listed in Appendix I or provided by your instructor. ◆

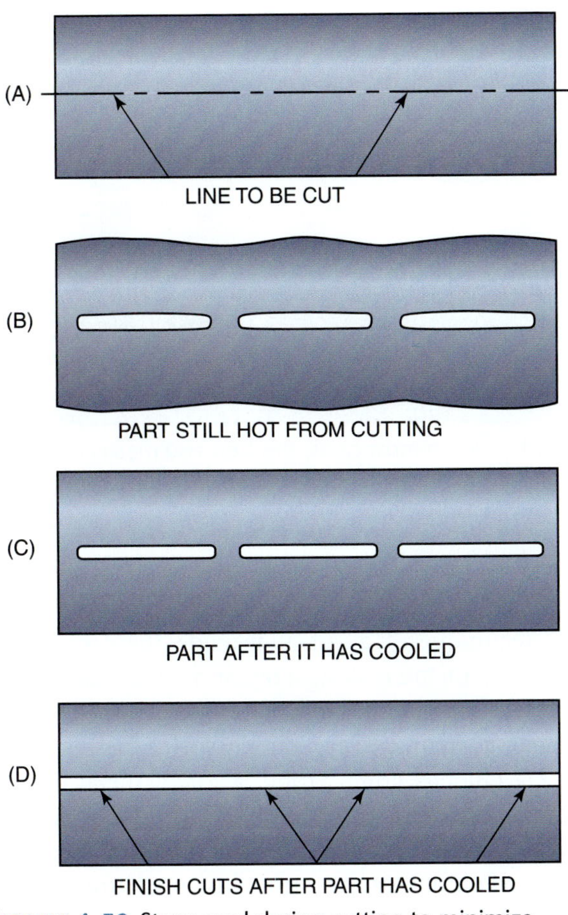

FIGURE 6-50 Steps used during cutting to minimize distortion.

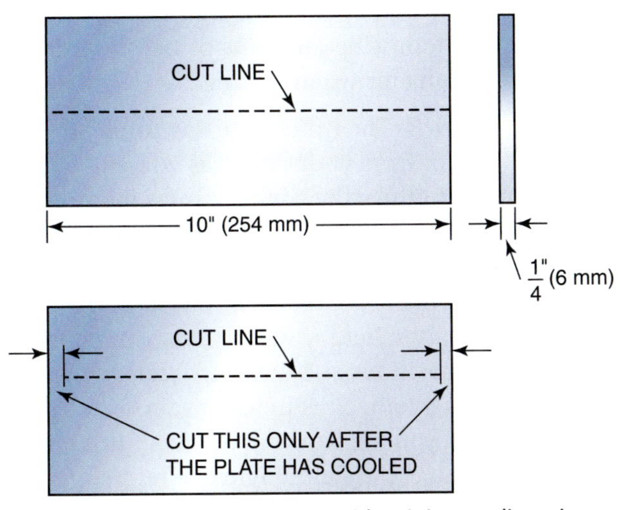

FIGURE 6-51 Making two cuts with minimum distortion. *Note:* Sizes of these and other cutting projects can be changed to fit available stock.

PRACTICE 6-9

Beveling a Plate

Use a properly lit and adjusted cutting torch, welding gloves, appropriate eye protection and clothing, and one piece of mild steel plate 6 in. (152 mm) long × 3/8 in. (10 mm) thick. You will make a 45° bevel down the length of the plate.

Mark the plate in strips 1/2 in. (13 mm) wide. Set the tip for beveling and cut a bevel. The bevel should be within ±3/32 in. (2 mm) of a straight line and ±5° of a 45° angle. There may be some soft slag, but no hard slag, on the beveled plate. Repeat this Practice until the cut can be made within tolerance. Turn off the cylinder valves, bleed the hoses, back out the pressure regulators, and clean up your work area when you are finished cutting.

Complete a copy of the "Student Welding Report" listed in Appendix I or provided by your instructor. ◆

PRACTICE 6-10

Vertical Straight Cut

For this Practice, you will need a properly lit and adjusted cutting torch, welding gloves, appropriate eye protection and clothing, and one piece of mild steel plate 6 in. (152 mm) long × 1/4 in. (6 mm) to 3/8 in. (10 mm) thick, marked in strips 1/2 in. (13 mm) wide and held in the vertical position. You will make a straight line cut. Make sure that the sparks do not cause a safety hazard and that the metal being cut off will not fall on any person or object.

Starting at the top, make one cut downward. Then, starting at the bottom, make the next cut upward. The cut must be free of hard slag and within ±3/32 in. (2 mm) of a straight line and ±5° of being square. Repeat these cuts until they can be made within tolerance. Turn off the cylinder valves, bleed the hoses, back out the pressure regulators, and clean up your work area when you are finished cutting.

Complete a copy of the "Student Welding Report" listed in Appendix I or provided by your instructor. ◆

PRACTICE 6-11

Overhead Straight Cut

Using a properly lit and adjusted cutting torch, welding gloves, appropriate eye protection and clothing, and one piece of mild steel plate 6 in. (152 mm) long × 1/4 in. (6 mm) to 3/8 in. (10 mm) thick, marked in strips 1/2 in. (13 mm) wide, you will make a cut in the overhead position. When making overhead cuts, it is important to be completely protected from the hot sparks. In addition to the standard safety clothing, you should wear a leather jacket, leather apron, cap, ear protection, and a full face shield.

The torch can be angled so that most of the sparks will be blown away. The metal should fall free when the cut is completed. The cut must be within 1/8 in.

FIGURE 6-52 Have plenty of water and a fire extinguisher available before starting to cut or weld. Courtesy of Larry Jeffus.

FIGURE 6-53 In most cases, this sign also means no welding or cutting outdoors. Courtesy of Larry Jeffus.

(3 mm) of a straight line and ±5° of being square. Repeat this practice until the cut can be made within tolerance. Turn off the cylinder valves, bleed the hoses, back out the pressure regulators, and clean up your work area when you are finished cutting.

Complete a copy of the "Student Welding Report" listed in Appendix I or provided by your instructor. ◆

Cutting Applications

It is important to know that not all the damage that can be done with a torch is to your machinery. Newspapers often have stories stating "the sparks from welders torches are thought to have started the fire". The torch they are referring to is the cutting torch. The torch throws out so many sparks that almost anything will catch fire. In the welding industry, the cutting torch is used in a shop or yard where there is almost nothing to catch fire. However in agriculture, we have dry crop stubble, wood chips, hay and almost every other thing that will burn lying around our work areas. When possible, move anything that might catch fire well away from your work because cutting sparks can easily travel 30 feet (10 m) or more. In some cases, such as working in a field or in the timber, it is impossible to remove all combustibles from around your work area. If you must work around combustibles wet the area first and keep a bucket of water and a fire extinguisher handy, **Figure 6-52**. Keeping someone as a fire watch will also help.

CAUTION

Never weld or cut in a dry or an area has been posted by the county or state as a fire zone, Figure 6-53. You have too great an investment in the land to accidentally start a fire while welding. Also you could be held legally responsible for a fire that might start if you ignore the warning.

Making practice cuts on a piece of metal that will only become scrap is a good way to learn the proper torch techniques. If a bad cut is made, there is no loss. But when you are making a repair on your equipment and make a bad cut, that can be disastrous. It is easy to do a lot of damage very quickly with a cutting torch. Once a cut has gone wrong, a lot of metal can be removed quickly which will take hours to repair. When you see a cut is going bad—and bad cuts happen to everyone—stop cutting. Do not assume the cut will get better. If you cannot make a good cut with a torch use a grinder or saw, or get help.

A number of factors that do not exist during practice cuts can affect your ability to make a quality cut on a part. The following are some of the things that can become problems when cutting:

- *Brace yourself:* The tip of a cutting torch is nearly a foot from your hands as you cut. This distance amplifies the slightest movement of your hand, which translates to grooves and notches in the cut. It is easier to brace yourself so you can make good cuts in the shop than in the field. In the field, you may be working in awkward positions and small spaces. One way of steadying yourself is to lean against the equipment you are working on. Even just resting your hand can help, **Figure 6-54**. Before even lighting the torch, practice how you can move across the cut and which way would be best to brace yourself.
- *Changing positions:* Often, parts are too large to be cut from one position, so you may have to move to complete the cut. Stopping and restarting a cut can result in a small flaw in the cut surface. To avoid this problem, always try to stop at corners if the cut cannot be completed without moving.

FIGURE 6-54 Bracing your hand on the tractor tongue will help keep you steady as you make the cut. Courtesy of Larry Jeffus.

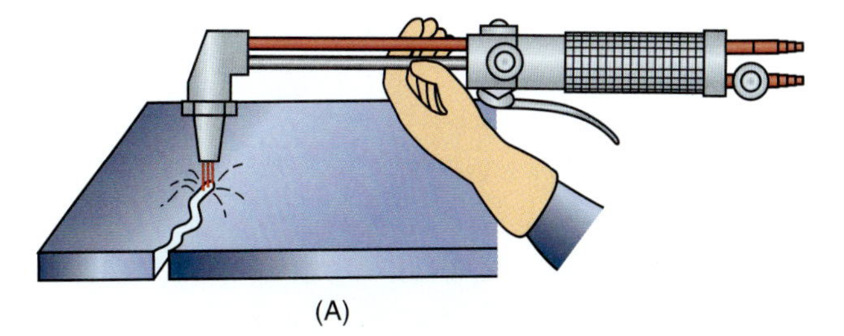

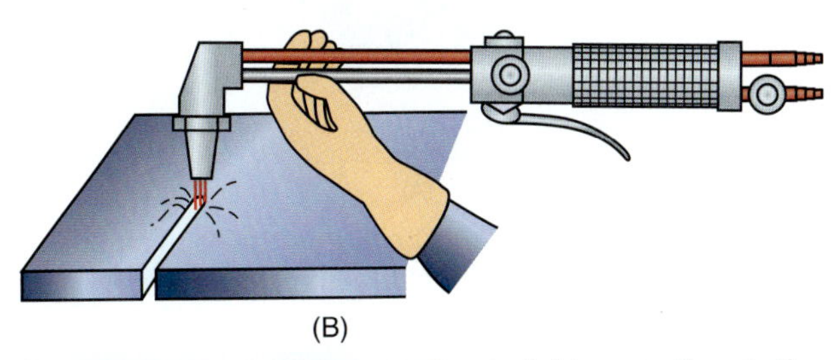

FIGURE 6-55 It is easier to make straight, smooth cuts if you can brace the torch closer to the tip, as in cut B.

- *Sparks:* All cuts create sparks that bounce around.You must always know where the sparks from a cut are being thrown. Make sure they are not being directed toward the fuel tanks, dead grass or straw, oil or other combustible materials. Sparks striking glass or mirrors can cause pitting. Cutting on equipment where the sparks hit another surface behind the one being cut makes the problem of bouncing sparks much worse. Sparks often find their way into your glove, under your arm, or to any other place that will become uncomfortable. Experienced welders will usually keep working if the sparks are not too large or too uncomfortable. With experience you will learn how to angle the torch, direct the cut, and position your body to minimize this problem. It is your responsibility to make sure the sparks do not start a fire or damage the equipment.
- *Hot surfaces:* As you continue making cuts to complete the part, it will begin to heat up. Depending on the size of the part, the number of cuts per part, and the number of parts being cut, this heat can become uncomfortable. You may find it necessary to hold the torch farther back from the tip, but this will affect the quality of your cuts, **Figure 6-55**. Sometimes you might be able to rest your hand on a block to keep it off the plate. Another problem with heat buildup is that it may become high enough to affect the equipments hydraulic or other systems. Planning your cutting sequence and allowing cooling time will help control this potential problem.
- *Tip cleaning:* The cutting tip will catch small sparks and become dirty or clogged. You must decide how dirty or clogged you will let the tip get before you stop to clean it. Time spent cleaning the tip reduces productivity, unfortunately. On the other hand, if you do not stop occasionally to clean up, the quality of the cut will become bad. It is your responsibility to decide when and how often to clean the tip.
- *Blow back:* As a cut progresses across the surface, it may cross parts underneath. During practice cuts this seldom if ever happens, but, depending on the part being cut, it will occur. If the part is small, the blowback may not cover you with sparks, plug the cutting tip, or cause a major flaw in the cut surface. If the part is large, then one or all of these events can occur. If you see that the blowback is not clearing quickly, it may be necessary to stop the cut. Stopping the cut halts the shower of sparks but leaves you with a problem restart.

PRACTICE 6-12

Cutting Out Internal and External Shapes

Using a properly lit and adjusted cutting torch, welding gloves, appropriate eye protection and clothing, and one piece of plate 1/4 in. (6 mm) to 3/8 in. (10 mm) thick, you may lay out and cut out one of the sample patterns shown in **Figure 6-56**, one of the projects in Chapter 17, or any other design available.

Choose the pattern that best fits the piece of metal you have and mark it using a center punch. The exact size and shape of the layout are not as important as the accuracy of the cut. The cut must

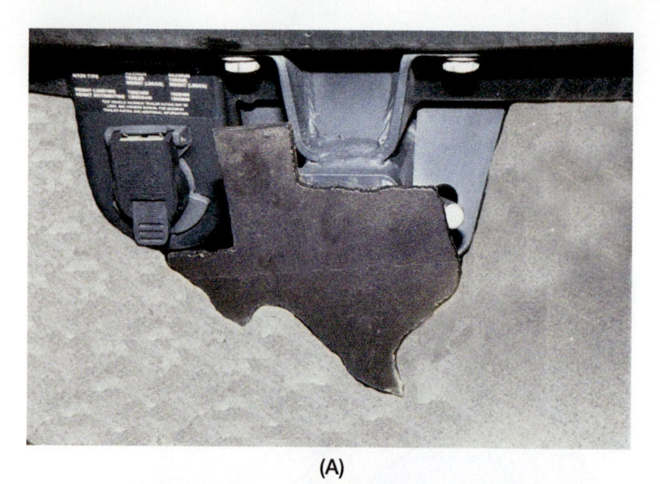
(A)

(B)

FIGURE 6-56 (A) You can cut out your state's silhouette and make it into a tractor hitch receiver cover. (B) Cut out all your family members' initials and put them in an entry gate arch. Courtesy of Larry Jeffus.

FIGURE 6-57 Beginning a cut with the torch concentrating the flame on the thin edge to speed starting. Courtesy of Larry Jeffus.

FIGURE 6-58 The torch is rotated to allow the preheating of the plate ahead of the cut. This speeds the cutting and also provides better visibility of the line being cut. Courtesy of Larry Jeffus.

be made so that the center-punched line is left on the part and so that there is no more than 1/8 in. (3 mm) between the cut edge and the line, **Figures 6-57** and **Figure 6-58**. Repeat this practice until the cut can be made within tolerance. Turn off the cylinder valves, bleed the hoses, back out the pressure regulators, and clean up your work area when you are finished cutting.

Complete a copy of the "Student Welding Report" listed in Appendix I or provided by your instructor. ◆

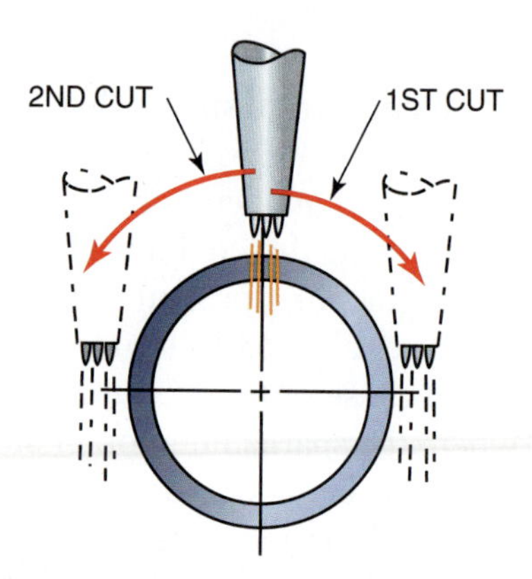

FIGURE 6-59 Small-diameter pipe can be cut without changing the angle of the torch. After the top is cut, roll the pipe to cut the bottom.

Pipe Cutting

Freehand pipe cutting may be done in one of two ways. On small diameter pipe, usually under 3 in. (76 mm), the torch tip is held straight up and down and moved from the center to each side, **Figure 6-59**. This technique can also be used successfully on larger pipe.

For large diameter pipe, 3 in. (76 mm) and larger, the torch tip is always pointed toward the center of the pipe, **Figure 6-60**. This technique is also used on all sizes of heavy-walled pipe and can be used on some smaller pipe sizes.

The torch body should be held so that it is parallel to the centerline of the pipe. Holding the torch parallel helps to keep the cut square.

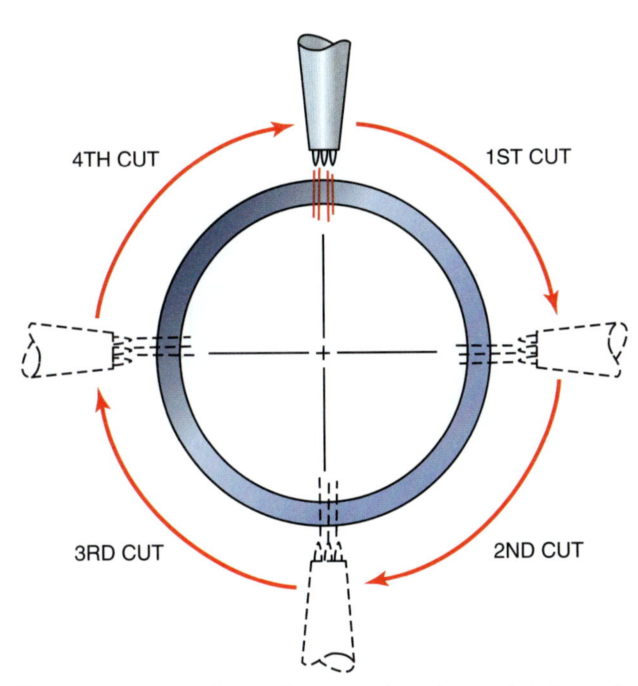

FIGURE 6-60 On large-diameter pipe, the torch is turned to keep it at a right angle to the pipe. The pipe should be cut as far as possible before stopping and turning it.

CAUTION

When cutting pipe, hot sparks can come out of the end of the pipe nearest you, causing severe burns. For protection from hot sparks, plug up the open end of the pipe nearest you, put up a barrier to the sparks, or stand to one side of the material being cut.

PRACTICE 6-13

Square Cut on Pipe, 1G (Horizontal Rolled) Position

Using a properly lit and adjusted cutting torch, welding gloves, appropriate eye protection and clothing, and one piece of schedule 40 steel pipe with a diameter of 3 in. (76 mm), you will cut off 1/2-in. (13-mm) -long rings.

Using a template and a piece of soapstone, mark several rings, each 1/2 in. (13 mm) wide, around the pipe. Place the pipe horizontally on the cutting table. Start the cut at the top of the pipe using the proper piercing technique. Move the torch backward along the line and then forward; this will keep slag out of the cut. If the end of the cut closes in with slag, this will cause the oxygen to gouge the edge of the pipe when the cut is continued. Keep the tip pointed straight down. When you have gone as far with the cut as you can comfortably, quickly flip the flame away from the pipe. Restart the cut at the top of the pipe and cut as far as possible in the other direction. Stop and turn the pipe so that the end of the cut is on top and the cut can be continued around the pipe. When the cut is completed, the ring must fall free. When the pipe is placed upright on a flat plate, the pipe must stand within 5° of vertical and have no gaps higher then 1/8 in. (3 mm) under the cut. Repeat this procedure until the cut can be made within tolerance. Turn off the cylinder valves, bleed the hoses, back out the pressure regulators, and clean up your work area when you are finished cutting.

Complete a copy of the "Student Welding Report" listed in Appendix I or provided by your instructor. ◆

PRACTICE 6-14

Square Cut on Pipe, 1G (Horizontal Rolled) Position

Using the same equipment, materials, and markings as described in Practice 6-13, you will cut off the 1/2 in. (13-mm) -long rings while keeping the tip pointed toward the center of the pipe.

Starting at the top, pierce the pipe. Move the torch backward to keep the slag out of the cut and then forward around the pipe, stopping when you have gone as far as you can comfortably. Restart the cut at the top and proceed with the cut in the other direction. Roll the pipe and continue the cut until the ring falls off freely. Stand the cut end of the pipe on a flat plate. The pipe must stand within 5° of vertical and have no gaps higher than 1/8 in. (3 mm). Repeat this practice until the cut can be made within tolerance. Turn off the cylinder valves, bleed the hoses, back out the pressure regulators, and clean up your work area when you are finished cutting.

Complete a copy of the "Student Welding Report" listed in Appendix I or provided by your instructor. ◆

PRACTICE 6-15

Square Cut on Pipe, 5G (Horizontal Fixed) Position

With the same equipment, materials, and markings as described in Practice 6-13, you will cut off 1/2-in. (13-mm) rings, using either technique, without rolling the pipe.

Start at the top and cut down both sides as far as you can comfortably. Reposition yourself and continue the cut under the pipe until the ring falls off freely. Stand the cut end of the pipe on a flat plate. The pipe must stand within 5° of vertical and have no gaps higher than 1/8 in. (3 mm). Repeat this practice until the cut can be made within tolerance. Turn

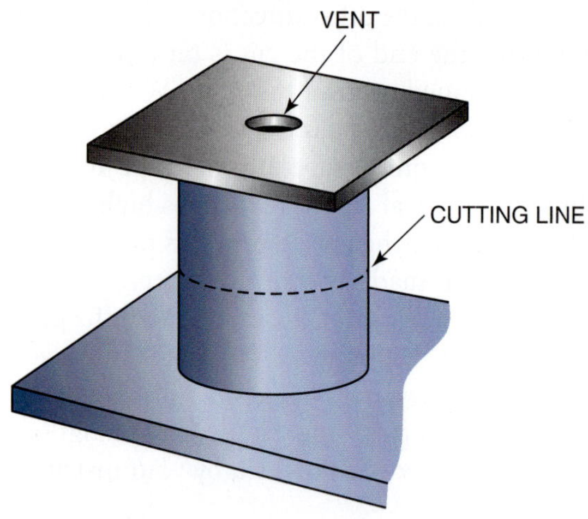

FIGURE 6-61 Place a plate on top of a short piece of pipe to keep the sparks from flying around the shop.

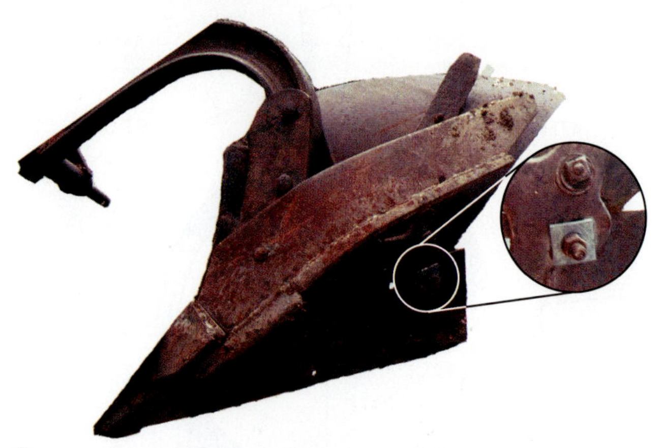

FIGURE 6-62 The nuts and bolts on this rusty plow could not be taken off without a torch. Coutesy of Larry Jeffus.

off the cylinder valves, bleed the hoses, back out the pressure regulators, and clean up your work area when you are finished cutting.

Complete a copy of the "Student Welding Report" listed in Appendix I or provided by your instructor. ◆

PRACTICE 6-16

Square Cut on Pipe, 2G (Vertical) Position

With the same equipment, materials, and markings as listed in Practice 6-13, you will cut off 1/2-in. (13-mm) rings, using either technique, from a pipe in the vertical position.

Place a flat piece of plate over the open top end of the pipe to keep the sparks contained, **Figure 6-61.** Start on one side and proceed around the pipe until the cut is completed. Because of slag, the ring may have to be tapped free. Stand the cut end of the pipe on a flat plate. The pipe must stand within 5° of vertical and have no gaps higher than 1/8 in. (3 mm). Repeat this practice until the cut can be made within tolerance. Turn off the cylinder valves, bleed the hoses, back out the pressure regulators, and clean up your work area when you are finished cutting.

Complete a copy of the "Student Welding Report" listed in Appendix I or provided by your instructor. ◆

Summary

The speed and accuracy with which an oxyacetylene torch can cut off damaged parts or cut out new parts makes it an indispensable tool on the farm or ranch. The torch's ability to heat and loosen rusty nuts or simply cut them off has earned it the nickname "smoke wrench." Frequently nuts and bolts on equipment that has been left out in the weather could not be removed without the torch, **Figure 6-62.** Anyone who has worked on agricultural equipment even stored in sheds knows how much time the smoke wrench can save on many jobs. It is even possible to cut a really rusty nut off without damaging the threads on the bolt. The thick layer of rust acts like an insulator keeping the bolt from getting heated as quickly as the nut. Because the bolt does not get heated to its kindling temperature, the threads will remain intact. You can even cut out a 1/2 in. (13-mm) rusty stud that is broken off in a cast-iron housing if you first drill a 1/4 in. (6-mm) hole through the stud. The hole allows the stud to be heated quickly and cut almost cleanly out. With good torch control this will work when an easy out would not.

Summary continued

In the hands of a skilled welder, a proper setup and cleaned oxyacetylene torch can make a cut that is nearly as accurate as can be made with a hacksaw, but with a lot less work. Taking your time to set up, clean, and adjust the torch before you begin working will save more time than it took to get the equipment set right to the cutting job.

Review

1. Which metal can be cut easily with an oxyacetylene torch?
   A. copper
   B. aluminum
   C. cast iron
   D. steel
2. Stainless steel can be cut easily with oxyacetylene.
   A. true
   B. false
3. Why must proper eye protection be worn when making oxyacetylene cuts?
   A. to protect your face from the heat
   B. because of the bright light and flying sparks
   C. to keep the smoke from the torch out of your eyes
   D. because the torch tip could explode
4. It is not safe to use sunglasses as gas cutting goggles.
   A. true
   B. false
5. Torches that use a mixing chamber are known as ___________.
   A. equal-pressure torches
   B. low-volume torches
   C. venturi torches
   D. combination torches
6. The location of the cutting lever is the same location on every torch.
   A. true
   B. false
7. What metal is used to make most cutting tips?
   A. stainless steel
   B. copper
   C. brass
   D. chrome
8. What is an easy way to determine the cutting thickness capacity of a torch tip?
   A. Count the number of holes in the tip.
   B. Measure the tip length.
   C. Look at the diameter of the center hole.
   D. Count the number of pieces that make up the tip.
9. All cutting tips are the same and can be used for any job.
   A. true
   B. false
10. Which of the following is not a reason that a cutting tip would need more preheat holes?
    A. skill of operator
    B. material thickness
    C. material shape
    D. surface condition
11. What is the most common fuel gas used with oxygen for the OFC process?
    A. MSP gases
    B. acetylene
    C. propane
    D. MAPP®
12. Why is it important to brace yourself when making a cut?
    A. For your own safety.
    B. To keep from falling.
    C. Because the torch is heavy and you might get tired.
    D. So the cut is smoother and straighter.

Review continued

13. The torch should be braced with the right hand if the welder is left-handed.
    A. true
    B. false
14. What is the name of the open space made in metal as it is cut through by a torch?
    A. drag lines
    B. groove
    C. gouge
    D. kerf
15. What tool can be used to layout a 10-ft (3-m) long straight line on metal for cutting?
    A. chalk line
    B. tape measure
    C. soap stone
    D. framing square
16. Each welding equipment manufacturer uses the same numbering system to designate the tip size.
    A. true
    B. false
17. What is the name for slag that is easy to remove from a cut?
    A. free form slag
    B. hard slag
    C. brittle slag
    D. soft slag
18. Stopping and starting a cut can result in a small flaw in the cut surface.
    A. true
    B. false
19. It is not all right to do cutting in the fields or woods when there is a fire danger.
    A. true
    B. false
20. What should you do if the cut stops going all the way through the metal and sparks begin to fly back on both you and the torch?
    A. travel faster
    B. mash the cutting lever down further
    C. stop
    D. adjust the flame higher

Chapter 7

Plasma Arc Cutting

OBJECTIVES

After completing this chapter, the student should be able to

- ☑ describe plasma and describe a plasma torch.
- ☑ explain how a plasma cutting torch works.
- ☑ list the advantages and disadvantages of using a plasma cutting torch.
- ☑ demonstrate an ability to set up and use a plasma cutting torch.

KEY TERMS

arc cutting
arc plasma
cup
dross
electrode setback
electrode tip
heat-affected zone
high-frequency alternating current
ionized gas
joules
kerf
nozzle
nozzle insulator
nozzle tip
pilot arc
plasma
plasma arc
plasma arc gouging
stack cutting
standoff distance
water shroud
water table

INTRODUCTION

The plasma cutting process (PAC) has become very popular as the result of the introduction of smaller, less expensive plasma equipment to the agricultural field, **Figure 7-1**. A typical portable agricultural plasma cutting system can cut mild steel up to 3/8 in. (9 mm) thick. Plasma cutters have the unique ability to cut metals without making them very hot. This means that there is less distortion and heat damage than would be caused with an oxyacetylene cutting torch. Very intricate shapes can be cut out without warping.

Plasma can cut sheet metal so easily that it has become popular to use it to cut out the smallest of decorations. Most often the smallest of animals, people, buildings, and scenery used

FIGURE 7-1 Plasma arc cutting machine. This unit can have additional power modules added to the base of its control module to give it more power. Courtesy of ESAB Welding & Cutting Products.

on gates, fences, barns, and so forth have been cut out using PAC, **Figure 7-2**.

Small plasma cutting machines can do many of the same cutting jobs that are done with an oxyacetylene torch, but without the expense of renting gas cylinders. Small agricultural plasma cutting machines can use 120V electrical power from any standard wall plug or auxiliary power plug on a portable welder, **Figure 7-3**.

Plasma machines can cut any type of metal including aluminum, stainless steel, and cast iron—all commonly found in agriculture. They can cut mild steel ranging from sheet metal up to about 3/8 in. (9 mm), which means that the plasma torch can do most of the cutting required on a farm or ranch. Larger, more powerful machines are available that can cut an inch (25 mm) or more, but their expense for most agricultural operations would be hard to justify.

Plasma

The word **plasma** has two meanings: it is the fluid portion of blood and it is a state of matter that is found in the region of an electrical discharge (arc). The plasma created by an arc is an **ionized gas** that has both electrons and positive ions whose charges are nearly equal to each other. For welding we use the electrical definition of *plasma*.

A plasma is present in any electrical discharge. It consists of charged particles that conduct the electrons across the gap. Both the glow of a neon tube and the bright fluorescent lightbulb are examples of low-temperature plasmas.

A plasma results when a gas is heated to a high enough temperature to convert into positive and negative ions, neutral atoms, and negative electrons. The temperature of an unrestricted arc is about 11,000°F, but the temperature created when the arc is concentrated to form a plasma is about 43,000°F, **Figure 7-4**. This is hot enough to rapidly melt any metal it comes in contact with.

FIGURE 7-2 Plasma cut sheet metal panels decorate this gate. Courtesy of Larry Jeffus.

FIGURE 7-3 Portable welder's auxiliary power plug can be used for plasma cutting. Courtesy of Larry Jeffus.

Arc Plasma

The term **arc plasma** is defined as gas that has been heated to at least a partially ionized condition, enabling it to conduct an electric current.[1] The term **plasma arc** is the term most often used in the welding industry when referring to the arc plasma used in welding and cutting processes. The plasma arc produces both the high temperature and intense light associated with all forms of arc welding and **arc cutting** processes.

[1]ANSI/AWS A3.0-89 An American National Standard, Standard Welding Terms and Definitions.

Plasma Torch

The plasma torch is a device that allows the creation and control of the plasma for cutting processes. The plasma is created in the cutting torch head. A plasma torch supplies electrical energy to a gas to change it into the high energy state of a plasma.

Torch Body The torch body is made of a special plastic that is resistant to high temperatures, ultraviolet light, and impact. It provides a good grip area and protects the cable and hose connections to the head. The torch body is available in a variety of lengths and sizes. Generally the longer, larger torches are used for the higher-capacity machines; however, sometimes you might want a longer or larger torch to give yourself better control or a longer reach.

Torch Head The torch head is attached to the torch body where the cables and hoses attach to the electrode tip, nozzle tip, and nozzle. The torch and head can be connected at any angle, such as 90°, 75°, or 180° (straight), or it can be flexible. The 75° and 90° angles are popular for manual operations, and the 180° straight torch heads are most often used for machine operations. Because of the heat in the head produced by the arc, some provisions for cooling the head and its internal parts must be made. This cooling for low-power torches may be either by air or water. Higher-power torches must be liquid cooled. It is possible to replace just the torch head on most torches if it becomes worn or damaged.

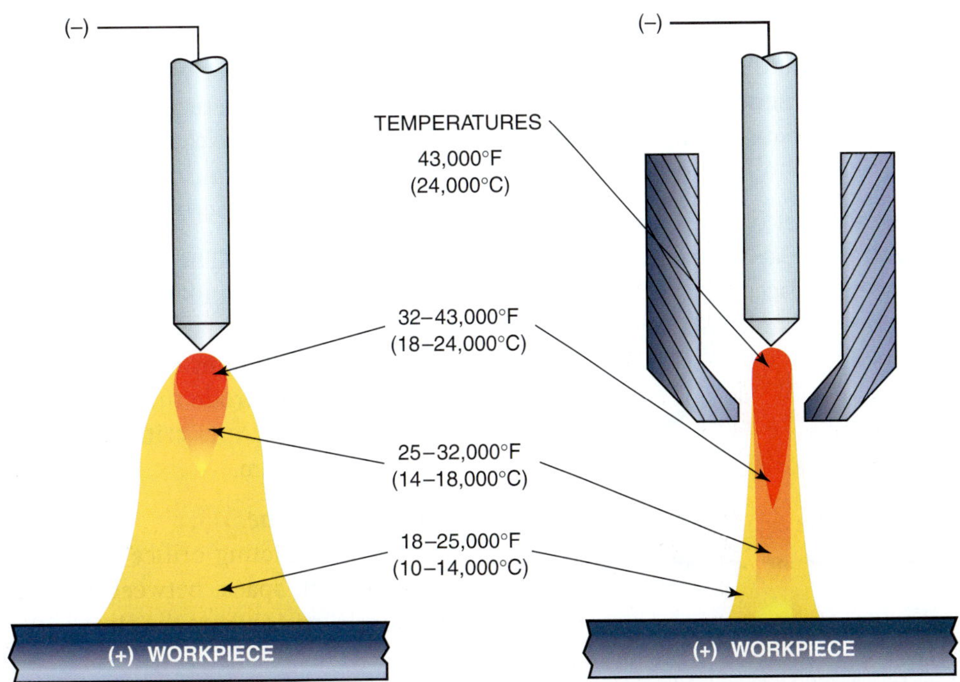

FIGURE 7-4 Approximate temperature differences between a standard arc and a plasma arc. Courtesy of the American Welding Society.

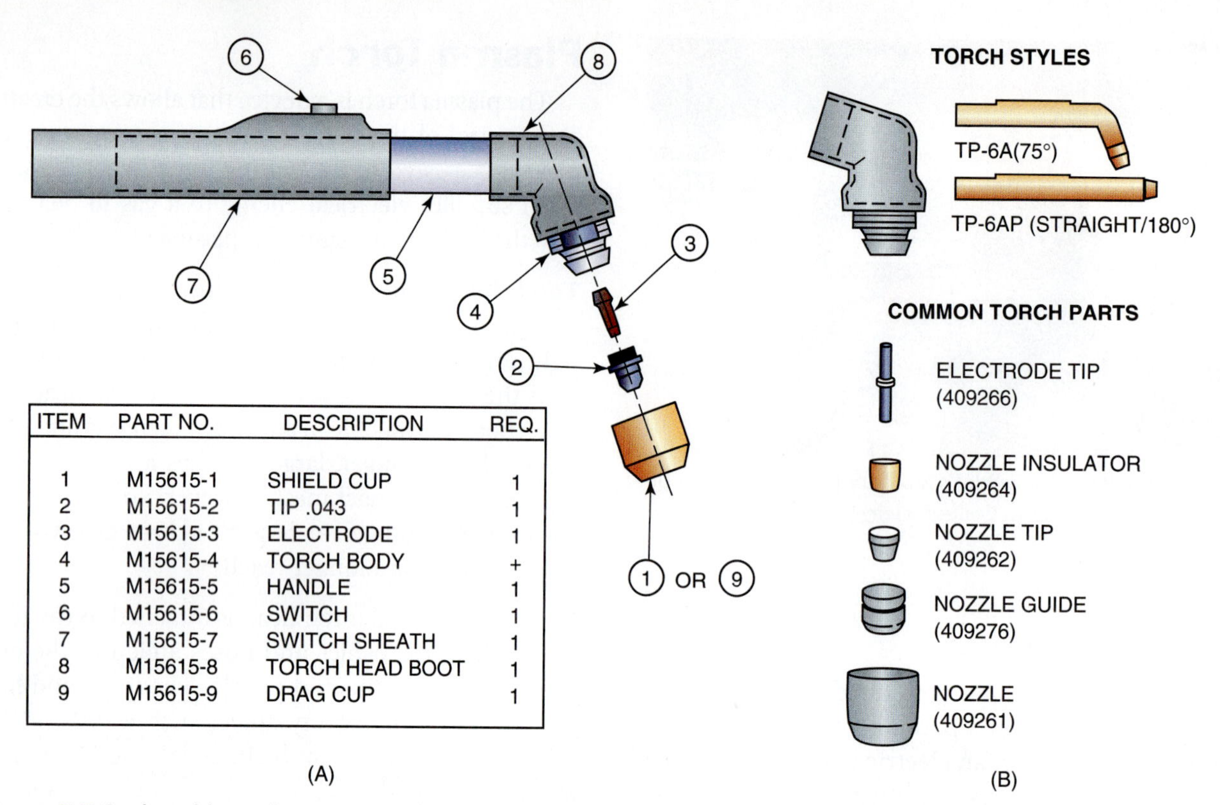

ITEM	PART NO.	DESCRIPTION	REQ.
1	M15615-1	SHIELD CUP	1
2	M15615-2	TIP .043	1
3	M15615-3	ELECTRODE	1
4	M15615-4	TORCH BODY	+
5	M15615-5	HANDLE	1
6	M15615-6	SWITCH	1
7	M15615-7	SWITCH SHEATH	1
8	M15615-8	TORCH HEAD BOOT	1
9	M15615-9	DRAG CUP	1

FIGURE 7-5 Replaceable torch parts. (A) Courtesy of Lincoln Electric Company. (B) Courtesy of Hobart Brothers Company.

Power Switch Most hand-held torches have a manual power switch that is used to start and stop the power source, gas, and cooling water (if used). The switch most often used is a thumb switch located on the torch body, but it may be a foot control or located on the panel for machine type equipment. The thumb switch may be molded into the torch body or it may be attached to the torch body with a strap clamp. The foot control must be rugged enough to withstand the welding shop environment. Some equipment has an automatic system that starts the plasma when the torch is brought close to the work.

Common Torch Parts The electrode tip, nozzle insulator, nozzle tip, nozzle guide, and nozzle are the parts of the torch that must be replaced periodically as they wear out or become damaged from use, **Figure 7-5**.

The metal parts are usually made out of copper, and they may be plated. The plating of copper parts will help them stay spatter free longer.

Electrode Tip The electrode tip is often made of copper electrode with a tungsten tip attached. The use of a copper/tungsten tip in the newer torches has improved the quality of work they can produce. By using copper, the heat generated at the tip can be conducted away faster. Keeping the tip as cool as possible lengthens the life of the tip and allows for better-quality cuts for a longer time. The newer-designed torches are a major improvement over earlier torches, some of which required the welder to accurately grind the tungsten electrode into shape. If you are using a torch that requires the grinding of the electrode tip, you must have a guide to ensure that the tungsten is properly prepared.

Nozzle Insulator The nozzle insulator is between the electrode tip and the nozzle tip. The nozzle insulator provides the critical gap spacing and the electrical separation of the parts. The spacing between the electrode tip and the nozzle tip, called electrode setback, is critical to the proper operation of the system.

Nozzle Tip The nozzle tip has a small, cone-shaped, constricting orifice in the center. The electrode setback space, between the electrode tip and the nozzle tip, is where the electric current forms the plasma. The preset close-fitting parts provide the restriction of the gas in the presence of the electric current so the plasma can be generated, **Figure 7-6**.

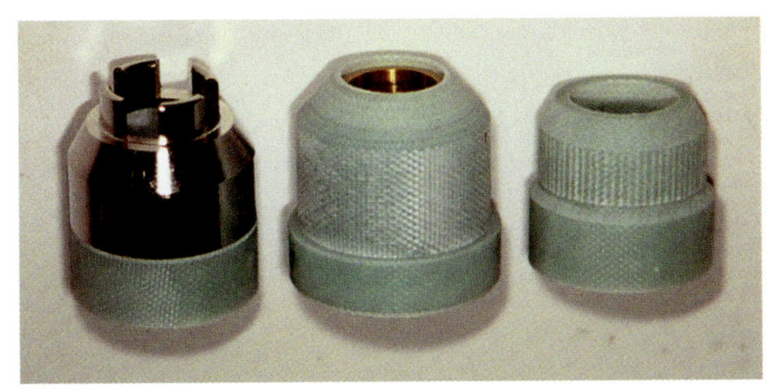

FIGURE 7-6 Different torches use different types of nozzle tips. Courtesy of Larry Jeffus.

FIGURE 7-7 Nozzles are available in a variety of shapes for different types of cutting jobs. Courtesy of Larry Jeffus.

The diameter of the constricting orifice and the electrode setback are major factors in the operation of the torch. As the diameter of the orifice changes, the plasma jet action will be affected. When the setback distance is changed, the arc voltage and current flow will change.

Nozzle The nozzle, sometimes called the cup, is made of ceramic or any other high-temperature-resistant substance. This helps protect the internal electrical parts from accidental shorting and provides control of the shielding gas or water injection if they are used, **Figure 7-7**.

Water Shroud A water shroud nozzle may be attached to some torches. The water surrounding the nozzle tip is used to control the potential hazards of light, fumes, noise, or other pollutants produced by the process.

Power and Gas Cables

A number of power and control cables and gas and cooling water hoses may be used to connect the power supply with the torch, **Figure 7-8**. This multipart cable is usually covered to provide some protection to the cables and hoses inside and to make handling the cable easier. This covering is heat resistant but will not prevent damage to the cables and hoses inside if it comes in contact with hot metal or is exposed directly to the cutting sparks.

Power Cable The power cable must have a high-voltage-rated insulation, and it is made of finely stranded copper wire to allow for maximum flexibility of the torch, **Figure 7-9**. For all nontransfer-type torches and those that use a high-frequency pilot arc, there are two power conductors, one positive (+) and one negative (−). The size and current carrying capacity of this cable is a controlling factor to the power range of the torch. As the capacity of the equipment increases, the cable must be made large enough to carry the increased current. The larger cables are less flexible and more difficult to manipulate. In order to make the cable smaller on water-cooled torches, the cable is run inside the cooling water return line. By putting the power cable inside the return water line, it allows a smaller cable to carry more current. The water prevents the cable from overheating.

Gas Hoses There may be two gas hoses running the torch. One hose carries the compressed air used to produce the plasma, and the other provides a shielding gas coverage. On some small-amperage cutting torches there is only one line for compressed air. The gas lines are made of a special heat-resistant, ultraviolet-light-resistant plastic. If it is necessary to replace the tubing because it is damaged, be sure to use the tubing provided by the manufacturer or a welding supplier. The tubing must be sized to carry the required gas flow rate within the pressure range of the torch, and it must be free from solvents and oils that might contaminate the gas. If the pressure of the gas supplied is excessive, the tubing may leak at the fittings or rupture. Check with the manufacturer for air pressure and flow rate requirements.

Control Wire The control wire is a two-conductor, low-voltage, stranded copper wire that connects the power switch to the power supply. This allows the welder to start and stop the plasma power and gas as needed during the cut or weld.

Water Tubing Medium- and high-amperage torches may be water cooled. The water for cooling early

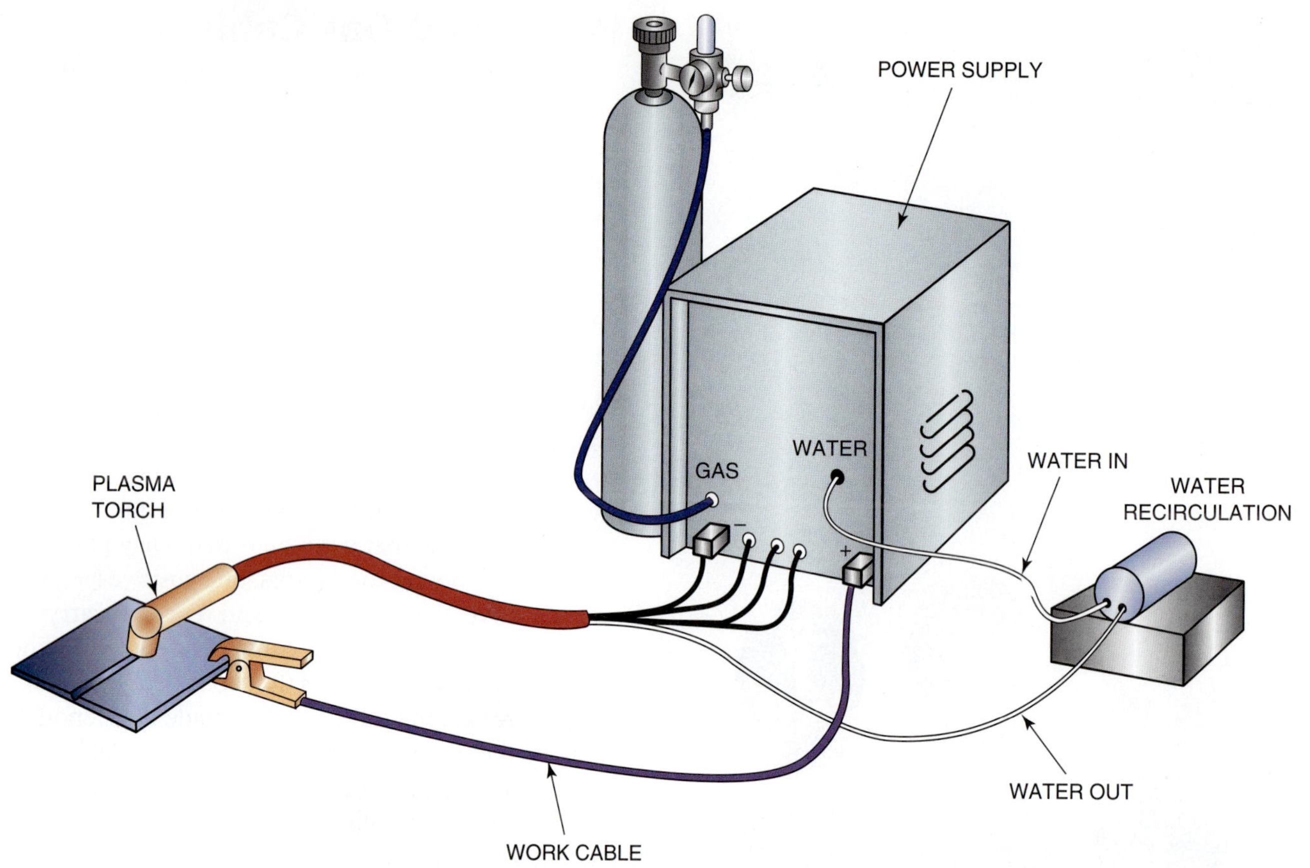

FIGURE 7-8 Typical manual plasma arc cutting setup.

FIGURE 7-9 Portable plasma arc cutting machine. Courtesy of Lincoln Electric Company.

model torches had to be deionized. Failure to use deionized water on these torches will result in the torch arcing out internally. This arcing may destroy or damage the torch's electrode tip and the nozzle tip. To see if your torch requires this special water, refer to the manufacturer's manual. If cooling water is required, it must be switched on and off at the same time as the plasma power. Allowing the water to circulate continuously might result in condensation in the torch. When the power is reapplied, the water will cause internal arcing damage.

Power Requirements

Voltage The production of the plasma requires a direct-current (DC), high-voltage, constant-current (drooping arc voltage) power supply. A constant-current-type machine allows for a rapid start of the plasma arc at the high open circuit voltage and a more controlled plasma arc as the voltage rapidly drops to the lower closed voltage level. The voltage required for most welding operations, such as shielded metal arc, gas metal arc, gas tungsten arc, and flux cored arc, ranges from 18 volts to 45 volts. The voltage for a plasma arc process ranges from 50 to 200 volts closed circuit and 150 to 400 volts open circuit. This higher electrical potential is required because the resistance of the gas increases as it is forced through a small orifice. The potential voltage of the power supplied must be high enough to overcome the resistance in the circuit in order for electrons to flow, **Figure 7-10**.

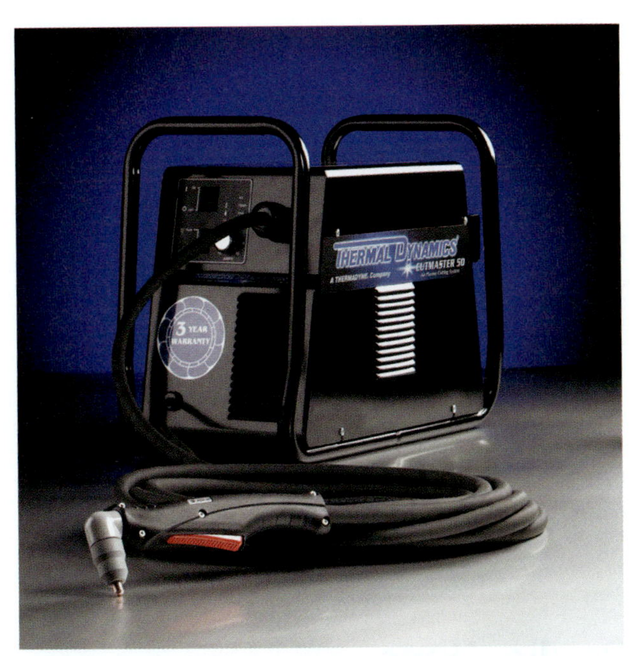

FIGURE 7-10 Inverter-type high voltage plasma arc cutting power supply. Courtesy of Thermal Dynamics®, a Thermadyne® Company.

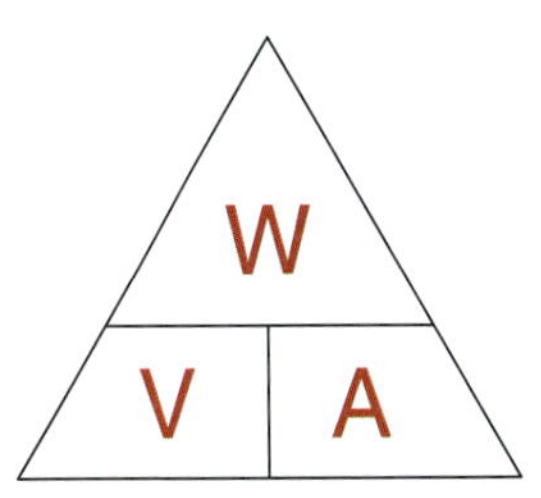

FIGURE 7-11 Ohm's Law.

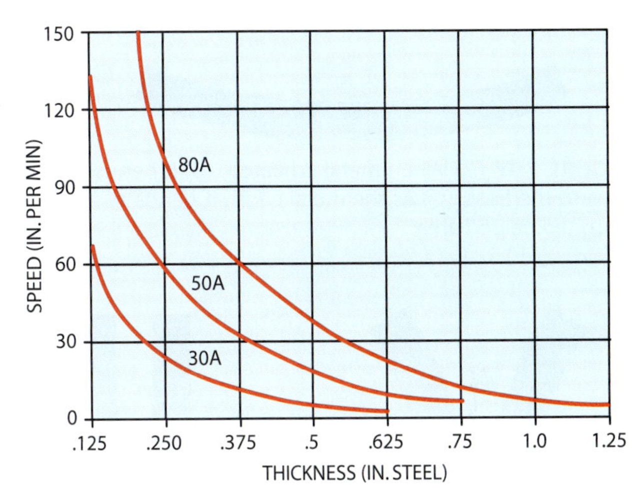

TABLE 7-1 Plasma Arc Cutting Parameters. Courtesy of ESAB Welding & Cutting Products.

Amperage Although the voltage is higher, the current (amperage) flow is much lower than it is with most other welding processes. Some low-powered PAC torches will operate with as low as 10 amps of current flow. High-powered plasma cutting machines can have amperages as high as 200 amps, and some very large automated cutting machines may have 1000-ampere capacities. The higher the amperage capacity the faster and thicker they will cut.

Watts The plasma process uses approximately the same amount of power, in watts, as a similar nonplasma process. Watts are the units of measure for electrical power. By determining the total watts used for both the nonplasma process and plasma operation, you can make a comparison. Watts used in a circuit are determined by multiplying the voltage times the amperage, **Figure 7-11.** For example, a 1/8-in.-diameter E6011 electrode will operate at 18 volts and 90 amperes. The total watts used would be

$$W = V \times A$$
$$W = 18 \times 90$$
$$W = 1620 \text{ watts of power}$$

A low-power PAC torch operating with only 20 amperes and 85 volts would be using a total of

$$W = V \times A$$
$$W = 85 \times 20$$
$$W = 1700 \text{ watts of power}$$

Heat Input

Although the total power used by both plasma and nonplasma processes is similar, the actual energy input into the work per linear foot is less with plasma. The very high temperatures of the plasma process allow much higher traveling rates so that the same amount of heat input is spread over a much larger area. This has the effect of lowering the joules per inch of heat the weld or cut will receive. **Table 7-1** shows the cutting performance of a typical plasma torch. Note the relationship among amperage, cutting speed, and metal thickness. The lower the amperage, the slower the cutting speed or the thinner the metal that can be cut.

A high travel speed with plasma cutting will result in a heat input that is much lower than that of the oxyfuel cutting process. A steel plate cut using the plasma process may have only a slight increase in temperature following the cut. It is often possible to pick up a part only moments after it is cut using plasma. The same part cut with oxyfuel would be much hotter and require a longer time to cool off.

Distortion

Anytime metal is heated in a localized zone or spot, it expands in that area and, after the metal cools, it is no longer straight or flat, **Figure 7-12.** If

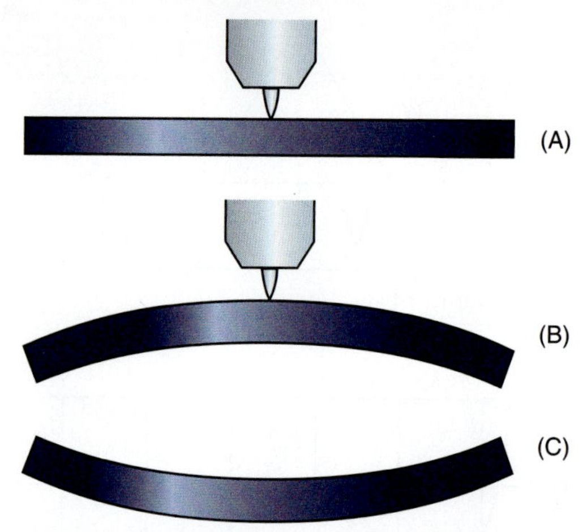

FIGURE 7-12 (A) When metal is heated, (B) it bends up toward the heat. (C) As the metal cools, it bends away from the heated area.

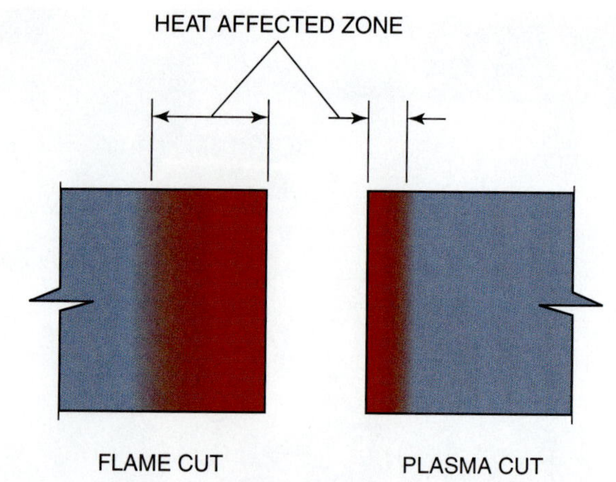

FIGURE 7-14 A smaller heat-affected zone will result in less hardness or weakening along the cut edge.

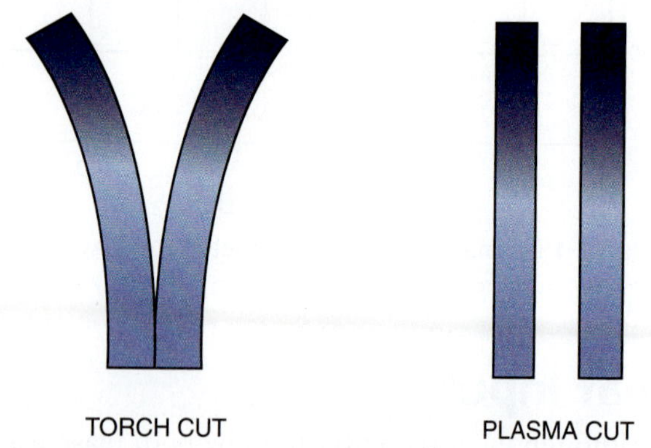

FIGURE 7-13 The heat of a torch cut causes metal to bend, but the plasma cut is so fast, little or no bending occurs.

a piece of metal is cut, there will be localized heating along the edge of the cut, and, unless special care is taken, the part will not be usable as a result of its distortion, **Figure 7-13.** This distortion is a much greater problem with thin metals. By using a plasma cutter, an auto body worker can cut the thin, low-alloy sheet metal of a damaged car with little problem from distortion.

On thicker sections, the hardness zone along the edge of a cut will be reduced so much that it is not a problem. When using oxyfuel cutting of thick plate, especially higher-alloyed metals, this hardness zone can cause cracking and failure if the metal is shaped after cutting, **Figure 7-14.** Often the plates must be preheated before they are cut using oxyfuel to reduce the heat-affected zone. This preheating adds greatly to the cost of fabrication both in time and fuel costs. By being able to make most cuts without preheating, the plasma process will greatly reduce fabrication cost.

Applications

Plasma cutting equipment is rapidly replacing oxyacetylene cutting equipment in agriculture. A major problem with keeping oxyacetylene cylinders is that they are usually rented. You have to pay for the cylinders month after month even if you're not using them. Plasma, however, does not need compressed gas cylinders so once you purchase the equipment there are no additional monthly charges. Agricultural plasma equipment can cut up to 3/8 in. (9 mm) thick metal. That is within the range for most small jobs, so plasma can do most metal cutting jobs around farms without the expense of renting cylinders. That is why plasma cutting is rapidly replacing the oxyacetylene torch on most agricultural operations.

Cutting Speed High cutting speeds are possible, up to 300 in./min; that is 25 ft/min, or about 1/4 mile an hour. The fastest oxyfuel cutting equipment could cut at only about one-fourth that speed, **Figure 7-15.**

Metals Any material that is conductive can be cut using the PAC process. In a few applications nonconductive materials can be coated with conductive material so they can be cut also. Although it is possible to make cuts in metal as thick as 7 inches, it is not cost effective. The most popular materials cut are carbon steel stainless steel, and aluminum sheet metal.

The PAC process is also used to cut expanded metals, screens, and other items that would require frequent starts and stops, **Figure 7-16.**

Standoff Distance The standoff distance is the distance from the nozzle tip to the work, **Figure 7-17.** This distance is very critical to producing quality plasma arc cuts. As the distance increases, the arc force is diminished and tends to spread out. This causes the kerf to be

FIGURE 7-15 Plasma arc cut in 2-in.-thick mild steel. Notice how smooth the machine cut edge is. Courtesy of Larry Jeffus.

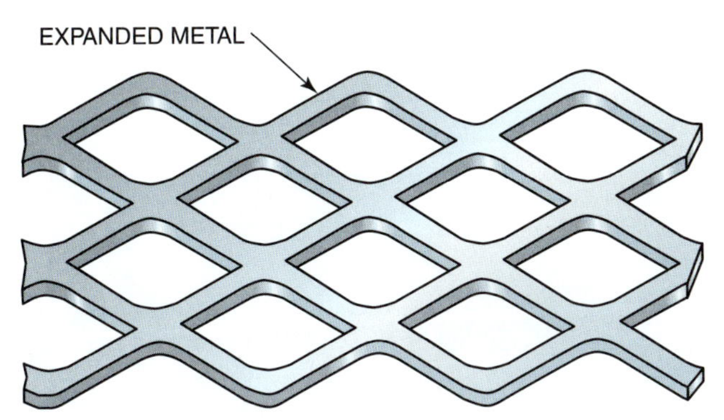

FIGURE 7-16 Expanded metal.

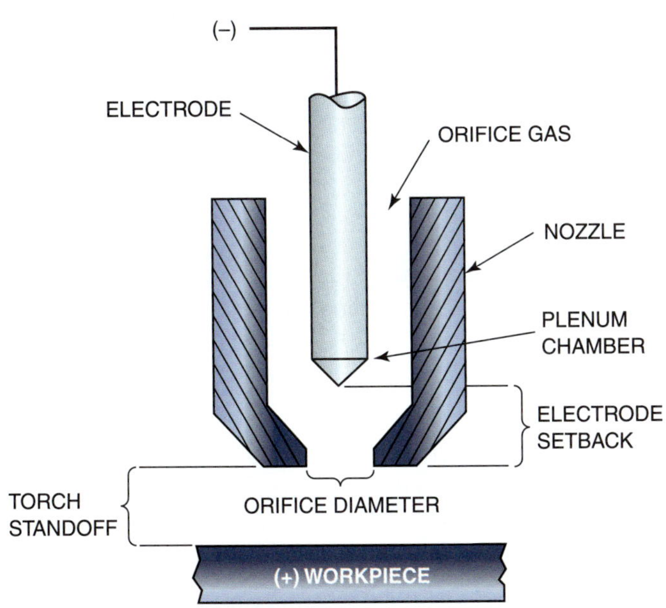

FIGURE 7-17 Conventional plasma arc terminology. Courtesy of the American Welding Society.

wider, the top edge of the plate to become rounded, and the formation of more dross on the bottom edge of the plate. However, if this distance becomes too close, the working life of the nozzle tip will be reduced. In some cases an arc can form between the nozzle tip and the metal that instantly destroys the tip.

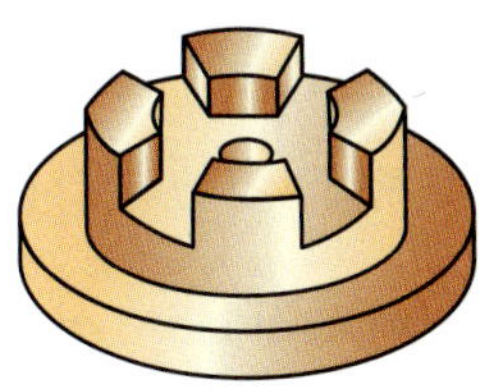

FIGURE 7-18 A castle nozzle tip can be used to allow the torch to be dragged across the surface.

On some torches, it is possible to drag the nozzle tip along the surface of the work without shorting it out. This is a large help when working on metal out of position or on thin sheet metal. Before you use your torch in this manner, you must check the owner's manual to see if it will operate in contact with the work, **Figure 7-18.** This technique will allow the nozzle tip orifice to become contaminated more quickly.

Starting Methods Because the electrode tip is located inside the nozzle tip, and a high initial resistance to current flow exists in the gas flow before the plasma is generated, it is necessary to have a specific starting method. Two methods are used to establish a current path through the gas.

The most common method uses a high-frequency alternating current carried through the conductor, the electrode, and back from the nozzle tip. This high-frequency current will ionize the gas and allow it to carry the initial current to establish a pilot arc, **Figure 7-19.** After the pilot arc has been started, the high-frequency starting circuit can be stopped. A pilot arc is an arc between the electrode tip and the nozzle tip within the torch head. This is a nontransfer arc, so the workpiece is not part of the current path. The low current of the pilot arc, although it is inside the torch, does not create enough heat to damage the torch parts.

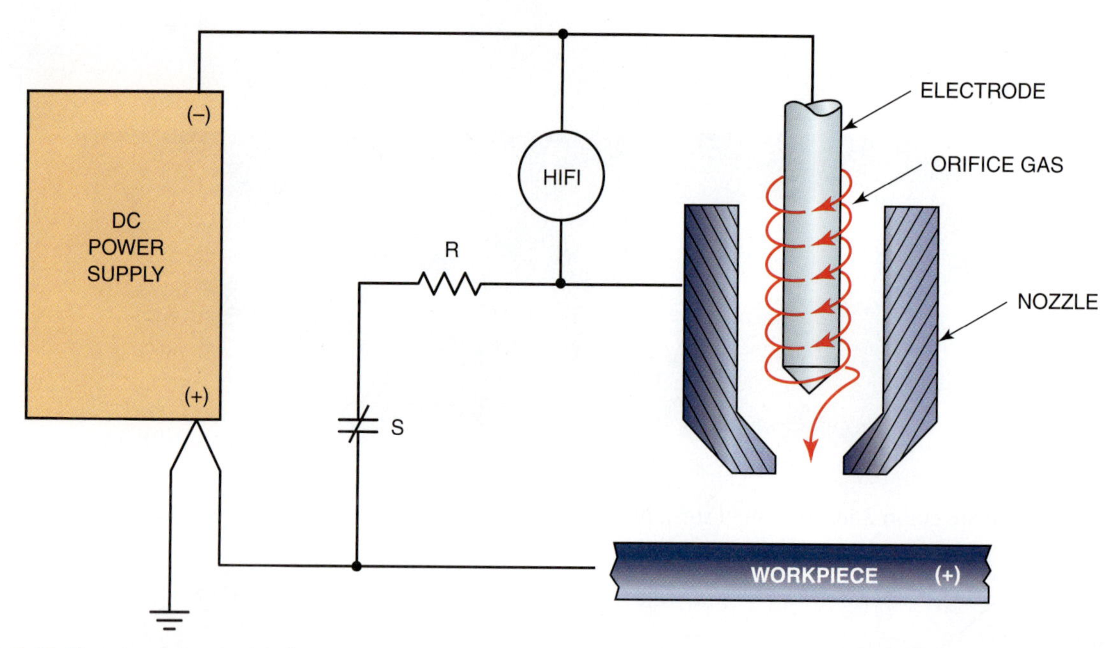

FIGURE 7-19 Plasma arc torch circuitry. Courtesy of the American Welding Society.

When the torch is brought close enough to the work, the primary arc will follow the pilot arc across the gap, and the main plasma is started. Once the main plasma is started, the pilot arc power can be shut off.

Kerf The kerf is the space left in the metal as the metal is removed during a cut. The width of a PAC kerf is often wider than that of an oxyfuel cut. Several factors will affect the width of the kerf. A few of the factors are as follows:

- standoff distance—The closer the torch nozzle tip is to the work, the narrower the kerf will be, **Figure 7-20**.
- orifice diameter—Keeping the diameter of the nozzle orifice as small as possible will keep the kerf smaller.
- power setting—Too high or too low a power setting will cause an increase in the kerf width.
- travel speed—As the travel speed is increased, the kerf width will decrease; however, the bevel on the sides and the dross formation will increase if the speeds are excessive.
- gas—The type of gas or gas mixture will affect the kerf width as the gas change affects travel speed, power, concentration of the plasma stream, and other factors.
- electrode and nozzle tip—As these parts begin to wear out from use or are damaged, the PAC quality and kerf width will be adversely affected.
- swirling of the plasma gas—On some torches, the gas is directed in a circular motion around the electrode before it enters the nozzle tip orifice. This swirling causes the plasma stream that is produced to be more dense with straighter sides. The result is an improved cut quality, including a narrow kerf, **Figure 7-21**.
- water injection—The injection of water into the plasma stream as it leaves the nozzle tip is not the same as the use of a water shroud. Water injection into the plasma stream will increase the swirl and further concentrate the plasma. This improves the cutting quality, lengthens the life of the nozzle tip, and makes a squarer, narrower kerf, **Figure 7-22**.

FIGURE 7-20 When the standoff distance is correct, as with this machine cut, almost no sparks bounce back on the cutting tip. Courtesy of ESAB Welding & Cutting Products.

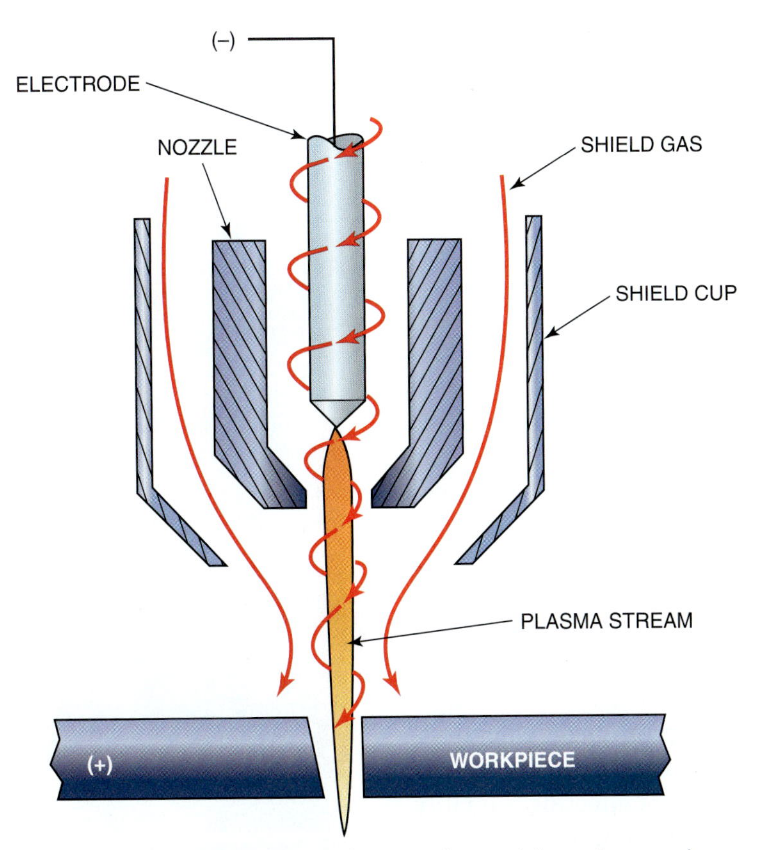

FIGURE 7-21 The cutting gas can swirl around the electrode to produce a tighter plasma column. Courtesy of the American Welding Society.

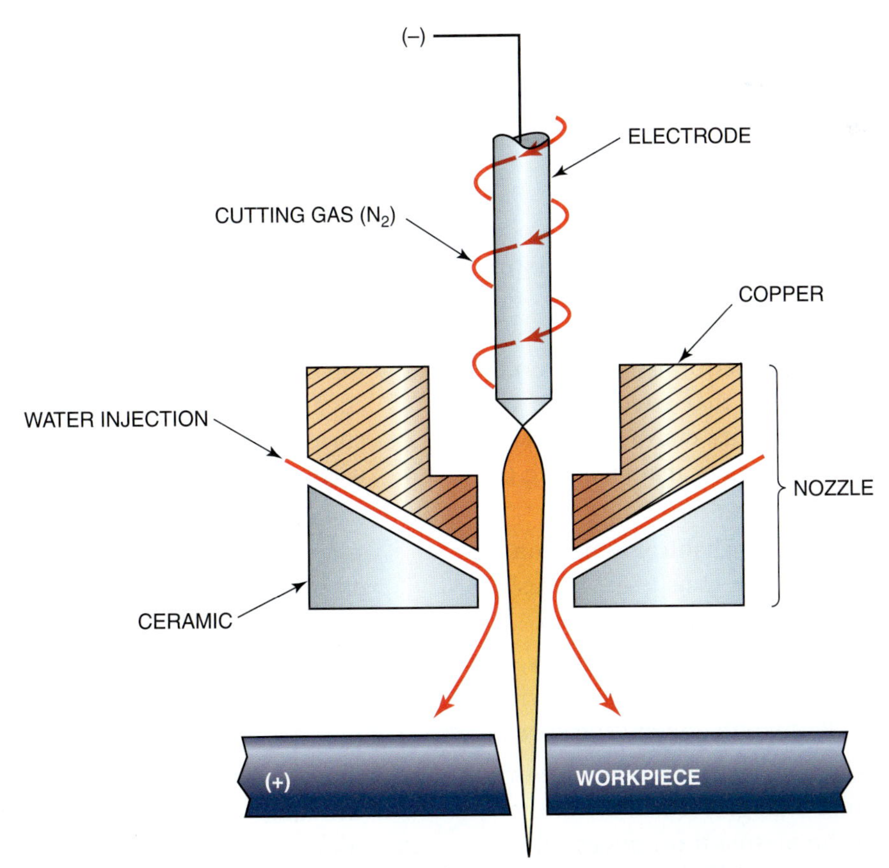

FIGURE 7-22 Water injection plasma arc cutting. Notice that the kerf is narrow, and one side is square. Courtesy of the American Welding Society.

Plate Thickness		Kerf Allowance	
in.	mm	in.	mm
1/8 to 1	3.2 to 25.4	+3/32	+2.4
1 to 2	25.4 to 51.0	+3/16	+4.8
2 to 5	51.0 to 127.0	+5/16	+8.0

TABLE 7-2 Standard Kerf Widths for Several Metal Thicknesses.

FIGURE 7-23 Plasma cutting. Courtesy of ESAB Welding & Cutting Products.

Table 7-2 lists some standard kerf widths for several metal thicknesses. These are to be used as a guide for nesting of parts on a plate to maximize the material used and minimize scrap. The kerf size may vary from this depending on a number of variables with your PAC system. You should make test cuts to verify the size of the kerf before starting any large production cuts.

Because the sides of the plasma stream are not parallel as they leave the nozzle tip, there is a bevel left on the sides of all plasma cuts. This bevel angle is from 1/2° to 3° depending on metal thickness, torch speed, type of gas, standoff distance, nozzle tip condition, and other factors affecting a quality cut. On thin metals, this bevel is undetectable and offers no problem in part fabrication or finishing.

The use of a plasma swirling type torch and the direction the cut is made can cause one side of the cut to be square and the scrap side to have all of the bevel, **Figure 7-23**. This technique is only effective provided that one side of the cut is to be scrap.

Gases

Almost any gas or gas mixture can be used today for the PAC process. Changing the gas or gas mixture is one method of controlling the plasma cut. Although the type of gas or gases used will have a major effect on the cutting performance, it is only one of a number of changes that a technician can make to help produce a quality cut. The following are some of the effects on the cut that changing the PAC gas(es) will have:

- force—The amount of mechanical impact on the material being cut; the density of the gas and its ability to disperse the molten metal
- central concentration—Some gases will have a more compact plasma stream. This factor will greatly affect the kerf width and cutting speed.
- heat content—As the electrical resistance of a gas or gas mixture changes, it will affect the heat content of the plasma it produces. The higher the resistance, the higher the heat produced by the plasma.
- kerf width—The ability of the plasma to remain in a tightly compact stream will produce a deeper cut with less of a bevel on the sides.
- dross formation—The dross that may be attached along the bottom edge of the cut can be controlled or eliminated.
- top edge rounding—The rounding of the top edge of the plate can often be eliminated by correctly selecting the gas(es) that are to be used.
- metal type—Because of the formation of undesirable compounds on the cut surface as the metal reacts to elements in the plasma, some metals may not be cut with specific gas(es).

Table 7-3 lists some of the popular gases and gas mixtures used for various PAC metals. The selection of a gas or gas mixture for a specific operation to maximize the system performance must be tested with the equipment and setup being used. With constant developments and improvements in the PAC system, new gases and gas mixtures are continuously being added to the list. In addition to the type of gas,

Metal	Gas
Carbon and low alloy steel	Nitrogen Argon with 0% to 35% Hydrogen Air
Stainless steel	Nitrogen Argon with 0% to 35% Hydrogen
Aluminum and aluminum alloys	Nitrogen Argon with 0% to 35% Hydrogen
All plasma arc gouging	Argon with 35% to 40% Hydrogen

TABLE 7-3 Gases for Plasma Arc Cutting and Gouging.

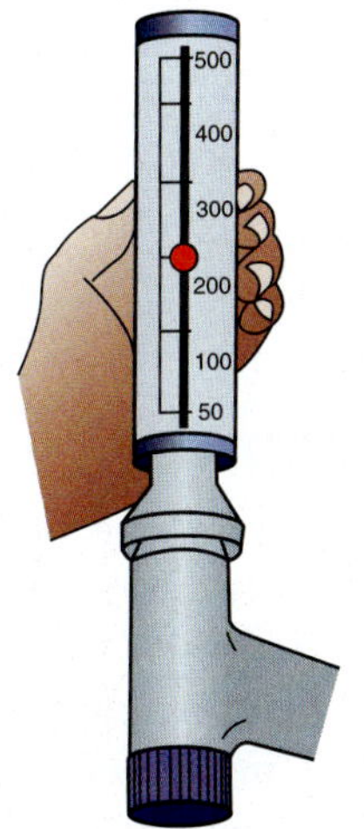

FIGURE 7-24 Plasma flow measuring kit. Courtesy of ESAB Welding & Cutting Products.

it is important to have the correct gas flow rate for the size tip, metal type, and thickness. Too low a gas flow will result in a cut having excessive dross and sharply beveled sizes. Too high a gas flow will produce a poor cut because of turbulence in the plasma stream and waste gas. A flow measuring kit can be used to test the flow at the plasma torch for more accurate adjustments, **Figure 7-24**.

Stack Cutting Because the PAC process does not rely on the thermal conductivity between stacked parts, like the oxyfuel process, thin sheets can be stacked and cut efficiently. With the oxyfuel stack cutting of sheets, it is important that there not be any air gaps between layers. Also, it is often necessary to make a weld along the side of the stack in order for the cut to start consistently.

The PAC process does not have these limitations. It is recommended that the sheets be held together for cutting, but this can be accomplished by using standard C-clamps. The clamping needs to be tight because, if the space between layers is excessive, the sheets may stick together. The only problem that will be encountered is that, because of the kerf bevel, the parts near the bottom might be slightly larger if the stack is very thick. This problem can be controlled by using the same techniques as described for making the kerf square.

Dross Dross is the metal compound that resolidifies and attaches itself to the bottom of a cut. This metal compound is made up mostly of unoxidized metal, metal oxides, and nitrides. It is possible to make cuts dross free if the PAC equipment is in good operating condition and the metal is not too thick for the size of torch being used. Because dross contains more unoxidized metal than most OFC slag, often it is much harder to remove if it sticks to the cut. The thickness that a dross-free cut can be made is dependent on a number of factors, including the gas(es) used for the cut, travel speed, standoff distance, nozzle tip orifice diameter, wear condition of electrode tip and nozzle tip, gas velocity, and plasma stream swirl.

Stainless steel and aluminum are easily cut dross free. Carbon steel, copper, and nickel-copper alloys are much more difficult to cut dross free.

Machine Cutting

Almost any of the plasma torches can be attached to some type of semiautomatic or automatic device to allow it to make machine cuts. The simplest devices are oxyfuel portable flame cutting machines that run on tracks, **Figure 7-25**. These portable machines are good for mostly straight or circular cuts. Complex shapes can be cut with a pattern cutter that uses a magnetic tracing system to follow the template's shape, **Figure 7-26**, or an automated robotic cutter, **Figure 7-27**.

Water Tables Machine cutting lends itself to the use of water cutting tables, although they can be used with most hand torches. The water table is used to reduce the noise level, control the plasma

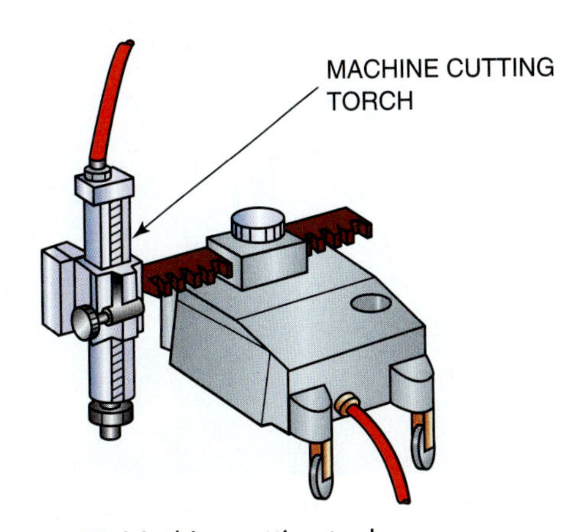

FIGURE 7-25 Machine cutting tool.

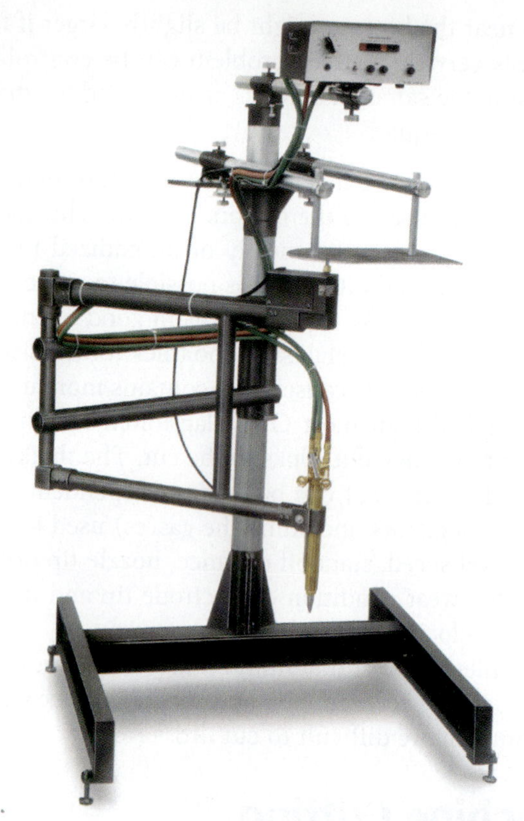

FIGURE 7-26 Portable pattern cutter can cut shapes, circles, and straight lines. Courtesy of ESAB Welding & Cutting Products.

FIGURE 7-27 Automated cutting. Courtesy of ESAB Welding & Cutting Products.

light, trap the sparks, eliminate most of the fume hazard, and reduce distortion.

Water tables either support the metal just above the surface of the water or they submerge the metal about 3 inches below the water's surface. Both types of water tables must have some method of removing the cut parts, scrap, and slag that build up in the bottom. Often the surface-type tables will have the PAC torch connected to a water shroud nozzle, **Figure 7-28**. By using a water shroud nozzle, the surface table will offer the same advantages to the PAC process as the submerged table offers. In most cases, the manufacturers of this type of equipment have made provisions for a special dye to be added to the water. This dye will help control the harmful light produced by the PAC. Check with the equipment's manufacturer for limitations and application of the use of dyes.

Manual Cutting

Manual plasma arc cutting is the most versatile of the PAC processes. It can be used in all positions, on almost any surface, and on most metals. This process is limited to low-power plasma machines; however, even these machines can cut up to 1 1/2-inch-thick metals. The limitation to low power, 100 amperes or less, is primarily for safety reasons. The higher-powered machines have extremely dangerous open circuit voltages that can kill a person if accidentally touched.

Setup The setup of most plasma equipment is similar, but do not ever attempt to set up a system without the manufacturer's owner's manual for the specific equipment.

Be sure all of the connections are tight and that there are no gaps in the insulation on any of the cables. Check the water and gas lines for leaks. Visually inspect the complete system for possible problems.

Before you touch the nozzle tip, be sure that the main power supply is off. The open circuit voltage on even low-powered plasma machines is high enough to kill a person. Replace all parts to the torch before the power is restored to the machine.

Plasma Arc Gouging

Plasma arc gouging is a recent introduction to the PAC processes. The process is similar to that of air carbon arc gouging in that a U-groove can be cut into the metal's surface. The removal of metal along a joint before the metal is welded or the removal of a

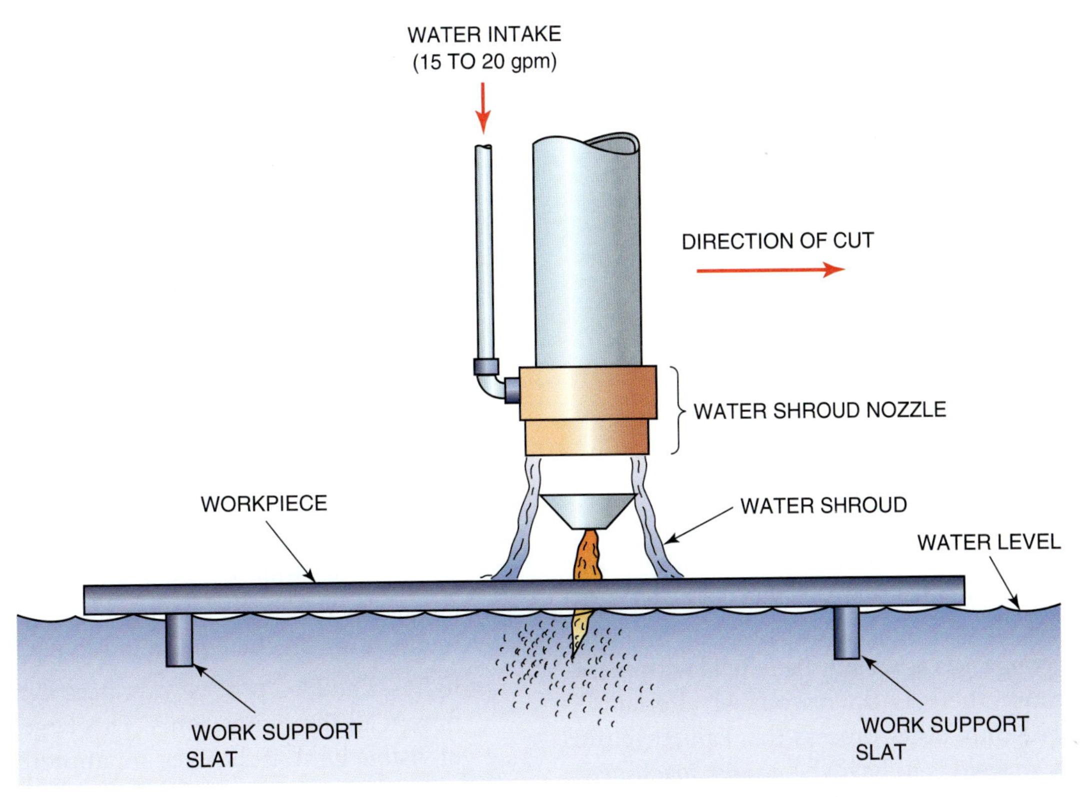

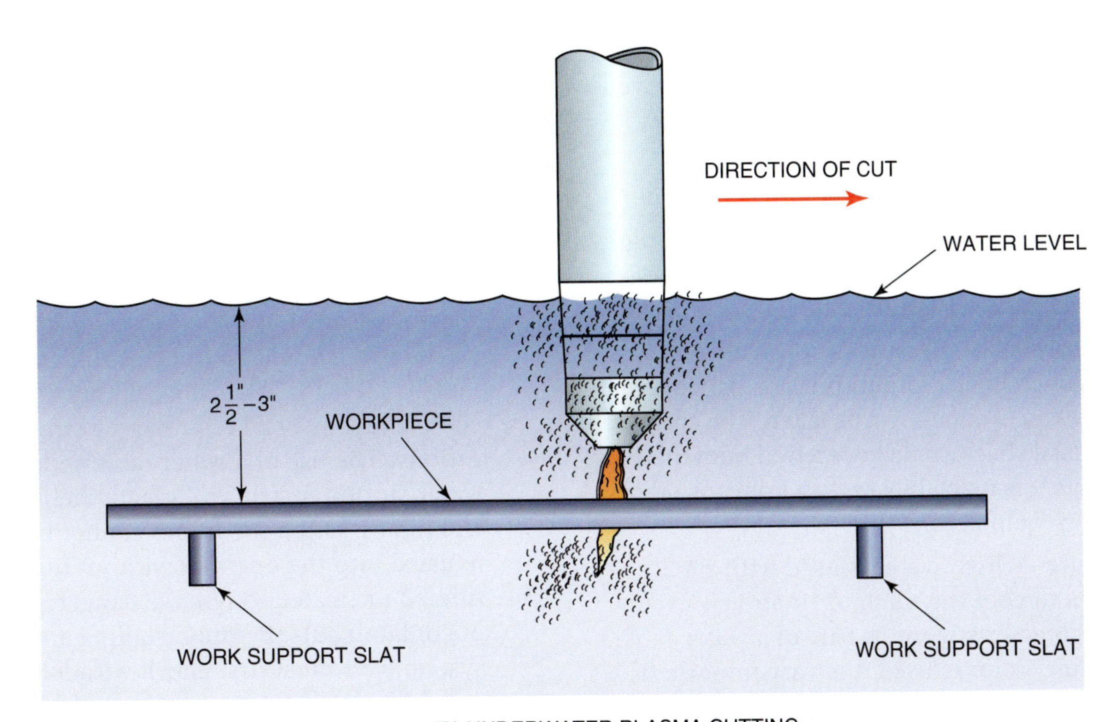

FIGURE 7-28 A water table can be used either with (A) a water shroud or (B) underwater torches. Courtesy of the American Welding Society.

defect for repairing can easily be done using this variation of PAC, **Figure 7-29**.

The torch is set up with a less-concentrated plasma stream. This will allow the washing away of the molten metal instead of thrusting it out to form a cut. The torch is held at approximately a 30° angle to the metal surface. Once the groove is started it can be controlled by the rate of travel, torch angle, and torch movement.

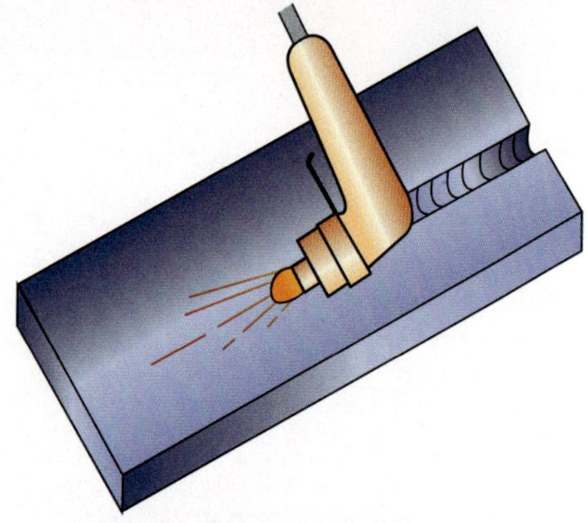

FIGURE 7-29 Plasma arc gouging a U-groove in a plate.

Plasma arc gouging is effective on most metals. Stainless steel and aluminum are especially good metals to gouge because there is almost no cleanup. The groove is clean, bright, and ready to be welded. Plasma arc gouging is especially beneficial with these metals because there is no reasonable alternative available. The only other process that can leave the metal ready to weld is to have the groove machined, and machining is slow and expensive compared to plasma arc gouging.

It is important not to try to remove too much metal in one pass. The process will work better if small amounts are removed at a time. If a deeper groove is required, multiple gouging passes can be used.

Safety

PAC has many of the same safety concerns as does most other electric welding or cutting processes. Some special concerns are specific to this process.

- electrical shock—Because the open circuit voltage is much higher for this process than for any other, extra caution must be taken. The chance that a fatal shock could be received from this equipment is much higher than from any other welding equipment.
- moisture—Often water is used with PAC torches to cool the torch or improve the cutting characteristic, or as part of a water table. Any time water is used it is very important that there be no leaks or splashes. The chance of electrical shock is greatly increased if there is moisture on the floor, cables, or equipment.
- noise—Because the plasma stream is passing through the nozzle orifice at a high speed, a loud sound is produced. The sound level increases as the power level increases. Even with low-power equipment the decibel (dB) level is above safety ranges. Some type of ear protection is required to prevent damage to operator and other people in the area of the PAC equipment when it is in operation. High levels of sound can have a cumulative effect on one's hearing. Over time, one's ability to hear will decrease unless proper precautions are taken. See the owner's manual for the recommendations for the equipment in use.
- light—The PAC process produces light radiation in all three spectrums. This large quantity of visible light, if the eyes are unprotected, will cause night blindness. The most dangerous of the lights is ultraviolet. Like other arc processes, this light can cause burns to the skin and eyes. The third light, infrared, can be felt as heat, and it is not as much of a hazard. Some type of eye protection must be worn when any PAC is in progress. **Table 7-4** lists the recommended lens shade numbers for various power-level machines.

Current Range A	Minimum Shade	Comfortable Shade
Less than 300	8	9
300 to 400	9	12
400 plus	10	14

TABLE 7-4 Recommended Shade Densities for Filter Lenses.

- fumes—This process produces a large quantity of fumes that are potentially hazardous. A specific means for removing them from the work space should be in place. A downdraft table is ideal for manual work, but some special pickups may be required for larger applications. The use of a water table and/or a water shroud nozzle will greatly help to control fumes. Often the fumes cannot be exhausted into the open air without first being filtered or treated to remove dangerous levels of contaminants. Before installing an exhaust system, you must first check with local, state, and federal officials to see if specific safeguards are required.
- gases—Some of the plasma gas mixtures include hydrogen; because this is a flammable gas, extra care must be taken to insure that the system is leak-proof.
- sparks—As with any process that produces sparks, the danger of an accidental fire is

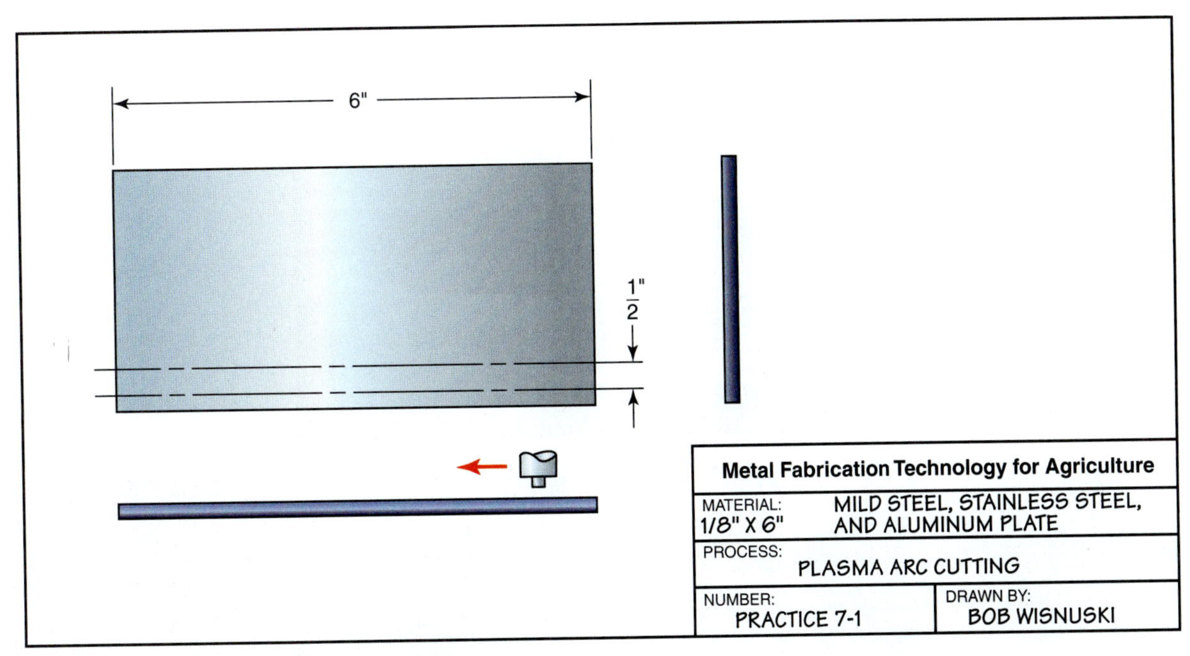

FIGURE 7-30 Straight square plasma arc cutting.

always present. This is a larger concern with PAC because the sparks are often thrown some distance from the work area and the operator's vision is restricted by a welding helmet. If there is any possibility that sparks will be thrown out of the immediate work area, a fire watch must be present. A fire watch is a person whose sole job is to watch for the possible starting of a fire. This person must know how to sound the alarm and have appropriate firefighting equipment handy. Never cut in the presence of combustible materials.

- operator check out—Never operate any PAC equipment until you have read the manufacturer's owner's and operator's manual for the specific equipment to be used. It is a good idea to have someone who is familiar with the equipment go through the operation after you have read the manual.

PRACTICE 7-1

Flat, Straight Cuts in Thin Plate

Using a properly set up and adjusted PAC machine, proper safety protection, one or more pieces of mild steel, stainless steel, and aluminum 6 in. (152 mm) long and 16 gauge and 1/8 in. (3 mm) thick, you will cut off 1/2-in.-wide strips, **Figure 7-30.**

- Starting at one end of the piece of metal that is 1/8 in. (3 mm) thick, hold the torch as close as possible to a 90° angle.
- Lower your hood and establish a plasma cutting stream.
- Move the torch in a straight line down the plate toward the other end, **Figure 7-31.**
- If the width of the kerf changes, speed up or slow down the travel rate to keep the kerf the same size for the entire length of the plate.

Repeat the cut using both thicknesses of all three types of metals until you can make consistently smooth cuts that are within ±3/32 inch of a straight line and ±5° of being square. Turn off the PAC equipment and clean up your work area when you are finished cutting.

Complete a copy of the "Student Welding Report" listed in Appendix I or provided by your instructor. ◆

PRACTICE 7-2

Flat, Straight Cuts in Thick Plate

Using a properly set up and adjusted PAC machine, proper safety protection, one or more pieces of mild steel, stainless steel, and aluminum 6 in. (152 mm) long and 1/4 in. and 1/2 in. thick, you will cut off 1/2-inch-wide strips. Follow the same procedure as outlined in Practice 7-1.

Repeat the cut using both thicknesses of all three types of metals until you can make consistently smooth cuts that are within ±3/32 inch of a straight line and ±5° of being square. Turn off the PAC

FIGURE 7-31 Starting at the edge of a plate, like starting an oxyfuel cut, move smoothly in a straight line toward the other end. Note the roughness left along the side of the plate from a previous cut. Courtesy of Larry Jeffus.

equipment and clean up your work area when you are finished cutting.

Complete a copy of the "Student Welding Report" listed in Appendix I or provided by your instructor. ◆

PRACTICE 7-3

Flat Cutting Holes

Using a properly set up and adjusted PAC machine, proper safety protection, one or more pieces of mild steel, stainless steel, and aluminum 16 gauge, 1/8 in., 1/4 in., and 1/2 in. thick, you will cut 1/2-in. and 1-in. holes.

- Starting with the piece of metal that is 1/8 in. (3 mm) thick, hold the torch as close as possible to a 90° angle.
- Lower your hood and establish a plasma cutting stream.
- Move the torch in an outward spiral until the hole is the desired size, **Figure 7-32**.

Repeat the hole-cutting process until both sizes of holes are made using all the thicknesses of all three types of metals until you can make consistently smooth cuts that are within ±3/32 inch of being round and ±5° of being square. Turn off the PAC equipment and clean up your work area when you are finished cutting.

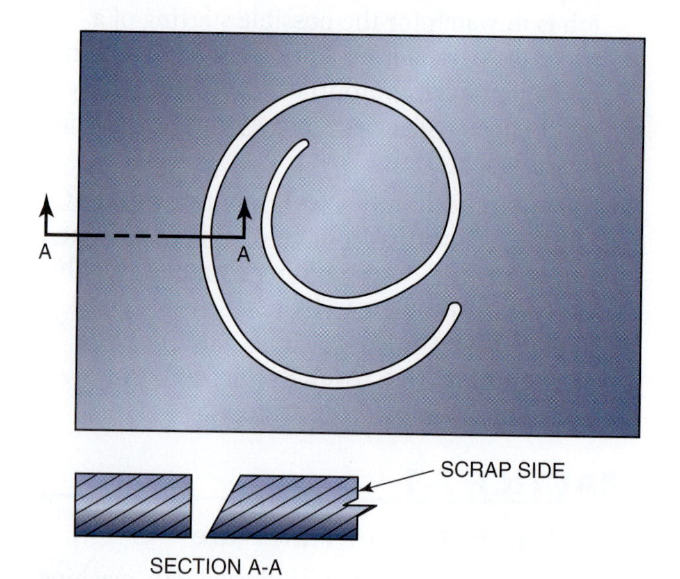

FIGURE 7-32 When cutting a hole, make a test to see which direction to make the cut so that the beveled side is on the scrap piece.

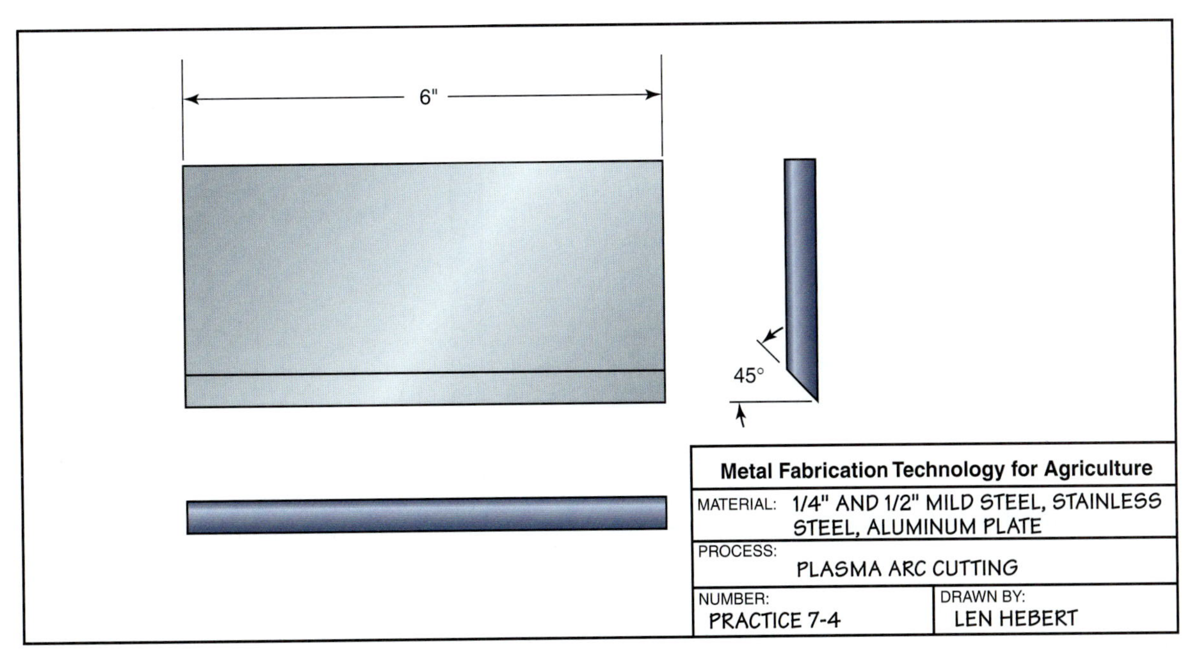

FIGURE 7-33 Beveled plasma arc cutting.

Complete a copy of the "Student Welding Report" listed in Appendix I or provided by your instructor. ◆

PRACTICE 7-4

Beveling of a Plate

Using a properly set up and adjusted PAC machine, proper safety protection, one or more pieces of mild steel, stainless steel, and aluminum 6 in. (152 mm) long and 1/4 in. and 1/2 in. thick, you will cut a 45° bevel down the length of the plate.

- Starting at one end of the piece of metal that is 1/4 in. thick, hold the torch as close as possible to a 45° angle.
- Lower your hood and establish a plasma cutting stream.
- Move the torch in a straight line down the plate toward the other end, **Figure 7-33.**

Repeat the cut using both thicknesses of all three types of metals until you can make consistently smooth cuts that are within ±3/32 inch of a straight line and ±5° of a 45° angle. Turn off the PAC equipment and clean up your work area when you are finished cutting.

Complete a copy of the "Student Welding Report" listed in Appendix I or provided by your instructor. ◆

PRACTICE 7-5

U-grooving of a Plate

Using a properly set up and adjusted PAC machine, proper safety protection, one or more pieces of mild steel, stainless steel, and aluminum 6 in. (152 mm) long and 1/4 in. or 1/2 in. thick, you will cut a U-groove down the length of the plate.

- Starting at one end of the piece of metal, hold the torch as close as possible to a 30° angle, **Figure 7-34.**
- Lower your hood and establish a plasma cutting stream.
- Move the torch in a straight line down the plate toward the other end.
- If the width of the U-groove changes, speed up or slow down the travel rate to keep the groove the same width and depth for the entire length of the plate.

Repeat the gouging of the U-groove using all three types of metals until you can make consistently smooth grooves that are within ±3/32 inch of a straight line and uniform in width and depth. Turn off the PAC equipment and clean up your work area when you are finished cutting.

Complete a copy of the "Student Welding Report" listed in Appendix I or provided by your instructor. ◆

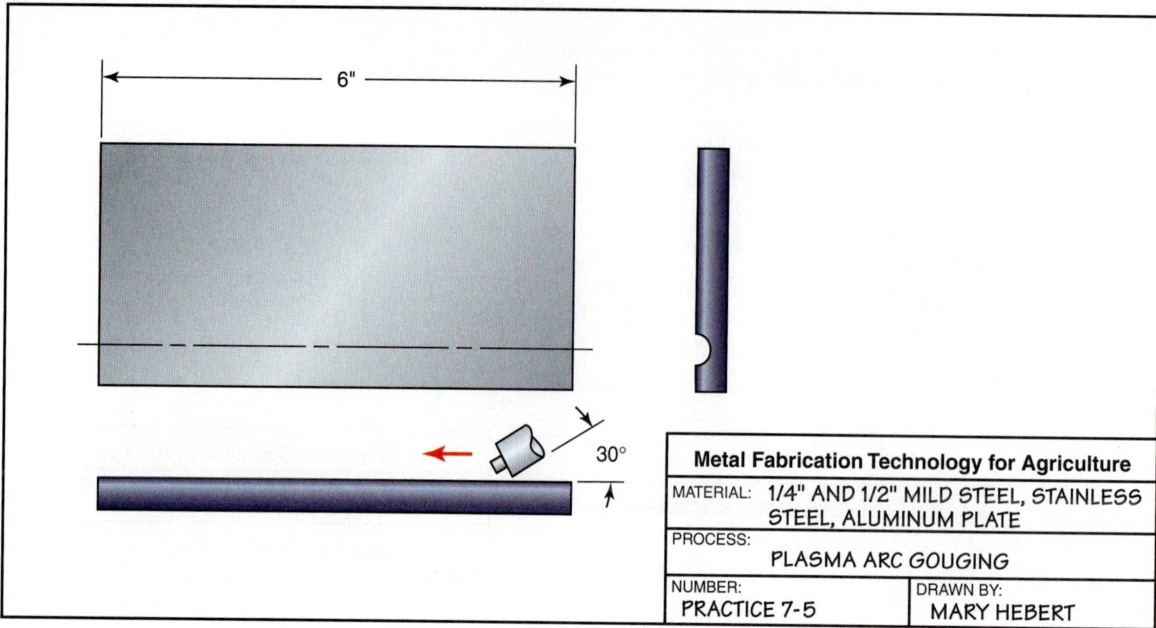

FIGURE 7-34 Plasma arc gouging.

Summary

A typical agricultural plasma cutting system power supply weighs about 30 lbs (13 k). It will be powered with 120V and/or 220V AC electrical power. It needs an external air supply capable of supplying approximately 65 psi (450 kPa) with a flow rate of 4.5 cubic feet per minute (CFM) (130L/min). In this small system you have a machine that can provide cutting for almost every project on your farm or ranch. Plasma cutting equipment is in a continuing state of development, which has provided significant breakthroughs in design. It is expected that this research and development will produce even smaller, more powerful, versatile, and portable systems.

One area that it is hoped research and development (R & D) will make advancements in is consumable supplies. R & D has already made a difference in the durability of the electrode, nozzle insulator, nozzle tip, and nozzle. Although in your training program you have probably seen that the plasma cutting torches consumable supplies have often been destroyed very quickly. Before the recent R & D, the failure rate of torch consumables, even with skilled welders, was very short. However, as you have developed better cutting skills, these torch consumables have lasted longer.

The need to make cuts on the farm or ranch will always exist. Plasma cutting will probably never completely replace oxyacetylene cutting in agriculture, but it certainly offers a great alternative.

Review

1. Most agricultural plasma cutting machines can cut mild steel up to ______ thickness.
   A. 1/16 in. (2 mm)
   B. 1/2 in. (13 mm)
   C. 1 in (25 mm)
   D. 3/8 in. (9 mm)
2. Which of the following materials cannot be cut with an agricultural plasma torch?
   A. aluminum
   B. glass
   C. cast iron
   D. stainless steel
3. Plasma for cutting is created in the __________.
   A. torch head
   B. torch handle
   C. power supply
   D. air compressor
4. Most hand-held torches have a manual power switch that is used to start and stop the power source.
   A. true
   B. false
5. The __________ of a plasma torch is often made of copper electrode with a tungsten tip attached.
   A. head
   B. body
   C. handle
   D. electrode tip
6. The __________ has a small, cone-shaped, constricting orifice in the center.
   A. nozzle tip
   B. electrode
   C. handle
   D. power supply
7. The production of the plasma requires a direct-current, high-voltage, constant current power supply.
   A. true
   B. false
8. __________ are the units of measure for electrical power.
   A. Volts
   B. Amps
   C. Ohms
   D. Watts
9. The high travel speed with plasma cutting will result in a heat input that is much higher than that of the oxyfuel cutting process.
   A. true
   B. false
10. The __________ is the distance from the nozzle tip to the work.
    A. cutting distance
    B. standoff distance
    C. work distance
    D. tip height
11. What type of current is used to start the plasma arc?
    A. DCEP
    B. DCSP
    C. high frequency
    D. open circuit
12. The __________ is the space left in the metal as the metal is removed during a cut.
    A. groove
    B. kerf
    C. gap
    D. crack
13. What gas is used to form the plasma?
    A. argon
    B. compressed air
    C. oxygen
    D. nitrogen
14. The heat affected zone on metal cut with a plasma torch is wider than that of metal cut with oxyacetylene.
    A. true
    B. false

Review continued

15. In order for a material to be cut with plasma, it only has to be conductive.
    A. true
    B. false
16. Why might a castle nozzle be used on a plasma torch?
    A. It allows the tip to be dragged across the surface.
    B. It directs the heat into the center of the castle.
    C. They are easier to grip with a gloved hand.
    D. You can see the line through the open ports.
17. What is it called when multiple layers of metal are cut at once?
    A. layer cutting
    B. pile cutting
    C. stack cutting
    D. group cutting
18. What is dross?
    A. It is the name for the sparks thrown from a PAC.
    B. It is the smell caused by cutting cast iron with PAC.
    C. It is the light that reflects from the cut onto the floor.
    D. It is the metal blown out of the kerf in a PAC.
19. The loud noise made by a PAC can be a safety hazard.
    A. true
    B. false
20. A fire watch is a person whose sole job is to watch for the possible starting of a fire.
    A. true
    B. false

Chapter 8

Shielded Metal Arc Equipment, Setup, and Operation

OBJECTIVES

After completing this chapter, the student should be able to

- ☑ explain the differences in welding with each of the three types of current.
- ☑ identify welding machines according to their type.
- ☑ demonstrate how to select and set the welding current.
- ☑ describe the proper maintenance of welding equipment.
- ☑ demonstrate how to safely set up an arc welding station.

KEY TERMS

amperage
anode
cathode
duty cycle
electrons
inverter
magnetic flux lines
open circuit voltage
operating voltage
output
rectifier
step-down transformer
voltage
wattage
welding cables
welding leads

INTRODUCTION

Shielded metal arc welding (SMAW) is a welding process that uses a flux covered metal electrode to carry an electrical current, **Figure 8-1**. The current forms an arc across the gap between the end of the electrode and the work. The electric arc creates sufficient heat to melt both the electrode and the work. Molten metal from the electrode travels across the arc to the molten pool on the base metal, where they mix together. The end of the electrode and molten pool of metal are surrounded, purified, and protected by a gaseous cloud and a covering of slag produced as the flux coating of the electrode burns or vaporizes. As the

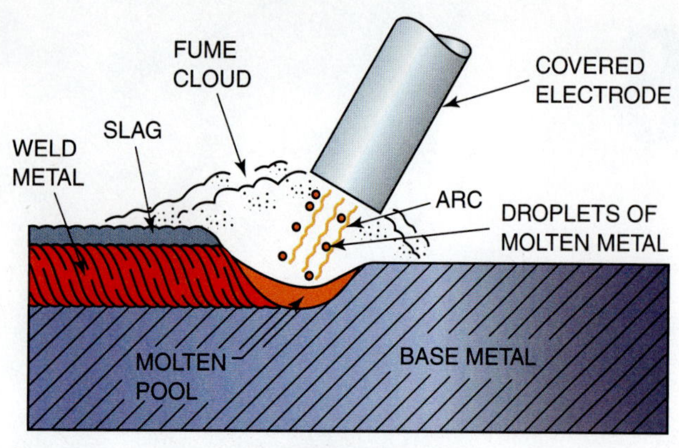

FIGURE 8-1 Shielded metal arc welding.

arc moves away, the mixture of molten electrode and base metal solidifies and becomes one piece.

SMAW is the most widely used agricultural welding process because of its low cost, flexibility, portability, and versatility. The machine and the electrodes are low cost. The machine itself can be as simple as a 110-volt, step-down transformer. The electrodes are available from a large number of manufacturers in packages from 1 lb (0.5 kg) to 50 lb (22 kg).

The SMAW process is very flexible in terms of the metal thicknesses that can be welded and the variety of positions in which it can be used. Metal as thin as 1/16 in. (2 mm) thick, or approximately 16 gauge, to several feet thick can be welded using the same machine with different settings. The flexibility of the process also allows metal in this thickness range to be welded in any position.

SMAW is a very portable process because engine-driven generator-type welders are available and a limited amount of equipment required for the process.

The process is versatile and is used to weld almost any metal or alloy, including cast iron, aluminum, stainless steel, and nickel.

Welding Current

The welding current is electric. An electric current is the flow of **electrons.** Electrons flow through a conductor from negative (−) to positive (+), **Figure 8-2.** Resistance to the flow of electrons (electricity) produces heat. The greater the resistance, the greater the heat. Air has a high resistance to current flow. As the electrons jump the air gap between the end of the electrode and the work, a great deal of heat is produced. Electrons flowing across an air gap produce an arc.

Measurement Three units are used to measure a welding current. They are voltage (V), amperage (A), and wattage (W). **Voltage,** or volts (V), is the measurement of electrical pressure, in the same way that pounds per square inch is a measurement of water pressure. Voltage controls the maximum gap the electrons can jump to form the arc. A higher voltage can jump a larger gap. **Amperage,** or amperes (A), is the measurement of the total number of electrons flowing, in the same way that gallons is a measurement of the amount of water flowing. Amperage controls the size of the arc. **Wattage,**

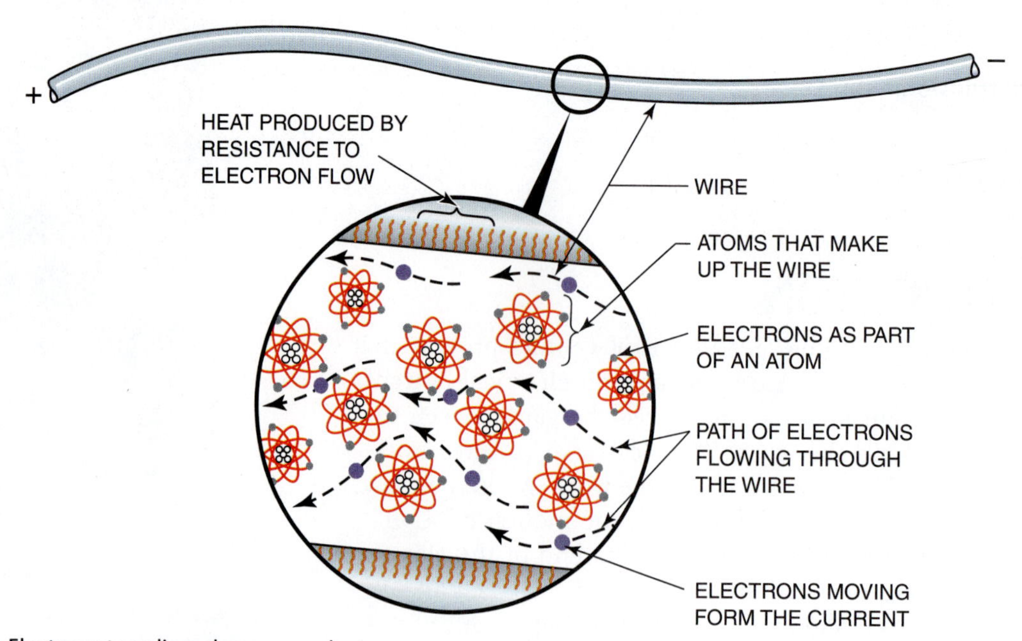

FIGURE 8-2 Electrons traveling along a conductor.

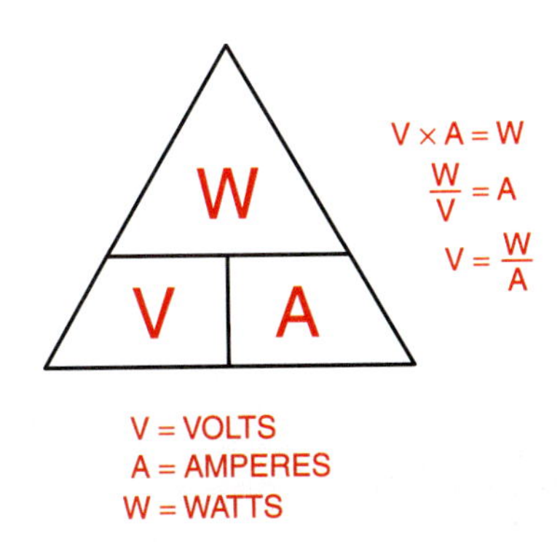

FIGURE 8-3 Ohm's Law.

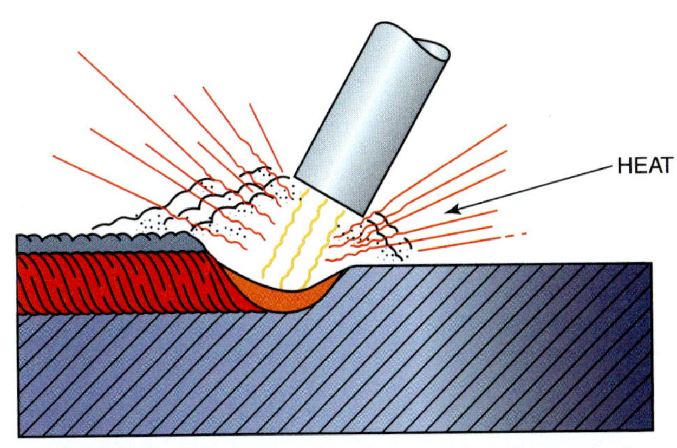

FIGURE 8-5 Energy is lost from the weld in the form of radiation and convection.

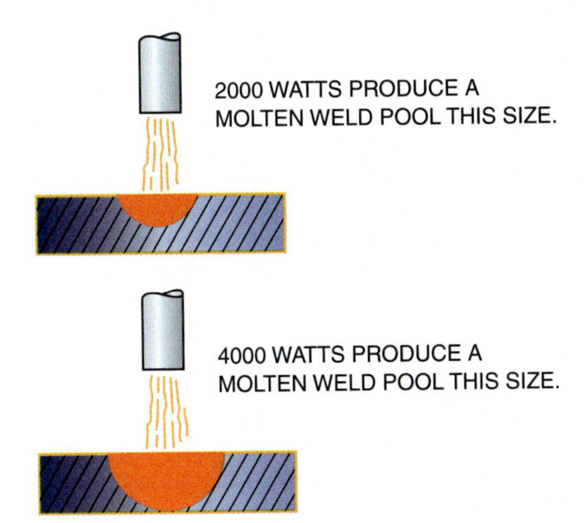

FIGURE 8-4 The molten weld pool size depends upon the energy (watts), the metal mass, and thermal conductivity.

or watts (W), are calculated by multiplying voltage (V) times amperes (A), **Figure 8-3**. Wattage is a measurement of the amount of electrical energy or power in the arc. The amount of watts being put into a weld per inch (cm) controls the width and depth of the weld bead, **Figure 8-4**.

Temperature The temperature of a welding arc exceeds 11,000°F (6000°C). The exact temperature depends on the resistance to the current flow. The resistance is affected by the arc length and the chemical composition of the gases formed as the electrode covering burns and vaporizes. As the arc lengthens, the resistance increases, thus causing a rise in the arc voltage and temperature. The shorter the arc, the lower the arc temperature produced.

Shielded metal arc welding electrodes have chemicals added to their coverings to stabilize the arc. These arc stabilizers reduce the arc resistance, making it easier to hold an arc. By lowering the resistance, the arc stabilizers also lower the arc temperature. Other chemicals within the gaseous cloud around the arc may raise or lower the resistance.

The amount of heat produced is determined by the size of the electrode and the amperage setting. Not all of the heat produced by an arc reaches the weld. Some of the heat is radiated away in the form of light and heat waves, **Figure 8-5**. Additional heat is carried away with the hot gases formed by the electrode covering. Heat also is lost through conduction in the work. In total, about 50% of all heat produced by an arc is missing from the weld.

The remaining 50% of the heat produced by the arc is not distributed evenly between both ends of the arc. This distribution depends on the composition of the electrode's coating, the type of welding current, and the polarity of the electrode's coating.

Currents The three different types of current used for welding are alternating current (AC), direct-current electrode negative (DCEN), and direct-current electrode positive (DCEP). The terms DCEN and DCEP have replaced the former terms *direct-current straight polarity (DCSP)* and *direct-current reverse polarity (DCRP)*. DCEN and DCSP are the same currents, and DCEP and DCRP are the same currents. Some electrodes can be used with only one type of current. Others can be used with two or more types of current. Each welding current has a different effect on the weld.

DCEN

In direct-current electrode negative, the electrode is negative, and the work is positive, **Figure 8-6**. DCEN welding current produces a high electrode melting rate.

DCEP

In direct-current electrode positive, the electrode is positive, and the work is negative, **Figure 8-7**. DCEP current produces the best welding arc characteristics.

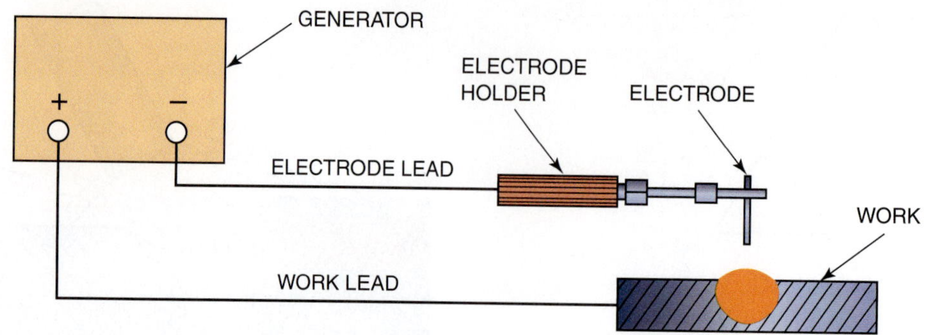

FIGURE 8-6 Electrode negative (DCEN), straight polarity (DCSP).

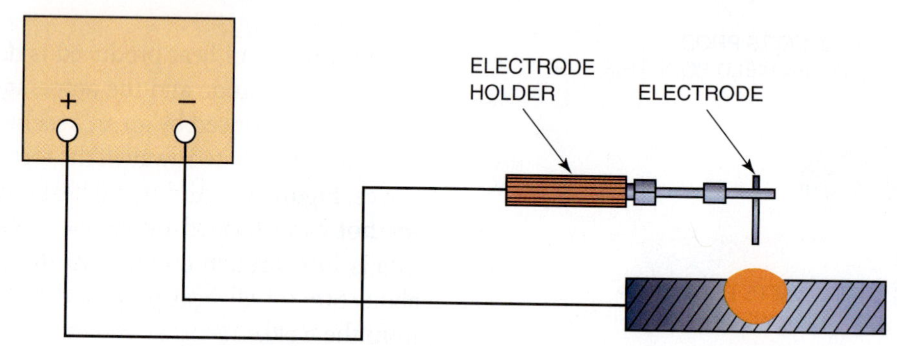

FIGURE 8-7 Electrode positive (DCEP), reverse polarity (DCRP).

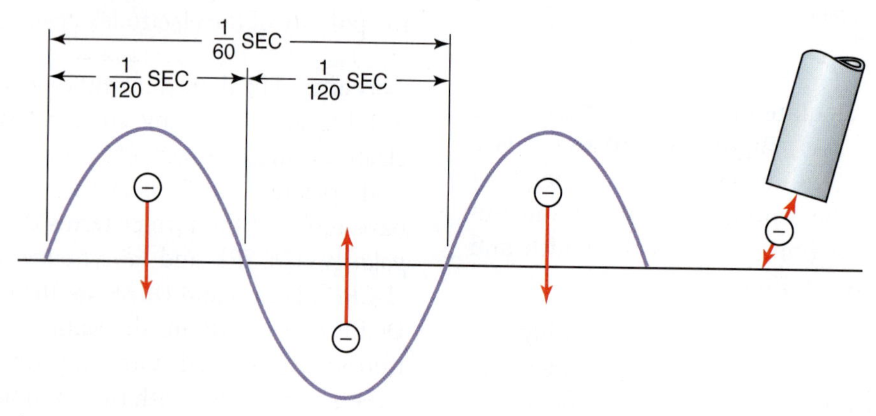

FIGURE 8-8 Alternating current (AC).

AC

In alternating current, the electrons change direction every 1/120 of a second so that the electrode and work alternate from anode to cathode, Figure 8-8. The rapid reversal of the current flow causes the welding heat to be evenly distributed on both the work and the electrode—that is, half on the work and half on the electrode. The even heating gives the weld bead a balance between penetration and buildup.

Types of Welding Power

Welding power can be supplied as

- Constant voltage (CV)—The arc voltage remains constant at the selected setting even if the arc length and amperage increase or decrease.
- Constant current (CC)—The total welding current (watts) remains the same. This type of

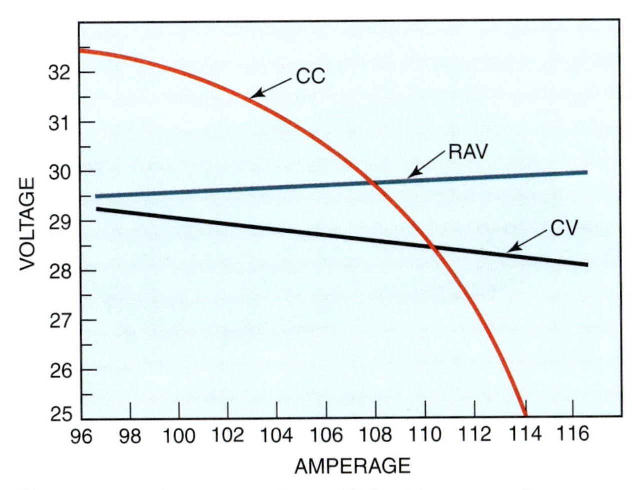

FIGURE 8-9 Constant voltage (CV), rising arc voltage (RAV), and constant current (CC).

power is also called drooping arc voltage (DAV), because the arc voltage decreases as the amperage increases.

The shielded metal arc welding (SMAW) process requires a constant current arc voltage characteristic, illustrated by the CC line in **Figure 8-9.** The shielded metal arc welding machine's voltage output decreases as current increases. This output power supply provides a reasonably high open circuit voltage before the arc is struck. The high open circuit voltage quickly stabilizes the arc. The arc voltage rapidly drops to the lower closed circuit level after the arc is struck. Following this short starting surge, the power (watts) remains almost constant despite the changes in arc length. With a constant voltage output, small changes in arc length would cause the power (watts) to make large swings. The welder would lose control of the weld.

Open Circuit Voltage

Open circuit voltage is the voltage at the electrode before striking an arc (with no current being drawn). This voltage is usually between 50 V and 80 V. The higher the open circuit voltage, the easier it is to strike an arc. The higher voltage also increases the chance of electrical shock.

Operating Voltage

Operating voltage, or closed circuit voltage, is the voltage at the arc during welding. This voltage will vary with arc length, type of electrode being used, type of current, and polarity. The voltage will be between 17 V and 40 V.

Arc Blow

When electrons flow they create lines of magnetic force that circle around the line of flow, **Figure 8-10.** Lines of magnetic force are referred to as magnetic flux lines. These lines space themselves evenly along a current-carrying wire. If the wire is bent, the flux lines on one side are compressed together, and those on the other side are stretched out, **Figure 8-11.** The unevenly spaced flux lines try to straighten the wire so that the lines can be evenly spaced once again. The force that they place on the wire is usually small. However, when welding with very high amperages, 600 amperes or more, the force may cause the wire to move.

The welding current flowing through a plate or any residual magnetic fields in the plate will result in uneven flux lines. These uneven flux lines can, in turn, cause an arc to move during a weld. This movement of the arc is called *arc blow.* Arc blow makes the arc drift as a string would drift in the wind. Arc blow is more noticeable in corners, at the ends of plates, and when the work lead is connected to only one side of a plate, **Figure 8-12.** If arc blow is a problem, it can be controlled by connecting the work lead to the end of the weld joint and making the weld in the direction

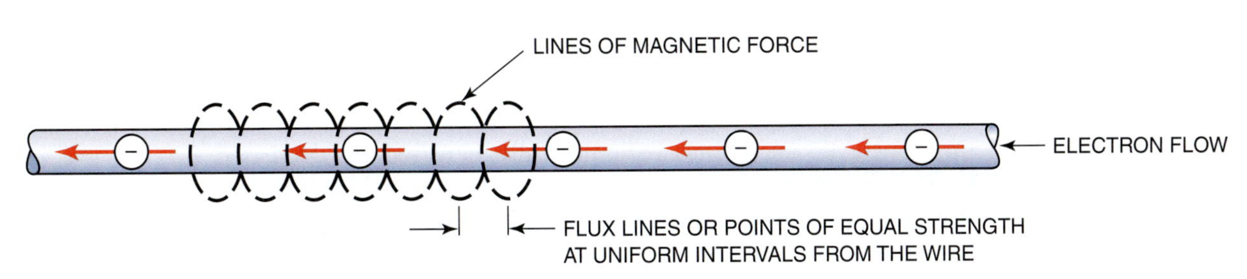

FIGURE 8-10 Magnetic force around a wire.

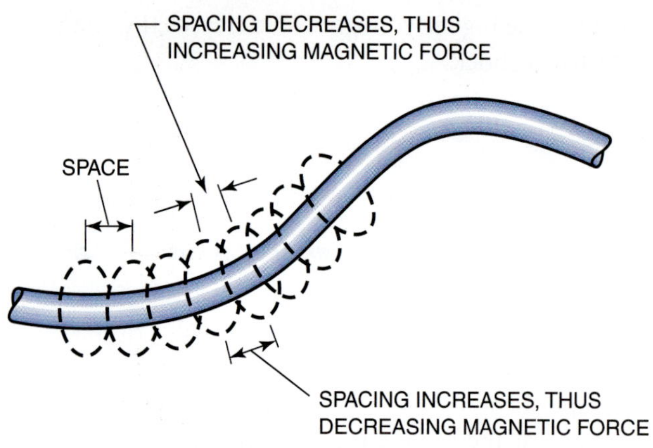

FIGURE 8-11 Magnetic forces concentrate around bends in wires.

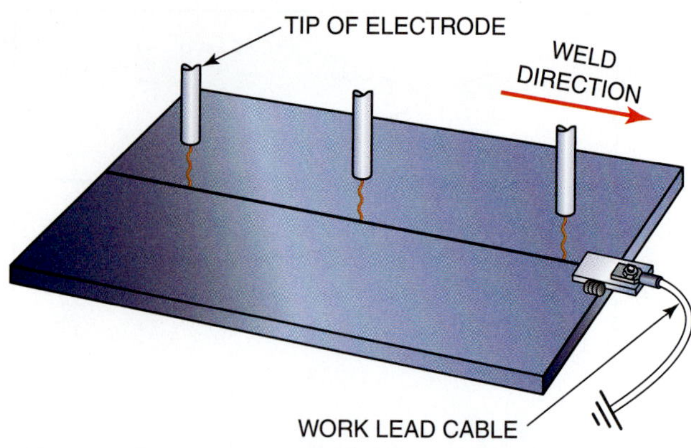

FIGURE 8-13 Correct current connections to control arc blow.

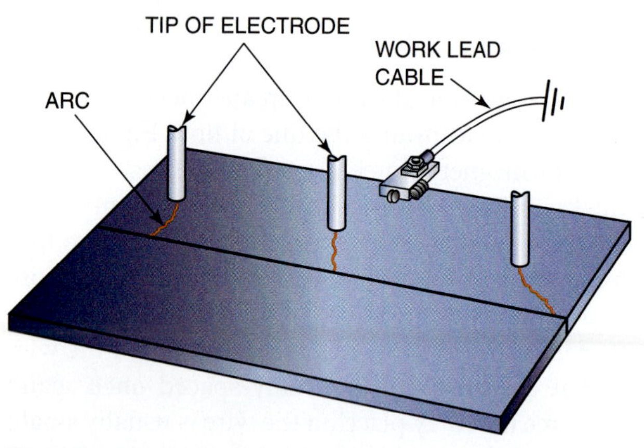

FIGURE 8-12 Arc blow.

toward the work lead, **Figure 8-13.** Another way of controlling arc blow is to use two work leads, one on each side of the weld. The best way to eliminate arc blow is to use alternating current. AC usually does not allow the flux lines to build long enough to bend the arc before the current changes direction. If it is impossible to move the work connection or to change to AC, a very short arc length can help control arc blow. A small tack weld or a change in the electrode angle can also help control arc blow.

Arc blow may not be a problem as you are learning to weld in the shop, because most welding tables are all steel. If you are using a pipe stand to hold your welding practice plates, arc blow can become a problem. In this case, try reclamping your practice plates.

Types of Power Sources

Two types of electrical devices can be used to produce the low-voltage, high-amperage current combination that arc welding requires. One type uses engines to drive alternators or generators. The other type uses **step-down transformers.** Because transformer-type welding machines are quieter, are more energy efficient, require less maintenance, and are less expensive, they are best for working in the shop or barn. Engine-powered generators are widely used for portable welding. They are best for working in the field or pasture.

Transformers A welding transformer uses the alternating current (AC) supplied to the shop at a high voltage to produce the low-voltage welding power. As electrons flow through a wire, they produce a magnetic field around the wire. If the wire is wound into a coil the weak magnetic field of each wire is concentrated to produce a much stronger central magnetic force. Because the current being used is alternating or reversing each 1/120 of a second, the magnetic field is constantly being built and allowed to collapse. By placing a second or secondary winding of wire in the magnetic field produced by the first or primary winding a current will be induced in the secondary winding. The placing of an iron core in the center of these coils will increase the concentration of the magnetic field, **Figure 8-14.**

A transformer with more turns of wire in the primary winding than in the secondary winding is known as a step-down transformer. A step-down transformer takes a high-voltage, low-amperage current and changes it into a low-voltage, high-amperage current. Except for some power lost by heat within a transformer, the power (watts) into a transformer equals the power (watts) out because the volts and amperes are mutually increased and decreased.

A transformer welder is a step-down transformer. It takes the high line voltage (110 V, 220 V, 440 V, etc.) and low-amperage current (30 A, 50 A, 60 A, etc.) and changes it into 17 V to 45 V at 190 A to 590 A.

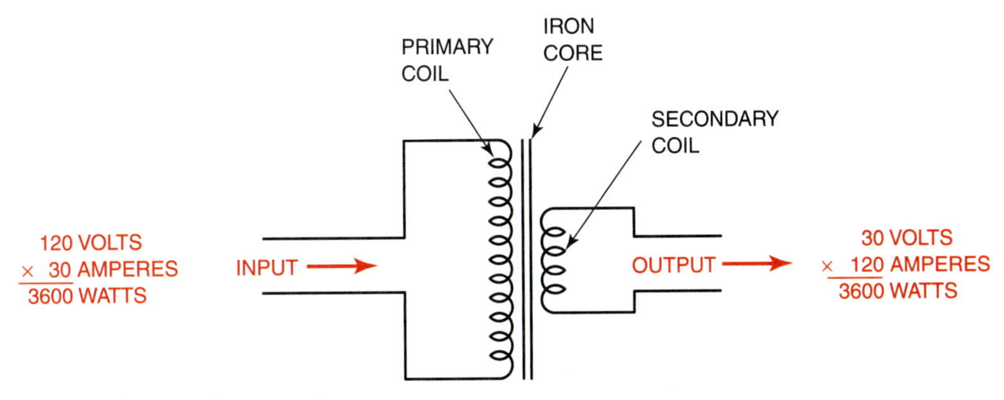

FIGURE 8-14 Diagram of a step-down transformer.

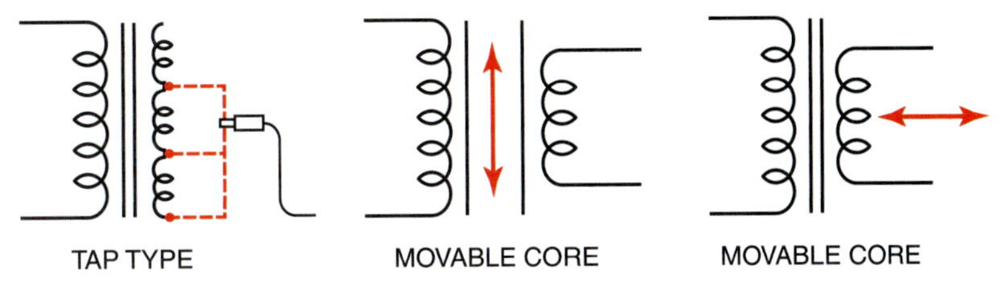

FIGURE 8-15 Major types of adjustable welding transformers.

FIGURE 8-16 Tap-type transformer welding machine.
Courtesy of Lincoln Electric Company.

Welding machines can be classified by the method by which they control or adjust the welding current. The major classifications are multiple-coil, called taps; movable coil; and movable core, **Figure 8-15**, and inverter type.

Multiple-coil The multiple-coil machine, or tap-type machine, allows the selection of different current settings by tapping into the secondary coil at a different turn value. The greater the number of turns, the higher is the amperage induced in the turns. These machines may have a large number of fixed amperes, **Figure 8-16**, or they may have two or more amperages that can be adjusted further with a fine adjusting knob. The fine adjusting knob may be marked in amperes, or it may be marked in tenths, hundredths, or any other unit.

EXPERIMENT 8-1

Estimating Amperages

Using a pencil and paper, you will prepare a rough estimate of the amperage setting of a welding machine. **Figure 8-17** shows a welding machine with low, medium, and high tap amperage ranges. A fine adjusting knob is marked with ten equal divisions, and each division is again divided by ten smaller lines.

The machine is set on the medium range, 50 to 250 amperes, and the fine adjusting knob is turned until it points to the line marked 5 (halfway between 0 and 10). This means that the amperage is halfway from 50 to 250, or 150 amperes. If the fine adjusting knob points between 2 and 3, the resulting amperage is 1/4 of the way from 50 to 250, or about 100 amperes. If the knob points between 7 and 8, the amperage is 3/4 of the way from 50 to 250, or about 200 amperes. If the knob points at 4, the amperage is more than 100 but a little

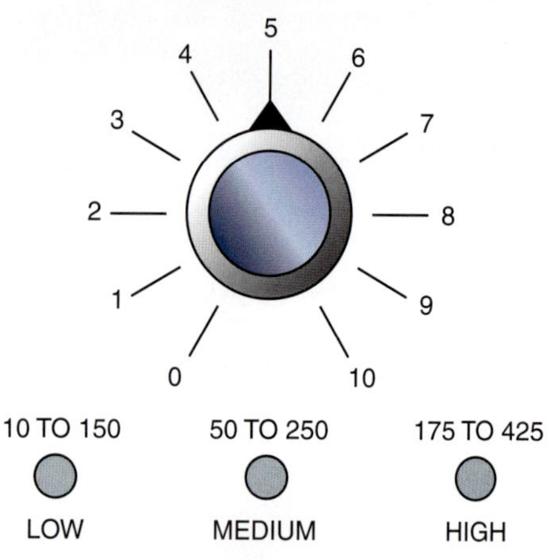

FIGURE 8-17 Tap-type welder knobs.

less than 150, or about 130 to 140 amperes. What is the amperage if the knob points at 6?

Since this is a method of estimating only, the amperage value obtained is close enough to allow an arc to be struck. The welder can then finish the fine adjusting to obtain a good weld. Complete a copy of the "Student Welding Report" listed in Appendix I or provided by your instructor. ◆

EXPERIMENT 8-2

Calculating the Amperage Setting

Using a pencil and paper or calculator, you will calculate the exact value for each space on the fine adjusting knob of a welding machine.

With the machine set on the medium range, from 50 to 250 amperes, first subtract the low amperage from the high amperage to get the amperage spread (250 − 50 = 200). Now divide the amperage spread by the number of units shown on the fine adjusting knob (200 ÷ 10 = 20). Each unit is equal to a 20-ampere increase, **Table 8-1.** When the knob points to 0, the amperage is 50; when the knob points to 1, the amperage is 70; and at 2, the amperage is 90, **Figure 8-18.** There are 100 small units on the fine adjusting knob. Dividing the amperage spread by the number of small units gives the amperage value for each unit (200 ÷ 100 = 2). Therefore, if the knob points to 6.1, the amperage is set at a value of 50 + 120 + 2 = 172 amperes. This method provides a good starting place for the current setting, but if the welding is to be made in accordance with a welding procedure's specific amperage setting it will be necessary to use a calibrated meter to make the correct setting.

Complete a copy of the "Student Welding Report" listed in Appendix I or provided by your instructor. ◆

Setting	Value in Amperes
0 = 50 + 0,	or 50 A
1 = 50 + 20,	or 70 A
2 = 50 + 40,	or 90 A
3 = 50 + 60,	or 110 A
4 = 50 + 80,	or 130 A
5 = 50 + 100,	or 150 A
6 = 50 + 120,	or 170 A
7 = 50 + 140,	or 190 A
8 = 50 + 160,	or 210 A
9 = 50 + 180,	or 230 A
10 = 50 + 200,	or 250 A

TABLE 8-1 Example of a Table Used to Calculate the Amperage Setting.

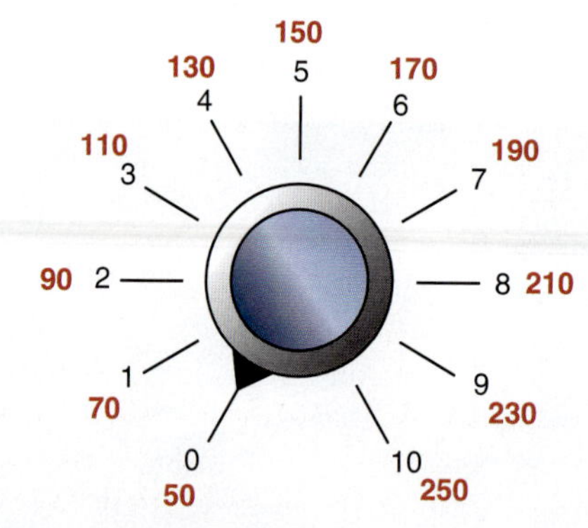

FIGURE 8-18 Fine adjusting knob.

PRACTICE 8-1

Estimating Amperages

Using a pencil and paper and the amperage ranges given in this practice (or from machines in the shop), you will estimate the amperage when the knob is at the 1/4, 1/2, and 3/4 settings, **Figure 8-19.**

Complete a copy of the "Student Welding Report" listed in Appendix I or provided by your instructor. ◆

PRACTICE 8-2

Calculating Amperages

Using a pencil and paper or a calculator, and the amperage ranges given in this practice (or from machines in the shop), you will calculate the amperages for each of the following knob settings: 1, 4, 7, 9, 2.3, 5.7, and 8.5.

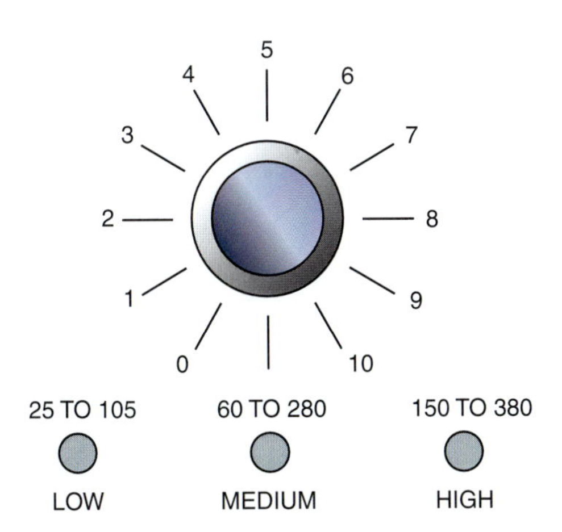

FIGURE 8-19 Practice 8-1.

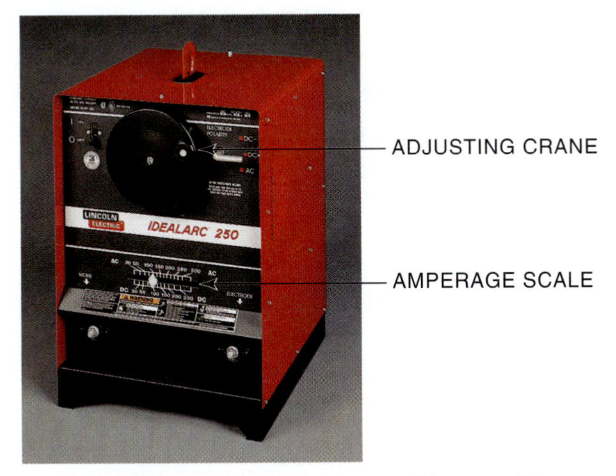

FIGURE 8-20 A movable, core-type welding machine.
Courtesy of Lincoln Electric Company.

Complete a copy of the "Student Welding Report" listed in Appendix I or provided by your instructor. ◆

Movable Coil or Core Movable coil or movable core machines are adjusted by turning a handwheel that moves the internal parts closer together or farther apart. The adjustment may also be made by moving a lever, **Figure 8-20.** These machines may have a high and low range, but they do not have a fine adjusting knob. The closer the primary and secondary coils are, the greater is the induced current; the greater the distance between the coils, the smaller is the induced current, **Figure 8-21.** Moving the core in concentrates more of the magnetic force on the secondary coil, thus increasing the current. Moving the core out allows the field to disperse, and the current is reduced, **Figure 8-22.**

Inverter Inverter welding machines will be much smaller than other types of machines of the same amperage range. This smaller size makes the welder

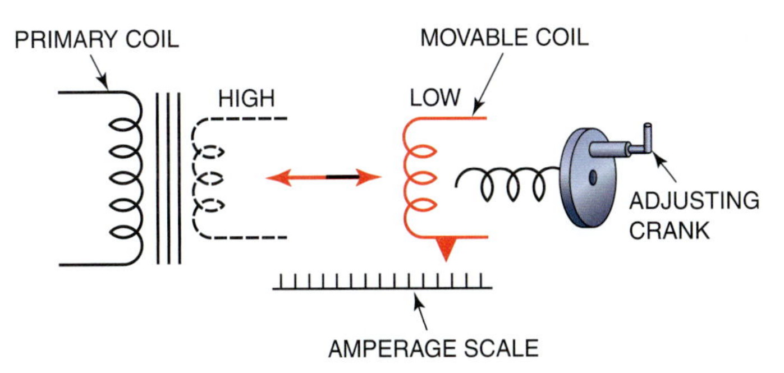

FIGURE 8-21 Movable coil.

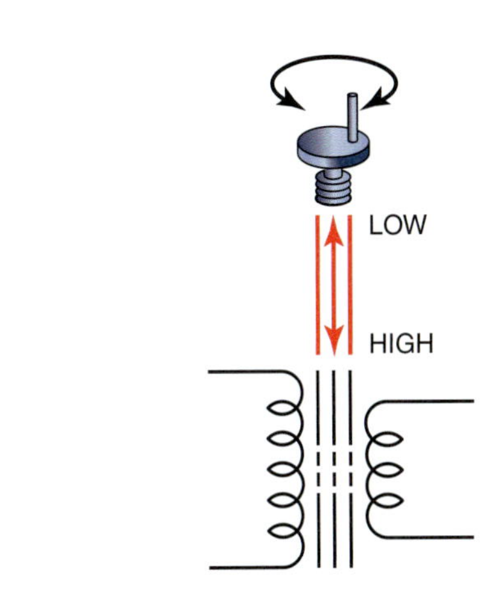

FIGURE 8-22 Movable coil.

much more portable as well as increasing the energy efficiency. In a standard welding transformer the iron core used to concentrate the magnetic field in the coils must be a specific size. The size of the iron core is determined by the length of time it takes for the magnetic field to build and collapse. By using solid state electronic parts the incoming power in an inverter welder is changed from 60 cycles a second to several thousand cycles a second. This higher frequency allows the use of a transformer that may be as light as 7 pounds and still do the work of a standard transformer weighing 100 pounds. Additional electronic parts remove the high frequency for the output welding power.

The use of electronics in the inverter-type welder allows it to produce any desired type of welding power. Before the invention of this machine, each type of welding required a separate machine. Now a single welding machine can produce the specific type of current needed for shielded metal arc welding, gas tungsten arc welding, gas metal arc welding, and plasma arc cutting. Because the machine can be light enough to be carried closer to work, shorter welding cables can be used. The welder does not have to walk as far to adjust the machine. Welding machine power wire

FIGURE 8-23 Engine generator welders make welding on a ranch easier. Courtesy of Larry Jeffus.

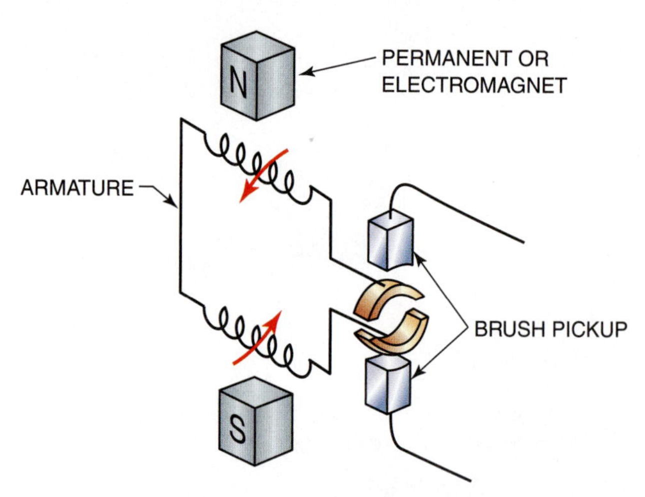

FIGURE 8-25 Diagram of a generator.

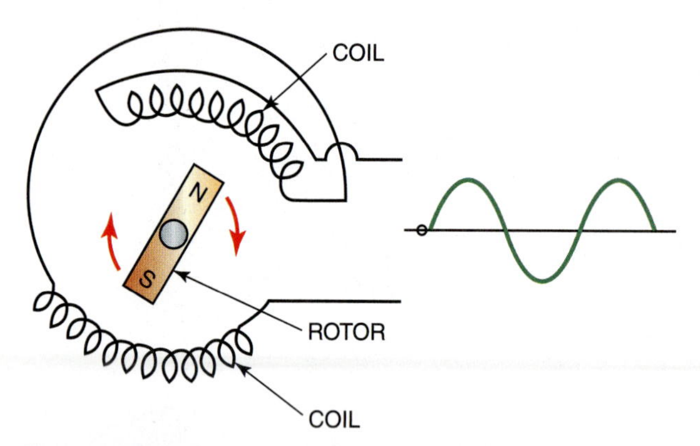

FIGURE 8-24 Schematic diagram of an alternator.

is cheaper than welding cables. Some manufacturers produce machines that can be stacked so that when you need a larger machine all you have to do is add another unit to your existing welder.

Generators and Alternators

Generators and alternators both produce welding electricity from a mechanical power source, **Figure 8-23.** Both devices have an armature that rotates and a stator that is stationary. As a wire moves through a magnetic force field, electrons in the wire are made to move, producing electricity.

In an alternator, magnetic lines of force rotate inside a coil of wire, **Figure 8-24.** An alternator can produce AC only. In a generator, a coil of wire rotates inside a magnetic force. A generator can produce AC or DC. It is possible for alternators and generators to both use diodes to change the AC to DC for welding. In generators, the welding current is produced on the armature and is picked up with brushes, **Figure 8-25.** In alternators, the welding current is produced on the stator, and only the small current for the electromagnetic force field goes across the brushes. Therefore, the brushes in an alternator are smaller and last longer. Alternators can be smaller in size and lighter in weight than generators and still produce the same amount of power.

Engine generators and alternators may run at the welding speed all the time, or they may have an option that reduces their speed to an idle when welding stops. This option saves fuel and reduces wear on the welding machine. To strike an arc when using this type of welder, stick the electrode to the work for a second. When you hear the welding machine (welder) pick up speed, remove the electrode from the work and strike an arc. In general, the voltage and amperage are too low to start a weld, so shorting the electrode to the work should not cause the electrode to stick. A timer can be set to control the length of time that the welder maintains speed after the arc is broken. The time should be set long enough to change electrodes without losing speed.

Portable welders often have 110-volt or 220-volt plug outlets, which can be used to run grinders, drills, lights, and other equipment. The power provided may be AC or DC. If DC is provided, only equipment with brush-type motors or tungsten light bulbs can be used. If the plug is not specifically labeled 110 volts AC, check the owner's manual before using it for such devices as radios or other electronic equipment. A typical portable welder is shown in **Figure 8-26.**

It is recommended that a routine maintenance schedule for portable welders be set up and followed. By checking the oil, coolant, battery, filters, fuel, and other parts, the life of the equipment can

(A)

(B)

FIGURE 8-26 Portable welders can be mounted in a (A) trailer or a (B) truck. Courtesy of Larry Jeffus.

be extended. A checklist can be posted on the welder, **Table 8-2**.

Rectifiers

Alternating welding current can be converted to direct current by using a series of rectifiers. A rectifier allows current to flow in one direction only, **Figure 8-27**.

If one rectifier is added, the welding power appears as shown in **Figure 8-28**. It would be difficult to weld with pulsating power such as this. A series of rectifiers, known as a bridge rectifier, can modify the alternating current so that it appears as shown in **Figure 8-29**.

Rectifiers become hot as they change AC to DC. They must be attached to a heat sink and cooled by having air blown over them. The heat produced by a rectifier reduces the power efficiency of the welding machine. **Figure 8-30** shows the amperage dial of a typical machine. Notice that at the same dial

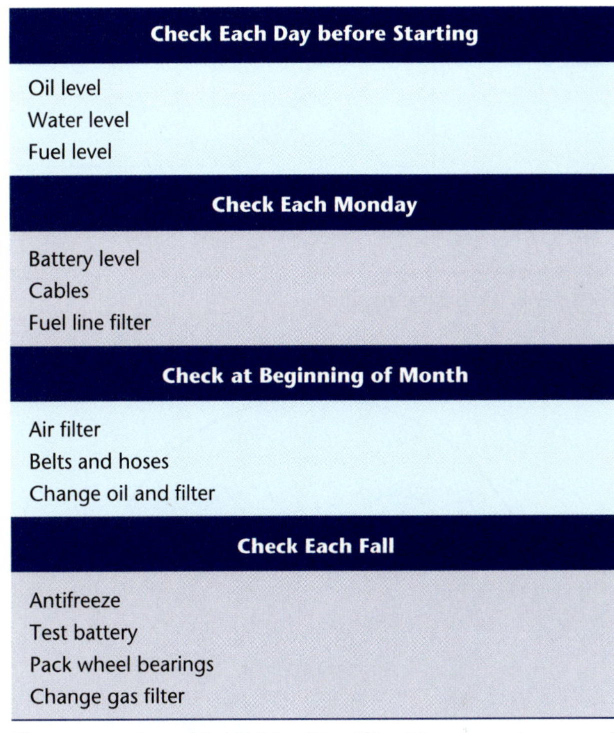

Check Each Day before Starting
Oil level
Water level
Fuel level
Check Each Monday
Battery level
Cables
Fuel line filter
Check at Beginning of Month
Air filter
Belts and hoses
Change oil and filter
Check Each Fall
Antifreeze
Test battery
Pack wheel bearings
Change gas filter

TABLE 8-2 Portable Welder Checklist. The owner's manual should be checked for any additional items that might need attention.

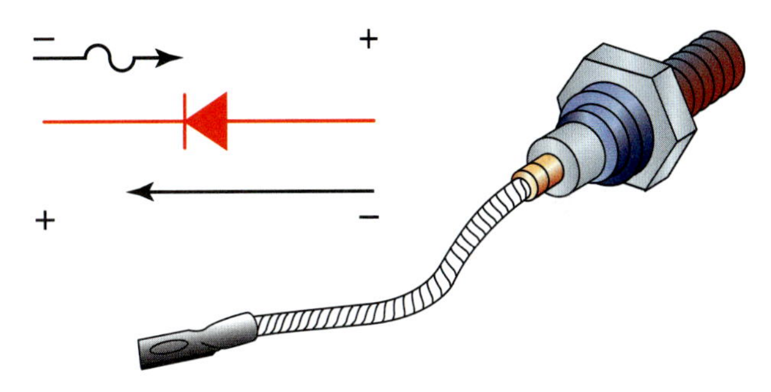

FIGURE 8-27 Rectifier.

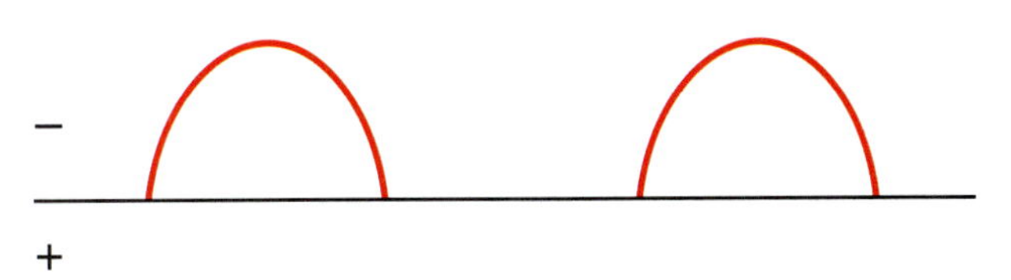

FIGURE 8-28 One rectifier in a welding power supply results in pulsating power.

settings for AC and DC, the DC is at a lower amperage. The difference in amperage (power) is due to heat lost in the rectifiers. The loss in power makes operation with AC more efficient and less expensive compared with DC.

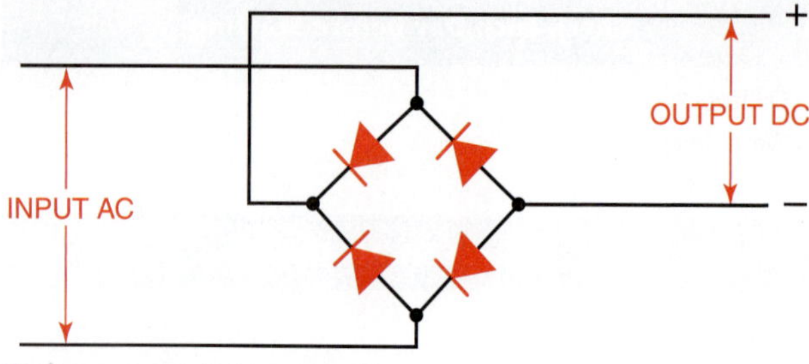

FIGURE 8-29 Bridge rectifier.

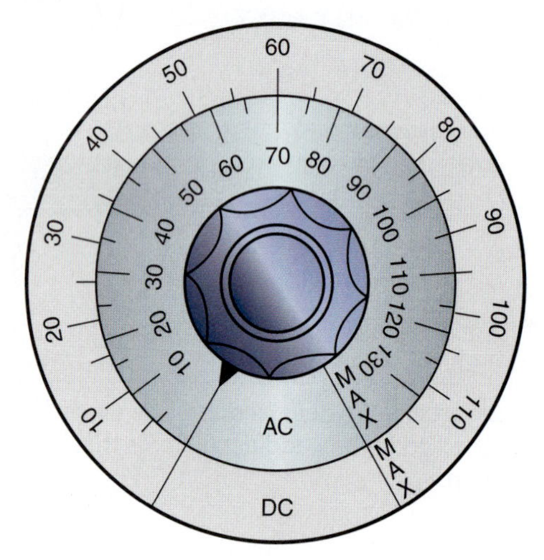

FIGURE 8-30 Typical dial on an AC-DC transformer rectifier welder.

A DC adapter for small AC machines is available from manufacturers. For some types of welding, AC does not work properly.

Duty Cycle

Welding machines produce internal heat at the same time they produce the welding current. Except for automatic welding machines, welders are rarely used every minute for long periods of time. The welder must take time to change electrodes, change positions, or change parts. Shielded metal arc welding never continues for long periods of time.

The duty cycle is the percentage of time a welding machine can be used continuously. A 60% duty cycle means that out of any ten minutes, the machine can be used for a total of six minutes at the maximum rated current. When providing power at this level, it must be cooled off for four minutes out of every ten minutes. The duty cycle increases as the amperage is lowered and decreases for higher amperages, **Figure 8-31.** Most welding machines weld at a 60% rate or less. Therefore, most manufacturers list the amperage rating for a 60% duty cycle on the nameplate that is attached to the machine. Other duty cycles are given on a graph in the owner's manual.

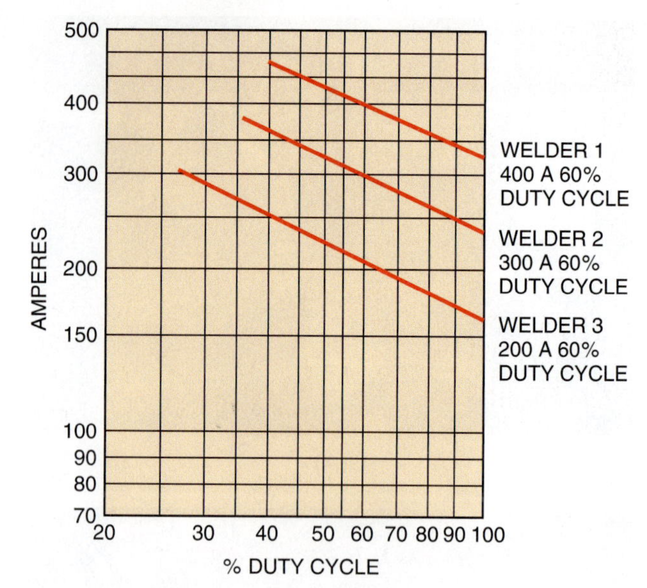

FIGURE 8-31 Duty cycle of a typical shielded metal arc welding machine.

The manufacturing cost of power supplies increases in proportion to their rated output and duty cycle. To reduce their price, it is necessary to reduce either their rating or their duty cycle. For this reason, some farm and ranch welding machines may have duty cycles as low as 20% even at a low welding setting of 90 to 100 amperes. For most agriculture repair and fabrication work, these welders perform very well. The duty cycle on these machines should never be exceeded because most agriculture welding can be completed in a few minutes of actual welding.

PRACTICE 8-3

Reading Duty Cycle Chart

Using a pencil and paper and the duty cycle chart in **Figure 8-31** (or one from machines in the shop), you will determine the following:

Welder 1: maximum welding amperage
percent duty cycle at maximum amperage

Welder 2: maximum welding amperage
percent duty cycle at maximum amperage

Welder 3: maximum welding amperage
percent duty cycle at maximum amperage

Length of Cable		Copper Welding Lead Sizes								
Amperes		**100**	**150**	**200**	**250**	**300**	**350**	**400**	**450**	**500**
ft	**m**									
50	15	2	2	2	2	1	1/0	1/0	2/0	2/0
75	23	2	2	1	1/0	2/0	2/0	3/0	3/0	4/0
100	30	2	1	1/0	2/0	3/0	4/0	4/0		
125	38	2	1/0	2/0	3/0	4/0				
150	46	1	2/0	3/0	4/0					
175	53	1/0	3/0	4/0						
200	61	1/0	3/0	4/0						
250	76	2/0	4/0							
300	91	3/0								
350	107	3/0								
400	122	4/0								

Length of Cable		Aluminum Welding Lead Sizes								
Amperes		**100**	**150**	**200**	**250**	**300**	**350**	**400**	**450**	**500**
ft	**m**									
50	15	2	2	1/0	2/0	2/0	3/0	4/0		
75	23	2	1/0	2/0	3/0	4/0				
100	30	1/0	2/0	4/0						
125	38	2/0	3/0							
150	46	2/0	3/0							
175	53	3/0								
200	61	4/0								
225	69	4/0								

TABLE 8-3 Copper and Aluminum Welding Lead Sizes.

Welder 1: maximum welding amperage at 100% duty cycle

Welder 2: maximum welding amperage at 100% duty cycle

Welder 3: maximum welding amperage at 100% duty cycle

Complete a copy of the "Student Welding Report" listed in Appendix I or provided by your instructor. ◆

Welding Cables

The terms welding cables and welding leads mean the same thing. Cables to be used for welding must be flexible, well insulated, and the correct size for the job. Most welding cables are made from stranded copper wire. The aluminum wires are lighter and less expensive than copper.

The insulation on welding cables will be exposed to weather, hot sparks, flames, grease, oils, sharp edges, impact, and other types of wear. To withstand such wear, only specially manufactured insulation should be used for welding cable. Several new types of insulation are available that will give longer service against these adverse conditions.

As electricity flows through a cable, the resistance to the flow causes the cable to heat up and increase the voltage drop. To minimize the loss of power and prevent overheating, the electrode cable and work cable must be the correct size. **Table 8-3** lists the minimum size cable that is required for each amperage and length. Large welding lead sizes make electrode manipulation difficult. Smaller cable can be spliced to the electrode end of a large cable to make it more flexible. This whip-end cable must not be over 10 ft (3 m) long.

CAUTION

A splice in a cable should not be within 10 ft (3 m) of the electrode because of the possibility of electrical shock.

PRACTICE 8-4

Determining Welding Lead Sizes

Using a pencil and paper and **Table 8-3**, Copper and aluminum welding lead sizes, you will determine the following:

1. The minimum copper welding lead size for a 200-amp welder with 100-ft (30-m) leads.

FIGURE 8-32 Keep leads stored so they will not be damaged. Courtesy of Larry Jeffus.

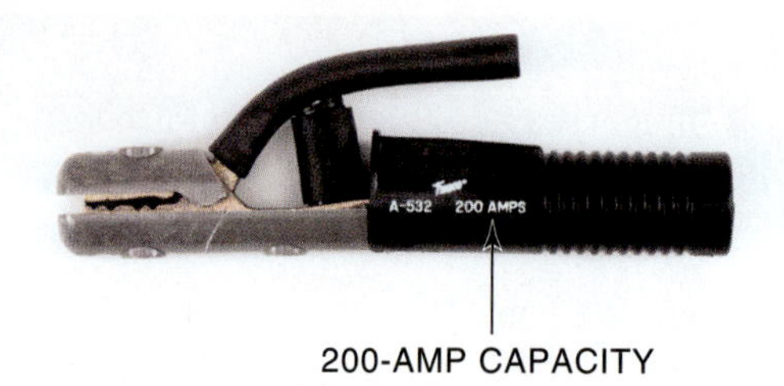

FIGURE 8-33 The amperage capacity of an electrode holder is often marked on its side. Courtesy of TWECO®, a Thermadyne® Company.

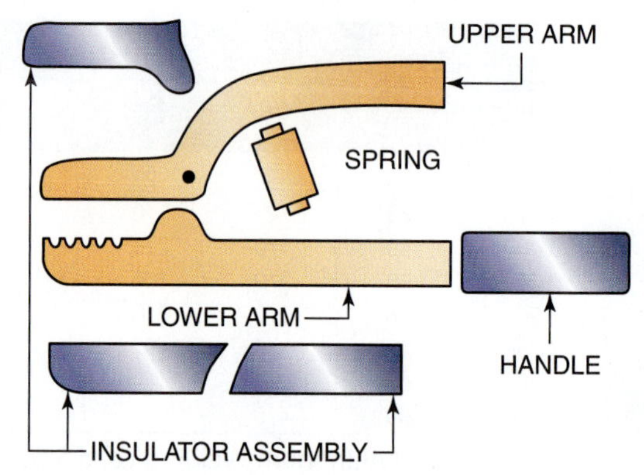

FIGURE 8-34 Replaceable parts of an electrode holder.

2. The minimum copper welding lead size for a 125-amp welder with 225-ft (69-m) leads.
3. The maximum length aluminum welding lead that can carry 300 amps.

Store welding leads so they will not get accidentally damaged, **Figure 8-32.**

Splices and end lugs are available from suppliers. Be sure that a good electrical connection is made whenever splices or lugs are used. A poor electrical connection will result in heat buildup, voltage drop, and poor service from the cable. Splices and end lugs must be well insulated against possible electrical shorting. Complete a copy of the "Student Welding Report" listed in Appendix I or provided by your instructor. ◆

Electrode Holders

The electrode holder should be of the proper amperage rating and in good repair for safe welding. Electrode holders are designed to be used at their maximum amperage rating or less. Higher amperage values will cause the holder to overheat and burn up. If the holder is too large for the amperage range being used, manipulation is hard, and operator fatigue increases. Make sure that the correct amperage holder is chosen, **Figure 8-33.**

CAUTION

Never dip a hot electrode holder in water to cool it off. The problem causing the holder to overheat should be repaired.

A properly sized electrode holder can overheat if the jaws are dirty or too loose, or if the cable is loose. If the holder heats up, welding power is being lost. In addition, a hot electrode holder is uncomfortable to work with.

Replacement springs, jaws, insulators, handles, screws, and other parts are available to keep the holder in good working order, **Figure 8-34.** To prevent excessive damage to the holder, welding electrodes should not be burned too short. A 2-in. (51-mm) electrode stub is short enough to minimize electrode waste and save the holder.

PRACTICE 8-5

Repairing Electrode Holders

Using the manufacturer's instructions for your type of electrode holder, required hand tools, and replacement parts, you will do the following:

CAUTION

Before starting any work, make sure that the power to the welder is off and locked off or the welding lead has been removed from the machine.

FIGURE 8-35 Make sure you have a good ground connection close to your welding. Courtesy of Larry Jeffus.

FIGURE 8-36 Slag, chips from grinding, and dust must be blown out occasionally so that they will not start a fire, or cause a short-out or other types of machine failure.

1. Remove the electrode holder from the welding cable.
2. Remove the jaw insulating covers.
3. Replace the jaw insulating covers.
4. Reconnect the electrode holder to the welding cable.
5. Turn on the welding power or reconnect the welding cable to the welder.
6. Make a weld to ensure that the repair was made correctly.

Complete a copy of the "Student Welding Report" listed in Appendix I or provided by your instructor. ◆

FIGURE 8-37 To prevent people from tripping, when cables must be placed in walkways, lay two blocks of wood beside the cables.

Work Clamps

The work clamp must be the correct size for the current being used, and it must clamp tightly to the material. Heat can build up in the work clamp, reducing welding efficiency, just as was previously described for the electrode holder. Power losses in the work clamp are often overlooked. The clamp should be touched occasionally to find out if it is getting hot.

In addition to power losses due to poor work lead clamping, a loose clamp may cause arcing that can damage a part. If the part is to be moved during welding, a swivel-type work clamp may be needed, **Figure 8-35**.

Setup

Arc welding machines should be located near the welding site, but far enough away so that they are not covered with spark showers. The machines may be stacked to save space, but there must be enough room between the machines to ensure that the air can circulate to keep the machines from overheating. The air that is circulated through the machine should be as free as possible of dust, oil, and metal filings. Even in a good location, the power should be turned off periodically and the machine blown out with compressed air, **Figure 8-36**.

The welding machine should be located away from cleaning tanks and any other sources of corrosive fumes that could be blown through it. Water leaks must be fixed and puddles cleaned up before a machine is used.

Power to the machine must be fused, and a power shut-off switch provided. The switch must be located so that it can be reached in an emergency without touching either the machine or the welding station. The machine case or frame must be grounded.

The welding cables should be sufficiently long to reach the work station but not so long that they must always be coiled. Cables must not be placed on the floor in aisles or walkways. If workers must cross a walkway, the cable must be installed overhead, or it must be protected by a ramp, **Figure 8-37**. The welding machine and its main power switch should be off while a person is installing or working on the cables.

The work station must be free of combustible materials. Screens should be provided to protect other workers from the arc light.

The welding cable should never be wrapped around arms, shoulders, waist, or any other part of the body. If the cable was caught by any moving equipment, such as a forklift, crane, or dolly, a welder could be pulled off balance or more seriously injured. If it is necessary to hold the weight off the cable so that the welding can more easily be done, a free hand can be used. The cable should be held so that if it is pulled it can be easily released.

CAUTION

The cable should never be tied to scaffolding or ladders. If the cable is caught by moving equipment, the scaffolding or ladder may be upset, causing serious personal injury.

Check the surroundings before starting to weld. If heavy materials are being moved in the area around you, there should be a safety watch. A safety watch can warn a person of danger while that person is welding.

Summary

Understanding the scientific theory of electricity and magnetism will aid you in understanding how the welding currents are produced and their reactions to changes in their physical surroundings. Understanding electromagnetic phenomena will aid you in controlling arc blow. Failure to control arc blow can result in load failures. In addition, understanding electricity will help you interpret information given on manufacturers' tables, charts, and equipment specifications.

Before starting to weld, be sure to check with the equipment's manufacturer's safety guidelines for proper operation and maintenance. Follow all recommended guidelines.

Keeping your work area clean and orderly will help prevent accidents.

Review

1. The __________ process is very flexible and the most commonly used process in farms and ranches.
   A. GTAW
   B. OFW
   C. RSW
   D. SMAW
2. An electric current is the flow of __________.
   A. atoms
   B. electrons
   C. molecules
   D. neutrons
3. __________ is the measurement of electrical pressure.
   A. Amperage
   B. Watts
   C. Voltage
   D. Amps

Review continued

4. __________ is calculated by multiplying voltage times amperes.
   A. Horse power
   B. Wattage
   C. Speed in RPM's
   D. Torque
5. The exact temperature of a welding arc depends on the resistance to the current flow.
   A. true
   B. false
6. The three types of currents used for welding are __________, direct-current electrode negative, and direct-current electrode positive.
   A. high amperage
   B. high power
   C. low power
   D. alternating current
7. In direct-current electrode negative, the electrode is positive and the work is negative.
   A. true
   B. false
8. With a constant current welding machine the voltage can __________ during a weld.
   A. stop
   B. change
   C. reverse
   D. not change
9. Open circuit voltage is the voltage at the electrode before striking an arc.
   A. true
   B. false
10. The maximum safe open circuit voltage for welders is __________ volts.
    A. 80
    B. 110
    C. 220
    D. 12
11. What is the term used to describe the voltage of the arc during welding?
    A. stepping voltage
    B. operating voltage
    C. maximum voltage
    D. line voltage
12. When electrons flow, they create lines of magnetic force that circle around the line of flow. These lines are referred to as magnetic flux lines.
    A. true
    B. false
13. A welding __________ uses the alternating current supplied to the shop at a high voltage to produce the low-voltage welding power.
    A. engine generator
    B. motor generator
    C. alternator
    D. transformer
14. A(n) __________ welding machine allows the selection of different current settings by tapping into the secondary coil at a different turn value.
    A. movable coil
    B. multiple-coil
    C. generator
    D. alternator
15. __________ welding machines are adjusted by turning a handwheel that moves the internal parts closer together or farther apart.
    A. Tap-type
    B. Movable coil
    C. Engine generator
    D. Portable
16. Inverter welding machines will be much smaller than other types of machines of the same amperage range.
    A. true
    B. false
17. In a(n) __________, magnetic lines of force rotate inside a coil of wire.
    A. tap-type
    B. transformer
    C. motor
    D. alternator

Review continued

18. An alternator can produce only DC.
    A. true
    B. false
19. A(n) __________allows current to flow in one direction only.
    A. transformer
    B. electrode
    C. rectifier
    D. ground
20. A(n) __________is the percentage of time a welding machine can be used continuously.
    A. duty cycle
    B. size
    C. amperage
    D. watts
21. The terms welding cables and welding leads refer to two different things.
    A. true
    B. false
22. Large heavy welding cable sizes make welding difficult.
    A. true
    B. false

Chapter 9

Shielded Metal Arc Welding of Plate

OBJECTIVES

After completing this chapter, the student should be able to

- ☑ set the welding amperage correctly.
- ☑ explain the effect of changing arc length on a weld.
- ☑ control weld bead contour during welding by using the proper weave pattern.
- ☑ demonstrate an ability to control undercut, overlap, porosity, and slag inclusions when welding.
- ☑ explain the effect of electrode angle on a weld.

KEY TERMS

amperage range
arc length
cellulose-based fluxes
chill plate
electrode angle
lap joint
mineral-based fluxes
rutile-based fluxes
square butt joint
stringer bead
tee joint
weave pattern

INTRODUCTION

Shielded metal arc welding (SMAW) or stick welding is the most commonly used agricultural welding process. Stick welding is so popular for a number of reasons. It can be used to make strong durable welds in a wide range of metal thicknesses and types. Transformer-type welders can set for years without use and still work when needed, and the welding rods have an almost unlimited storage life, as long as they are kept dry. These factors mean that the equipment is there when you need it to make a repair or build a project.

In addition to the standard rods that can be used to make welds in steel 1/8-in. (3 mm) and thicker, there are a wide variety of specialty electrodes. These specialty rods allow you to use the same welding machine to weld on cast iron, stainless steel, and aluminum. There are rods for cutting that will cut cast iron and stainless, metals that cannot be cut with oxygen and acetylene. Wear-resistant and buildup rods can be used to repair scraper and front end loader blades and buckets.

The basic stick-welding skills you will learn in this chapter are used for almost every agricultural welding project, such as replacing the jack on a horse trailer **Figure 9-1**. Welding is a skill that takes practice to perfect. The repetition in the chapter practices is designed to give you the opportunity to develop your skills. The more time you spend welding, the better your welding skills will become.

FIGURE 9-1 Stick welding is an excellent process to be used for most farm or ranch repair work. Courtesy of Larry Jeffus.

PRACTICE 9-1

Shielded Metal Arc Welding Safety

Using a welding work station, welding machine, welding electrodes, welding helmet, eye and ear protection, welding gloves, proper work clothing, and any special protective clothing that may be required, demonstrate, to your instructor and other students, the safe way to prepare yourself and the welding work station for welding. Include in your demonstration appropriate references to burn protection, eye and ear protection, material specification data sheets, ventilation, electrical safety, general work clothing, special protective clothing, and area clean-up.

Complete a copy of the "Student Welding Report" listed in Appendix I or provided by your instructor. ◆

EXPERIMENT 9-1

Striking the Arc

Using a properly set up and adjusted arc welding machine, the proper safety protection, as demonstrated in Practice 9-1, E6011 welding electrodes having a 1/8-in. (3-mm) diameter, and one piece of mild steel plate, 1/4-in. (6-mm) thick, you will practice striking an arc, **Figure 9-2**.

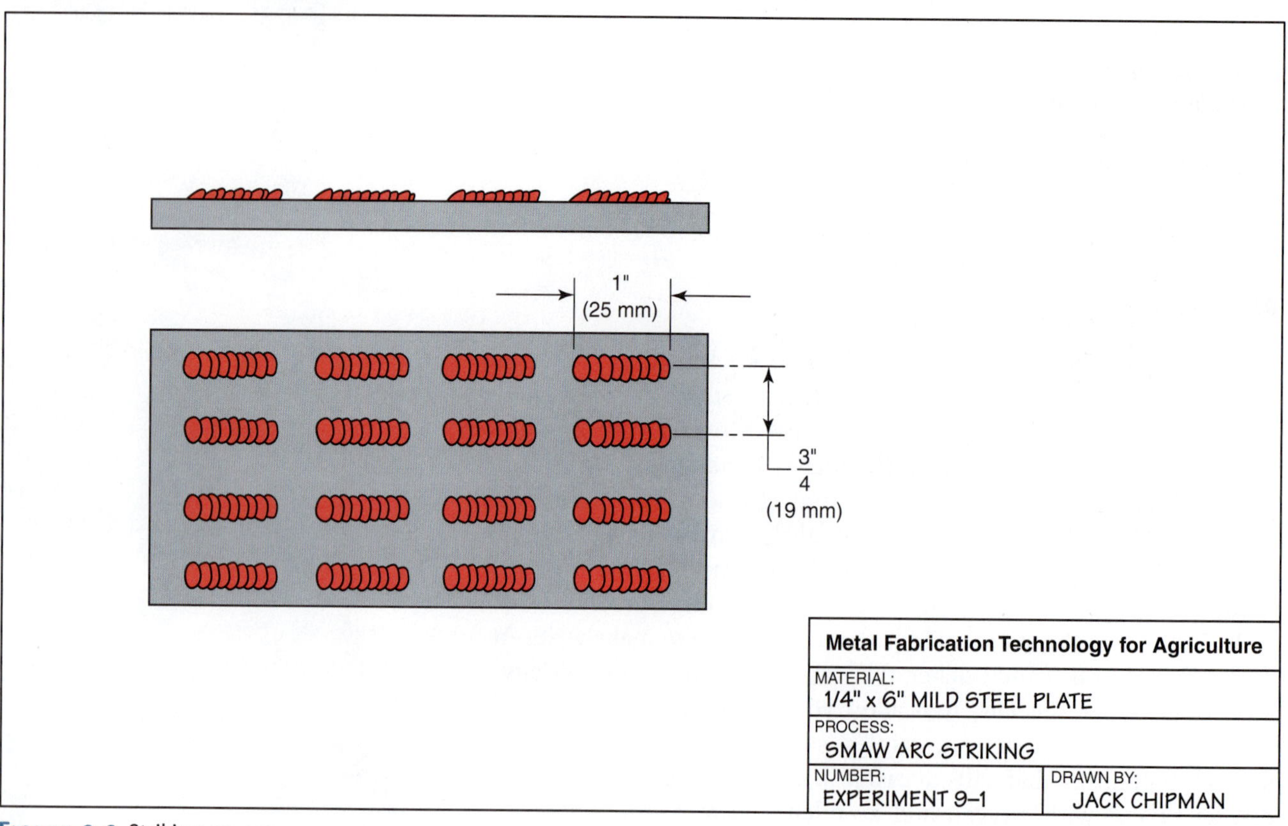

FIGURE 9-2 Striking an arc.

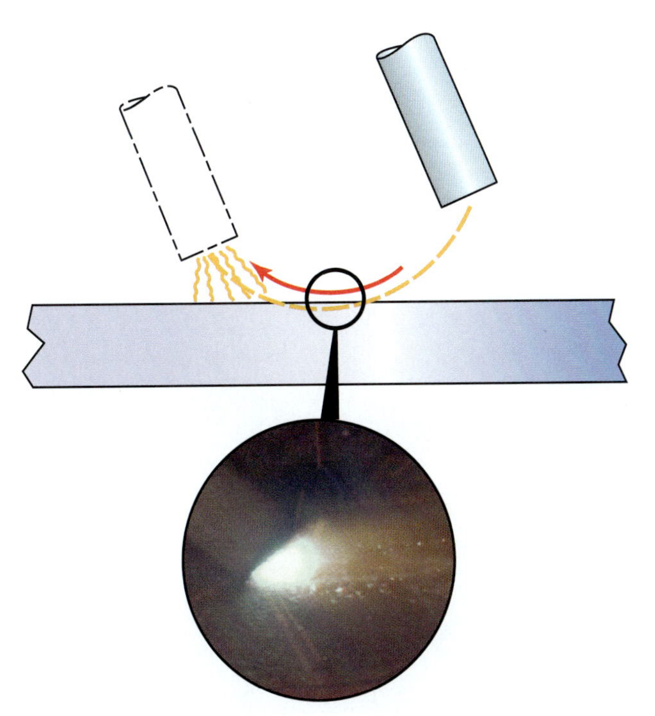

FIGURE 9-3 Striking the arc. Photo courtesy of Larry Jeffus.

FIGURE 9-4 If the flux is broken off the end completely or on one side, the arc can be erratic or forced to the side. Courtesy of Larry Jeffus.

With the electrode held over the plate, lower your helmet. Scratch the electrode across the plate (like striking a large match), **Figure 9-3.** As the arc is established, slightly raise the electrode to the desired arc length. Hold the arc in one place until the molten weld pool builds to the desired size. Slowly lower the electrode as it burns off and move it forward to start the bead.

If the electrode sticks to the plate, quickly squeeze the electrode holder lever to release the electrode. Break the electrode free by bending it back and forth a few times. Do not touch the electrode without gloves, because it will still be hot. If the flux breaks away from the end of the electrode, throw out the electrode because restarting the arc will be very difficult, **Figure 9-4.**

Break the arc by rapidly raising the electrode after completing a 1-in. (25-mm) weld bead. Restart the arc as you did before, and make another short weld. Repeat this process until you can easily start the arc each time. Turn off the welding machine and clean up your work area when you are finished welding.

Complete a copy of the "Student Welding Report" listed in Appendix I or provided by your instructor. ◆

EXPERIMENT 9-2

Striking the Arc Accurately

Using the same materials and setup as described in Experiment 9-1, you will start the arc at a specific spot in order to prevent damage to the surrounding plate.

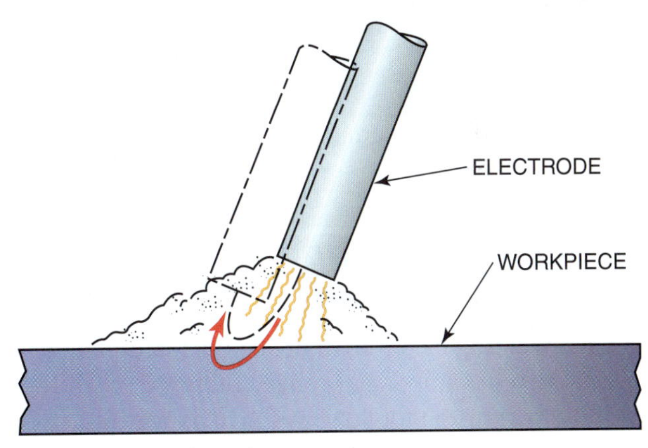

FIGURE 9-5 Striking the arc on a spot.

Hold the electrode over the desired starting point. After lowering your helmet, swiftly bounce the electrode against the plate, **Figure 9-5.** A lot of practice is required to develop the speed and skill needed to prevent the electrode from sticking to the plate.

A more accurate method of starting the arc involves holding the electrode steady by resting it on your free hand like a pool cue. The electrode is rapidly pushed forward so that it strikes the metal exactly where it should. This is an excellent method of striking an arc. Striking an arc in an incorrect spot may cause damage to the base metal.

Practice starting the arc until you can start it within 1/4 in. (6 mm) of the desired location. Turn off the welding machine and clean up your work area when you are finished welding.

Complete a copy of the "Student Welding Report" listed in Appendix I or provided by your instructor. ◆

Electrode	Classification					
Size	E6010	E6011	E6012	E6013	E7016	E7018
3/32 in. (2.4 mm)	40–80	50–70	40–90	40–85	75–105	70–110
1/8 in. (3.2 mm)	70–130	85–125	75–130	70–120	100–150	90–165
5/32 in. (4 mm)	110–165	130–160	120–200	130–160	140–190	125–220

TABLE 9-1 Welding amperage range.

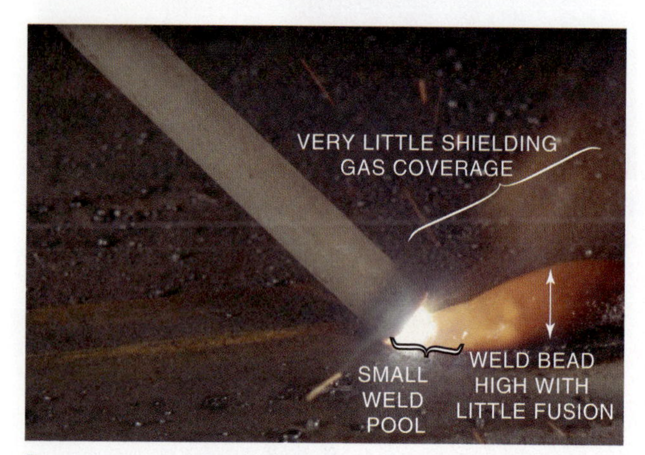

FIGURE 9-6 Welding with the amperage set too low.
Courtesy of Larry Jeffus.

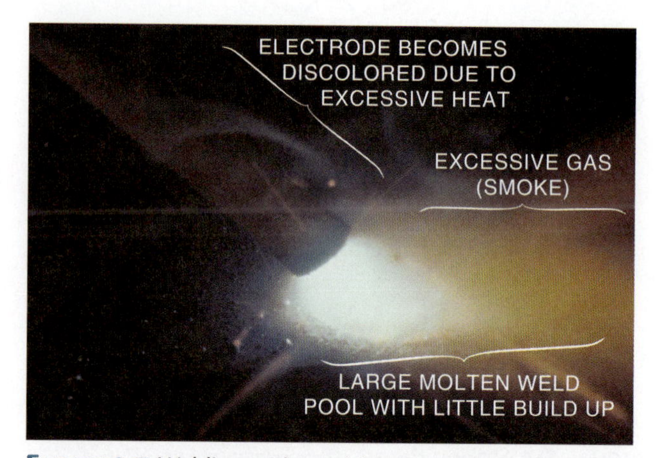

FIGURE 9-7 Welding with too high an amperage.
Courtesy of Larry Jeffus.

Effect of Too High or Too Low Current Settings

Each welding electrode must be operated in a particular current (amperage) range, **Table 9-1**. Welding with the current set too low results in poor fusion and poor arc stability, **Figure 9-6**. The weld may have slag or gas inclusions because the molten weld pool was not fluid long enough for the flux to react. Little or no penetration of the weld into the base plate may also be evident. With the current set too low, the arc length is very short. A very short arc length results in frequent shortening and sticking of the electrode.

The core wire of the welding electrode is limited in the amount of current it can carry. As the current is increased, the wire heats up because of electrical resistance. This preheating of the wire causes some of the chemicals in the covering to be burned out too early, **Figure 9-7**. The loss of the proper balance of elements causes poor arc stability. This condition leads to spatter, porosity, and slag inclusions.

An increase in the amount of spatter is also caused by a longer arc. The weld bead made at a high amperage setting is wide and flat with deep penetration. The spatter is excessive and is mostly hard. The spatter is called hard because it fuses to the base plate and is difficult to remove, **Figure 9-8**. The electrode covering is discolored more than 1/8 in. (3 mm) to 1/4 in. (6 mm) from the end of the electrode. Extremely high settings may also cause the electrode to discolor, crack, glow red, or burn.

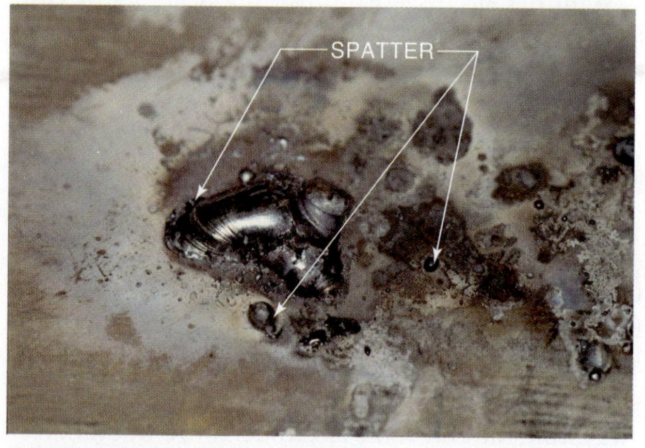

FIGURE 9-8 Hard weld spatter fused to base metal.
Courtesy of Larry Jeffus.

EXPERIMENT 9-3

Effects of Amperage Changes on a Weld Bead

For this experiment, you will need an arc welding machine, welding gloves, safety glasses, welding helmet, appropriate clothing, E6011 welding electrodes having a 1/8-in. (3-mm) diameter, and one piece of mild steel plate, 1/4 in. (6 mm) to 1/2 in. (13 mm) thick. You will observe what happens to the weld bead when the amperage settings are raised and lowered.

Starting with the machine set at approximately 90 A AC or DCRP, strike an arc and make a weld 1 in. (25 mm) long. Break the arc. Raise the current setting by 10 A, strike an arc, and make another weld

1 in. (25 mm) long. Repeat this procedure until the machine amperage is set at the maximum value.

Replace the electrode and reset the machine to 90 A. Make a weld 1 in. (25 mm) long. Stop and lower the current setting by 10 A. Repeat this procedure until the machine amperage is set at a minimum value.

Cool and chip the plate, comparing the different welds for width, buildup, molten weld pool size, spatter, slag removal, and penetration, **Figure 9-9 (A)** and **(B)**. In addition, compare the electrode stubs. Turn off the welding machine and clean up your work area when you are finished welding.

Complete a copy of the "Student Welding Report" listed in Appendix I or provided by your instructor. ◆

CAUTION

Do not change the current settings during welding. A change in the setting may cause arcing inside the machine, resulting in damage to the machine.

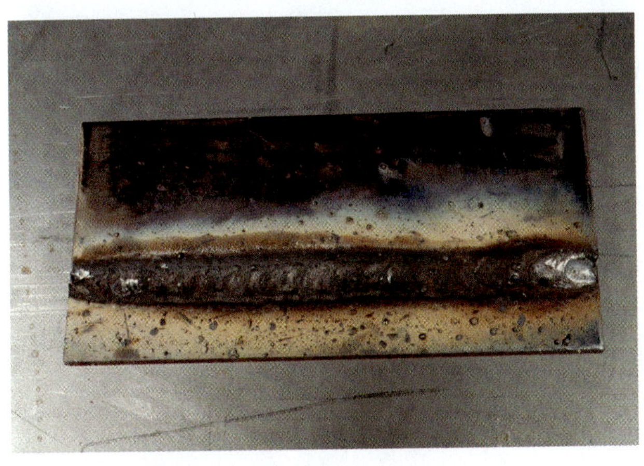

(A) WELD BEFORE CLEANING

(B) WELD AFTER CLEANING

FIGURE 9-9 Weld before cleaning and after cleaning.
Courtesy of Larry Jeffus.

Electrode Size and Heat

The selection of the correct size of welding electrode for a weld is determined by the skill of the welder, the thickness of the metal to be welded, and the size of the metal. The 1/8-in. (3 mm) electrode is the most commonly used size for agriculture welding. It can be used to make welds on thin metal up to thick plates. Using small diameter electrodes requires less skill than using large diameter electrodes. The deposition rate, or the rate that weld metal is added to the weld, is slower when small diameter electrodes are used. Small diameter electrodes will make acceptable welds on thick plate, but more time is required to make the weld.

Large diameter electrodes may overheat the metal if they are used with thin or small pieces of metal. To determine if a weld is too hot, watch the shape of the trailing edge of the molten weld pool, **Figure 9-10**. Rounded ripples indicate the weld is cooling uniformly and that the heat is not excessive. If the ripples are pointed, the weld is cooling too slowly because of excessive heat. Extreme overheating can cause a burn-through, which is hard to repair.

To correct an overheating problem, a welder can turn down the amperage, use a shorter arc, travel at a faster rate, use a chill plate (a large piece of metal used to absorb excessive heat), or use a smaller electrode at a lower current setting.

EXPERIMENT 9-4

Excessive Heat

Using a properly set up and adjusted arc welding machine, the proper safety protection, E6011 welding electrodes having a 1/8-in. (3-mm) diameter, and three pieces of mild steel plate, 1/8 in. (3 mm), 3/16 in. (4.8 mm), and 1/4 in. (6 mm) thick, you will observe the effects of overheating on the weld.

AMOUNT OF HEAT DIRECTED AT WELD	WELD POOL
TOO LOW	
CORRECT	
TOO HOT	

FIGURE 9-10 The effect on the shape of the molten weld pool caused by the heat input.

Make a stringer weld on each of the three plates using the same amperage setting, travel rate, and arc length for each weld. Cool and chip the welds. Then compare the weld beads for width, reinforcement, and appearance.

Using the same amperage settings, make additional welds on the 1/8 in. (3 mm) and 3/16 in. (4.8-mm) plates. Vary the arc lengths and travel speeds for these welds. Cool and chip each weld and compare the beads for width, reinforcement, and appearance. Make additional welds on the 1/8 in. (3 mm) and 3/16 in. (4.8-mm) plates, using the same arc length and travel speed as in the earlier part of this experiment, but at a lower amperage setting. Cool and chip the welds and compare the beads for width, reinforcement, and appearance.

The plates should be cooled between each weld so that the heat from the previous weld does not affect the test results. Turn off the welding machine and clean up your work area when you are finished welding.

Complete a copy of the "Student Welding Report" listed in Appendix I or provided by your instructor. ◆

Arc Length

The **arc length** is the distance the arc must jump from the end of the electrode to the plate or weld pool surface. As the weld progresses, the electrode becomes shorter as it is consumed. To maintain a constant arc length, the electrode must be lowered continuously. Maintaining a constant arc length is important, as too great a change in the arc length will adversely affect the weld.

As the arc length is shortened, metal transferring across the gap may short out the electrode, causing it to stick to the plate. The weld that results is narrow and has a high buildup, **Figure 9-11**.

FIGURE 9-11 Welding with too short an arc length.
Courtesy of Larry Jeffus.

Long arc lengths produce more spatter because the metal being transferred may drop outside of the molten weld pool. The weld is wider and has little buildup, **Figure 9-12**.

There is a narrow range for the arc length in which it is stable, metal transfer is smooth, and the bead shape is controlled. Factors affecting the length are the type of electrode, joint design, metal thickness, and current setting.

Some welding electrodes, such as E7024, have a thick flux covering. The rate at which the covering melts is slow enough to permit the electrode coating to be rested against the plate. The arc burns back inside the covering as the electrode is dragged along touching the joint, **Figure 9-13**. For this type of welding electrode, the arc length is maintained by the electrode covering. E7024 electrodes require very little welding skill to use. Because of the size of the molten weld pool, they are usually only used in the flat position on thick metal.

An arc will jump to the closest metal conductor. On joints that are deep or narrow, the arc is pulled to

FIGURE 9-12 Welding with too long an arc length.
Courtesy of Larry Jeffus.

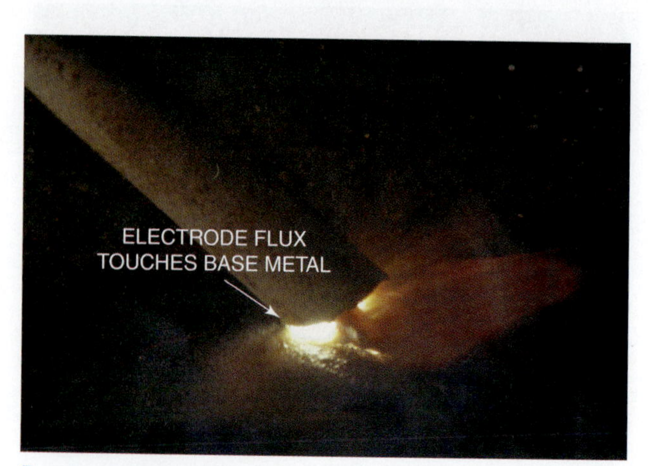

FIGURE 9-13 Welding with a drag technique.
Courtesy of Larry Jeffus.

one side and not to the root, **Figure 9-14**. As a result, the root fusion is reduced or may be nonexistent, thus causing a poor weld. If a very short arc is used, the arc is forced into the root for better fusion.

Because shorter arcs produce less heat and penetration, they are best suited for use on thin metal or thin-to-thick metal joints. Using this technique, metal as thin as 16 gauge can be arc welded easily. Higher amperage settings are required to maintain a short arc that gives good fusion with a minimum of slag inclusions. The higher settings, however, must be within the amperage range for the specific electrode.

Finding the correct arc length often requires some trial and adjustment. Most welding jobs require an arc length of 1/8 in. (3 mm) to 3/8 in. (10 mm), but this distance varies. It may be necessary to change the arc length when welding to adjust for varying welding conditions.

EXPERIMENT 9-5

Effect of Changing the Arc Length on a Weld

Using an arc welding machine, welding gloves, safety glasses, welding helmet, appropriate clothing, E6011 welding electrodes having a 1/8-in. (3-mm) diameter, and one piece of mild steel plate, 1/4 in. (6 mm) to 1/2 in. (13 mm) thick, you will observe the effect of changing the arc length on a weld.

Starting with the welding machine set at approximately 90 A AC or DCRP, strike an arc and make a weld 1 in. (25 mm) long. Continue welding while slowly increasing the arc length until the arc is broken. Restart the arc and make another weld 1 in. (25 mm) long. Welding should again be continued while slowly shortening the arc length until the arc stops. Quickly break the electrode free from the plate, or release the electrode by squeezing the lever on the electrode holder.

Cool and chip both welds. Compare the welding beads for width, reinforcement, uniformity, spatter, and appearance. Turn off the welding machine and clean up your work area when you are finished welding.

Complete a copy of the "Student Welding Report" listed in Appendix I or provided by your instructor. ◆

Electrode Angle

The electrode angle is measured from the electrode to the surface of the metal. The term used to identify the electrode angle is affected by the direction of travel, generally leading or trailing, **Figure 9-15**. The relative angle is important because there is a jetting force blowing the metal and flux from the end of the electrode to the plate.

Leading Angle A leading electrode angle pushes molten metal and slag ahead of the weld, **Figure 9-16**. When welding in the flat position, caution must

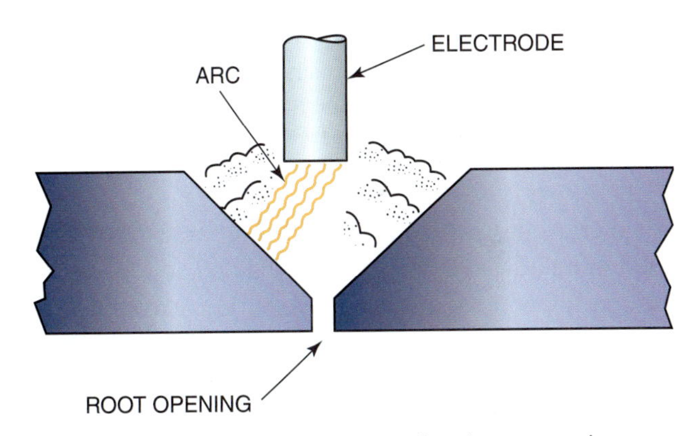

FIGURE 9-14 The arc may jump to the closest metal, reducing root penetration.

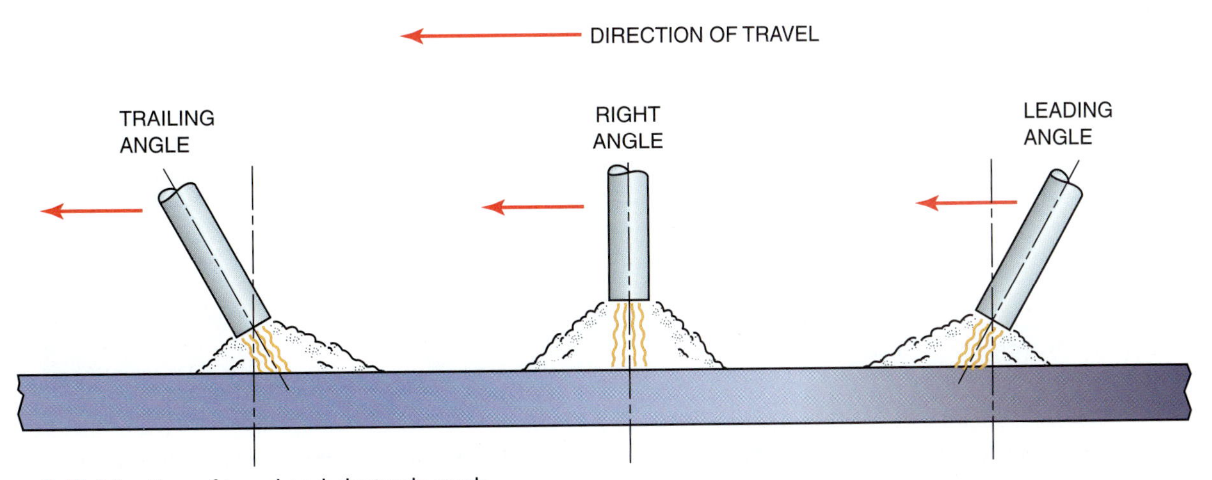

FIGURE 9-15 Direction of travel and electrode angle.

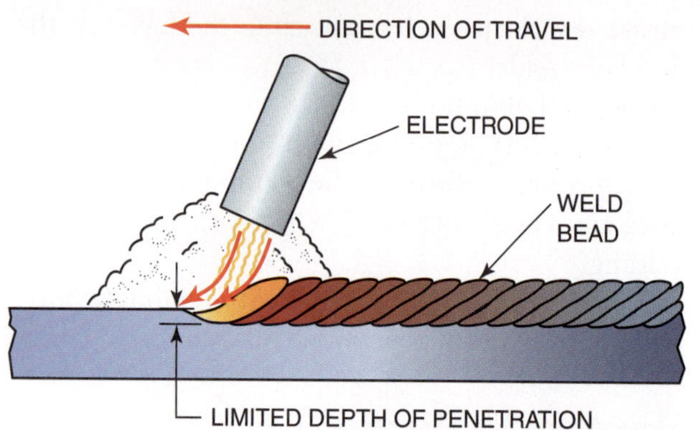

FIGURE 9-16 Leading, lag, or pushing electrode angle.

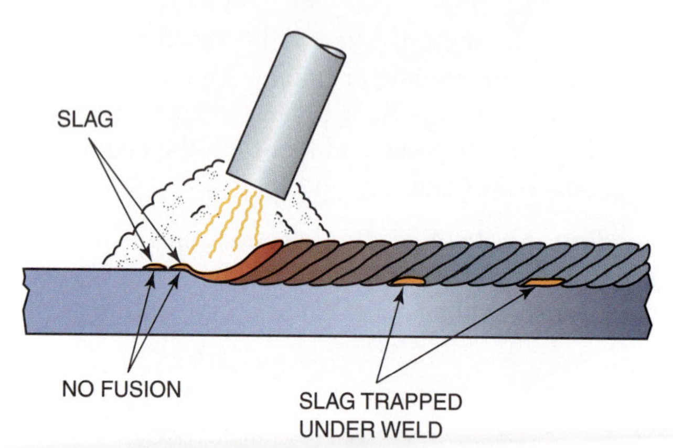

FIGURE 9-17 Some electrodes, such as E7018, may not remove the deposits ahead of the molten weld pool, resulting in discontinuities within the weld.

be taken to prevent cold lap and slag inclusions. The solid metal ahead of the weld cools and solidifies the molten filler metal and slag before they can melt the solid metal. This rapid cooling prevents the metals from fusing together, **Figure 9-17**. As the weld passes

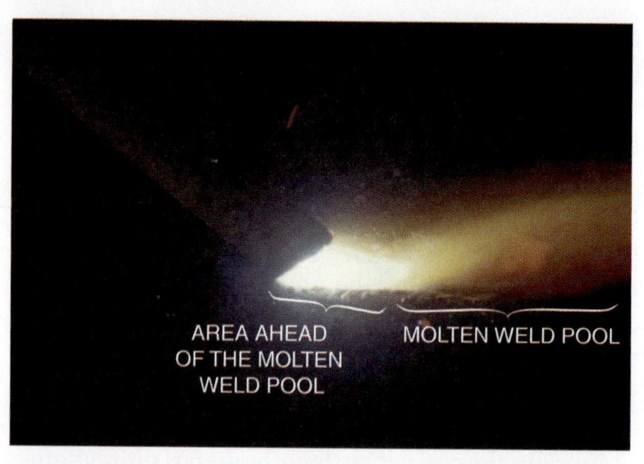

FIGURE 9-18 Metal being melted ahead of the molten weld pool helps to ensure good weld fusion. Courtesy of Larry Jeffus.

over this area, heat from the arc may not melt it. As a result, some cold lap and slag inclusions are left.

The following are suggestions for preventing cold lap and slag inclusions:

- Use as little leading angle as possible.
- Ensure that the arc melts the base metal completely, **Figure 9-18**.
- Use a penetrating-type electrode that causes little buildup.
- Move the arc back and forth across the molten weld pool to fuse both edges.

A leading angle can be used to minimize penetration or to help hold molten metal in place for vertical welds, **Figure 9-19**.

Trailing Angle A trailing electrode angle pushes the molten metal away from the leading edge of the molten weld pool toward the back where it solidifies,

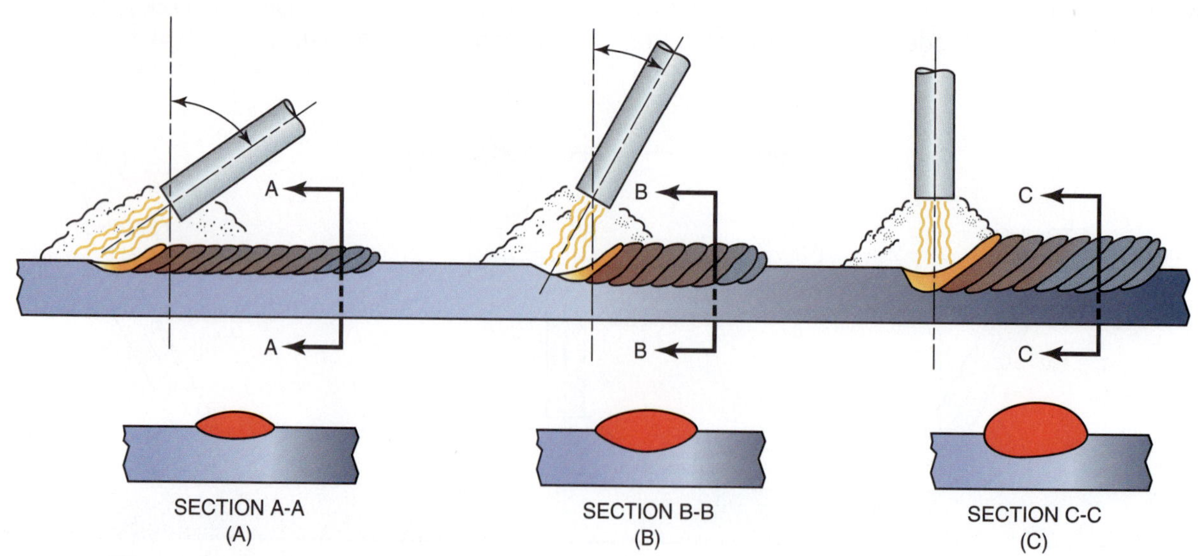

FIGURE 9-19 Effect of a leading angle on weld bead buildup, width, and penetration. As the angle increases toward the vertical position (C), penetration increases.

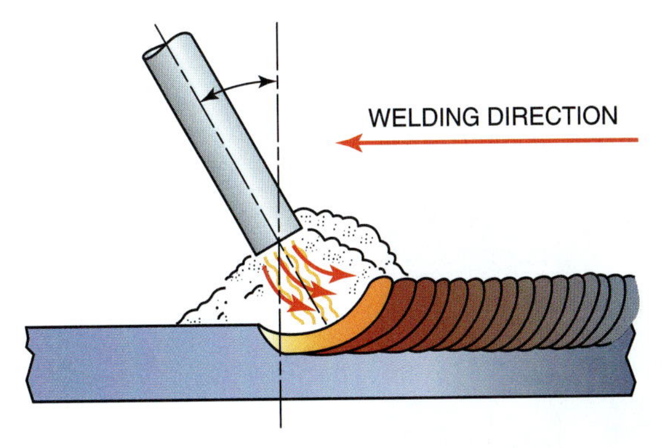

FIGURE 9-20 Trailing electrode angle.

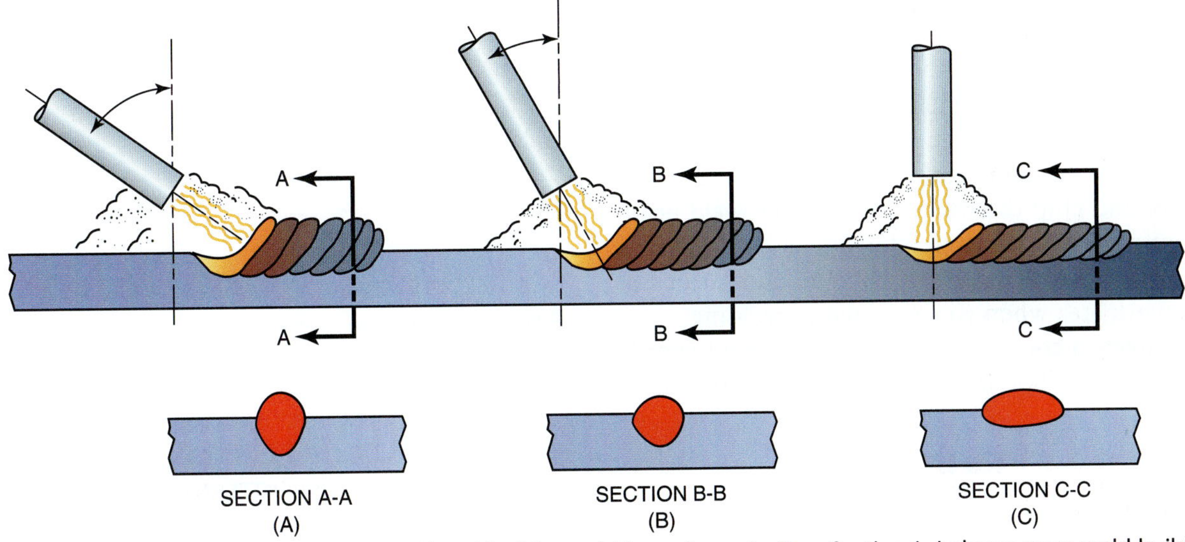

FIGURE 9-21 Effect of a trailing angle on weld bead buildup, width, and penetration. Section A-A shows more weld buildup due to a greater angle of the electrode.

Figure 9-20. As the molten metal is forced away from the bottom of the weld, the arc melts more of the base metal, which results in deeper penetration. The molten metal pushed to the back of the weld solidifies and forms reinforcement for the weld, **Figure 9-21**.

EXPERIMENT 9-6

Effect of Changing the Electrode Angle on a Weld

Using a properly set up and adjusted arc welding machine, the proper safety protection, E6011 welding electrodes having a 1/8-in. (3-mm) diameter, and one piece of mild steel plate, 1/4 in. (6 mm) to 1/2 in. (13 mm) thick, you will observe the effect of changes in the electrode angle on a weld.

Start welding with a sharp trailing angle. Make a weld about 1 in. (25 mm) long. Closely observe the molten weld pool at the points shown in **Figure 9-22**. Slowly increase the electrode angle and continue to observe the weld.

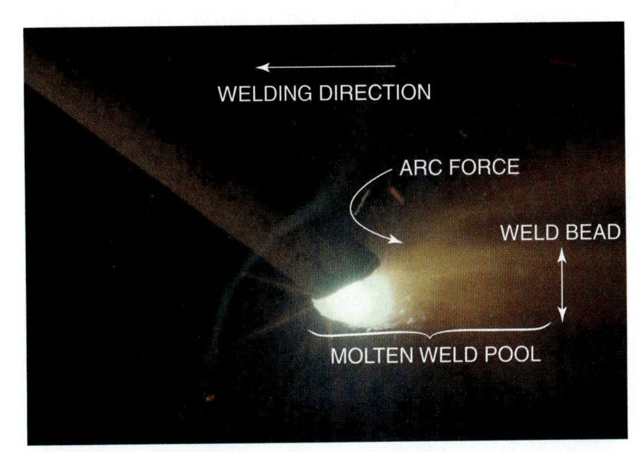

FIGURE 9-22 Welding with a trailing angle. Courtesy of Larry Jeffus.

When you reach a 90° electrode angle, make a weld about 1 in. (25 mm) long. Observe the parts of the weld molten weld pool as shown in **Figure 9-22**.

Continue welding and change the electrode angle to a sharp leading angle. Observe the weld molten weld pool at the points shown in **Figure 9-23**.

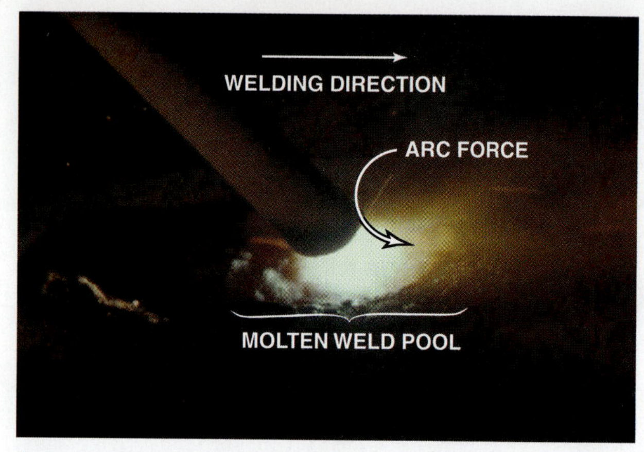

FIGURE 9-23 Welding with a leading angle.
Courtesy of Larry Jeffus.

During this experiment, you must maintain a constant arc length, travel speed, and weave pattern if the observations and results are to be accurate.

Cool and chip the weld. Compare the weld bead for uniformity in width, reinforcement, and appearance. Turn off the welding machine and clean up your work area when you are finished welding.

Complete a copy of the "Student Welding Report" listed in Appendix I or provided by your instructor. ◆

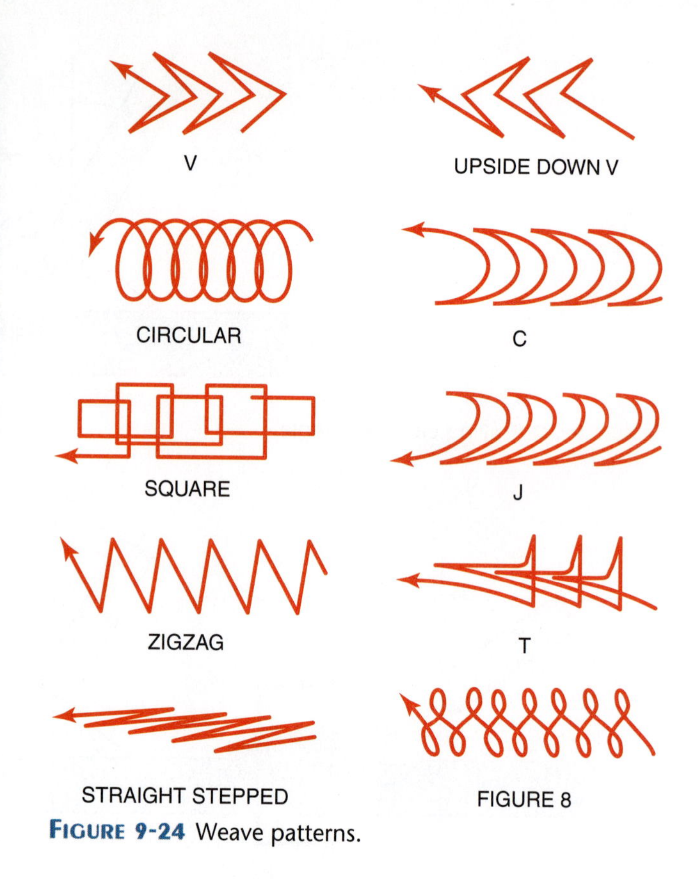

FIGURE 9-24 Weave patterns.

Electrode Manipulation

The movement or weaving of the welding electrode can control the following characteristics of the weld bead: penetration, buildup, width, porosity, undercut, overlap, and slag inclusions. The exact **weave pattern** for each weld is often the personal choice of the welder. However, some patterns are especially helpful for specific welding situations. The pattern selected for a flat (1G) butt joint is not as critical as is the pattern selection for other joints and other positions.

Many weave patterns are available for the welder to use. **Figure 9-24** shows ten different patterns that can be used for most welding conditions.

The circular pattern is often used for flat position welds on butt, tee, and outside corner joints, and for buildup or surfacing applications. The circle can be made wider or longer to change the bead width or penetration, **Figure 9-25**.

The "C" and square patterns are both good for most 1G (flat) welds, but can also be used for vertical (3G) positions. These patterns can also be used if there is a large gap to be filled when both pieces of metal are nearly the same size and thickness.

The "J" pattern works well on flat (1F) lap joints, all vertical (3G) joints, and horizontal (2G) butt and lap (2F) welds. This pattern allows the heat to be concentrated on the thicker plate, **Figure 9-26**. It also allows the reinforcement to be built up on the metal deposited during the first part of the pattern. As a result, a uniform bead contour is maintained during out-of-position welds.

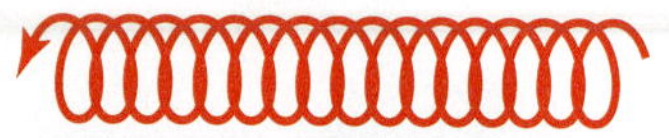
THIS WEAVE PATTERN RESULTS IN A NARROW BEAD WITH DEEP PENETRATION.

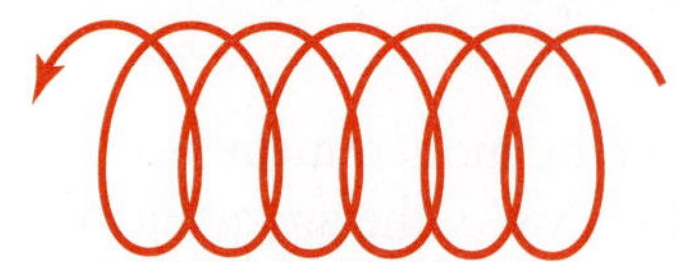
THIS WEAVE PATTERN RESULTS IN A WIDE BEAD WITH SHALLOW PENETRATION.

FIGURE 9-25 Changing the weave pattern width to change the weld bead characteristics.

The "T" pattern works well with fillet welds in the vertical (3F) and overhead (4F) positions, **Figure 9-27**. It also can be used for deep groove welds for the hot pass. The top of the "T" can be used to fill in the toe of the weld to prevent undercutting.

The straight step pattern can be used for stringer beads, root pass welds, and multiple pass welds in all positions. For this pattern, the smallest quantity of

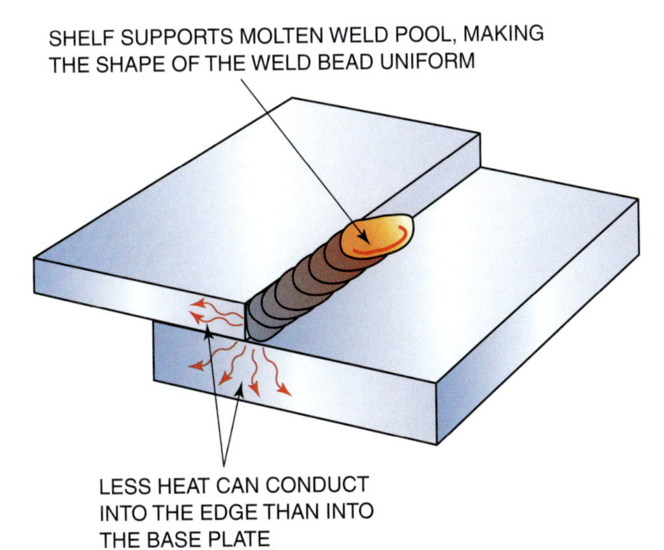

FIGURE 9-26 The "J" pattern allows the heat to be concentrated on the thicker plate.

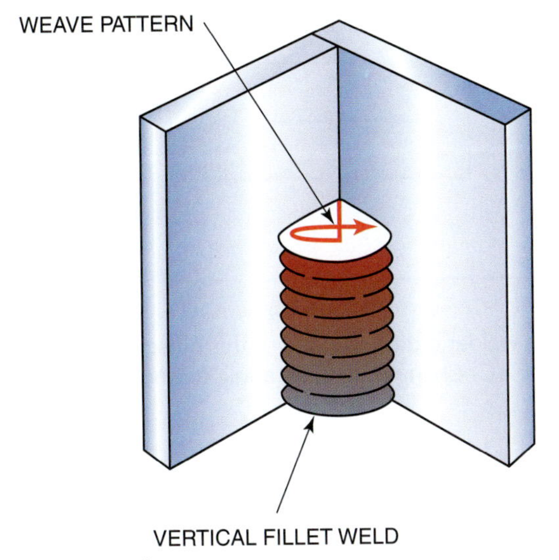

FIGURE 9-27 "T" pattern.

metal is molten at one time as compared to other patterns. Therefore, the weld is more easily controlled. At the same time that the electrode is stepped forward, the arc length is increased so that no metal is deposited ahead of the molten weld pool, **Figure 9-28** and **Figure 9-29**. This action allows the molten weld pool to cool to a controllable size. In addition, the arc burns off any paint, oil, or dirt from the metal before it can contaminate the weld.

The figure 8 pattern and the zigzag pattern are used as cover passes in the flat and vertical positions. Do not weave more than 2 1/2 times the width of the electrode. These patterns deposit a large quantity of metal at one time. A shelf can be used to support the molten weld pool when making vertical welds using either of these patterns, **Figure 9-30**.

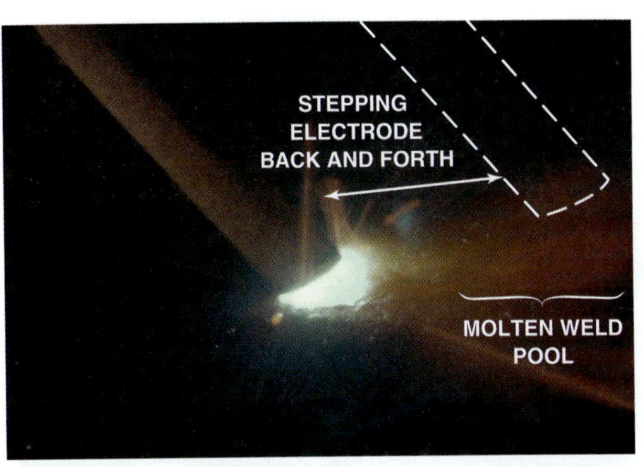

FIGURE 9-28 The electrode is moved slightly forward and then returned to the weld pool. Courtesy of Larry Jeffus.

FIGURE 9-29 The electrode does not deposit metal or melt the base metal. Courtesy of Larry Jeffus.

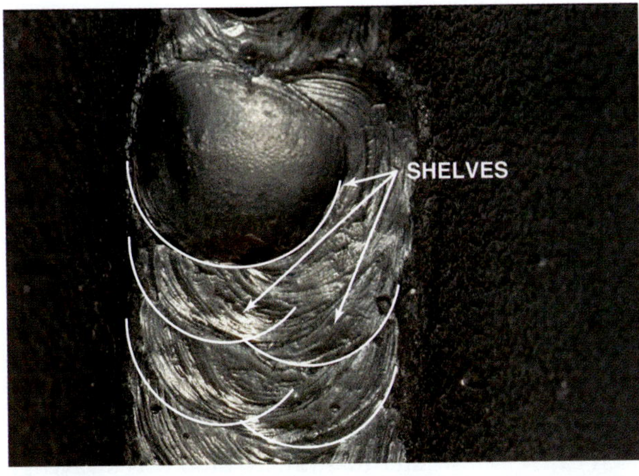

FIGURE 9-30 Using the shelf to support the molten pool for vertical welds. Courtesy of Larry Jeffus.

Positioning of the Welder and the Plate

The welder should be in a relaxed, comfortable position before starting to weld. A good position is important for both the comfort of the welder and the

(A)

(B)

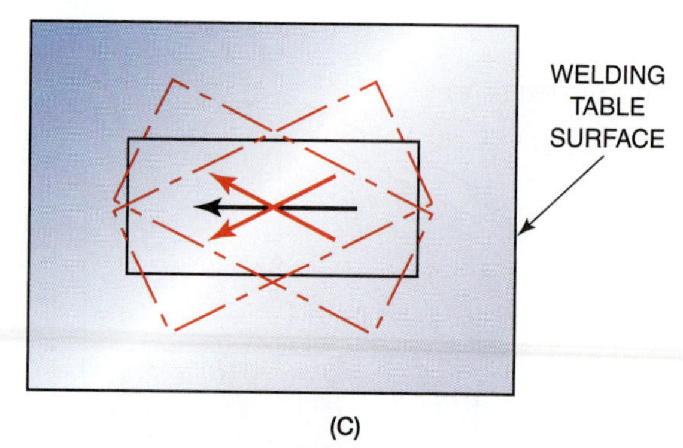

(C)

FIGURE 9-31 (A) Brace yourself by leaning up against something like the welder did on the side of this horse trailer, or (B) lean your arm over like this ranch hand has done while building his fence. (C) Change the plate angle to find the most comfortable welding position. Photos courtesy of Larry Jeffus.

quality of the welds. Welding in an awkward position can cause welder fatigue, which leads to poor welder coordination and poor-quality welds. Welders must have enough freedom of movement so that they do not need to change position during a weld. Body position changes should be made only during electrode changes.

When the welding helmet is down, the welder is blind to the surroundings. Due to the arc, the field of vision of the welder is also very limited. These factors often cause the welder to sway. To stop this swaying, the welder should lean against or hold on to a stable object. When welding, even if a welder is seated, touching a stable object will make that welder more stable and will make welding more relaxing.

Welding is easier if the welder can find the most comfortable angle. The welder should be in either a seated or a standing position in front of the welding table. The welding machine should be turned off. With an electrode in place in the electrode holder, the welder can draw a straight line along the plate to be welded. By turning the plate to several different angles, the welder should be able to determine which angle is most comfortable for welding, **Figure 9-31A**, **Figure 9-31B**, and **Figure 9-31C**.

Practice Welds

Practice welds are grouped according to the type of joint and the type of welding electrode. The welder or instructor should select the order in which the welds are made. The stringer beads should be practiced first in each position before the welder tries the different joints in each position. Some time can be saved by starting with the stringer beads. If this is done, it is not necessary to cut or tack the plate together, and a number of beads can be made on the same plate.

Students will find it easier to start with butt joints. The lap, tee, and outside corner joints are all about the same level of difficulty.

Starting with the flat position allows the welder to build skills slowly, so that out-of-position welds become easier to do. The horizontal tee and lap welds are almost as easy to make as the flat welds. Overhead welds are as simple to make as vertical welds, but they are harder to position. Horizontal butt welds are more difficult to perform than most other welds.

Electrodes Arc welding electrodes used for practice welds are grouped into three filler metal (F number) classes according to their major welding characteristics. The groups are E6010 and E6011, E6012 and E6013, and E7016 and E7018.

F3 E6010 and E6011 Electrodes

Both of these electrodes have cellulose-based fluxes. As a result, these electrodes have a forceful arc with little slag left on the weld bead. E6010 and E6011 are the most utilitarian welding electrodes for use around the farm or ranch. They can be used on metal that has a little rust, oil, or dirt without seriously affecting the weld's strength. The E6010 electrodes can only weld with direct current (DC) welding machines. Because E6011 electrodes can be used with AC, small agricultural transformer-type welders that put out only AC welding current can be used.

F2 E6012 and E6013 Electrodes

These electrodes have rutile-based fluxes, giving a smooth, easy arc with a thick slag left on the weld bead. Both E6012 and E6013 are easy electrodes to use. They do not have forceful arcs, so they can be used on thinner metals such as some thicker sheet metal gauges that are used as guards on equipment or mower decks.

F4 E7016 and E7018 Electrodes

Both of these electrodes have a mineral-based flux. The resulting arc is smooth and easy, with a very heavy slag left on the weld bead. Of these two electrodes, E718 is the one used most often to make high-strength welds on agriculture equipment. Store these electrodes in a dry place. If they get wet, they will still weld, but the welds will not be as strong.

The cellulose- and rutile-based groups of electrodes have characteristics that make them the best electrodes for starting specific welds. The electrodes with the cellulose-based fluxes do not have heavy slags that may interfere with the welder's view of the weld. This feature is an advantage for flat tee and lap joints. Electrodes with the rutile-based fluxes (giving an easy arc with low spatter) are easier to control and are used for flat stringer beads and butt joints.

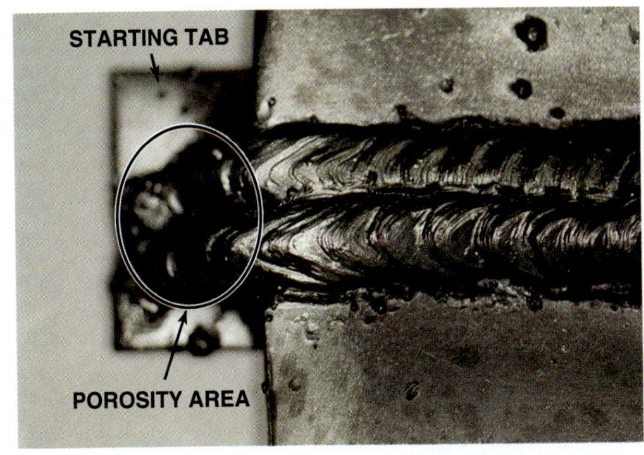

FIGURE 9-32 Porosity is found on the starting tab where it will not affect the weld. Courtesy of Larry Jeffus.

FIGURE 9-33 Stringer bead. Courtesy of Larry Jeffus.

Unless a specific electrode is required for a job, welders can select what they consider to be the best electrode for a specific weld. Welders often have favorite electrodes to use on specific jobs.

Electrodes with mineral-based fluxes should be the last choice. Welds with a good appearance are more easily made with these electrodes, but strong welds are hard to obtain. Without special care being taken during the start of the weld, porosity will be formed in the weld. **Figure 9-32** shows a starting tab used to prevent this porosity from becoming part of the finished weld. More information on electrode selection can be found in Chapter 17.

Stringer Beads

A straight weld bead on the surface of a plate, with little or no side-to-side electrode movement, is known as a stringer bead. Stringer beads are used by students to practice maintaining arc length, weave patterns, and electrode angle so that their welds will be straight, uniform, and free from defects. Stringer beads, **Figure 9-33**, are also used to set the machine amperage and for buildup or surfacing applications.

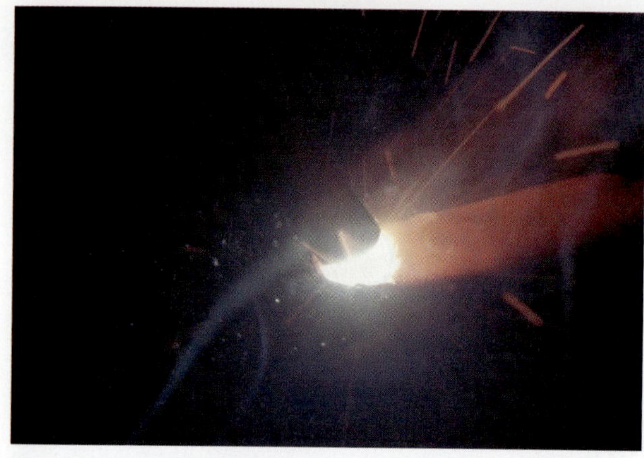

FIGURE 9-34 New welders frequently see only the arc and sparks from the electrode. Courtesy of Larry Jeffus.

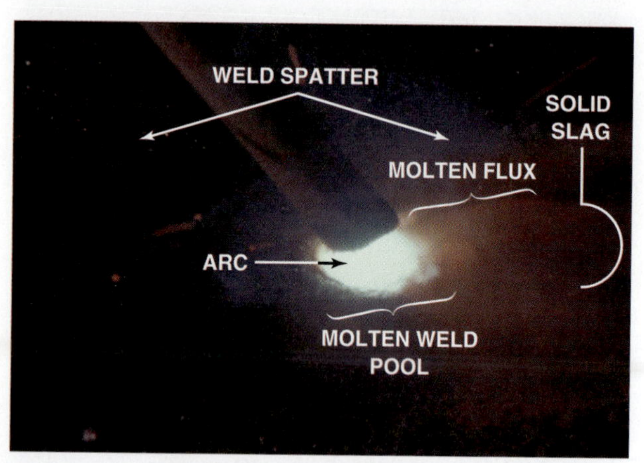

FIGURE 9-35 More experienced welders can see the molten pool, metal being transferred across the arc, and penetration into the base metal. Courtesy of Larry Jeffus.

The stringer bead should be straight. A beginning welder needs time to develop the skill of viewing the entire welding area. At first, the welder sees only the arc, **Figure 9-34**. With practice, the welder begins to see parts of the molten weld pool. After much practice, the welder will see the molten weld pool (front, back, and both sides), slag, buildup, and the surrounding plate, **Figure 9-35**. Often, at this skill level, the welder may not even notice the arc.

A straight weld is easily made once the welder develops the ability to view the entire welding zone. The welder will occasionally glance around to ensure that the weld is straight. In addition, it can be noted if the weld is uniform and free from defects. The ability of the welder to view the entire weld area is demonstrated by making consistently straight and uniform stringer beads.

After making practice stringer beads, a variety of weave bead patterns should be practiced to gain the ability to control the molten weld pool when welding out-of-position.

PRACTICE 9-2

Straight Stringer Beads in the Flat Position Using E6010 or E6011 Electrodes, E6012 or E6013 Electrodes, and E7016 or E7018 Electrodes

Using a properly set up and adjusted arc welding machine, proper safety protection, as demonstrated in Practice 9-1, arc welding electrodes with a 1/8-in. (3mm) diameter, and one piece of mild steel plate, 6 in. (152 mm) long × 1/4 in. (6 mm) thick, you will make straight stringer beads.

- Starting at one end of the plate, make a straight weld the full length of the plate.
- Watch the molten weld pool at this point, not the end of the electrode. As you become more skillful, it is easier to watch the molten weld pool.
- Repeat the beads with all three (F) groups of electrodes until you have consistently good beads.
- Cool, chip, and inspect the bead for defects after completing it. Turn off the welding machine and clean up your work area when you are finished welding.

Complete a copy of the "Student Welding Report" listed in Appendix I or provided by your instructor. ◆

PRACTICE 9-3

Stringer Beads in the Vertical Up Position Using E6010 or E6011 Electrodes, E6012 or E6013 Electrodes, and E7016 or E7018 Electrodes

Using the same setup, materials, and electrodes as listed in Practice 9-2, you will make vertical up stringer beads. Start with the plate at a 45° angle.

This technique is the same as that used to make a vertical weld. However, a lower level of skill is required at 45°, and it is easier to develop your skill. After the welder masters the 45° angle, the angle is increased successively until a vertical position is reached, **Figure 9-36**.

Before the molten metal drips down the bead, the back of the molten weld pool will start to bulge, **Figure 9-37**. When this happens, increase the speed of travel and the weave pattern.

Cool, chip, and inspect each completed weld for defects. Repeat the beads as necessary with all three (F) groups of electrodes until consistently good beads are obtained in this position. Turn off the welding

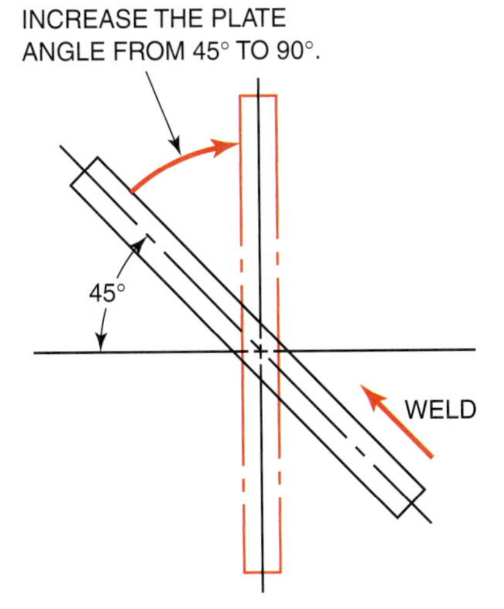

FIGURE 9-36 Once the 45° angle is mastered, the plate angle is increased successively until a vertical position (90°) is reached.

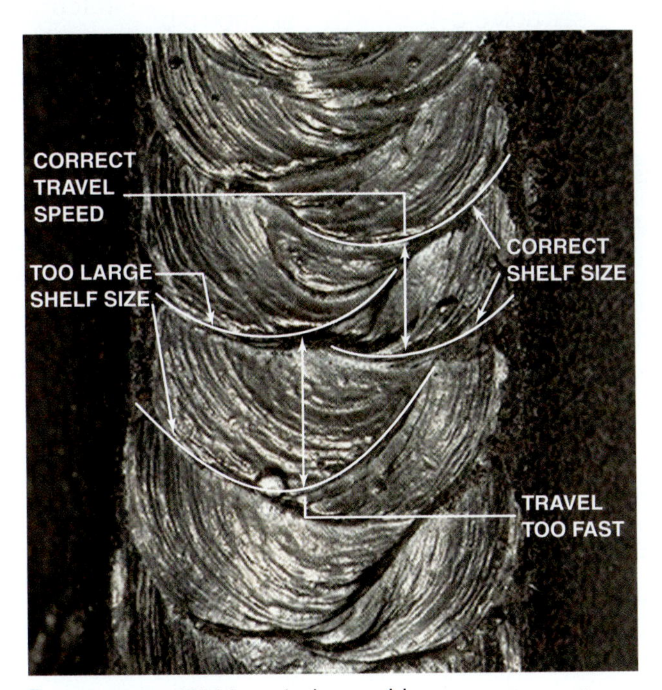

FIGURE 9-37 E7018 vertical up weld. Courtesy of Larry Jeffus.

machine and clean up your work area when you are finished welding.

Complete a copy of the "Student Welding Report" listed in Appendix I or provided by your instructor. ◆

PRACTICE 9-4

Horizontal Stringer Beads Using E6010 or E6011 Electrodes, E6012 or E6013 Electrodes, and E7016 or E7018 Electrodes

Using the same setup, materials, and electrodes as listed in Practice 9-2, you will make horizontal stringer beads on a plate.

When the welder begins to practice the horizontal stringer bead, the plate may be reclined slightly,

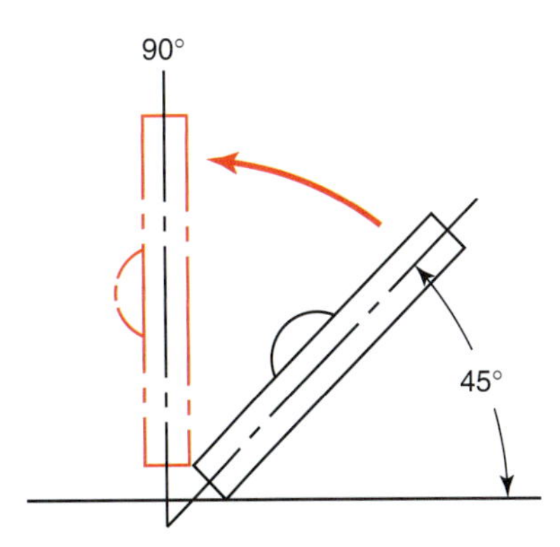

FIGURE 9-38 Change the plate angle as welding skill improves.

Figure 9-38. This placement allows the welder to build the required skill by practicing the correct techniques successfully. The "J" weave pattern is suggested for this practice. As the electrode is drawn along the straight back of the "J," metal is deposited. This metal supports the molten weld pool, resulting in a bead with a uniform contour, **Figure 9-39.**

Angling the electrode up and back toward the weld causes more metal to be deposited along the top edge of the weld. Keeping the bead small allows the surface tension to hold the molten weld pool in place.

Gradually increase the angle of the plate until it is vertical and the stringer bead is horizontal. Repeat the beads as needed with all three (F) groups of electrodes until consistently good beads are obtained in this position. Turn off the welding machine and clean up your work area when you are finished welding.

Complete a copy of the "Student Welding Report" listed in Appendix I or provided by your instructor. ◆

Square Butt Joint

The **square butt joint** is made by tack welding two flat pieces of plate together, **Figure 9-40.** The space between the plates is called the root opening or root gap. Changes in the root opening will affect penetration. As the space increases, the weld penetration also increases. The root opening for most butt welds will vary from 0 in. (0 mm) to 1/8 in. (3 mm). Excessively large openings can cause burn-through or a cold lap at the weld root, **Figure 9-41.**

After a butt weld is completed, the plate can be cut apart so it can be used for rewelding. The strips for butt welding should be no smaller than 1 in. (25 mm) wide. If they are too narrow, there will be a problem with heat buildup.

If the plate strips are no longer flat after the weld has been cut out, they can be tack welded together and flattened with a hammer, **Figure 9-42.**

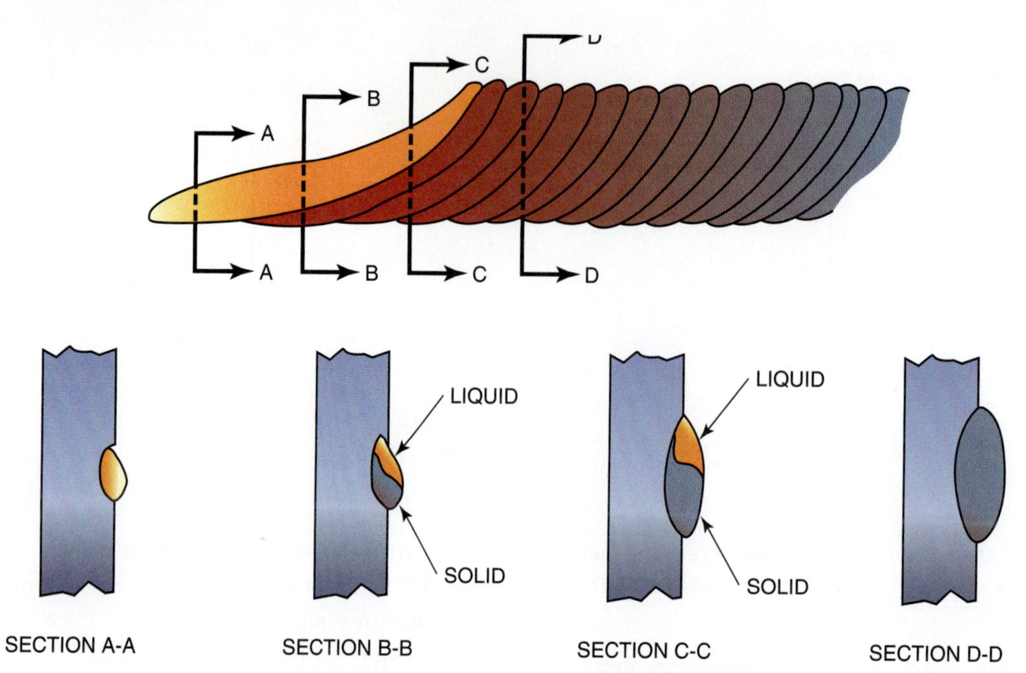

FIGURE 9-39 The progression of a horizontal bead.

FIGURE 9-40 The tack weld should be small and uniform to minimize its effect on the final weld. Courtesy of Larry Jeffus.

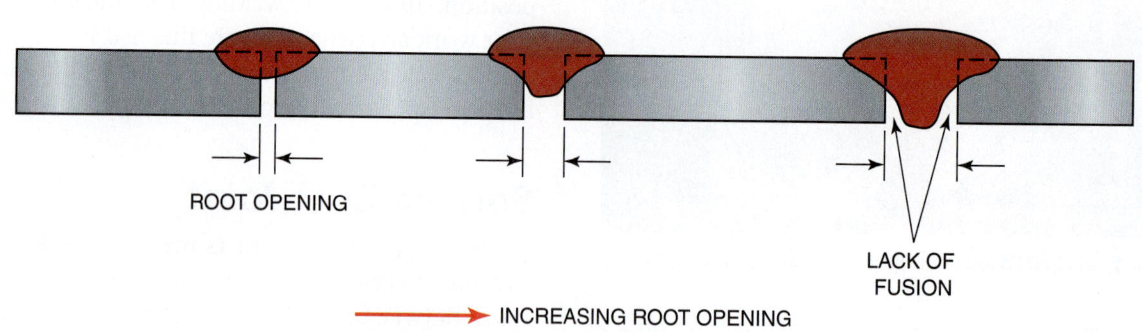

FIGURE 9-41 Effect of root opening on weld penetration.

PRACTICE 9-5

Welded Square Butt Joint in the Flat Position (1G) Using E6010 or E6011 Electrodes, E6012 or E6013 Electrodes, and E7016 or E7018 Electrodes

Using a properly set up and adjusted arc welding machine, proper safety protection, arc welding electrodes having a 1/8-in. (3-mm) diameter, and two or more pieces of mild steel plate, 6 in. (152 mm) long × 1/4 in. (6 mm) thick, you will make a welded square butt joint in the flat position, **Figure 9-43**.

Tack weld the plates together and place them flat on the welding table. Starting at one end, establish a molten weld pool on both plates. Hold the electrode in the molten weld pool until it flows together, **Figure 9-44**. After the gap is bridged by the molten weld pool, start weaving the electrode slowly back and forth across the joint. Moving the electrode too

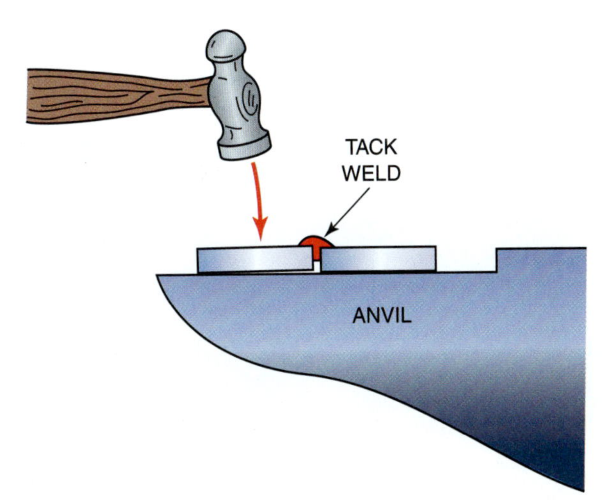

FIGURE 9-42 After the plates are tack welded together, they can be forced into alignment by striking them with a hammer.

quickly from side to side may result in slag being trapped in the joint, **Figure 9-45**.

Continue the weld along the 6-in. (152-mm) length of the joint. Normally, deep penetration is not required for this type of weld. If full plate penetration is required, the edges of the butt joint should be beveled or a larger than normal root gap should be used. Cool, chip, and inspect the weld for uniformity and soundness. Repeat the welds as needed to master all three (F) groups of electrodes in this position. Turn off the welding machine and clean up your work area when you are finished welding.

Complete a copy of the "Student Welding Report" listed in Appendix I or provided by your instructor. ◆

PRACTICE 9-6

Vertical (3G) Up-Welded Square Butt Weld Using E6010 or E6011 Electrodes, E6012 or E6013 Electrodes, and E7016 or E7018 Electrodes

Using the same setup, materials, and electrodes as listed in Practice 9-5, you will make vertical up-welded square butt joints.

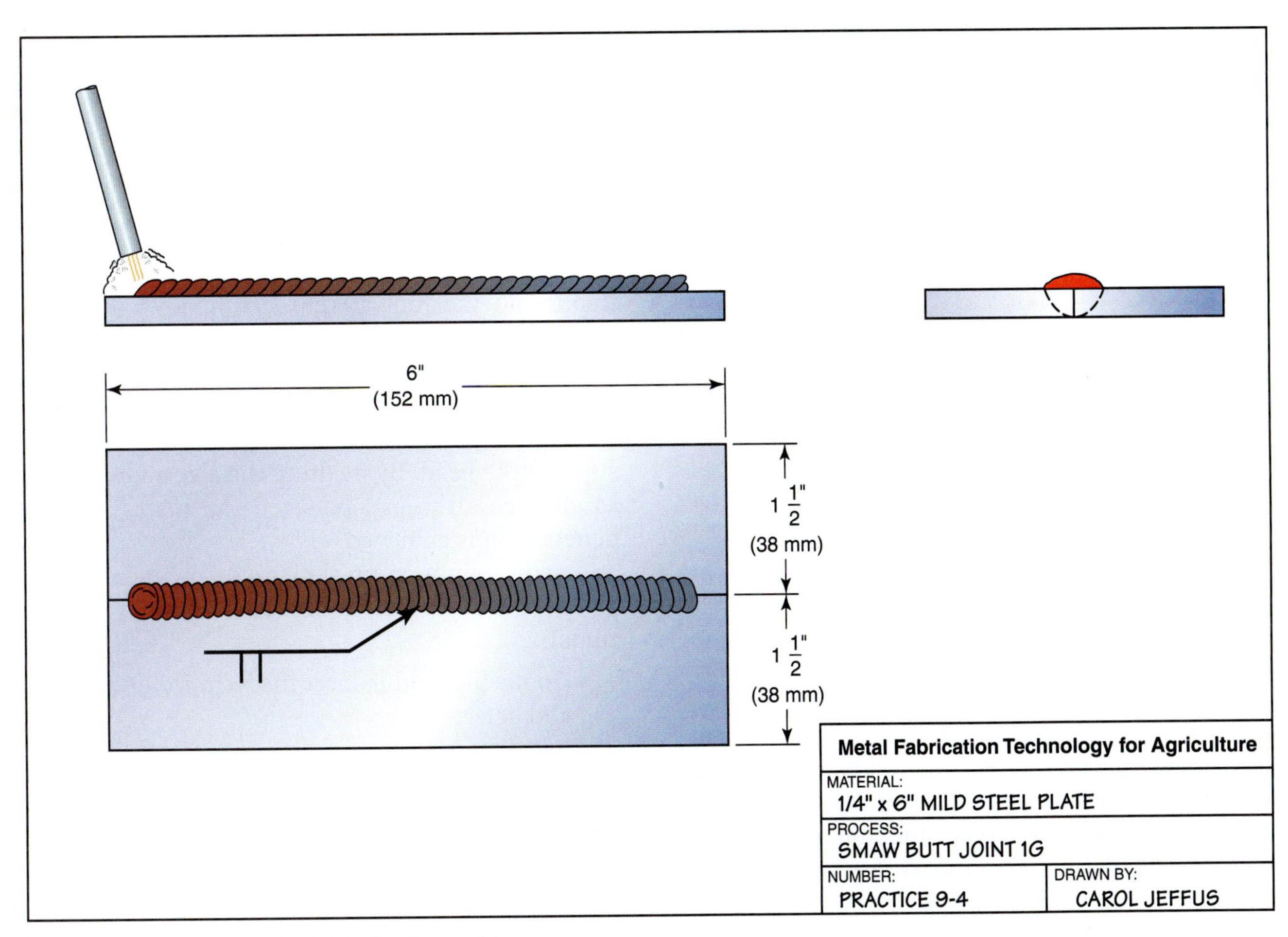

FIGURE 9-43 Square butt joint in the flat position.

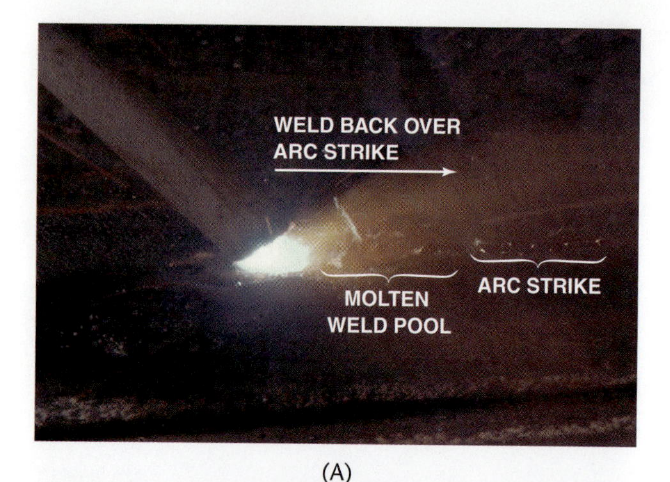

(A)

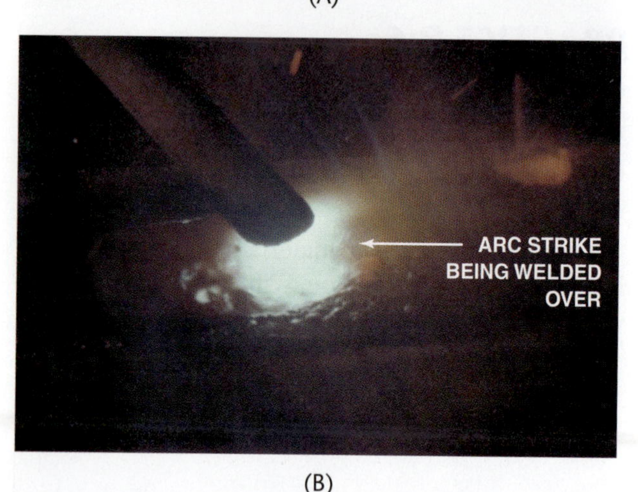

(B)

FIGURE 9-44 (A) After the arc is established, hold it in one area long enough to establish the size of molten weld pool desired. (B) Weld back over the arc strike to melt it into the weld. Courtesy of Larry Jeffus.

With the plates at a 45° angle, start at the bottom and make the molten weld pool bridge the gap between the plates, **Figure 9-46**. Build the bead size slowly so that the molten weld pool has a shelf for support. The "C," "J," or square weave pattern works well for this joint.

As the electrode is moved up the weld, the arc is lengthened slightly so that little or no metal is deposited ahead of the molten weld pool. When the electrode is brought back into the molten weld pool, it should be lowered to deposit metal, **Figure 9-47**.

As skill is developed, increase the plate angle until it is vertical. Cool, chip, and inspect the weld for uniformity and defects. Repeat the welds with all three (F) groups of electrodes until you can consistently make welds free of defects. Turn off the welding machine and clean up your work area when you are finished welding.

Complete a copy of the "Student Welding Report" listed in Appendix I or provided by your instructor. ◆

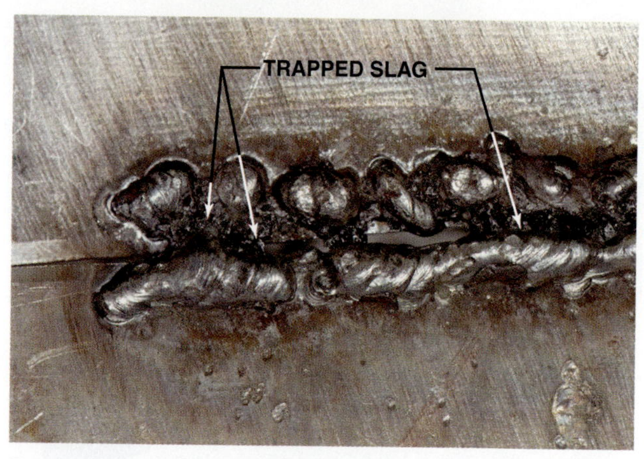

FIGURE 9-45 Moving the electrode from side to side too quickly can result in slag being trapped between the plates. Courtesy of Larry Jeffus.

PRACTICE 9-7

Welded Horizontal (2G) Square Butt Weld Using E6010 or E6011 Electrodes, E6012 or E6013 Electrodes, and E7016 or E7018 Electrodes

Using the same setup, materials, and electrodes as described in Practice 9-5, you will make a welded horizontal square butt joint.

- Start practicing these welds with the plate at a slight angle.
- Strike the arc on the bottom plate and build the molten weld pool until it bridges the gap.

If the weld is started on the top plate, slag will be trapped in the root at the beginning of the weld because of poor initial penetration. The slag may cause the weld to crack when it is placed in service.

The "J" weave pattern is recommended in order to deposit metal on the lower plate so that it can support the bead. By pushing the electrode inward as you cross the gap between the plates, deeper penetration is achieved.

As you acquire more skill, gradually increase the plate angle until it is vertical and the weld is horizontal.

- Cool, chip, and inspect the weld for uniformity and defects.
- Repeat the welds with all three (F) groups of electrodes until you can consistently make welds free of defects. Turn off the welding machine and clean up your work area when you are finished welding.

Complete a copy of the "Student Welding Report" listed in Appendix I or provided by your instructor. ◆

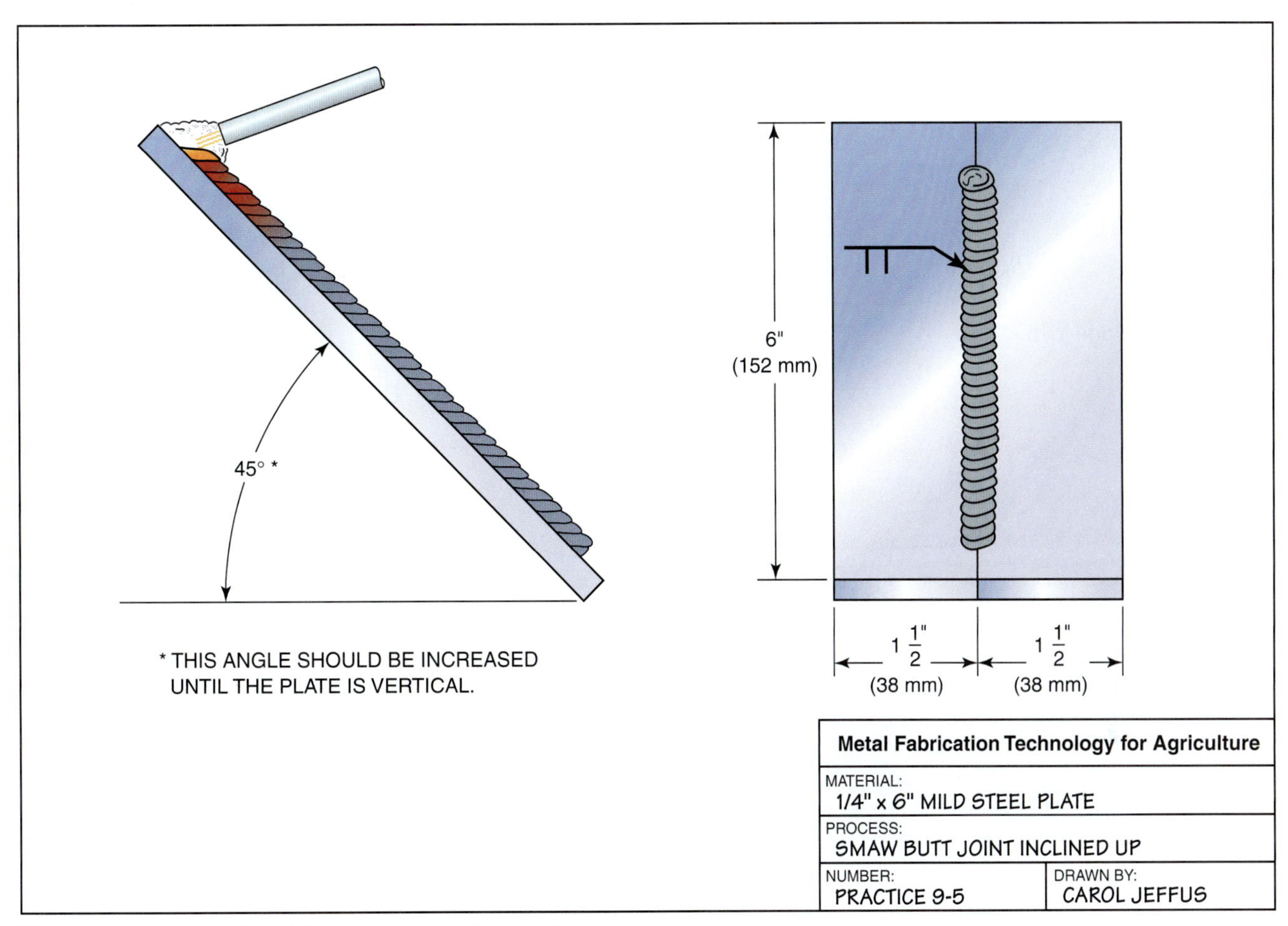

FIGURE 9-46 Square butt joint in the vertical up position.

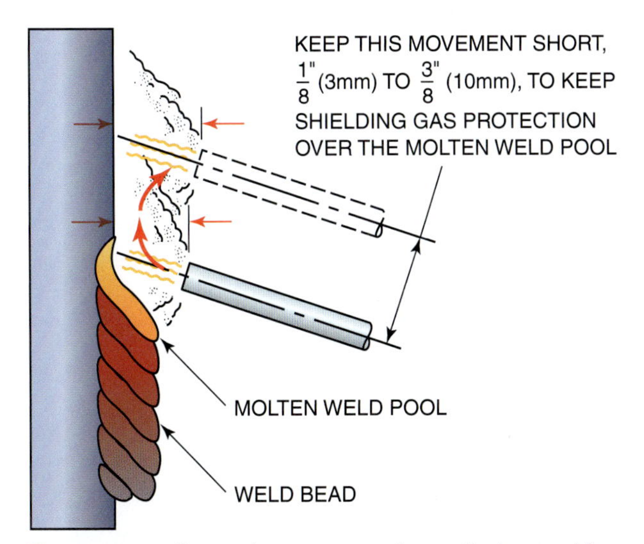

FIGURE 9-47 Electrode movement for vertical up welds.

Edge Weld

An edge weld joint is made by placing the edges of the plate evenly, **Figure 9-48**. When assembling the edge joint the plates should be clamped tightly together; there should not be any gap between the plates. Both edges of the plate assembly can be welded.

Make the tack welds to hold the plates together along the ends of the joint, **Figure 9-49**.

The size of the weld should equal the thickness of the plate being joined. A good indication the weld is being made large enough is when the weld bead width is equal to the width of the joint, **Figure 9-50**. The weld bead should also have a slight buildup.

PRACTICE 9-8

Edge Weld in the Flat Position Using E6010 or E6011 Electrodes, E6012 or E6013 Electrodes, and E7016 or E7018 Electrodes

Using a properly set up and adjusted arc welding machine, proper safety protection, as demonstrated in Practice 9-1, arc welding electrodes with a 1/8-in. (3-mm) diameter, and two pieces of mild steel plate, 6 in. (152 mm) long × 1/4 in. (6 mm) thick, you will make a weld on an edge joint, **Figure 9-51**.

- Clamp the plates flat together and make a tack weld along each end of the plates.
- Starting at one end of the plate, make a straight weld the full length of the plate. Make the weld bead as wide as the width of the edge joint.

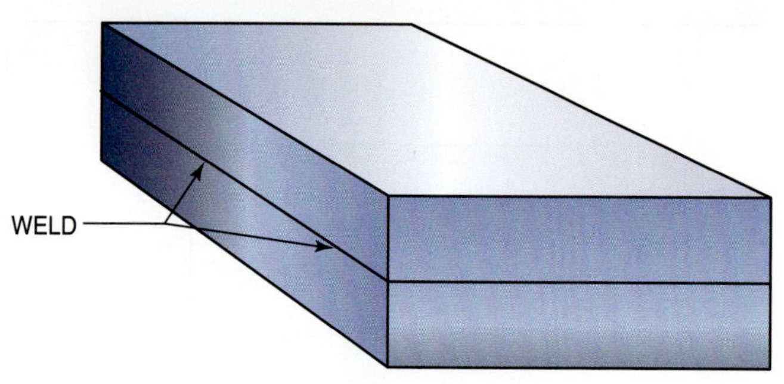

FIGURE 9-48 Edge joint.

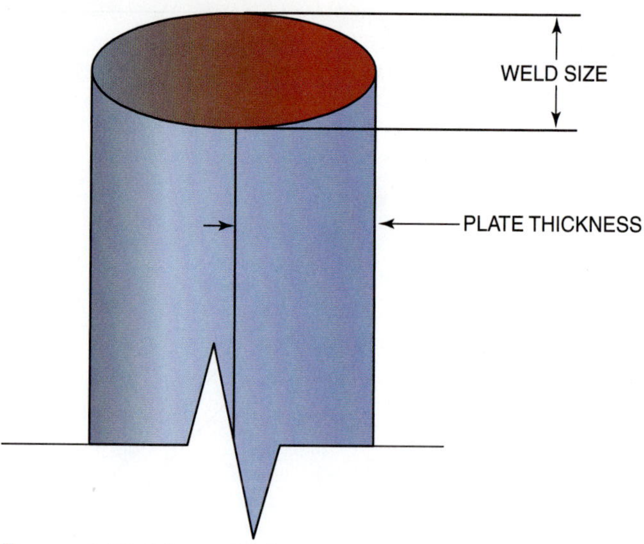

FIGURE 9-50 Edge weld size.

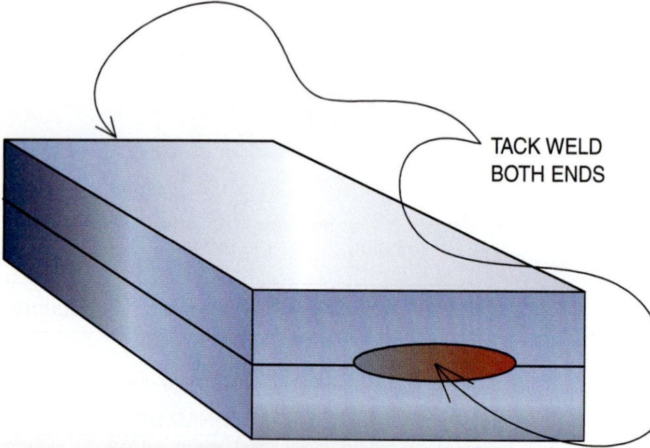

FIGURE 9-49 Make tack welds at the ends of the joint.

- Watch the molten weld pool, not the end of the electrode.
- Cool, chip, and inspect the weld for uniformity and defects.
- Repeat the welds as needed with all three (F) groups of electrodes until you can consistently make welds free of defects.

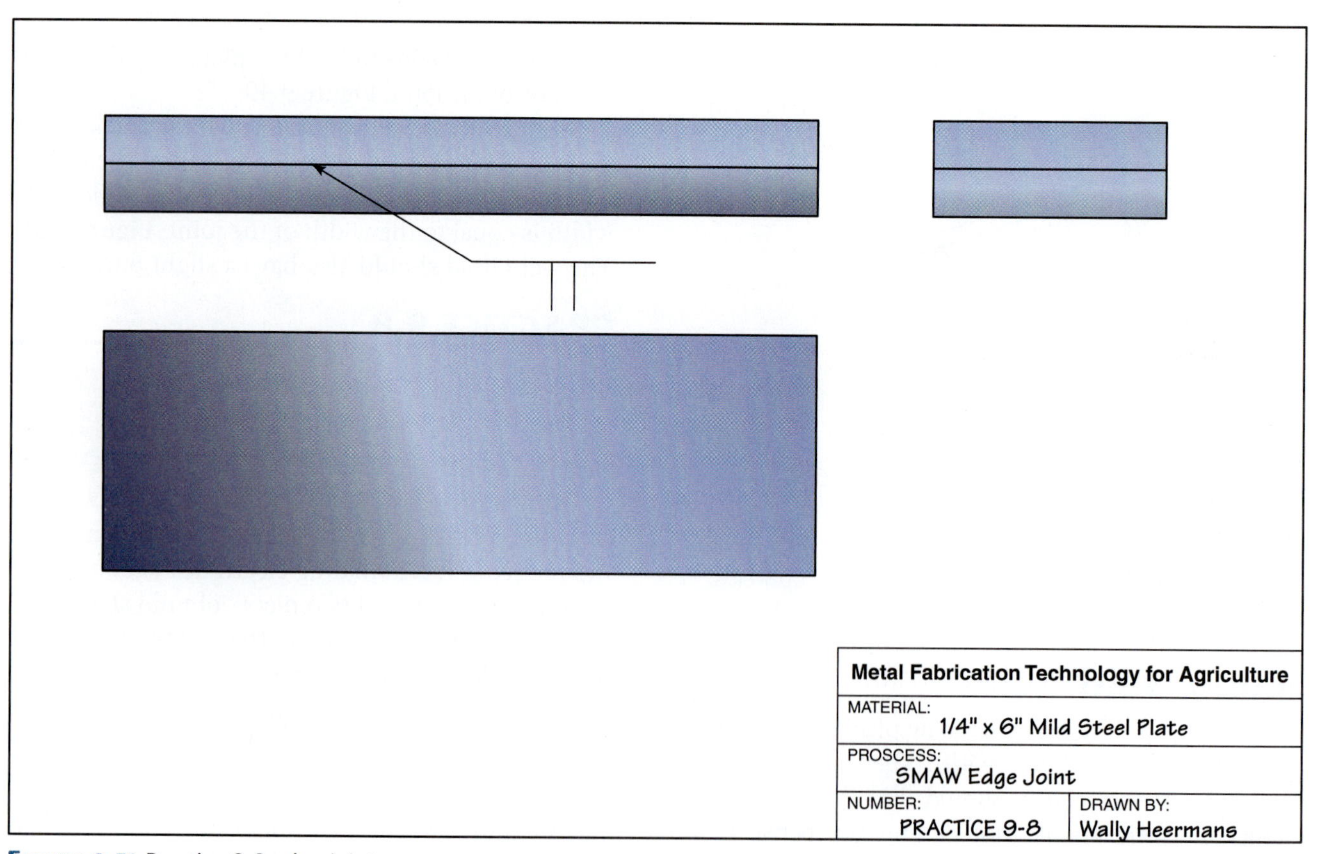

FIGURE 9-51 Practice 9-8 edge joint.

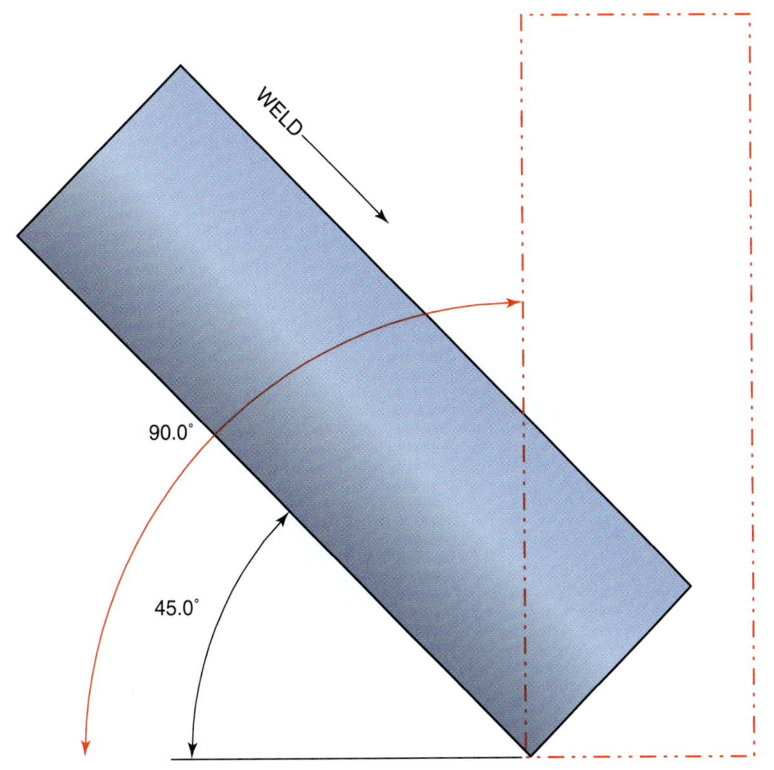

FIGURE 9-52 Vertical down.

- Turn off the welding machine and clean up your work area when you are finished welding.

Complete a copy of the "Student Welding Report" listed in Appendix I or provided by your instructor. ◆

PRACTICE 9-9

Edge Joint in the Vertical Down Position Using E6010 or E6011 Electrodes, E6012 or E6013 Electrodes, and E7016 or E7018 Electrodes

Using the same setup, materials, and electrodes as listed in Practice 9-8, you will make a vertical down weld on an edge joint. Start with the plates at a 45° angle.

This technique is the same as that used to make vertical down welds. However, a lower level of skill is required at 45°, and it is easier to develop your skill. After you master the 45° angle, the angle is increased successively until a vertical position is reached, **Figure 9-52.**

- Make the weld bead as wide as the joint. Controlling a weld bead this size is more difficult, but you must develop the skill required to control this larger molten weld pool.
- Cool, chip, and inspect the weld for uniformity and defects.
- Repeat the welds as needed with all three (F) groups of electrodes until you can consistently make welds free of defects. Turn off the welding machine and clean up your work area when you are finished welding.

Complete a copy of the "Student Welding Report" listed in Appendix I or provided by your instructor. ◆

PRACTICE 9-10

Edge Joint in the Vertical Up Position Using E6010 or E6011 Electrodes, E6012 or E6013 Electrodes, and E7016 or E7018 Electrodes

Using the same setup, materials, and electrodes as listed in Practice 9-8, you will make a vertical up weld on an edge joint. Start with the plates at a 45° angle.

This technique is the same as that used to make vertical up welds. However, a lower level of skill is required at 45°, and it is easier to develop your skill. After you master the 45° angle, the angle is increased successively until a vertical position is reached, **Figure 9-53.**

Before the molten metal drips down the bead, the back of the molten weld pool will start to bulge, **Figure 9-54.** When this happens, increase the speed of travel and the weave pattern.

- Cool, chip, and inspect the weld for uniformity and defects.
- Repeat the welds as needed with all three (F) groups of electrodes until you can consistently

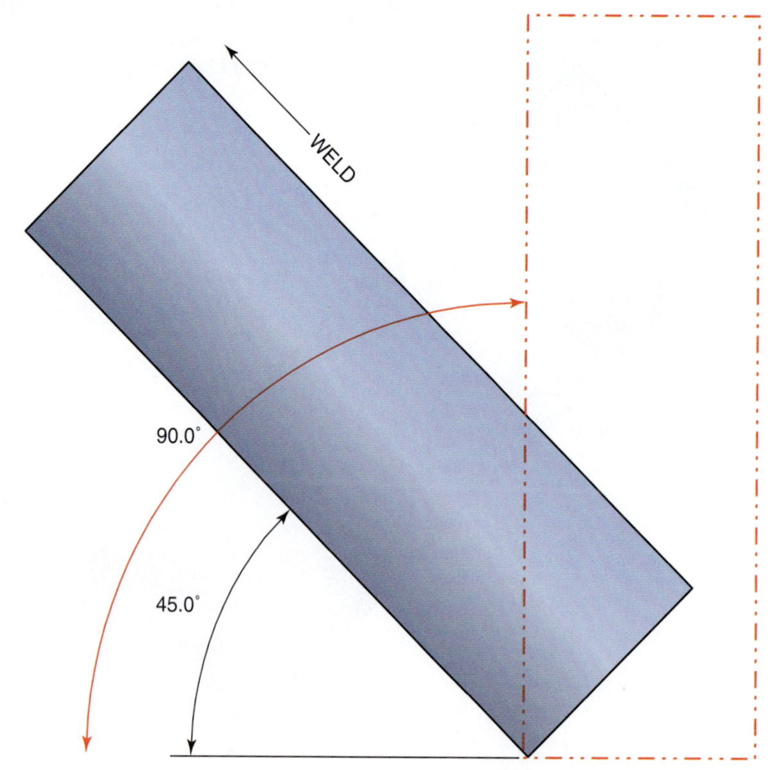

FIGURE 9-53 Vertical up.

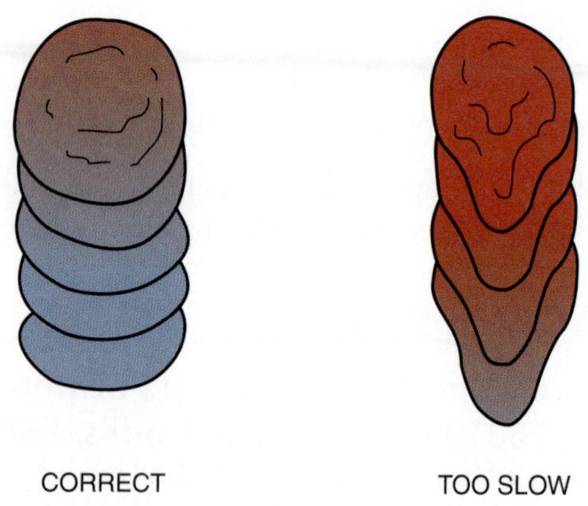

FIGURE 9-54 Watch the trailing edge of the weld pool to judge the correct travel speed.

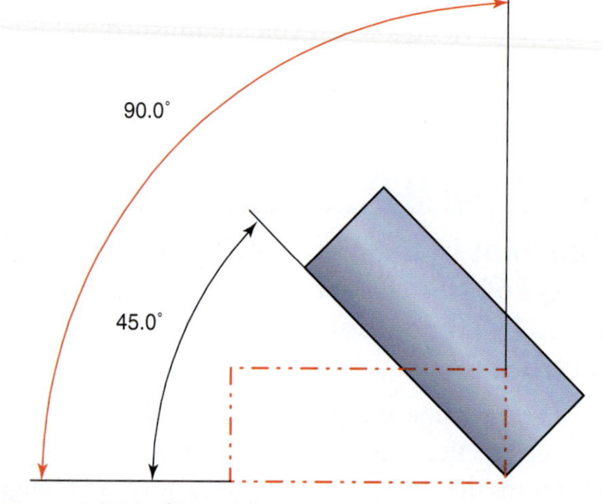

FIGURE 9-55 Incline angle.

make welds free of defects. Turn off the welding machine and clean up your work area when you are finished welding.

Complete a copy of the "Student Welding Report" listed in Appendix I or provided by your instructor. ◆

PRACTICE 9-11

Edge Joint in the Horizontal Position Using E6010 or E6011 Electrodes, E6012 or E6013 Electrodes, and E7016 or E7018 Electrodes

Using the same setup, materials, and electrodes as listed in Practice 9-8, you will make a horizontal weld on an edge joint. When you begin to practice the horizontal weld, the plate may be reclined slightly, **Figure 9-55.** This placement allows the welder to build the required skill by practicing the correct techniques successfully. The "J" weave or stepped pattern is suggested for this practice. As the electrode is drawn back to the back edge of the weld pool, metal is deposited. Use the metal being deposited to support the molten weld pool.

Angling the electrode up and back toward the weld causes more metal to be deposited along the top edge of the weld. Keeping the bead small allows

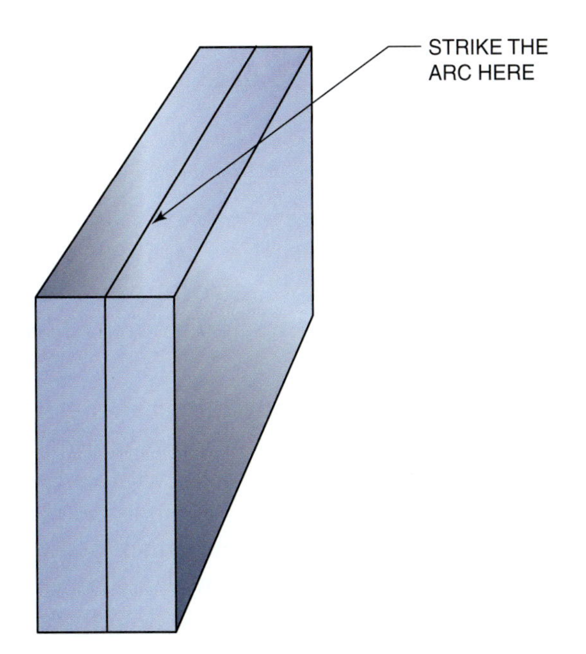

FIGURE 9-56 Strike the arc in the joint.

the surface tension to hold the molten weld pool in place.

Gradually increase the angle of the plate until it and the weld bead are horizontal.

- Cool, chip, and inspect the weld for uniformity and defects.
- Repeat the welds as needed with all three (F) groups of electrodes until you can consistently make welds free of defects. Turn off the welding machine and clean up your work area when you are finished welding.

Complete a copy of the "Student Welding Report" listed in Appendix I or provided by your instructor. ◆

PRACTICE 9-12

Edge Joint in the Overhead Position Using E6010 or E6011 Electrodes, E6012 or E6013 Electrodes, and E7016 or E7018 Electrodes

Using the same setup, materials, and electrodes as listed in Practice 9-8, you will make an overhead weld on an edge joint.

- With the electrode pointed in a slightly trailing angle, **Figure 9-56**, strike the arc in the joint.
- Keep a very short arc length.
- Use the stepped pattern and move the electrode forward slightly when the molten weld pool grows to the correct size, **Figure 9-57**.

FIGURE 9-57 Step the electrode. Courtesy of Larry Jeffus.

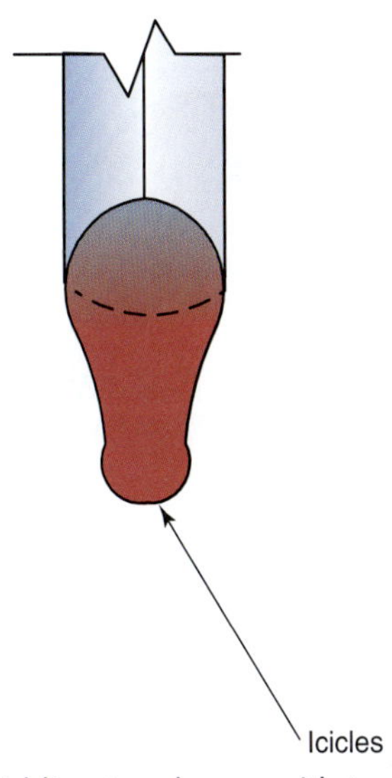

FIGURE 9-58 Welding too slow or with too high an amperage setting will result in the weld metal dripping down like icicles.

As the molten weld pool gets larger it has a tendency to quickly become convex. If you keep the arc in the molten weld pool once the joint is filled and the weld face is flat it will quickly overfill and become convex. This can result in the weld face forming drips of metal hanging from the weld like icicles, **Figure 9-58**.

- When the molten weld pool cools and begins to shrink, move the arc back near the center of the weld.
- Hold the arc in this new location until the molten weld pool again grows to the correct size.
- Step the electrode forward again and keep repeating this pattern until the weld progresses along the entire weld joint length.
- Cool, chip, and inspect the weld for uniformity and defects.
- Repeat the welds as needed with all three (F) groups of electrodes until you can consistently make welds free of defects. Turn off the welding machine and clean up your work area when you are finished welding.

Complete a copy of the "Student Welding Report" listed in Appendix I or provided by your instructor. ◆

Outside Corner Joint

An outside corner joint is made by placing the plates at a 90° angle to each other, with the edges forming a V groove, **Figure 9-59**. There may or may not be a slight root opening left between the plate edges. Small tack welds should be made approximately 1/2 in. (13 mm) from each end of the joint.

The weld bead should completely fill the V groove formed by the plates and may have a slightly convex surface buildup. The back side of an outside corner joint can be used to practice fillet welds, or four plates can be made into a box tube shape, **Figure 9-60**.

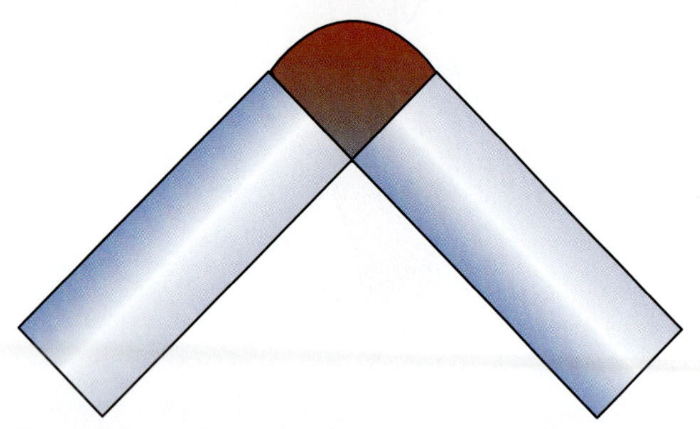

FIGURE 9-59 V formed by an outside corner joint.

PRACTICE 9-13

Outside Corner Joint in the Flat Position Using E6010 or E6011 Electrodes, E6012 or E6013 Electrodes, and E7016 or E7018 Electrodes

Using a properly set up and adjusted arc welding machine, proper safety protection, as demonstrated in Practice 9-1, arc welding electrodes with a 1/8-in. (3-mm) diameter, and two pieces of mild steel plate 6 in. (152 mm) long × 1/4 in. (6 mm) thick, you will make a weld on an outside corner joint.

- Starting at one end of the plate, make a straight weld the full length of the plate.
- Watch the molten weld pool at this point, not the end of the electrode. As you become more skillful, it is easier to watch the molten weld pool.
- Cool, chip, and inspect the weld for uniformity and defects.
- Repeat the welds as needed with all three (F) groups of electrodes until you can consistently make welds free of defects. Turn off the weld-

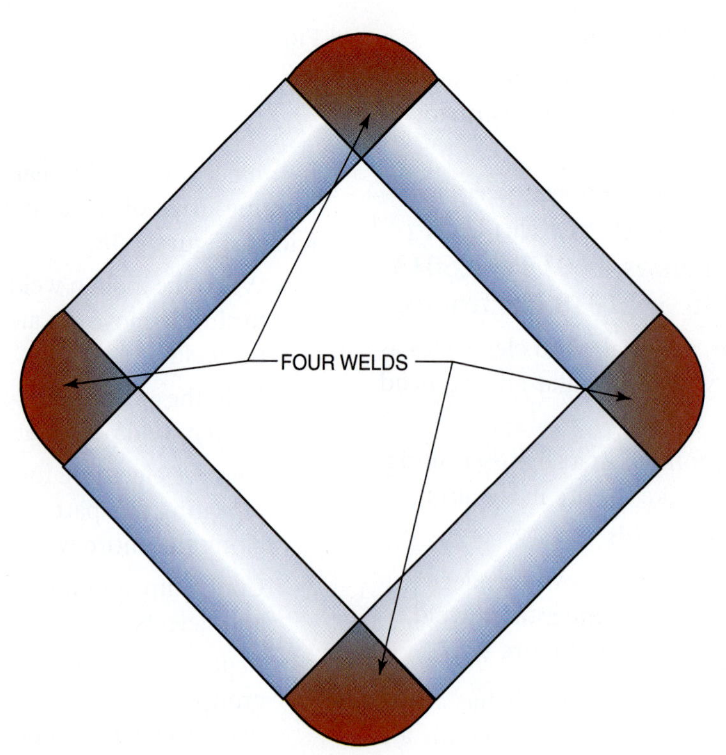

FIGURE 9-60 Box tube made from four outside corner joint welds.

ing machine and clean up your work area when you are finished welding.

Complete a copy of the "Student Welding Report" listed in Appendix I or provided by your instructor. ◆

PRACTICE 9-14

Outside Corner Joint in the Vertical Down Position Using E6010 or E6011 Electrodes, E6012 or E6013 Electrodes, and E7016 or E7018 Electrodes

Using the same setup, materials, and electrodes as listed in Practice 9-13, you will make a vertical down weld on an outside corner joint. Start with the plate at a 45° angle.

This technique is the same as that used to make vertical down welds. However, a lower level of skill is required at 45°, and it is easier to develop your skill. After you master the 45° angle, the angle is increased successively until a vertical position is reached, **Figure 9-61.**

- Cool, chip, and inspect the weld for uniformity and defects.
- Repeat the welds as needed with all three (F) groups of electrodes until you can consistently make welds free of defects. Turn off the welding machine and clean up your work area when you are finished welding.

Complete a copy of the "Student Welding Report" listed in Appendix I or provided by your instructor. ◆

PRACTICE 9-15

Outside Corner Joint in the Vertical Up Position Using E6010 or E6011 Electrodes, E6012 or E6013 Electrodes, and E7016 or E7018 Electrodes

Using the same setup, materials, and electrodes as listed in Practice 9-13, you will make a vertical up weld on an outside corner joint. Start with the plate at a 45° angle.

This technique is the same as that used to make vertical up welds. However, a lower level of skill is required at 45°, and it is easier to develop your skill. After the welder masters the 45° angle, the angle is increased successively until a vertical position is reached, **Figure 9-62.**

Before the molten metal drips down the bead, the back of the molten weld pool will start to bulge, **Figure 9-63.** When this happens, increase the speed of travel and the weave pattern.

- Cool, chip, and inspect the weld for uniformity and defects.
- Repeat the welds as needed with all three (F) groups of electrodes until you can consistently make welds free of defects. Turn off the welding machine and clean up your work area when you are finished welding.

Complete a copy of the "Student Welding Report" listed in Appendix I or provided by your instructor. ◆

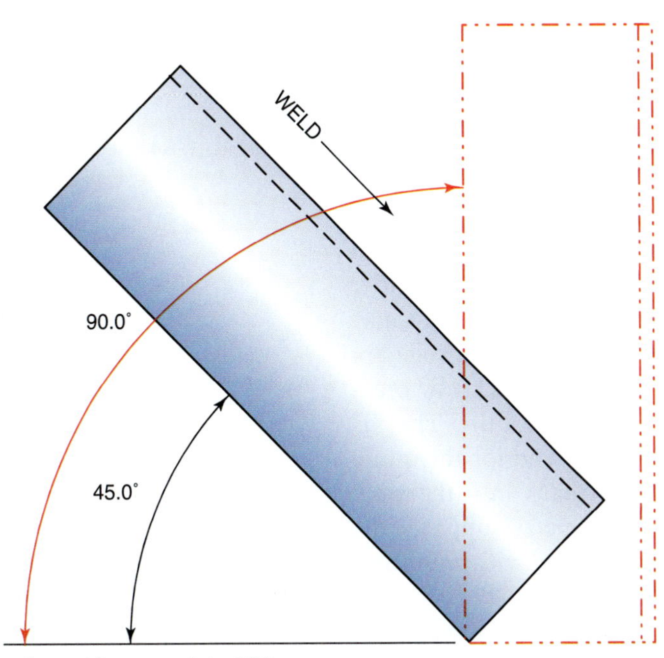

FIGURE 9-61 Start with a 45° angle and increase it to 90°.

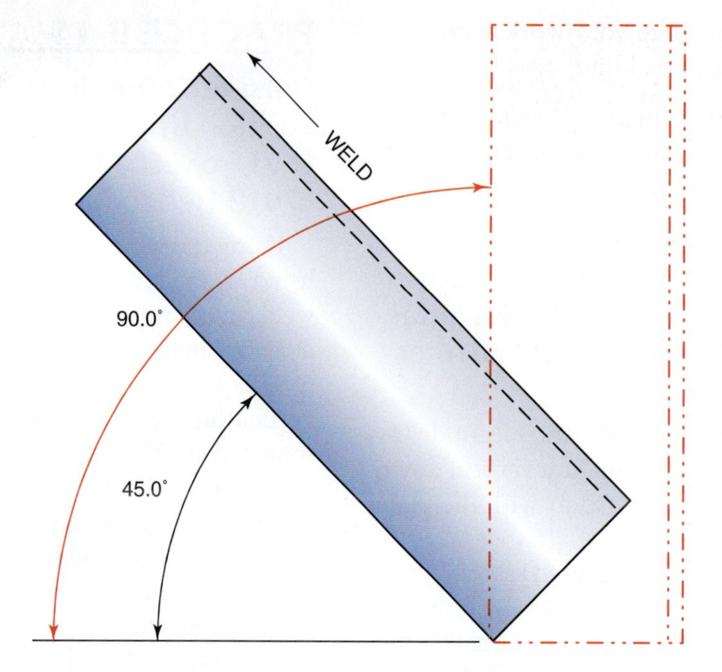

FIGURE 9-62 Vertical up.

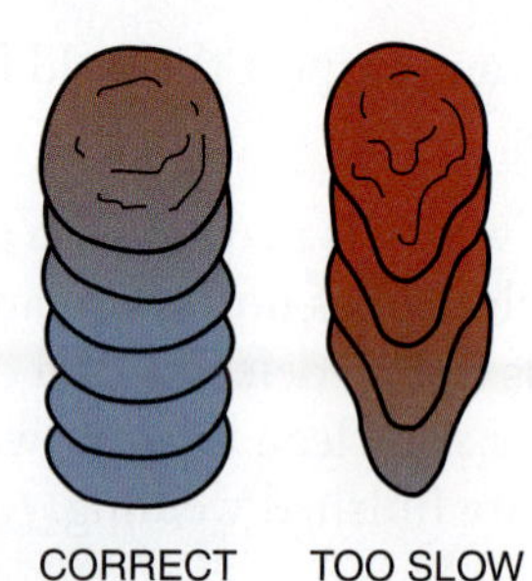

FIGURE 9-63 Watch the trailing edge of the weld pool to judge the correct travel speed.

PRACTICE 9-16

Outside Corner Joint in the Horizontal Position Using E6010 or E6011 Electrodes, E6012 or E6013 Electrodes, and E7016 or E7018 Electrodes

Using the same setup, materials, and electrodes as listed in Practice 9-13, you will make a horizontal weld on an outside corner joint. When the welder begins to practice the horizontal weld, the joint may be reclined slightly, **Figure 9-64**. This placement allows the welder to build the required skill by practicing the correct techniques successfully. The "J" weave or stepped pattern is suggested for this practice. As the electrode is drawn back into the weld pool, metal is deposited. This metal supports the molten weld pool, resulting in a bead with a uniform contour, **Figure 9-65**.

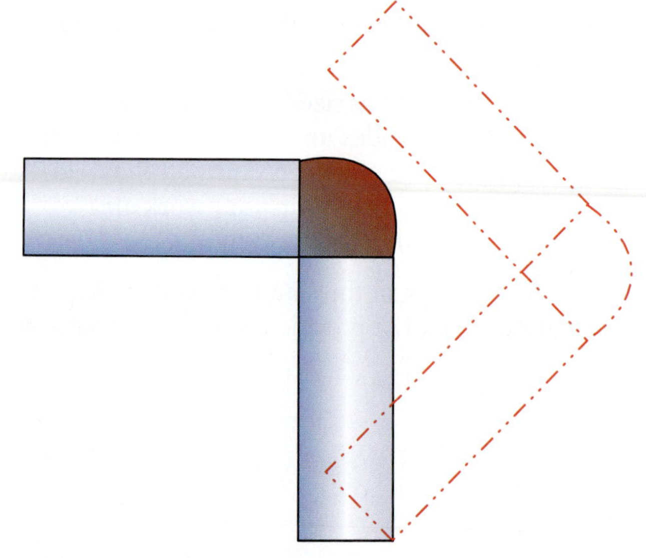

FIGURE 9-64 Incline angle.

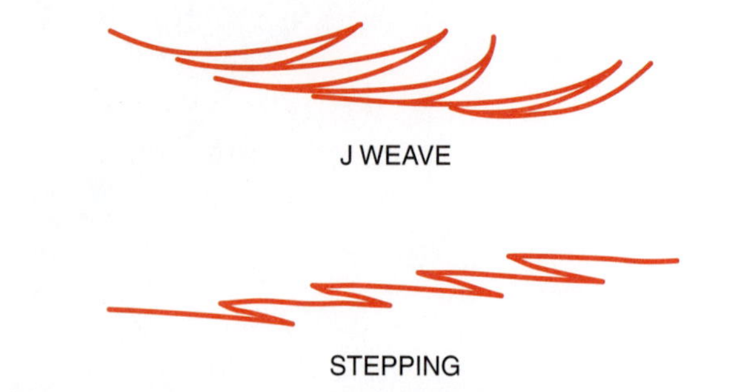

FIGURE 9-65 "J" weave or stepping.

Angling the electrode up and back toward the weld causes more metal to be deposited along the top edge of the weld. Keeping the bead small allows the surface tension to hold the molten weld pool in place.

Gradually increase the angle of the plate until it is vertical and the weld bead is horizontal.

- Cool, chip, and inspect the weld for uniformity and defects.
- Repeat the welds as needed with all three (F) groups of electrodes until you can consistently make welds free of defects. Turn off the welding machine and clean up your work area when you are finished welding.

Complete a copy of the "Student Welding Report" listed in Appendix I or provided by your instructor. ◆

PRACTICE 9-17

Outside Corner Joint in the Overhead Position Using E6010 or E6011 Electrodes, E6012 or E6013 Electrodes, and E7016 or E7018 Electrodes

Using the same setup, materials, and electrodes as listed in Practice 9-13, you will make an overhead welded outside corner joint.

- With the electrode pointed slightly into the joint, **Figure 9-66**, strike the arc in the joint.
- Keep a very short arc length.
- Use the stepped pattern and move the electrode forward slightly when the molten weld pool grows to the correct size, **Figure 9-67**.

As the molten weld pool gets larger it has a tendency to quickly become convex. If you keep the arc in the molten weld pool once the joint is filled and the weld face is flat it will quickly overfill and become convex. This can result in the weld face forming drips of metal hanging from the weld like icicles, **Figure 9-68**.

- When the molten weld pool cools and begins to shrink, move the arc back near the center of the weld.
- Hold the arc in this new location until the molten weld pool again grows to the correct size.
- Step the electrode forward again and keep repeating this pattern until the weld progresses along the entire weld joint length.
- Cool, chip, and inspect the weld for uniformity and defects.
- Repeat the welds as needed with all three (F) groups of electrodes until you can consistently make welds free of defects. Turn off the welding machine and clean up your work area when you are finished welding.

Complete a copy of the "Student Welding Report" listed in Appendix I or provided by your instructor. ◆

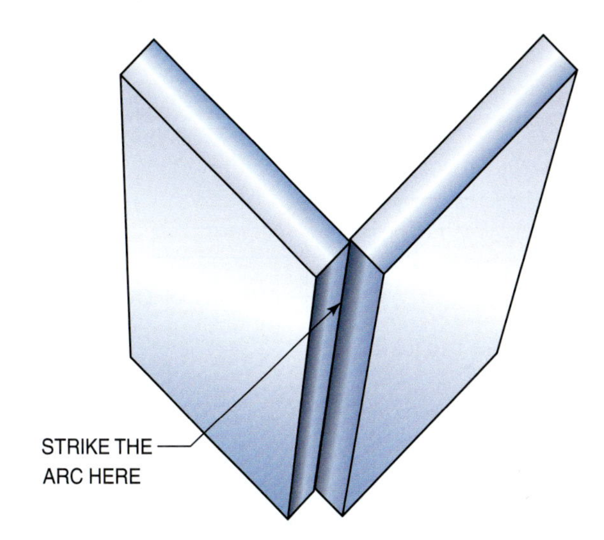

FIGURE 9-66 Strike arc in the joint.

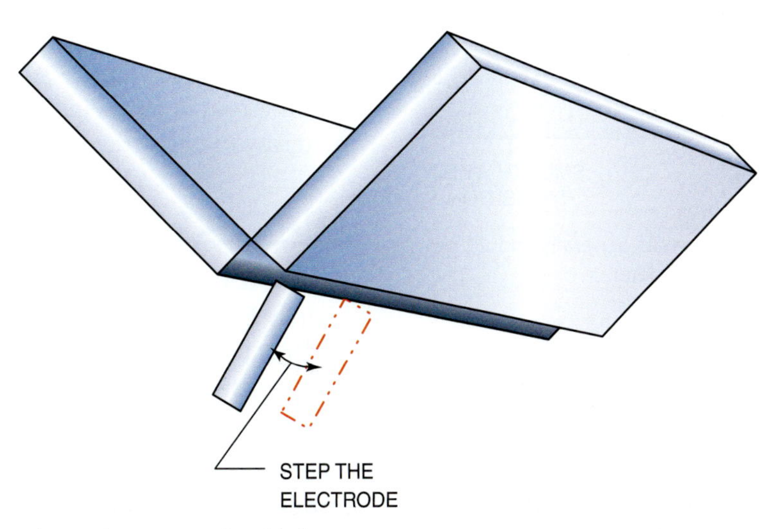

FIGURE 9-67 Stepping the electrode to control weld size.

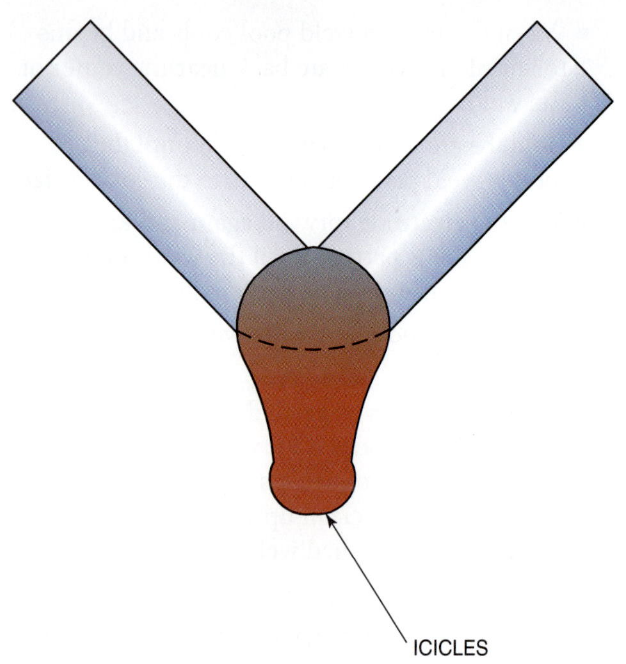

FIGURE 9-68 Welding too slowly or with too high of an amperage setting will result in the weld metal dripping down like icicles.

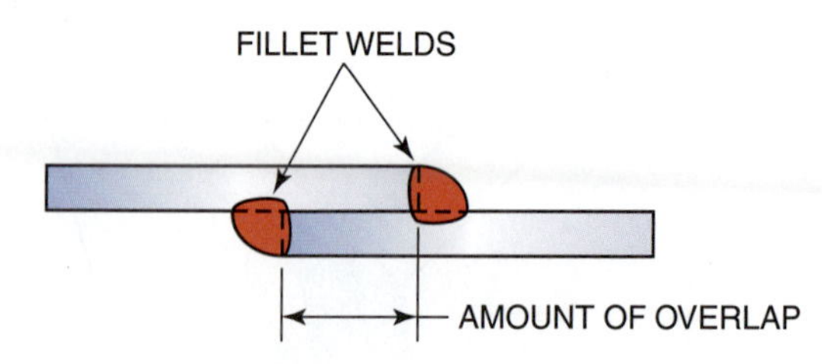

FIGURE 9-69 Lap joint.

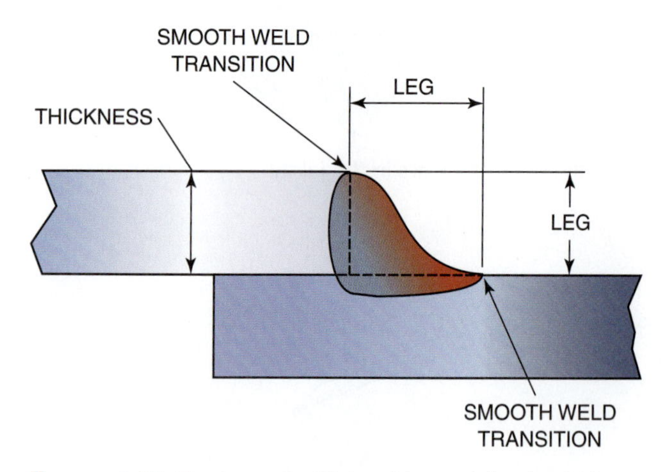

FIGURE 9-70 The legs of a fillet weld generally should be equal to the thickness of the base metal.

Lap Joint

A lap joint is made by overlapping the edges of the two plates, **Figure 9-69.** The joint can be welded on one side or both sides with a fillet weld. In Practice 9-7, both sides should be welded unless otherwise noted.

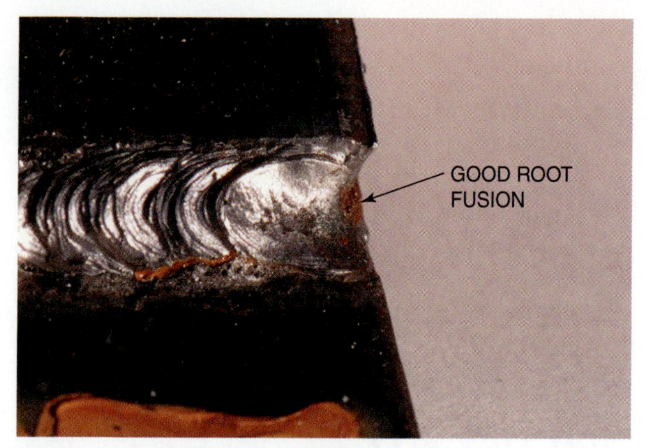

FIGURE 9-71 Watch the root of the weld bead to be sure there is complete fusion. Courtesy of Larry Jeffus.

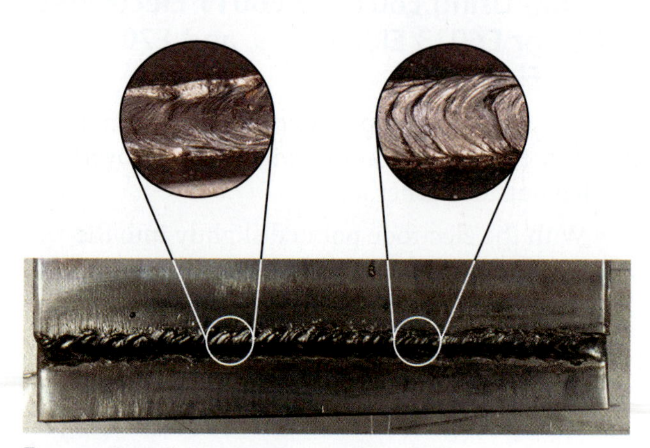

FIGURE 9-72 Lap joint. Courtesy of Larry Jeffus.

As the fillet weld is made on the lap joint, the buildup should equal the thickness of the plate, **Figure 9-70.** A good weld will have a smooth transition from the plate surface to the weld. If this transition is abrupt, it can cause stresses that will weaken the joint.

Penetration for lap joints does not improve their strength; complete fusion is required. The root of fillet welds must be melted to ensure a completely fused joint. If the molten weld pool shows a notch during the weld, **Figure 9-71**, this is an indication that the root is not being fused together. The weave pattern will help prevent this problem, **Figure 9-72.**

PRACTICE 9-18

Welded Lap Joint in the Flat Position (1F) Using E6010 or E6011 Electrodes, E6012 or E6013 Electrodes, and E7016 or E7018 Electrodes

Using a properly set up and adjusted arc welding machine, proper safety protection, arc welding electrodes having a 1/8-in. (3-mm) diameter, and two or more pieces of mild steel plate, 6 in. (152 mm) long

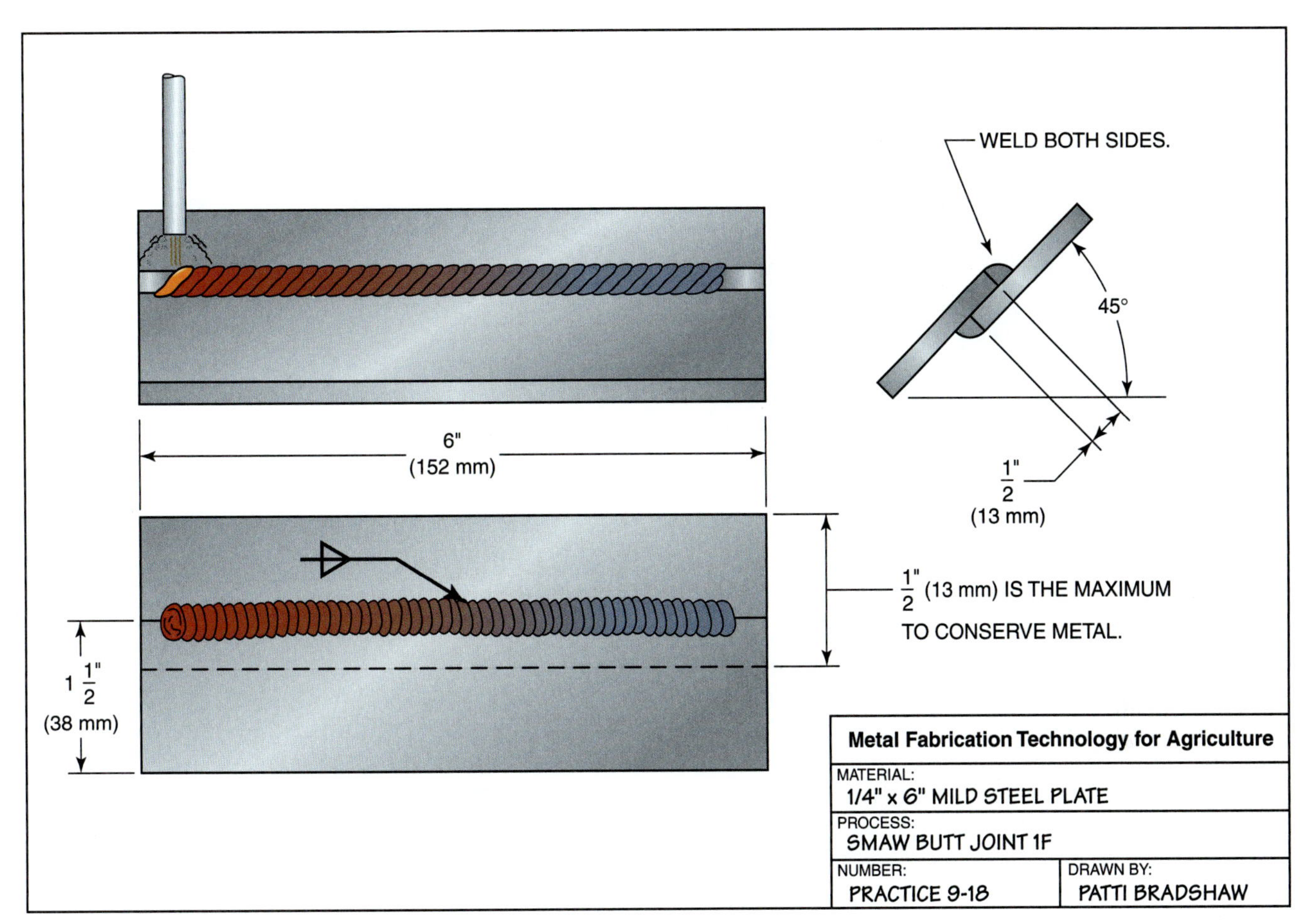

FIGURE 9-73 Lap joint in the flat position.

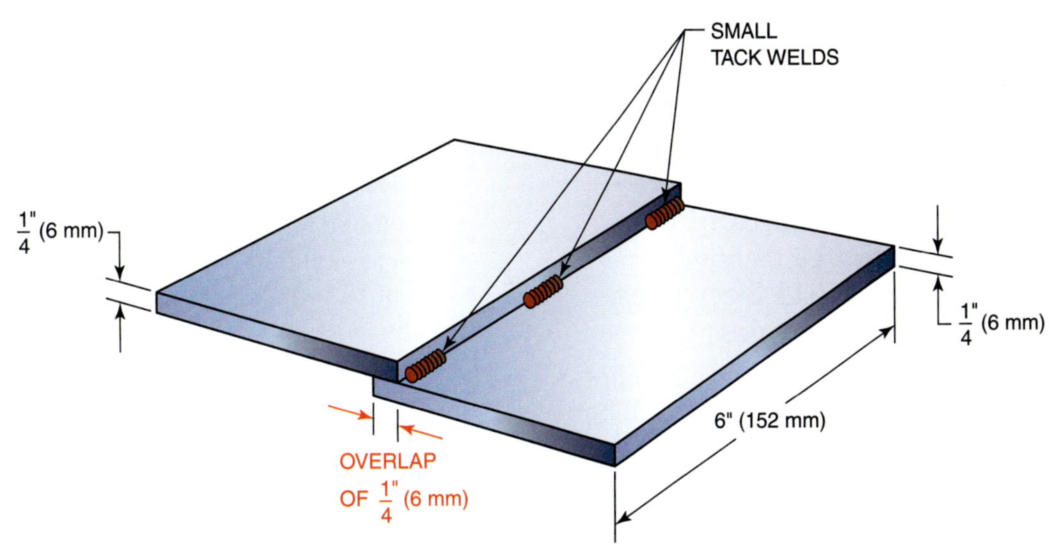

FIGURE 9-74 Tack welding the plates together.

× 1/4 in. (6 mm) thick, you will make a welded lap joint in the flat position, **Figure 9-73.**

Hold the plates together tightly with an overlap of no more than 1/4 in. (6 mm). Tack weld the plates together. A small tack weld may be added in the center to prevent distortion during welding, **Figure 9-74.** Chip the tacks before you start to weld.

The "J," "C," or zigzag weave pattern works well on this joint. Strike the arc and establish a molten pool directly in the joint. Move the electrode out on the bottom plate and then onto the weld to the top edge of the top plate, **Figure 9-75.** Follow the surface of the plates with the arc. Do not follow the trailing edge of the weld bead. Following the molten

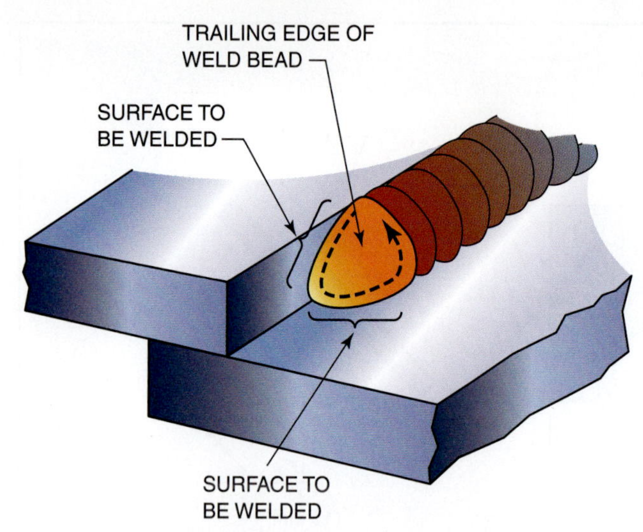

FIGURE 9-75 Follow the surface of the plate to ensure good fusion.

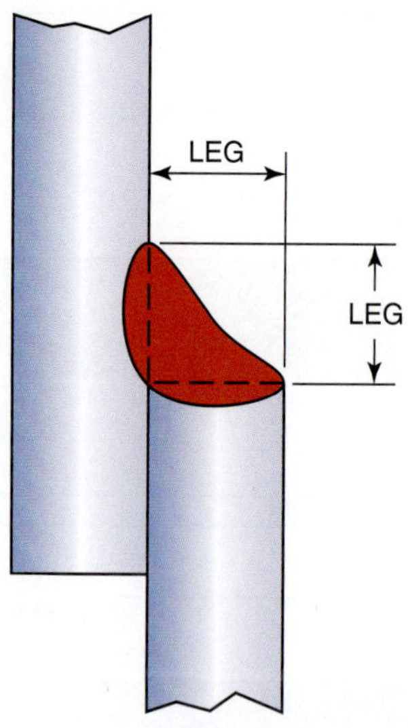

FIGURE 9-76 The horizontal lap joint should have a fillet weld that is equal on both plates.

weld pool will not allow for good root fusion and will cause slag to collect in the root. If slag does collect, a good weld is not possible. Stop the weld and chip the slag to remove it before the weld is completed. Cool, chip, and inspect the weld for uniformity and defects. Repeat the welds with all three (F) groups of electrodes until you can consistently make welds free of defects. Turn off the welding machine and clean up your work area when you are finished welding.

Complete a copy of the "Student Welding Report" listed in Appendix I or provided by your instructor. ◆

PRACTICE 9-19

Welded Lap Joint in the Horizontal Position (2F) Using E6010 or E6011 Electrodes, E6012 or E6013 Electrodes, and E7016 or E7018 Electrodes

Using the same setup, materials, and electrodes as listed in Practice 9-18, you will make a welded horizontal lap joint.

The horizontal lap joint and the flat lap joint require nearly the same technique and skill to achieve a proper weld, **Figure 9-76**. Use the "J," "C," or zigzag weave pattern to make the weld. Do not allow slag to collect in the root. The fillet must be equally divided between both plates for good strength. After completing the weld, cool, chip, and inspect the weld for uniformity and defects. Repeat the welds using all three (F) groups of electrodes until you can consistently make welds free of defects. Turn off the welding machine and clean up your work area when you are finished welding.

Complete a copy of the "Student Welding Report" listed in Appendix I or provided by your instructor. ◆

PRACTICE 9-20

Lap Joint in the Vertical Position (3F) Using E6010 or E6011 Electrodes, E6012 or E6013 Electrodes, and E7016 or E7018 Electrodes

Using the same setup, materials, and electrodes as listed in Practice 9-18, you will make a vertical up welded lap joint.

- Start practicing this weld with the plate at a 45° angle.
- Gradually increase the angle of the plate to vertical as skill is gained in welding this joint. The "J" or "T" weave pattern works well on this joint.
- Establish a molten weld pool in the root of the joint.
- Use the "T" pattern to step ahead of the molten weld pool, allowing it to cool slightly. Do not deposit metal ahead of the molten weld pool.
- As the molten weld pool size starts to decrease, move the electrode back down into the molten weld pool.
- Quickly move the electrode from side to side in the molten weld pool, filling up the joint.
- Cool, chip, and inspect the weld for uniformity and defects.
- Repeat the welds as necessary with all three (F) groups of electrodes until you can consistently

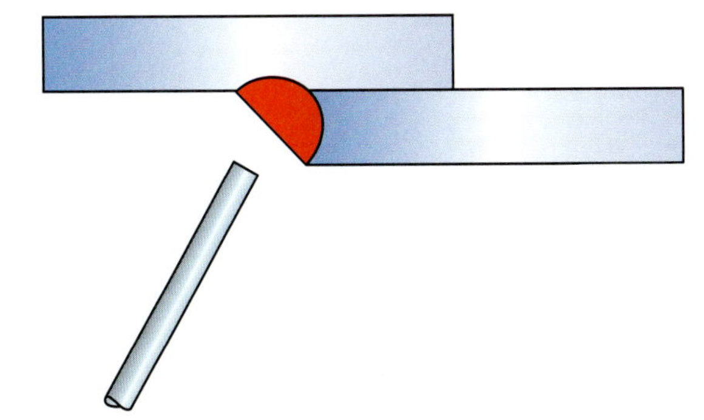

FIGURE 9-77 Point the electrode slightly toward the root of the joint.

make welds free of defects. Turn off the welding machine and clean up your work area when you are finished welding.

Complete a copy of the "Student Welding Report" listed in Appendix I or provided by your instructor. ◆

PRACTICE 9-21

Lap Joint in the Overhead Position (4F) Using E6010 or E6011 Electrodes, E6012 or E6013 Electrodes, and E7016 or E7018 Electrodes

Using the same setup, materials, and electrodes as listed in Practice 9-18, you will make an overhead welded lap joint.

- With the electrode pointed slightly into the joint, **Figure 9-77**, strike the arc in the inside corner of the lap joint.
- Keep a very short arc length.
- Use the stepped pattern and move the electrode forward slightly when the molten weld pool grows to the correct size, **Figure 9-78**.

As the molten weld pool gets larger it has a tendency to quickly become convex. If you keep the arc in the molten weld pool once the joint is filled and the weld face is flat it will quickly overfill and become convex. This can result in the weld face forming drips of metal hanging from the weld like icicles, **Figure 9-79**.

- When the molten weld pool cools and begins to shrink, move the arc back near the center of the weld.
- Hold the arc in this new location until the molten weld pool again grows to the correct size.
- Step the electrode forward again and keep repeating this pattern until the weld progresses along the entire weld joint length.

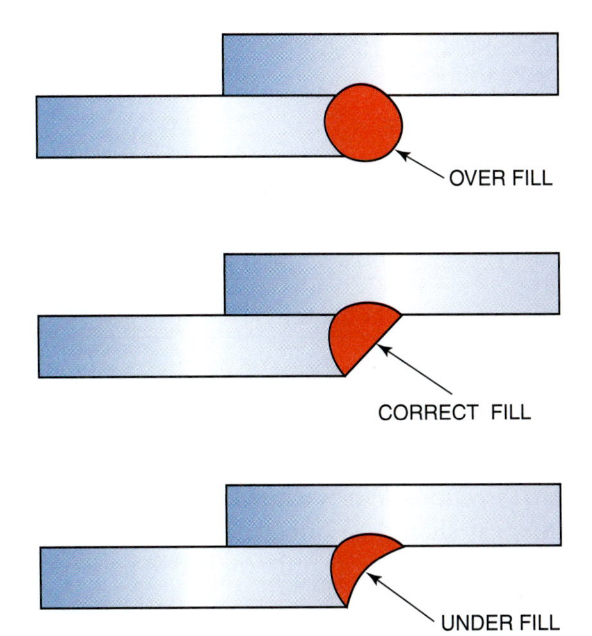

FIGURE 9-78 Correct fillet weld size for overhead welds.

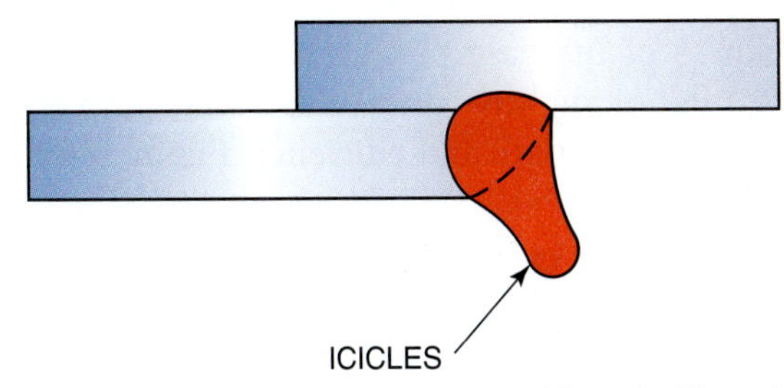

FIGURE 9-79 Overfilling the molten weld pool will result in drips of metal called icicles.

- Cool, chip, and inspect the weld for uniformity and defects.
- Repeat the welds as needed with all three (F) groups of electrodes until you can consistently make welds free of defects. Turn off the welding machine and clean up your work area when you are finished welding.

Complete a copy of the "Student Welding Report" listed in Appendix I or provided by your instructor. ◆

Tee Joint

The tee joint is made by tack welding one piece of metal on another piece of metal at a right angle, **Figure 9-80**. After the joint is tack welded together, the slag is chipped from the tack welds. If the slag is not removed, it will cause a slag inclusion in the final weld.

The heat is not distributed uniformly between both plates during a tee weld. Because the plate that forms the stem of the tee can conduct heat away from the arc in only one direction, it will heat up faster

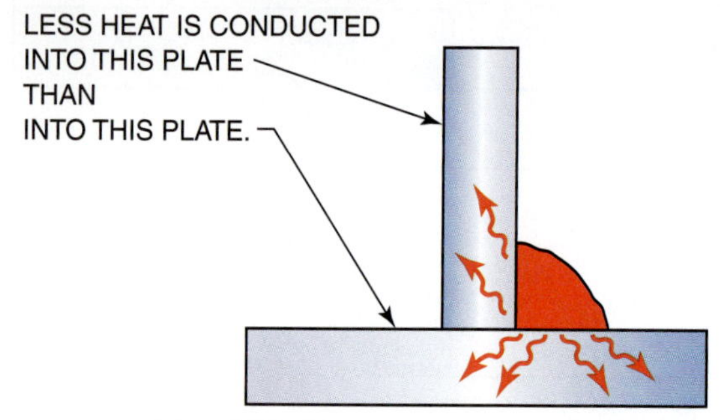

FIGURE 9-80 Tee joint.

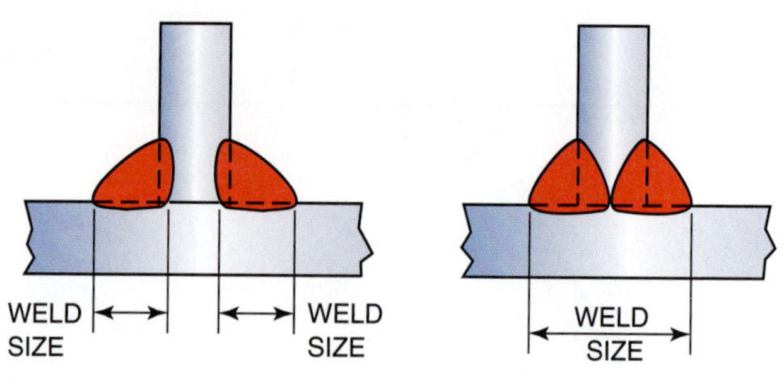

FIGURE 9-81 If the total weld sizes are equal, then both tee joints would have equal strength.

than the base plate. Heat escapes into the base plate in two directions. When using a weave pattern, most of the heat should be directed to the base plate to keep the weld size more uniform and to help prevent undercut

A welded tee joint can be strong if it is welded on both sides, even without having deep penetration, **Figure 9-81**. The weld will be as strong as the base plate if the size of the two welds equals the total thickness of the base plate. The weld bead should have a flat or slightly concave appearance to ensure the greatest strength and efficiency, **Figure 9-82**.

PRACTICE 9-22

Tee Joint in the Flat Position (1F) Using E6010 or 6011 Electrodes, E6012 or E6013 Electrodes, and E7016 or E7018 Electrodes

Using a properly set up and adjusted arc welding machine, proper safety protection, arc welding electrodes having a 1/8-in. (3-mm) diameter, and two or more pieces of mild steel plate, 6 in. (152 mm) long × 1/4 in. (6 mm) thick, you will make a welded tee joint in the flat position, **Figure 9-83**.

After the plates are tack welded together, place them on the welding table so the weld will be flat. Start at one end and establish a molten weld pool on both plates. Allow the molten weld pool to flow together before starting the bead. Any of the weave patterns will work well on this joint. To prevent slag inclusions, use a slightly higher than normal amperage setting.

When the 6-in. (152-mm) -long weld is completed, cool, chip, and inspect it for uniformity and soundness. Repeat the welds as needed for all these groups of electrodes until you can consistently make welds free of defects. Turn off the welding machine and clean up your work area when you are finished welding.

Complete a copy of the "Student Welding Report" listed in Appendix I or provided by your instructor. ◆

PRACTICE 9-23

Tee Joint in the Horizontal Position (2F) Using E6010 or E6011 Electrodes, E6012 or E6013 Electrodes, and E7016 or E7018 Electrodes

Using the same setup, materials, and electrodes as listed in Practice 9-22, you will make a welded tee joint in the horizontal position.

Place the tack welded tee plates flat on the welding table so that the weld is horizontal and the plates are flat and vertical, **Figure 9-84**. Start the arc on the flat plate and establish a molten weld pool in the root on both plates. Using the "J" or "C" weave pattern, push the arc into the root and

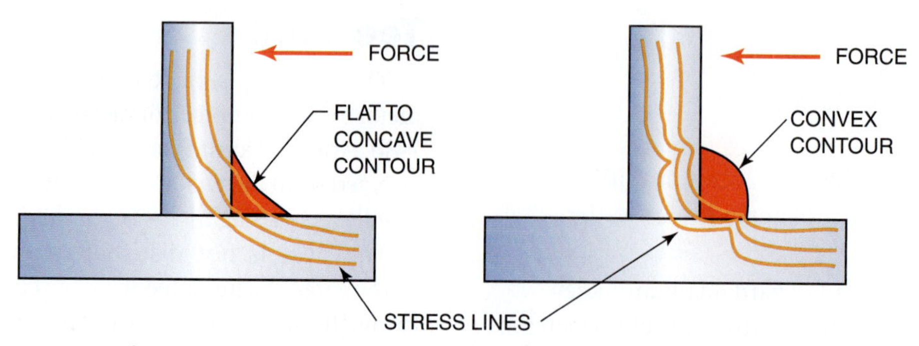

FIGURE 9-82 The stresses are distributed more uniformly through a flat or concave fillet weld.

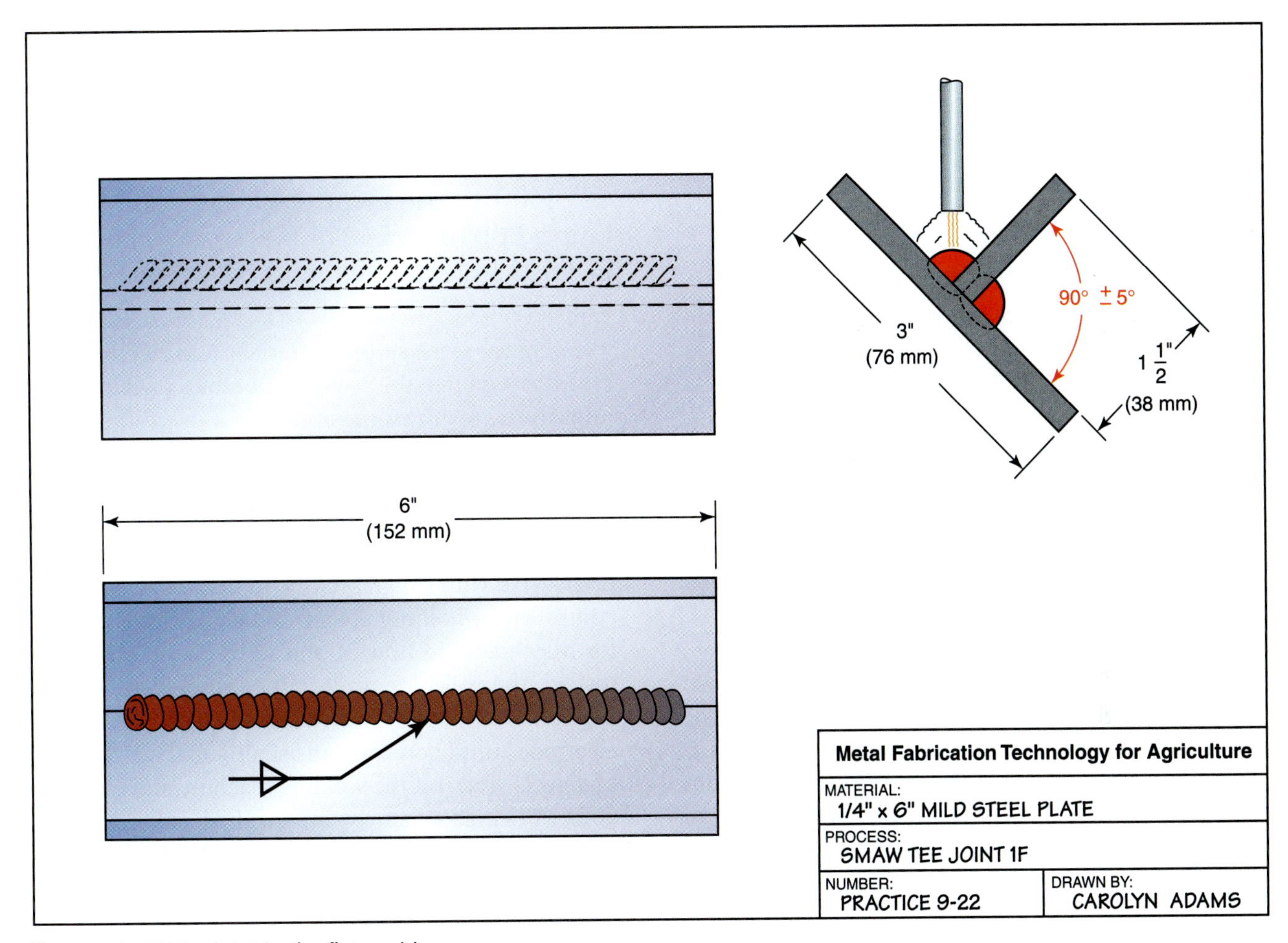

FIGURE 9-83 Tee joint in the flat position.

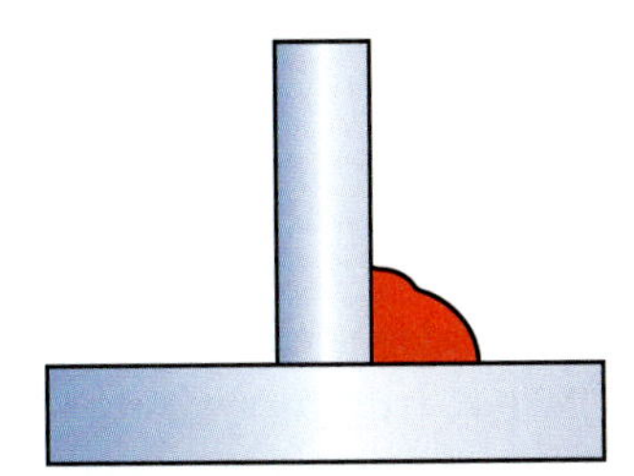

FIGURE 9-84 Horizontal tee.

slightly up the vertical plate. You must keep the root of the joint fusing together with the weld metal. If the metal does not fuse, a notch will appear on the leading edge of the weld bead. Poor or incomplete root fusion will cause the weld to be weak and easily cracked under a load.

When the weld is completed, cool, chip, and inspect it for uniformity and defects. Undercut on the vertical plate is the most common defect. Repeat the welds with all three (F) groups of electrodes until you can consistently make welds free of defects. Turn off the welding machine and clean up your work area when you are finished welding.

Complete a copy of the "Student Welding Report" listed in Appendix I or provided by your instructor. ◆

PRACTICE 9-24

Tee Joint in the Vertical Position (3F) Using E6010 or E6011 Electrodes, E6012 or E6013 Electrodes, and E7016 or E7018 Electrodes

Using the same setup, materials, and electrodes as listed in Practice 9-22, you will make a welded tee joint in the vertical position.

Practice this weld with the plate at a 45° angle. This position will allow you to develop your skill for the vertical position. Start the arc and molten weld pool deep in the root of the joint. Build a shelf large enough to support the bead as it progresses up the joint. The square, "J," or "C" pattern can be used, but the "T" or stepped pattern will allow deeper root penetration.

For this weld, undercut is a problem on both sides of the weld. It can be controlled by holding the arc

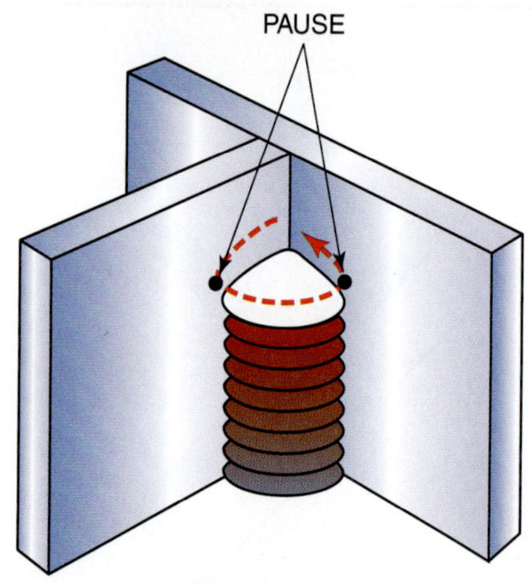

FIGURE 9-85 Pausing just above the undercut will fill it. This action also causes undercut, but that will be filled on the next cycle.

on the side long enough for filler metal to flow down and fill it, **Figure 9-85**. Cool, chip, and inspect the weld for uniformity and defects. Repeat the welds as necessary with all three (F) groups of electrodes until you can consistently make welds free of defects. Turn off the welding machine and clean up your work area when you are finished welding.

Complete a copy of the "Student Welding Report" listed in Appendix I or provided by your instructor. ◆

PRACTICE 9-25

Tee Joint in the Overhead Position (4F) Using E6010 or E6011 Electrodes, E6012 or E6013 Electrodes, and E7016 or E7018 Electrodes

Using the same setup, materials, and electrodes as listed in Practice 9-12, you will make a welded tee joint in the overhead position.

Start the arc and molten weld pool deep in the root of the joint. Keep a very short arc length. The stepped pattern will allow deeper root penetration.

For this weld, undercut is a problem on both sides of the weld with a high buildup in the center. It can be controlled by holding the arc on the side long enough for filler metal to flow in and fill it. Cool, chip, and inspect the weld for uniformity and defects. Repeat the welds as necessary with all three (F) groups of electrodes until you can consistently make welds free of defects. Turn off the welding machine and clean up your work area when you are finished welding.

Complete a copy of the "Student Welding Report" listed in Appendix I or provided by your instructor. ◆

Summary

The ability to make strong, high-quality stick welds will be of great benefit to you in agriculture. It will allow you to make repairs and build projects in metal that will benefit your operation and result in cost savings. You will not have to wait for a welder to find your place and repair a vital piece of equipment during planting, harvest, or any other critical time. The versatility of the stick-welding process means that you can repair almost anything that breaks.

As you have learned to make good welds, you now know that the adage "the bigger the glob the better the job" is not true in welding. Making the correct-size weld is important; a too large weld can cause a part to fail as easily as one that is too small.

Welding is like any other skill. You need to keep in practice. Welding at every opportunity will help you keep your skills sharp.

Review

1. ________ equipment has as one of its advantages the fact that it can be left for long periods of time without developing a problem.
   A. FCAQ
   B. OFW
   C. GTAW
   D. SMAW
2. Welding with a current that is too low results in poor fusion and poor arc stability.
   A. true
   B. false
3. The _________conducts the electricity in a stick electrode.
   A. outside covering
   B. gas cloud
   C. core wire
   D. flux
4. Extremely high amperage settings may cause the electrode to ______.
   A. burn off slower
   B. discolor
   C. create less spatter
   D. catch on fire
5. A _______ is a large piece of metal used to absorb excessive heat.
   A. chill plate
   B. sheet metal disk
   C. fin
   D. damp-off
6. The _______ is the distance the arc must jump from the end of the electrode to the plate or the weld pool surface.
   A. voltage drop
   B. electrode diameter
   C. electrode spacing
   D. arc length
7. Maintaining a constant arc length is not important to making good welds.
   A. true
   B. false
8. _____ arc lengths produce more spatter because the metal being transferred may drop outside of the molten weld pool.
   A. Short
   B. Angled
   C. Long
   D. Little
9. A _______ electrode angle pushes molten metal and slag ahead of the weld.
   A. leading
   B. trailing
   C. right angled
   D. both leading and right angled
10. What is the name for a welding bead with little or no side-to-side movement?
    A. weave bead
    B. vertical bead
    C. outside corner joint
    D. stringer bead
11. Which welding position can be done with almost any electrode weave pattern?
    A. vertical up
    B. flat
    C. horizontal
    D. overhead
12. Which weave pattern would be best for building up a surface?
    A. stepping
    B. J
    C. circular
    D. T
13. Before striking an arc you should_______ yourself.
    A. clean
    B. brace
    C. ground
    D. insulate
14. What type of joint is formed when two plates are laid side by side on a flat surface?
    A. lap joint
    B. square butt joint
    C. tee joint
    D. outside corner joint
15. Which change will not affect weld penetration?
    A. root spacing
    B. arc length
    C. amperage
    D. length of weld

Review continued

16. A good indication that the weld is being made large enough is when the weld bead width is equal to the width of the joint.
    A. true
    B. false
17. A(n) ____________is made by placing two plates at a 90° angle to each other, with the edges forming a V groove.
    A. V groove joint
    B. outside corner joint
    C. lap joint
    D. square butt joint
18. A lap joint cannot be made with two fillet welds.
    A. true
    B. false
19. As a fillet weld is made on the lap joint, the buildup should equal the thickness of the plate.
    A. true
    B. false
20. After the joint is tack welded together, it is not necessary to chip the slag before welding the joint.
    A. true
    B. false

Chapter 10

Shielded Metal Arc Welding of Pipe

OBJECTIVES

After completing this chapter, the student should be able to

- ☑ explain the difference between low-, medium-, and high-pressure pipe systems.
- ☑ describe the differences between pipe and tubing.
- ☑ list the different weld passes and explain the purposes of each one.
- ☑ demonstrate skill in welding pipe in all positions.

KEY TERMS

concave root surface
horizontal fixed pipe position
horizontal rolled pipe position
land
pipe
pressure range
root face
root gap
root suck back
tubing
vertical fixed pipe position
welding uphill or downhill

INTRODUCTION

Stick welding is very well suited to the fabrication and repair of piping systems, and equipment and structures made of pipe or tubing. Welded pipe and tubing is used for many things around the farm or ranch, such as carrying water for irrigation and watering livestock; building fences, round bale hauling equipment, and pole barns; and much more, **Figure 10-1**.

Pipe and tubing are excellent construction materials because they are strong. All of the other structural shapes, such as angle iron, channel iron, I-beams, and square tubing have a stronger and weaker side, **Figure 10-2**. They cannot withstand the same amount of bending or twisting forces in all directions but because pipe is round, it can. This makes pipe the best choice for most structures. The disadvantage to using pipe is that it is more difficult to lay out, fabricate, and weld than most other shapes. In this chapter, you will learn how to make strong pipe joints.

(A)

(B)

FIGURE 10-1 (A) Constructing a pipe corral. (B) Round bale hauling implement. Courtesy of Larry Jeffus.

The purpose for which pipe will be used largely determines how it is welded. This text groups pipe welds into the following three general categories:

- Low-pressure or light structural service
- Medium-pressure or medium structural service
- High-pressure or heavy structural service

Low-pressure, or noncritical, piping systems may be used to carry low-pressure water, natural gas, and non-hazardous materials used in agriculture. These types of noncritical piping assemblies are also found on such structural items as handrails, truck racks, columns for barns, and other light-duty products. These pipe joints must be free from pinholes, undercut, slag inclusions, or any other defect that may cause the joints to leak or break prematurely. The weld does not require 100% penetration, although penetration should be uniform. Much of the strength of these pipe joints comes from the reinforcement. These welds should always be located so that they are easily repairable if necessary.

Medium-pressure piping is used for medium-pressure water, corrosive chemicals, and waste disposal. Medium- to heavy-service welded structural items include such things as signs, equipment, railings or light posts, trailer axles, and equipment frames or stands. These pipe joints must withstand heavy loads, but their failure will not be disastrous. Much of their strength is due to weld reinforcement, but there should be 100% root penetration around most of the joint. The root, filler, and reinforcement passes are usually welded with E7018 electrodes for added strength.

High-pressure, or critical, piping systems are used for high-pressure steam, radioactive materials, the Alaskan pipeline, fired or unfired boilers, refinery reactor lines, aircraft air frames, motorcycle frames, race car roll cages, truck axles, and several other critical, heavy-duty applications. The welds on critical piping systems must be as strong or stronger than the pipe itself. Often, the pipe used for these applications is extra heavy-duty pipe, with heavier wall thicknesses.

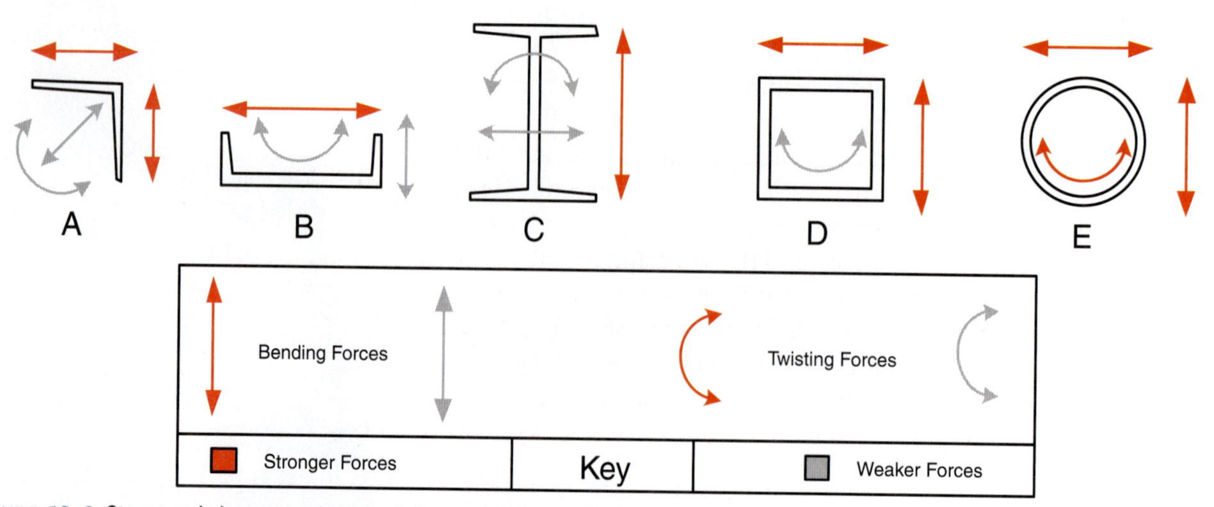

FIGURE 10-2 Structural shapes and their ability to resist bending and twisting forces.

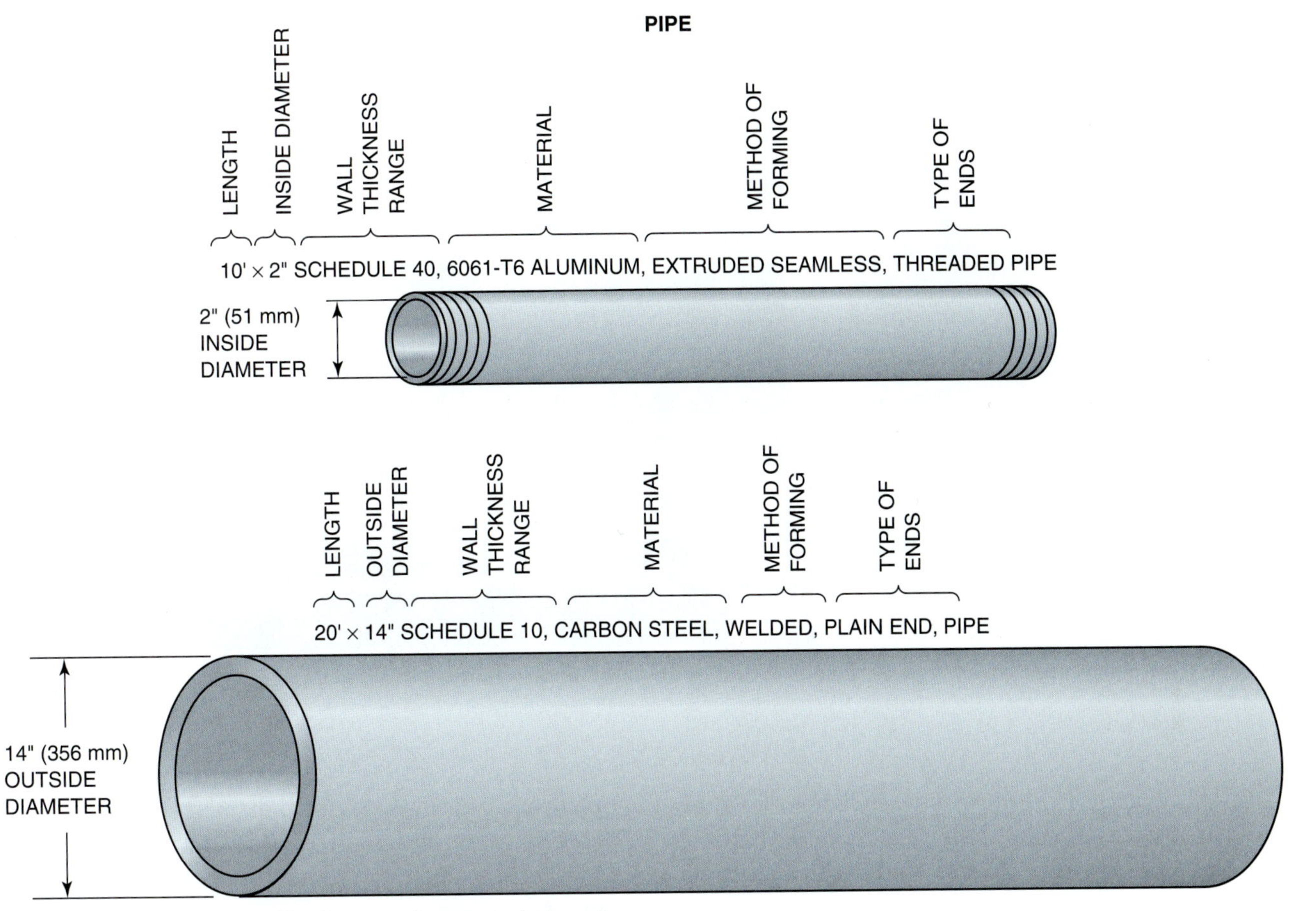

FIGURE 10-3 Typical specifications used when ordering pipe.

The weld must have 100% root penetration over 100% of the joint. Root, filler, and reinforcement weld passes are made with an E7018 or stronger electrode. The welds are usually tested, and defects are repaired.

Pipe and Tubing

Although pipe and tubing are similar in some aspects, they have different types of specifications and uses. They are only sometimes interchangeable.

The specifications for pipe sizes are given as the inside diameter for pipe 12 in. (305 mm) in diameter or smaller, and as the outside diameter for pipe larger than 12 in. (305 mm) in diameter. Tubing sizes are always given as the outside diameter. The desired shape of tubing, such as square, round, or rectangular, must also be listed with the ordering information.

The wall thickness of tubing is measured in inches (millimeters) or as U.S. standard sheet metal gauge thickness. The wall thickness for pipe is determined by its schedule, or pressure range. The larger the diameter of the pipe, the greater its area. As the area increases, so must the wall thickness in order for the wall to withstand the same pressure range, **Figure 10-3**.

The strength of pipe is given as a schedule. Schedules 10 through 180 are available; schedule 40 is often considered a standard strength. Tubing strength is the ability of tubing to withstand compression, bending, or twisting loads. Tubing should also be specified as rigid or flexible.

Pipe and tubing are both available as welded (seamed) or extruded (seamless).

Most pipe that will be welded into a system is used to carry liquids or gases from one place to another. These systems may be designed to carry large or small quantities or materials having a wide range of pressures. In agriculture, pipe is used for structural applications.

Small diameter flexible tubing is commonly used to carry pressurized liquids or gases. Ridged tubing is normally used for structural applications. Some tubing is designed for specific purposes, such as electrical mechanical tubing (EMT), which is used to protect electrical wiring. Tubing can be used to replace some standard structural shapes such as I-beams, channels, and angles for buildings. Tubing is also available in sizes that will slide one inside the other to be used in places where telescoping tubing is required, **Figure 10-4**.

FIGURE 10-4 Space shuttle launch tower is constructed using round and rectangular tubing. Courtesy of NASA.

In this chapter, the term *pipe* will refer to pipe only. However, it should be understood that the welding sequence, procedures, and skill can also be used on thick-wall round tubing.

Advantages of Welded Pipe Most pipe 1 1/2 in. (38 mm) in diameter and all steel pipe 2 in. (51 mm) and larger are generally arc welded. Welded piping systems, compared to pipe joined by any other method, are stronger, require less maintenance, last for longer periods of time, allow smoother flow, and weigh less.

Strength

The thickness of the pipe and fitting is the same when they are welded together. Threaded pipe is weakened because the threads reduce the wall thickness of the pipe, **Figure 10-5**.

Less Maintenance Required

Over much time and use, welded pipe joints are resistant to leaks.

Longer Lasting

Welded pipe joints resist corrosion caused by electrochemical reactions because all the parts are made of the same types of metal. Small cracks between the threads on threaded pipe are likely spots for corrosion to start.

Smoother Flow

The inside of a welded fitting is the same size as the pipe itself. As material flows through the pipe, less turbulence is caused by unequal diameters, **Figure 10-6**. Large irrigation systems may be several miles in length. Lowered resistance to product flow can save on operating energy costs.

Lighter Weight

Threaded fittings are larger and weigh more than welded fittings. It takes much more time to thread and assemble a joint of pipe than it does to make a welded pipe joint. Pipe fittings are also expensive and not always available.

Other advantages of welded pipe include the following:

- An ability to make specially angled fittings by cutting existing fittings, **Figure 10-7**.
- Odd-shaped parts can be fabricated.
- A lot of highly specialized equipment is not required for each different size of pipe.
- Alignment of parts is easier. It is not necessary to overtighten or undertighten fittings so that they will line up.
- Removing, replacing, or changing parts is easy because special connections are not needed to remove the parts.

Preparation and Fitup

The ends of pipe must be beveled for maximum penetration and high joint strength. The end can be beveled by flame cutting, machining, grinding, or a combination process. When maximum strength is

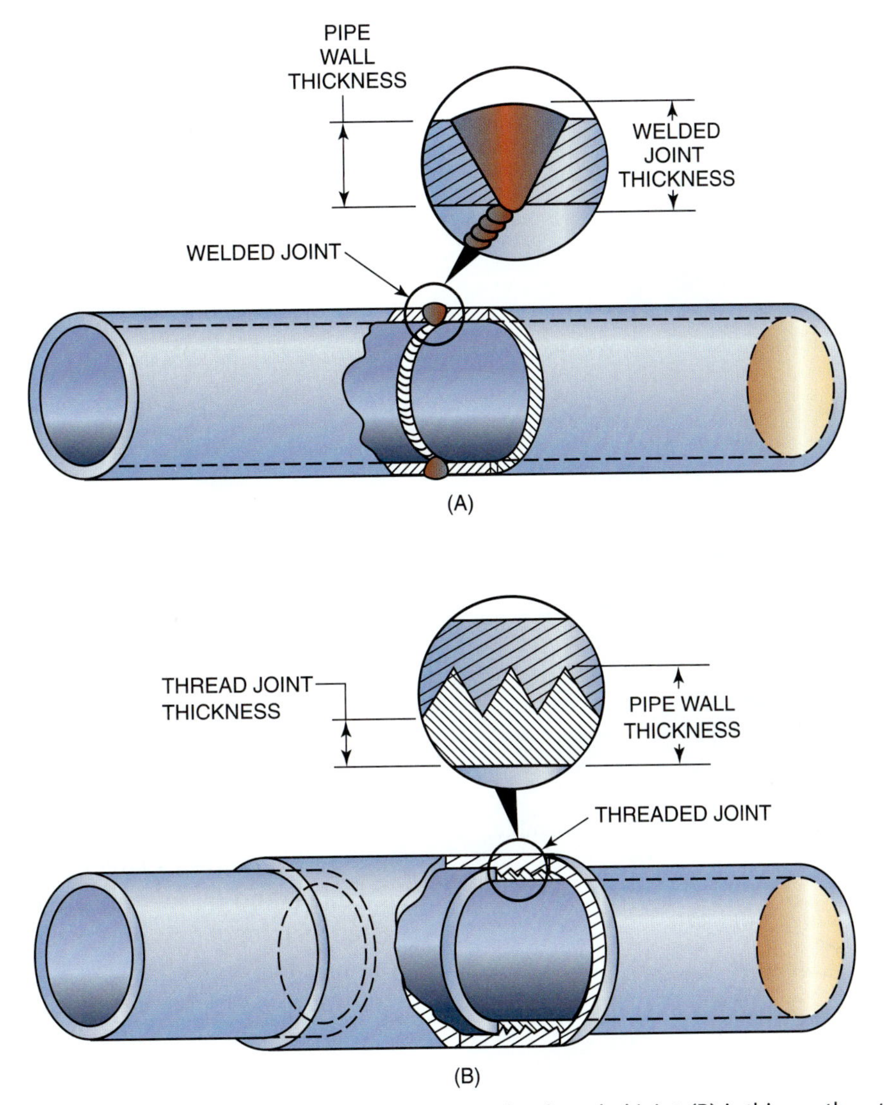

FIGURE 10-5 The welded joint (A) is thicker than the original pipe; the threaded joint (B) is thinner than the original pipe.

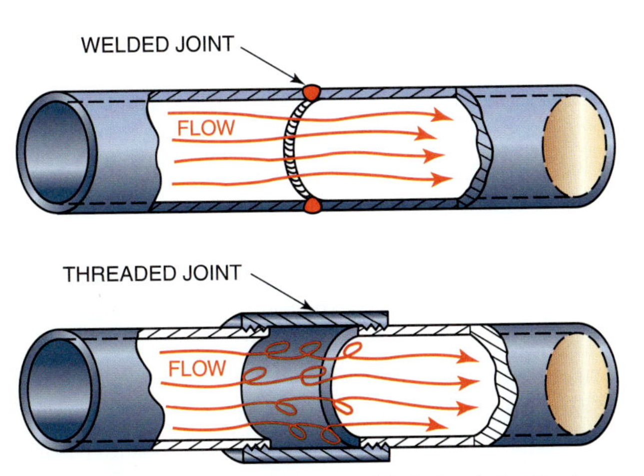

FIGURE 10-6 The flow along a welded pipe is less turbulent than that in a threaded pipe.

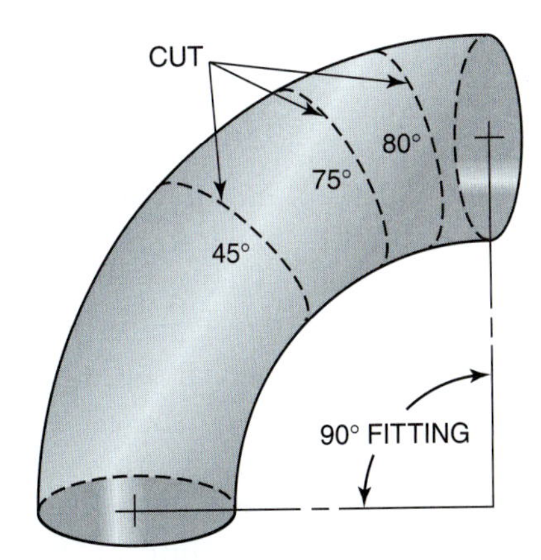

FIGURE 10-7 A standard 90° fitting can be cut to any special angle that is needed.

required, it is important that the bevel be at the correct angle, about 37 1/2°. The sharp or featured inner edge of the bevel should be ground flat forming a chamfer. This area is called a root face or land. Final shaping should be done with a grinder so that the root gap will be uniform.

The bevel on the end of the pipe is usually cut by a hand-held torch and ground smooth. Chapter 6

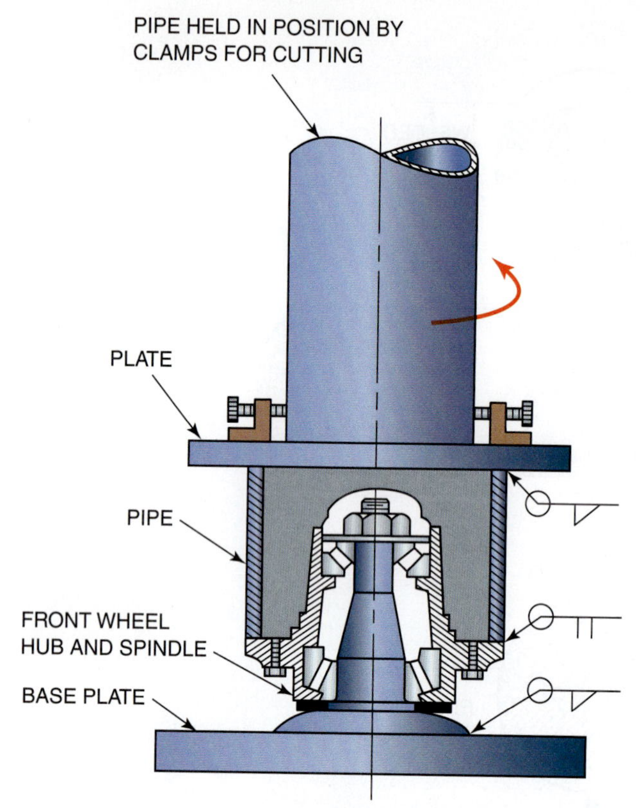

FIGURE 10-8 Turntable built from a front wheel assembly.

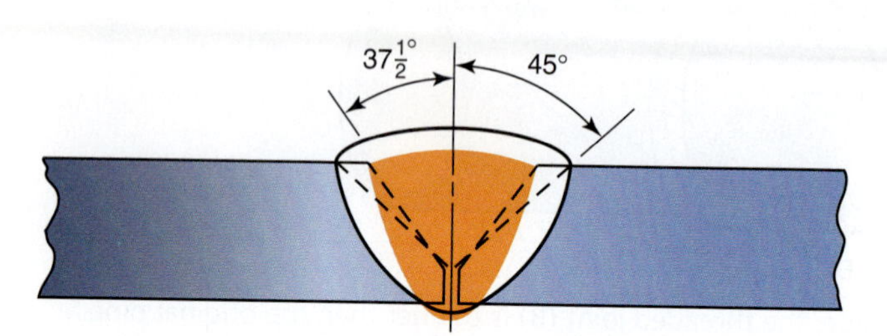

FIGURE 10-9 The 37 1/2° angled joint may use nearly 50% less filler metal, time, and heat, as compared with the 45° angled joint.

describes how to set up and operate flame-cutting equipment. A turntable, similar to the one shown in **Figure 10-8**, can be made in the ag shop and used for beveling short pieces of pipe. The turntable can be used vertically or horizontally. By turning the table slowly with pipe held between the clamps, a hand torch can be used to produce smooth pipe bevels.

The 37 1/2° angle allows easy access for the electrode with a minimum amount of filler metal required to fill the groove, **Figure 10-9**.

The root face will help a welder control both penetration and root suck back. Penetration control is improved because there is more metal near the edge to absorb excessive arc heat. This makes machine adjustments less critical by allowing the molten weld pool to be quickly cooled between each electrode movement. **Root suck back** is caused by the surface

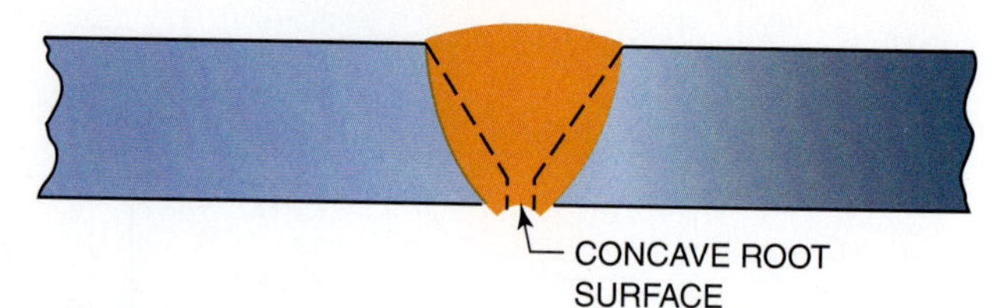

FIGURE 10-10 Root surface concavity.

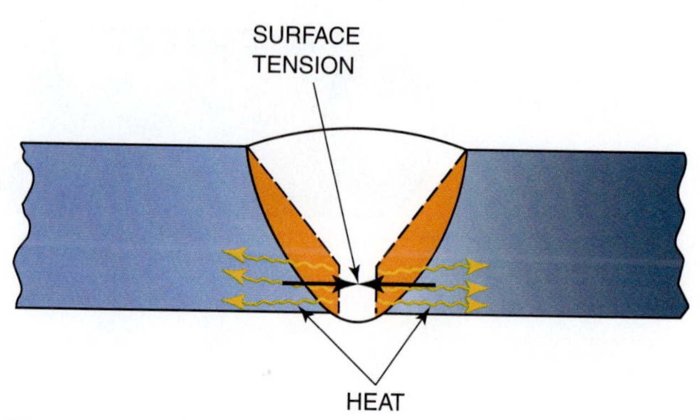

FIGURE 10-11 Heat is drawn out of the molten weld pool, and surface tension holds the pool in place.

tension of the molten metal trying to pull itself into a ball, forming a **concave root surface**, **Figure 10-10**. The root face allows a larger molten weld pool to be controlled, and because of the increased size of the molten weld pool, it is not so affected by surface tension, **Figure 10-11**.

Fitting pipe together and holding it in place for welding become more difficult as the diameter of the pipe gets larger. Devices for clamping and holding pipe in place are available, or a series of wedges and dogs can be used, **Figure 10-12**. In the Practices for this chapter, the pieces of pipe the welder will be using are about 1 1/2 in. (38 mm) wide. However, when welding on larger diameter pipe sizes, the weld specimens must be larger than 1 1/2 in. (38 mm). Welds on larger pipe sizes need more metal to help absorb the higher heat required to make these welds.

A welder can use a vise to hold the pipes in place for tack welding. If the pipe is not round and does not align properly, first tack weld the pipe together and then quickly hit the tack while holding the pipe over the horn on an anvil, **Figure 10-13**. This action will force the pipe into alignment. For pipe that is too distorted to be forced into alignment in this manner, a welder must grind down the high points to ensure a good fit.

Practice Welds

One of the major problems to be overcome in pipe welding is learning how to make the transition from one position to another. The rate of change in welding position is slower with large diameter pipes, but

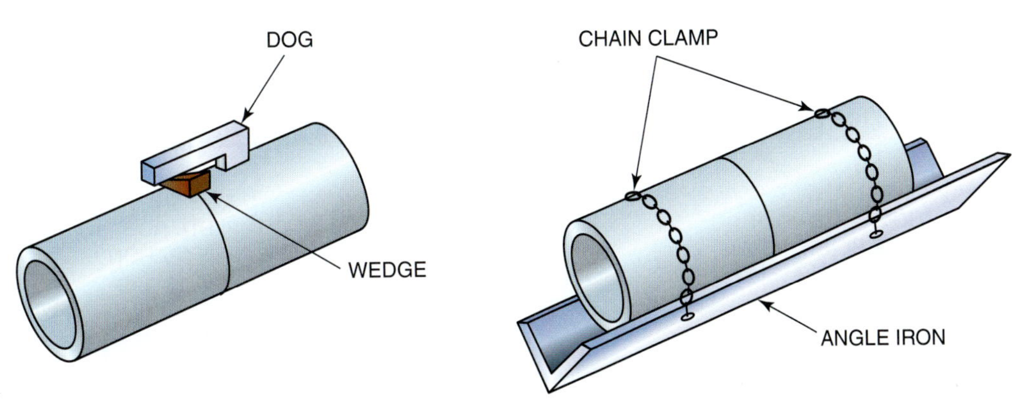

FIGURE 10-12 Shop fabrications used to align pipe joints.

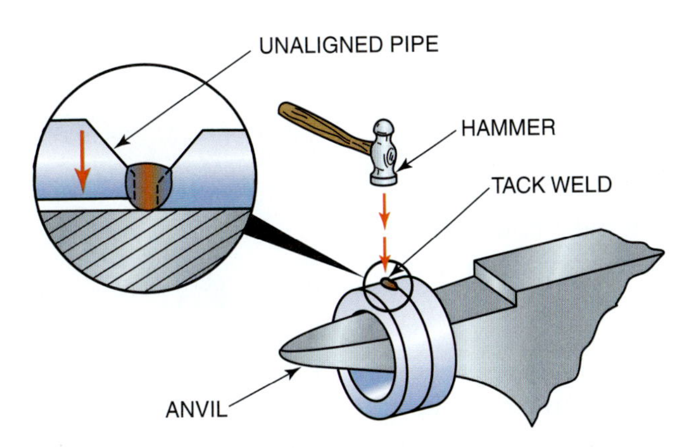

FIGURE 10-13 Hitting a hot tack weld can align a pipe joint.

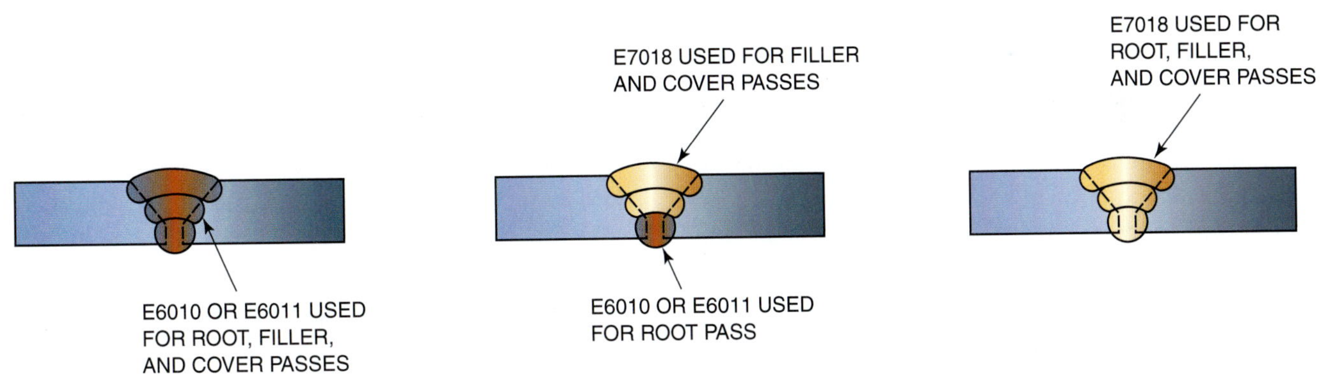

FIGURE 10-14 Single or multiple types of electrodes may be used when producing a pipe weld. The electrode selected is most often controlled by a code or specification.

the large diameter pipes require more time to weld. When a welder first starts welding, a large diameter pipe should be used in order to make learning this transition easier. As welders develop skill and the technique of pipe welding, they can change to the small diameter pipe sizes. Pipe as small as 3 in. (76 mm) can be welded quickly. It is large enough for the welder to be able to cut out test specimens.

Pipe used for these practice welds should be no shorter than 1 1/2 in. (38 mm). Pipe that is shorter than 1 1/2 in. (38 mm) rapidly becomes overheated, making welding more difficult.

To progress more quickly with pipe welds, a welder should master grooved plate welds. Once plate welding is mastered in all positions, pipe skills are faster and easier to develop.

Pipe welding is either performed with E6010 or E6011 electrodes for the complete weld, or these electrodes are used for the root pass, and E7018 electrodes are used to complete the joint. Pipe welding can also be done using the E7018 electrode for the entire weld, **Figure 10-14.**

The practice pieces of pipe used in the shop are much shorter than the pieces of pipe used in industry.

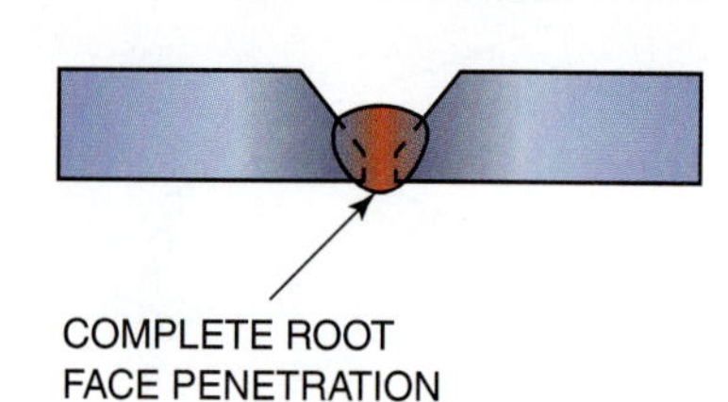

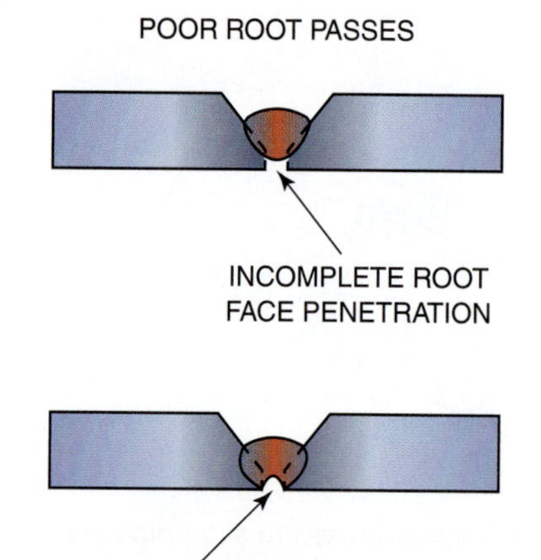

FIGURE 10-15 Root pass.

When learning to weld on short pipe, it is a good idea to avoid positioning oneself where longer pipe would eventually be located. Often it is easier to stand at the end of a pipe rather than to one side; however, this cannot be done if the weld is being made on a full length of pipe.

Weld Standards The surface of the pipe on both sides of the weld is important. No arc strikes should be made on this surface. Arc strikes outside of the weld groove may cause cracking on truck axles or other heavily loaded pipe joints. Arc strikes form small hardness spots, which, if not remelted by the weld, will crack as the pipe expands and contracts with pressure changes. Because of the importance of not having arc strikes outside the weld groove on pipe welds, you should try to avoid them from the beginning. In Chapter 9, Experiment 9-2, several techniques are described to avoid arc strikes outside of the welding zone. You may want to refer back to this section if you have difficulty in making arc starts accurately.

Root Weld A root weld is the first weld in a joint, **Figure 10-15**. It is part of a series of welds that make up a multiple pass weld. The root weld is used to establish the contour and depth of penetration. The most important part of a root weld is the internal root face, or, in the case of pipe, the inside surface, (**Figure 10-16**). The face, or outside shape, or contour of the root weld is not so important.

The face of a root weld is not important if the root surface is clean, smooth, and uniform. A grinder is used to remove excessive buildup and reshape the face of the root pass. This grinding removes slag along the sides of the weld bead and makes it easier to add the next pass. Not all root passes are ground. Pipe that is to be used in low- and medium-pressure systems is not usually ground. Grinding each root pass takes extra time and does not give the welder the experience of using a hot pass. Most slag must be completely removed by chipping before the hot pass is used.

FIGURE 10-16 The root face must be uniform. Courtesy of Larry Jeffus.

If you need more experience or practice in making an open root weld, refer to Chapter 9.

Hot Pass The hot pass is used to quickly burn out small amounts of slag trapped along the edge of the root pass. This is slag that cannot be removed easily by chipping or wire brushing. The hot pass can also be used to reshape the root pass by using high current settings and a faster than normal travel speed.

Slag is mostly composed of silicon dioxide, which melts at about 3100°F (1705°C). Steel melts at approximately 2600°F (1440°C). A temperature of more than 500°F (270°C) hotter than the surrounding metal is required to melt slag. The slag can be floated to the surface by melting the surrounding metal. A high current will quickly melt enough surface to allow the slag to float free; a fast travel speed will prevent burnthrough. The fast travel speed forms a concave weld bead that is easy to clean for the welds that will follow.

Filler Pass After thoroughly removing slag from the weld groove by chipping, wire brushing, or grinding, it is ready to be filled. The filler pass(es) may be either a series of stringer beads, **Figure 10-17**, or a weave bead, **Figure 10-18.** Stringer beads require less welder skill because of the small amount of metal that is molten at one time. Stringer beads are as strong or stronger than weave beads.

The weld bead crater must be cleaned before the next electrode is started. Failure to clean the crater will result in slag inclusions. On high-strength, high-pressure pipe welds, the crater should be slightly ground to ensure its cleanliness, **Figure 10-19.** When the bead has gone completely around the pipe, it should continue past the starting point so that good fusion is ensured, **Figure 10-20.** The locations of starting and stopping spots for each weld pass must be staggered. The weld groove should be filled level with these beads so that it is ready for the cover pass.

Cover Pass The final covering on a weld is referred to as the cover pass or cap. It may be a weave or stringer bead. The cover pass should not be too wide or have too much reinforcement, **Figure 10-21.** Cover

FIGURE 10-17 Filler pass using stringer beads. Courtesy of Larry Jeffus.

ELECTRODE DIAMETER	BEAD WIDTH
$\frac{1"}{8}$ (3 mm)	$\frac{1"}{4}$ (6 mm)
$\frac{5"}{32}$ (4 mm)	$\frac{5"}{16}$ (8 mm)
$\frac{3"}{16}$ (4.8 mm)	$\frac{3"}{8}$ (10 mm)

FIGURE 10-18 Filler pass using weave bead. The bead width should not be more than two times the rod diameter. Courtesy of Larry Jeffus.

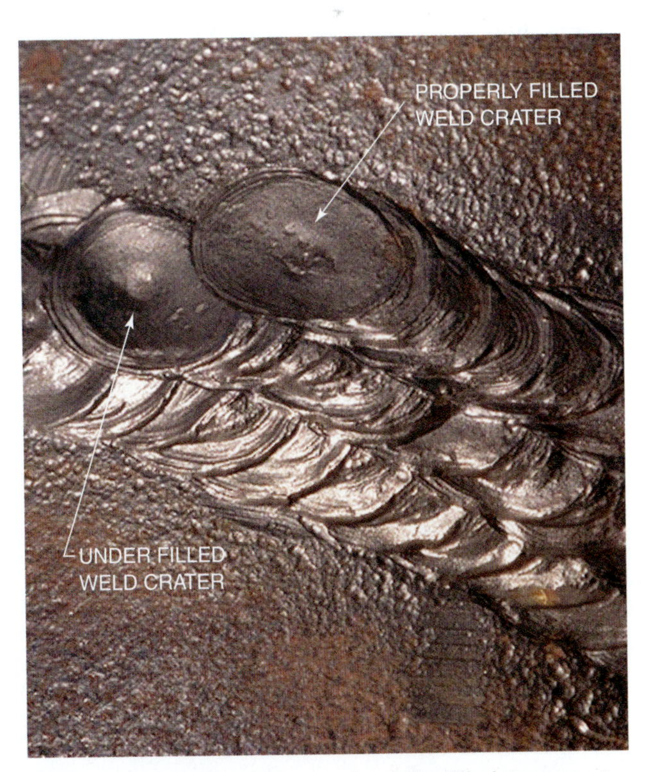

FIGURE 10-19 The weld crater should be filled to prevent cracking and cleaned of slag before restarting the arc. Courtesy of Larry Jeffus.

FIGURE 10-20 Avoid starting and stopping all weld passes in the same area. Courtesy of Larry Jeffus.

passes that are excessively large will reduce the pipe's strength, not increase it. A large cover pass will cause the stresses in the pipe to be concentrated at the sides of the weld. An oversized weld will not allow the pipe to expand and contract uniformly along its length. This concentration is similar to the restriction a rubber band would have on an inflated balloon if it were put around its center.

The cover pass should be kept as uniform and as neat-looking as possible, **Figure 10-22.** A visual check is often all that low- and medium-pressure welds receive, and a nice-looking cover will pass testing each time. A good cover pass, during a visual inspection, is used to indicate that the weld underneath is sound.

1G Horizontal Rolled Position

The horizontal rolled pipe position is commonly used in shops where structures or small systems can be positioned for the convenience of the welder, **Figure 10-23.** The penetration and buildup of the weld are controlled more easily with the pipe in this position. Weld visibility and welder comfort are improved so that welder fatigue is less of a problem. The pipe can be rolled continuously with some types of positioners, and the weld can be made in one continuous bead.

PRACTICE 10-1

Beading, 1G Position, Using E6010 or E6011 Electrodes and E7018 Electrodes

Using a properly set up and adjusted arc welding machine, proper safety protection, E6010 or E6011 and E7018 arc welding electrodes, having a 1/8-in. (3-mm) diameter, schedule 40 mild steel pipe 3 in. (76 mm) or larger in diameter, you will make a straight stringer bead around a horizontally rolled pipe.

Place the pipe horizontally on the welding table in a vee block made of angle iron, **Figure 10-24.** The vee block will hold the pipe steady and allow it to be moved easily between each bead. Strike an arc on the pipe at the 11 o'clock position. Make a stringer bead over the 12 o'clock position, stopping at the 1 o'clock position, **Figure 10-25.** Roll the pipe until the end of the weld is at the 11 o'clock position. Clean the weld crater by chipping and wire brushing.

Strike the arc again and establish a molten weld pool at the leading edge of the weld crater. With the molten weld pool reestablished, move the electrode back on the weld bead just short of the last full ripple,

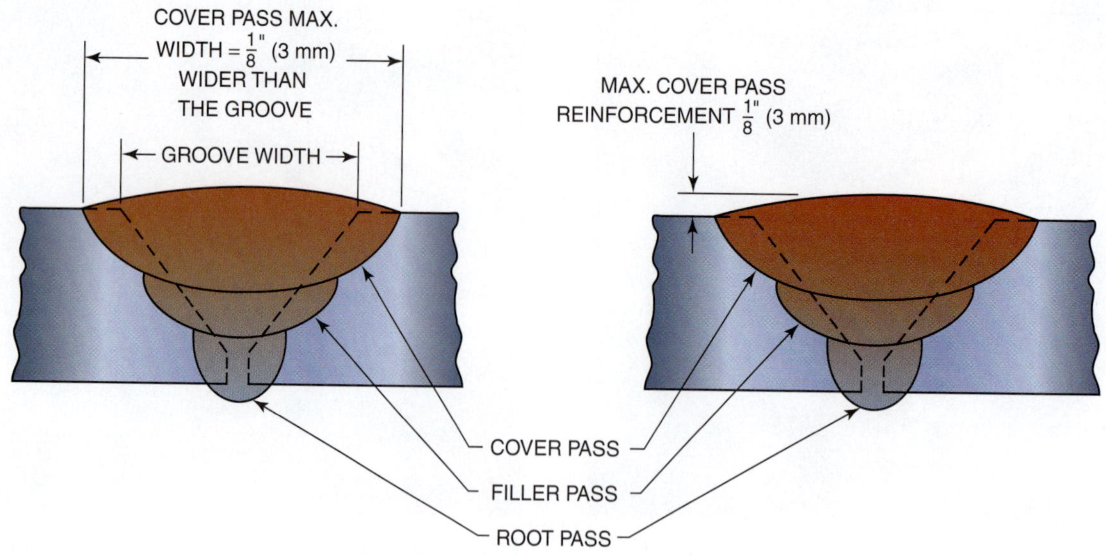

FIGURE 10-21 Excessively wide or built-up welds restrict pipe expansion at the joint, which may cause premature failure. Check the appropriate code or standard for exact specifications.

FIGURE 10-22 Uniformity in each pass shows a high degree of welder skill and increases the probability the weld will pass testing. Courtesy of Larry Jeffus.

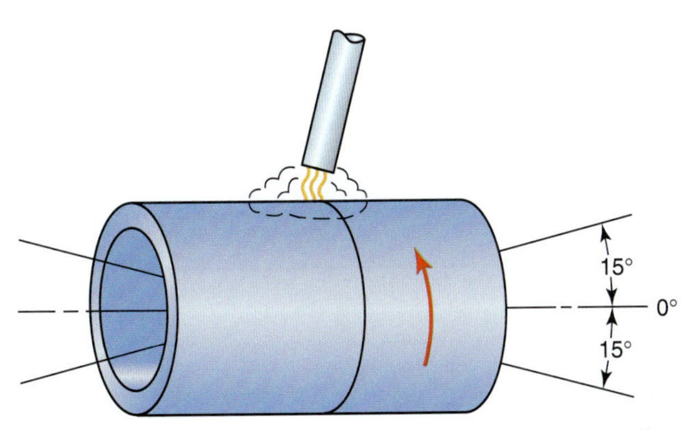

FIGURE 10-23 1G position. The pipe is rolled horizontally.

Figure 10-26. This action will both reestablish good fusion and keep the weld bead size uniform. Now that the new weld bead is tied into the old weld, continue welding to the 1 o'clock position again. Stop welding, roll the pipe, clean the crater, and resume welding. Keep repeating this procedure until the weld is completely around the pipe. Before the last weld is started, clean the end of the first weld so that the end and beginning beads can be tied together smoothly. When you reach the beginning bead, swing your electrode around on both sides of the weld bead. A poor beginning of a weld bead is always high and narrow and has little penetration, **Figure 10-27**. By swinging the weave pattern (the "C" pattern is best) on both sides of the bead, you can make the bead correctly so the width is uniform. The added heat will give deeper penetration at the starting point. Hold the arc in the crater for a moment until it is built up but do not overfill the crater.

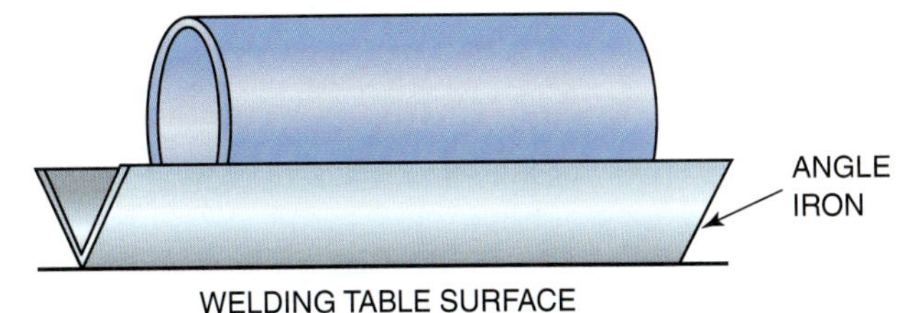

FIGURE 10-24 Angle iron pipe support.

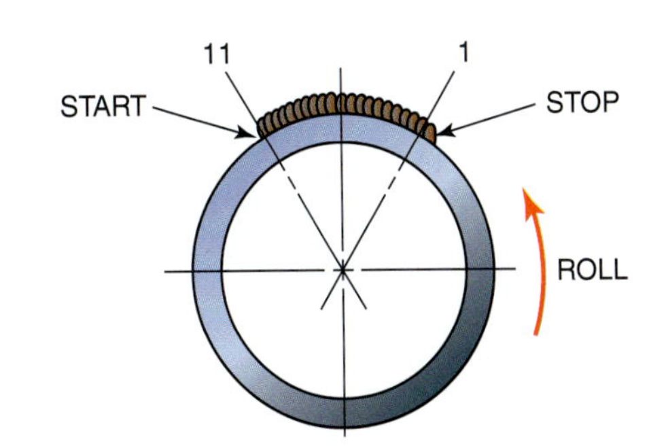

FIGURE 10-25 1G pipe welding area.

Cool, chip, and inspect the bead for defects. Repeat the beads as needed until they are mastered. Turn off the welding machine and clean up your work area when you are finished welding.

Complete a copy of the "Student Welding Report" listed in Appendix I or provided by your instructor. ◆

PRACTICE 10-2

Butt Joint, 1G Position, Using E6010 or E6011 Electrodes

Using a properly set up and adjusted arc welding machine, proper safety protection, E6011 or E6011 arc welding electrodes having a 1/8-in. (3-mm) diameter, and two or more pieces of schedule 40 mild steel pipe 3 in. (76 mm) or larger in diameter, you will make a pipe butt joint in the 1G horizontal rolled position, **Figure 10-28**.

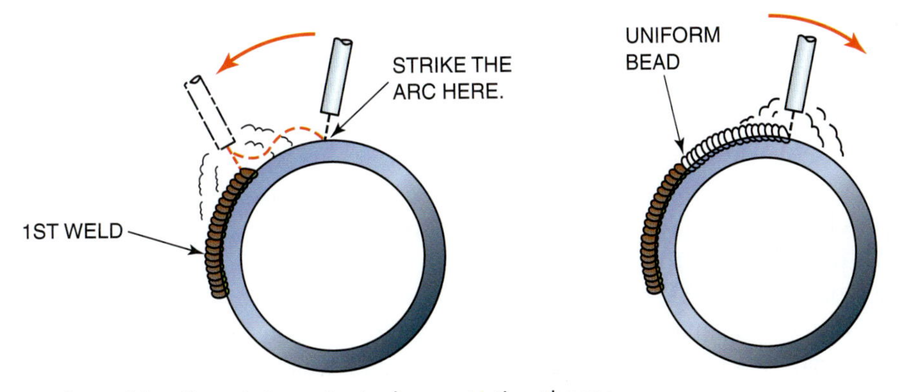

FIGURE 10-26 Keeping the weld uniform is important when restarting the arc.

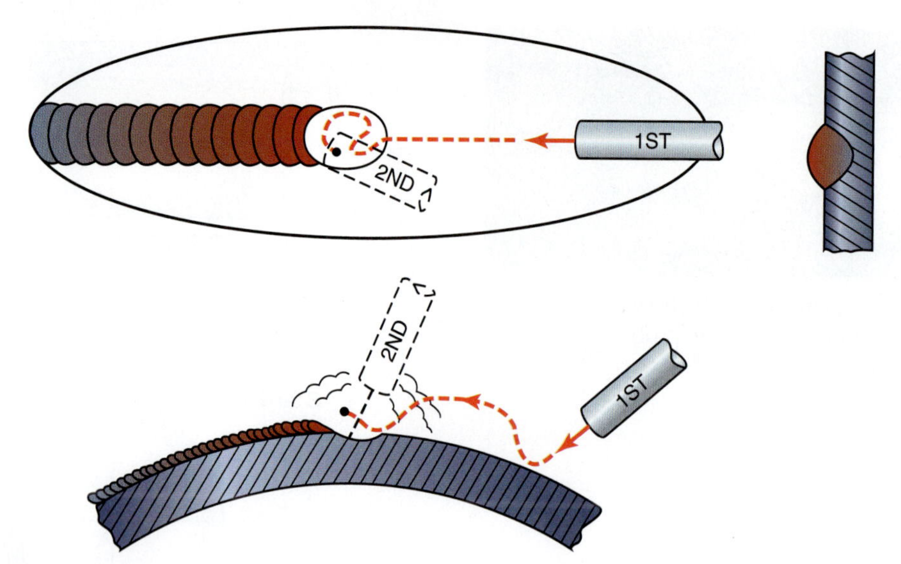

FIGURE 10-27 Restarting the weld.

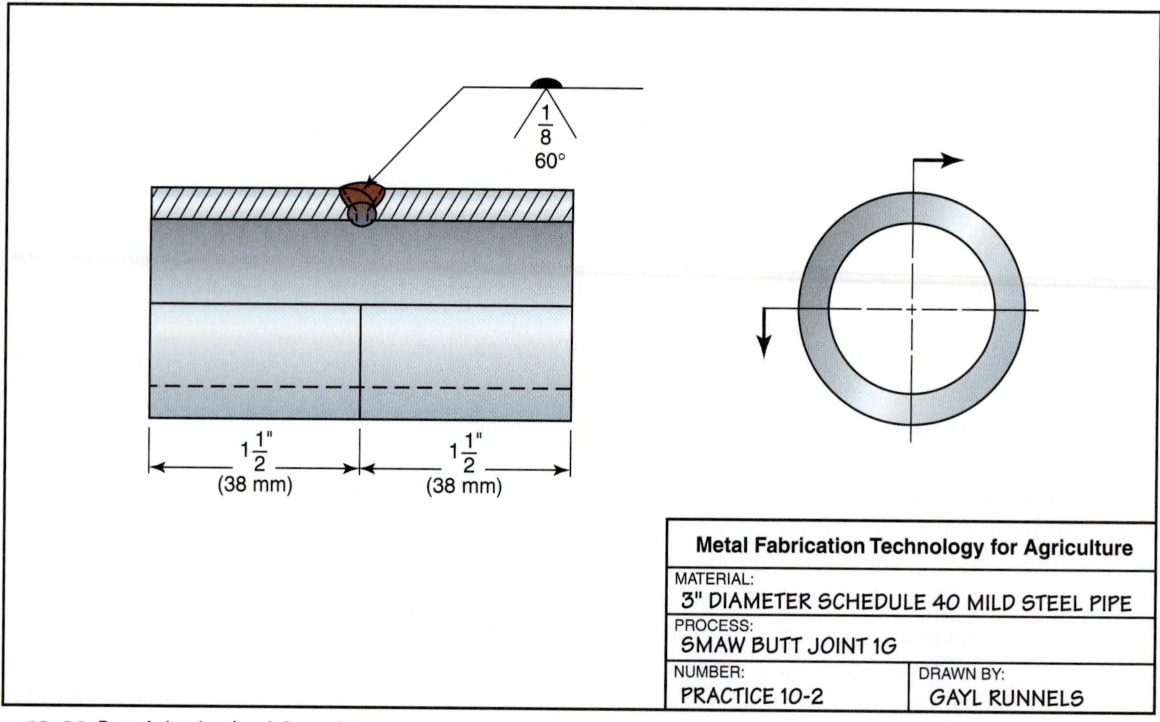

FIGURE 10-28 Butt joint in the 1G position.

Tack weld two pieces of pipe together as shown in **Figure 10-29**. Place the pipe horizontally in a vee block on the welding table. Start the root pass at the 11 o'clock position. Using a very short arc and high current setting, weld toward the 1 o'clock position. Stop and roll the pipe, chip the slag, and repeat the weld until you have completed the root pass.

Clean the root pass by chipping and wire brushing. The root pass should not be ground this time. Replace the pipe in the vee block on the table so that the hot pass can be done. Turn up the machine amperage, enough to remelt the root weld surface, for the hot pass. Use a stepped electrode pattern, moving forward each time the molten weld pool washes out the slag, and returning each time the molten weld pool is nearly all solid, **Figure 10-30**. Weld from the 11 o'clock position to the 1 o'clock position before stopping, rolling, and chipping the weld. Repeat this procedure until the hot pass is complete.

The filler pass and cover pass may be the same pass on this joint. Turn down the machine amperage. Use a "T," "J," "C," or zigzag pattern for this weld. Start the weld at the 10 o'clock position and stop at the 12 o'clock position. Sweep the electrode so that the molten weld pool melts out any slag trapped by the hot pass. Watch the back edge of the bead to see that the

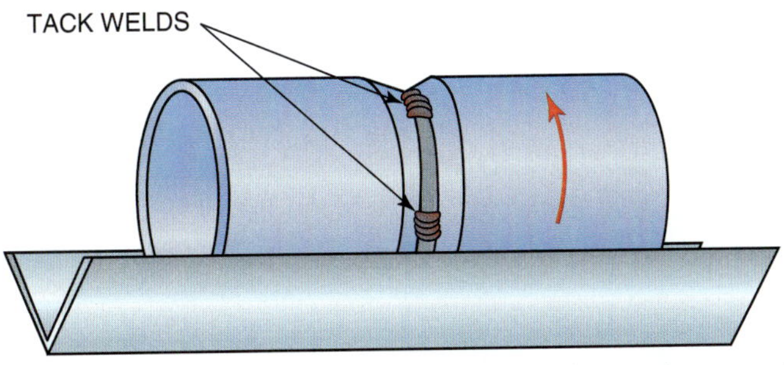

FIGURE 10-29 The tack welds are to be evenly spaced around the pipe. Use four tacks on small diameter pipe and six or more on large diameter pipe.

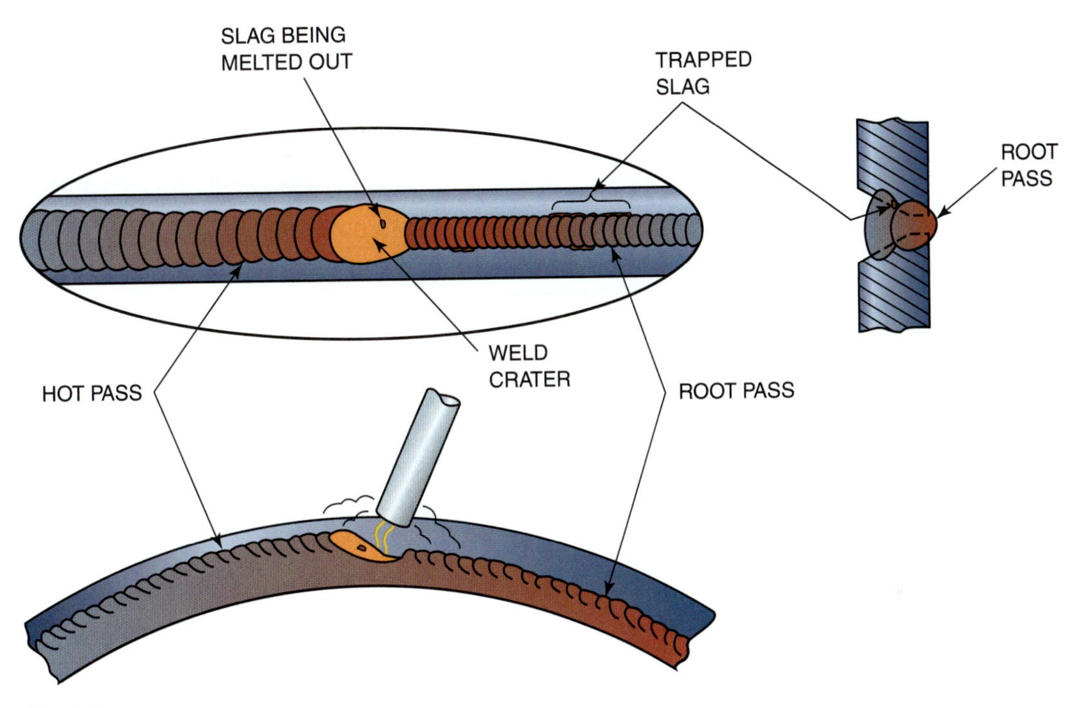

FIGURE 10-30 Hot pass.

molten weld pool is filling the groove completely. Turn, chip, and continue the bead until the weld is complete. Repeat this weld until you can consistently make welds free of defects. Turn off the welding machine and clean up your work area when you are finished welding.

Complete a copy of the "Student Welding Report" listed in Appendix I or provided by your instructor. ◆

PRACTICE 10-3

Butt Joint, 1G Position, Using E6010 or E6011 Electrodes for the Root Pass with E7018 Electrodes for the Filler and Cover Passes

Using the same setup, materials, and procedures as described in Practice 10-2, you will make a horizontal rolled butt joint in pipe, **Figure 10-31**. The root pass is to have 100% penetration over 80% or more of the length of the weld.

Set the pipe in the vee block on the welding table and make the root pass as explained in Practice 10-2. Watch for 100% penetration with no icicles. A hot pass or grinder can be used to clean the face of the root pass. Use an E7018 electrode for the filler and cover passes. The E7018 electrode should not be weaved more than 2 1/2 times the diameter of the electrode. Excessively wide weaving will allow the molten weld pool to become contaminated, **Figure 10-32.**

After the weld is completed, visually inspect it for 100% penetration around 80% of the root length. Check the weld for uniformity and visual defects on the cover pass. Repeat this weld until you can consistently make welds free of defects. Turn off the welding machine and clean up your work area when you are finished welding.

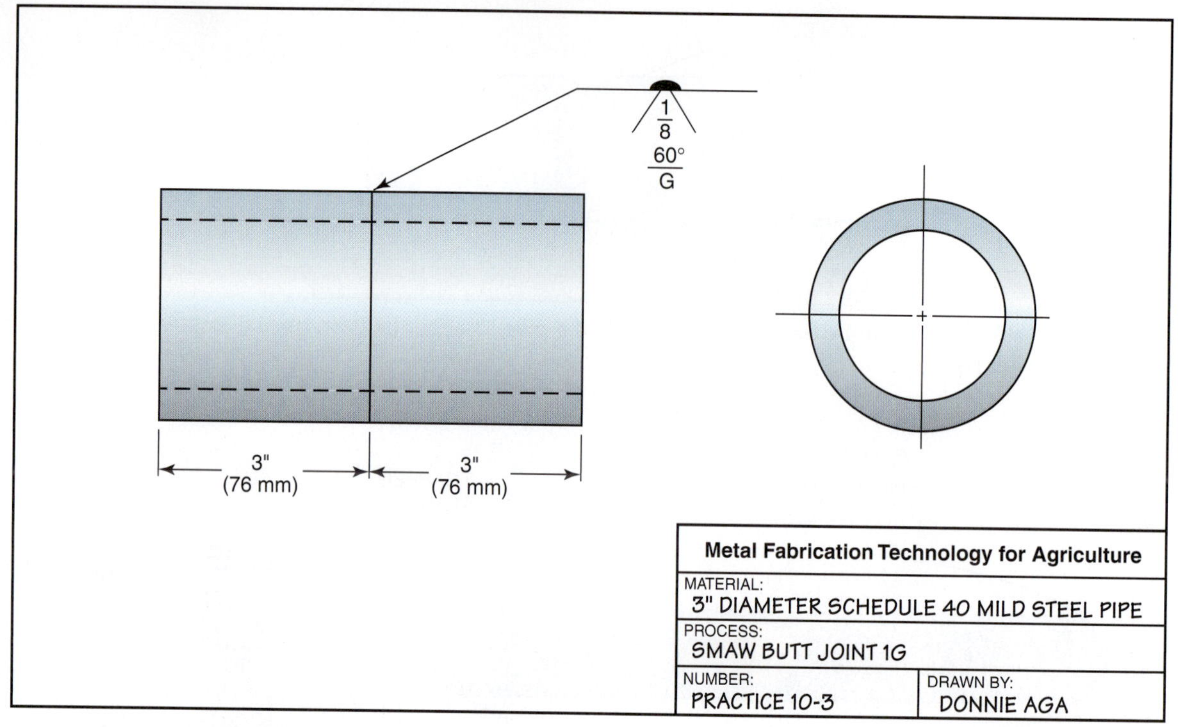

FIGURE 10-31 Butt joint in the 1G position to be tested.

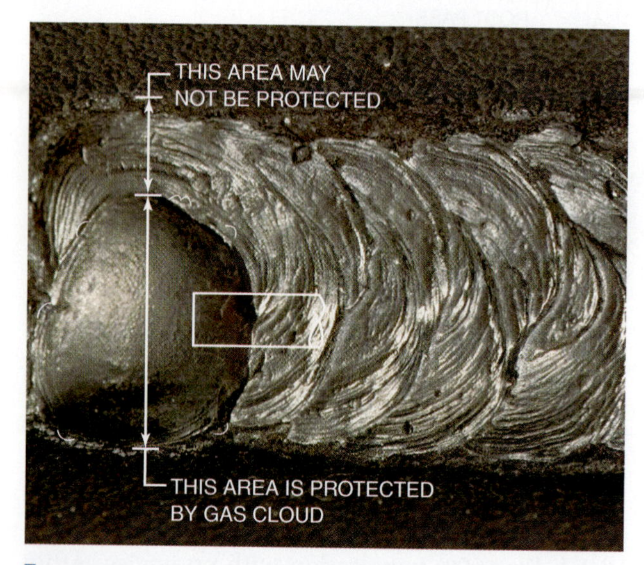

FIGURE 10-32 Weave beads more than two and one half times the diameter of the electrode may be nice looking, but the atmosphere may contaminate the unprotected part of the molten weld pool. Courtesy of Larry Jeffus.

Complete a copy of the "Student Welding Report" listed in Appendix I or provided by your instructor. ◆

2G Vertical Fixed Position

In the 2G vertical fixed pipe position, the pipe is vertical and the weld is horizontal, **Figure 10-33**. With these welds, the welder does not need to change welding positions constantly. The major problem that faces welders when welding pipe in this position is that the area to be welded is often located in corners. Because of this location, reaching the back side of the weld is difficult. In the practices that follow, you may turn the pipe between welds. As a welder gains more experience, welds in tight places will become easier.

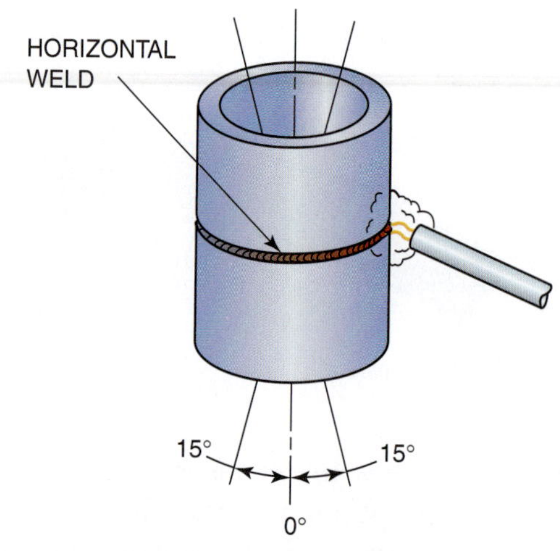

FIGURE 10-33 2G position. The pipe is fixed vertically, and the weld is made horizontally around it.

The welds must be completed in the correct sequence, **Figure 10-34**. The root pass goes in as with other joints. To reduce the sagging of the bottom of the weld, increase the electrode to work angle. As long as the weld is burned in well and does not have cold lap on the bottom, the weld is correct. Each of the filler and cover welds that will follow must be supported by the previous weld bead.

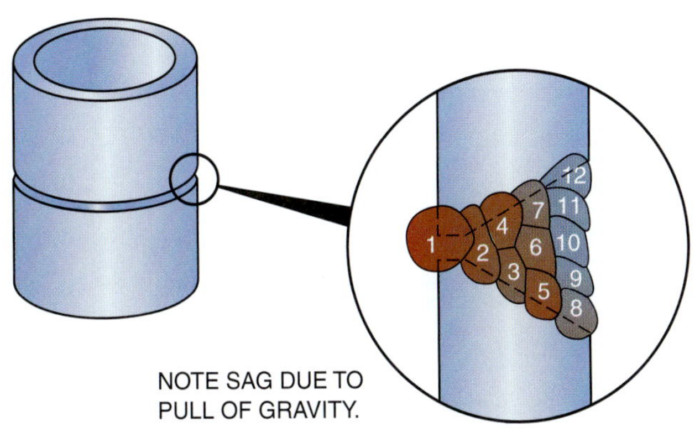

FIGURE 10-34 2G pipe welding position.

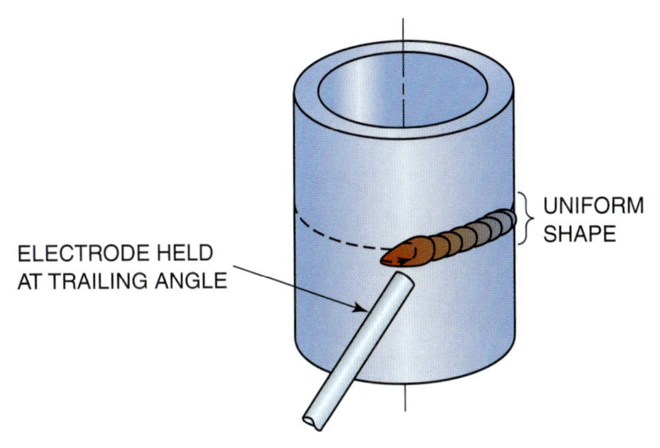

FIGURE 10-35 Electrode position and weave pattern for a weld on vertical pipe.

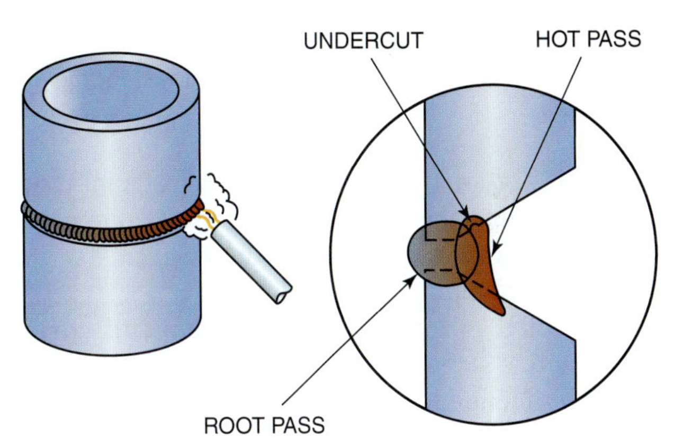

FIGURE 10-36 Hot pass.

PRACTICE 10-4

Stringer Bead, 2G Position, Using E6010 or E6011 Electrodes and E7018 Electrodes

Using the same setup, materials, and electrodes as listed in Practice 10-1, you will make straight stringer beads on a pipe that is in the vertical position.

The "J" weave pattern should be used so that the molten weld pool will be supported by the lower edge of the solidified metal, **Figure 10-35**. Keep the electrode at an upward and trailing angle so the arc force will help to keep the weld in place.

Repeat these stringer beads as needed, with both groups of electrodes, until you can consistently make welds free of defects. Turn off the welding machine and clean up your work area when you are finished welding.

Complete a copy of the "Student Welding Report" listed in Appendix I or provided by your instructor. ◆

PRACTICE 10-5

Butt Joint, 2G Position, Using E6010 or E6011 Electrodes

Using a properly set up and adjusted arc welding machine, proper safety protection, E6010 or E6011 arc welding electrodes having a 1/8-in. (3-mm) diameter, and two or more pieces of schedule 40 mild steel pipe 3 in. (76 mm) or larger in diameter, you will make a butt joint on a pipe that is in the vertical position.

Place the pipe on the arc welding table. Strike an arc and make a root weld that is as long as possible. If the root gap is uniform, a step pattern must be used. After completing and cleaning the root pass, make a hot pass. The hot pass need only burn the root pass clean, **Figure 10-36**. Undercut on the top pipe is acceptable.

The filler and cover passes should be stringer beads. By keeping the molten weld pool size small, control is easier. Cool, chip, and inspect the completed weld for uniformity and defects. Repeat this weld until you can consistently make welds free of defects. Turn off the welding machine and clean up your work area when you are finished welding.

Complete a copy of the "Student Welding Report" listed in Appendix I or provided by your instructor. ◆

PRACTICE 10-6

Butt Joint, 2G Position, Using E6010 or E6011 Electrodes for the Root Pass and E7018 Electrodes for the Filler and Cover Passes

Using the same setup, materials, and procedures as described in Practice 10-4, you will make a vertical fixed pipe weld. The root pass is to have 100% penetration over 80% or more of the length of the weld.

Place the pipe vertically on the welding table. Hold the electrode at a 90° angle to the pipe axis and with a slight trailing angle, **Figure 10-37**. The electrode should be held tightly into the joint. If a burnthrough occurs, quickly push the electrode back over the burnthrough while increasing the trailing

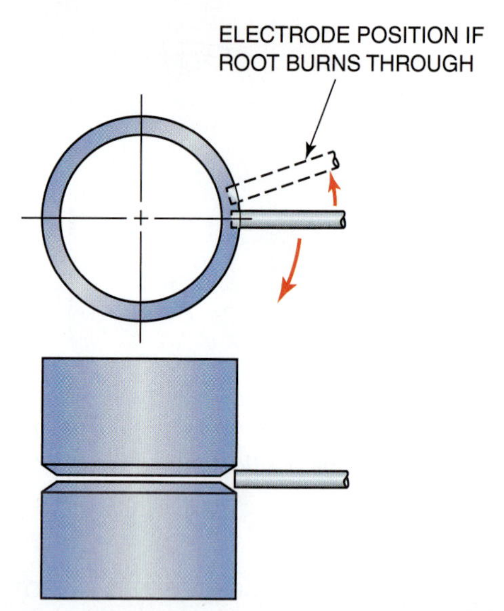

FIGURE 10-37 Electrode position and movement for the root pass.

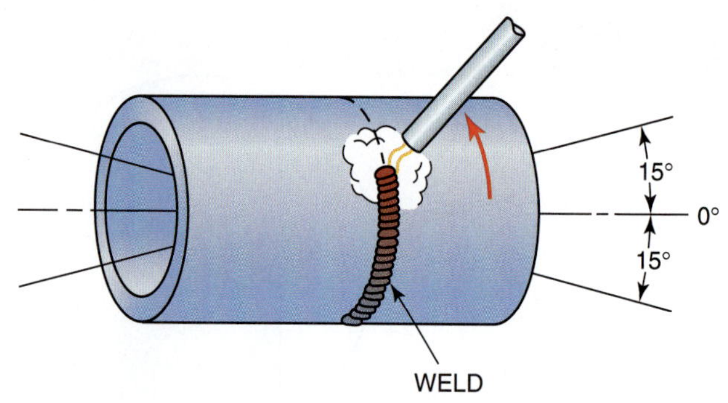

FIGURE 10-38 5G horizontal fixed position.

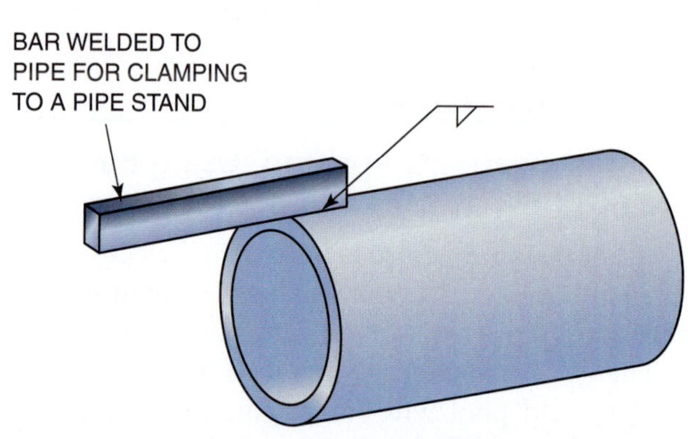

FIGURE 10-39 Holding the pipe in place by welding a piece of flat stock to the pipe and then clamping the flat stock to a pipe stand.

angle. This action forces the weld metal back into the opening. When the root pass is complete, chip the surface slag and then clean out the trapped slag by grinding or chipping, or use a hot pass.

Use E7018 electrodes for the filler and cover passes with a stringer pattern. The weave beads are not recommended with this electrode and position because they tend to undercut the top and overlap the bottom edge. After the weld is completed, visually inspect it for 100% penetration around 80% of the root length. Check the weld for uniformity and visual defects on the cover pass. Repeat the weld until you can consistently make welds free of defects. Turn off the welding machine and clean up your work area when you are finished welding.

Complete a copy of the "Student Welding Report" listed in Appendix I or provided by your instructor. ◆

5G Horizontal Fixed Position

The 5G horizontal fixed pipe position is the most often used pipe welding position. Welds produced in flat, vertical up or vertical down, and overhead positions must be uniform in appearance and of high quality.

When practicing these welds, mark the top of the pipe for future reference. Moving the pipe will make welding easier, but the same side must stay on the top at all times, **Figure 10-38.**

The root pass can be performed by welding uphill or downhill. In industry, the method used to weld the root pass is determined by established weld procedures. If there are no procedures requiring a specific direction, the choice is usually made based upon fitup. A close parallel root opening can be welded uphill or downhill. A root opening that is wide or uneven must be welded uphill. In the following Practices, the welder can make the choice of direction, but both directions should be tried.

The pipe may be removed from the welding position for chipping, wire brushing, or grinding. The pipe can be held in place by welding a piece of flat stock to it and clamping the flat stock to a pipe stand, **Figure 10-39.**

The electrode angle should always be upward, **Figure 10-40.** Changing the angle toward the top and bottom will help control the bead shape. The bead, if welded downhill, should start before the 12 o'clock position and continue past the 6 o'clock position to ensure good fusion and tie-in of the welds. The arc must always be struck inside the joint preparation groove.

PRACTICE 10-7

Stringer Bead, 5G Position, Using E6010 or E6011 Electrodes and E7018 Electrodes

Using the same setup, materials, and electrodes as listed in Practice 10-1, you will make straight stringer beads in the horizontal fixed 5G position using both groups of electrodes.

Clamp the pipe horizontally between waist and chest level. Starting at the 11 o'clock position, make a downhill straight stringer bead through the 12 o'clock

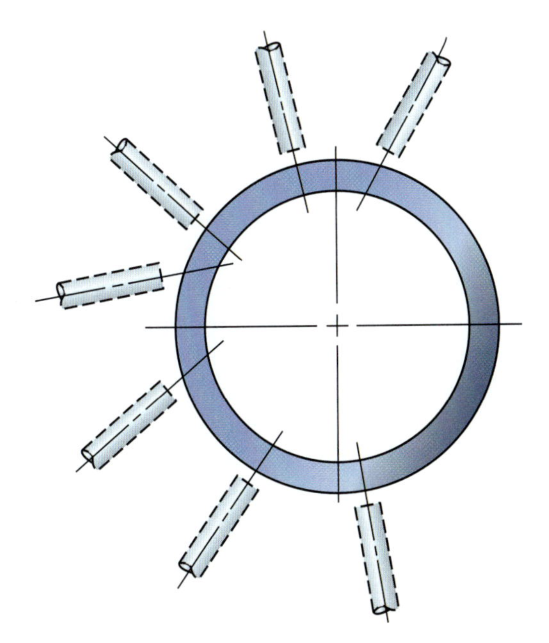

FIGURE 10-40 Electrode angle.

and 6 o'clock positions. Stop at the 7 o'clock position, **Figure 10-41**. Using a new electrode, start at the 5 o'clock position and make an uphill straight stringer bead through the 6 o'clock and 12 o'clock positions. Stop at the 1 o'clock position. Change the electrode angle to control the molten weld pool.

Repeat these stringer beads as needed, with each group of electrodes, until you can consistently make welds free of defects. Turn off the welding machine and clean up your work area when you are finished welding.

Complete a copy of the "Student Welding Report" listed in Appendix I or provided by your instructor. ◆

PRACTICE 10-8

Butt Joint, 5G Position, Using E6010 or E6011 Electrodes for the Root Pass and E7018 Electrodes for the Filler and Cover Passes

Using the same setup, materials, and procedures as listed in Practice 10-2, you will make a horizontal fixed 5G pipe weld. The root pass is to have 100% penetration over 80% or more of the length of the weld.

Mark the top of the pipe and mount it horizontally between waist and chest level. Weld the root pass uphill or downhill using E6010 or E6011 electrodes. Either grind the root pass or use a hot pass to clean out trapped slag.

Use E7018 electrodes for the filler and cover passes with stringer or weave patterns. When the weld is completed, visually inspect it for 100% penetration around 80% of the root length. Check the weld for uniformity and visual defects on the cover pass. Repeat the weld until you can consistently make welds free of

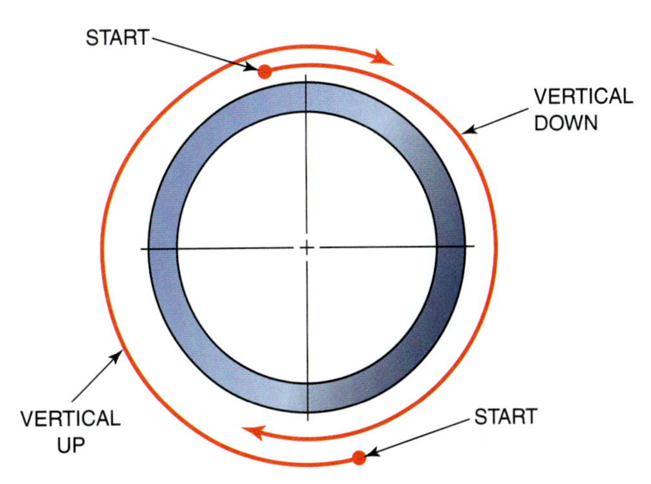

FIGURE 10-41 Stop at the 7 o'clock position.

defects. Turn off the welding machine and clean up your work area when you are finished welding.

Complete a copy of the "Student Welding Report" listed in Appendix I or provided by your instructor. ◆

PRACTICE 10-9

Butt Joint, 5G Position, Using E6010 or E6011 Electrodes

Using a properly set up and adjusted arc welding machine, proper safety protection, E6010 or E6011 arc welding electrodes having a 1/8-in. (3-mm) diameter, and two or more pieces of schedule 40 mild steel pipe 3 in. (76 mm) or larger in diameter, you will make a butt joint on a horizontally fixed 5G pipe.

Mark the top of the pipe and mount it between waist and chest level. Depending upon the root gap, make a root weld uphill or downhill using E6010 or E6011 electrodes. Check the root penetration to determine if it is better in one area than in another area. Chip and wire brush the weld and set the machine for a hot pass. Start the hot pass at the bottom and weld upward on both sides. The bead should be kept uniform with little buildup.

Using stringer or weave beads, make the filler and cover welds. If stringer beads are used, downhill welds can be made. Cool, chip, and inspect the weld for uniformity and defects. Repeat this weld until you can consistently make welds free of defects. Turn off the welding machine and clean up your work area when you are finished welding.

Complete a copy of the "Student Welding Report" listed in Appendix I or provided by your instructor. ◆

Pipe Tee Joints

The tee joint is the second most common pipe joint, used in the construction of almost every pipe structure from fences to barns, **Figure 10-42.** The round shape of the pipe makes it more difficult to fit together than other structural shapes. The two parts of a pipe tee joint are the straight pipe, called the top, and the intersecting pipe called the saddle. The most difficult part to fit is the saddle.

There are several ways the tee joint can be made to fit the top pipe. One way is to bend the pipe; another way involves cutting a saddle. The easiest way to get small diameter pipes to fit together is to collapse the end of the tee pipe's sides slightly inward, **Figure 10-43.** This joint is easier to make but offers less strength than cutting the saddle to fit around the pipe. Saddled tee joints can be cut into any diameter pipe. Laying out and cutting the saddle to shape will take practice.

FIGURE 10-42 Pipe barn structure. Courtesy of Larry Jeffus.

PRACTICE 10-10

Tee Joint 1G Position in Small Diameter Pipe, Using E6010 or E6011 Electrodes

Using the same setup, materials, and electrodes as listed in practice 10-1, you will make a fillet weld on small diameter pipe fitted in a 90° joint, **Figure 10-44.**

The end of the tee pipe is to be collapsed inward slightly. Pipe that is 1 in. (25mm) or less in diameter can be sheared off, automatically forming the collapse. On larger pipe, up to 2 in. (50mm), the end can be beaten flat using a hammer and anvil. Heating the end before the pipe is bent can be done on any diameter pipe, but it is required on pipe larger than 2 in. (50 mm) in diameter.

If the end of the pipe is not collapsed completely, it will make a stronger joint. Leaving an opening about 1/4 of the original pipe diameter will make the joint easy to weld and will give it adequate strength for most applications, like the fence in **Figure 10-45.**

Lay the pipes on a flat surface and use a square to set them at a 90° angle, and then make a small track weld in three places, **Figure 10-44.** The gap along the side of the joint will be wider than normal, **Figure 10-46.** To make a weld across this gap you can use the technique of repeatedly striking and breaking the arc. Hold the arc only long enough to deposit a small amount of metal. Stop the arc and watch the molten weld pool cool. Once it has cooled to a dull red, re-strike the arc. Repeat this process as often as needed until the molten weld pool bridges the gap. Turn the pipe over and put a tack weld on the other side of the joint.

With the pipe laying flat, strike an arc at the edge of the joint and make a weld along the one side. The side of the pipe can absorb more heat than the end of the tee joint, so using a J weave will keep much of the heat off the end of the tee joint, **Figure 10-47.** If the weld bead begins to sag or drops through, stop welding con-

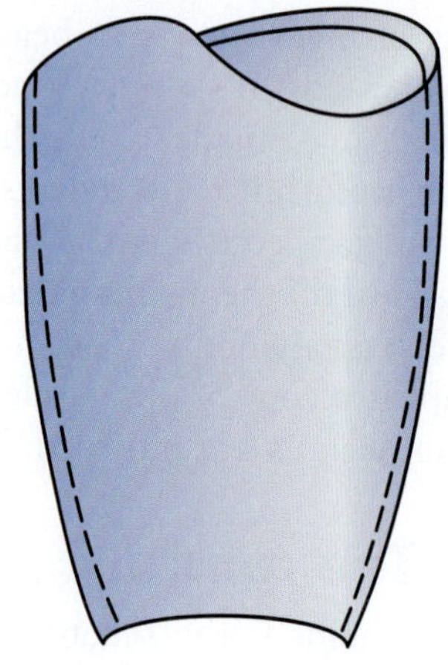

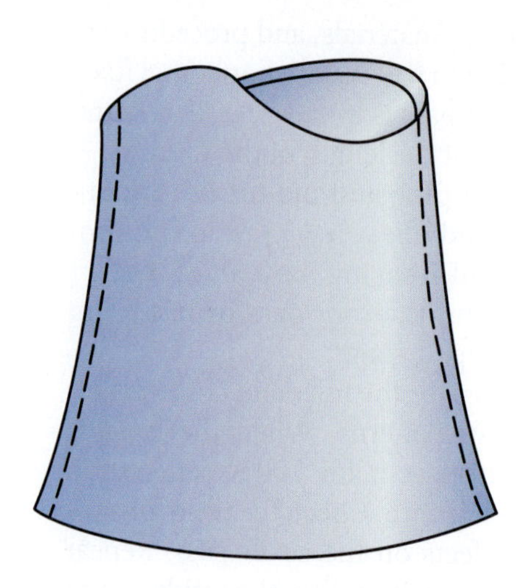

FIGURE 10-43 Flattened pipe end for T-joint.

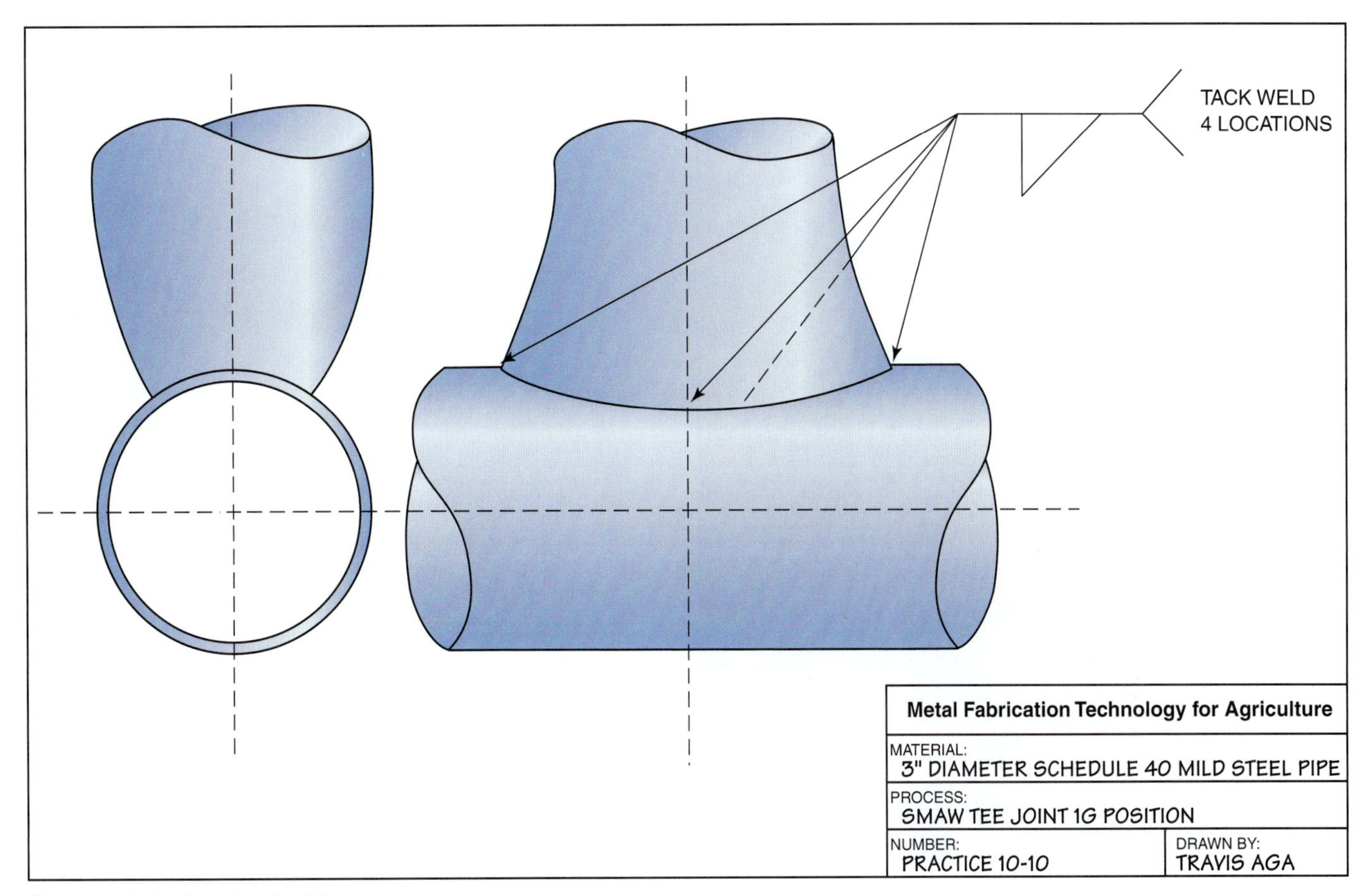

FIGURE 10-44 Practice 10-10.

tinuously and use the same starting and stopping technique used for the tack weld. If the welding becomes very hot because of the slow travel speed with the wide gap, stop welding and start a new bead on the other end. Intermittently swap welding from one end to the other so the pipe cools. Continue doing this back and forth until the weld bead meets in the middle.

Turn the pipe over and make a weld on the other side. Check the ends of the tee to see that there is no hole or gap that water could get into. If necessary, stand the pipe up and make a weld across the end.

After the weld is complete, visually inspect it for uniformity and visual defects. Repeat this weld until you can consistently make welds that are free of defects. Turn off the welding machine and clean up your work area when you are finished welding.

Complete a copy of the "Student Welding Report" listed in Appendix I or provided by your instructor. ◆

Laying Out a Saddle Pipe Tee Joint

The saddle end can be laid out using a pipe-layout tool or a template. The layout tool shown in **Figure 10-48** is the fastest and easiest to use. It also can be used on a wide range of pipe diameters. Templates can be purchased or shop made; see Chapter 17 for a detailed description of how to create shop-made pipe templates.

Lay out the saddle on the pipe using a template or layout tool. Make sure the layout is straight by checking to see that the high and low lobes are opposed to each other. Also check to see that they are the same distance from the end of the pipe, **Figure 10-49.**

Place the pipe horizontally. Put on all required safety equipment, including gloves and cutting goggles. Follow all safety rules and use a properly lit and adjusted oxyacetylene cutting torch to cut out the saddle. A detailed description of flame cutting pipe is located in Chapter 6. Cool the pipe and clean off any slag from the cut.

Check the fit of the saddle to the pipe. For pipe that is being used for fences, barns, and other structures the joint gap at the saddle can be much wider than that acceptable for high-pressure piping. For most agricultural application the gap on pipe 2 in. (50 mm) and smaller should be no more than 1/4 in. (6 mm), **Figure 10-50A.** Larger diameter pipe can have a gap of approximately 3/8 in. (9 mm). If necessary, mark the high spots and use a hand grinder to remove them so the proper fit can be obtained, **Figure 10-50B.**

FIGURE 10-45 Pipe end bent to form a rail in a stock pen. Courtesy of Larry Jeffus.

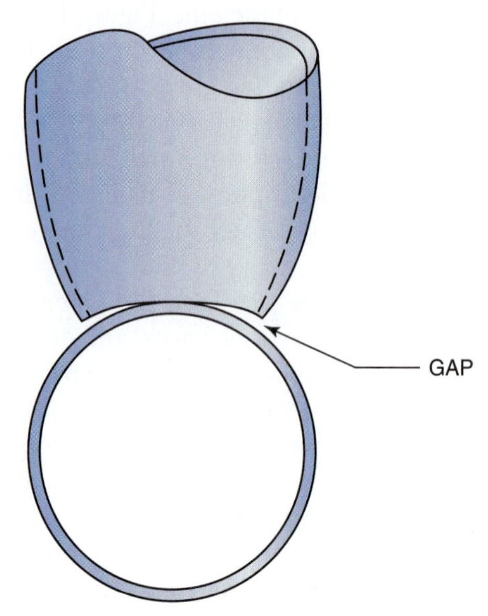

FIGURE 10-46 Large gap may be formed on the side of the T-joint.

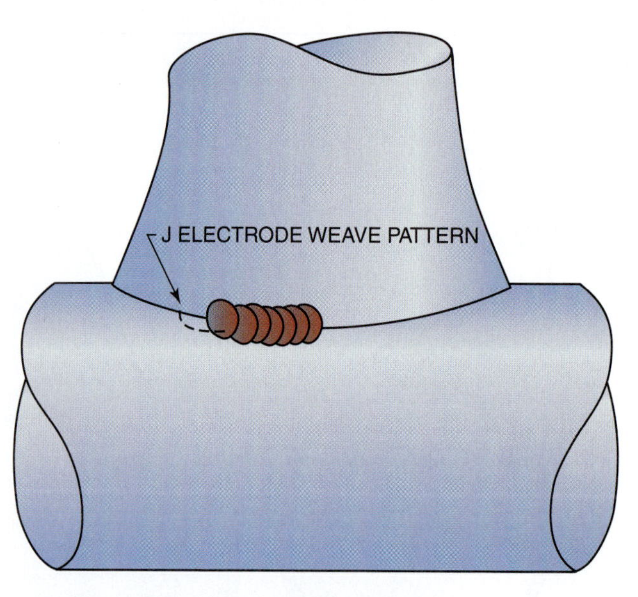

FIGURE 10-47 Use a J electrode weave pattern to keep excessive heat from the edge of the pipe.

FIGURE 10-48 Pipe layout tool. Courtesy of Larry Jeffus.

CAUTION

Always use proper eye protection and other safeguards when using a grinder.

Using a saddle as a template, draw a soapstone line around the straight pipe to locate the hole, **Figure 10-51**. The line drawn on the pipe will be the outside diameter of the saddle pipe. The hole needs to be cut slightly smaller to the outside diameter of the saddle pipe. Use a tape measure to locate several setback points around the outside line. Sketch a line through the setback points to show the size of the cutout.

Place the pipe horizontally and use the same safety equipment, including gloves and cutting goggles, and properly lit and adjusted oxyacetylene cutting torch. Following all safety rules, cut out the saddle hole. A detailed description of flame cutting pipe

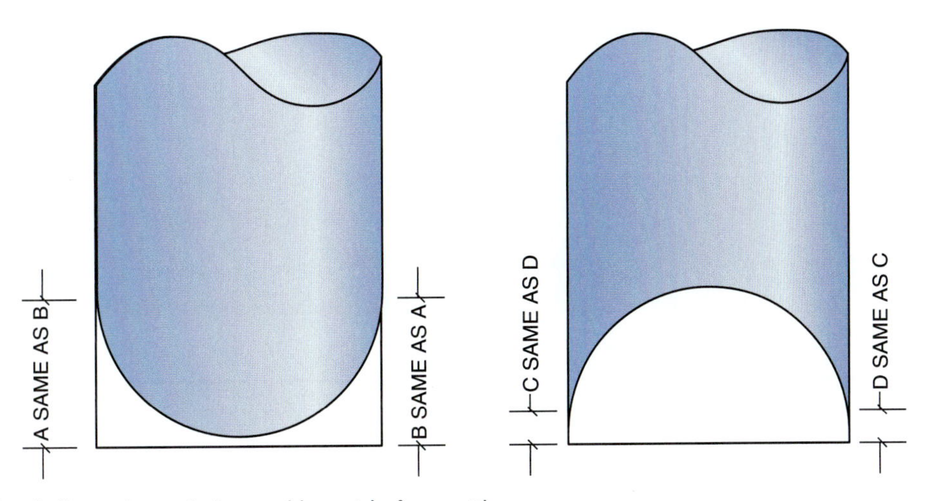

FIGURE 10-49 Check dimensions of pipe end layout before cutting.

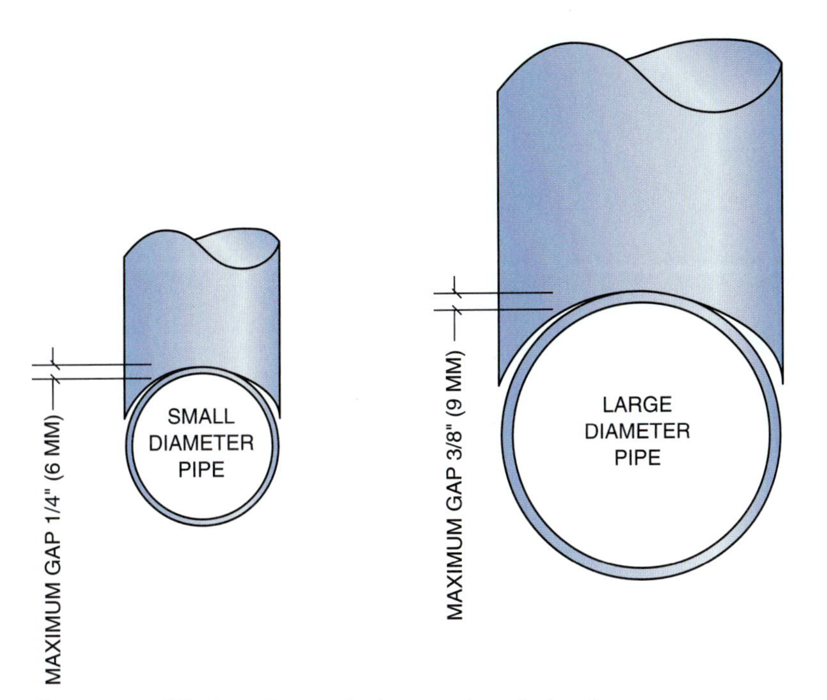

FIGURE 10-50(A) Gap tolerance for large and small pipe sizes.

FIGURE 10-50(B) Pipe joint cut and fitted ready for clean up and welding. Photo courtesy of Larry Jeffus.

is in Chapter 6. Cool the pipe and clean off any slag from the cut.

NOTE: Pipe joints being used for structural applications will be stronger if the hole is not cut.

PRACTICE 10-11

Saddled Tee Joint 5G Position in Pipe, Using E6010 or E6011 Electrodes

Using the same setup, materials, and electrodes as listed in Practice 10-1, you will make a fillet weld on a saddled pipe fitting, **Figure 10-52.**

Lay the pipes on a flat surface and use a square to set them at a 90° angle. Then make a small tack weld in three places. Turn the pipe over and put a tack weld on the other side of the joint.

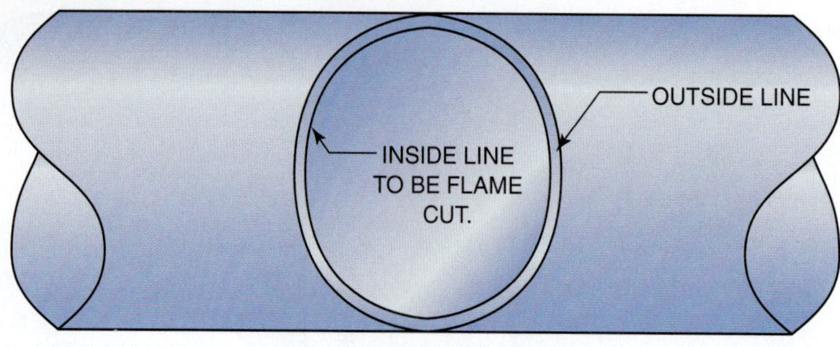

FIGURE 10-51 Laying out the hole to be flame cut for a pipe T joint.

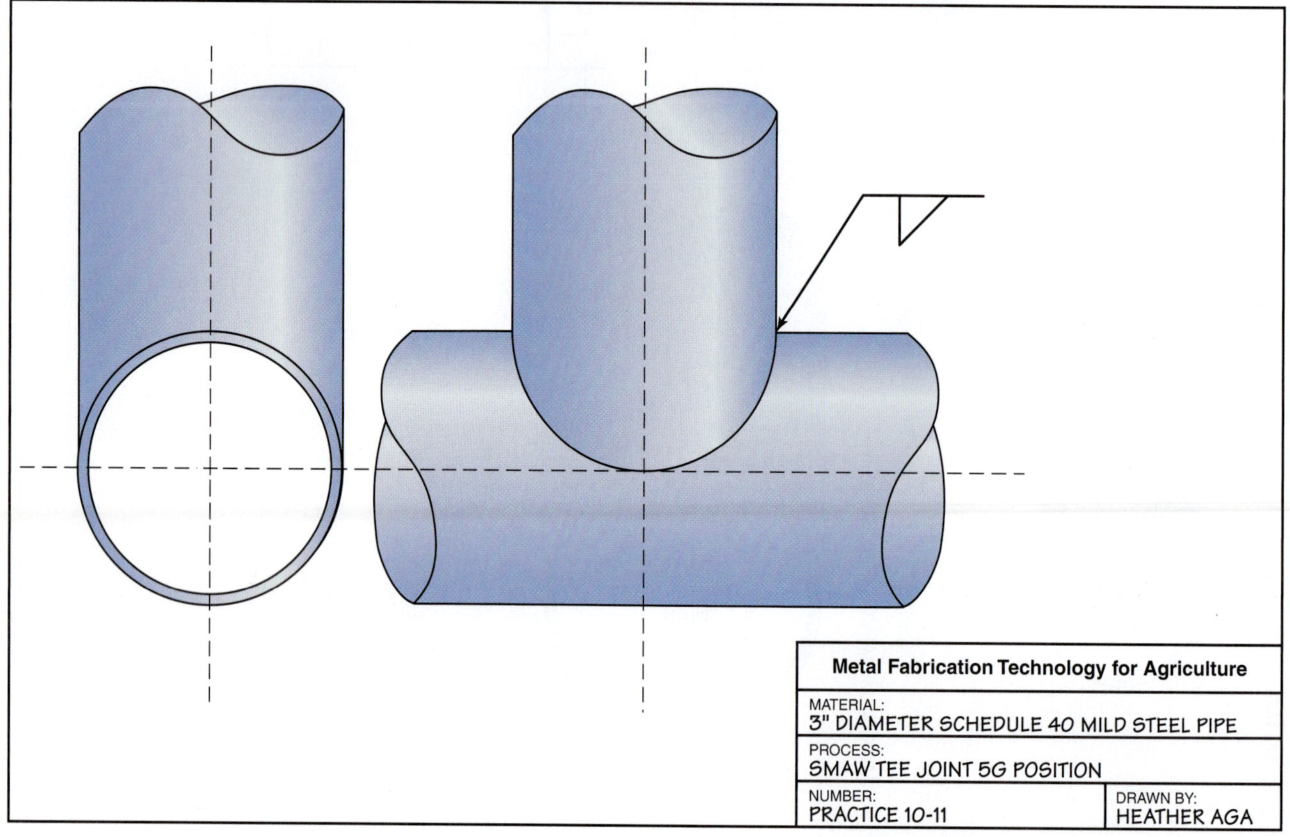

FIGURE 10-52 Practice 10-11.

With the pipe horizontal and the tee sticking straight up, strike an arc at the edge of the saddle where it meets the centerline of the side of the pipe. Make a weld 1/4 of the way around the pipe, stopping at the top of the horizontal pipe. The side of the pipe can absorb more heat than the end of the saddle so using a J weave will keep much of the heat off the end of the tee. Staggering the start and stop points will give the joint greater strength, **Figure 10-53**. Repeat this process on the other side of the saddle.

After the weld is complete, visually inspect it for uniformity and visual defects. Repeat this weld until you can consistently make welds that are free of defects. Turn off the welding machine and clean up your work area when you are finished welding.

Complete a copy of the "Student Welding Report" listed in Appendix I or provided by your instructor. ◆

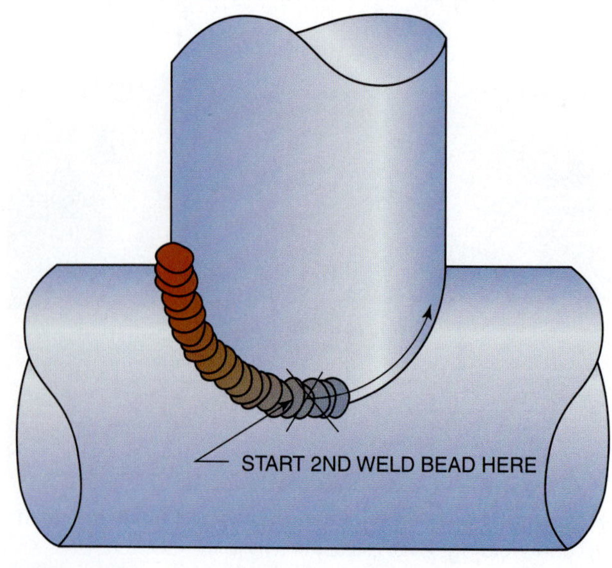

FIGURE 10-53 Overlapping the weld start points increases joint strength.

Summary

Pipe is a very versatile construction material, which is why it is so popular with farmers and ranchers. It has many uses, some functional and some ornamental, like the cactus mailbox post in **Figure 10-54**. The only limit on the uses for pipe is the welder's imagination. As you develop your pipe welding and layout skills, you will be able to build the things you need and want.

Not all pipe welds have to be nuclear quality. Some welds, like the ones used to build fences, can be less than perfect and can, with a little grinding, look great. In addition, when items built out of pipe are no longer needed, the pipe can easily be recycled for other projects.

FIGURE 10-54 Three inch pipe cactus mail box post.
Courtesy of Larry Jeffus.

Review

1. The size of pipe and tubing are given in the same way.
   A. true
   B. false
2. Tubing sizes are always given as inside diameter.
   A. true
   B. false
3. The wall thickness for pipe is determined __________.
   A. by the method of manufacturing
   B. by schedule
   C. as welded or seamless
   D. by location and code
4. The terms *pipe* and *tubing* are interchangeable.
   A. true
   B. false
5. Why are the ends of pipes beveled before welding?
   A. greater strength
   B. remove trash
   C. get down to clean metal
   D. remove paint
6. The __________ angle allows easy access for the electrode with a minimum amount of filler metal required to fill the groove.
   A. 45°
   B. 22 1/2°

Review continued

C. 90°
D. 37 1/2°

7. The most important surface of a root weld is the __________.
   A. root face
   B. face
   C. right side
   D. outside
8. Which of the following is not a way that the end of a pipe can be beveled?
   A. sanding
   B. grinding
   C. machining
   D. flame cutting
9. One of the major problems to overcome in pipe welding is learning how to make the transition from one position to another.
   A. true
   B. false
10. Dogs and wedges are used to __________.
    A. tighten the work table
    B. align the ends of pipe
    C. mount the pipe in position for welding
    D. level the welder
11. Striking the arc outside of the weld groove can __________.
    A. make it easier to find the groove through the welding lens
    B. help warm the electrode so the weld starts better
    C. cause the paint not to stick to the finished job
    D. cause hard spots
12. A __________ is the first weld in a joint and is used to establish the contour and depth of penetration.
    A. filler weld
    B. root weld
    C. cap weld
    D. cover weld
13. What can be used to remove excessive buildup and reshape the face of the root pass?
    A. chipping hammer
    B. sandpaper
    C. buffing wheel
    D. grinder
14. The __________ is used to quickly burn out small amounts of slag trapped along the edge of the root pass.
    A. cover pass
    B. hot pass
    C. filler pass
    D. capping pass
15. What is the primary function of the filler passes in pipe welding?
    A. Keep the pipe hot so it does not cool fast and get brittle.
    B. Make the inside surface look smooth.
    C. Fill up the weld groove.
    D. Burn out trapped slag.
16. An advantage of making stringer beads is that __________.
    A. they take longer, which is good when you're paid by the hour
    B. some types of electrodes can only be used for stringer welds
    C. they look nicer than weave beads
    D. they take less skill than wider weld beads
17. What is the name of the last weld bead put on a pipe joint?
    A. cover pass
    B. root pass
    C. filler pass
    D. hot pass
18. The consistent high quality and quantity of welds produced in this position make it desirable for welders.
    A. horizontal fixed position
    B. horizontal roller pipe position
    C. vertical fixed position
    D. overhead position
19. The penetration and the buildup of the weld are hard to control in the 1G horizontal rolled pipe position.
    A. true
    B. false
20. What is the major difference between making a tee joint in a pipe by bending the end of the pipe and by cutting a saddle?
    A. The saddle is faster and takes less skills.
    B. The saddle is stronger.
    C. The bent end fits better with less gap.
    D. The bent end can only be welded in the flat position.

Chapter 11

Gas Metal Arc Welding Equipment, Setup, and Operation

OBJECTIVES

After completing this chapter, the student should be able to

- ☑ describe the various methods of metal transfer.
- ☑ explain the effect of slope and inductance on gas metal arc welding.
- ☑ list four variables used to control the gas metal arc welding bead.
- ☑ describe the different electrode feed methods.
- ☑ name the parts of a gas metal arc welding setup.
- ☑ list the advantages of gas metal arc spot welding.

KEY TERMS

axial spray metal transfer
buried-arc transfer
electrode extension (stickout)
globular transfer
pinch effect
short-circuiting transfer
slope
transition current

INTRODUCTION

The *gas metal arc welding process* (GMAW) is referred to by a number of commonly used names; MIG, which is short for *metal inert gas welding*, is the most popular. Other names include MAG, for *metal active gas welding*, and *wire welding*, which describes the electrode used in the process. In this process a continuous feed wire electrode enters the arc and is melted. The molten metal transfers across the arc and mixes with the molten base metal to form a weld. The arc, molten electrode droplets and molten base metal are all protected from contamination from air by shielding gas, **Figure 11-1**.

One of the major advantages of MIG welding is that a single sized electrode wire can be used to make welds on a wide range of metal thicknesses. Adjusting voltage and amperage settings, and changing shielding gas or gases allows this single wire size to make welds on 16-gauge sheet metal and 1 in. (25 mm) plate. Additionally it is easy to change the type of wire and shielding gas to weld on stainless steel and aluminum. This makes this

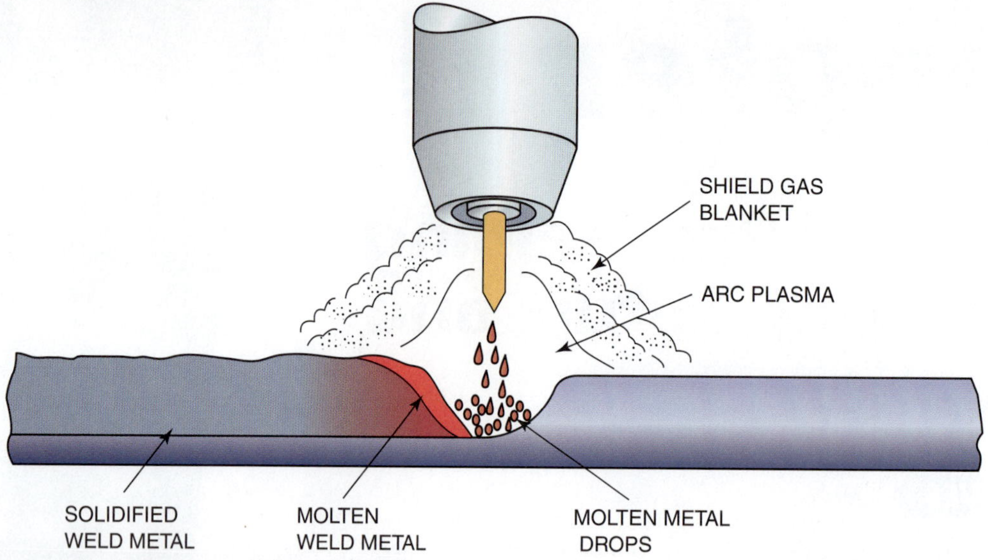

FIGURE 11-1 Gas shielded metal arc welding (GMAW).

Function	Manual (MA) (Example: SMAW)	Semiautomatic (SA) (Example: GMAW)	Machine (ME) (Example: GMAW)	Automatic (AU) (Example: GMAW)
Maintain the arc	Welder	Machine	Machine	Machine
Feed the filler metal	Welder	Machine	Machine	Machine
Provide the joint travel	Welder	Welder	Machine	Machine
Provide the joint guidance	Welder	Welder	Welder	Machine

TABLE 11-1 Methods of Performing Welding Processes.

piece of welding equipment one of the most flexible and versatile to be found on the ranch or farm.

The GMAW process may be performed as semiautomatic (SA), machine (ME), or automatic (AU) welding, **Table 11-1.** In agriculture the GMA welding process is performed as a semiautomatic process and is often mistakenly referred to as "semiautomatic welding," **Figure 11-2.**

In this chapter, the semiautomatic GMA welding process will be covered. The skill required to set up and operate this process is basic to the understanding and operation of other wire-feed processes. The reaction of the weld to changes in voltage, amperage, feed speed, stickout, and gas is similar to that of most wire-feed processes.

Metal Transfer

When first introduced, the GMA process was used with argon as a shielding gas to weld aluminum. Even though argon (Ar) was then expensive, the process was accepted immediately because it was much more productive than TIG or GTA. It produced higher-quality welds than SMA. This new arc welding process required very little postweld cleanup because it was slag and spatter free.

FIGURE 11-2 Semiautomatic GMA welding setup. Courtesy of Lincoln Electric Company.

Axial Spray Metal Transfer The freedom from spatter associated with the argon-shielded GMAW process results from a unique mode of metal trans-

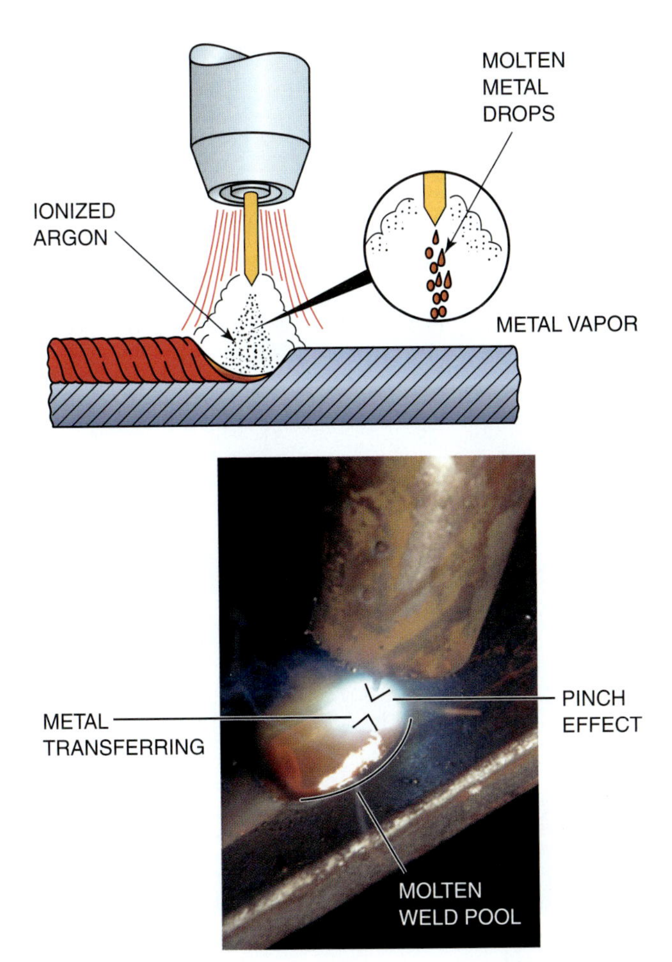

FIGURE 11-3 Axial spray metal transfer. Note the pinch effect of filler wire and the symmetrical metal transfer column. Photo courtesy of Larry Jeffus.

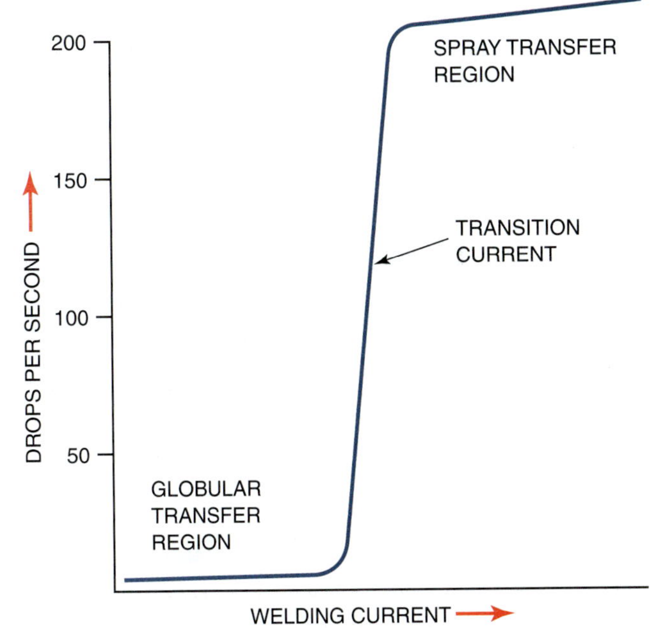

FIGURE 11-4 Desirable spray transfer shown schematically.

fer called axial spray metal transfer, **Figure 11-3**. This process is identified by the pointing of the wire tip from which very small drops are projected axially across the arc gap to the molten weld pool. There are hundreds of drops per second crossing from the wire to the base metal. These drops are propelled by arc forces at high velocity in the direction the wire is pointing. This projection of drops enables welding in the vertical and overhead positions without losing control of transfer. In many cases, the molten weld pool may be too large to be controlled in these positions. Because the drops are very small and directed at the molten weld pool, the process is spatter free.

This spray transfer process requires three conditions: argon shielding (or argon-rich shielding gas mixtures), DCEP polarity, and a current level above a critical amount called the transition current. The shielding gas is usually a mixture of 95% to 98% argon and 2% to 5% oxygen. The added percentage of oxygen allows greater weld penetration. **Figure 11-4** illustrates how the rate of drops transferred changes in relationship to the welding current. At low currents, the drops are large and are transferred at rates below 10 per second. These drops move slowly, falling from the electrode tip as gravity pulls them down. They tend to bridge the gap between the electrode tip end and molten weld pool. This produces a momentary short circuit that throws off spatter. However, the mode of transfer changes very abruptly above the critical current, producing the desirable spray. This change in the rate of transfer as related to current is shown schematically in **Figure 11-4**.

The transition current depends on the alloy being welded. It also is proportional to the wire diameter, meaning that higher currents are needed with larger diameter wires. The need for high current density imposes some restrictions on the process. The high current hinders welding sheet metal because the high heat cuts through sheet metal. High current also limits its use to the flat, vertical down, and horizontal welding positions. Weld control in the vertical up or overhead position is very difficult to impossible. **Table 11-2** lists the welding parameters for a variety of gases, wire sizes, and metal thicknesses for GMA welding of mild steel.

Globular Transfer The globular transfer process is rarely used by itself because it transfers the molten metal across the arc in much larger droplets. It is used in combination with pulsed spray transfer.

Globular transfer can be used on thin materials and at a very low current range. It can be used with higher current but is not as effective as other welding modes of metal transfer.

Mild Steel	Wire-feed speed, in./min		Voltage, V				
Base-Material Thickness, in.	0.035-in.	0.045-in.	CO_2	75 Ar-25 CO_2	Ar	98 Ar-2 O_2	Current A
0.036	105–115	–	18	16	–	–	50–60
0.048	140–160	70	19	17	–	–	70–80
0.060	180–220	90–110	20	17.7	–	–	90–110
0.075	240–260	120–130	20.7	18	20	–	120–130
1/8	280–300	140–150	21.5	18.5	20.5	–	140–150
3/16	320–340	160–175	22	19	21.5	23.5	160–170
1/4	360–380	185–195	22.7	19.5	22.5	24.5	180–190
5/16	400–420	210–220	23.5	20.5	23.5	25	200–210
3/8	420–520	220–270	25	22	25	26.5	220–250
1/2 and up	–	375	28	26	29	31	300

TABLE 11-2 GMA Welding Parameters for Mild Steel.

CAUTION

The heat produced during axial spray welding using large diameter wire or high current may be intense enough to cause the filter lens in a welding helmet to shatter. Be sure the helmet is equipped with a clear plastic lens on the inside of the filter lens. Avoid getting your face too close to the intense heat.

Buried-arc Transfer Carbon dioxide was one of the first gases studied during the development of the GMAW process. It was abandoned temporarily because of excessive spatter and porosity in the weld. After argon was accepted for shielding, further work with carbon dioxide demonstrated that the spatter was associated with globular metal transfer. The large drops are partially supported by arc forces, **Figure 11-5**. As they become heavy enough to overcome those forces and drop into the pool, they bridge the gap between the wire and the weld pool, producing explosive short circuits and spatter.

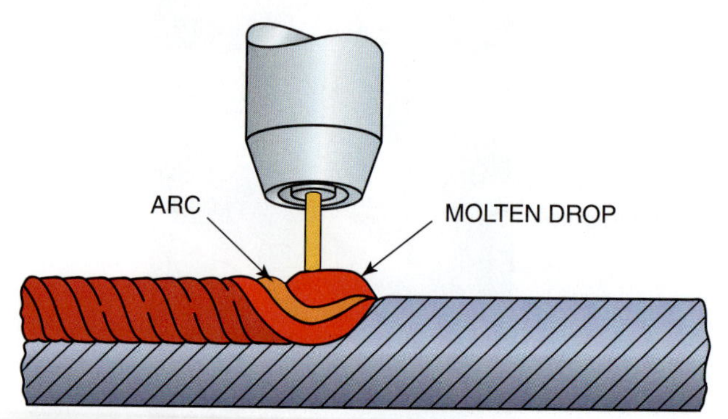

FIGURE 11-5 Globular metal transfer. Large drop is supported by arc forces.

Additional work showed that the arc in carbon dioxide was very forceful. Because of this, the wire tip could be driven below the surface of the molten weld pool. With the shorter arcs, the drop size is reduced, and any spatter produced as the result of short circuits was trapped in the cavity produced by the arc—hence the name buried-arc transfer, **Figure 11-6**. The resultant welds tend to be more highly crowned than those produced with open arcs, but they are relatively free of spatter and offer a decided advantage of welding speed. These characteristics make the buried-arc process useful for high-speed mechanized welding of thin sections, such as that found in compressor domes for hermetic air-conditioning and refrigeration equipment or for automotive components.

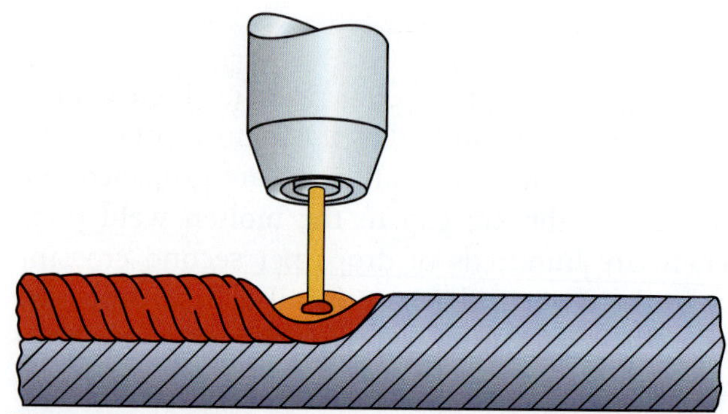

FIGURE 11-6 Buried-arc transfer. Wire tip is with the weld crater. Spatter is trapped.

Because carbon dioxide is an oxidizing gas, its applications to welding carbon steels are restricted. It cannot be used to fabricate most nonferrous materials. Neither should it be used to weld stainless steels because carbon corrodes the weld metal.

Carbon dioxide and helium are similar in that metal transfer in both gases is globular. Helium, too, can be used with the buried-arc technique. It has the advantage of inertness, potentially making it useful for the same types of applications as carbon dioxide but in nonferrous alloys.

Short-circuiting Transfer GMAW-S Low currents allow the liquid metal at the electrode tip to

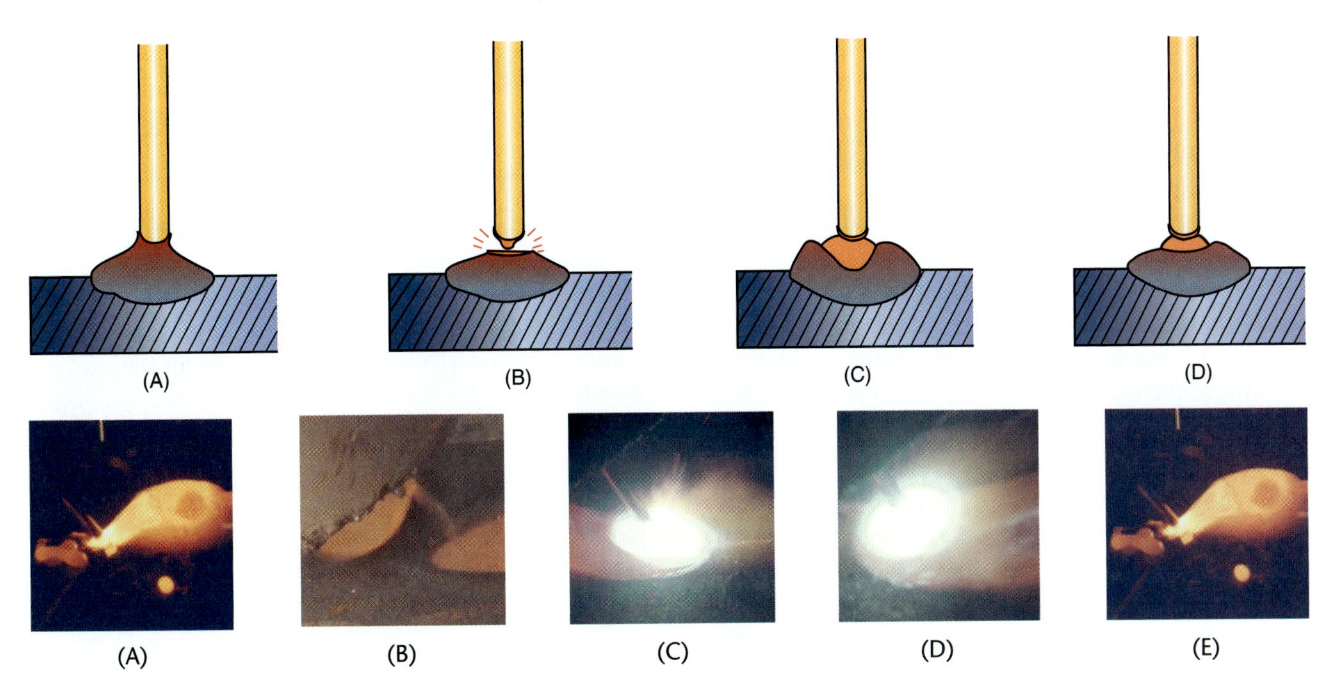

FIGURE 11-7 Schematic of short-circuiting transfer. Photos courtesy of Larry Jeffus.

be transferred by direct contact with the molten weld pool. This process requires close interaction between the wire feeder and the power supply. This technique is called the short-circuiting transfer.

The short-circuiting mode of transfer is the most common process used with GMA welding

- on thin or properly prepared thick sections of material
- on a combination of thick to thin materials
- with a wide range of electrode diameters
- with a wide range of shielding gases

The 0.023, 0.030, 0.035, and 0.045 wire electrodes are the recommended diameters for the short-circuiting mode. Shielding gas used on carbon steel is carbon dioxide (CO_2) or a combination of 25% CO_2 and 75% argon (Ar). The amperage range may be as low as 35 for materials of 24 gauge or as high as 225 for materials up to 1/8 inch in thickness on square groove weld joints. Thicker base metals can be welded if the edges are beveled to accept a complete joint weld penetration.

The transfer mechanisms in this process are quite simple and straightforward, as shown schematically in **Figure 11-7**. To start, the wire is in direct contact with the molten weld pool (**Figure 11-7A**). Once the electrode touches the molten weld pool, the arc and its resistance are removed. Without the arc resistance, the welding amperage quickly rises as it begins to flow freely through the tip of the wire into the molten weld pool. The resistance to current flow is highest at the point where the electrode touches the molten weld pool. The resistance is high because both the electrode tip and weld pool are very hot. The higher the temperature the higher the resistance to current flow. A combination of high current flow and high resistance causes a rapid rise in the temperature of the electrode tip.

As the current flow increases, the interface between the wire and molten weld pool is heated until it explodes into a vapor (**Figure 11-7B**), establishing an arc. This small explosion produces sufficient force to depress the molten weld pool. A gap between the electrode tip and the molten weld pool (**Figure 11-9C**) immediately opens. With the resistance of the arc reestablished, the voltage increases as the current decreases.

The low current flow is insufficient to continue melting the electrode tip off as fast as it is being fed into the arc. As a result, the arc length rapidly decreases (**Figure 11-7D**) until the electrode tip contacts the molten weld pool (**Figure 11-7A**). The liquid formed at the wire tip during the arc-on interval is transferred by surface tension to the molten weld pool, and the cycle begins again with another short circuit.

If the system is properly tuned, the rate of short circuiting can be repeated from approximately 20 to 200 times per second, causing a characteristic buzzing sound. The spatter is low and the process easy to use. The low heat produced by GMAW-S makes the system easy to use in all positions on sheet metal, low carbon steel, low alloy steel, and stainless steel ranging in thickness from 25 gauge (0.02 in.; 0.5 mm) to 12 gauge (0.1 in.; 2.6 mm). The short-circuiting process does not produce enough heat to make quality welds in sections much thicker than 1/4 in. (6 mm) unless it is used for the root pass on a grooved weld or to fill gaps in joints. Although this technique is highly effective, lack-of-fusion defects can occur unless the process is perfectly tuned and the welder is highly skilled, especially on thicker metal.

Carbon dioxide works well with this short-circuiting process because it produces the forceful arc needed during the arc-on interval to displace the weld pool. Helium can be used as well. Pure argon is not as effective because its arc tends to be sluggish and not very fluid. However, a mixture of 25% carbon dioxide and 75% argon produces a less harsh arc and a flatter, more fluid and desirable weld profile. Although more costly, this gas mixture is preferred.

New technology in wire manufacturing has allowed smaller wire diameters to be produced. These smaller diameters have become the preferred size even though they are more expensive. The short-circuiting process works better with a short electrode stickout.

The power supply is most critical. It must have a constant potential output and sufficient inductance to slow the time rate of current increase during the short-circuit interval. Too little inductance causes spatter due to high current surges. Too much inductance causes the system to become sluggish. The short-circuiting rate decreases enough to make the process difficult to use. Also, the power supply must sustain an arc long enough to premelt the electrode tip in anticipation of the transfer at recontact with the weld pool.

Base Metal Type	AWS Filler Metal Specification
Aluminum and aluminum alloys	A5.10
Copper and copper alloys	A5.6
Magnesium alloys	A5.19
Nickel and nickel alloys	A5.14
Stainless steel (austenitic)	A5.9
Steel (carbon)	A5.18
Titanium and titanium alloys	A5.16

TABLE 11-3 AWS Filler Metal Specifications for Different Base Metals.

Base Metal	Electrode Diameter		Amperage
	Inch	Millimeter	Range
Carbon steel	0.023	0.6	35–190
	0.030	0.8	40–220
	0.035	0.9	60–280
	0.045	1.2	125–380
	1/16	1.6	275–450
Stainless steel	0.023	0.6	40–150
	0.030	0.8	60–160
	0.035	0.9	70–210
	0.045	1.2	140–310
	1/16	1.6	280–450

TABLE 11-4 Filler Metal Diameters and Amperage Ranges.

Filler Metal Specifications

GMA welding filler metals are available for a variety of metals, **Table 11-3.** The most frequently used filler metals are AWS specification A5.18 for carbon steel and AWS specification A5.9 for stainless steel. These filler metals are available in diameter sizes ranging from 0.023 in. (0.6 mm) to 1/8 in. (3.2 mm). **Table 11-4** lists the most common sizes and the amperage ranges for these electrodes. The amperage will vary depending on the method of metal transfer, type of shielding gas, and base metal thickness.

For more information on each of the GMA welding electrodes, refer to Chapter 15, Filler Metal Selection.

Wire Melting and Deposition Rates

The wire melting rates, deposition rates, and wire-feed speeds of the consumable wire welding processes are affected by the same variables. Before discussing them, however, these terms need to be defined. The wire melting rate, measured in inches per minute (mm/sec) or pounds (kg/hr), is the rate at which the arc consumes the wire. The deposition rate, the measure of weld metal deposited, is nearly always less than the melting rate because not all of the wire is converted to weld metal. Some is lost as slag, spatter, or fume. The amount of weld metal deposited in ratio to the wire used is called the deposition efficiency.

Deposition efficiencies depend on the process, on the gas used, and even on how the welder sets welding conditions. With efficiencies of approximately 98%, solid wires with argon shields are best. Some of the self-shielded cored wires are poorest with efficiencies as low as 80%.

Welders can control the deposition rate by changing the current, electrode extension, and diameter of the wire. To obtain higher melting rates, they can increase the current or wire extension or decrease the wire diameter. Knowing the precise constants is unimportant. However, it is important to know that current greatly affects melting rate and that the extension must be controlled if results are to be reproducible.

Welding Power Supplies

To better understand the terms used to describe the different welding power supplies, you need to know the following electrical terms:

- Voltage, or volts (V), is a measurement of electrical pressure, in the same way that pounds per square inch is a measurement of water pressure.
- Electrical potential means the same thing as voltage and is usually expressed by using the term *potential* (*P*). The terms *voltage, volts,* and *potential* can all be interchanged when referring to electrical pressure.
- Amperage, or amps (A), is the measurement of the total number of electrons flowing, in the same way that gallons is a measurement of the amount of water flowing.
- Electrical current means the same thing as amperage and is usually expressed by using the term *current* (*C*). The terms *amperage, amps,* and *current* can all be interchanged when referring to electrical flow.

GMAW power supplies are the constant-voltage, constant-potential (CV, CP)-type machines, unlike SMAW power supplies, which are the constant-current (CC)-type machines and are sometimes called drooping arc voltage (DAV). It is impossible to make acceptable welds using the wrong type of power supply. Constant-voltage power supplies are available as transformer-rectifiers or as motor-generators, **Figure 11-8.** Some newer machines use electronics, enabling them to supply both types of power at the flip of a switch.

The relationships between current and voltage with different combinations of arc length or wire-feed speeds are called volt-ampere characteristics. The volt-ampere characteristics of arcs in argon with constant arc lengths or constant wire-feed speeds are shown in **Figure 11-9.** To maintain a constant arc length while increasing current, it is necessary to increase voltage. For example, with a 1/8-in. (3-mm) arc length, increasing current from 150 to 300 amperes requires a voltage increase from about 26 to 31 volts. The current increase illustrated here results from increasing the wire-feed speed from 200 to 500 inches per minute.

Speed of the Wire Electrode The wire-feed speed is generally recommended by the electrode manufacturer and is selected in inches per minute (ipm), or how fast the wire exits the contact tube. The welder uses a wire speed control dial on the wire-feed unit to control ipm. It can be advanced or slowed to control the burn-off rate, or how fast the electrode transfers into the weld pool, to meet the welder's skill in controlling the weld pool, **Table 11-5.**

To accurately measure wire-feed ipm, snip off the wire at the contact tube. Squeeze the trigger for fifteen seconds; release and snip off the wire electrode. Measure the number of inches of wire that was fed out

FIGURE 11-8 Transformer-rectifier welding power supply. Courtesy of ESAB Welding & Cutting Products.

in the fifteen seconds. Now using basic shop math, multiply its total length in inches by 4. The result is how many inches of wire were fed per minute.

Power Supplies for Short-circuiting Transfer Although the GMA power source is said to have a constant potential (CP), it is not perfectly constant. The graph in **Figure 11-10** shows that there is a slight decrease in voltage as the amperage increases within the working range. The rate of decrease is known as **slope.** It is expressed as the voltage decrease per 100-ampere increase—for example, 10 V/100 A. For short-circuiting welding, some are equipped to allow changes in the slope by steps or continuous adjustment.

The slope, which is called the *volt-ampere curve,* is often drawn as a straight line because it is fairly straight within the working range of the machine. Whether it is drawn as a curve or a straight line, the slope can be found by finding two points. The first point is the set voltage as read from the voltmeter when the gun switch is activated but no welding is being done. This is referred to as the open circuit voltage. The second point is the voltage and amperage as read during a weld. The voltage control is not adjusted during the test but the amperage can be changed. The slope is the voltage difference between the first and second readings. The difference can be

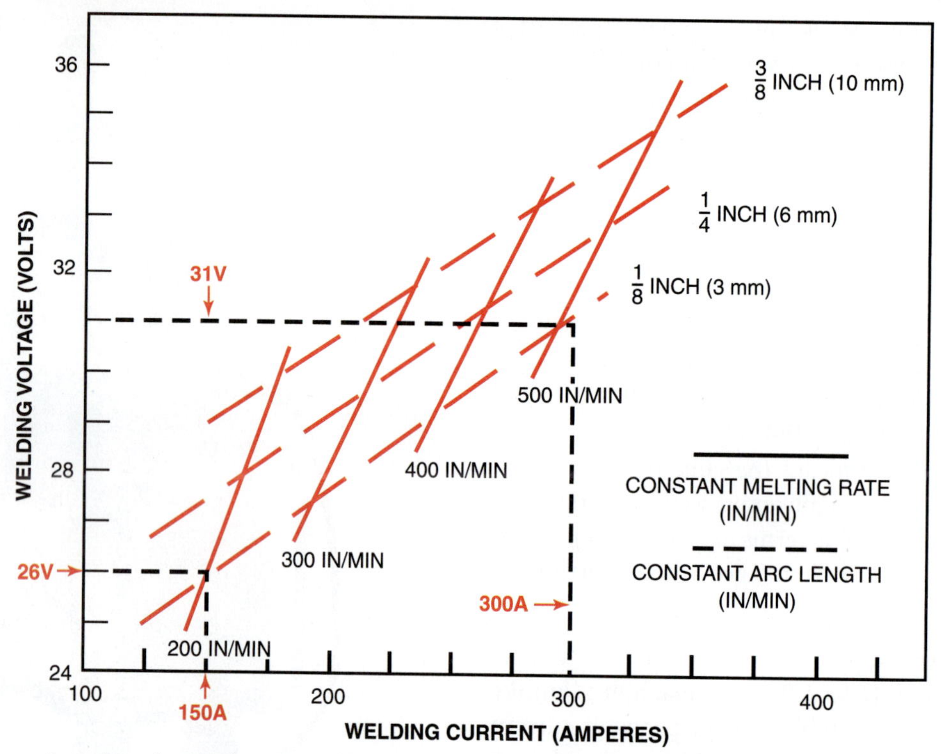

FIGURE 11-9 The arc length and arc voltage are affected by the welding current and wire-feed speed (0.045-in. [1.43-mm] wire; 1-in. [25-mm] electrode extension).

Wire-Feed Speed* in./min (m/min)	Wire Diameter Amperages			
	.030 in. (0.8 mm)	.035 in. (0.9 mm)	.045 in. (1.2 mm)	.062 in. (1.6 mm)
100 (2.5)	40	65	120	190
200 (5.0)	80	120	200	330
300 (7.6)	130	170	260	425
400 (10.2)	160	210	320	490
500 (12.7)	180	245	365	—
600 (15.2)	200	265	400	—
700 (17.8)	215	280	430	—

*To check feed speed, run out wire for one minute and then measure its length.

TABLE 11-5 Typical Amperages for Carbon Steel.

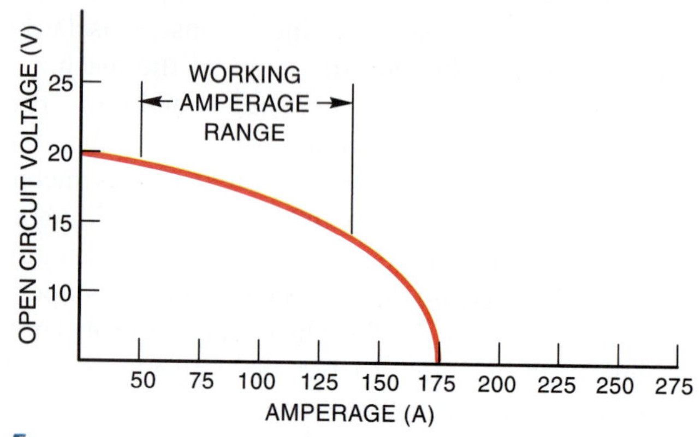

FIGURE 11-10 Constant potential welder slope.

found by subtracting the second voltage from the first voltage. Therefore, for settings over 100 amperes, it is easier to calculate the slope by adjusting the wire feed so that you are welding with 100 amperes, 200 amperes, 300 amperes, and so on. In other words, the voltage difference can be simply divided by 1 for 100 amperes, 2 for 200 amperes, and so forth.

The machine slope is affected by circuit resistance. Circuit resistance may result from a number of factors, including poor connections, long leads, or a dirty contact tube. A higher resistance means a steeper slope. In short-circuiting machines, increasing the inductance increases the slope. This increase slows the current's rate of change during short circuiting and the arcing intervals, **Figure 11-11.** Therefore, slope and inductance become synonymous in this discussion. As the slope increases, both the short-circuit current and pinch effect are reduced. A flat slope has both an increased short-circuit current and a greater pinch effect.

The machine slope affects the short-circuiting metal transfer mode more than it does the other modes. Too much current and pinch effect from a flat slope cause a violent short and arc restart cycle, which results in increased spatter. Too little current and pinch effect from a steep slope result in the short circuit not being cleared as the wire freezes in the molten pool and piles up on the work, **Table 11-6.**

The slope should be adjusted so that a proper spatter-free metal transfer occurs. On machines that

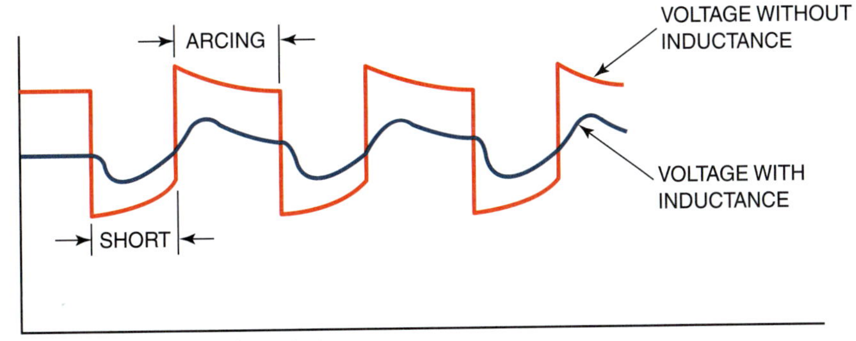

FIGURE 11-11 Voltage pattern with and without inductance.

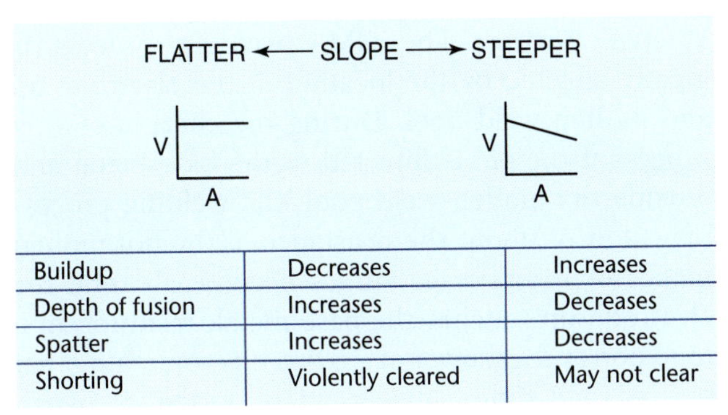

Buildup	Decreases	Increases
Depth of fusion	Increases	Decreases
Spatter	Increases	Decreases
Shorting	Violently cleared	May not clear

TABLE 11-6 Effect of Slope.

have adjustable slopes, this is easily set. Experiment 12-5 describes a method of adjusting the circuit resistance to change the slope on machines that have a fixed slope. This is done by varying the contact tube-to-work distance. The GMA filler wire is much too small to carry the welding current and heats up due to its resistance to the current flow. The greater the tube-to-work distance, the greater the circuit resistance and the steeper the slope. By increasing or decreasing this distance, a proper slope can be obtained so that the short circuiting is smoother with less spatter.

Molten Weld Pool Control

The GMAW molten weld pool can be controlled by varying the following factors: shielding gas, power settings, weave pattern, travel speed, electrode extension, and gun angle.

Shielding Gas The shielding gas selected for a weld has a definite effect on the weld produced. The properties that can be affected include the method of metal transfer, welding speed, weld contour, arc cleaning effect, and fluidity of the molten weld pool.

In addition to the effects on the weld itself, the metal to be welded must be considered in selecting a shielding gas. Some metals must be welded with an inert gas such as argon or helium or mixtures of argon and helium. Other metals weld more favorably with reactive gases such as carbon dioxide or with mixtures of inert gases and reactive gases such as argon and oxygen or argon and carbon dioxide, **Table 11-7**. The most commonly used shielding gases are 75% argon + 25% CO_2, argon + 1% to 5% oxygen, and carbon dioxide, **Figure 11-12**.

- *75% argon + 25% CO_2*: This mixture is used for the short-circuiting metal transfer process on

Metal	Shielding Gas	Chemical Reaction
Aluminum	Argon	Inert
	Argon + helium	Inert
Copper and	Argon	Inert
Copper alloys	Argon + helium	Inert
Magnesium	Argon	Inert
	Argon + helium	Inert
Nickel and	Argon	Inert
nickel alloys	Argon + helium	Inert
Steel, carbon	Argon + oxygen	Slightly oxidizing
	Argon + CO_2	Oxidizing
	CO_2	Oxidizing
Steel, low alloy	Argon + oxygen	Slightly oxidizing
	Argon + helium + CO_2	Slightly oxidizing
	Argon + CO_2	Oxidizing
Steel, stainless	Argon + oxygen	Slightly oxidizing
	Argon + helium + CO_2	Slightly oxidizing
Titanium	Argon	Inert

TABLE 11-7 GMAW Shielding Gases and Base Metals.

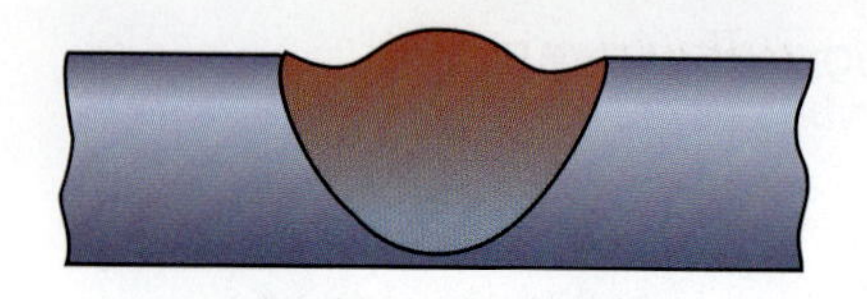

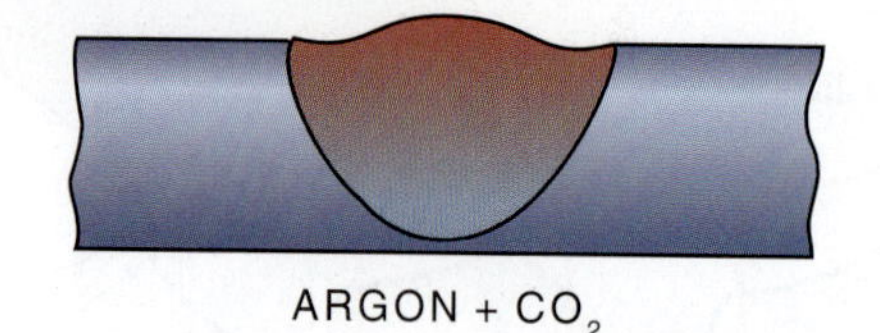

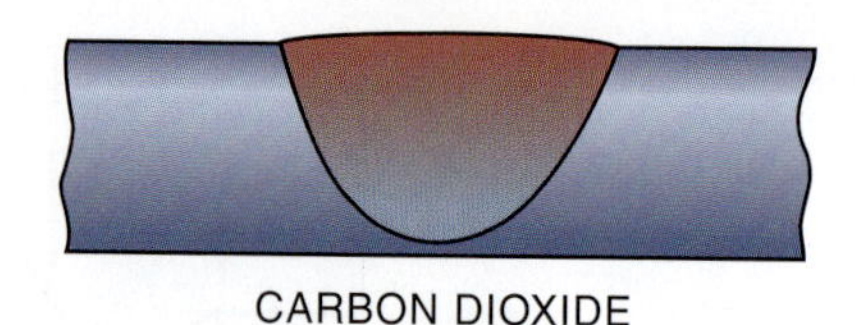

FIGURE 11-12 Effect of shielding gas on weld bead shape.

carbon and low alloy steels. It produces welds with good wetting characteristics, little spatter, high welding speeds, and low distortion.

- *Argon + 1% to 5% oxygen:* This mixture is used for the axial spray transfer method. It produces welds with good wetting, arc stability, little undercut, high welding speeds, and minimum distortion. As the percentage of the oxygen increases, the tendency for oxidation increases. For this reason, only 1% to 2% oxygen is used on stainless steels and low alloy steels but up to 5% oxygen can be used for carbon steels.
- *Carbon dioxide (CO_2):* This gas is used for the short-circuiting metal transfer process on carbon steel. It produces welds with deep penetration, high welding speeds, and noticeable spatter.

Power Settings As the power settings, voltage, and amperage are adjusted, the weld bead is affected. Making an acceptable weld requires a balancing of the voltage and amperage. If either or both are set too high or too low, the weld penetration can decrease. A GMA welding machine has no direct amperage settings. Instead, the amperage at the arc is adjusted by changing the wire-feed speed. As a result of the welding machine's maintaining a constant voltage when the wire-feed speed increases, more amperage flows across the arc. This higher amperage is required to melt the wire so that the same arc voltage can be maintained. The higher amperage is used to melt the filler wire and does not increase the penetration. In fact, the weld penetration may decrease significantly.

Increasing and decreasing the voltage changes the arc length but may not put more heat into the weld. Like changes in the amperage, these voltage changes may decrease weld penetration.

Weave Pattern The GMA welding process is greatly affected by the location of the electrode tip and molten weld pool. During the short-circuiting process if the arc is directed to the base metal and outside the molten weld pool, the welding process may stop. Without the resistance of the hot molten metal, high amperage surges occur each time the electrode tip touches the base metal, resulting in a loud pop and a shower of sparks. It is something that occurs each time a new weld is started. So when making the weave pattern, you must keep the arc and electrode tip directed into the molten weld pool. Other than the sensitivity to arc location, most of the SMAW weave pattern can be used for GMA welds.

Travel Speed Because the location of the arc inside the molten weld pool is important, the welding travel speed cannot exceed the ability of the arc to melt the base metal. Too high a travel speed can result in overrunning of the weld pool and an uncontrollable arc. Fusion between the base metal and filler metal can completely stop if the travel rate is too fast. If the travel rate is too slow and the weld pool size increases excessively, it can also restrict fusion to the base plate.

Electrode Extension The **electrode extension (stickout)** is the distance from the contact tube to the arc measured along the wire. Adjustments in this distance cause a change in the wire resistance and the resulting weld bead, **Figure 11-13**.

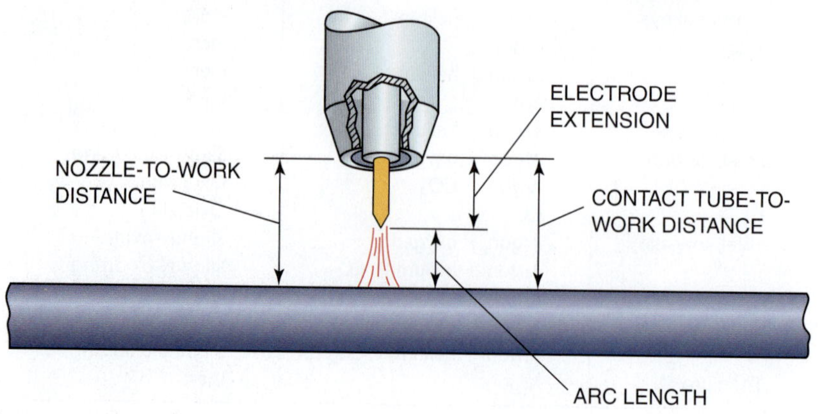

FIGURE 11-13 Electrode-to-work distances.

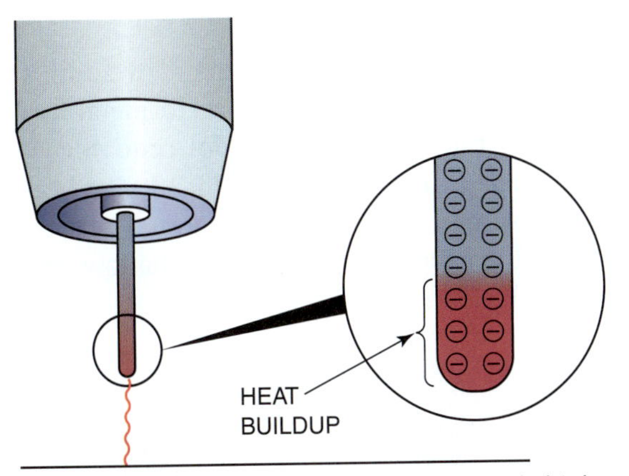

FIGURE 11-14 Heat buildup due to the extremely high current for the small conductor (electrode).

GMA welding currents are relatively high for the wire sizes, even for the low current values used in short-circuiting arc metal transfer, **Figure 11-14.** As the length of wire extending from the contact tube to the work increases, the voltage, too, should increase. Since this change is impossible with a constant-voltage power supply, the system compensates by reducing the current. In other words, by increasing the electrode extension and maintaining the same wire-feed speed, the current has to change to provide the same resistance drop. This situation leads to a reduction in weld heat, penetration, and fusion, and an increase in buildup. On the other hand, as the electrode extension distance is shortened, the weld heats up, penetrates more, and builds up less, **Figure 11-15.**

Experiment 12-3 explains the technique of using varying extension lengths to change the weld characteristics. Using this technique, a welder can make acceptable welds on metal ranging in thickness from 16 gauge to 1/4 in. (6 mm) or more without changing the machine settings. When using this technique, the nozzle-to-work distance should be kept the same so that enough shielding gas coverage is provided. Some nozzles can be extended to provide coverage. Others must be exchanged with the correct-length nozzle, **Figure 11-16.**

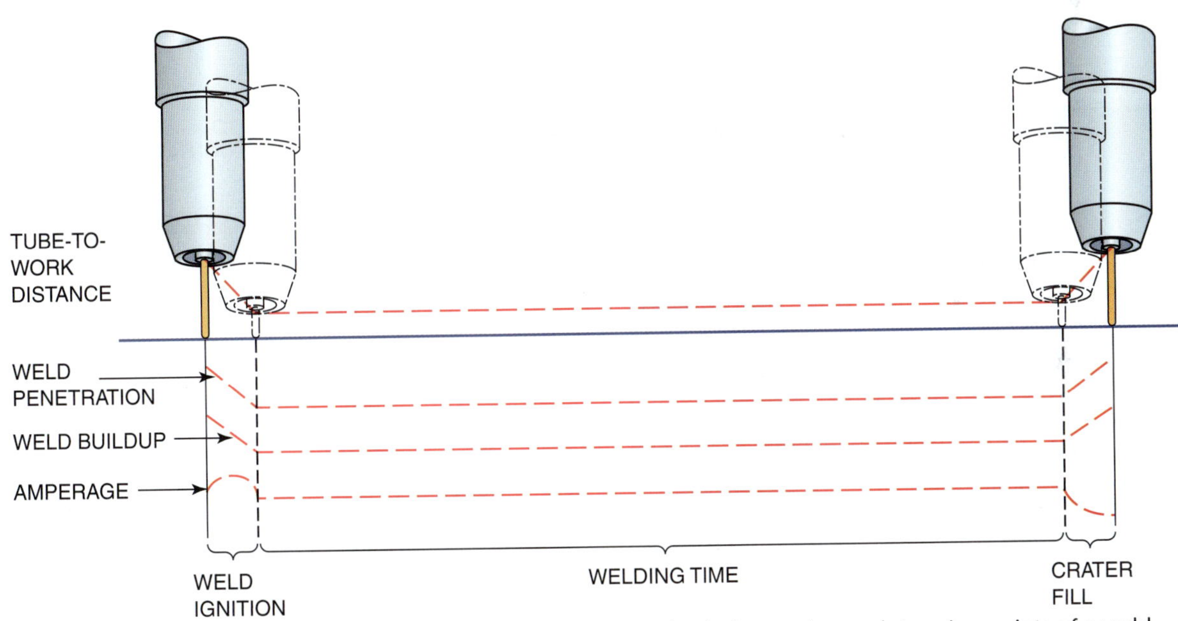

FIGURE 11-15 Using the changing tube-to-work distance to improve both the starting and stopping points of a weld.

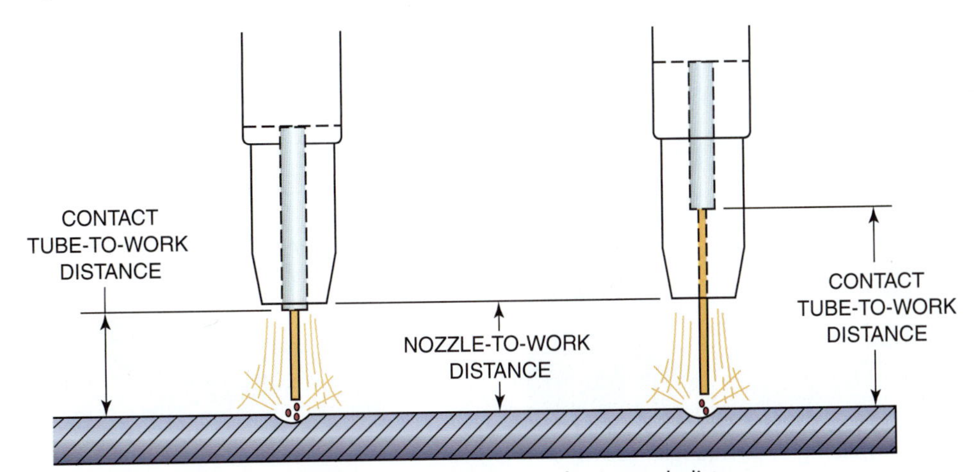

FIGURE 11-16 Nozzle-to-work distance can differ from the contact tube-to-work distance.

Gun Angle The GMA welding gun may be held so that the relative angle among the gun, work, and welding bead being made is either vertical or has a drag angle or a push angle. Changes in this angle will affect the weld bead. The effect is most noticeable during the short-circuiting arc and globular transfer modes.

Backhand welding is the welding technique that uses a drag angle, **Figure 11-17**. The welding technique that uses a push angle is known as forehand welding, **Figure 11-18**.

Backhand Welding

A dragging angle, or backhand, welding technique directs the arc force into the molten weld pool of metal. This action, in turn, forces the molten metal back onto the trailing edge of the molten weld pool and exposes more of the unmelted base metal, **Figure 11-19**. The digging action pushes the penetration deeper into the base metal while building up the weld head. If the weld is sectioned, the profile of the bead is narrow and deeply penetrated, with high buildup.

FIGURE 11-17 Backhand welding or drag angle. Courtesy of Larry Jeffus.

FIGURE 11-18 Forehand welding or push angle. Courtesy of Larry Jeffus.

Forehand Welding

Using a push angle, or forehand, welding technique, the arc force pushes the weld metal forward and out of the molten weld pool onto the cooler metal ahead of the weld, **Figure 11-20**. The heat and metal are spread out over a wider area. The sectional profile of the bead is wide, showing shallow penetration with little buildup.

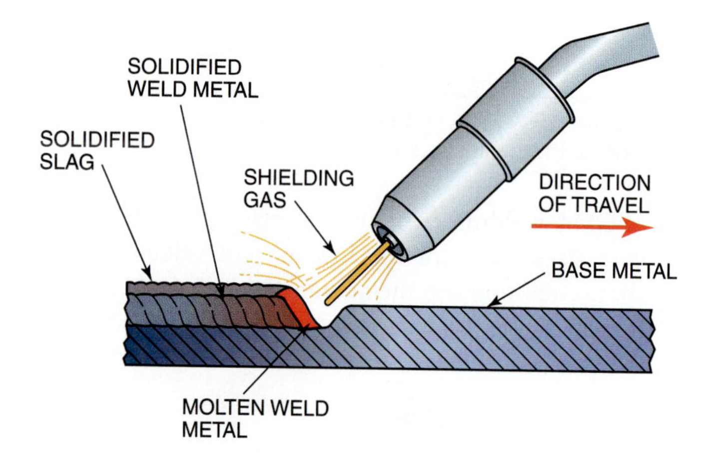

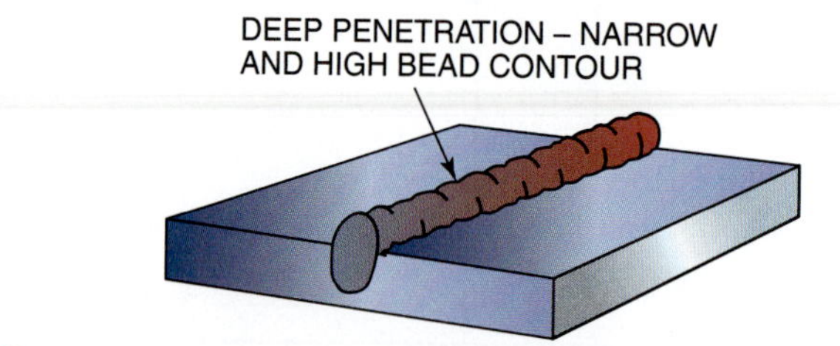

FIGURE 11-19 Backhand welding or dragging angle.

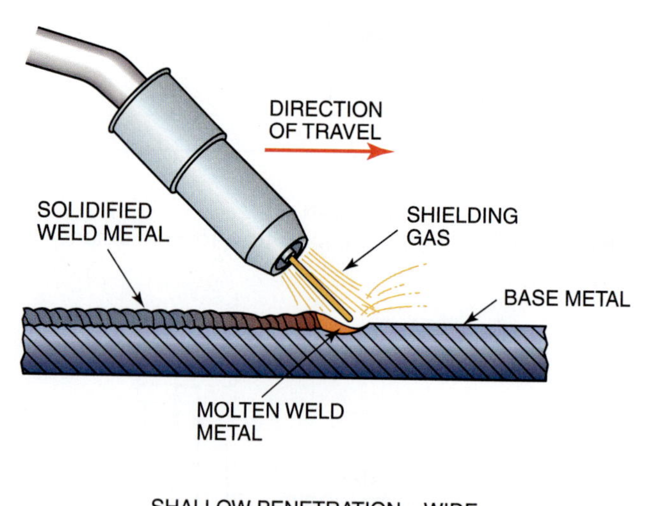

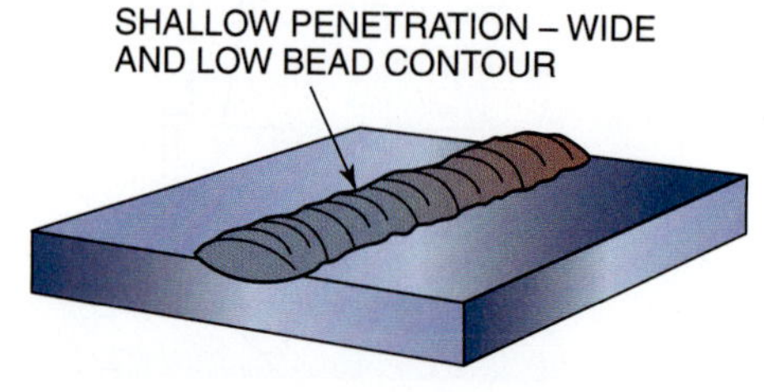

FIGURE 11-20 Forehand welding or pushing angle.

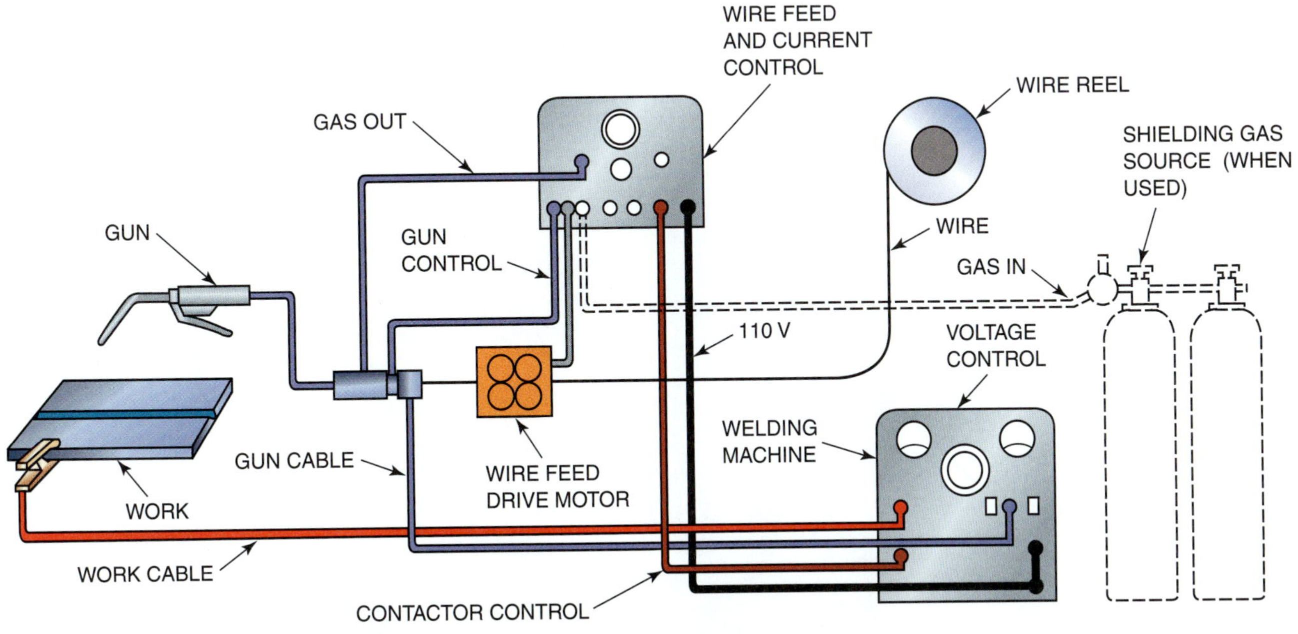

FIGURE 11-21 Schematic of equipment setup for GMA welding. Courtesy of Hobart Brothers Company.

FIGURE 11-22 Small 110V GMA welder. Courtesy of Thermal Arc®, a Thermadyne Company.

The greater the angle, the more defined is the effect on the weld. As the angle approaches vertical, the effect is reduced. This allows the welder to change the weld bead as effectively as the changes resulting from adjusting the machine current settings.

Equipment

The basic GMAW equipment consists of the gun, electrode (wire) feed unit, electrode (wire) supply, power source, shielding gas supply with flowmeter/regulator, control circuit, and related hoses, liners, and cables, **Figures 11-21** and **Figure 11-22**. Larger, more complex systems may have water for cooling, solenoids for controlling gas flow, and carriages for moving the work or the gun or both, **Figure 11-23**.

FIGURE 11-23 Robot welding. Courtesy of ESAB Welding & Cutting Products.

Power Source The power source may be either a transformer rectifier or generator-type. The transformers are stationary and commonly require a three-phase power source. Engine generators are ideal for portable use or where sufficient power is not available.

The welding machine produces a DC welding current ranging from 40 amperes to 600 amperes with

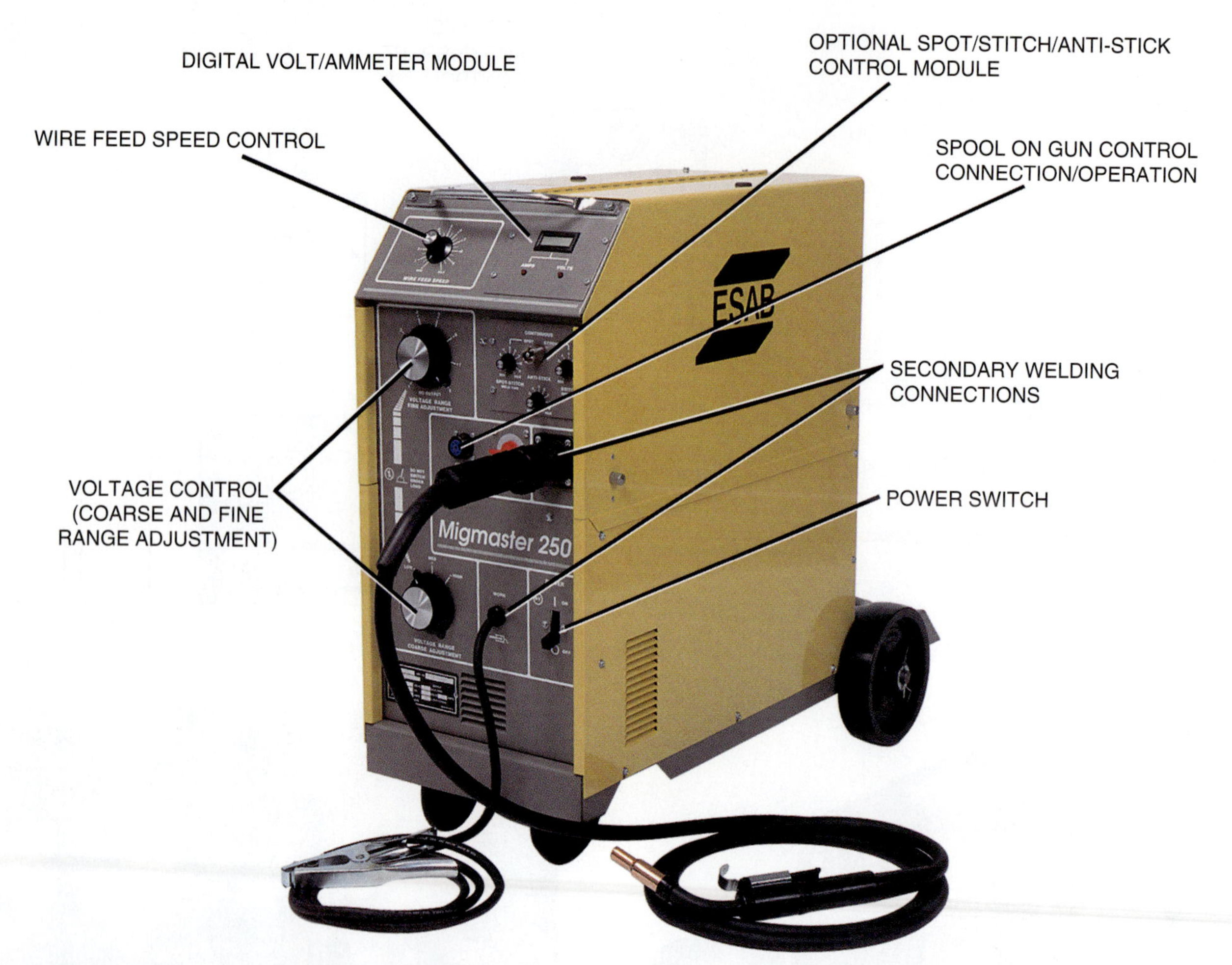

FIGURE 11-24 A 200-ampere constant-voltage power supply for multipurpose GMAW applications. Courtesy of ESAB Welding & Cutting Products.

10 volts to 40 volts, depending upon the machine. In the past, some GMA processes used AC welding current, but DCRP is used almost exclusively now. Typical power supplies are shown in **Figure 11-24**.

Because of the long periods of continuous use, GMA welding machines have a 100% duty cycle. This allows the machine to be run continuously without damage.

Electrode (Wire) Feed Unit The purpose of the electrode feeder is to provide a steady and reliable supply of wire to the weld. Slight changes in the rate at which the wire is fed have distinct effects on the weld.

The motor used in a feed unit is usually a DC-type that can be continuously adjusted over the desired range. **Figure 11-25** shows typical wire-feed units and accessories.

Push-type Feed System The wire rollers are clamped securely against the wire to provide the necessary friction to push the wire through the conduit to the gun. The pressure applied on the wire can be adjusted. A groove is provided in the roller to aid in alignment and to lessen the chance of slippage. Most manufacturers provide rollers with smooth or knurled U-shaped or V-shaped grooves, **Figure 11-26**. Knurling (a series of ridges cut into the groove) helps grip larger diameter wires so that they can be pushed along more easily. Soft wires, such as aluminum, are easy to damage if knurled rollers are used. Soft wires are best used with U-grooved rollers. Even V-grooved rollers can distort the surface of the wire, causing problems. V-grooved rollers are best suited for hard wires, such as mild steel and stainless steel. It is also important to use the correct-size grooves in the rollers.

Variations of the push-type electrode wire feeder include the pull-type and push-pull-type. The difference is in the size and location of the drive rollers. In the push-type system, the electrode must have enough strength to be pushed through the conduit without kinking. Mild steel and stainless steel can be readily pushed 15 ft (4 m) to 20 ft (6 m), but aluminum is much harder to push over 10 ft (3 m).

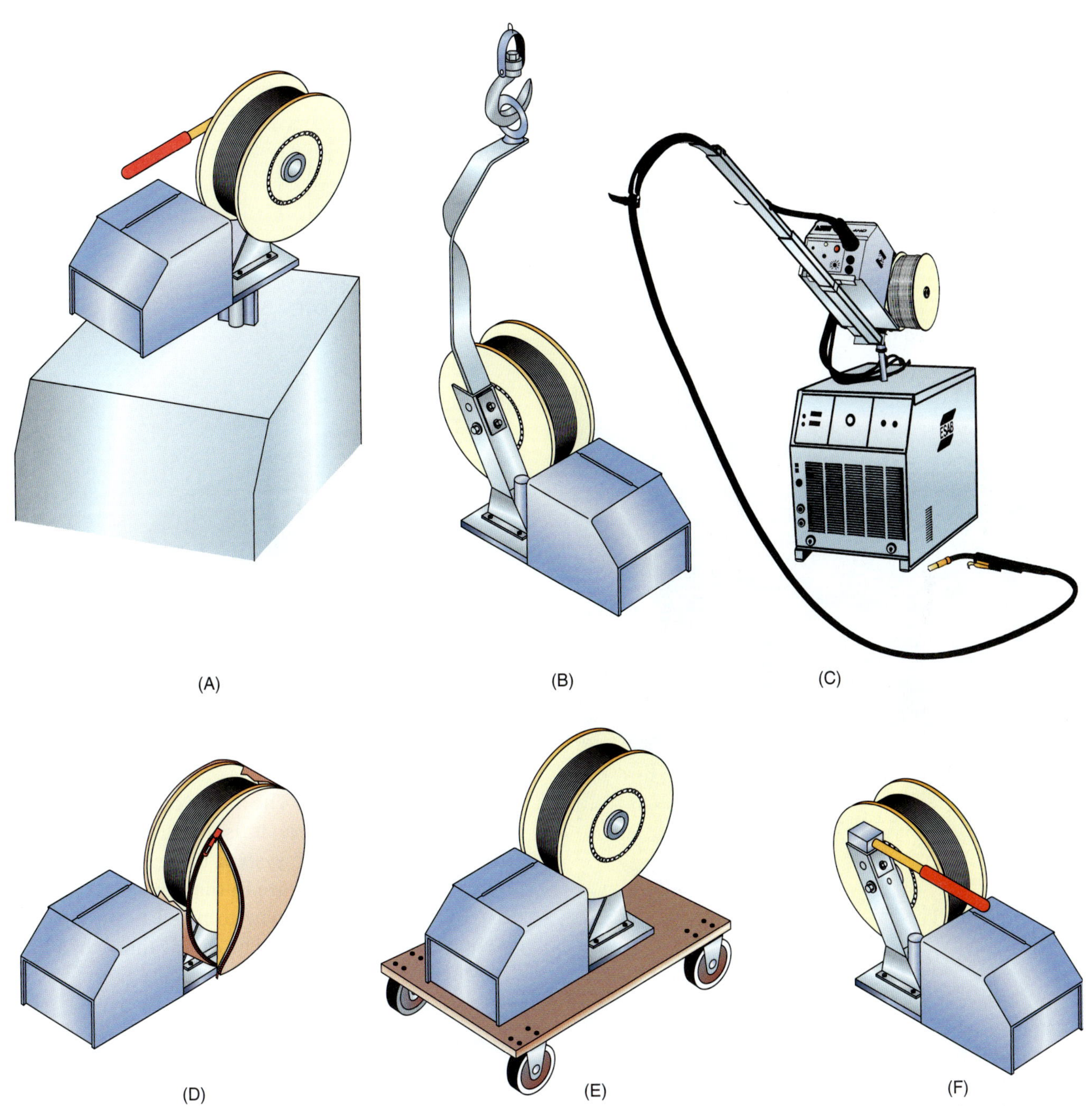

FIGURE 11-25 A variety of accessories are available for most electrode feed systems: (A) swivel post, (B) boom hanging bracket, (C) counterbalance mini-boom, (D) spool cover, (E) wire feeder wheel cart, and (F) carrying handle.

Pull-type Feed System

In pull-type systems, a smaller but higher-speed motor is located in the gun to pull the wire through the conduit. Using this system, it is possible to move even soft wire over great distances. The disadvantages are that the gun is heavier and more difficult to use, rethreading the wire takes more time, and the operating life of the motor is shorter.

Push-pull-type Feed System

Push-pull-type feed systems use a synchronized system with feed motors located at both ends of the electrode conduit, **Figure 11-27**. This system can be used to move any type of wire over long distances by periodically installing a feed roller into the electrode conduit. Compared to the pull-type system, the advantages of this system include the ability to move wire over longer distances, faster rethreading, and increased motor life due to the reduced load. A disadvantage is that the system is more expensive.

Linear Electrode Feed System Linear electrode feed systems use a different method to move the wire and change the feed speed. Standard systems use rollers that pinch the wire between the rollers.

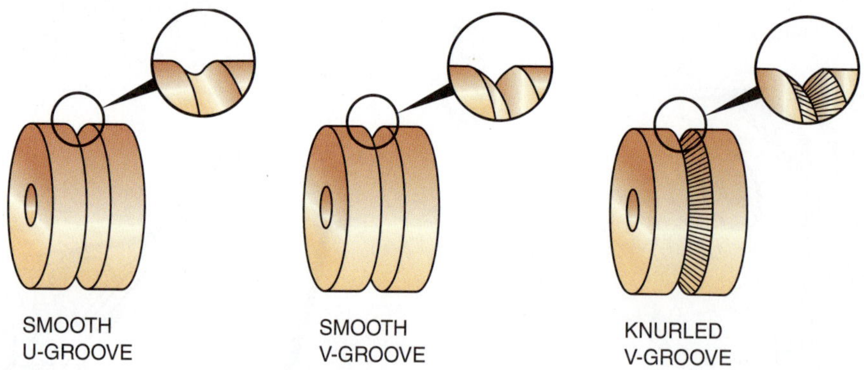

FIGURE 11-26 Feed rollers.

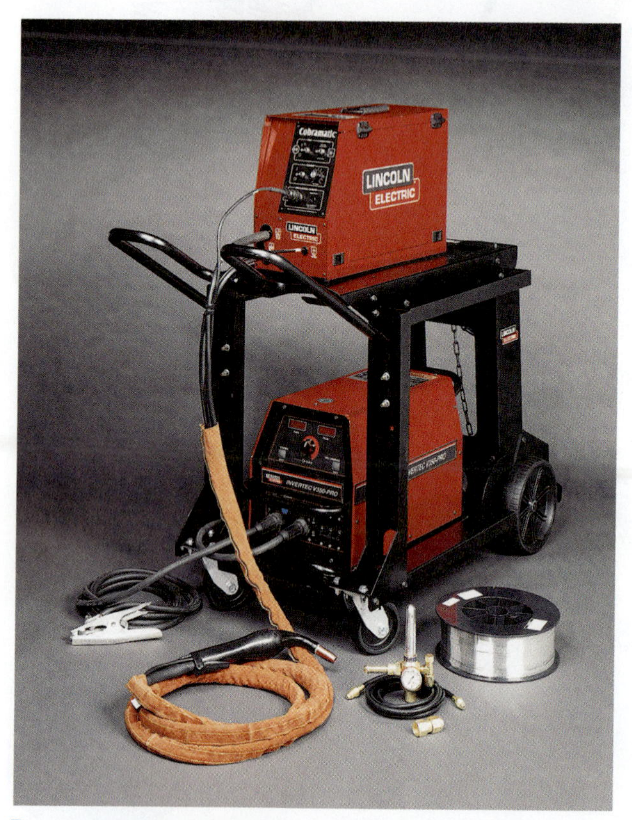

FIGURE 11-27 Wire-feed system that enables the wire to be moved through a longer cable. Courtesy of Lincoln Electric Company.

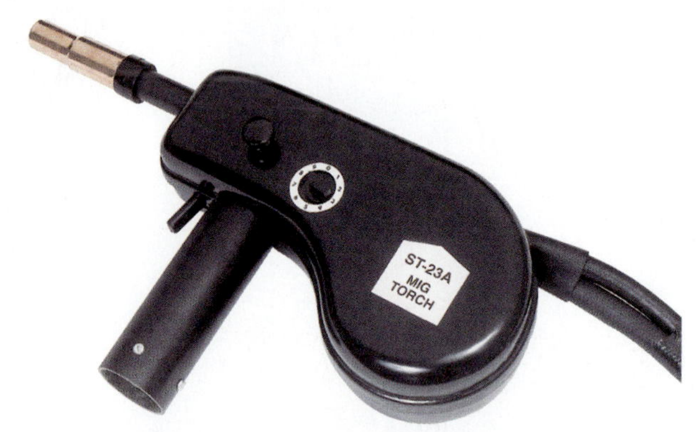

FIGURE 11-28 Feeder/gun for GMA welding. Courtesy of ESAB Welding & Cutting Products.

A system of gears is used between the motor and rollers to provide roller speed within the desired range. The linear feed system does not have gears or conventional-type rollers.

The linear feed system uses a small motor with a hollow armature shaft through which the wire is fed. The rollers are attached so that they move around the wire. Changing the roller pitch (angle) changes the speed at which the wire is moved without changing the motor speed. This system works in the same way that changing the pitch on a screw, either coarse threads or fine threads, affects the rate that the screw will move through a spinning nut.

The advantage of a linear system is that the bulky system of gears is eliminated, thus reducing weight, size, and wasted power. The motor operates at a constant high speed where it is more efficient. The reduced size allows the system to be housed in the gun or within an enclosure in the cable. Several linear wire feeders can be synchronized to provide an extended operating range. The disadvantage of a linear system is that the wire may become twisted as it is moved through the feeder.

Spool Gun A spool gun is a compact, self-contained system consisting of a small drive system and a wire supply, **Figure 11-28.** This system allows the welder to move freely around a job with only a power lead and shielding gas hose to manage. The major control system is usually mounted on the welder. The feed rollers and motor are found in the gun just behind the nozzle and contact tube. Because of the short distance the wire must be moved, very soft wires (aluminum) can be used. A small spool of welding wire is located just behind the feed rollers. The small spools of wire required in these guns are often very expensive. Although the guns are small, they feel heavy when being used.

Electrode Conduit The electrode conduit or liner guides the welding wire from the feed rollers to the gun. It may be encased in a lead that contains the shielding gas.

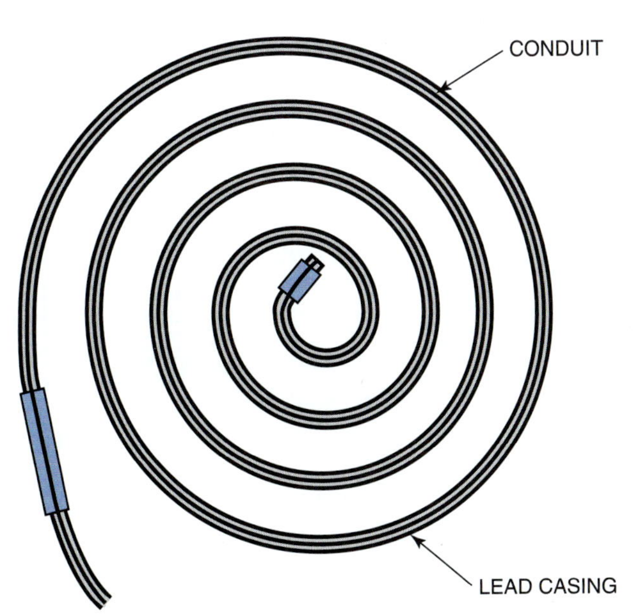

FIGURE 11-29 Tightly coiled lead casing will force the liner out of the gun.

FIGURE 11-30 A typical GMA welding gun used for most welding processes with a heat shield attached to protect the welder's gloved hand from intense heat generated when welding with high amperages. Courtesy of Tweco®, a Thermadyne® Company.

Power cable and gun switch circuit wires are contained in a conduit that is made of a tightly wound coil having the needed flexibility and strength. The steel conduit may have a nylon or Teflon® liner to protect soft, easily scratched metals, such as aluminum, as they are fed.

If the conduit is not an integral part of the lead, it must be firmly attached to both ends of the lead. Failure to attach the conduit can result in misalignment, which causes additional drag or makes the wire jam completely. If the conduit does not extend through the lead casing to make a connection, it can be drawn out by tightly coiling the lead, **Figure 11-29**. Coiling will force the conduit out so that it can be connected. If the conduit is too long for the lead, it should be cut off and filed smooth. Too long a lead will bend and twist inside the conduit, which may cause feed problems.

Welding Gun The welding gun attaches to the end of the power cable, electrode conduit, and shielding gas hose, **Figure 11-30**. It is used by the welder to produce the weld. A trigger switch is used to start and stop the weld cycle. The gun also has a contact tube, which is used to transfer welding current to the electrode moving through the gun, and a gas nozzle, which directs the shielding gas onto the weld, **Figure 11-31**.

Spot Welding

GMA can be used to make high-quality arc spot welds. Welds can be made using standard or specialized equipment, **Figure 11-32**. The arc spot weld produced by GMAW differs from electric resistance spot welding. The GMAW spot weld starts on one surface of one member and burns through to the other member, **Figure 11-33**. Fusion between the members occurs, and a small nugget is left on the metal surface.

GMA spot welding has some advantages such as the following: (1) welds can be made in thin-to-thick materials; (2) the weld can be made when only one side of the materials to be welded is accessible; and (3) the weld can be made when there is paint on the interfacing surfaces. The arc spot weld can also be used to assemble parts for welding to be done at a later time.

Thin metal can be attached to thicker sections using an arc spot weld. If a thin-to-thick butt, lap, or tee joint is to be welded with complete joint penetration, often the thin material will burn back, leaving a hole, or there will not be enough heat to melt the thick section. With an arc spot weld, the burning back of the thin material allows the thicker metal to be melted. As more metal is added to the weld, the burn-through is filled, **Figure 11-33**.

The GMA spot weld is produced from only one side. Therefore, it can be used on awkward shapes and in cases where the other side of the surface being welded should not be damaged. This makes it an excellent process for auto body repair. In addition, because the metals are melted and the molten weld pool is agitated, thin films of paint between the members being joined need not be removed. This is an added benefit for auto body repair work.

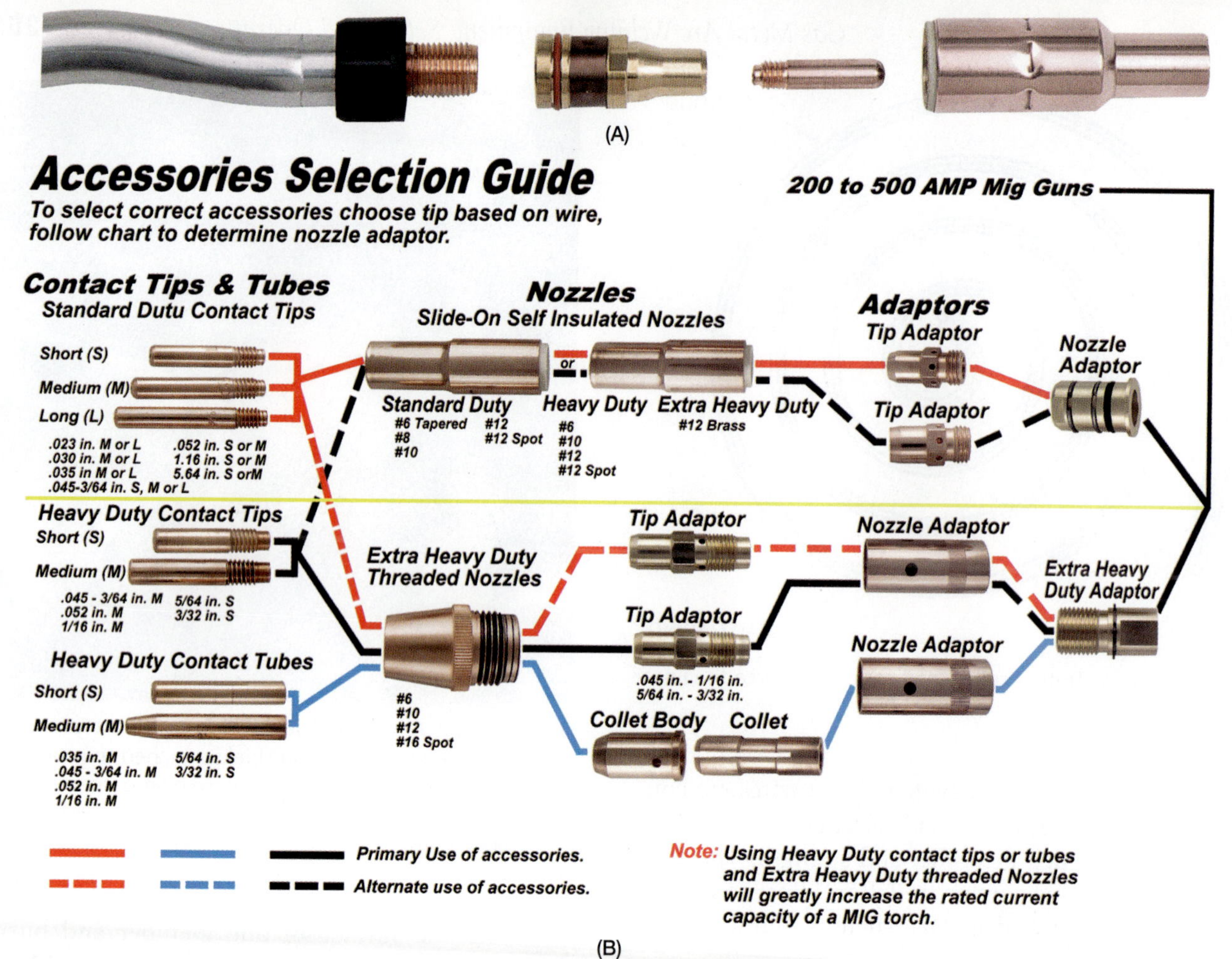

FIGURE 11-31 (A) Typical replaceable parts of a GMA welding gun. (B) Accessories and parts selection guide for a GMA welding gun. Courtesy of ESAB Welding & Cutting Products.

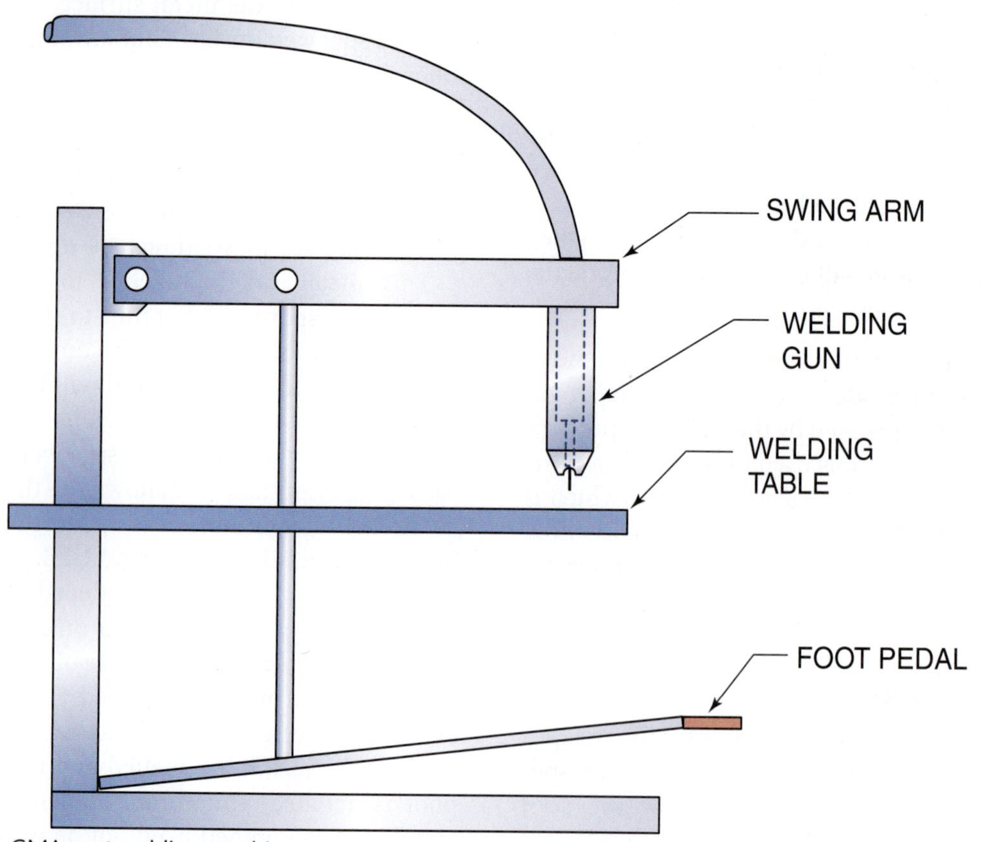

FIGURE 11-32 GMA spot welding machine.

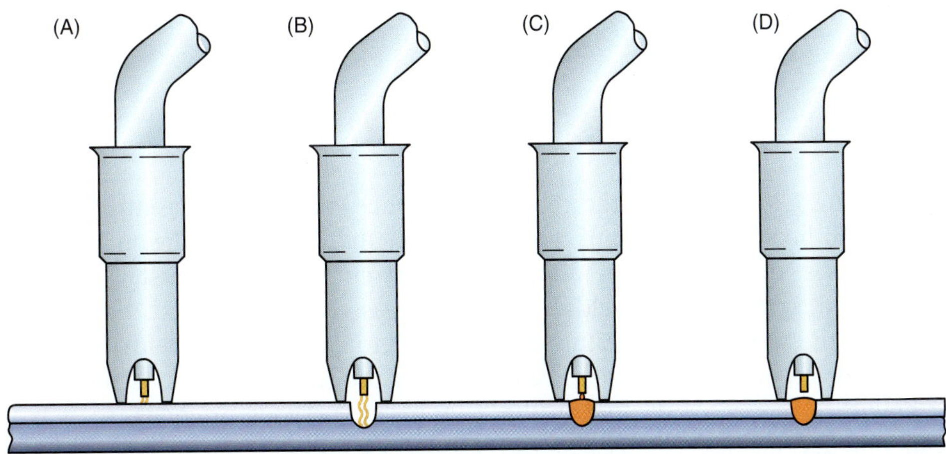

FIGURE 11-33 GMA spot weld: (A) the arc starts, (B) a hole is burned through the first plate, (C) the hole is filled with weld metal, and (D) the wire feed stops and the arc burns the electrode back.

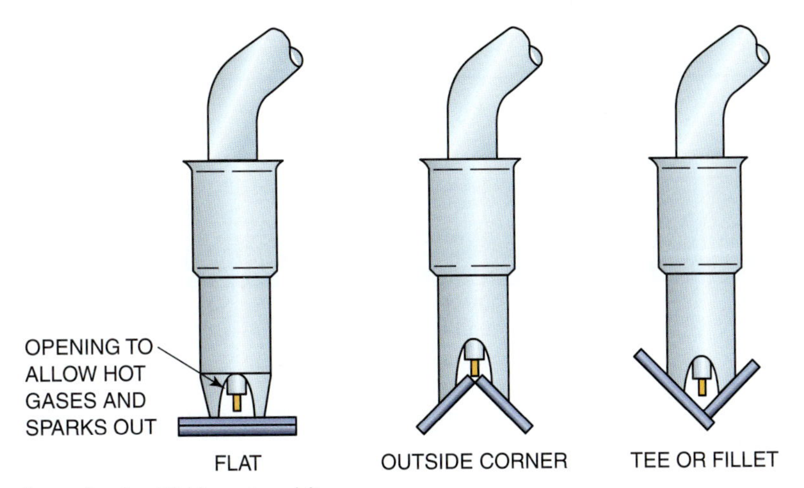

FIGURE 11-34 Specialized nozzles for GMA spot welding.

CAUTION

Safety glasses and/or flash glasses must be worn to protect the eyes from flying sparks.

Specially designed nozzles provide flash protection, part alignment, and arc alignment, **Figure 11-34**. As a result, for some small jobs it is possible to perform the weld with only safety glasses. Welders can shut their eyes and turn their head during the weld.

The optional control timer provides weld time and burn-back time. To make a weld, the amperage, voltage, and length of welding time must be set correctly. The burn-back time is a short period at the end of the weld when the wire feed stops but the current does not. This allows the wire to be burned back so it does not stick in the weld, **Figure 11-33.**

CAUTION

This is not advisable for any work requiring more than just a few spot welds. Prolonged exposure to the reflected ultraviolet light will cause skin burns.

Summary

The key to making GMA welds is the setup. Once you understand how the volts, amps, tube-to-work distance, and travel direction adjustments affect the weld, making great welds only requires following the joint. Recognizing the effects these changes have on welding takes time but knowing how they interact is the key. In the shop, you can practice making changes in the machine setting and observing how each change affects the weld. Then, when you are on your own in the field you will be able to set up the GMA welding machine and make welds.

Once you master the setup for one metal transfer method, start working on mastering the others. Short circuit metal transfer, or short arc, is the process most often used; however, it cannot be used to make every weld. Sometimes you will need the higher heat or faster fill that the other transfer methods provides, so it is important to practice and develop your skills with each of the processes. **Table 11-8** lists many of the factors that affect GMA welding but without shop practice it will be difficult or impossible to know which adjustment to make to correct a problem.

Troubleshooting for GMAW	
Problem	**Correction**
Arc Blow	1. Change gun angle. 2. Move ground clamp. 3. Use backup bars, brass or copper. 4. Demagnetize part.
Cracked Welds	1. Check filler wire compatibility with base metal. 2. Use pre- & postheat on weldment. 3. Use a convex weld bead. 4. Check design of root opening. 5. Change welding speed. 6. Change shielding gas.
Dirty Welds	1. Decrease gun angle. 2. Hold gun nozzle closer to work. 3. Increase gas flow. 4. Clean weld joint area, gas flow. 5. Check for draft that may be blowing shielding gas away. 6. Check gun nozzle for damaged or worn parts. 7. Center contact tip in gun nozzle. 8. Clean filler wire before it enters wire drive. 9. Check cables and gun for air or water leaks. 10. Keep unused filler wire in shipping containers.
Wide Weld Bead	1. Increase welding speed. 2. Reduce current. 3. Use a different welding technique. 4. Shorten arc length.
Incomplete Penetration	1. Increase current. 2. Reduce welding speed. 3. Shorten arc length. 4. Increase root opening. 5. Change gun angle.
Irregular Arc Start	1. Use wire cutters to cut off the end of the filler wire before starting new weld. 2. Check ground. 3. Check contact tip. 4. Check polarity. 5. Check for drafts. 6. Increase gas flow.

TABLE 11-8 Troubleshooting Guide for GMA Welding.

Troubleshooting for GMAW	
Problem	**Correction**
Irregular Wirefeed Burn-back	1. Check contact tip. 2. Check wire feed speed. 3. Increase drive roll pressure. 4. Check voltage. 5. Check polarity. 6. Check wire spool for kinks or bends. 7. Clean or replace worn conduit liner.
Welding Cables Overheating	1. Check for loose cable connections. 2. Use larger cables. 3. Use shorter cables. 4. Decrease welding time.
Porosity	1. Check for drafts. 2. Check shielding gas. 3. Increase gas flow. 4. Decrease gun angle. 5. Hold nozzle close to work. 6. Do not weld if metal is wet. 7. Clean weld joint area. 8. Center contact tip with gun nozzle. 9. Check gun nozzle for damage. 10. Check gun and cables for air or water leaks.
Spatter	1. Change gun angle. 2. Shorten arc length. 3. Decrease wire speed. 4. Check for draft.
Undercutting	1. Reduce current. 2. Change gun angle. 3. Use different welding technique. 4. Reduce welding speed. 5. Shorten arc length.
Incomplete Fusion	1. Increase current. 2. Change welding technique. 3. Shorten arc length. 4. Check joint preparation. 5. Clean weld joint area.
Unstable Arc	1. Clean weld area. 2. Check contact tip. 3. Check for loose cable connections.

TABLE 11-8 Troubleshooting Guide for GMA Welding. *(continued)*

Review

1. Which of the following is not another name for GMAW?
   A. wire welding
   B. MAG
   C. FCAW
   D. MIG
2. GMA welding as performed for most agricultural welding is a __________ process.
   A. semiautomatic
   B. automatic
   C. machine
   D. manual
3. In the axial spray metal transfer, metal is transferred across the arc as __________.
   A. fire spray of vaporized metal
   B. large globules
   C. hundreds of small droplets
   D. direct contact
4. According to Table 11-2, what would the wire-feed speed be for 1/8 in. thick base metal using 0.035-in. wire?
   A. 140–160
   B. 360–380

Review continued

C. 280–300
D. 420–520

5. Which of the following wire diameters is not recommended for the short-circuiting metal transfer mode?
A. 0.023 in.
B. 0.035 in.
C. 0.045 in.
D. 0.065 in.

6. According to Table 11-4 the amperage range for a 0.035 electrode would be __________ amps for welding on carbon steel.
A. 60–280
B. 35–190
C. 275–450
D. 70–210

7. During the short-circuiting metal transfer mode, metal is transferred to the weld pool by __________.
A. fine spray of vaporized metal
B. large globules
C. hundreds of small droplets
D. direct contact

8. The short-circuiting mode of transfer is the most common process used with GMA welding.
A. true
B. false

9. __________ is/are a measurement of electrical pressure in the same way that pounds per square inch is a measurement of water pressure.
A. Watts
B. Amperage
C. Ohms
D. Voltage

10. __________ is/are the measurement of the total number of electrons flowing in the same way that gallons is a measurement of water flowing.
A. Watts
B. Amperage
C. Ohms
D. Voltage

11. In GMA welding you adjust the __________ to change the welding amperage.
A. wire feed speed
B. voltage
C. gun angle
D. joint gap

12. As the slope increases, both the short-circuit current and the pinch effect are increased.
A. true
B. false

13. __________ is used for short circuiting metal transfer process on carbon steel.
A. 100% argon
B. 100% nitrogen
C. 100% oxygen
D. 100% carbon dioxide

14. The shielding gas selected for a weld has a definite effect on the weld produced.
A. true
B. false

15. The distance from the contact tip to the arc at the end of the electrode is called __________.
A. electrode extension
B. arc length
C. stick out
D. contact tube-to-work distance

16. As the tube-to-work distance increases the weld penetration increases.
A. true
B. false

17. Forehand welding produces a weld bead that had __________ penetration and __________ buildup.
A. deep/wide
B. shallow/narrow
C. shallow/wide
D. deep/narrow

18. The purpose of the electrode feeder is to provide a steady and reliable supply of wire to the weld.
A. true
B. false

19. Why are some wire feed rollers knurled?
A. To put bumps in the wire so it makes better contact in the contact tip.
B. So the wire can bend easier as it is fed through the welding cable.
C. To help control the size of the molten weld metal drops.
D. To help feed large diameter wires.

20. GMA welding can be used to make spot welds.
A. true
B. false

Chapter 12

Gas Metal Arc Welding

OBJECTIVES

After completing this chapter, the student should be able to

- ☑ set up a constant-potential semiautomatic arc welding unit.
- ☑ make satisfactory welds in all positions using the short-circuiting metal transfer method.
- ☑ make satisfactory welds in the 1F, 2F, and 1G positions using the pulsed-arc metal transfer method.
- ☑ make satisfactory welds in the 1F and 1G positions using the axial spray metal transfer method.

KEY TERMS

bird-nesting
cast
conduit liner
contact tube
feed rollers
flow rate
spool drag
wire-feed speed

INTRODUCTION

The most important part of MIG (GMA welding) is setting up the welder. You must be able to follow the weld joint with uniform speed and manipulation to make good MIG welds, but the setup affects MIG welding more than it does most other welding processes. Changes in voltage, wire feed speed (amperage), electrode extension, and welding gun angle all have dramatic effects on the weld. At times, even slight changes in any one of these can mean the difference between a weld that fuses the metal together and one that just lays on the surface. A change in electrode extension, for example, caused this weld not to stick, **Figure 12-1.** The weld bead has a smooth uniform shape and good width, but it did not even fuse to the surface of the base metal. Before the electrode extension was increased, this vertical down weld had great fusion. This is an extreme example of lack of fusion, but it shows how important welder setup is in MIG welding.

Although the example shown in Figure 12-1 is extreme, the lack of fusion in MIG welds is often a problem. It has been so common that some agricultural workers think the MIG process cannot be trusted for heavily loaded welds. However when done correctly, MIG can make welds that fully penetrate metal up to 1/4 in. (6 mm) thick. In this chapter you will learn, through experiments and practices, how to set up and make welds with 100% reliability.

FIGURE 12-1 Weld separated from the plate; there is no fusion between the weld and plate. Courtesy of Larry Jeffus.

Setup

The same equipment may be used for semiautomatic GMAW, FCAW, and SAW. The basic GMAW installation consists of the following: welding gun, gun switch circuit, electrode conduit-welding contractor control, electrode feed unit, electrode supply, power source, shielding gas supply, shielding gas flowmeter regulator, shielding gas hoses, and both power and work cables. Typical water-cooled and air-cooled guns are shown in **Figure 12-2**. The equipment setup in this chapter is similar to equipment built by other manufacturers, which means that any skills developed can be transferred easily to other equipment.

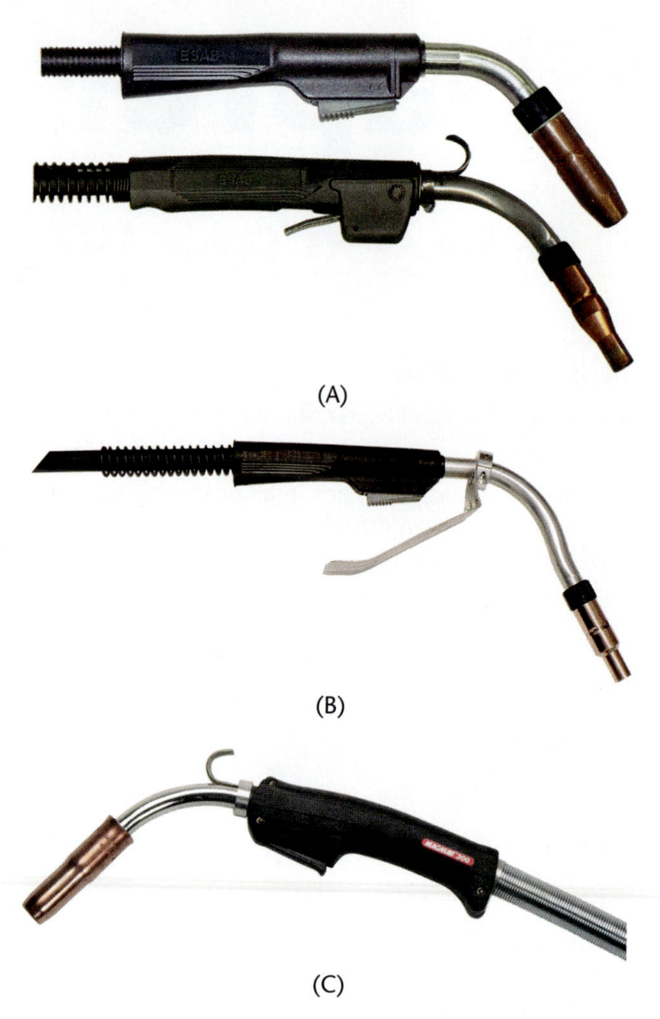

FIGURE 12-2 (A) Air-cooled GMA welding gun, (B) water-cooled GMA welding gun, and (C) water-cooled GMA welding nozzle. Courtesy of ESAB Welding & Cutting Products.

PRACTICE 12-1

GMAW Equipment Setup

For this practice, you will need a GMAW power source, a welding gun, an electrode feed unit, an electrode supply, a shielding gas supply, a shielding gas flowmeter regulator, electrode conduit, power and work leads, shielding gas hoses, assorted hand tools, spare parts, and any other required materials. In this practice, you will properly set up a GMA welding installation.

If the shielding gas supply is a cylinder, it must be chained securely in place before the valve protection cap is removed, **Figure 12-3**. Standing to one side of the cylinder, quickly crack the valve to blow out any dirt in the valve before the flowmeter regulator is attached, **Figure 12-4**. Attach the correct hose from the regulator to the "gas-in" connection on the electrode feed unit or machine.

Install the reel of electrode (welding wire) on the holder and secure it, **Figure 12-5**. Check the

FIGURE 12-3 Make sure the gas cylinder is chained securely in place before removing the safety cap. Courtesy of Larry Jeffus.

FIGURE 12-4 Attach the flowmeter regulator. Be sure the tube is vertical. Courtesy of Larry Jeffus.

FIGURE 12-5 When installing the spool of wire, check the label to be sure that the wire is the correct type and size. Courtesy of Larry Jeffus.

feed roller size to ensure that it matches the wire size, **Figure 12-6**. The conduit liner size should be checked to be sure that it is compatible with the wire size. Connect the conduit to the feed unit. The conduit or an extension should be aligned with the groove in the roller and set as close to the roller as possible without touching, **Figure 12-7**. Misalignment at this point can contribute to a bird's nest, **Figure 12-8**. Bird-nesting of the electrode wire results when the feed roller pushes the wire into a tangled ball because the wire would not go through the outfeed side conduit, and appears to look like a bird's nest.

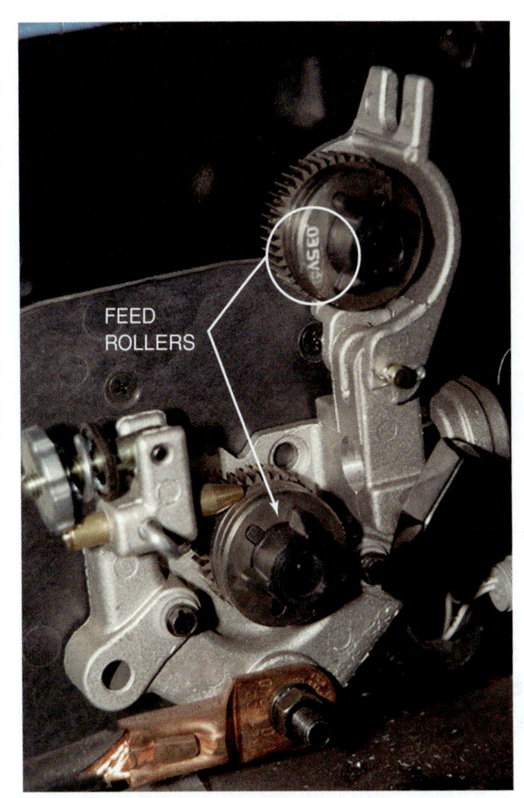

FIGURE 12-6 Check to be certain that the feed rollers are the correct size for the wire being used. Courtesy of Larry Jeffus.

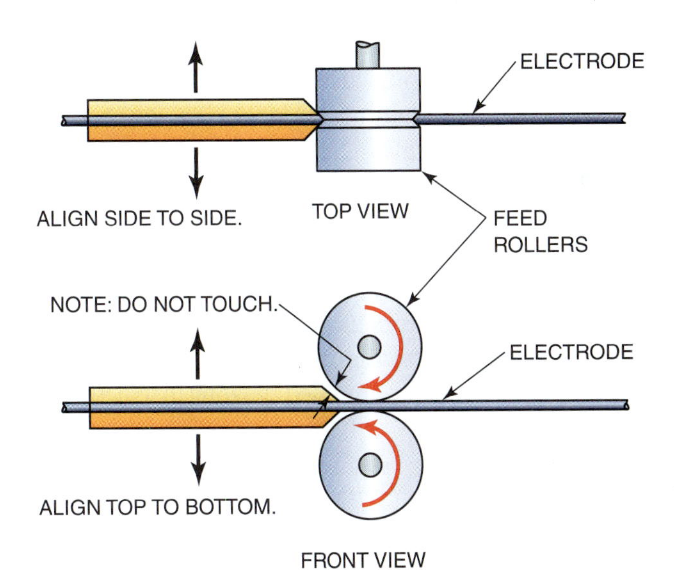

FIGURE 12-7 Feed roller and conduit alignment.

Be sure the power is off before attaching the welding cables. The electrode and work leads should be attached to the proper terminals. The electrode lead should be attached to electrode or positive (+). If necessary, it is also attached to the power cable part of the gun lead. The work lead should be attached to work or negative (−).

The shielding "gas-out" side of the solenoid is then also attached to the gun lead. If a separate splice is

FIGURE 12-8 "Bird's nest" in the filler wire at the feed rollers. Courtesy of Larry Jeffus.

required from the gun switch circuit to the feed unit, it should be connected at this time, **Figure 12-9A, 12-9B,** and **12-9C**. Check to see that the welding contractor circuit is connected from the feed unit to the power source.

The welding gun should be securely attached to the main lead cable and conduit, **Figure 12-10**. There should be a gas diffuser attached to the end of the conduit liner to ensure proper alignment. A contact tube (tip) of the correct size to match the electrode wire size being used should be installed, **Figure 12-11**. A shielding gas nozzle is attached to complete the assembly.

Recheck all fittings and connections for tightness. Loose fittings can leak; loose connections can cause added resistance, reducing the welding efficiency. Some manufacturers include detailed setup instructions with their equipment, **Figure 12-12**.

Complete a copy of the "Student Welding Report" listed in Appendix I or provided by your instructor. ◆

PRACTICE 12-2

Threading GMAW Wire

Using the GMAW machine that was properly assembled in Practice 12-1, you will turn the machine on and thread the electrode wire through the system.

Check to see that the unit is assembled correctly according to the manufacturer's specifications. Switch on the power and check the gun switch circuit by depressing the switch. The power source relays, feed relays, gas solenoid, and feed motor should all activate.

Cut the end of the electrode wire free. Hold it tightly so that it does not unwind. The wire has a natural curve that is known as its cast. The cast is mea-

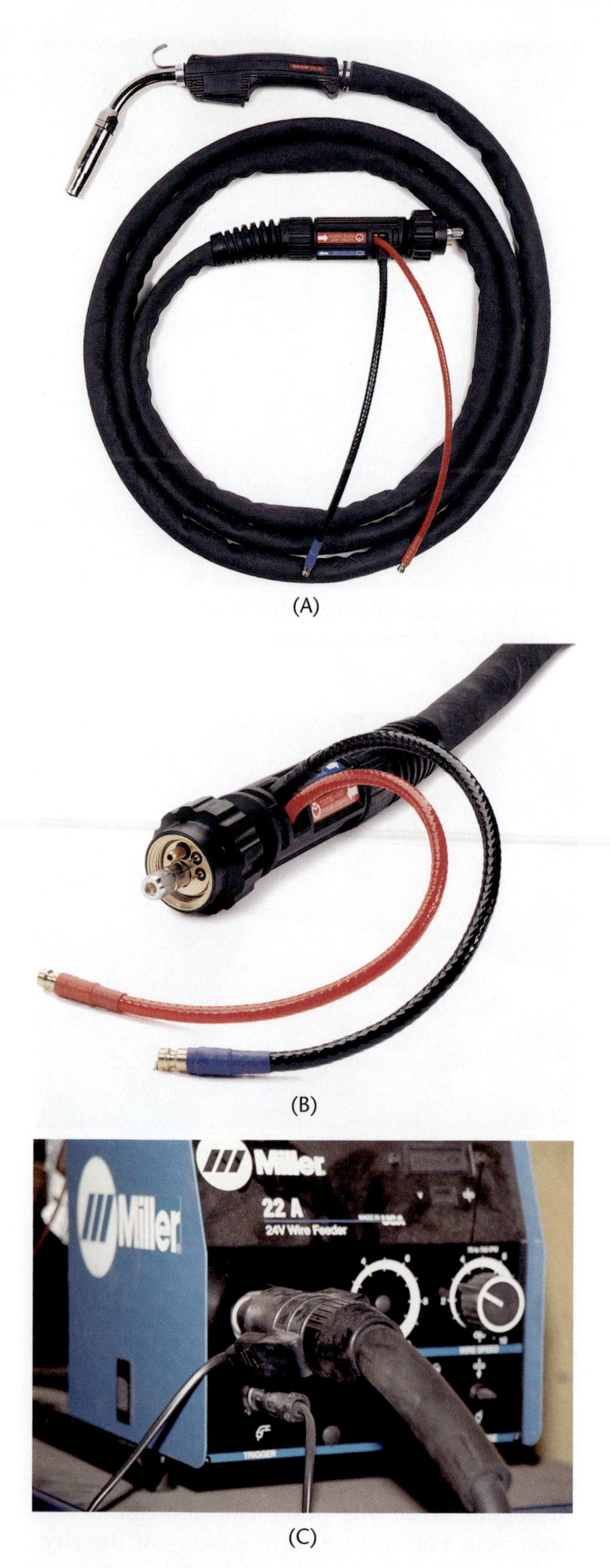

FIGURE 12-9 Connect the leads and other lines as shown in the owner's manual. (A & B) Courtesy of ESAB Welding & Cutting Products. (C) Courtesy of Larry Jeffus.

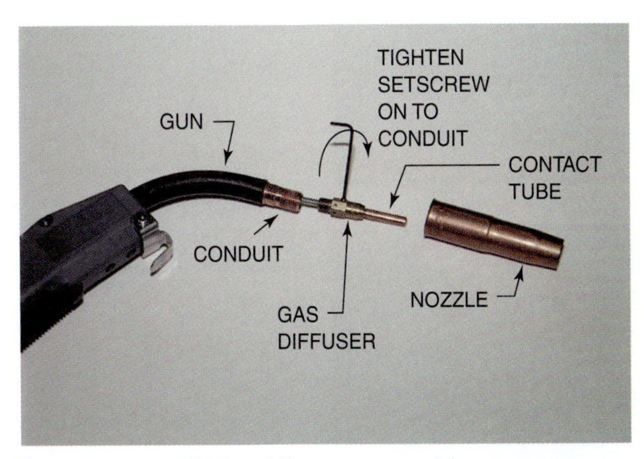

FIGURE 12-10 GMA welding gun assembly. Courtesy of Larry Jeffus.

sured by the diameter of the circle that the wire would make if it were loosely laid on a flat surface, **Figure 12-13.** The cast helps the wire make a good electrical contact as it passes through the contact tube, **Figure 12-14.** However, the cast can be a problem when threading the system. To make threading easier, straighten about 12 in. (305 mm) of the end of the wire and cut any kinks off.

Separate the wire **feed rollers** and push the wire first through the guides, then between the rollers, and finally into the **conduit liner,** **Figure 12-15.** Reset the rollers so there is a slight amount of compression on the wire, **Figure 12-16.** Set the **wire-feed speed** control to a slow speed. Hold the welding gun

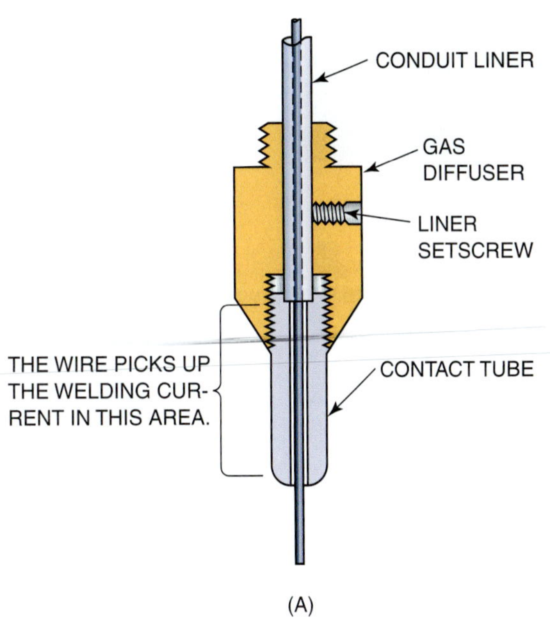

(A)

(B)

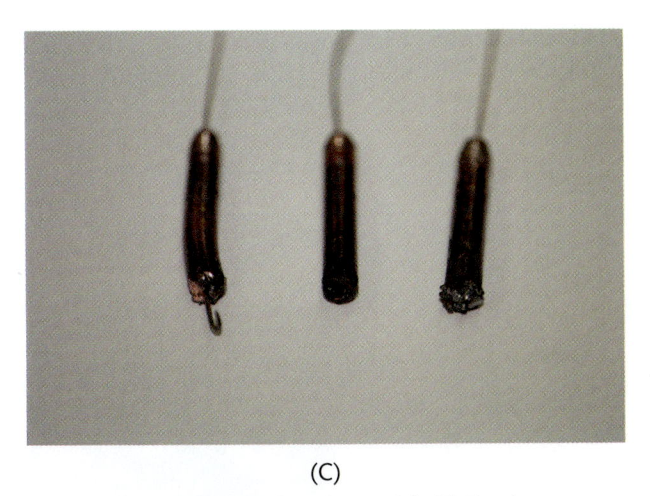

(C)

FIGURE 12-11 The contact tube must be the correct size. (A) Too small a contact tube will cause the wire to stick. (B) Too large a contact tube can cause arcing to occur between the wire and tube. (C) Heat from the arcing can damage the tube. (B) Courtesy of Brett V. Hahn (C) Courtesy of Larry Jeffus.

Open the side cover.

Remove the empty wire spool.

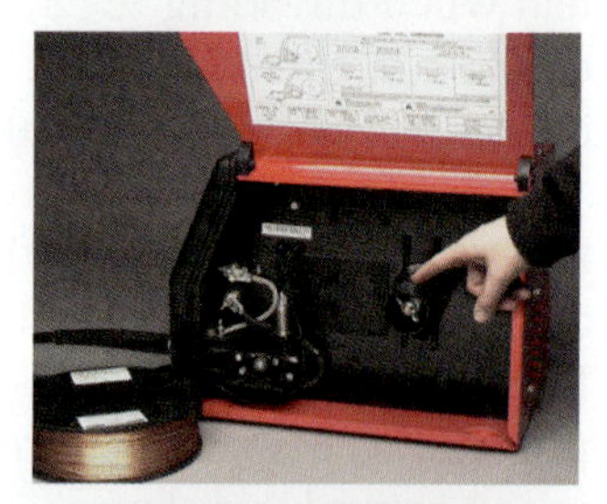

Release upper feed roller.

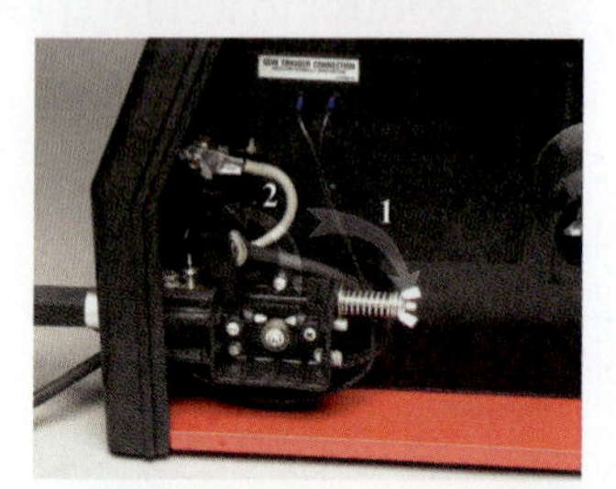

Reload the wire spool with the free end unreeling from the bottom.

Thread wire through guide between rollers and into wire cable.

Set the polarity as DCEP from GMA welding.

Turn the input switch on.

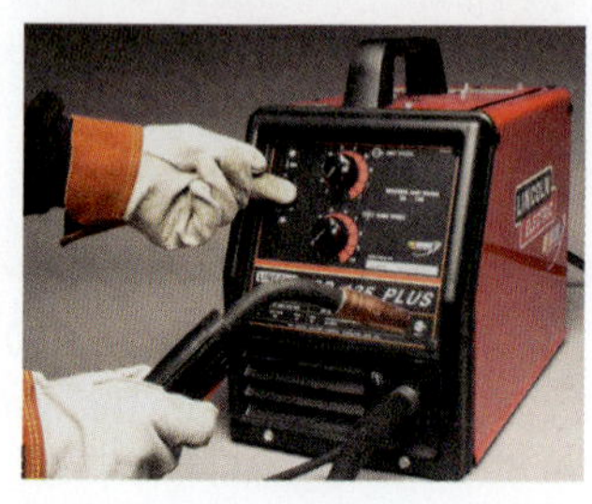

With the gun trigger pressed, adjust the feed roller tension.

Check the setting guide inside the machine door.

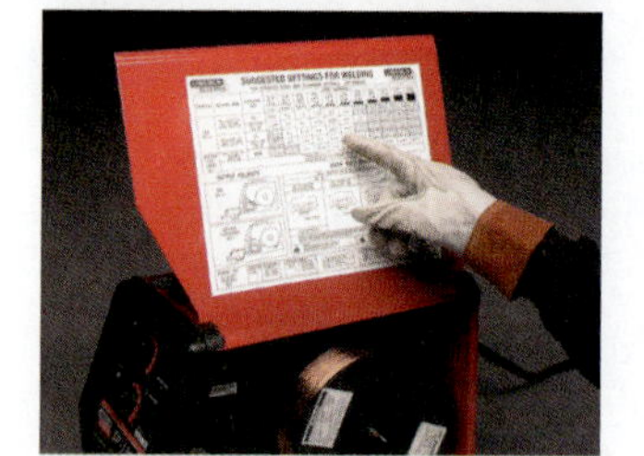

Set the voltage and wire feed for the metal you are going to be welding.

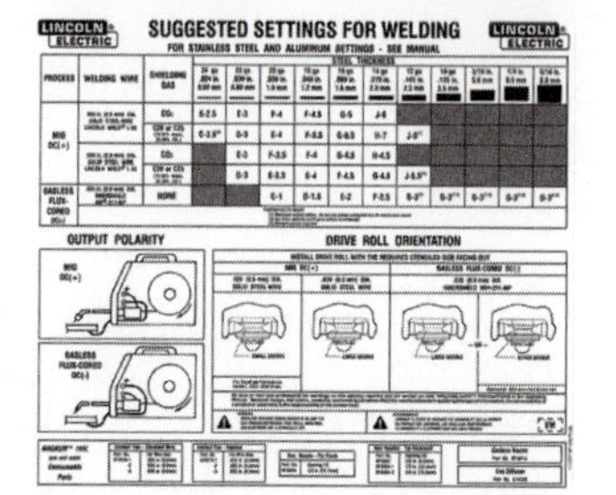

Attach work cable clamp to work to be welded.

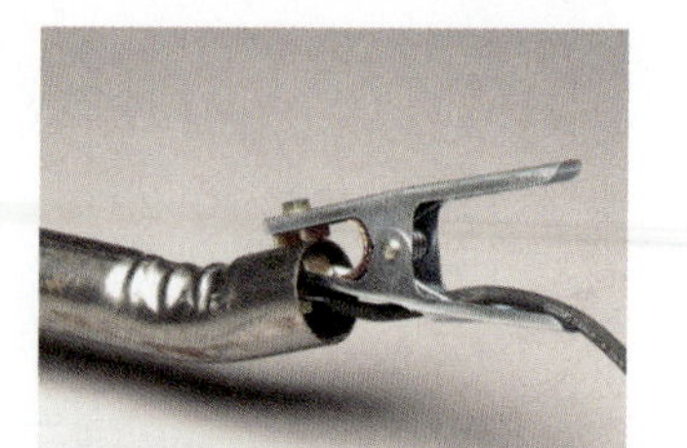

Connect gas to coupling at rear of case and turn on shielding gas.

ALWAYS WEAR PROPER SAFETY EQUIPMENT. Pull trigger and weld.

FIGURE 12-12 Example of manufacturer's setup instructions. Courtesy of Lincoln Electric Company.

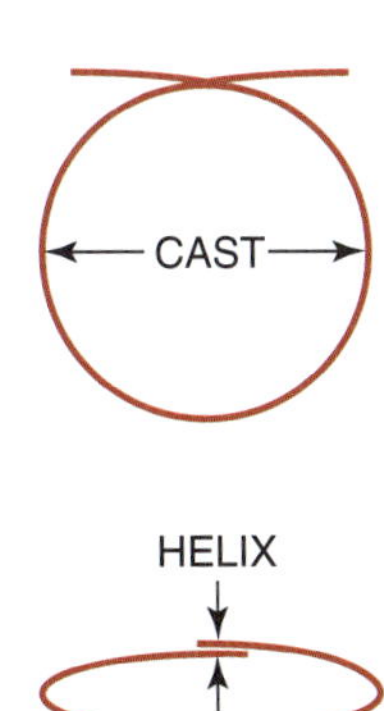

FIGURE 12-13 Cast and helix of GMA welding wire.

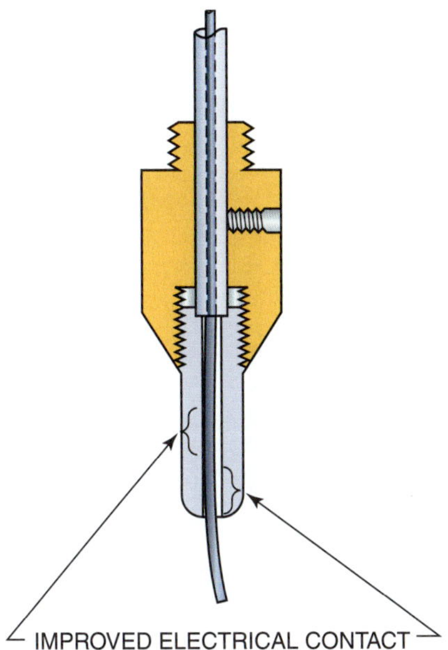

FIGURE 12-14 Cast forces the wire to make better electrical contact with the tube.

FIGURE 12-15 Push the wire through the guides by hand. Courtesy of Larry Jeffus.

FIGURE 12-16 Adjust the wire feed tensioner. Courtesy of Larry Jeffus.

so that the electrode conduit and cable are as straight as possible.

Press the gun switch. The wire should start feeding into the liner. Watch to make certain that the wire feeds smoothly and release the gun switch as soon as the end comes through the contact tube. ◆

CAUTION

If the wire stops feeding before it reaches the end of the contact tube, stop and check the system. If no obvious problem can be found, mark the wire with tape and remove it from the gun. It then can be held next to the system to determine the location of the problem.

With the wire feed running, adjust the feed roller compression so that the wire reel can be stopped easily by a slight pressure. Too light a roller pressure will cause the wire to feed erratically. Too high a pressure can turn a minor problem into a major disaster. If the wire jams at a high roller pressure, the feed rollers keep feeding the wire, causing it to bird nest and possibly short out. With a light pressure, the wire can stop, preventing bird nesting. This is very important with soft wires. The other advantage of a light pressure is that the feed will stop if something like clothing or gas hoses are caught in the reel.

With the feed running, adjust the spool drag so that the reel stops when the feed stops. The reel should not coast to a stop because the wire can be snagged easily. Also, when the feed restarts, a jolt occurs when the slack in the wire is taken up. This jolt can be enough to momentarily stop the wire, possibly causing a discontinuity in the weld.

When the test runs are completed, the wire can either be rewound or cut off. Some wire feed units have a retract button. This allows the feed driver to reverse and retract the wire automatically. To rewind the wire on units without this retract feature, release the rollers and turn them backward by hand. If the machine will not allow the feed rollers to be released without upsetting the tension, you must cut the wire.

Complete a copy of the "Student Welding Report" listed in Appendix I or provided by your instructor. ◆

Do not discard pieces of wire on the floor. They present a hazard to safe movement around the machine. In addition, a small piece of wire can work its way into a filter screen on the welding power source. If the piece of wire shorts out inside the machine, it could become charged with high voltage, which could cause injury or death. Always wind the wire tightly into a ball or cut it into short lengths before discarding it in the proper waste container.

Gas Density and Flow Rates

Density is the chief determinant of how effective a gas is for arc shielding. The lower the density of a gas, the higher will be the flow rate required for equal arc protection. Flow rates, however, are not in proportion to the densities. Helium, with about one-tenth the density of argon, requires only twice the flow for equal protection.

EXPERIMENT 12-1

Setting Gas Flow Rate

Using the equipment setup as described in Practice 12-1, and the threaded machine as described in Practice 12-2, you will set the shielding gas flow rate.

The exact flow rate required for a certain job will vary depending upon welding conditions. This experiment will help you determine how those conditions affect the flow rate. You will start by setting the shielding gas flow rate at 35 cfh (16 L/min).

Turn on the shielding gas supply valve. If the supply is a cylinder, the valve is opened all the way. With the machine power on and the welding gun switch depressed, you are ready to set the flow rate. Slowly turn in the adjusting screw and watch the float ball as it rises in a tube on a column of gas. The faster the gas flows, the higher the ball will float. A scale on the tube allows you to read the flow rate. Different scales are used with each type of gas being used. Since various gases have different densities (weights), the ball will float at varying levels even though the flow rates are the same, **Figure 12-17**. The line corresponding to the flow rate may be read as it compares to the top, center, or bottom of the ball, depending upon the manufacturer's instructions. There should be some marking or instruction on the tube or regulator to tell a person how it should be read, **Figure 12-18**.

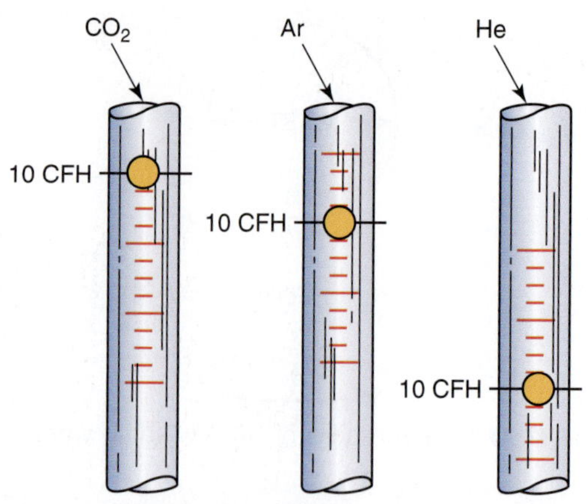

FIGURE 12-17 Each of these gases is flowing at the same cfh (L/min) rate. Because helium (He) is less dense, its indicator ball is the lowest. Be sure that you are reading the correct scale for the gas being used.

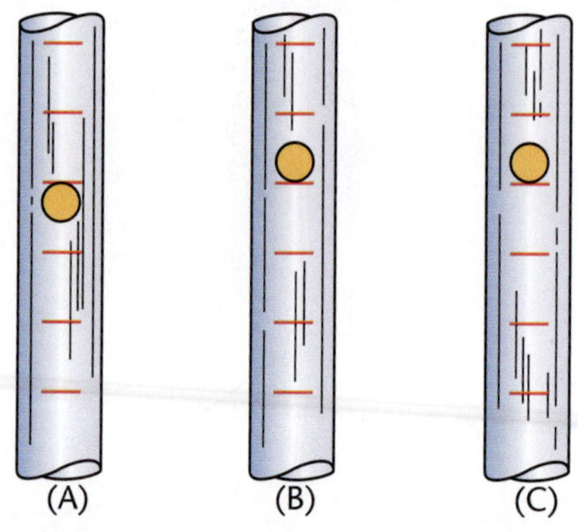

FIGURE 12-18 Three methods of reading a flowmeter: (A) top of ball, (B) center of ball, and (C) bottom of ball.

Release the welding gun switch, and the gas flow should stop. Turn off the power and spray the hose fittings with a leak-detecting solution.

When stopping for a period of time, the shielding gas supply valve should be closed and the hose pressure released.

Complete a copy of the "Student Welding Report" listed in Appendix I or provided by your instructor. ◆

Arc-voltage and Amperage Characteristics

The arc-voltage and amperage characteristics of GMA welding are different from most other welding processes. The voltage is set on the welder, and the amperage is set by changing the wire-feed speed. At any one voltage setting the amperage required to melt the wire must change as it is fed into the weld. It requires more amperage to melt the wire the faster it is fed, and less the slower it is fed.

Wire Feed Speed* in./min (m/min)	Wire Diameter			
	.030 in. (0.8 mm)	.035 in. (0.9 mm)	.045 in. (1.2 mm)	.062 in. (1.6 mm)
100 (2.5)	40	65	120	190
200 (5.0)	80	120	200	330
300 (7.6)	130	170	260	425
400 (10.2)	160	210	320	490
500 (12.7)	180	245	365	–
600 (15.2)	200	265	400	–
700 (17.8)	215	280	430	–

*To check feed speed, run out wire for one minute and then measure its length.

TABLE 12-1 Typical Amperages for Carbon Steel.

Because changes in the wire-feed speed directly change the amperage, it is possible to set the amperage by using a chart and measuring the length of wire fed per minute, **Table 12-1.** The voltage and amperage required for a specific metal transfer method differ for various wire sizes, shielding gases, and metals.

The voltage and amperage setting will be specified for all welding done according to a welding procedure specification (WPS) or other codes and standards. However, most welding—like that done in small production shops, as maintenance welding, for repair work, in farm shops, and the like—is not done to specific code or standard and therefore no specific setting exists. For that reason, it is important to learn to make the adjustments necessary to allow you to produce quality welds.

EXPERIMENT 12-2

Setting the Current

Using a properly assembled GMA welding machine, proper safety protection, and one piece of mild steel plate approximately 12 in. (305 mm) long × 1/4 in. (6 mm) thick, you will change the current settings and observe the effect on GMAW.

On a scale of 0 to 10, set the wire-feed speed control dial at 5, or halfway between the low and high settings of the unit. The voltage is also set at a point halfway between the low and high settings. The shielding gas can be CO_2, argon, or a mixture. The gas flow should be adjusted to a rate of 35 cfh (16 L/min).

Hold the welding gun at a comfortable angle, lower your welding hood, and pull the trigger. As the wire feeds and contacts the plate, the weld will begin. Move the gun slowly along the plate. Note the following welding conditions as the weld progresses: voltage, amperage, weld direction, metal transfer, spatter, molten weld pool size, and penetration. Stop and record your observations in **Table 12-2.** Evaluate the quality of the weld as acceptable or unacceptable.

Reduce the voltage somewhat and make another weld, keeping all other weld variables (travel speed, stickout, direction, amperage) the same. Observe the weld and upon stopping, record the results. Repeat this procedure until the voltage has been lowered to the minimum value indicated on the machine. Near the lower end the wire may stick, jump, or simply no longer weld.

Return the voltage indicator to the original starting position and make a short test weld. Stop and compare the results to those first observed. Then slightly increase the voltage setting and make another weld. Repeat the procedure of observing and recording the results as the voltage is increased in steps until the maximum machine capability is obtained. Near the maximum setting the spatter may become excessive if CO_2 shielding gas is used. Care must be taken to prevent the wire from fusing to the contact tube.

Return the voltage indicator to the original starting position and make a short test weld. Compare the results observed with those previously obtained.

Weld Acceptability	Voltage	Amperage	Spatter	Molten Pool Size	Penetration
Good	20	75	Light	Small	Little

Electrode diameter .035 in. (0.9 mm)
Shielding gas CO_2
Welding direction Backhand

TABLE 12-2 Setting the Current.

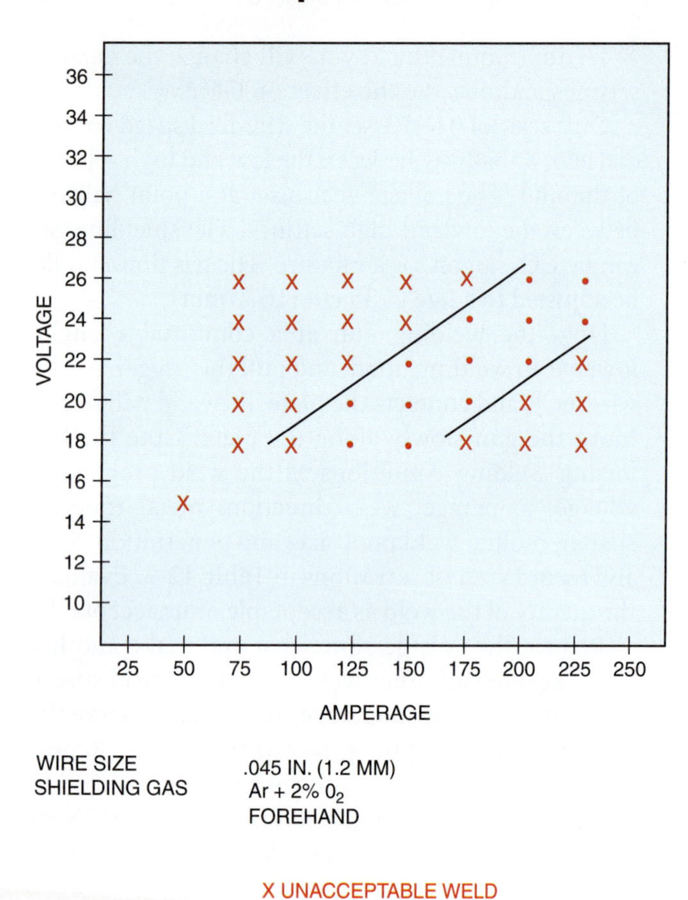

TABLE 12-3 Graph for GMAW Machine Settings.

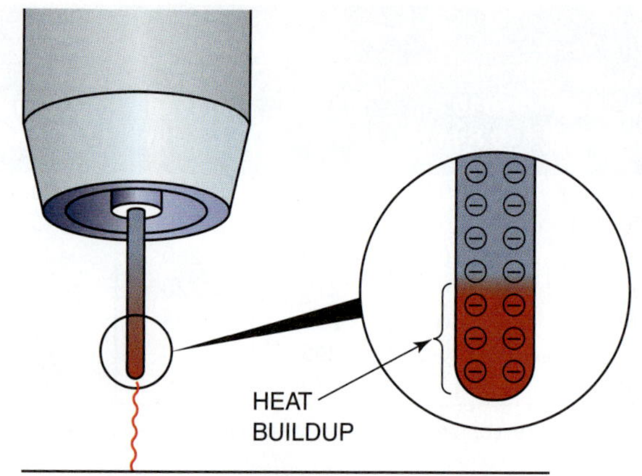

FIGURE 12-19 Heat buildup due to the extremely high current for the small conductor (electrode).

Lower the wire-feed speed setting slightly and use the same procedure as before. First lower and then raise the voltage through a complete range and record your observations. After a complete set of test results is obtained from this amperage setting, again lower the wire-feed speed for a new series of tests. Repeat this procedure until the amperage is at the minimum setting shown on the machine. At low amperages and high voltage settings, the wire may tend to pop violently as a result of the uncontrolled arc.

Return the wire-feed speed and voltages to the original settings. Make a test weld and compare the results with the original tests. Slightly raise the wire speed and again run a set of tests as the voltage is changed in small steps. After each series, return the voltage setting to the starting point and increase the wire-feed speed. Make a new set of tests.

All of the test data can be gathered into an operational graph for the machine, wire type, size, and shielding gas. Use **Table 12-3** to plot the graph. The acceptable welds should be marked on the lines that extend from the appropriate voltages and amperages. Upon completion, the graph will give you the optimum settings for the operation of this particular GMAW setup. The optimum settings are along a line in the center of the acceptable welds.

Experienced welders will follow a much shorter version of this type of procedure anytime they are starting to work on a new machine or testing for a new job. This experiment can be repeated using different types of wire, wire sizes, shielding gases, and weld directions. Turn off the welding machine and shielding gas and clean up your work area when you are finished welding.

Complete a copy of the "Student Welding Report" listed in Appendix I or provided by your instructor. ◆

Electrode Extension

Because of the constant-potential (CP) power supply, the welding current will change as the distance between the contact tube and the work changes. Although this change is slight, it is enough to affect the weld being produced. The longer the electrode extension, the greater the resistance to the welding current flowing through the small welding wire. This results in some of the welding current being changed to heat at the tip of the electrode, **Figure 12-19.** With a standard SMA welding CC power supply this would also reduce the arc voltage, but with a CP power supply the voltage remains constant and the amperage increases. If the electrode extension is shortened, the welding current decreases.

The increase in current does not result in an increase in penetration, because the current is being used to heat the electrode tip and not being transferred to the weld metal. Penetration is reduced and buildup is increased as the electrode extension is lengthened. Penetration is increased and buildup decreased as the electrode extension is shortened. Controlling the weld penetration and buildup by changing the electrode will help maintain weld bead shape during welding. It will also help you better

Weld Acceptability	Voltage	Amperage	Electrode Extension	Contact Tube to Work Distance	Bead Shape
Poor	20	100	1 in. (25 mm)	1 1/4 in. (31 mm)	Narrow, high, with little penetration

Electrode diameter .035 in. (0.9 mm)
Shielding gas CO_2
Welding direction Forehand

TABLE 12-4 Electrode Extension.

understand what may be happening if a weld starts out correctly but begins to change as it progresses along the joint. You may be changing the electrode extension without noticing the change.

EXPERIMENT 12-3

Electrode Extension

Using a properly assembled GMA welding machine, proper safety protection, and a few pieces of mild steel, each about 12 in. (305 mm) long and ranging in thickness from 16 gauge to 1/2 in. (13 mm), you will observe the effect of changing electrode extension on the weld.

Start at a low current setting. Using the graph developed in Experiment 12-2, set both the voltage and amperage. The settings should be equal to those on the optimum line established for the wire type and size being used with the same shielding gas.

Holding the welding gun at a comfortable angle and height, lower your helmet and start to weld. Make a weld approximately 2 in. (51 mm) long. Then reduce the distance from the gun to the work while continuing to weld. After a few inches, again shorten the electrode extension. Keep doing this in steps until the nozzle is as close as possible to the work. Stop and return the gun to the original starting distance.

Repeat the process just described but now increase the electrode extension in steps of a few inches each. Keep increasing the electrode extension until the weld will no longer fuse or the wire becomes impossible to control.

Change the plate thickness and repeat the procedure. When the series has been completed with each plate thickness, raise the voltage and amperage to a medium setting and repeat the process. Upon completing this series of tests, adjust the voltage and amperage upward to a high setting. Make a full series of tests using the same procedures as before.

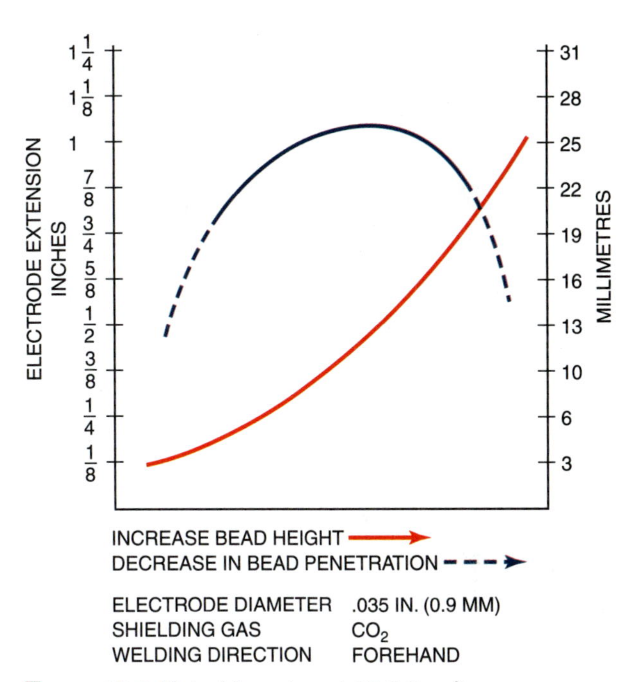

TABLE 12-5 Plot of Experiment 12-3 Results.

Record the results in **Table 12-4** after each series of tests. The final results can be plotted on a graph, as was done in **Table 12-5**, to establish the optimum electrode extension for each thickness, voltage, and amperage. Turn off the welding machine and shielding gas and clean up your work area when you are finished welding.

Complete a copy of the "Student Welding Report" listed in Appendix I or provided by your instructor. ◆

Welding Gun Angle

The term *welding gun angle* refers to the angle between the GMA welding gun and the work as it relates to the direction of travel. Backhand welding, or dragging angle, **Figure 12-20**, produces a weld with deep penetration and higher buildup. Forehand welding, or pushing angle, **Figure 12-21**, produces a weld with shallow penetration and little buildup.

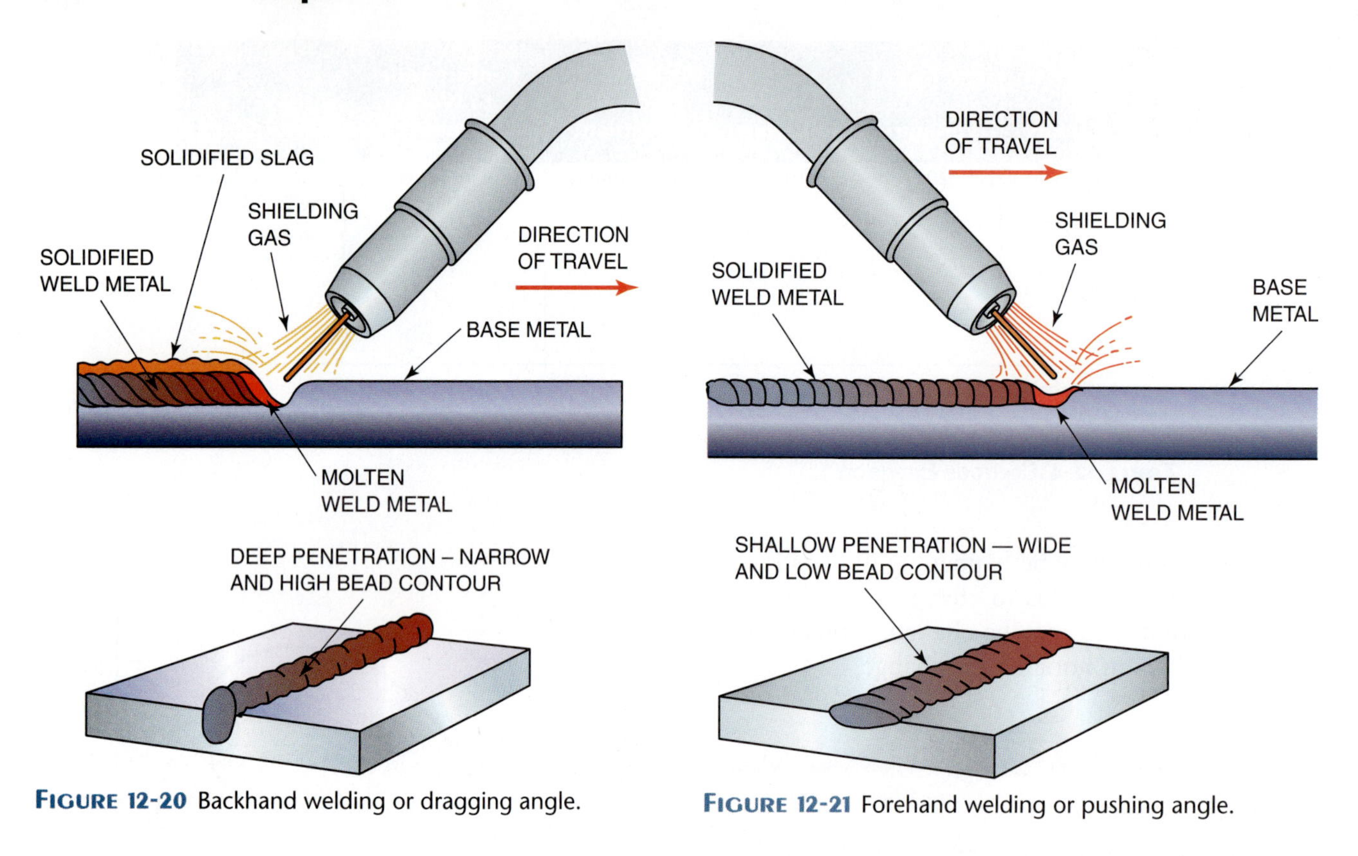

FIGURE 12-20 Backhand welding or dragging angle.

FIGURE 12-21 Forehand welding or pushing angle.

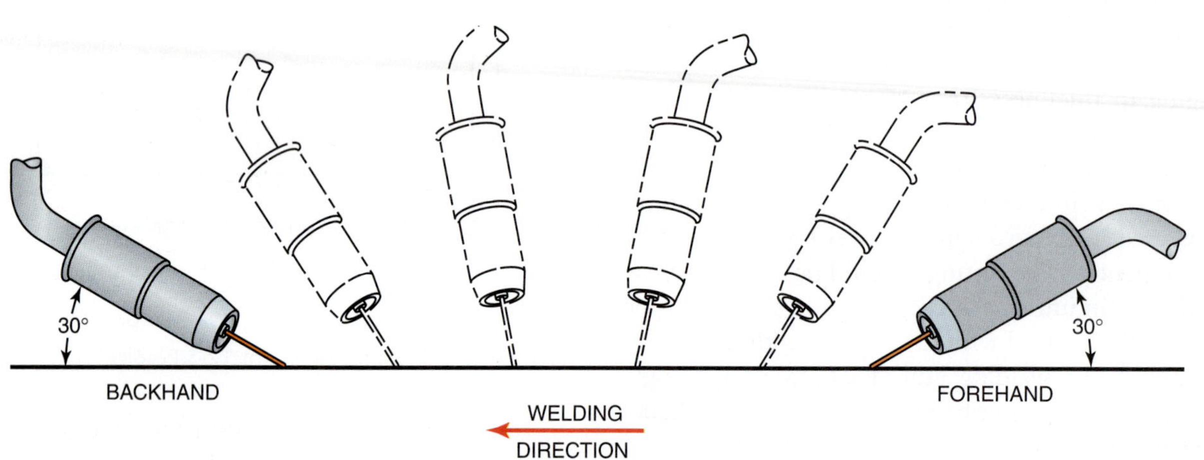

FIGURE 12-22 Welding gun angle.

Slight changes in the welding gun angle can be used to control the weld as the groove spacing changes. A narrow gap may require more penetration, but as the gap spacing increases a weld with less penetration may be required. Changing the electrode extension and welding gun angle at the same time can result in a quality weld being made with less than ideal conditions.

EXPERIMENT 12-4

Welding Gun Angle

Using a properly assembled GMA welding machine, proper safety protection, and some pieces of mild steel, each approximately 12 in. (305 mm) long and ranging in thickness from 16 gauge to 1/2 in. (13 mm), you will observe the effect of changing the welding gun angle on the weld bead.

Starting with a medium current setting and a plate that is 1/4 in. (6 mm) thick, hold the welding gun at a 30° angle to the plate in the direction of the weld, **Figure 12-22.** Lower your welding hood and depress the trigger. When the weld starts, move in a straight line and slowly pivot the gun angle as the weld progresses. Keep the travel speed, electrode extension, and weave pattern (if used) constant so that any change in the weld bead is caused by the angle change.

The pivot should be completed in the 12 in. (305 mm) of the weld. You will proceed from a 30° push-

Shielding Gas	Chemical Behavior	Uses and Usage Notes
1. Argon	Inert	Welding virtually all metals except steel
2. Helium	Inert	Al and Cu alloys for greater heat and to minimize porosity
3. Ar and He (20% to 80% to 50% to 50%)	Inert	Al and Cu alloys for greater heat and to minimize porosity but with quieter, more readily controlled arc action
4. N_2	Reducing	On Cu, very powerful arc
5. Ar + 25% to 30% N_2	Reducing	On Cu, powerful but smoother operating, more readily controlled arc than with N_2
6. Ar + 1% to 2% O_2	Oxidizing	Stainless and alloy steels, also for some deoxidized copper alloys
7. Ar + 3% to 5% O_2	Oxidizing	Plain carbon, alloy, and stainless steels (generally requires highly deoxidized wire)
8. Ar + 3% to 5% O_2	Oxidizing	Various steels using deoxidized wire
9. Ar + 20% to 30% O_2	Oxidizing	Various steels, chiefly with short-circuiting arc
10. Ar + 5% O_2 + 15% CO_2	Oxidizing	Various steels using deoxidized wire
11. CO_2	Oxidizing	Plain-carbon and low alloy steels, deoxidized wire essential
12. CO_2 + 3% to 10% O_2	Oxidizing	Various steels using deoxidized wire
13. CO_2 + 20% O_2	Oxidizing	Steels

TABLE 12-6 Shielding Gases and Gas Mixtures Used for Gas Metal Arc Welding.

ing angle to a 30° dragging angle. Repeat this procedure using different welding currents and plate thicknesses.

After the welds are complete, note the differences in width and reinforcement along the welds. Turn off the welding machine and shielding gas and clean up your work area when you are finished welding.

Complete a copy of the "Student Welding Report" listed in Appendix I or provided by your instructor. ◆

Effect of Shielding Gas on Welding

Shielding gases in the gas metal arc process are used primarily to protect the molten metal from oxidation and contamination. Other factors must be considered, how-ever, in selecting the right gas for a particular application. Shielding gas can influence arc and metal transfer characteristics, weld penetration, width of fusion zone, surface shape patterns, welding speed, and undercut tendency, **Table 12-6**. Inert gases such as argon and helium provide the necessary shielding because they do not form compounds with any other substance and are insoluble in molten metal. When used as pure gases for welding ferrous metals, argon and helium may produce an erratic arc action, promote undercutting, and result in other flaws.

It is therefore usually necessary to add controlled quantities of reactive gases to achieve good arc action and metal transfer with these materials. Adding oxygen or carbon dioxide to the inert gas tends to stabilize the arc, promote favorable metal transfer, and minimize spatter. As a result, the penetration pattern is improved and undercutting is reduced or eliminated.

Oxygen or carbon dioxide is often added to argon. The amount of reactive gas required to produce the desired effects is quite small. As little as 0.5% of oxygen will produce a noticeable change; 1% to 5% of oxygen is more common. Carbon dioxide may be added to argon in the 20% to 30% range. Mixtures of argon with less than 10% carbon dioxide may not have enough arc voltage to give the desired results.

Adding oxygen or carbon dioxide to an inert gas causes the shielding gas to become oxidizing. This in turn may cause porosity in some ferrous metals. In this case, a filler wire containing suitable deoxidizers should be used. The presence of oxygen in the shielding gas can also cause some loss of certain alloying elements, such as chromium, vanadium, aluminum, titanium, manganese, and silicon. Again, the addition of a deoxidizer to the filler wire is necessary.

Pure carbon dioxide has become widely used as a shielding gas for GMA welding of steels. It allows higher welding speed, better penetration, and good mechanical properties, and it costs less than the inert gases. The chief drawback in the use of carbon dioxide is the less-steady-arc characteristics and considerable weld-metal-spatter losses. The spatter can be kept to a minimum by maintaining a very short, uniform arc length. Consistently sound welds can be produced using carbon dioxide shielding, provided that a filler wire having the proper deoxidizing additives is used.

Practices

The practices in this chapter are grouped according to those requiring similar techniques and setups. To make acceptable GMA welds consistently, the major skill required is the ability to set up the equipment and weldment. Changes such as variations in material thickness, position, and type of joint require changes both in technique and setup. A correctly set up GMA welding station can, in many cases, be operated with minimum skill. Often the only difference between a

Deoxidizing Element	Strength
Aluminum (Al)	Very strong
Manganese (Mn)	Weak
Silicon (Si)	Weak
Titanium (Ti)	Very strong
Zirconium (Zr)	Very strong

Table 12-7 Sufficient Deoxidizing Elements Must Be Added to the Filler Wire to Minimize Porosity in the Molten Weld Pool.

welder earning a minimum wage and one earning the maximum wage is the ability to correct machine setups.

Ideally, only a few tests would be needed for the welder to make the necessary adjustments in setup and manipulation techniques to achieve a good weld. The previous welding experiments should have given the welder a graphic set of comparisons to help that welder make the correct changes. In addition to keeping the test data, you may want to keep the test plates for a more accurate comparison.

The grouping of practices in this chapter will keep the number of variables in the setup to a minimum. Often, the only change required before going on to the next weld is to adjust the power settings.

Figures that are given in some of the practices will give the welder general operating conditions, such as voltage, amperage, and shielding gas and/or gas mixture. These are general values, so the welder will have to make some fine adjustments. Differences in the type of machine being used and the material surface condition will affect the settings. For this reason, it is preferable to use the settings developed during the experiments.

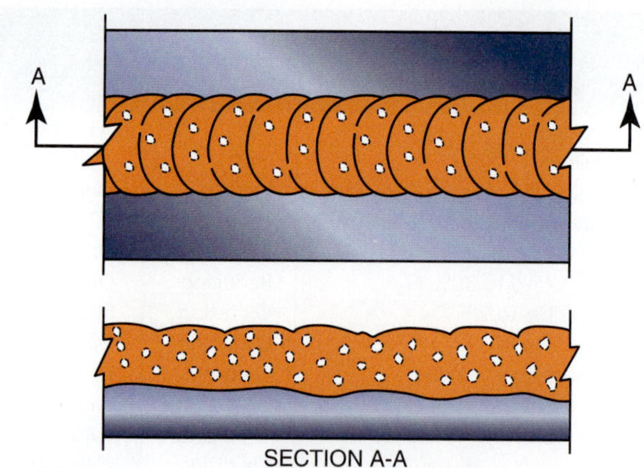

Figure 12-23 Uniformly scattered porosities.

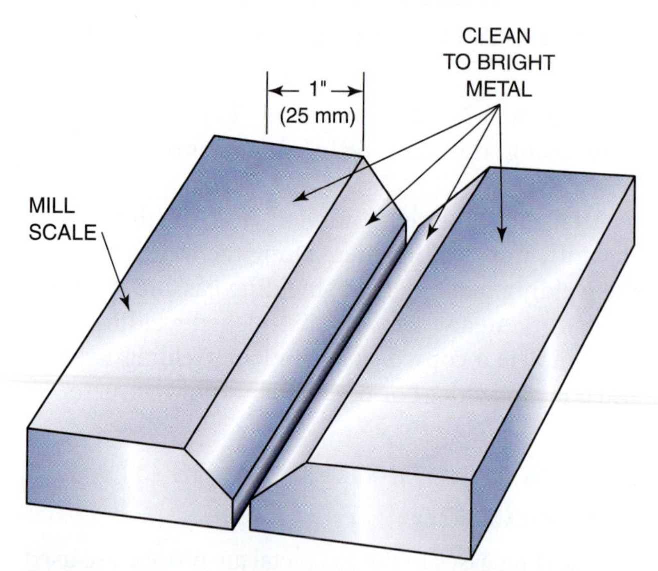

Figure 12-24 Clean all surfaces to bright metal before welding.

Metal Preparation

All hot-rolled steel has an oxide layer, which is formed during the rolling process, called mill scale. *Mill scale* is a thin layer of dark gray or black iron oxide. Some hot-rolled steels that have had this layer removed either mechanically or chemically can be purchased. However, almost all of the hot-rolled steel used today still has this layer because it offers some protection from rusting.

Mill scale is not removed for noncode welding, because it does not prevent most welds from being suitable for service. For practice welds that will be visually inspected mill scale can usually be left on the plate. Filler metals and fluxes usually have deoxidizers added to them so that the adverse effects of the mill scale are reduced or eliminated, **Table 12-7.** But with GMA welding wire it is difficult to add enough deoxidizers to remove all effects of mill scale. The porosity that mill scale causes is most often confined to the interior of the weld and is not visible on the surface, **Figure 12-23.** Because it is not visible on the surface, it usually goes unnoticed and the weld passes visual inspection.

If the practices are going to be destructively tested, then all welding surfaces within the weld groove and the surrounding surfaces within 1 in. (25 mm) must be cleaned to bright metal, **Figure 12-24.** Cleaning may be either grinding, filing, sanding, or blasting.

Flat Position, 1G and 1F Positions

PRACTICE 12-3

Stringer Beads Using the Short-Circuiting Metal Transfer Method in the Flat Position

Using a properly set up and adjusted GMA welding machine, **Table 12-8,** proper safety protection, .035-in. and/or .045-in. (0.9-mm and/or 1.2-mm) diameter

Process	Wire Diameter	Amperage Range (Optimum)	Voltage Range (Optimum)	Shielding Gas
Short-circuiting	0.030	60 (100) 140	14 (15) 16	100% CO_2 75% Ar + 25% CO_2 98% Ar + 2% O
	0.035	90 (130) 150	16 (17) 20	

TABLE 12-8 Typical Welding Current Settings for Short-Circuiting Metal Transfer for Mild Steel.

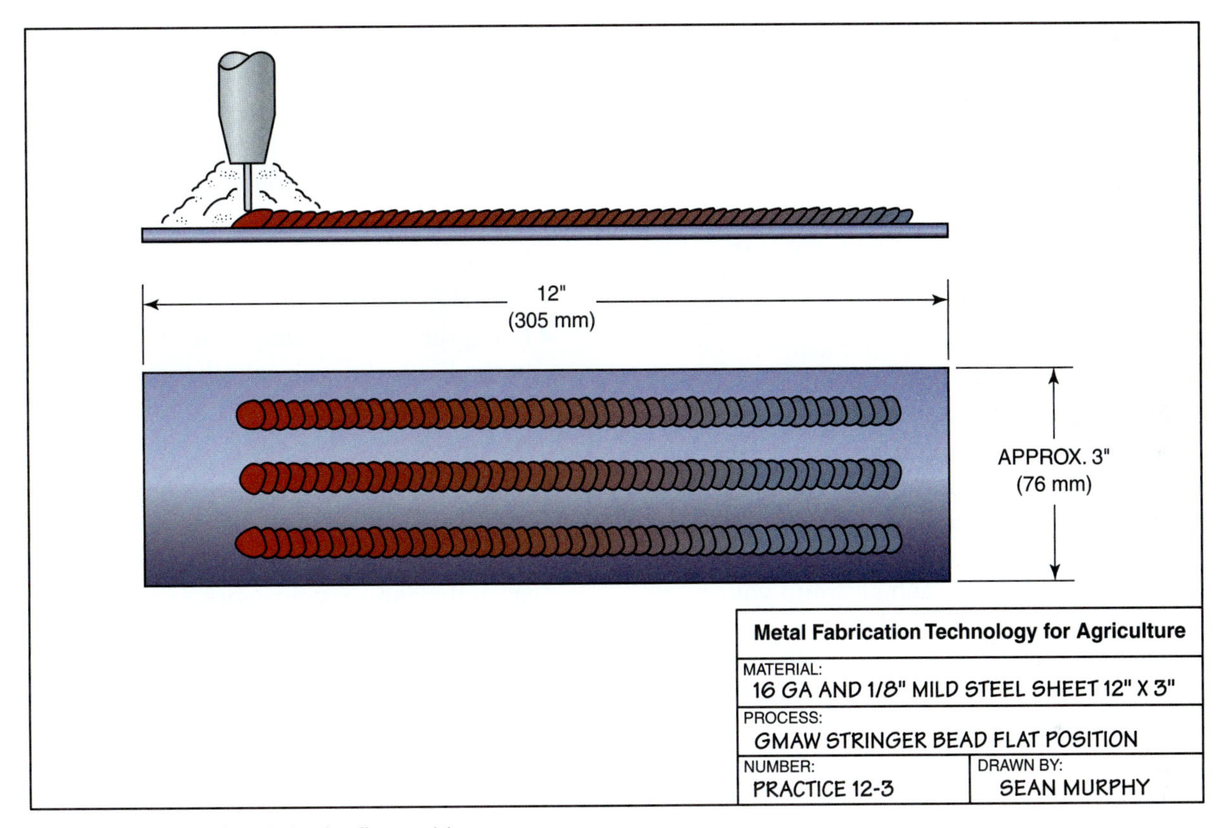

FIGURE 12-25 Stringer beads in the flat position.

wire, and two or more pieces of mild steel sheet 12 in. (305 mm) long and 16 gauge and 1/8 in. (3 mm) thick, you will make a stringer bead weld in the flat position, **Figure 12-25**.

Starting at one end of the plate and using either a pushing or dragging technique, make a weld bead along the entire 12-in. (305-mm) length of the metal. After the weld is complete, check its appearance. Make any needed changes to correct the weld (refer to **Table 12-3** and **Table 12-5**). Repeat the weld and make additional adjustments. After the machine is set, start to work on improving the straightness and uniformity of the weld.

Keeping the bead straight and uniform can be hard because of the limited visibility due to the small amount of light and the size of the molten weld pool. The welder's view is further restricted by the shielding gas nozzle, **Figure 12-26**. Even with limited visibility, it is possible to make a satisfactory weld by watching the edge of the molten weld pool, the sparks, and the

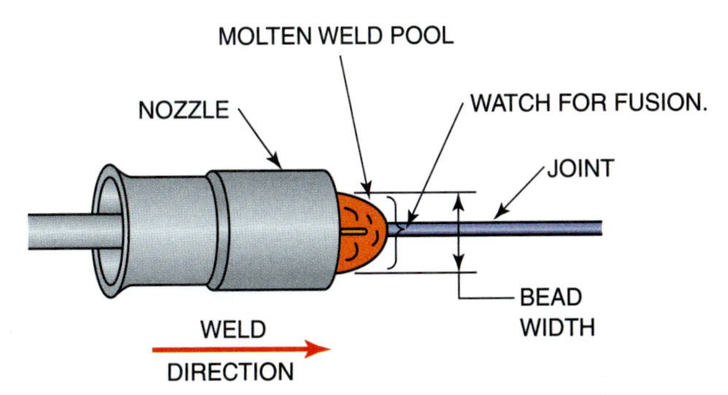

FIGURE 12-26 The shielding gas nozzle restricts the welder's view.

weld bead produced. Watching the leading edge of the molten weld pool (forehand welding, pushing technique) will show you the molten weld pool fusion and width. Watching the trailing edge of the molten weld pool (backhand welding, dragging technique) will show you the amount of buildup and the relative heat input, **Figure 12-27**. The quantity and size of sparks

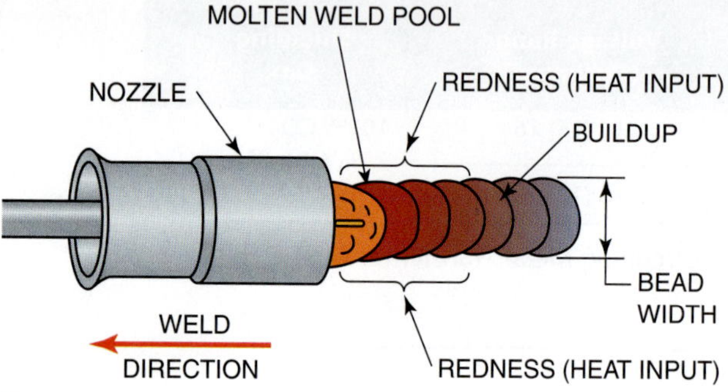

FIGURE 12-27 Watch the trailing edge of the molten weld pool.

produced can indicate the relative location of the filler wire in the molten weld pool. The number of sparks will increase as the wire strikes the solid metal ahead of the molten weld pool. The gun itself will begin to vibrate or bump as the wire momentarily pushes against the cooler, unmelted base metal before it melts. Changes in weld width, buildup, and proper joint tracking can be seen by watching the bead as it appears from behind the shielding gas nozzle.

Repeat each type of bead as needed until consistently good beads are obtained. Turn off the welding machine and shielding gas and clean up your work area when you are finished welding.

Complete a copy of the "Student Welding Report" listed in Appendix I or provided by your instructor. ◆

PRACTICE 12-4

Flat Position Butt Joint, Lap Joint, and Tee Joint

Using the same equipment, materials, and procedures listed in Practice 12-3, make welded butt joints, lap joints, and tee joints in the flat position, **Figure 12-28A**, **Figure 12-28B**, and **Figure 12-28C**.

- Tack weld the sheets together and place them flat on the welding table, **Figure 12-29.**
- Starting at one end, run a bead along the joint. Watch the molten weld pool and bead for signs that a change in technique may be required.
- Make any needed changes as the weld progresses. By the time the weld is complete, you should be making the weld nearly perfectly.
- Using the same technique that was established in the last weld, make another weld. This time, the entire 12 in. (305 mm) of weld should be flawless.

Repeat each type of joint with both thicknesses of metal as needed until consistently good beads are obtained. Turn off the welding machine and shielding gas and clean up your work area when you are finished welding.

Complete a copy of the "Student Welding Report" listed in Appendix I or provided by your instructor. ◆

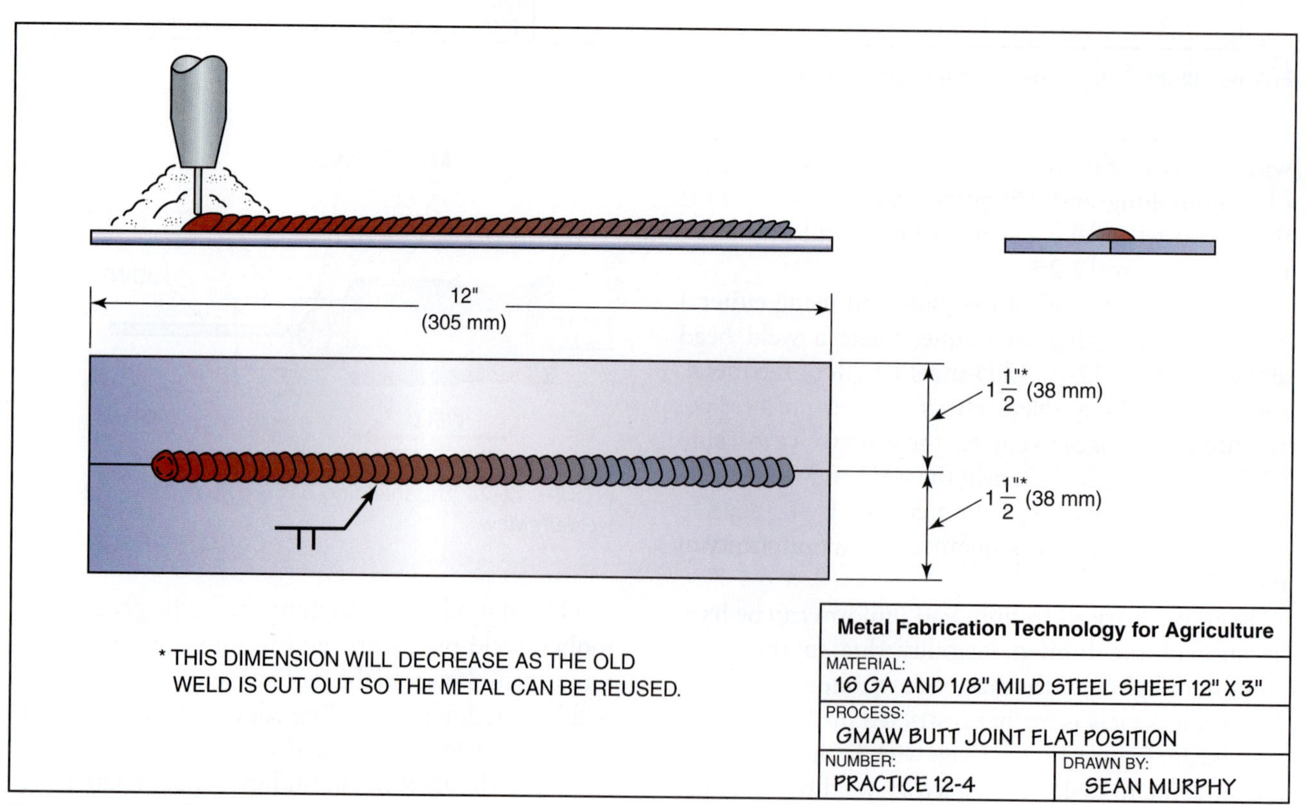

FIGURE 12-28(A) Butt joint in the flat position.

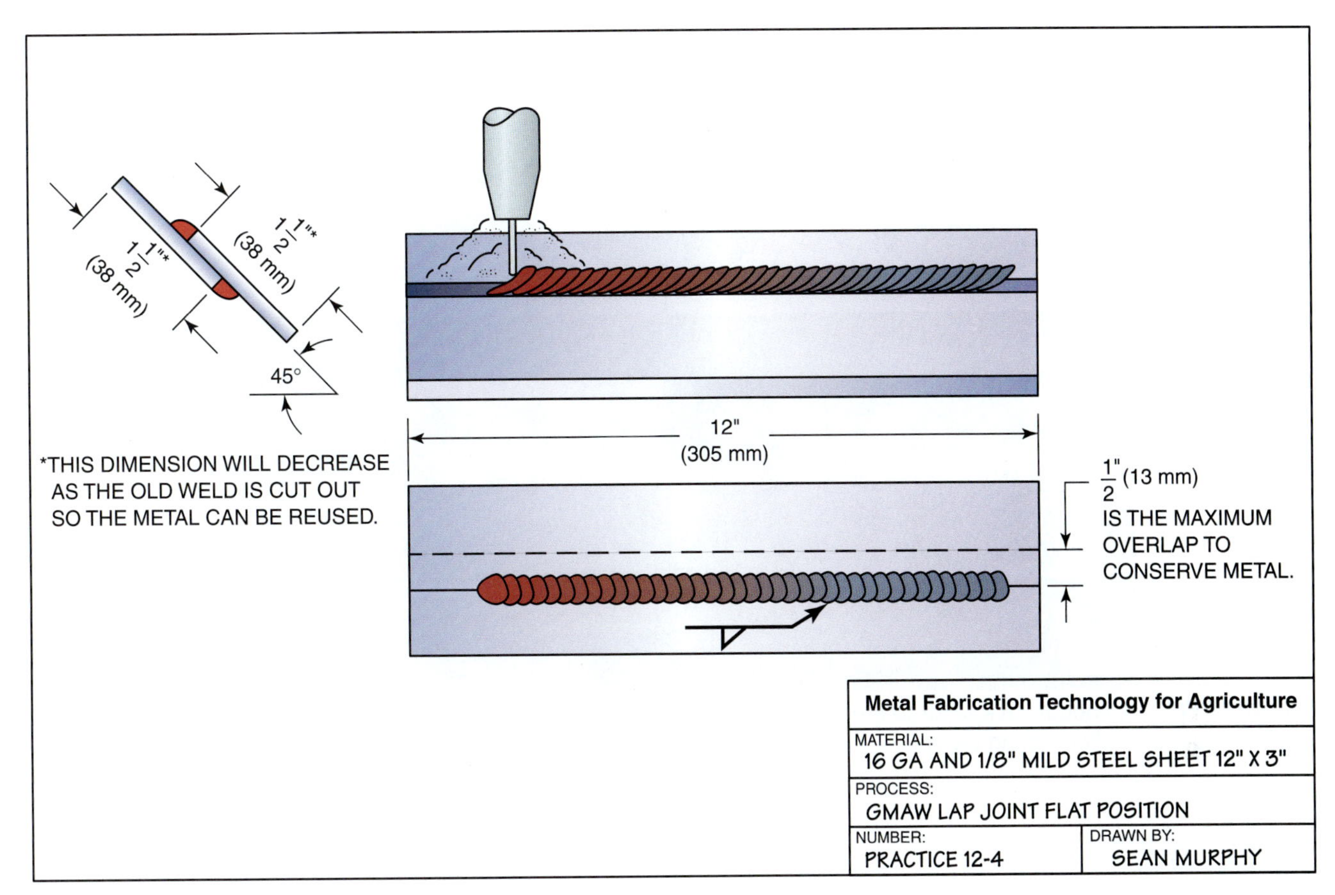

FIGURE 12-28(B) Lap joint in the flat position.

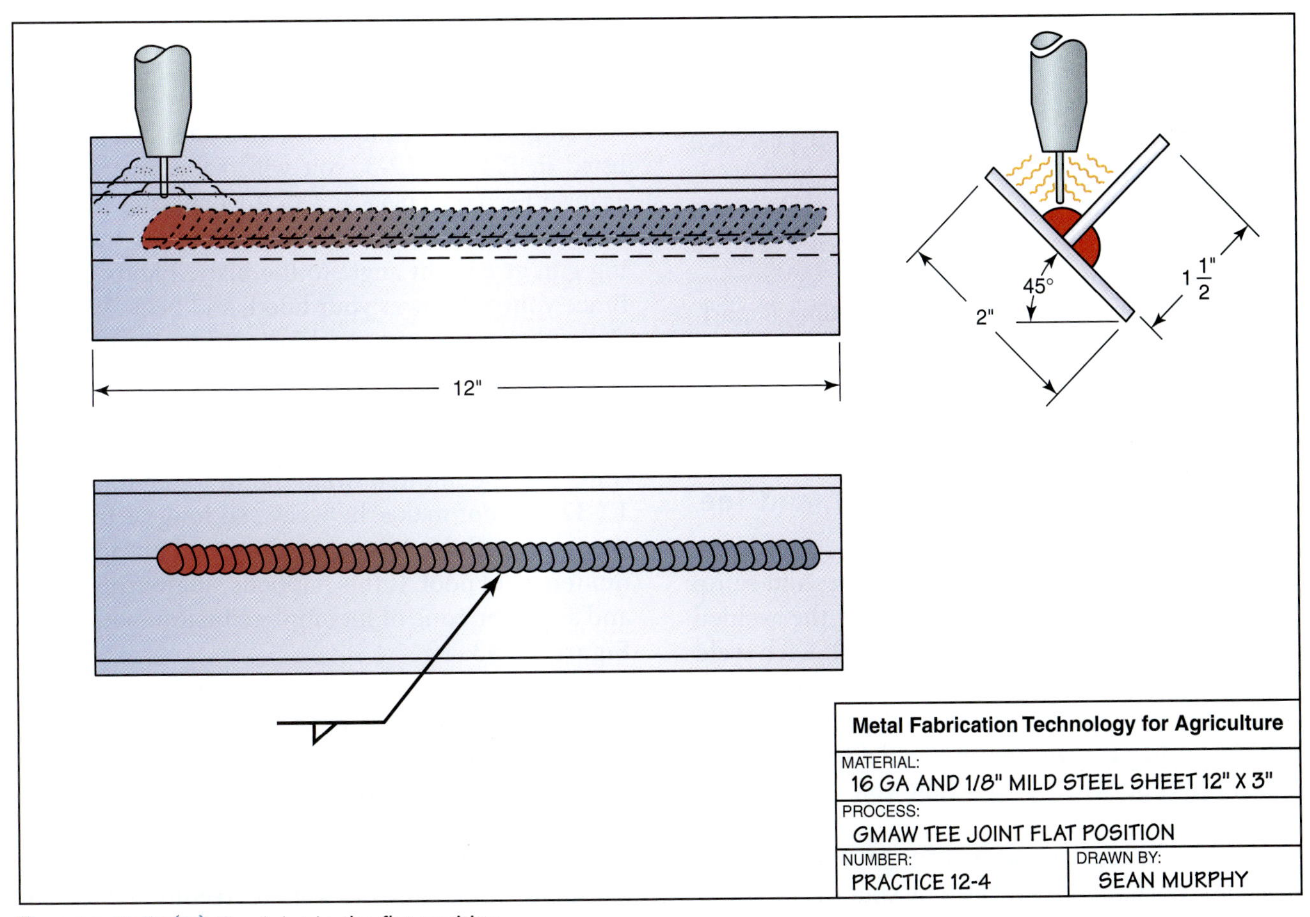

FIGURE 12-28(C) Tee joint in the flat position.

FIGURE 12-29 Use enough tack welds to keep the joint in alignment during welding. Small tack welds are easier to weld over without adversely affecting the weld. Courtesy of Larry Jeffus.

PRACTICE 12-5

Flat Position Butt Joint, Lap Joint, and Tee Joint, All with 100% Penetration

Using the same equipment, materials, and setup listed in Practice 12-3, make a welded joint in the flat position with 100% penetration, along the entire 12-in. (305-mm) length of the welded joint. Repeat each type of joint as needed until consistently good beads are obtained. Turn off the welding machine and shielding gas and clean up your work area when you are finished welding.

Complete a copy of the "Student Welding Report" listed in Appendix I or provided by your instructor. ◆

PRACTICE 12-6

Flat Position Butt Joint, Lap Joint, and Tee Joint, All Welds to Be Tested

Using the same equipment, materials, and setup listed in Practice 12-3, make each of the welded joints in the flat position, **Figure 12-30**. Each weld joint must pass the bend test. Repeat each type of weld joint until all pass the guided bend test. Turn off the welding machine and shielding gas and clean up your work area when you are finished welding.

Complete a copy of the "Student Welding Report" listed in Appendix I or provided by your instructor. ◆

Vertical Up 3G and 3F Positions

PRACTICE 12-7

Stringer Bead at a 45° Vertical Up Angle

Using the same equipment, materials, and setup as listed in Practice 12-3, you will make a vertical up stringer bead on a plate at a 45° inclined angle.

Start at the bottom of the plate and hold the welding gun at a slight angle to the plate, **Figure 12-31.** Brace yourself, lower your hood, and begin to weld. Depending upon the machine settings and type of shielding gas used, you will make a weave pattern.

If the molten weld pool is large and fluid (hot), use a "C" or "J" weave pattern to allow a longer time for the molten weld pool to cool, **Figure 12-32.** Do not make the weave so long or fast that the wire is allowed to strike the metal ahead of the molten weld pool. If this happens, spatter increases and a spot or zone of incomplete fusion may occur, **Figure 12-33.**

If the molten weld pool is small and controllable, use a small "C," zigzag, or "J" weave pattern to control the width and buildup of the weld. A slower speed can also be used. Watch for complete fusion along the leading edge of the molten weld pool.

A weld that is high and has little or no fusion is too "cold." Changing the welding technique will not

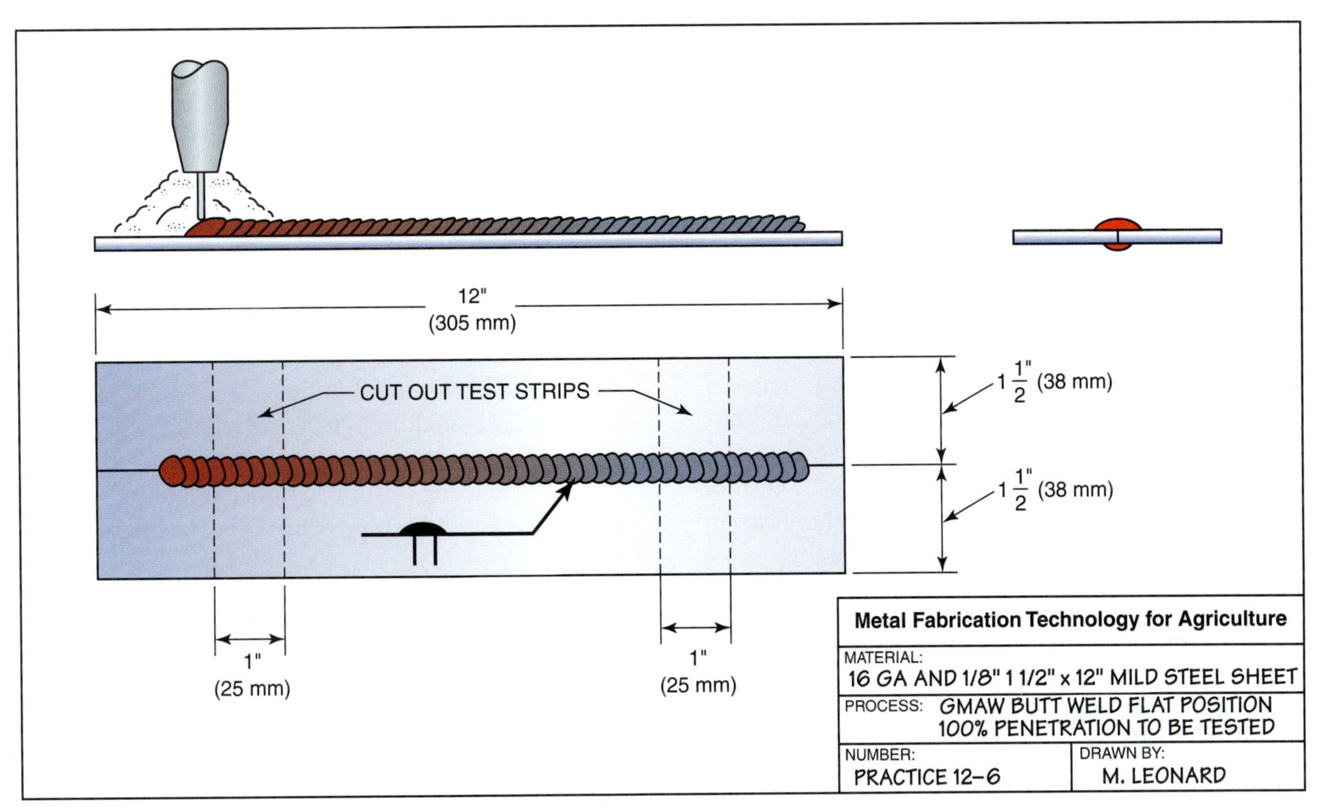

FIGURE 12-30 Butt joint in the flat position to be tested.

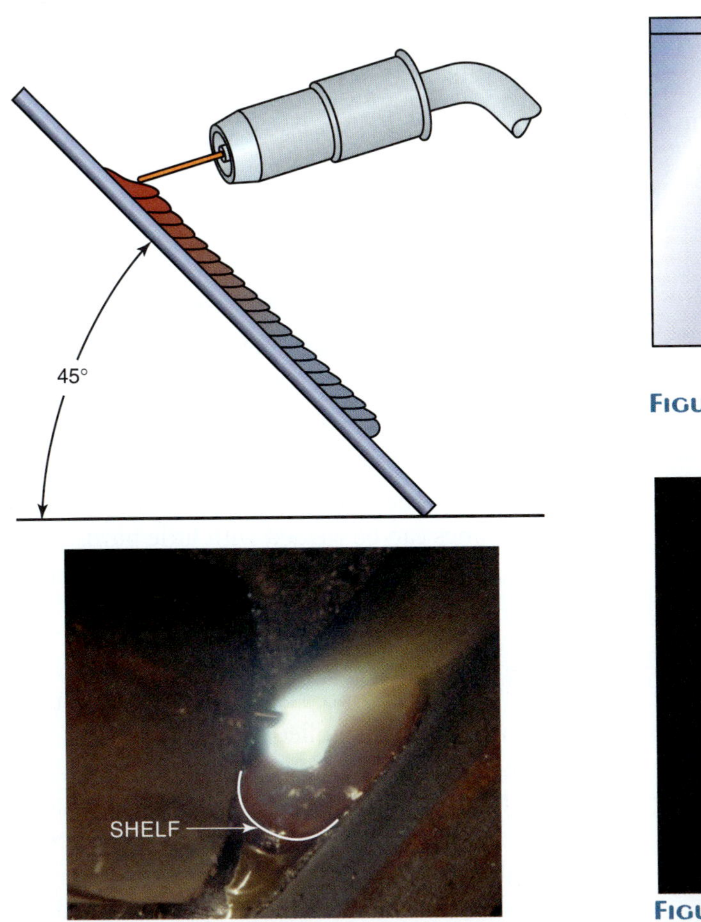

FIGURE 12-31 Vertical up position. Photo courtesy of Larry Jeffus.

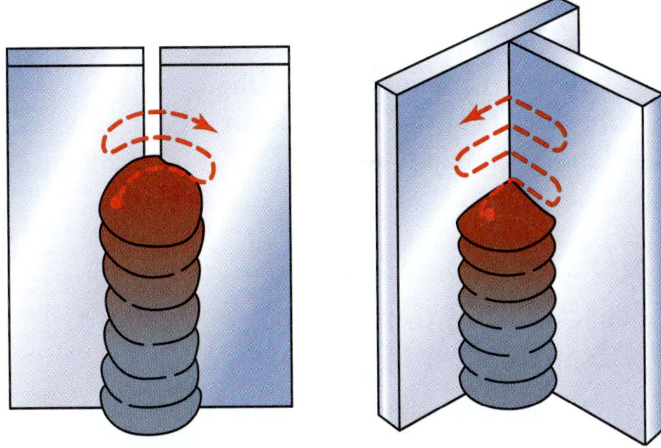

FIGURE 12-32 Vertical up weld patterns.

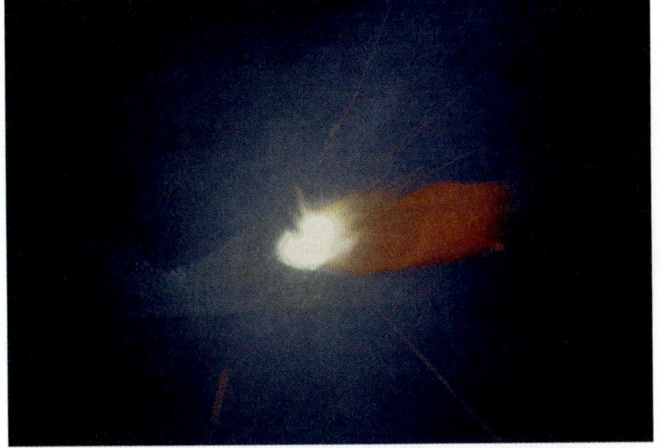

FIGURE 12-33 Burst of spatter caused by incorrect electrode contact with base metal. Courtesy of Larry Jeffus.

correct this problem. The welder must stop welding and make the needed adjustments.

As the weld progresses up the plate, the back or trailing edge of the molten weld pool will cool, forming a shelf to support the molten metal. Watch the shelf to be sure that molten metal does not run over, forming a drip. When it appears that the metal may flow over the shelf, either increase the weave lengths or stop and start the current for brief moments to allow the weld to cool. Stopping for brief moments will not allow the shielding gas to be lost.

Continue to weld along the entire 12-in. (305-mm) length of plate. Repeat this weld as needed until a straight and uniform weld bead is produced. Turn off the welding machine and shielding gas and clean up your work area when you are finished welding.

Complete a copy of the "Student Welding Report" listed in Appendix I or provided by your instructor. ◆

PRACTICE 12-8

Stringer Bead in the Vertical Up Position

Repeat Practice 12-7 and increase the angle until you have mastered a straight and uniform weld bead in the vertical up position. Turn off the welding machine and shielding gas and clean up your work area when you are finished welding.

Complete a copy of the "Student Welding Report" listed in Appendix I or provided by your instructor. ◆

PRACTICE 12-9

Butt Joint, Lap Joint, and Tee Joint in the Vertical Up Position at a 45° Angle

Using the same equipment, materials, and setup as listed in Practice 12-3, you will make vertical up welded joints on a plate at a 45° inclined angle.

Tack weld the metal pieces together and brace them in position. Check to see that you have free movement along the entire joint to prevent stopping and restarting during the weld. Avoiding stops and starts both speeds up the welding time and eliminates discontinuities.

The weave pattern should allow for adequate fusion on both edges of the joint. Watch the edges to be sure that they are being melted so that adequate fusion and penetration occur.

Repeat each type of joint as needed until consistently good beads are obtained. Turn off the welding machine and shielding gas and clean up your work area when you are finished welding.

Complete a copy of the "Student Welding Report" listed in Appendix I or provided by your instructor. ◆

PRACTICE 12-10

Butt Joint, Lap Joint, and Tee Joint in the Vertical Up Position with 100% Penetration

Using the same equipment, materials, and setup as listed in Practice 12-3, you will increase the plate angle gradually as you develop skill until you are making satisfactory welds in the vertical up position.

Repeat each type of joint as needed until consistently good beads are obtained. Turn off the welding machine and shielding gas and clean up your work area when you are finished welding.

Complete a copy of the "Student Welding Report" listed in Appendix I or provided by your instructor. ◆

PRACTICE 12-11

Butt Joint, Lap Joint, and Tee Joint in the Vertical Up Position, All Welds to Be Tested

Using the same equipment, materials, and setup as listed in Practice 12-3, you will make the welded joints in the vertical up position. Each weld must pass the bend test. Repeat each type of weld joint until all pass the bend test. Turn off the welding machine and shielding gas and clean up your work area when you are finished welding.

Complete a copy of the "Student Welding Report" listed in Appendix I or provided by your instructor. ◆

Vertical Down 3G and 3F Positions

The vertical down welding technique can be useful when making some types of welds. The major advantages of this technique are the following:

- Speed—Very high rates of travel are possible.
- Shallow penetration—Thin sections or root openings can be welded with little burn-through.
- Good bead appearance—The weld has a nice width-to-height ratio and is uniform.

Vertical down welds are often used on thin sheet metals or in the root pass in grooved joints. The combination of controlled penetration and higher welding speeds makes vertical down the best choice for such welds. The ease with which welds having a good appearance can be made is deceiving. Generally, more skill is required to make sound welds with this technique than in the vertical up position. The most common problem with these welds is lack of fusion or overlap. To prevent these problems, the arc must be kept at or near the leading edge of the molten weld pool.

PRACTICE 12-12

Stringer Bead at a 45° Vertical Down Angle

Using the same equipment, materials, and setup as listed in Practice 12-3, you will make a vertical downing stringer bead on a plate at a 45° inclined angle.

Holding the welding gun at the top of the plate with a slight dragging angle, **Figure 12-34**, will help to increase penetration, hold back the molten weld pool, and improve visibility of the weld. Be sure that your movements along the 12-in. (305-mm) length of plate are unrestricted.

Lower your hood and start the weld. Watch both the leading edge and sides of the molten weld pool for fusion. The leading edge should flow into the base metal, not curl over it. The sides of the molten weld pool should also show fusion into the base metal and not be flashed (ragged) along the edges.

The weld may be made with or without a weave pattern. If a weave pattern is used, it should be a "C" pattern. The "C" should follow the leading edge of the weld. Some changes on the gun angle may help to increase penetration. Experiment with the gun angle as the weld progresses.

FIGURE 12-34 Vertical down position. Courtesy of Larry Jeffus.

Repeat these welds until you have established a rhythm and technique that work well for you. The welds must be straight and uniform and have complete fusion. Turn off the welding machine and shielding gas and clean up your work area when you are finished welding.

Complete a copy of the "Student Welding Report" listed in Appendix I or provided by your instructor. ◆

PRACTICE 12-13

Stringer Bead in the Vertical Down Position

Repeat Practice 12-12 and increase the angle of the plate until you have developed the skill to repeatedly make good welds in the vertical down position. The weld bead must be straight and uniform and have complete fusion. Turn off the welding machine and shielding gas and clean up your work area when you are finished welding.

Complete a copy of the "Student Welding Report" listed in Appendix I or provided by your instructor. ◆

PRACTICE 12-14

Butt Joint, Lap Joint, and Tee Joint in the Vertical Down Position

Using the same equipment, materials, and setup as listed in Practice 12-3, you will make vertical down welded joints.

Tack weld the pieces of metal together and brace them in position. Using the same technique developed in Practice 12-11, start at the top of the joint and weld down the length of the joint. When the weld is complete, inspect it for discontinuities and make any necessary changes in your technique. Repeat each type of joint as needed until consistently good welds are obtained. Turn off the welding machine and shielding gas and clean up your work area when you are finished welding.

Complete a copy of the "Student Welding Report" listed in Appendix I or provided by your instructor. ◆

PRACTICE 12-15

Butt Joint and Tee Joint in the Vertical Down Position with 100% Penetration

Using the same equipment, materials, and setup as listed in Practice 12-3, you will make welded joints with 100% weld penetration.

It may be necessary to adjust the root opening to meet the penetration requirements. The lap joint was omitted from this practice because little additional skill can be developed with it that is not already acquired with the tee joint. Repeat each type of joint as needed until consistently good welds are obtained. Turn off the welding machine and shielding gas and clean up your work area when you are finished welding.

Complete a copy of the "Student Welding Report" listed in Appendix I or provided by your instructor. ◆

PRACTICE 12-16

Butt Joint and Tee Joint in the Vertical Down Position, Welds to Be Tested

Using the same equipment, materials, and setup as listed in Practice 12-3, you will make the welded joints in the vertical down position. Each weld must pass the bend test. Repeat each type of weld joint until both pass the bend test. Turn off the welding machine and shielding gas and clean up your work area when you are finished welding.

Complete a copy of the "Student Welding Report" listed in Appendix I or provided by your instructor. ◆

Horizontal 2G and 2F Positions

PRACTICE 12-17

Horizontal Stringer Bead at a 45° Angle

Using the same equipment, materials, and setup as listed in Practice 12-3, you will make a horizontal stringer bead on a plate at a 45° reclined angle.

Start at one end with the gun pointed in a slightly upward direction, **Figure 12-35.** You may use a pushing or a dragging leading or a trailing gun angle, depending upon the current setting and penetration desired. Undercutting along the top edge and overlap along the bottom edge are problems with both gun angles. Careful attention must be paid to the manipulation "weave" technique used to overcome these problems.

The most successful weave patterns are the "C" and "J" patterns. The "J" pattern is the most frequently used. The "J" pattern allows weld metal to be deposited along a shelf created by the previous weave, **Figure 12-36.** The length of the "J" can be changed to control the weld bead size. Smaller weld beads are easier to control than large ones.

Repeat these welds until you have established the rhythm and technique that work well for you. The weld must be straight and uniform and have complete fusion. Turn off the welding machine and shielding gas and clean up your work area when you are finished welding.

Complete a copy of the "Student Welding Report" listed in Appendix I or provided by your instructor. ◆

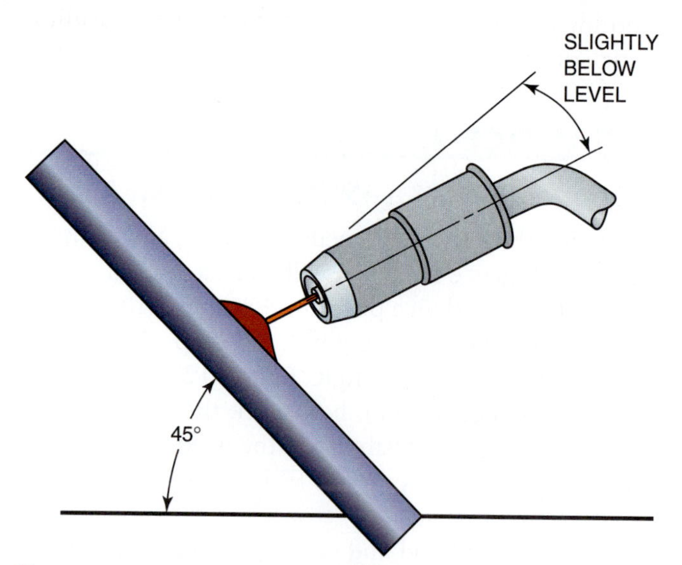

FIGURE 12-35 45° horizontal position.

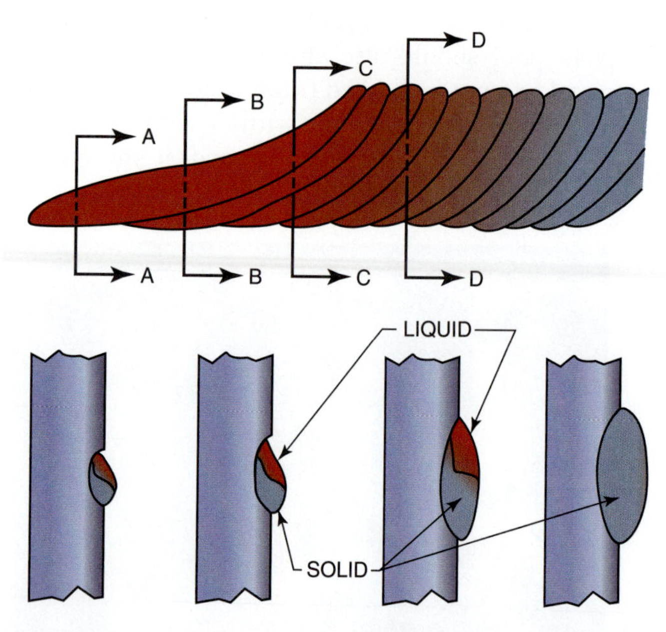

FIGURE 12-36 The actual size of the molten weld pool remains small along the weld.

PRACTICE 12-18

Stringer Bead in the Horizontal Position

Repeat Practice 12-17 and increase the angle of the plate until you have developed the skill to repeatedly make good horizontal welds on a vertical surface. The weld bead must be straight and uniform and have complete fusion. Turn off the welding machine and shielding gas and clean up your work area when you are finished welding.

Complete a copy of the "Student Welding Report" listed in Appendix I or provided by your instructor. ◆

PRACTICE 12-19

Butt Joint, Lap Joint, and Tee Joint in the Horizontal Position

Using the same equipment, materials, and setup listed in Practice 12-3, you will make horizontal welded joints.

Tack weld the pieces of metal together and brace them in position using the same skills developed in Practice 12-17. Starting at one end, make a weld along the entire length of the joint. When making the butt or lap joints, it may help to recline the plates at a 45° angle until you have developed the technique required. Repeat each type of joint as needed until consistently good welds are obtained. Turn off the welding machine and shielding gas and clean up your work area when you are finished welding.

Complete a copy of the "Student Welding Report" listed in Appendix I or provided by your instructor. ◆

PRACTICE 12-20

Butt Joint and Tee Joint in the Horizontal Position with 100% Penetration

Using the same equipment, materials, and setup as listed in Practice 12-3, you will make overhead joints having 100% penetration in the horizontal position.

It may be necessary to adjust the root opening to meet the penetration requirements. Repeat each type of joint as needed until consistently good welds are obtained. Turn off the welding machine and shielding gas and clean up your work area when you are finished welding.

Complete a copy of the "Student Welding Report" listed in Appendix I or provided by your instructor. ◆

PRACTICE 12-21

Butt Joint and Tee Joint in the Horizontal Position, Welds to Be Tested

Using the same equipment, materials, and setup as listed in Practice 12-3, you will make the welded joints in the horizontal position. Each weld must pass the bend test. Repeat each type of weld joint until both pass the bend test. Turn off the welding

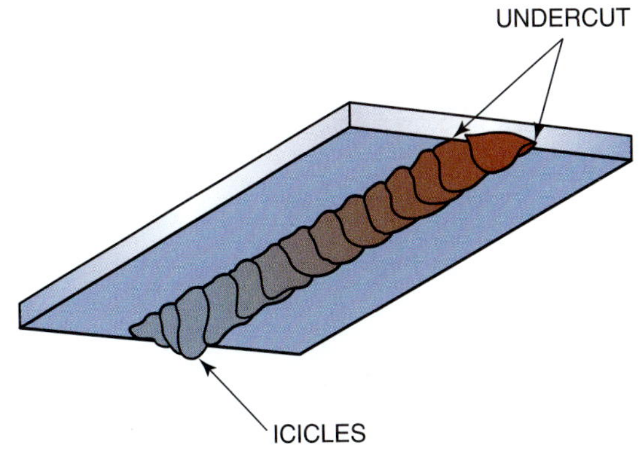

FIGURE 12-37 Overhead weld.

machine and shielding gas and clean up your work area when you are finished welding.

Complete a copy of the "Student Welding Report" listed in Appendix I or provided by your instructor. ◆

Overhead 4G and 4F Positions

There are several advantages to the use of short-circuiting arc metal transfer in the overhead position, including

- Small molten weld pool size—The smaller size of the molten weld pool allows surface tension to hold it in place. Less molten weld pool sag results in improved bead contour with less undercut and fewer icicles, **Figure 12-37.**
- Direct metal transfer—The direct metal transfer method does not rely on other forces to get the filler metal into the molten weld pool. This results in efficient metal transfer and less spatter and loss of filler metal.

PRACTICE 12-22

Stringer Bead Overhead Position

Using the same equipment, materials, and setup as listed in Practice 12-3, you will make a welded stringer bead in the overhead position.

The molten weld pool should be kept as small as possible for easier control. A small molten weld pool can be achieved by using lower current settings, by traveling faster, or by pushing the molten weld pool. The technique used is the welder's choice. Often a combination of techniques can be used with excellent results.

Lower current settings require closer control of gun manipulation to ensure that the wire is fed into the molten weld pool just behind the leading edge. The low power will cause overlap and more spatter if this wire-to-molten weld pool contact position is not closely maintained.

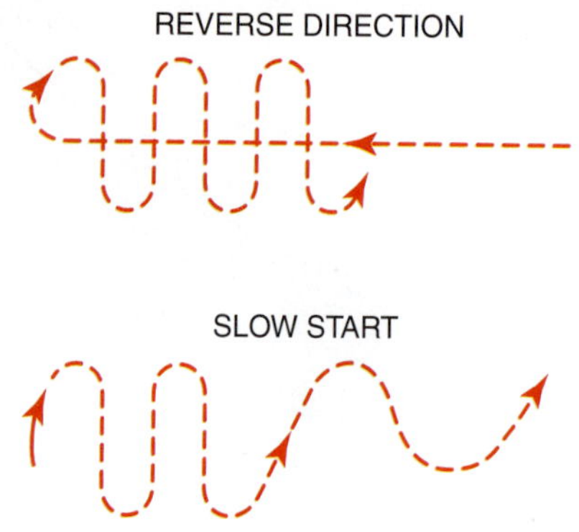

FIGURE 12-38 Two methods of concentrating heat at the beginning of a weld bead to aid in penetration depth.

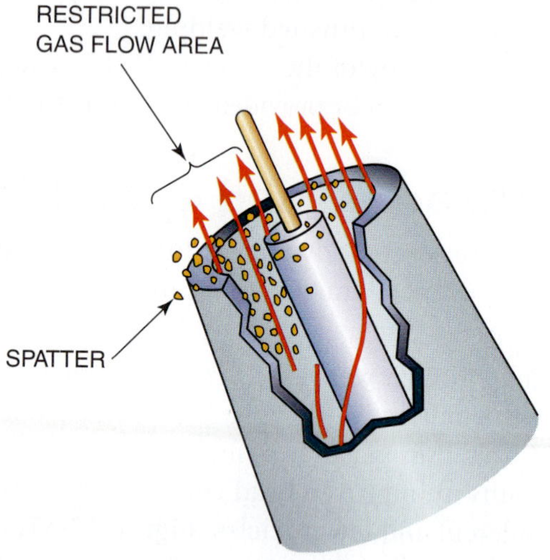

FIGURE 12-39 Shielding gas flow affected by excessive weld spatter in nozzle.

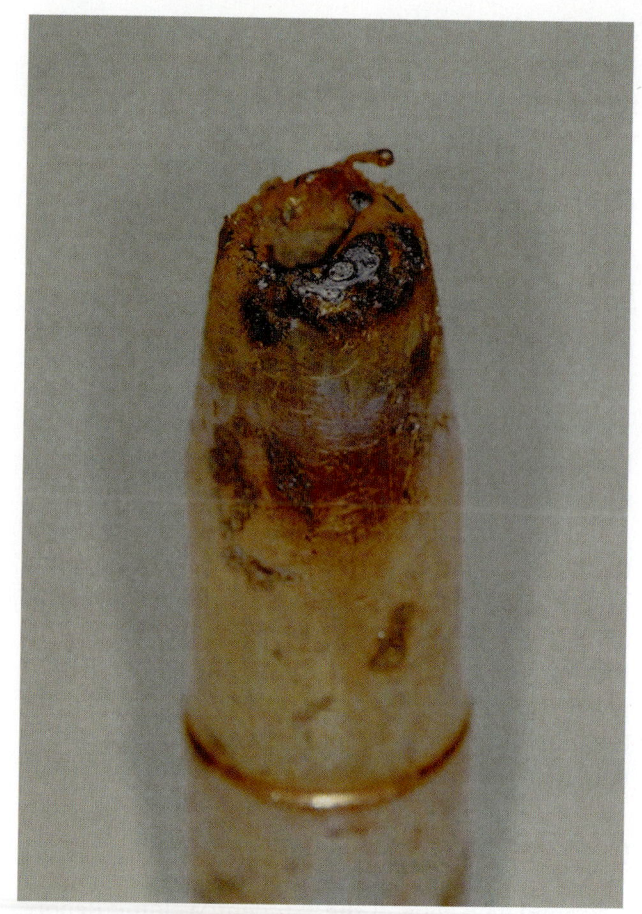

FIGURE 12-40 Gas nozzle damaged after shorting out against the work. Courtesy of Larry Jeffus.

Faster travel speeds allow the welder to maintain a high production rate even if multiple passes are required to complete the weld. Weld penetration into the base metal at the start of the bead can be obtained by using a slow start or quickly reversing the weld direction. Both the slow start and reversal of weld direction put more heat into the start to increase penetration, **Figure 12-38**. The higher speed also reduces the amount of weld distortion by reducing the amount of time that heat is applied to a joint.

The pushing or trailing gun angle forces the bead to be flatter by spreading it out over a wider area as compared to the bead resulting from a dragging or backhand gun angle. The wider, shallow molten weld pool cools faster, resulting in less time for sagging and the formation of icicles.

When welding overhead, extra personal protection is required to reduce the danger of burns. Leather sleeves or leather jackets should be worn.

Much of the spatter created during overhead welding falls into the shielding gas nozzle. The effectiveness of the shielding gas is reduced, **Figure 12-39**, and the contact tube may short out to the gas nozzle, **Figure 12-40**. Turbulence caused by the spatter obstructing the gas may lead to weld contamination. The shorted gas nozzle may arc to the work, causing damage both to the nozzle and to the plate. To control the amount of spatter, a longer stickout and/or a sharper gun-to-plate angle is required to allow most of the spatter to fall clear of the gas nozzle. The nozzle can be dipped, sprayed, or injected automatically, **Figure 12-41**, with antispatter to help prevent the spatter from sticking. Applying antispatter will not stop the spatter from building up, but it does make its removal much easier.

Make several short weld beads using various techniques to establish the method that is most successful and most comfortable for you. After each weld, stop and evaluate it before making a change. When you have decided on the technique to be used, make a welded stringer bead that is 12 in. (305 mm) long.

Repeat the weld until it can be made straight, uniform, and free from any visual defects. Turn off the welding machine and shielding gas and clean up your work area when you are finished welding.

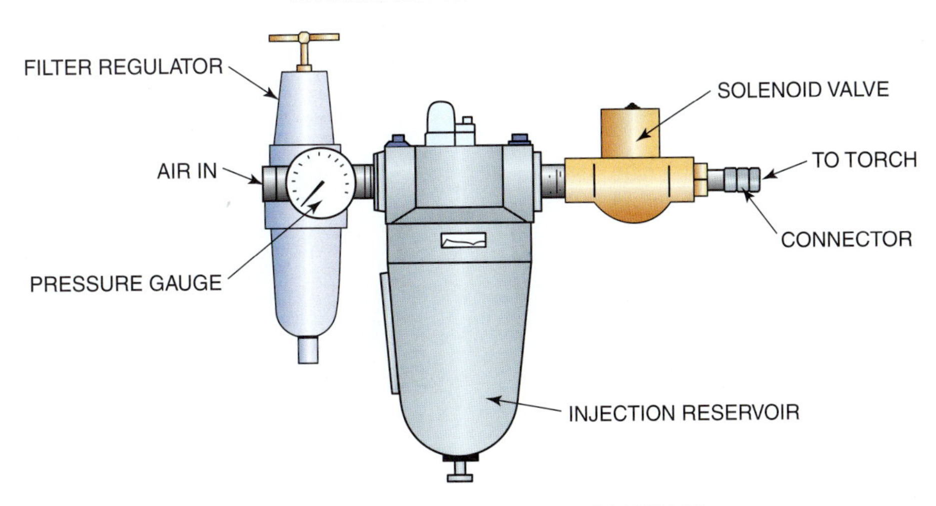

FIGURE 12-41 Automatic anti-spatter system that can be added to a GMA welding gun.

Complete a copy of the "Student Welding Report" listed in Appendix I or provided by your instructor. ◆

PRACTICE 12-23

Butt Joint, Lap Joint, and Tee Joint in the Overhead Position

Using the same equipment, materials, and setup as listed in Practice 12-3, you will make an overhead welded joint.

Tack weld the pieces of metal together and secure them in the overhead position. Be sure you have an unrestricted view and freedom of movement along the joint. Start at one end and make a weld along the joint. Use the same technique developed in Practice 12-22.

Repeat the weld until it can be made straight, uniform, and free from any visual defects. Turn off the welding machine and shielding gas and clean up your work area when you are finished welding.

Complete a copy of the "Student Welding Report" listed in Appendix I or provided by your instructor. ◆

PRACTICE 12-24

Butt Joint and Tee Joint in the Overhead Position with 100% Penetration

Using the same equipment, materials, and setup as listed in Practice 12-3, you will make overhead welded joints having 100% penetration.

Tack weld the metal together. It may be necessary to adjust the root opening to allow 100% weld metal penetration. During these welds, it may be necessary to use a dragging or backhand torch angle. When used with a "C" or "J" weave pattern, this torch angle

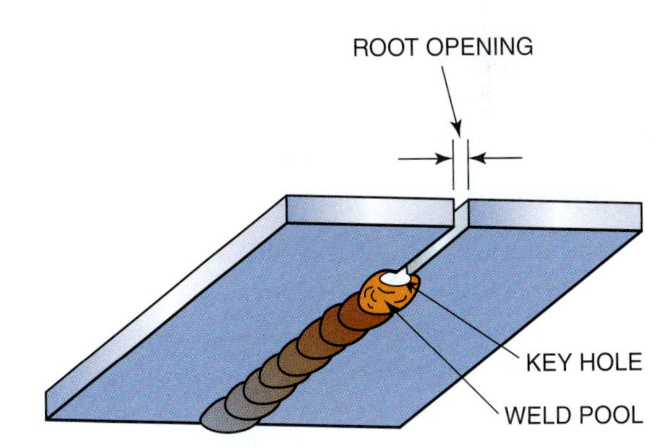

FIGURE 12-42 Overhead welding.

helps to achieve the desired depth of penetration. A key hole just ahead of the molten weld pool is a good sign that the metal is being penetrated, **Figure 12-42.**

Repeat the weld until it can be made straight, uniform, and free from any visual defects. Turn off the welding machine and shielding gas and clean up your work area when you are finished welding.

Complete a copy of the "Student Welding Report" listed in Appendix I or provided by your instructor. ◆

PRACTICE 12-25

Butt Joint and Tee Joint in the Overhead Position, Welds to Be Tested

Using the same equipment, materials, and setup as listed in Practice 12-3, you will make a welded joint in the overhead position. Each weld must pass the bend test. Repeat each type of weld joint until both pass the bend test. Turn off the welding machine and shielding gas and clean up your work area when you are finished welding.

FIGURE 12-43 Weld bead made with GMAW globular metal transfer mode. Courtesy of Larry Jeffus.

FIGURE 12-44 Molten globule of metal being transferred across the arc. Courtesy of Larry Jeffus.

Complete a copy of the "Student Welding Report" listed in Appendix I or provided by your instructor. ◆

Pulsed-arc Metal Transfer, 1G Position

PRACTICE 12-26

Stringer Bead

Using a properly set up and adjusted GMA welding machine (see **Table 12-8**), proper safety protection, .035-in. and/or .045-in. (0.9-mm and/or 1.2-mm) -diameter wire, and two or more pieces of mild steel plate 12 in. (305 mm) long × 1/4 in. (6 mm) thick, make a stringer bead weld in the flat position.

- Start at one end of the plate and use either a push or drag technique to make a weld bead along the entire 12-in. (305-mm) length of the metal.
- After completing the weld, check its appearance and make any changes needed to correct the weld, **Figure 12-43.**
- Repeat the weld and make additional adjustments as required in the frequency, amplitude, or pulse width.
- After the machine is set, start working on improving the straightness and uniformity of the weld.

The location of the arc in the molten weld pool is not as critical in this method of metal transfer as it is in the short-circuiting arc method. If the arc is too far back on the molten weld pool, however, fusion along the leading edge may be reduced, **Figure 12-44.** Moving the arc to the cool metal ahead of the molten weld pool often causes an increase in spatter.

The weave pattern selected should allow the arc to follow the leading edge of the molten weld pool if deep penetration is needed. A pattern that moves the arc back from the leading edge completely or in part results in reduced penetration.

Repeat the weld until it can be made straight, uniform, and free from any visual defects. Turn off the welding machine and shielding gas and clean up your work area when you are finished welding.

Complete a copy of the "Student Welding Report" listed in Appendix I or provided by your instructor. ◆

PRACTICE 12-27

Butt Joint

Using the same equipment, materials, and setup as listed in Practice 12-26, you will make a welded joint in the flat position, **Figure 12-45.**

The heat produced during GMA welding is not enough to force 100% penetration consistently through metal thicker than 3/16 in. (4.8 mm). On occasion, or with extra effort, it is possible to obtain 100% penetration. However, the method generally is not reliable enough for most industrial applications. To ensure that the joint is completely fused, it is prepared for welding with a groove. Any one of several groove designs can be used, **Figure 12-46.** The V-groove is most frequently used because of the ease with which it can be produced.

The V-groove should be made with a 45° inclined angle, **Figure 12-47.** The small diameter filler metal allows the usual 60% inclined angle for SMAW to be reduced. This reduced groove size requires less filler metal and can be welded in less time, resulting in lower cost.

After the metal has been grooved, tack weld it together. The root opening between the plates should be 1/8 in. (3 mm) or less. Starting at one end, make a single pass weld along the entire joint. A weave pattern that follows the contour of the groove should be used to ensure complete fusion, **Figure 12-48.**

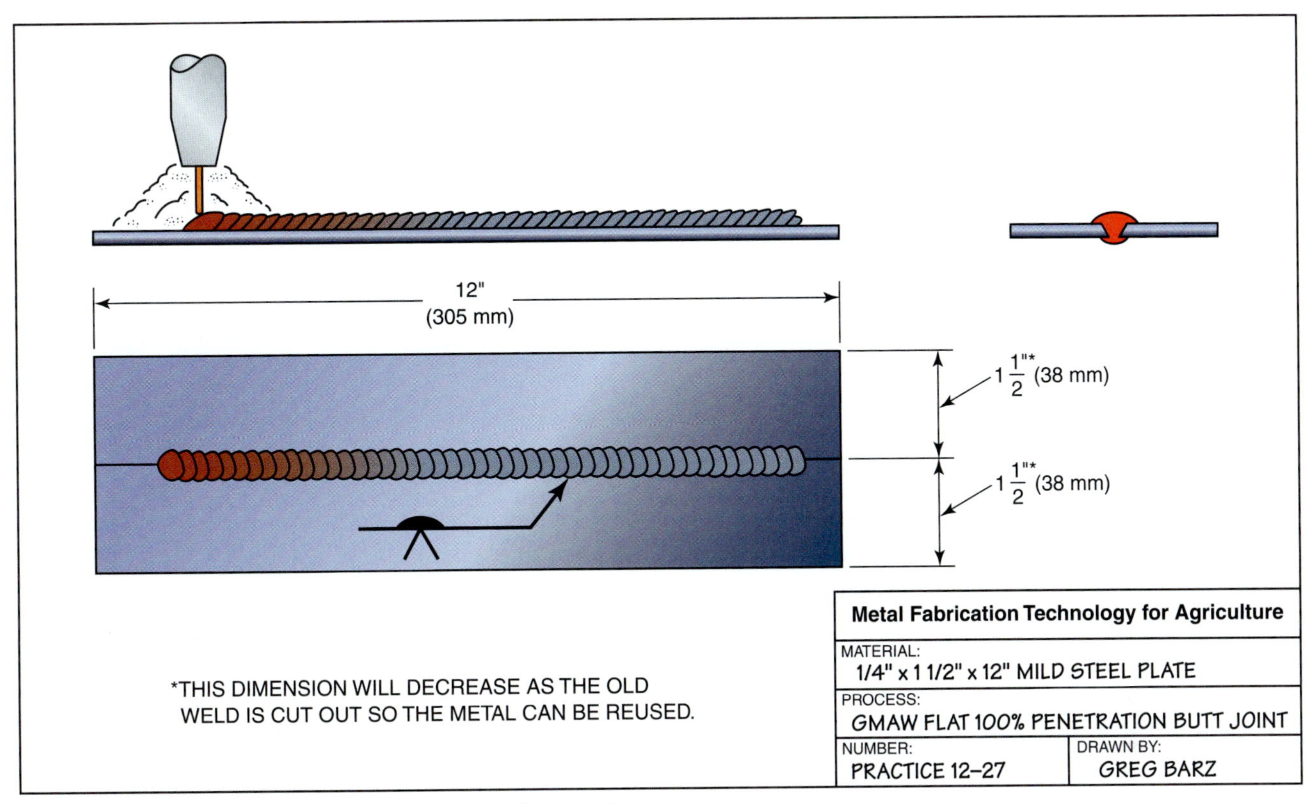

FIGURE 12-45 Single V-groove butt joint in the flat position.

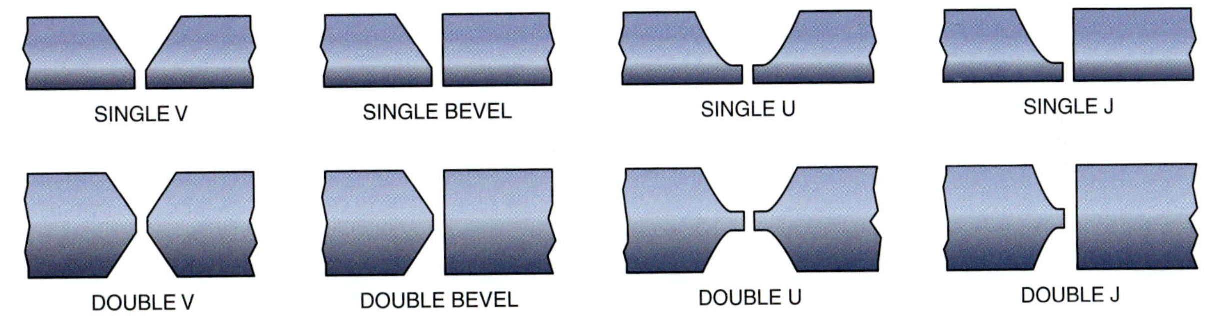

FIGURE 12-46 Typical groove designs.

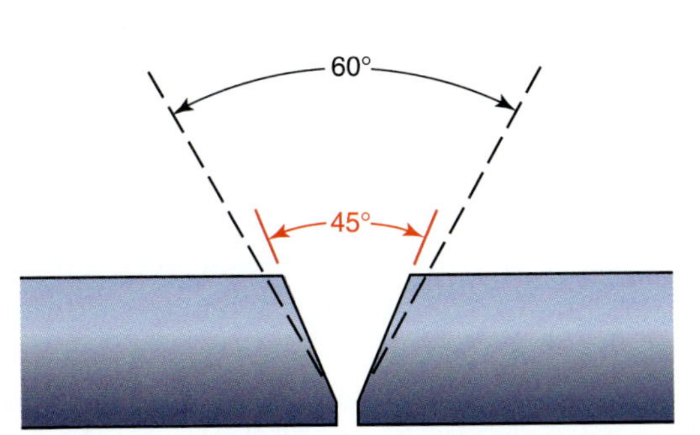

FIGURE 12-47 A smaller groove angle reduces both weld time and filler metal required to make the weld.

FIGURE 12-48 Uniform weld produced with a weave pattern. Courtesy of Larry Jeffus.

The completed weld should be uniform in width and reinforcement and have no visual defects. Repeat the weld until it can be made straight, uniform, and free from any visual defects. Turn off the welding machine and shielding gas and clean up your work area when you are finished welding.

Complete a copy of the "Student Welding Report" listed in Appendix I or provided by your instructor. ◆

PRACTICE 12-28

Butt Joint with 100% Penetration

Using the same equipment, materials, and setup as listed in Practice 12-16, you will make a groove weld having 100% penetration.

Using the technique developed in Practice 12-27 and making the necessary adjustments in the root gap, the weld metal must fuse 100% of the plate thickness. Watch the molten weld pool. If it appears to sink or does not increase in size, you are probably burning through. This action will cause excessive root reinforcement. To correct this problem, speed up the travel rate and/or increase the contact tube-to-work distance.

After the weld is complete, inspect the root surface for the proper appearance. Repeat the weld until it can be made straight, uniform, and free from any visual defects. Turn off the welding machine and shielding gas and clean up your work area when you are finished welding.

Complete a copy of the "Student Welding Report" listed in Appendix I or provided by your instructor. ◆

PRACTICE 12-29

Butt Joint to Be Tested

Using the same equipment, materials, and setup as listed in Practice 12-26, you will make a flat groove weld. The weld must pass the bend test. Repeat the weld until the test can be passed. Turn off the welding machine and shielding gas and clean up your work area when you are finished welding.

Complete a copy of the "Student Welding Report" listed in Appendix I or provided by your instructor. ◆

PRACTICE 12-30

Tee Joint and Lap Joint in the 1F Position

Using the same equipment, materials, and setup as listed in Practice 12-26, you will make a fillet weld in the flat position.

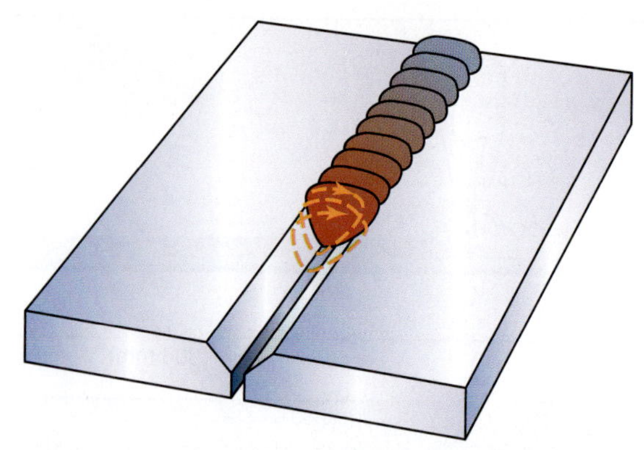

FIGURE 12-49 The electrode should be moved along the groove contour.

It is not necessary to groove the plates used for a tee or lap joint to obtain a sound weld. Fillet weld strength can be obtained by making the weld the proper size for the plate thickness.

The face or surface of a fillet weld should be as flat as possible. Welds with excessive buildup waste metal. Welds with too little buildup will be weak, flat, or concave. Fillet weld beads have fewer stress points to cause weld failure during cyclical loading.

A weave pattern that follows the contour of the joint should be used to ensure adequate fusion, **Figure 12-49**. Repeat the weld until it can be made straight, uniform, and free from any visual defects. Turn off the welding machine and shielding gas and clean up your work area when you are finished welding.

Complete a copy of the "Student Welding Report" listed in Appendix I or provided by your instructor. ◆

PRACTICE 12-31

Tee Joint and Lap Joint in the 2F Position

Using the same equipment, materials, and setup as listed in Practice 12-26, you will make a fillet weld in the horizontal position.

Tack weld the metal together and place the assembly in the horizontal position. The pulsed-arc metal transfer method can be used for a horizontal weld. However, care must be taken to ensure that the legs of the fillet are equal. Because of the size and fluidity of the molten weld pool, undercutting along the top edge and overlap along the bottom edge can also be problems.

To control or eliminate these defects, the beads must be small and quickly made. In addition, a proper weave pattern must be established. The pattern must follow the plate surfaces and establish a shelf to support the weld. After the weld is com-

Process	Wire Diameter	Amperage Range	Voltage Range	Shielding Gas
Axial-spray	0.030	115–200	15–27	75% Ar + 25% CO_2 98% Ar + 2% O
	0.035	165–300	18–32	
	0.045	200–450	20–34	

Table 12-9 Typical Welding Current Settings for Axial Spray Metal Transfer for Mild Steel.

plete, inspect it for defects and measure it for uniformity. Repeat the weld until it can be made straight, uniform, and free from any visual defects. Turn off the welding machine and shielding gas and clean up your work area when you are finished welding.

Complete a copy of the "Student Welding Report" listed in Appendix I or provided by your instructor. ◆

Axial Spray

PRACTICE 12-32

Stringer Bead, 1G Position

Using a properly set up and adjusted GMA welding machine (see **Table 12-9**), proper safety protection, .035-in. and/or .045-in. (0.9-mm and/or 1.2-mm) -diameter wire, and two or more pieces of mild steel plate 12 in. (305 mm) long × 1/4 in. (6 mm) thick, you will make a welded stringer bead in the flat position.

Start at one end of the plate and use either a push or drag technique to make a weld bead along the entire 12-in. (305-mm) length of the metal. After the weld is complete, check its appearance and make any changes needed to correct the weld, **Figure 12-50**. Repeat the weld and make any additional adjustments required. After the machine is set, start working on improving the straightness and uniformity of the weld. Turn off the welding machine and shielding gas and clean up your work area when you are finished welding.

Complete a copy of the "Student Welding Report" listed in Appendix I or provided by your instructor. ◆

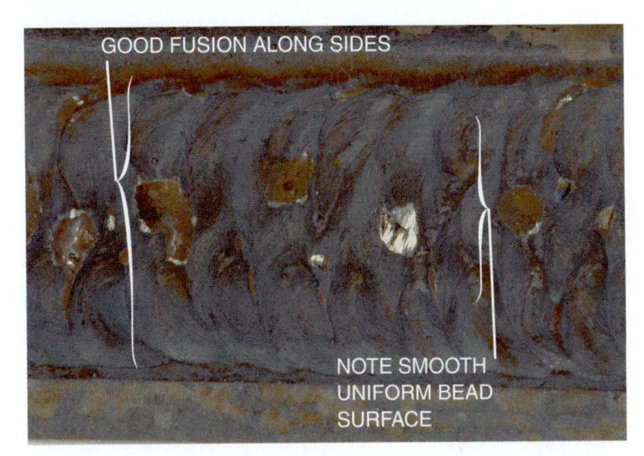

Figure 12-50 Weld bead made with GMAW axial spray metal transfer. Courtesy of Larry Jeffus.

PRACTICE 12-33

Butt Joint, Lap Joint, and Tee Joint Using the Axial Spray Method

Using the same equipment, materials, and setup as listed in Practice 12-32, you will make a flat and horizontal weld using axial spray metal transfer, **Figure 12-51**.

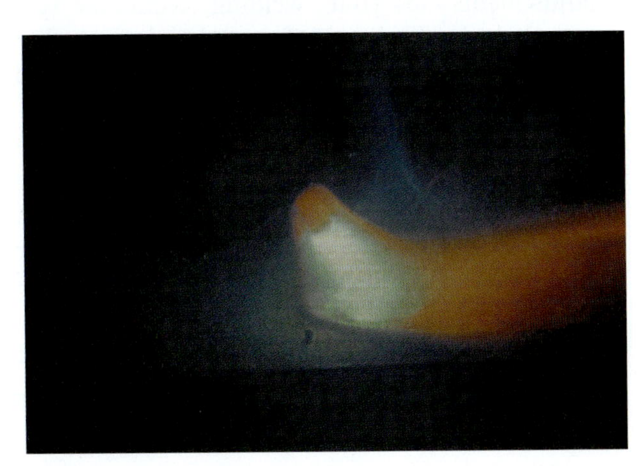

Figure 12-51 GMAW axial spray metal transfer. Courtesy of Larry Jeffus.

Tack weld the metal together and place the assembly in the flat position on the welding table. Start at one end and make a uniform weld along the entire 12-in. (305-mm) length of the joint. Watch the sides of the fillet weld for signs of undercutting.

Repeat the weld until it can be made straight, uniform, and free from any visual defects. Turn off the welding machine and shielding gas and clean up your work area when you are finished welding.

Complete a copy of the "Student Welding Report" listed in Appendix I or provided by your instructor. ◆

PRACTICE 12-34

Butt Joint and Tee Joint

Using the same equipment, materials, and setup as listed in Practice 12-32, you will make a flat weld using axial spray metal transfer. Each weld must pass the guided bend test. Repeat each type of weld joint as needed until the bend test can be passed. Turn off the welding machine and shielding gas and clean up your work area when you are finished welding.

Complete a copy of the "Student Welding Report" listed in Appendix I or provided by your instructor. ◆

Summary

Slight changes in welding gun angle and electrode extension can make significant differences in the quality of the weld produced. As a new welder you might find it difficult to tell the effect of these changes if they are slight. Therefore, as you start to learn this process it is a good idea to make more radical changes so it is easier for you to see their effects on the weld. Later as you develop your skills you can use these slight changes to aid in controlling the weld's quality and appearance as it progresses along the joint. Small adjustments in your welding technique are required to compensate for slight changes that occur along a welding joint, such as joint gap and the increasing temperature of the base metal.

Variations in conditions can significantly affect welding setup for the MIG process. Before starting an actual weld in the field you should practice to test your setup. Practice on scrap metal of a similar thickness and type of metal to be welded. A practice weld before you begin welding can significantly reduce the chances that your weld will not meet standards or specifications. Making these sample or test welds is more important when you are welding in the field, since welds outside of the shop are more difficult to control and anticipate. Think of this much as an athlete warms up before competing.

You will find it beneficial when you are initially setting up your welder to have someone assist you, so that he or she can make changes in the welding machine's settings as you are welding. This teamwork can significantly increase your set up accuracy and reduce set up time. Later on in the field, having developed a keen eye for watching the weld, you can then make these adjustments for yourself more rapidly and accurately. Working with another student in a group effort like this will also give you a better understanding of how other individuals' set up preferences affect their welds. Welding is an art, and therefore each welder may have slight differences in preference for voltage, amperage, gas flow, and other set up variables. This gives you an opportunity to learn more from others.

Review

1. Lack of fusion is when the weld metal does not fuse to the base metal.
   A. true
   B. false
2. What is one of the most common problems with MIG welds?
   A. lack of fusion
   B. excessive penetration
   C. under fill
   D. too many sparks
3. Which of the following welding processes does not use fire feeders like those used in MIG?

Review continued

A. GMAW
B. FCAW
C. GTAW
D. SAW

4. Which of the following pieces of equipment is not needed for MIG welding?
A. oxygen regulator
B. shielding gas supply
C. electrode feed unit
D. welding gun

5. What must be done to the shielding gas cylinder before it is connected to the flowmeter regulator?
A. cleaned
B. chained securely in place
C. painted the correct color
D. turned so the valve will be pointing in a direction that makes the gauge easy to read.

6. What angle should the flowmeter regulator be set at?
A. 45°
B. 22 1/2°
C. 75°
D. 90°

7. When the electrode wire gets knotted up at the feed rollers it is called __________.
A. birds nesting
B. tangling
D. winding
D. de-spooling

8. What may happen if the wire end is cut loose from the spool without holding it?
A. unwind
B. get dirty
C. become bent
D. re-coil

9. The spool drag should be set so the spool does not roll freely when the feed rollers stop.
A. true
B. false

10. Short cut pieces of welding wire should not be discarded on the floor because __________.
A. they can be reused
B. it wastes wire that can be used for gas welding
C. they make the shop look dirty
D. they might get caught inside a welding machine, shorting out the electrical power

11. Use Table 12-1 to determine what the amperage would be if .035 in. (0.9 mm) wire is being fed at a rate of 400 in./min (10.2 m/min).
A. 210 amps
B. 400 amps
C. 130 amps
D. 430 amps

12. As the electrode extension increases, the weld's penetration __________ and buildup __________.
A. increases/increases
B. decreases/decreases
C. increases/decreases
D. decreases/increases

13. When welding with a backhand or dragging angle, the weld's penetration __________ and the buildup __________.
A. increases/increases
B. decreases/decreases
C. increases/decreases
D. decreases/increases

14. When the welding gun is pointed in the direction of the weld, it is referred to as being at a __________.
A. dragging angle
B. backhand angle
C. forehand angle
D. right angle

15. According to Table 12-6 helium is an inert gas.
A. true
B. false

16. How far back from the joint should the surface of the metal next to a groove joint be?
A. 1 in. (25 mm)
B. 2 in. (50 mm)
C. 1/2 in. (13 mm)
D. 1 1/2 in. (38 mm)

17. Which metal transfer method would be best for welding on 16 gauge sheet metal?
A. short circuiting metal transfer
B. globular metal transfer
C. axial spray transfer
D. pulsed globular transfer

Review continued

18. What is an advantage of vertical down welding?
    A. deep penetration
    B. fast groove filling
    C. thick sections
    D. very high rates of travel
19. What is the best method of metal transfer for overhead welding?
    A. direct metal transfer
    B. axial spray transfer
    C. globular transfer
    D. arc transfer
20. During overhead welding, keeping the nozzle directly below the arc will help prevent spatter from blocking the gas flow.
    A. true
    B. false

Chapter 13

Flux Cored Arc Welding Equipment, Setup, and Operation

OBJECTIVES

After completing this chapter, the student should be able to

- ☑ explain the effect of flux on the weld.
- ☑ explain the difference between self shielding and gas shielding.
- ☑ describe the different ways electrodes are made.
- ☑ name the parts of a flux cored arc welding setup.
- ☑ list the advantages of flux cored arc spot welding.
- ☑ list the major limitations of the flux cored arc welding process.

KEY TERMS

air-cooled
backhand
carbon
coils
deoxidizers
dual shield
FCA welding
flux cored arc welding (FCAW)
forehand
lime-based flux
rutile-based flux
self-shielding
slag
spools
water-cooled

INTRODUCTION

Flux cored arc welding (FCAW) process has become extremely popular for agricultural welding. Its popularity is due in part to its portability, versatility, ease of use, and lower equipment cost. It is very portable; some small welders weigh less than 30 pounds (66 k). They can be plugged into a standard wall outlet or portable engine generator. They are versatile and can be used to make welds in any position in metals ranging from 24-gauge sheet metal up to an inch or more in thickness with the same diameter electrode wire. This means that you might only need to have a single wire size to do any job on the farm or ranch. It is easy to use, because once it is set up correctly, the welder in most cases only has to follow the

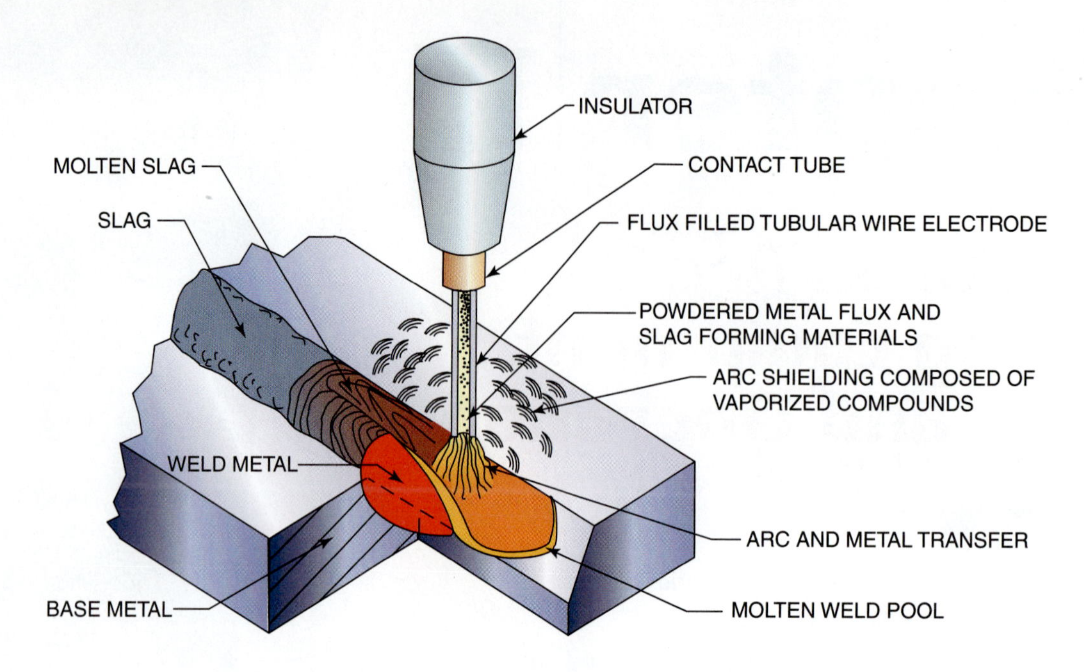

(A) SELF SHIELDED FLUX CORED ARC WELDING (FCAW-S)

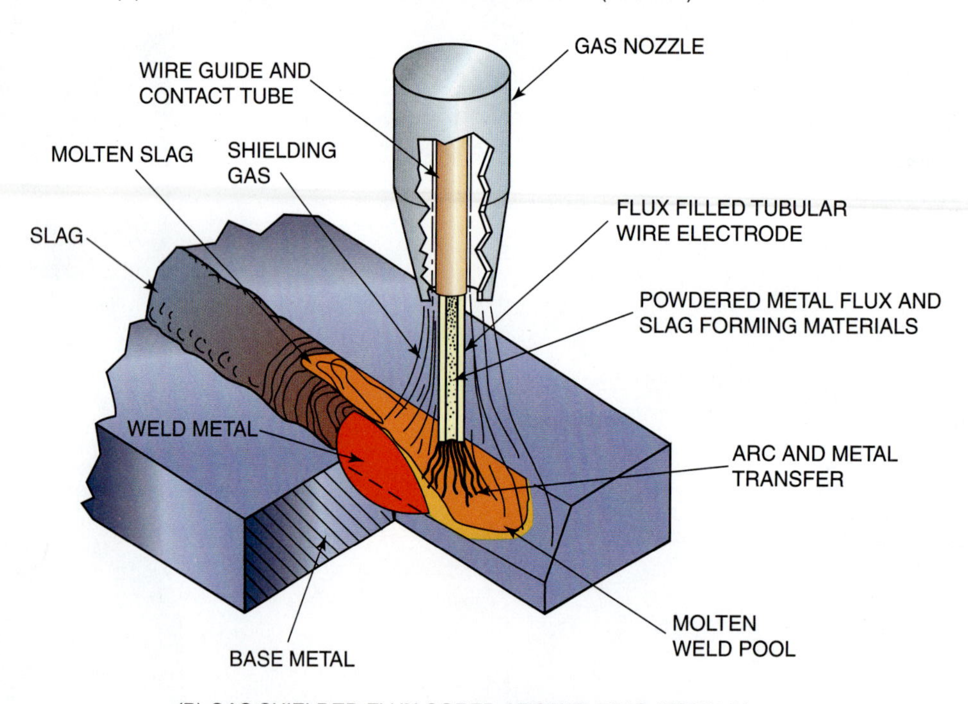

(B) GAS SHIELDED FLUX CORED ARC WELDING (FCAW-G)

FIGURE 13-1 (A) Self shielded flux cored arc welding; (B) gas shielded flux cored arc welding. Courtesy of the American Welding Society.

joint being welded. Proper setup of the equipment is the key to making good FCA welds. The popularity of the equipment has resulted in lower prices and greater accessibility to the equipment and supplies. Retail stores are now selling FCA welding equipment, with some of the larger stores and supply houses even carrying more than one brand.

In FCAW an arc between a continuously fed tubular wire electrode and the work produces welds. Heat from the arc melts the end of the wire electrode and the surface of the base metal. Molten droplets form as the wire electrode is melted across the arc and mixed with the molten base metal forming a weld. The molten weld metal is protected from contamination by the gases formed as the flux core of the wire electrode vaporizes. An additional shielding gas is added for some types of welding. When FCA welding is being done without a gas shield, it is called self shielding, **Figure 13-1A.** When a shielding gas is added, it is called dual shielded, **Figure 13-1B.**

Almost all FCA welding in agriculture is done with self-shielding wire electrodes.

The equipment and setup for FCA welding is very similar to that of GMA welding. The major difference in the equipment is that GMA welding always requires a shielding gas, but FCA welding can be done with or without a shielding gas. FCA welding equipment is easier to carry around the ranch or farm than GMA welding because no heavy shielding gas cylinder is required. Unlike most other pieces of welding equipment, an FCA welder can easily be carried to the barnyard or field, or lifted to the top of large equipment to make repairs or onto sheds and barns for construction, **Figure 13-2.**

FIGURE 13-2 FCA Welder setting on a wood frame so it can be lifted to the top of a combine to make a welded repair. Courtesy of Larry Jeffus.

Principles of Operation

FCA welding is similar in a number of ways to the operation of GMA welding, **Figure 13-3.** Both processes use a constant-potential (CP) or constant-voltage (CV) power supply. CP power supplies provide a controlled voltage (potential) to the welding electrode. The amperage (current) varies with the speed that the electrode is being fed into the molten weld pool. Just like GMA welding, higher electrode feed speeds produce higher currents and slower feed speeds result in lower currents, assuming all other conditions remain constant.

The effects on the weld of electrode extension, gun angle, welding direction, travel speed, and other welder manipulations are similar to those experienced in GMA welding. Like GMA welding, having a correctly set welder does not ensure a good weld. The skill of the welder is an important factor in producing high-quality welds.

The flux inside the electrode provides the molten weld pool with protection from the atmosphere,

(A)

(B)

FIGURE 13-3 FCA welding (A) without smoke extraction and (B) with smoke extraction. Courtesy of Lincoln Electric Company.

improves strengths through chemical reactions and alloys, and improves the weld shape.

Atmospheric contamination of molten weld metal occurs as it travels across the arc gap and within the pool before it solidifies. The major atmospheric contaminations come from oxygen and nitrogen, the major elements in air. The addition of fluxing and gas forming elements to the core electrode reduce or eliminate their effects.

Improved strength and other physical or corrosive properties of the finished weld are improved by the flux. Small additions of alloying elements, deoxidizers, and slag agents all can improve the desired weld properties. Carbon, chromium, and vanadium can be added to improve hardness, strength, creep resistance, and corrosion resistance. Aluminum, silicon, and titanium all help remove oxides and/or nitrides in the weld. Potassium, sodium, and zirconium are added to form slag.

A slag covering of the weld is useful for several reasons. Slag helps the weld by protecting the hot metal from the effects of the atmosphere, controlling the bead shape by serving as a dam or mold, and serving as a blanket to slow the weld's cooling rate, which improves its physical properties, **Figure 13-4**.

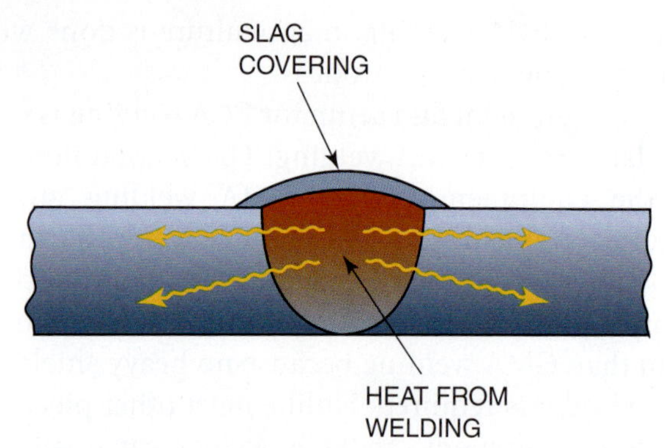

FIGURE 13-4 The slag covering keeps the welding heat from escaping quickly, thus slowing the cooling rate.

Equipment

Power Supply The FCA welding power supply is the same type that is required for GMAW, called constant-potential, constant-voltage (CP, CV). The words *potential* and *voltage* have the same electrical meaning and are used interchangeably.

Guns FCA welding guns are available water-cooled or air-cooled. Although most of the FCA welding guns that you will find in agriculture are air-cooled, some industry needs water-cooled guns because of the higher heat caused by longer welds made at higher currents. The water-cooled FCA welding gun is more efficient than an air-cooled gun at removing waste heat. The air-cooled gun is more portable because it has fewer hoses, and it may be easier to manipulate than the water-cooled gun.

Also, the water-cooled gun requires a water reservoir or another system to give the needed cooling. There are two major ways that water can be supplied to the gun for cooling. Cooling water can be supplied directly from the building's water system, or it can be supplied from a recirculation system.

Smoke Extraction Nozzles Because of the large quantity of smoke that can be generated during FCA welding, systems for smoke extraction that fit on the gun have been designed, **Figure 13-3B.** These systems use a vacuum blower to pull the smoke back into a specially designed smoke extraction nozzle on the welding gun. The disadvantage of having a slightly heavier gun is offset by the system's advantages. The advantages of the system are as follows:

- Cleaner air for the welder to breathe because the smoke is removed before it rises to the welder's face.
- Reduced heating and cooling cost because the smoke is concentrated, so less shop air must be removed with the smoke.

Electrode Feed Electrode feed systems are similar to those used for GMAW. Electrode feed systems are sometimes called wire feeders. They use rollers to push the wire though the welding cable assembly to the gun. Small feed systems may have only one drive roller, but larger systems have two or more drive rollers, **Figure 13- 5.**

Advantages

FCA welding offers the welding industry a number of important advantages.

High Deposition Rate High rates of weld buildup are possible. FCA welding deposition rates of more than 25 lb/hr (12 kg/hr) of weld metal are possible. This compares to about 10 lb/hr (6 kg/hr) for SMA welding using a very large diameter electrode of 1/4 in. (6 mm).

Portability Small, portable FCA welding machines can be carried almost anywhere and plugged into a standard wall outlet. Their low input power requirement means they can easily be used with extension cords. Some even have a two-level input power switch so they can be used on either a 15 amp or a 20 amp, 120 volt circuit found in barns, shops, or garages. Even with their low input

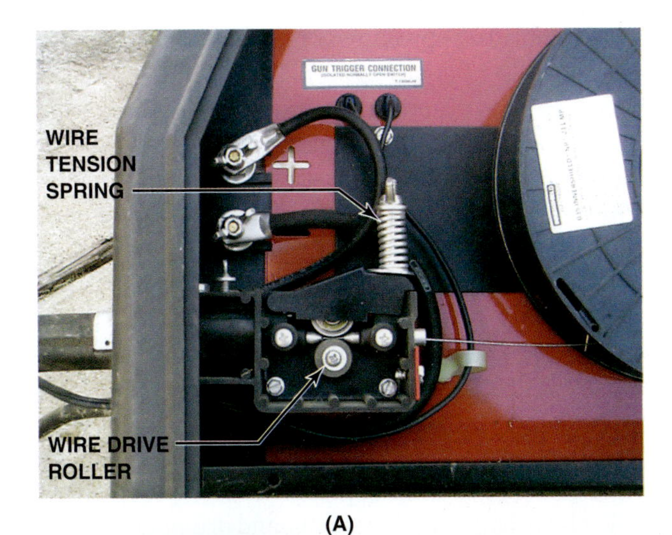

(A)

(B)

FIGURE 13-5 The drive roller (A) must have the wire tension roller (B) properly set with the tension spring to push the wire to the welding gun. Courtesy of Larry Jeffus.

power requirement, they can make single pass welds on metal as thick as 3/8 in. (9 mm) and multiple pass welds on metal up to 1 in. (25 mm) and thicker.

Minimum Electrode Waste The FCA method makes efficient use of filler metal; from 75% to 90% of the weight of the FCA electrode is metal, the remainder being flux. SMA electrodes have a maximum of 75% filler metal; some SMA electrodes have much less. Also, a stub must be left at the end of each SMA welding electrode. The stub will average 2 in. (51 mm) in length, resulting in a loss of 11% or more of the SMAW filler electrode purchased. FCA welding has no stub loss, so nearly 100% of the FCAW electrode purchased is used.

Narrow Groove Angle Because of the deep penetration characteristic, no edge beveling preparation is required on some joints. When bevels are cut, the

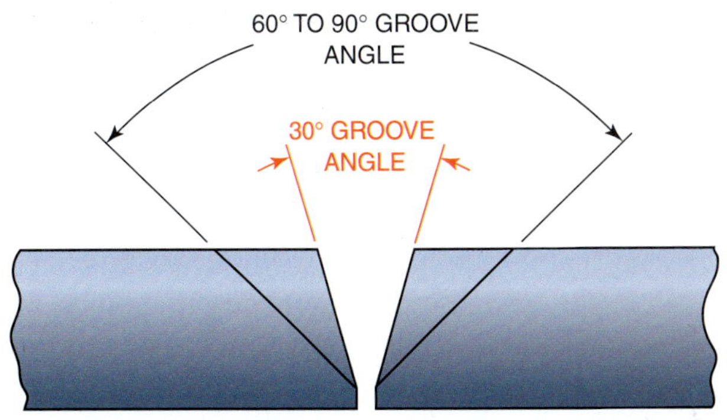

FIGURE 13-6 The narrower groove angle for FCAW saves on filler metal, welding time, and heat input into the part.

joint included angle can be reduced to as small as 35°, **Figure 13-6.** The reduced groove angle results in a smaller-sized weld. This can save 50% of filler metal with about the same savings in time and weld power used.

Minimum Precleaning The addition of deoxidizers and other fluxing agents permits high-quality welds to be made on plates with light surface rust and mill scale. This eliminates most of the precleaning required. Often it is possible to make excellent welds on plates in the "as cut" condition; no grinding is needed.

All-position Welding Small diameter electrode sizes in combination with special fluxes allow excellent welds in all positions. The slags produced assist in supporting the weld metal. This process is easy to use, and, when properly adjusted, it is much easier to use than other all-position arc welding processes.

Flexibility Changes in power settings can permit welding to be made on thin-gauge sheet metals or thicker plates using the same electrode size. Multipass welds allow joining unlimited thickness metals. This, too, is attainable with one size of electrode.

High Quality The addition of the flux gives the process the high level of reliability needed for welding on trailers, axles, tow bars, and structural steel.

Excellent Control The molten weld pool is more easily controlled with FCAW than with GMAW. The surface appearance is smooth and uniform even with less operator skill. Visibility is improved by removing the nozzle when using self-shielded electrodes.

Limitations

The main limitation of flux cored arc welding is that it is confined to steel. Generally, all low and medium carbon steels and some low alloy steels, cast irons, and a limited number of stainless steels are presently weldable using FCAW.

The equipment and electrodes used for the FCAW process are more expensive. However, the cost is quickly recoverable through higher productivity.

The flux must be removed before the weldment is finished (painted) to prevent crevice corrosion.

With the increased welding output comes an increase in smoke and fume generation. The existing ventilation system in a shop might need to be increased to handle the added volume.

Electrodes

Methods of Manufacturing The electrodes have flux tightly packed inside. FCA filler metal is packaged in a number of forms for purchase by the end user, **Figure 13-7.** The AWS has a standard for the size of each of the package units. Although the dimensions of the packages are standard, the weight of filler wire is not standard. More of the smaller diameter wire can fit into the same space than a larger diameter wire, so a package of .030 in. (0.8-mm) wire weighs more than the same-sized package of 3/32 in. (2.3-mm) wire. The standard packing units for FCA wires are spools, coils, reels, and drums, **Table 13-1.**

Spools are made of plastic or fiber board and are disposable. They are completely self-contained and are available in approximate weights from 1 lb up to around 50 lb (.5 kg to 25 kg). The smaller spools, 4-in. and 8-in. (102-mm and 203-mm), weighing from 1 lb to 7 lbs, are most often used for agriculture or for home/hobby use; 12-in. and 14-in. (305-mm and 356-mm) spools are often used in schools and agricultural shops.

Coils come wrapped and/or wire tied together. They are unmounted, so they must be supported on a frame on the wire feeder in order to be used. Coils are available in weights around 60 lb (27 kg). Because FCA wires on coils do not have the expense of a disposable core, these wires cost a little less per pound, so they are more desirable for higher-production shops.

Reels are large wooden spools, and drums are shaped like barrels. Both reels and drums are used for high-production jobs. Both can contain approximately 300 lb to 1000 lb (136 kg to 454 kg) of FCA wire. Because of their size, they are used primarily at fixed welding stations. Such stations are often associated with some form of automation, such as turntables or robotics.

Electrode Cast and Helix Feed out 10 ft of wire electrode and cut it off. Lay it on the floor and observe that it forms a circle. The diameter of the circle is known as the cast of the wire, **Figure 13-8.**

Note that the wire electrode does not lay flat. One end is slightly higher than the other. This height is the helix of the wire.

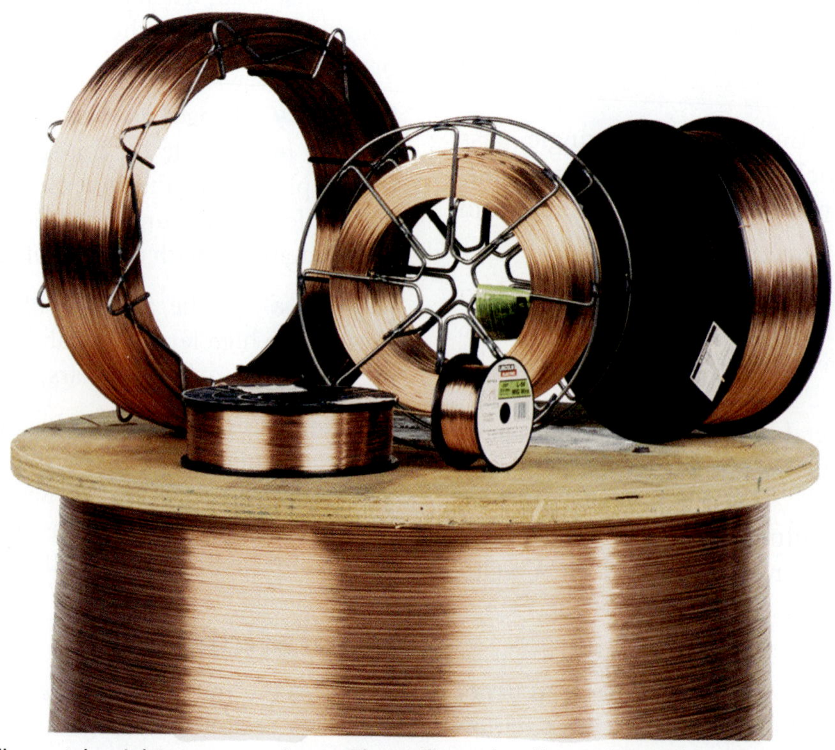

FIGURE 13-7 FCAW filler metal weights are approximate. They will vary by alloy and manufacturer. Courtesy of Lincoln Electric Company.

Packaging	Outside Diameter	Width	Arbor (Hole) Diameter
Spools	4 in. (102 mm)	1 3/4 in. (44.5 mm)	5/8 in. (16 mm)
	8 in. (203 mm)	2 1/4 in. (57 mm)	2 1/16 in. (52.3 mm)
	12 in. (305 mm)	4 in. (102 mm)	2 1/16 in. (52.3 mm)
	14 in. (356 mm)	4 in. (102 mm)	2 1/16 in. (52.3 mm)
Reels	22 in. (559 mm)	12 1/2 in. (318 mm)	1 5/16 in. (33.3 mm)
	30 in. (762 mm)	16 in. (406 mm)	1 5/16 in. (33.3 mm)
Coils	16 1/4 in. (413 mm)	4 in. (102 mm)	12 in. (305 mm)
	Outside Diameter	**Inside Diameter**	**Height**
Drums	23 in. (584 mm)	16 in. (406 mm)	34 in. (864 mm)

TABLE 13-1 Packaging Size Specification for Commonly Used FCA Filler Wire.

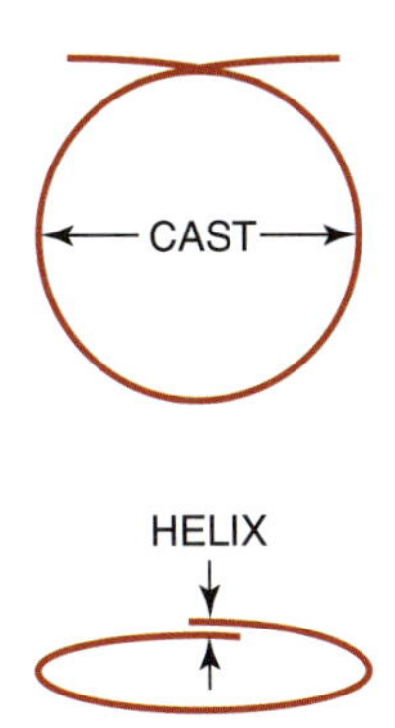

FIGURE 13-8 Method of measuring cast and helix of FCAW filler wire.

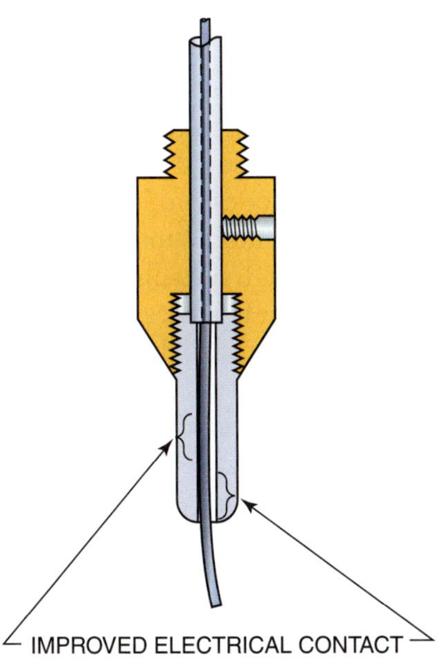

FIGURE 13-9 Cast forces the wire to make better electrical contact with the tube.

There are specifications for both cast and helix for all FCA welding wires.

The cast and helix causes the wire to rub on the inside of the contact tube, **Figure 13-9.** The slight bend in the electrode wire ensures a positive electrical contact between the contact tube and filler wire.

Deoxidizing Element	Strength
Aluminum (Al)	Very strong
Manganese (Mn)	Weak
Silicon (Si)	Weak
Titanium (Ti)	Very strong
Zirconium (Zr)	Very strong

TABLE 13-2 Deoxidizing Elements Added to Filler Wire (to Minimize Porosity in the Molten Weld Pool).

Flux

The fluxes used are mainly rutile or lime based. The purpose of the fluxes is the same as in the SMAW process. That is, they can provide all or part of the following to the weld:

- *Deoxidizers:* Oxygen that is present in the welding zone has two forms. It can exist as free oxygen from the atmosphere surrounding the weld. Oxygen also can exist as part of a compound such as an iron oxide or carbon dioxide (CO_2). In either case it can cause porosity in the weld if it is not removed or controlled. Chemicals are added that react to the presence of oxygen in either form and combine to form a harmless compound, **Table 13-2.** The new compound can become part of the slag that solidifies on top of the weld, or some of it may stay in the weld as very small inclusions. Both methods result in a weld with better mechanical properties with less porosity.
- *Slag formers:* Slag serves several vital functions for the weld. It can react with the molten weld metal chemically, and it can affect the weld bead physically. In the molten state it moves through the molten weld pool and acts as a magnet or sponge to chemically combine with impurities in the metal and remove them, **Figure 13-10.** Slags can be refractory, become solid at a high temperature, and solidify over the weld molten, helping it hold its shape and slowing its cooling rate.

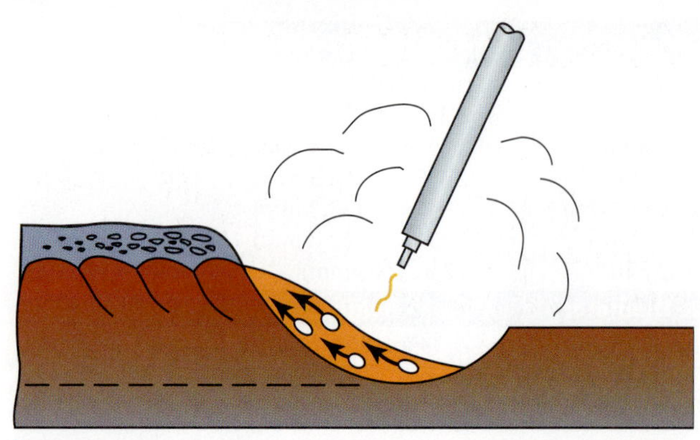

FIGURE 13-10 Impurities being floated to the surface by slag.

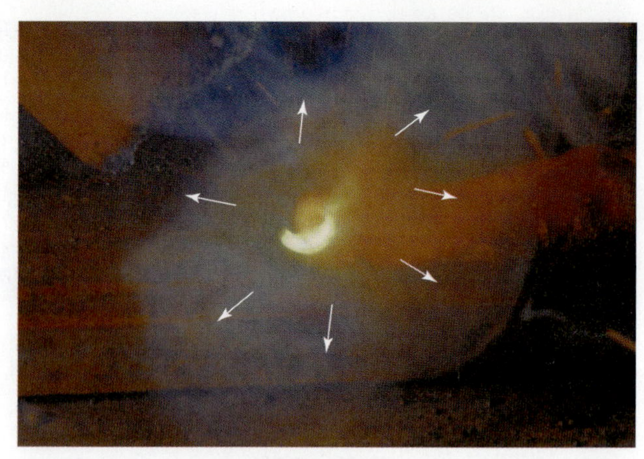

FIGURE 13-11 Rapidly expanding gas cloud. Courtesy of Larry Jeffus.

- *Fluxing agents:* Molten weld metal tends to have a high surface tension, which prevents it from flowing outward toward the edges of the weld. This causes undercutting along the junction of the weld and the base metal. Fluxing agents make the weld more fluid and allow it to flow outward, filling the undercut.
- *Arc stabilizers:* Chemicals in the flux affect the arc resistance. As the resistance is lowered, the arc voltage drops and penetration is reduced. When the arc resistance is increased, the arc voltage increases and weld penetration is increased. Although the resistance within the ionized arc stream may change, the arc is more stable and easier to control. It also improves the metal transfer by reducing spatter caused by an erratic arc.
- *Alloying elements:* Because of the difference in the mechanical properties of metal that is formed by rolling or forging and metal that is melted to form a cast weld nugget, the metallurgical requirements of the two also differ. Some elements change the weld's strength, ductility, hardness, brittleness, toughness, and corrosion resistance. Other alloying elements in the form of powder metal can be added to both alloys and add to the deposition rate.
- *Shielding gas:* As elements in the flux are heated by the arc some of them vaporize and form voluminous gaseous clouds hundreds of times larger than their original volume. This rapidly expanding cloud forces the air around the weld zone away from the molten weld metal, **Figure 13-11.** Without the protection this process affords the molten metal, it would rapidly oxidize. Such oxidization would severely affect the weld's mechanical properties, rendering it unfit for service.

All FCAW fluxes are divided into two groups based on the acid or basic chemical reactivity of the slag. The AWS classifies T-1 as acid and T-5 as basic.

The **rutile-based flux** is acidic, T-1. It produces a smooth, stable arc and a refractive high-temperature slag for out-of-position welding. These electrodes produce a fine drop transfer, a relatively low fume, and an easily removed slag. The main limitation of the rutile fluxes is that their fluxing elements do not produce as high a quality deposit as do the T-5 systems.

The **lime-based flux** is basic, T-5. It is very good at removing certain impurities from the weld metal, but its low melting temperature slag is fluid, which makes it generally unsuitable for out-of-position welding. These electrodes produce a more globular transfer, more spatter, more fume, and a more adherent slag than do the T-1 systems. These characteristics are tolerated when it is necessary to deposit very tough weld metal and for welding materials having a low tolerance for hydrogen.

Some rutile-based electrodes allow the addition of a shielding gas. With the weld being protected partially by the shielding gas, more elements can be added to the flux, which produces welds with the best of both flux systems, high-quality welds in all positions.

Some fluxes can be used on both single and multiple pass welds, and others are limited to single pass welds only. Using a single pass welding electrode for multipass welds may result in an excessive amount of manganese. The manganese is necessary to retain strength when making large, single pass welds. However, with the lower dilution associated with multipass techniques, it can strengthen the weld metal too much and reduce its ductility. In some cases, small welds that deeply penetrate the base metal can help control this problem.

Classifications		Comments		Shielding Gas
T-1		Requires clean surfaces and produces little spatter. It can be used for single and multiple pass welds in the flat (1G and 1F) and horizontal (2F) positions.		Carbon dioxide (CO_2)
T-2		Requires clean surfaces and produces little spatter. It can be used for single pass welds in the flat (1G and 1F) and horizontal (2F) positions only.		Carbon dioxide (CO_2)
T-3		Used on thin-gauge steel for single pass welds in the flat (1G and 1F) and horizontal (2F) positions only.		None
T-4		Low penetration and moderate tendency to crack for single and multiple pass welds in the flat (1G and 1F) and horizontal (2F) positions.		None
T-5		Low penetration and a thin, easily removed slag, used for single and multiple pass welds in the flat (1G and 1F) position only.		With or without carbon dioxide (CO_2)
T-6		Similar to T-5 without externally applied shielding gas.		None
T-G		The composition and classification of this electrode is not given in the preceding classes. It may be used for single or multiple pass welds.		With or without shielding

TABLE 13-3 Welding Characteristics of Seven Flux Classifications.

Table 13-3 lists the shielding and polarity for the flux classifications of mild steel FCAW electrodes. The letter *G* is used to indicate an unspecified classification. The *G* means that the electrode has not been classified by the American Welding Society. Often the exact composition of fluxes are kept as a manufacturer's trade secret. Therefore, only limited information about the electrode's composition will be given. The only information often supplied is current, type of shielding required, and some strength characteristics.

As a result of the relatively rapid cooling of the weld metal, the weld may tend to become hard and brittle. This factor can be controlled by both adding elements to the flux and the slag formed by the flux, **Table 13-4**. Ferrite is the softer, more ductal form of iron. The addition of ferrite-forming elements can control the hardness and brittleness of a weld. Refractory fluxes are sometimes called "fast-freeze" because they solidify at a higher temperature than the weld metal. By becoming solid first, this slag can cradle the molten weld pool and control its shape. This property is very important for out-of-position welds.

The impurities in the weld pool can be metallic or nonmetallic compounds. Metallic elements that are added to the metal during the manufacturing process in small quantities may be concentrated in the weld. These elements improve the grain structure, strength, hardness, resistance to corrosion, or other mechanical properties in the metal's as-rolled or formed state. But weld nugget is a small casting, and some alloys adversely affect the properties of this casting (weld metal). Nonmetallic compounds are primarily slag inclusions left in the metal from the fluxes used during manufacturing. The welding fluxes form slags that are less dense than the weld metal so that they will float to the surface before the weld solidifies.

Element	Reaction in Weld
Silicon (Si)	Ferrite former and deoxidizer
Chromium (Cr)	Ferrite and carbide former
Molybdenum (Mo)	Ferrite and carbide former
Columbium (Cb)	Strong ferrite former
Aluminum (Al)	Ferrite former and deoxidizer

TABLE 13-4 Ferrite-Forming Elements Used in FCA Welding Fluxes.

Electrode Classification The American Welding Society classifies FCA welding electrodes as tubular wire. **Table 13-5** lists the AWS specifications for filler metals for flux cored arc welding.

Mild Steel

In the AWS classification for mild steel, FCA electrode starts with the letter *E*, which stands for *electrode*, **Figure 13-12**. The *E* is followed by a single one- or two-digit number to indicate the minimum tensile strength, in pounds per square inch

Metal	AWS Filler Metal Classification
Mild steel	A5.20
Stainless steel	A5.22
Chromium–molybdenum	A5.29

TABLE 13-5 Filler Metal Classification Numbers.

(psi), of a good weld. The actual strength is obtained by adding four zeros to the right of the number given. For example, E6xT-x is 60,000 psi, and E11xT-x is 110,000 psi.

The next number, 0 or 1, indicates the welding position. Ex0T is to be used in a horizontal or flat position only. Ex1T is an all-position filler metal.

The *T* located to the right of the tensile strength and weld position numbers indicates that this is a tubular, flux cored wire. The last number—1, 2, 3, 4, 5, 6, 7, 8, 10, 11—or the letter *G* or *GS* is used to indicate the type of flux and whether the filler metal can be used for single or multiple pass welds. The electrodes with the numbers ExxT-2, ExxT-3, ExxT-10, and ExxT-GS are intended for single pass welds only.

Stainless Steel Electrodes

The AWS classification for stainless steel for FCA electrodes starts with the letter *E* as its prefix. Following the *E* prefix, the American Iron and Steel Institute's (AISI) three-digit stainless steel number is used. This number indicates the type of stainless steel in the filler metal.

To the right of the AISI number, the AWS adds a dash followed by a suffix number. The number 1 is used to indicate an all-position filler metal, and the number 3 is used to indicate an electrode to be used in the flat and horizontal positions only.

Care of Flux Core Electrodes Wire electrodes may be wrapped in sealed plastic bags for protection from the elements. Others may be wrapped in a special paper, and some are shipped in cans or cardboard boxes.

A small paper bag of a moisture-absorbing material, crystal desiccant, is sometimes placed in the shipping containers. It is enclosed to protect wire electrodes from moisture. Some wire electrodes require storage in an electric rod oven to prevent contamination from excessive moisture. Read the manufacturer's recommendations located in or on the electrode shipping container for information on use and storage.

Weather conditions affect your ability to make high-quality welds. Humidity increases the chance of moisture entering the weld zone. Water (H_2O), which consists of two parts hydrogen and one part oxygen, separates in the weld pool. Hydrogen is trapped and causes undesirable hydrogen entrapment, when only one part of hydrogen is expelled. Hydrogen entrapment can cause weld beads to crack or become brittle. The evaporating moisture will also cause porosity.

To prevent hydrogen entrapment, porosity, and atmospheric contamination, it may be necessary to preheat the base metal to drive out moisture. Storing the wire electrode in a dry location is recommended. The electrode may develop restrictions due to the tangling of the wire or become oxidized with excessive rusting if the wire electrode package is mishandled, thrown, dropped, or stored in a damp location.

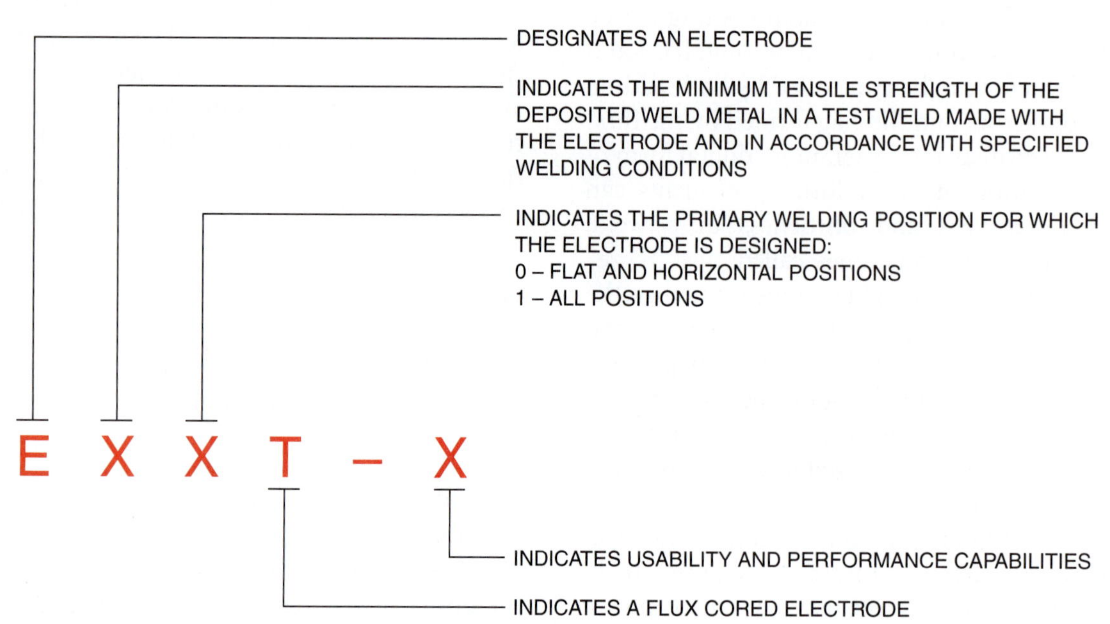

FIGURE 13-12 Identification system for mild steel FCAW electrodes. Courtesy of the American Welding Society.

CAUTION

Always keep the wire electrode dry and handle it as you would any important tool or piece of equipment.

Shielding Gas

FCA welding wire can be manufactured so that all of the required shielding of the molten weld pool is provided by the vaporization of some of the flux within the tubular electrode. When the electrode provides all of the shielding, it is called self-shielding. Other FCA welding wire must use an externally supplied shielding gas to provide the needed protection of the molten weld pool. When a shielding gas is added, it is called dual shield.

Care must be taken to use the cored electrodes with the recommended gases or not to use gas at all with the self-shielded electrodes. Using a self-shielding flux cored electrode with a shielding gas may produce a defective weld. The shielding gas will prevent the proper disintegration of much of the deoxidizers. This results in the transfer of these materials across the arc to the weld. In high concentrations, the deoxidizers can produce slags that become trapped in the welds, causing undesirable defects. Lower concentrations may cause brittleness only. In either case, the chance of weld failure is increased. If these electrodes are used correctly, there is no problem.

The selection of a shielding gas will affect the arc and weld properties. The weld bead width, buildup, penetration, spatter, chemical composition, and mechanical properties are all affected as a result of the shielding gas selection.

Shielding gas comes in high-pressure cylinders. These cylinders are supplied with 2000 psi of pressure. Because of this high pressure, it is important that the cylinders be handled and stored safely. See Chapter 2 for specific cylinder safety instructions.

Gases used for FCA welding include CO_2 and mixtures of argon and CO_2. Argon gas is easily ionized by the arc. Ionization results in a highly concentrated path from the electrode to the weld. This concentration results in a smaller droplet size that is associated with the axial spray mode of metal transfer, **Figure 13-13.** A smooth stable arc results and there is a minimum of spatter. This transfer mode continues as CO_2 is added to the argon until the mixture contains more than 25% of CO_2.

As the percentage of CO_2 increases in the argon mixture, weld penetration increases. This increase in penetration continues until a 100% CO_2 shielding gas is reached. But as the percentage of CO_2 is increased the arc stability decreases. The less stable arc causes an increase in spatter. A mixture of 75% argon and 25% CO_2 works best for jobs requiring a mixed gas. This mixture is sometimes called C-25.

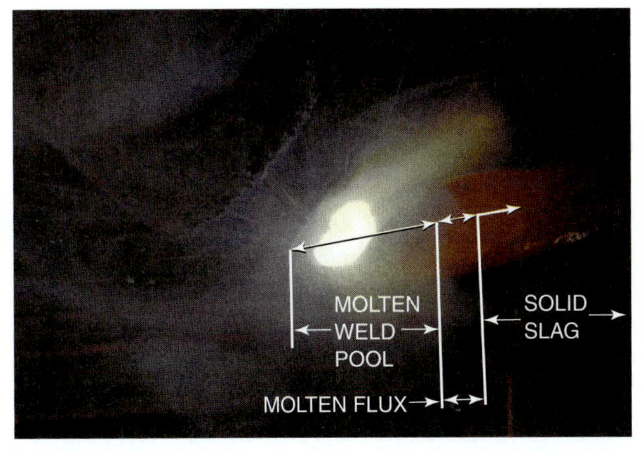

FIGURE 13-13 Axial spray transfer mode. Courtesy of Larry Jeffus.

Straight CO_2 is used for some welding. But the CO_2 gas molecule is easily broken down in the welding arc. It forms carbon monoxide (CO) and free oxygen (O). Both gases are reactive to some alloys in the electrode. As these alloys travel from the electrode to the molten weld pool, some of them form oxides. Silicon and manganese are the primary alloys that become oxidized and lost from the weld metal.

Most FCA welding electrodes are specifically designed to be used with or without shielding gas and for a specific shielding gas or percentage mixture. For example, an electrode designed specifically for use with 100% CO_2 will have higher levels of silicon and manganese to compensate for the losses to oxidization. But if 100% argon or a mixture of argon and CO_2 is used, the weld will have an excessive amount of silicon and manganese. The weld will not have the desired mechanical or metallurgical properties. Although the weld may look satisfactory, it will probably fail prematurely.

CAUTION

Never use an FCA welding electrode with a shielding gas it is not designated to be used with. The weld it produces may be unsafe.

Welding Techniques

A welder can control weld beads made by FCA welding by making changes in the techniques used.

The following explains how changing specific welding techniques will affect the weld produced.

Gun Angle *Gun angle, work angle,* and *travel angle* are names used to refer to the relation of the gun to the work surface. The gun angle can be used to control the weld pool. The electric arc produces an electrical force known as the arc force. The arc force can be used to counteract the gravitational pull that tends to make the liquid weld pool sag or run ahead of the arc. By manipulating the electrode travel angle for the flat and horizontal position of welding to a 20° to 45° angle from the vertical, the weld pool can be controlled. A 40° to 50° angle from the vertical plate is recommended for fillet welds.

Shallower angles are needed when welding thinner materials to prevent burn-through. Steeper angles are used for thicker materials.

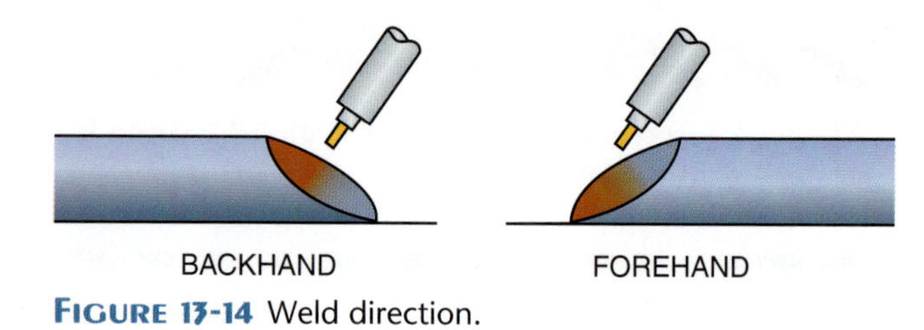

FIGURE 13-14 Weld direction.

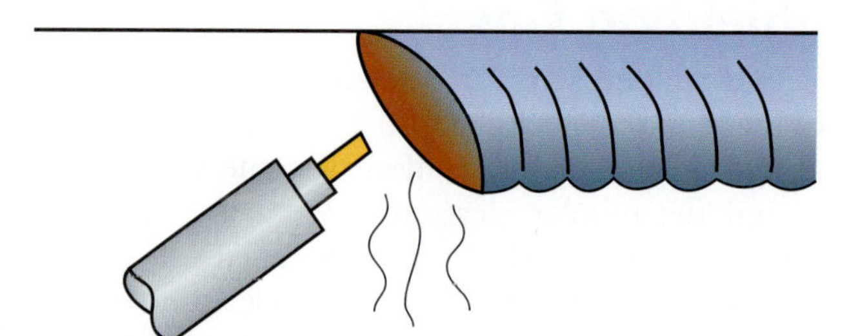
FIGURE 13-15 Weld gun position to control spatter buildup.

Forehand/Backhand Techniques Both the longitudinal and transverse angles remain the same whether you are pushing or pulling the weld bead. Pushing a weld bead is the forehand welding technique and pulling, or dragging, is the backhand welding technique, **Figure 13-14**.

Advantages of the Backhand Technique

By using the backhand technique you can readily see the bead as it progresses. The gun position is easier to maintain, ensuring a consistent depth of penetration. The backhand method is the preferred method to weld with FCAW in the flat or horizontal position.

This method may result in a slightly slower weld progression due to the amount of weld reinforcement required. The slower-moving backhand method allows a slower progression of the weld. This longer welding time at the base metal provides a preheat time in the weld zone as the weld bead progresses. The preheat time allows the weld bead to flow easily into the weld pool and allows deeper weld bead penetration.

A gun angle of approximately 90° or slightly backhand works best for overhead welds, **Figure 13-15**.

Disadvantages of the Backhand Technique

The weld bead may have a more pronounced weld face when you use the backhand technique. Because of the convex, or raised or rounded, shape of the weld bead, more work is required if the product has to be finished by grinding with a power tool. The surface of the weld bead is ground flush or blended into the surface of the base metal so the welded joint is not noticeable. If all variables are met and the travel speed is correct, these disadvantages are easily overcome.

When using the backhand technique, the weld joint is harder to follow because of the welder's hand position and the FCAW gun position. With these obstructions it is easier to wander from the seam of the weld joint. An inexperienced welder sometimes directs the wire too far back into the weld pool, causing the wire to build up in the face of the weld pool. Loss of depth of penetration, or depth that fusion extends into the base metal during FCAW, then occurs. Weld reinforcement will be excessive, which increases welding time and wastes welding electrode, or the high heat and arc force of the FCAW process will blow a hole through the base metal, making a mess that must be fixed.

Advantages of the Forehand Technique

An advantage of the forehand technique is that you can readily see the joint where the bead will be deposited. The contact tube is easier to see, ensuring a more consistent wire electrode extension. The forehand technique works well on vertical up and overhead joints for better control of the weld pool, **Figure 13-16**.

Disadvantages of the Forehand Technique

Less weld reinforcement is applied to the weld joint at the leading edge of the weld pool, where you are able to see new metal. This is metal that has not yet been welded but is in the direct path of the weld pool as the weld progresses. Some spattered slag can be thrown in front of the weld bead, **Figure 13-17**. If this spatter becomes excessive it can interfere with the weld's fusion to the base metal.

As the hot weld pool moves into the unpreheated new metal (base metal), a rougher-looking surface may appear on the weld bead face, and/or the weld bead ripple may not be as smooth as one produced using the backhand method.

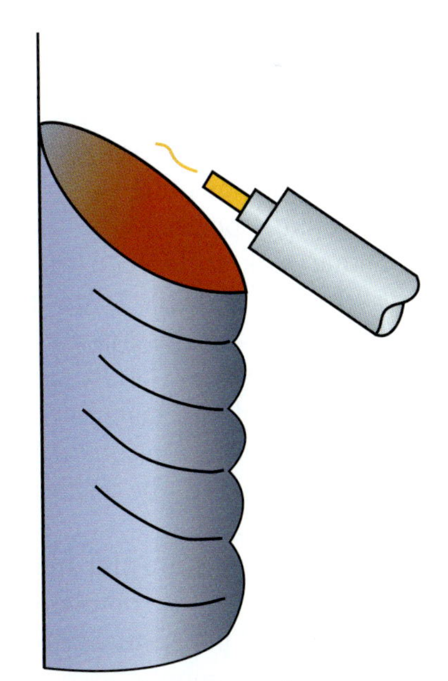

FIGURE 13-16 Vertical up gun angle.

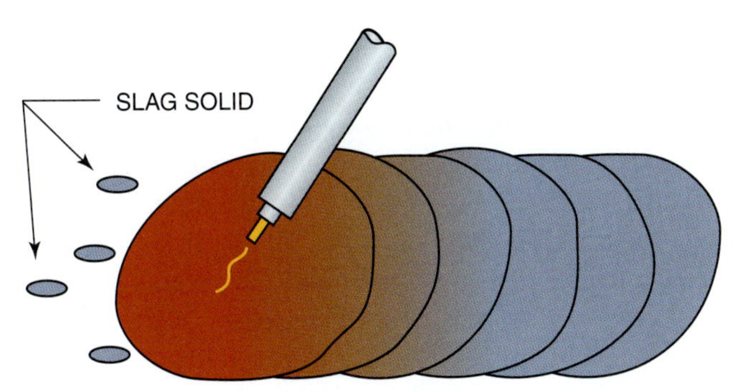

FIGURE 13-17 Large quantities of solid slag in front of a weld can cause slag to be trapped under the weld bead.

Depth of penetration is not as great with the forehand method as it is with the backhand method. The weld progresses so quickly that the arc does not preheat the weld zone as well as the slower-moving backhand method does. Some of the heavy slag cover may become entrapped.

When the flux cored wire electrode is pushed, some electrodes may cause a condition in a weld called honey combing, in which the weld shows no external sign of an existing problem but, when cut open, is full of cavities, resembling the interior of a honey comb.

The amount of spatter may be slightly increased with the forehand technique, depending on the electrode.

Discuss the pros and cons of the two welding techniques with your instructor.

Travel Speed The American Welding Society defines travel speed as the linear rate at which the arc is moved along the weld joint. Fast travel speeds deposit less filler metal. If the rate of travel increases, the filler metal cannot be deposited fast enough to adequately fill the path melted by the arc. This causes the weld bead to have a groove melted into the base metal next to the weld and left unfilled by the weld. This condition is known as undercut.

Undercut occurs along the edges or toes of the weld bead. Slower travel speeds will, at first, increase penetration and increase the filler weld metal deposited. As the filler metal increases, the weld bead will build up in the weld pool. Because of the deep penetration of flux cored wire, the angle at which you hold the gun is very important for a successful weld.

If all welding conditions are correct and remain constant, the preferred rate of travel for maximum weld penetration is a travel speed that allows you to stay within the selected welding variables and still control the fluidity of the weld pool. This is an intermediate travel speed, or progression, which is not too fast or too slow.

Another way to figure out correct travel speed is to consult the manufacturer's recommendations chart for the ipm burn-off rate for the selected electrode.

Mode of Metal Transfer The mode of metal transfer is used to describe how the molten weld metal is transferred across the arc to the base metal. The mode of metal transfer that is selected, the shape of the completed weld bead, and the depth of weld penetration depend on the welding power source, wire electrode size, type and thickness of material, type of shielding gas used, and best welding position for the task.

Spray Transfer—FCAW-G

The spray transfer mode is the most common process used with gas shielded FCAW, **Figure 13-13.**

As the gun trigger is depressed, the shielding gas automatically flows and the electrode bridges the distance from the contact tube to the base metal, making contact with the base metal to complete a circuit. The electrode shorts and becomes so hot that the base metal melts and forms a weld pool. The electrode melts into the weld pool and burns back toward the contact tube. A combination of high amperage and the shielding gas along with the electrode size produces a pinching effect on the molten electrode wire, causing the end of the electrode wire to spray across the arc.

The characteristic of spray-type transfer is a smooth arc, through which hundreds of small droplets per second are transferred through the arc from the electrode to the weld pool. At that moment a transfer of metal is taking place. Spray transfer can produce a high quantity of metal droplets, up to approximately 250 per second above the transition current, or critical current. This means the current is dependent on the electrode size, composition of the electrode, and shielding gas so a spray transfer can take place. Below the transition current (critical current) globular transfer takes place.

In order to achieve a spray transfer, high current and larger diameter electrode wire is needed. A shielding gas of carbon dioxide (CO_2), a mixture of carbon dioxide (CO_2) and argon (Ar), or an argon (Ar) oxygen (O_2) mixture is needed. FCAW-G is a welding process that, with the correct variables, can be used

- on thin or properly prepared thick sections of material,
- on a combination of thick to thin materials,
- with small or large electrode diameters, and
- with a combination of shielding gases.

Globular Transfer—FCAW-G

Globular transfer occurs when the welding current is below the transition current, **Figure 13-18.** The electrode forms a molten ball at its end that grows in size to approximately two to three times the original electrode diameter. These large molten balls are then transferred across the arc at the rate of several drops per second.

The arc becomes unstable because of the gravitational pull from the weight of these large drops. A spinning effect caused by a natural phenomenon takes place when argon gas is introduced to a large ball of molten metal electrode. This causes a spinning motion as the molten ball transfers across the arc to the base metal. This unstable globular transfer can produce excessive spatter.

Both FCAW-S and FCAW-G use DCEN when welding on thin-gauge materials to keep the heat in the base metal and the small diameter electrode at a controllable burn-off rate. The electrode can then be stabilized, and it is easier to manipulate and control the weld pool in all weld positions. Larger diameter electrodes are welded with DCEP because the larger diameters can keep up with the burn-off rates.

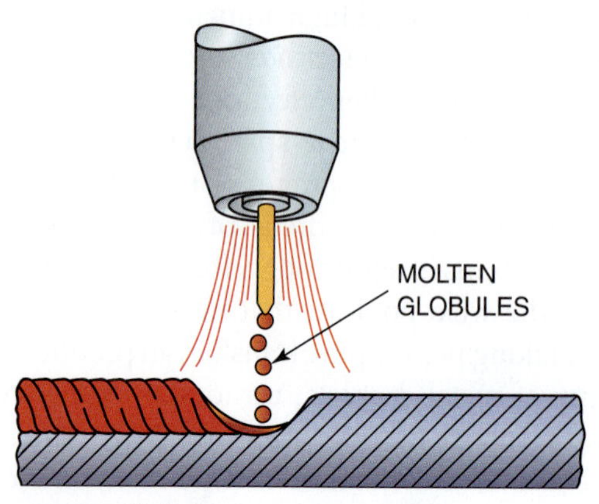

FIGURE 13-18 Globular transfer method.

The recommended weld position means the position in which the workpiece is placed for welding. All welding positions use either spray or globular transfer, but for now we will concentrate on the flat and horizontal welding positions.

In the flat welding position the workpiece is placed flat on the work surface. In the horizontal welding position the workpiece is positioned perpendicular to the work- bench surface.

The amperage range may be from 30 to 400 amperes or more for welding materials from gauge thickness up to 1 1/2 inches. On square groove weld joints, thicker base metals can be welded with little or no edge preparation. This is one of the great advantages of FCAW. If edges are prepared and cut at an angle (beveled) to accept a complete joint weld penetration, the depth of penetration will be greatly increased. FCAW is commonly used for general repairs to mild steel in the horizontal, vertical, and overhead welding positions, sometimes referred to as out-of-position welding.

Electrode Extension The electrode extension is measured from the end of the electrode contact tube to the point the arc begins at the end of the electrode, **Figure 13-19.** Compared to GMA welding, the electrode extension required for FCAW is much greater. The longer extension is required for several reasons. The electrical resistance of the wire causes the wire to heat up, which can drive out moisture from the flux. This preheating of the wire also results in a smoother arc with less spatter.

Porosity FCA welding can produce high-quality welds in all positions, although porosity in the weld can be a persistent problem. Porosity can be caused by moisture in the flux, improper gun manipulation, or surface contamination.

The flux used in the FCA welding electrode is subject to picking up moisture from the surrounding atmosphere, so the electrodes must be stored in a dry area. Once the flux becomes contaminated with moisture, it is very difficult to remove. Water (H_2O) breaks down into free hydrogen and oxygen in the presence of an arc, **Figure 13-20.** The hydrogen can be absorbed into the molten weld metal, where it can cause postweld cracking. The oxygen is absorbed into the weld metal also, but it forms bubbles of porosity as the weld begins to solidify.

If a shielding gas is being used, the FCA welding gun gas nozzle must be close enough to the weld to provide adequate shielding gas coverage. If there is a wind or if the nozzle-to-work distance is excessive, the shielding will be inadequate and cause weld porosity. If welding is to be done outside or in an area subject to drafts, the

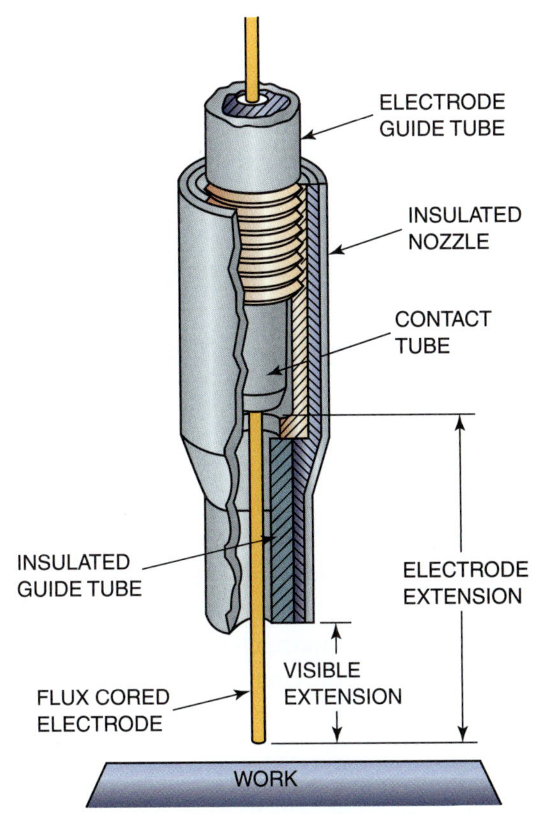

FIGURE 13-19 Self-shielded electrode nozzle. Courtesy of the American Welding Society.

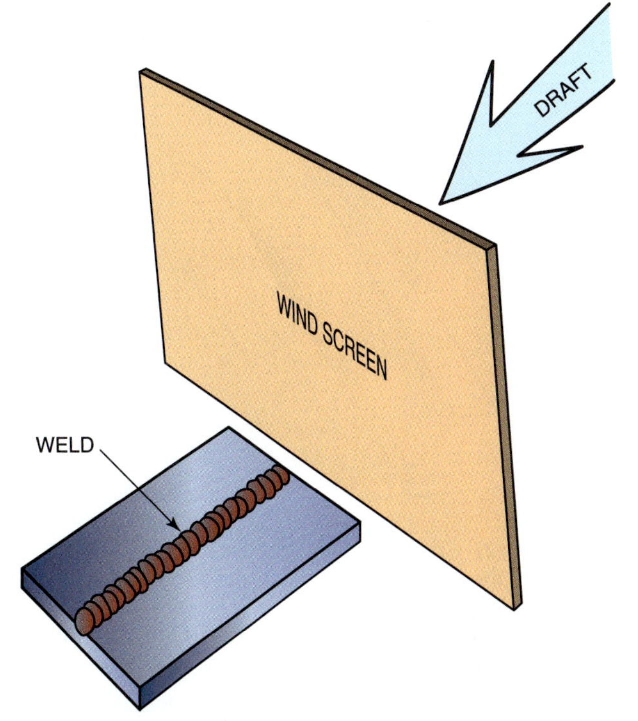

FIGURE 13-21 A wind screen can keep the welding shielding from being blown away.

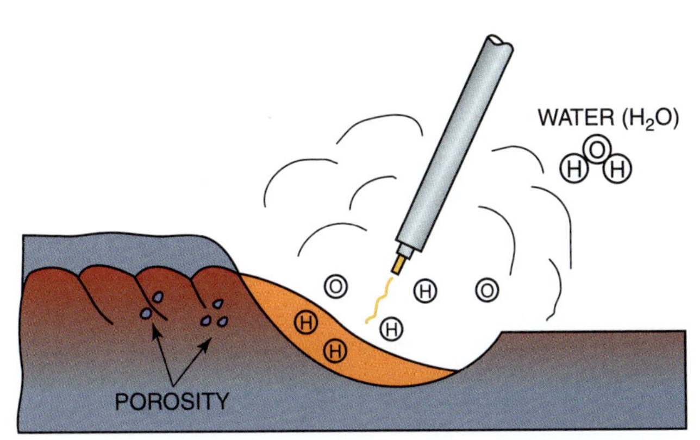

FIGURE 13-20 Water (H_2O) breaks down in the presence of the arc and the hydrogen (H) is dissolved in the molten weld metal.

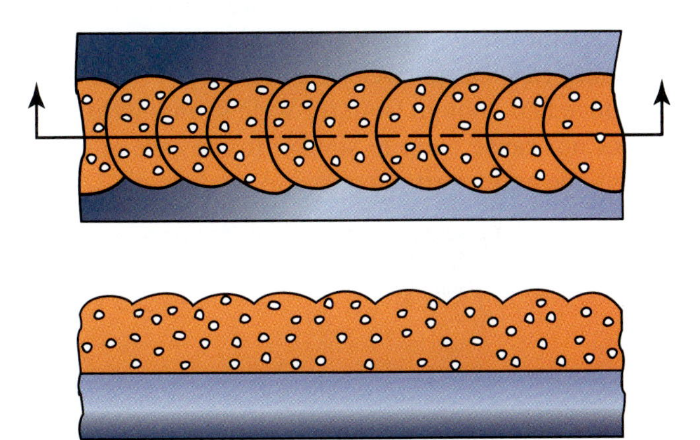

FIGURE 13-22 Uniformly scattered porosity.

gas flow rate must be increased or a wind shield must be placed to protect the weld, **Figure 13-21.**

A common misconception is that the flux within the electrode will either remove or control weld quality problems caused by surface contaminations. That is not true. The addition of flux makes FCA welding more tolerant to surface conditions than GMA welding, although it still is adversely affected by such contaminations.

New hot-rolled steel has a layer of dark gray or black iron oxide called mill scale. Although this layer is very thin, it may provide a source of enough oxygen to cause porosity in the weld. If mill scale causes porosity, it is usually uniformly scattered through the weld, **Figure 13-22.** Unless severe, uniformly scattered porosity is usually not visible in the finished weld. It is trapped under the surface as the weld cools.

Because porosity is under the weld surface, nondestructive testing methods, including X ray, magnetic particle, and ultrasound, must be used to locate it in a weld. It can be detected by mechanical testing such as guided-bend, free-bend, and nick-break testing for establishing weld parameters. Often it is better to remove the mill scale before welding rather than risking the production of porosity.

All welding surfaces within the weld groove and the surrounding surfaces within 1 in. (25 mm) must be cleaned to bright metal, **Figure 13-23.** Cleaning may be either grinding, filing, sanding, or blasting.

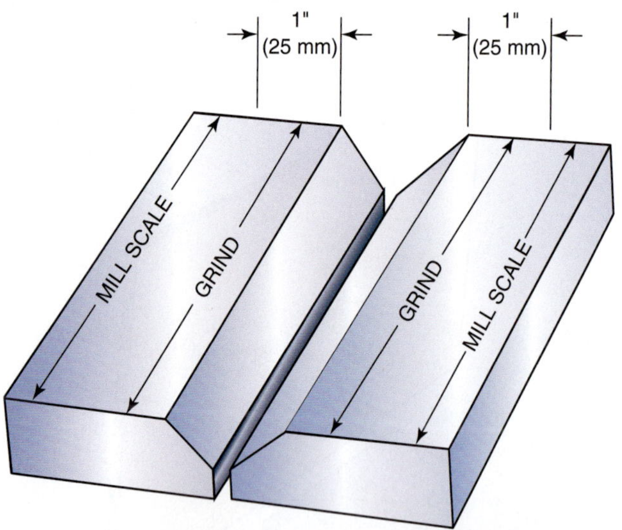

FIGURE 13-23 Grind mill scales off plates within 1 in. (25mm) of the groove.

Any time FCA welds are to be made on metals that are dirty, oily, rusty, or wet or that have been painted, the surface must be precleaned. Cleaning can be done chemically or mechanically.

One advantage of chemically cleaning oil and paint is that it is easier to clean larger areas. Both oil and paint smoke easily when heated, and such smoke can cause weld defects. They must be removed far enough from the weld so that weld heat does not cause them to smoke. In the case of small parts the entire part may need to be cleaned.

Chemically cleaning oil and paint off metal must be done according to the cleaner manufacturer's directions. The work must be done in an appropriate, approved area. The metal must be dry, and all residues of the cleaner must be removed before welding begins.

Troubleshooting FCA Welding

Troubleshooting FCA welding problems is often a trial and error process, wherein you make one adjustment or change and then make a trial weld to see if the problem has improved. If it has not improved or if the problem is worse, reset the machine and try another adjustment. Keep doing this until the problem is resolved, making only one adjustment or change at a time. Making two or more changes at the same time can result in one improving the weld and the others causing new problems.

The most common causes of FCA welding problems are equipment setup. However in the field, worn and dirty parts will, from time to time, cause problems similar to those caused by improper setup. If the cause is misdiagnosed good parts might be replaced. To avoid this, use the list of common FCA welding problems in **Table 13-6** to try to solve the weld problem before replacing parts.

Problem	Cause
Gun nozzle arcs to work	1. Weld spatter buildup in nozzle 2. Contact tube bent and touching nozzle
Wire feed but no arc	1. Bad or missing work clamp (ground) 2. Loose jumper lead in welder 3. Electrode not contacting bare metal
Arc burns off wire at contact tube end	1. Feed rollers tension too loose 2. Wrong sized feed rollers 3. Wire welded to contact tube 4. Wire liner worn or damaged 5. Out of wire
Wire feeds erratically	1. Feed rollers tension loose 2. Dirty liner 3. Worn or dirty contact tube
Wire pops and gun jerks	Too high a wire feed speed or too low a voltage
Wire burns back and large globules of metal cross the arc	1. Too high a voltage or too low a wire feed speed 2. Wire slipping in feed rollers, not feeding smoothly
Weld does not burn into base metal	1. Too long an electrode extinction 2. Too low voltage and amperage settings
Weld burns through the base metal	1. Too short an electrode extinction 2. Too high voltage and amperage settings
Porosity in weld	1. Poor shielding gas coverage on duel-shield wire 2. Wrong shielding gas type being used 3. Shielding gas used on self-shielding wire 4. Single pass electrode used for multi pass weld
Poor shielding gas coverage	1. Plugged or dirty gas diffuser 2. Too high or too low shielding gas flow rate 3. Shielding gas cylinder near empty

TABLE 13-6 Troubleshooting for FCA Welding.

Because of the similarities between FCA and GMA welding processes, the GMA welding troubleshooting chart in Table 11-8 on page 288 can also be used to resolve FCA welding problems. Often the equipment manufacturer will include a list of troubleshooting tips in the instruction booklet.

FIGURE 13-24 Although this is not the manufacturer's recommended way of storing a FCA welder, it is unfortunately how they are treated on farms and ranches. Courtesy of Larry Jeffus.

Summary

Because of its flexibility, portability, and range of metal thicknesses, flux cored arc welding is quickly becoming the most commonly used arc welding process on many farms and ranches. Most small agricultural FCA welders have their wire feed drive wheels and wire spools located inside the machine under a cover where they are protected from most dust and dirt, **Figure 13-24.** FCA welders can be low maintenance as long as they are stored in a dry place.

Setting up the welder is often the most difficult part of making a good FCA weld. The manufacturer's list of the machine settings located inside most welders' wire feed covers is very helpful. When similar welds are repeated periodically, some farmers and ranchers mark their favorite settings on the machine dial. This is especially helpful when you are not welding very often. Do not mark the machine dial at school.

Although FCA welding can be done with or without a shielding gas, most agricultural FCA welding is done without. There is very little difference in the welding techniques required to make welds with or without shielding gas, so whichever way you learned FCA welding in school, you should not have a problem making welds in the field either way.

Review

1. FCA welding can be used on metal ranging in thickness from __________ to __________ .
   A. 1/4 in. to 3/4 in.
   B. 24-gauge sheet metal to more than 1 in.
   C. 1/16 in. to 1/2 in.
   D. 32-gauge sheet metal to 16-gauge sheet metal
2. What type of power supply provides a controlled voltage to the welding electrode?
   A. AC
   B. CC
   C. DC
   D. CP
3. As with GMA welding, having a correctly set welder does not ensure a good weld.
   A. true
   B. false
4. The __________ inside the electrode wire provides the molten weld pool with protection from the atmosphere, improves strengths through chemical reactions and alloys, and improves the weld shape.

A. shielding gas
B. flux
C. filler metal
D. hot gases

5. What does slag do for the hot weld metal?
A. Protects it from the atmosphere while it cools.
B. Keeps it dry.
C. Adds filler metal.
D. Provides different colored sparks.

6. What is it about the FCA welding electrode that provides alloys to the weld?
A. slag
B. shielding gas
C. flux
D. welding power supply

7. What gases are major atmospheric contaminants to molten weld metal?
A. oxygen and acetylene
B. argon and carbon dioxide
C. nitrogen and carbon monoxide
D. oxygen and nitrogen

8. Because of the __________ characteristic, no edge beveling preparation is required on some joints in metal up to 1/2 in.
A. wire feed system
B. deep penetration
C. welder's power supply
D. shielding gas

9. Small agricultural FCA welding wire feed systems usually have __________ feed rollers.
A. 1
B. 2
C. 3
D. 4

10. The main limitation of flux cored arc welding is that it is confined to ferrous metals and nickel-based alloys.
A. true
B. false

11. What input voltage do most small agricultural FCA welders need?
A. 240V
B. 208V
C. 90V
D. 120V

12. Where in FCA welding can an "arc stabilizer" be found?
A. welding power supply
B. welding gun
C. flux
D. wire feed control

13. According to Table 13-3, which of the following FCA welding electrodes can be used with or without a shielding gas?
A. T-4
B. T-2
C. T-3
D. T-5

14. An E70T-1 mild steel FCA welding electrode would produce welds with tensile strength of approximately __________ psi.
A. 70
B. 700
C. 7,000
D. 70,000

15. Which of the following can cause problems with FCA welding wire that is stored improperly?
A. CO_2
B. H_2O
C. argon
D. C-25

16. When the electrode FCA welding electrodes provide all of the shielding, they are called duel shielded.
A. true
B. false

17. Which welding position is preferred for welding in the flat position?
A. backhand
B. forehand
C. leading angle
D. 90° angle

18. Depth of penetration is not as great with the backhand method as it is with the forehand method.
A. true
B. false

19. The __________ is measured for the end of the electrode contact tube to the point the arc begins at the end of the electrode.
A. work distance
B. weld arc length
C. height of arc
D. electrode extension

20. __________ can be caused by moisture in the flux, improper gun manipulation, or surface contamination.
A. Smoke
B. Porosity
C. Spatter
D. Electric shocks

Chapter 14

Flux Cored Arc Welding

OBJECTIVES

After completing this chapter, the student should be able to

- ☑ set up a constant potential, semiautomatic FCA welding system.
- ☑ explain the effects that changing from a self-shielded to a dual-shielded electrodes system has on welding.
- ☑ control weld bead contour during welding by using the proper weave pattern.
- ☑ demonstrate an ability to control undercut, overlap, porosity, and slag inclusions when welding in all positions.

KEY TERMS

amperage range
conduit liner
contact tube
critical weld
feed rollers
lap joint
root face
stringer bead
tee joint
voltage range
weave bead
wire-feed speed

INTRODUCTION

Setup of the FCA weld station is the key to making quality welds. It may be possible, using a poorly set up FCA welder, to make an acceptable weld in the flat position. The FCA welding process is often forgiving; thus welds can often be made even when the welder is not set correctly. However, such welds will have major defects such as excessive spatter, undercut, overlap, porosity, slag inclusions, and poor weld bead contours. Setup becomes even more important for out-of-position welds. Making vertical and overhead welds can be difficult for a student welder with a properly set up system, but it becomes impossible with a system that is out of adjustment.

Learning to set up and properly adjust the FCA welding system will allow you to produce high-quality welds at a high level of productivity.

FCAW is set up and manipulated in a manner similar to that of GMAW. The results of changes in electrode extension, voltage, amperage, and torch angle are essentially the same.

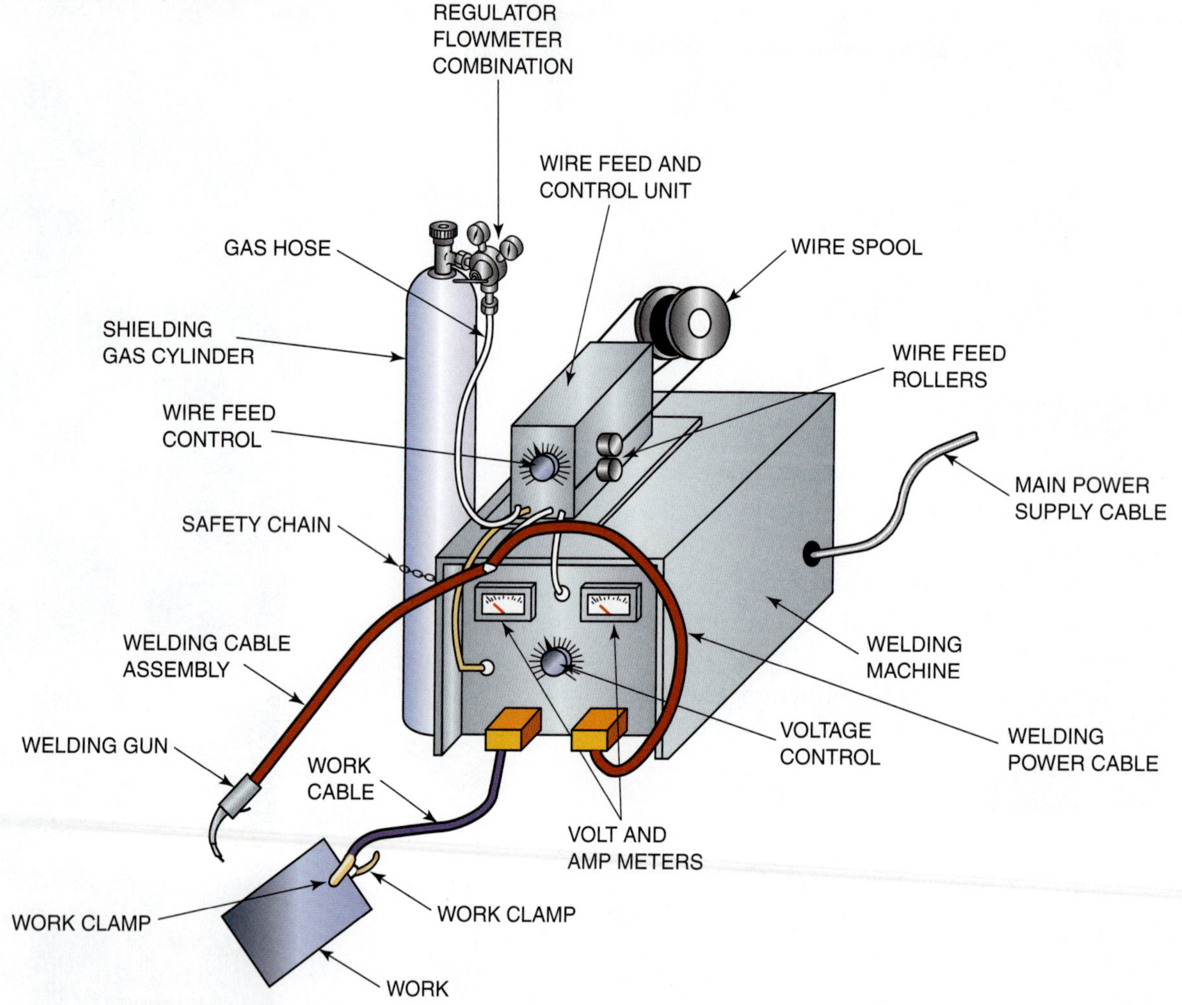

FIGURE 14-1 Typical FCA welding station and equipment identification.

Although every manufacturer's FCA welding equipment is designed differently, all equipment is set up in a similar manner. It is always best to follow the specific manufacturer's recommendations regarding setup as provided in its equipment literature. You will find, however, that, in the field, manufacturers' literature is not always available for the equipment you are asked to use. It is therefore important to have a good general knowledge and understanding of the setup procedure for FCA welding equipment. **Figure 14-1** shows all of the various components that make up an FCA welding station.

CAUTION

FCA welding produces a lot of ultraviolet light, heat, sparks, slag, and welding fumes. Proper personal, protective clothing and special protective clothing must be worn to prevent burns from the ultraviolet light and hot weld metal. Eye protection must be worn to prevent injury from flying sparks and slag. Forced ventilation and possibly a respirator must be used to prevent fume-related injuries. Refer to the safety precautions provided by the equipment and electrode manufacturers and to Chapter 2, "Safety in Welding and Fabrication," for additional safety help.

Practices

The practices in this chapter are grouped according to those requiring similar techniques and setups. Plate welds are covered first, then sheet metal. The practices start with 1/4-in. (6-mm) mild steel plates; they are used because they require the least preparation times. The thicker 3/8-in. (9.5-mm) plates provide the basics of practicing groove welding. Sheet metal is grouped together because it presents a unique set of learning skills.

The major skill required for making consistently acceptable FCA welds is the ability to set up the welding system. Changes such as variations in material thickness, position, and type of joint require changes both in technique and setup. A correctly set up FCA welding station can, in many cases, be operated by a less-skilled welder. Often the only difference between a welder earning a minimum wage and one earning the maximum wage is the ability to correct machine setups.

For several reasons the FCA welding practice plates will be larger than most other practice plates. Welding heat and welding speed are the major factors that necessitate this increased size. FCA welding is both high energy and fast, and the welding energy (heat) input is so great that small practice plates may glow red by the end of a single weld pass. This would seriously affect the weld quality. To prevent this from happening, wider plates are used. Because of the higher welding speeds, longer plates are usually used.

Plates less than 1/2 in. (13 mm) will be 12 in. (305 mm) long for most practices. In addition to controlling the heat buildup, the longer plates are needed to give the welder enough time to practice welding. Learning to make longer welds is a skill that must also be practiced, because the FCA welding process is used in industry to make long production welds.

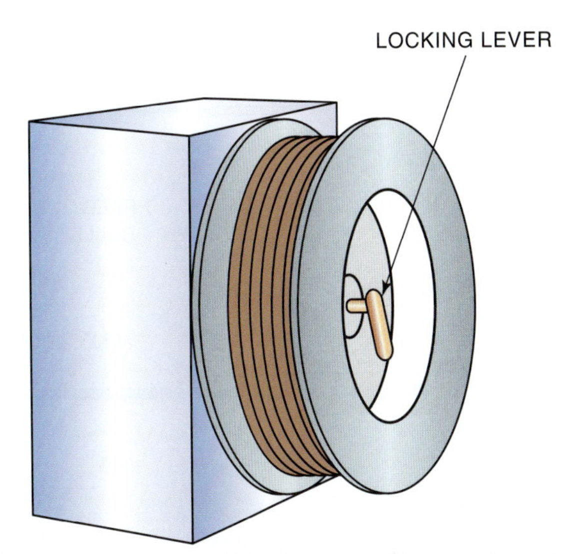

FIGURE 14-2 Wire reel may be secured by a center nut or locking lever.

FIGURE 14-3 Check to be certain that the feed rollers are the correct size for the wire being used. Courtesy of Larry Jeffus.

PRACTICE 14-1

FCAW Equipment Setup

For this practice, you will need a semiautomatic welding power source approved for FCA welding, welding gun, electrode feed unit, electrode supply, electrode conduit, power and work leads, assorted hand tools, spare parts, and any other required materials. In this practice, you will demonstrate to a group of students and your instructor how to properly set up an FCA welding station. Some manufacturers include detailed set up instructions with their equipment. If such instructions are available for your equipment, follow them. Otherwise, use the following instructions.

For FCA duel-shielded welding equipment setup, follow the setup procedures outlined in the GMA welding Practice 12-1 on page 292.

Install the reel of electrode (welding wire) on the holder and secure it, **Figure 14-2.** Check the feed roller size to ensure that it matches the wire size, **Figure 14-3.** The conduit liner size should be checked for compatibility with the wire size. Connect the conduit to the feed unit. The conduit or an extension should be aligned with the groove in the roller and set as close to the roller as possible without touching, **Figure 14-4.** Misalignment at this point can contribute to a bird's nest, **Figure 14-5.** Bird-nesting of the electrode wire, so-called because it looks like a bird's nest, results when the feed roller pushes the wire into a tangled ball because the wire would not go through the outfeed side conduit.

Be sure the power is off before attaching the welding cables. The electrode and work leads should be attached to the proper terminals. The electrode lead should be attached to electrode or positive (+). If necessary, it is also attached to the power cable part of the gun lead. The work lead should be attached to work or negative (−).

The welding cable liner or wire conduit must be securely attached to the gas diffusers and contact tube

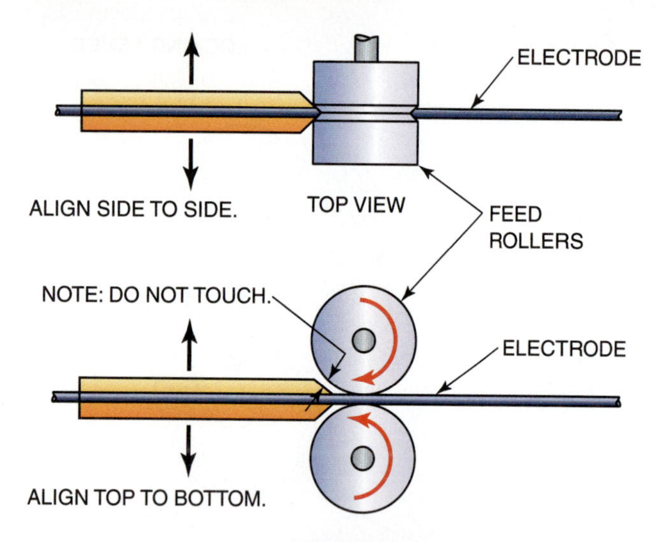

FIGURE 14-4 Feed roller and conduit alignment.

FIGURE 14-5 "Bird's nest" in the filler wire at the feed rollers. Courtesy of Larry Jeffus.

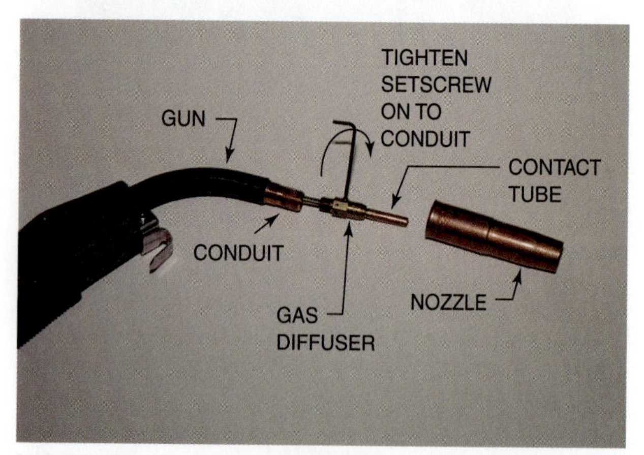

FIGURE 14-6 Securely attach conduit to gas diffuser and contact tube to prevent wire jams caused by misalignment. Courtesy of Larry Jeffus.

(tip), **Figure 14-6.** The contact tube must be the correct size to match the electrode wire size being used. If a shielding gas is to be used, a gas nozzle would be attached to complete the assembly. Even if a gas nozzle is not needed for a shielding gas, it is still installed.

FIGURE 14-7 Push the wire through the guides by hand. Courtesy of Larry Jeffus.

Because it is easy for a student to touch the work with the contact tube during welding, an electrical short may occur. This short-out of the contact tube will immediately destroy the tube. Although the gas nozzle may interfere with some visibility, it may be worth the trouble for a new welder. FCA welding is more sensitive to changes in arc voltage than is (stick) SMAW welding. Such variations in FCA welding voltage can dramatically and adversely affect your ability to maintain weld bead control. A loose or poor connection will result in increased circuit resistance and a loss of welding voltage. To be sure that you have a good work connection, remove any dirt, rust, oil, or other surface contamination at the point the work clamp is connected to the weldment.

Complete a copy of the "Student Welding Report" listed in Appendix I or provided by your instructor. ◆

PRACTICE 14-2

Threading FCAW Wire

Using the FCAW machine that was properly assembled in Practice 12-1, you will turn the machine on and thread the electrode wire through the system.

Check that the unit is assembled correctly according to the manufacturer's specifications. Switch on the power and check the gun switch circuit by depressing the switch. The power source relays, feed relays, gas solenoid, and feed motor should all activate.

Cut off the end of the electrode wire if it is bent. When working with the wire, be sure to hold it tightly. The wire will become tangled if it is released. The wire has a natural curl known as *cast*. Straighten out about 12 in. (300 mm) of the curl to make threading easier.

Separate the wire feed rollers and push the wire first through the guides, then between the rollers, and finally into the conduit liner, **Figure 14-7.** Reset the rollers so there is a slight amount of compression on

FIGURE 14-8 Adjust the wire feed tensioner. Courtesy of Larry Jeffus.

the wire, **Figure 14-8**. Set the wire-feed speed control to a slow speed. Hold the welding gun so that the electrode conduit and cable are as straight as possible.

Press the gun switch. Pressing the gun switch to start the wire feeder is called triggering the gun. The wire should start feeding into the liner. Watch to make certain that the wire feeds smoothly and release the gun switch as soon as the end comes through the contact tube.

If the wire stops feeding before it reaches the end of the contact tube, stop and check the system. If no obvious problem can be found, mark the wire with tape and remove it from the gun. It then can be held next to the system to determine the location of the problem.

With the wire feed running, adjust the feed roller compression so that the wire reel can be stopped easily by a slight pressure. Too light a roller pressure will cause the wire to feed erratically. Too high a pressure can crush some wires, causing some flux to be dropped inside the wire liner. If this happens, you will have a continual problem with the wire not feeding smoothly or jamming.

With the feed running, adjust the spool drag so that the reel stops when the feed stops. The reel should not coast to a stop, because the wire can be snagged easily. Also, when the feed restarts, a jolt occurs when the slack in the wire is taken up. This jolt can be enough to momentarily stop the wire, possibly causing a discontinuity in the weld.

When the test runs are completed, the wire can either be rewound or cut off. Some wire-feed units have a retract button. This allows the feed driver to reverse and retract the wire automatically. To rewind the wire on units without this retract feature, release the rollers and turn them backward by hand. If the machine will not allow the feed rollers to be released without upsetting the tension, you must cut the wire. Some wire reels have covers to prevent the collection of dust, dirt, and metal rings on the wire, **Figure 14-9**.

(A)

(B)

FIGURE 14-9 (A) Covered wire reel. (B) Wire cover on a dual wire feed system. Courtesy of Lincoln Electric Company.

Complete a copy of the "Student Welding Report" listed in Appendix I or provided by your instructor. ◆

Flat-position Welds

PRACTICE 14-3

Stringer Beads Flat Position

Using a properly set up and adjusted FCA welding machine, **Table 14-1**, proper safety protection, 0.035-in. and/or 0.045-in. (0.9-mm and/or 1.2-mm) -diameter E70T-1 and/or E70T-5 electrodes, and one or more pieces of mild steel plate, 12 in. (305 mm) long and 1/4 in. (6 mm) or thicker, you will make a stringer bead weld in the flat position, **Figure 14-10.**

Starting at one end of the plate and using a dragging technique, make a weld bead along the entire 12-in. (305-mm) length of the metal. After the weld is complete, check its appearance. Make any needed changes to correct the weld. Repeat the weld and make additional adjustments. After the machine is set, start to work on improving the straightness and

Electrode		Welding Power			Shielding Gas		Base Metal	
Type	Size	Amps	Wire Feed Speed IPM (cm/min)	Volts	Type	Flow	Type	Thickness
E70T-1	0.035 in. (0.9 mm)	130 to 150	288 to 380 (732 to 975)	22 to 25	None	n/a	Low-carbon steel	1/4 in. to 1/2 in. (6 mm to 13 mm)
E70T-1	0.045 in. (1.2 mm)	150 to 210	200 to 300 (508 to 762)	28 to 29	None	n/a	Low-carbon steel	1/4 in. to 1/2 in. (6 mm to 13 mm)
E70-5	0.035 in. (0.9 mm)	130 to 200	288 to 576 (732 to 1,463)	20 to 28	75% argon 25% CO_2	30 cfh	Low-carbon steel	1/4 in. to 1/2 in. (6 mm to 13 mm)
E70T-5	0.045 in. (1.2 mm)	150 to 250	200 to 400 (508 to 1,016)	23 to 29	75% argon 25% CO_2	35 cfh	Low-carbon steel	1/4 in. to 1/2 in. (6 mm to 13 mm)

TABLE 14-1 FCA Welding Parameters for Use if Specific Settings Are Unavailable from Electrode Manufacturer.

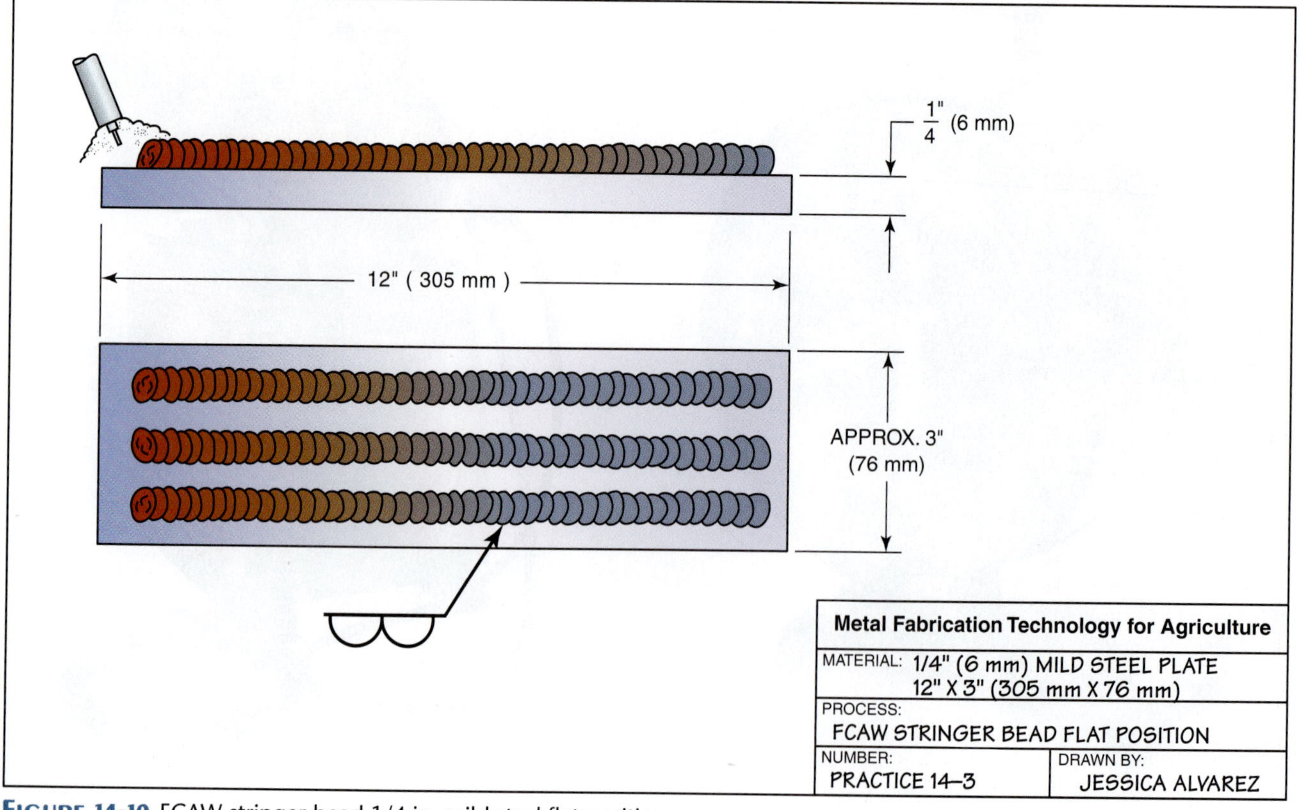

FIGURE 14-10 FCAW stringer bead 1/4 in. mild steel flat position.

uniformity of the weld. Use weave patterns of different widths and straight stringers without weaving.

Repeat with both classifications of electrodes as needed until beads can be made straight, uniform, and free from any visual defects. Turn off the welding machine and shielding gas and clean up your work area when you are finished welding.

Complete a copy of the "Student Welding Report" listed in Appendix I or provided by your instructor. ◆

Square-groove Welds

One advantage of FCA welding is the ability to make 100%-joint-penetrating welds without beveling the edges of the plates. These full-joint-penetrating welds can be made in plates that are 1/4 in. (6 mm) or less in thickness. Welding on thicker plates risks the possibility of a lack of fusion on both sides of the root face, **Figure 14-11.**

There are several disadvantages of having to bevel a plate before welding:

- Beveling the edge of a plate adds an operation to the fabrication process. Unbeveled plates can be sheared or thermally cut (OFC, PAC, PAC, LBC, etc.) to size, assembled, and welded; but beveled plates must first be cut to size, then beveled by grinding, machining, or thermally, assembled, and welded.
- Both more filler metal and welding time are required to fill a beveled joint than are required to make a square jointed weld.
- Beveled joints have more heat from the thermal beveling and additional welding required to fill the groove. The lower heat input to the square joint means less distortion.

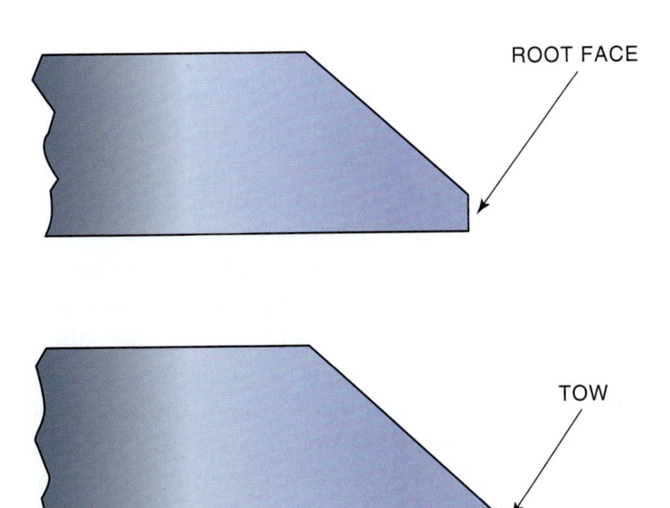

FIGURE 14-11 A beveled joint may or may not have a flat surface, called a root face.

The major disadvantage of making square jointed welds is that as the plate thickness approaches 1/4 in. (6 mm) or the weld is out of position a much higher level of skill is required. The skill required to make quality square welds can be acquired by practicing on thinner metal. It is much easier to make this type of weld in 1/8-in. (3-mm) -thick metal and then move up in thickness as your skills improve.

PRACTICE 14-4

Butt Joint 1G

Using a properly set up and adjusted FCA welding machine, proper safety protection, 0.035-in. and/or 0.045-in. (0.9-mm and/or 1.2-mm) -diameter E70T-1 and/or E70T-5 electrodes, and one or more pieces of mild steel plate, 12 in. (305 mm) long and 1/4 in. (6 mm) or less in thickness, you will make a groove weld in the flat position, **Figure 14-12.**

- Tack weld the plates together and place them in position to be welded.
- Starting at one end, run a bead along the joint. Watch the molten weld pool and bead for signs that a change in technique may be required.
- Make any needed changes as the weld progresses in order to produce a uniform weld.

Repeat with both classifications of electrodes as needed until defect-free welds can consistently be made in the 1/4-in. (6-mm) -thick plate. Turn off the welding machine and shielding gas and clean up your work area when you are finished welding.

Complete a copy of the "Student Welding Report" listed in Appendix I or provided by your instructor. ◆

V-Groove and Bevel-Groove Welds

Although for speed and economy engineers try to avoid specifying welds that require beveling the edges of plates, it is not always possible. Anytime the metal being welded is thicker than 1/4 in. (6 mm) and a 100% joint penetration weld is required, the edges of the plate must be prepared with a bevel. Fortunately, FCA welding allows a narrower groove to be made and still achieve a thorough thickness weld, **Figure 14-13.**

All FCA groove welds are made using three different types of weld passes, **Figure 14-14.**

- *Root pass:* The first weld bead of a multiple pass weld. The root pass fuses the two parts together and establishes the depth of weld metal penetration.

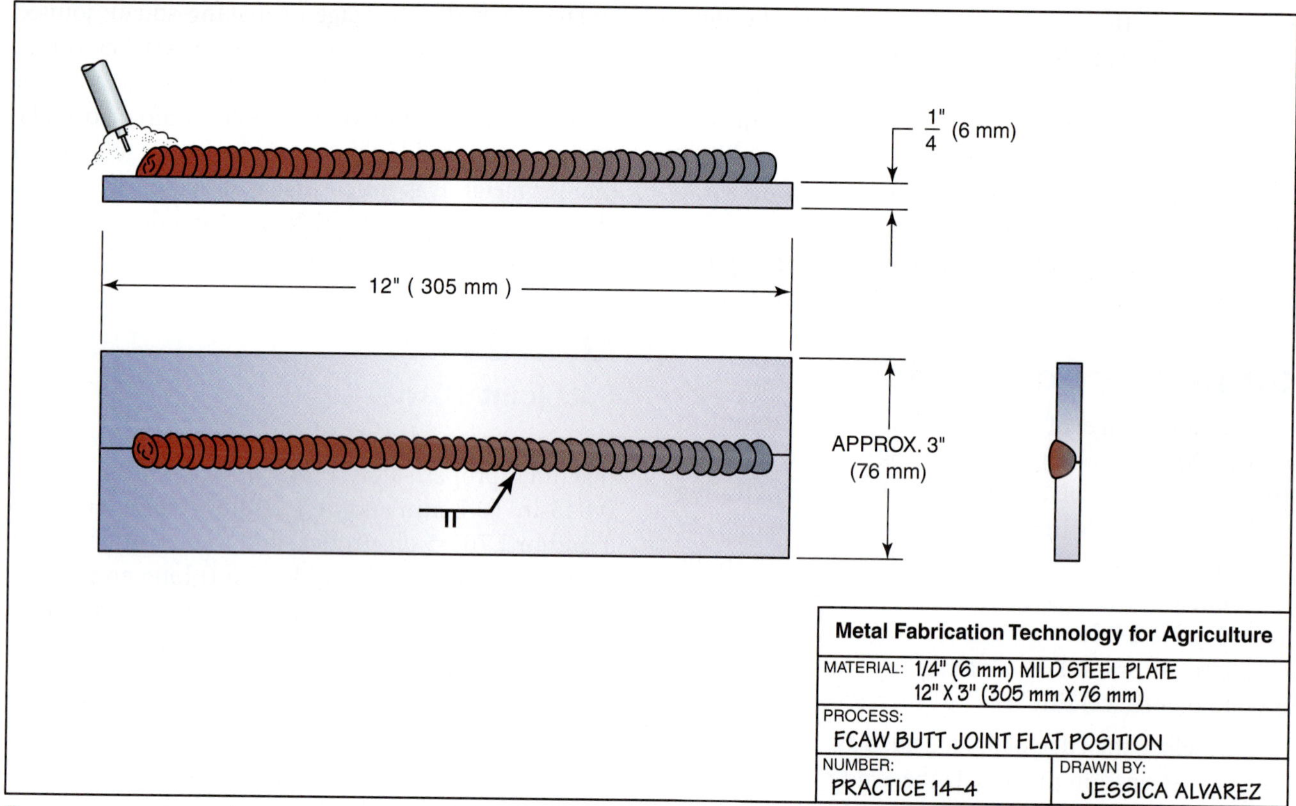

FIGURE 14-12 FCAW butt joint 1/4 in. mild steel flat position.

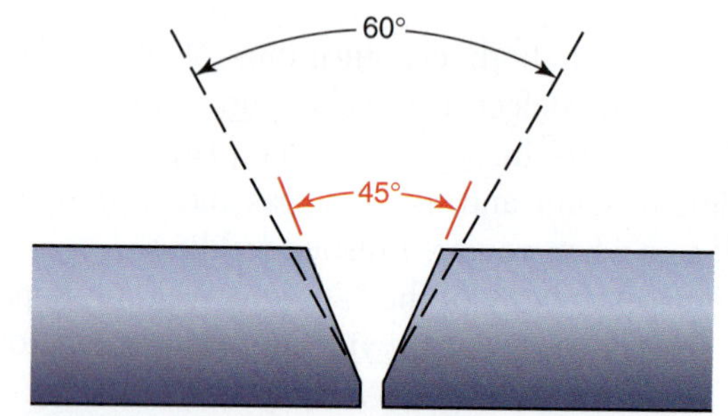

FIGURE 14-13 A smaller groove angle reduces both weld time and filler metal required to make the weld.

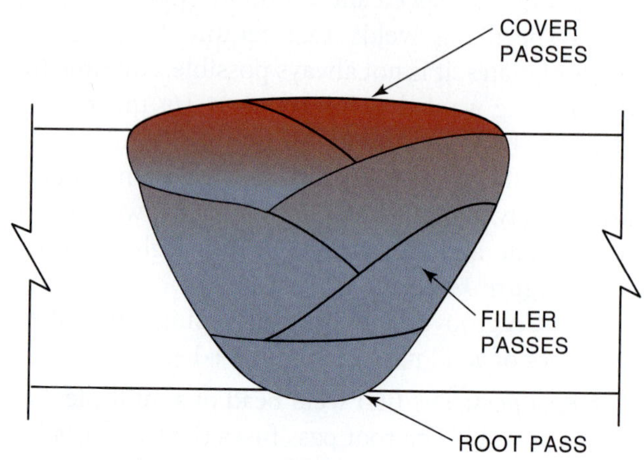

FIGURE 14-14 The three different types of weld passes that make up a weld.

- *Filler pass:* Made after the root pass is completed and used to fill the groove with weld metal. More than one pass is often required.
- *Cover pass:* The last weld pass on a multipass weld. The cover pass may be made with one or more welds. It must be uniform in width, reinforcement, and appearance.

Root Pass A good root pass is needed in order to obtain a sound weld. The root may be either open or closed and made using a backing strip, **Figure 14-15.**

The backing strips are usually made from a piece of 1/4-in. (6-mm) -thick, 1-in. (25-mm) -wide metal that should be 2 in. (50 mm) longer than the base plates. The strip is attached to the plate by tack welds made on the sides of the strip, **Figure 14-16.**

Most production welds do not use backing strips, so they are made as open root welds. Because of the difficulty in controlling FCA weld's root weld face contours, however, open-root joints are often avoided on critical welds. If an open-root weld is needed because of weldment design, the root pass may be put in with an SMA electrode or the root face of the FCA weld can be retouched by grinding and/or back welding.

Care must be taken with any root pass not to have the weld face too convex, **Figure 14-17.** Convex weld

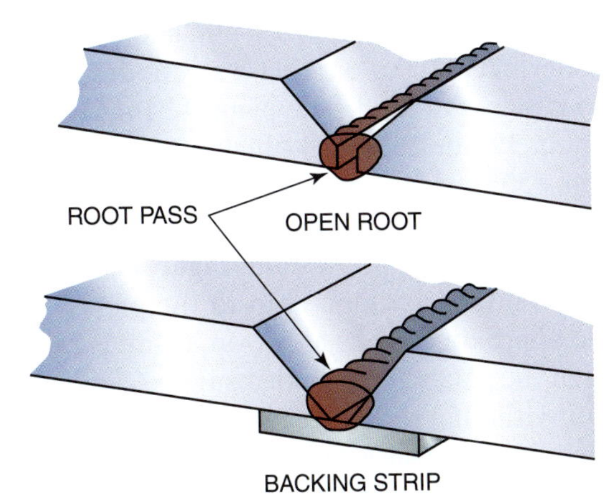

FIGURE 14-15 Root pass maximum deposit 1/4 in. (6mm) thick.

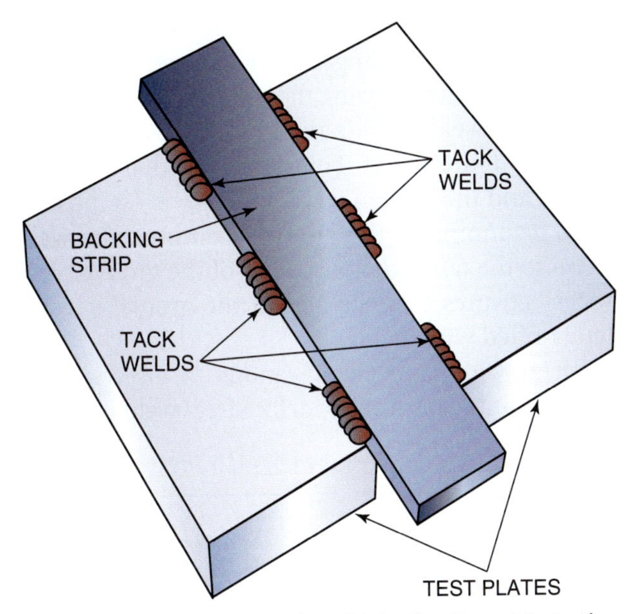

FIGURE 14-16 Securely tack weld the backing strip to the test plates.

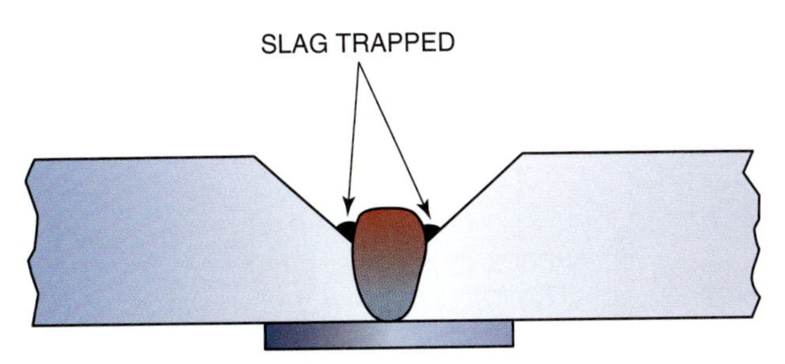

FIGURE 14-17 Slag trapped beside weld bead is hard to remove.

faces tend to trap slag along the tow of the weld. FCA weld slag can be extremely difficult to remove in this area, especially if there is any undercutting. To avoid this, adjust the welding power settings, speed, and weave pattern so that a flat or slightly concave weld face is produced, **Figure 14-18**.

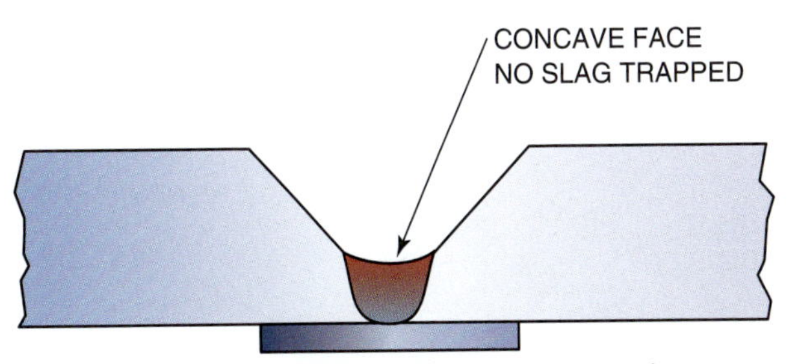

FIGURE 14-18 Flat or concave weld faces are easier to clean off.

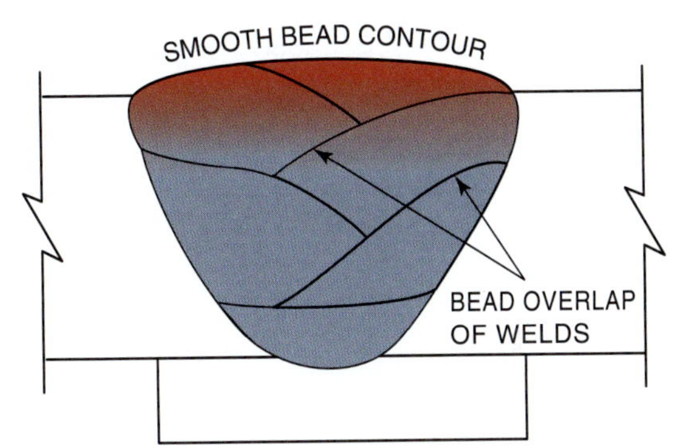

FIGURE 14-19 The surface of a multipass weld should be as smooth as if it were made by one weld.

Filler Pass Filler passes are made with either stringer beads or weave beads for flat or vertically positioned welds, but stringer beads work best for horizontal and overhead-positioned welds. When multiple-pass filler welds are required, each weld bead must overlap the others along the edges. Edges should overlap smoothly enough so that the finished bead is uniform, **Figure 14-19**. Stringer beads usually overlap about 25% to 50%, and weave beads overlap approximately 10% to 25%.

Each weld bead must be cleaned before the next bead is started. The filler pass ends when the groove has been filled to a level just below the plate surface.

Cover Pass The cover pass may or may not simply be a continuation of the weld beads used to make the filler pass(es). The major difference between the filler pass and the cover pass is the weld face importance. Keeping the face and tow of the cover pass uniform in width, reinforcement, and appearance and defect free is essential. Most welds are not tested beyond a visual inspection. For that reason the appearance might be the only factor used for accepting or rejecting welds.

The cover pass must meet a strict visual inspection standard. The visual inspection looks to see that the weld is uniform in width and reinforcement. There should be no arc strikes on the plate other than those on the weld itself. The weld must be free of both incomplete fusion and cracks. The weld must

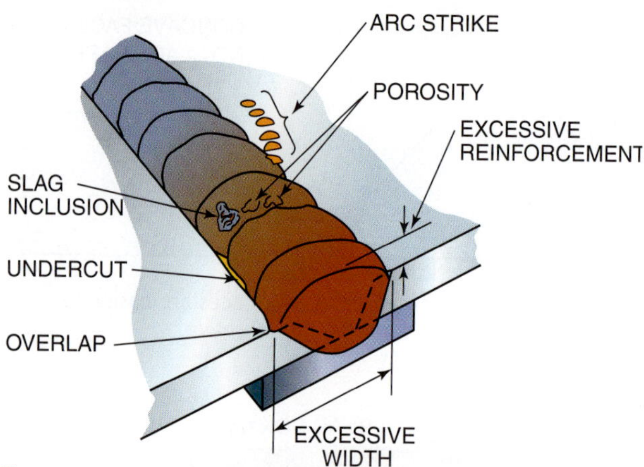

FIGURE 14-20 Common discontinuities found during a visual examination.

be free of overlap, and undercut must not exceed either 10% of the base metal or 1/32 in. (0.8 mm), whichever is less. Reinforcement must have a smooth transition with the base plate and be no higher than 1/8 in. (3 mm), **Figure 14-20.**

PRACTICE 14-5

Butt Joint 1G

Using a properly set up and adjusted FCA welding machine, Table 14-1, proper safety protection, 0.035-in. and/or 0.045-in. (0.9-mm and/or 1.2-mm) -diameter E70T-1 and/or E70T-5 electrodes, and one or more pieces of mild steel plate, 12-in. (305-mm) -long and 3/8-in. (9.5-mm) -thick beveled plate, and a 14-in. (355-mm) -long, 1-in. (25-mm) -wide, and 1/4-in. (6-mm) -thick backing strip, you will make a groove weld in the flat position, **Figure 14-21.**

Tack weld the backing strip to the plates. There should approximately be an 1/8-in. (3-mm) root gap between the plates. The beveled surface can be made with or without a root face, **Figure 14-22.**

Place the test plates in position at a comfortable height and location. Be sure that you have complete and free movement along the full length of the weld joint. It is often a good idea to make a practice pass along the joint with the welding gun without power to make sure nothing will interfere with your making the weld. Be sure the welding cable is free and will not get caught on anything during the weld.

Start the weld outside the groove on the backing strip tab, **Figure 14-23.** This is done so that the arc is smooth and the molten weld pool size is established at the beginning of the groove. Continue the weld out on to the tab at the outer end of the groove. This process ensures that the end of the groove is completely filled with weld.

Repeat with both classifications of electrodes as needed until consistently defect-free welds can be

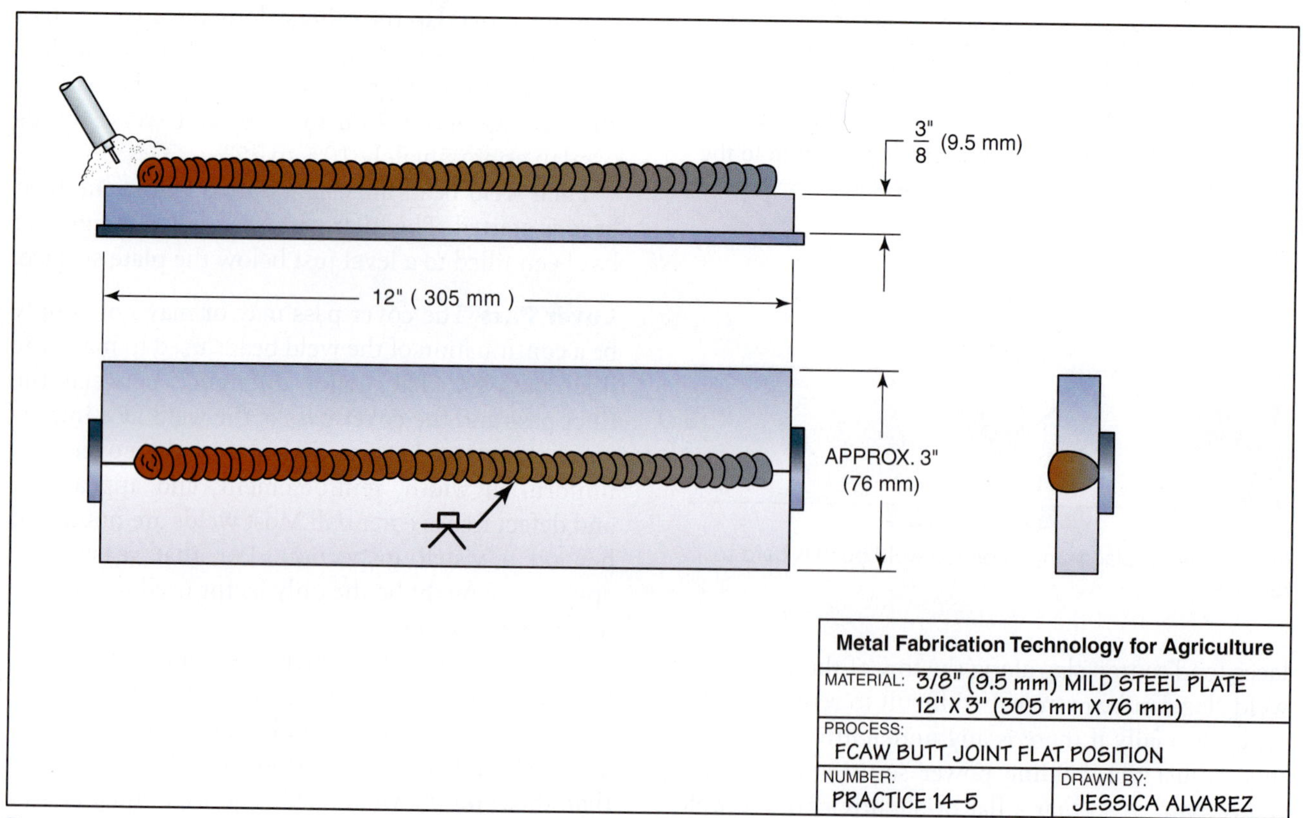

FIGURE 14-21 FCAW butt joint 3/8 in. mild steel flat position.

made. Turn off the welding machine and shielding gas and clean up your work area when you are finished welding.

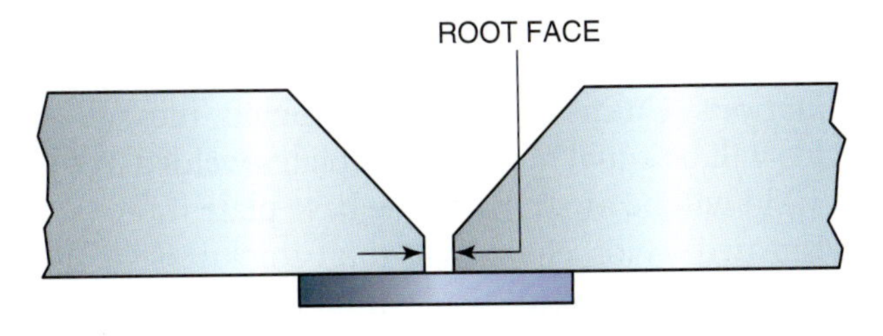

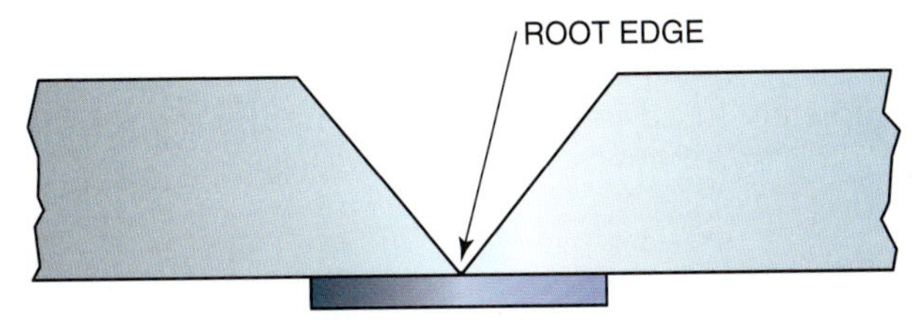

FIGURE 14-22 Groove layout with and without a root face.

FIGURE 14-23 Run-off tabs help control possible underfill or burn-back at the starting and stopping points of a groove weld. Courtesy of Larry Jeffus.

Complete a copy of the "Student Welding Report" listed in Appendix I or provided by your instructor. ◆

Fillet Welds

A fillet weld is the type of weld made on the lap joint and tee joint. It should be built up equal to the thickness of the plate, **Figure 14-24**. On thick plates the fillet must be made up of several passes as with a groove weld. The difference with a fillet weld is that a smooth transition from the plate surface to the weld is required. If this transition is abrupt, it can cause stresses that will weaken the joint.

The lap joint is made by overlapping the edges of the plates. They should be held together tightly before tack welding them together. A small tack weld may be added in the center to prevent distortion during welding, **Figure 14-25**. Chip the tacks before you start to weld.

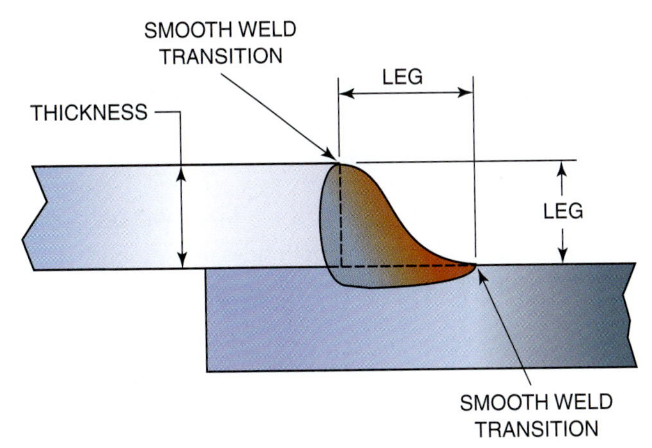

FIGURE 14-24 The legs of a fillet weld should generally be equal to the thickness of the base metal.

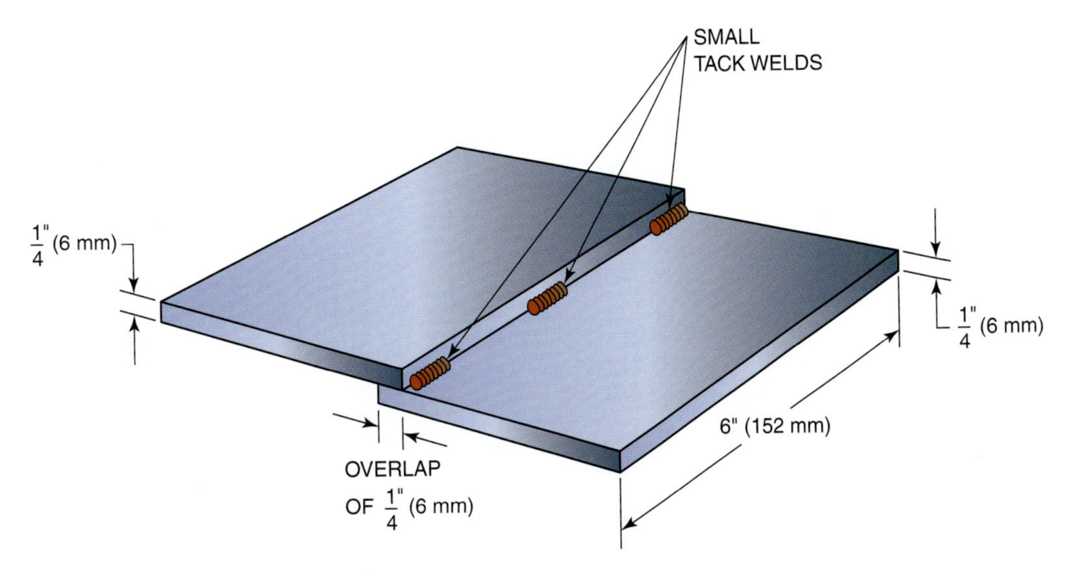

FIGURE 14-25 Tack welding the plates together.

The tee joint is made by tack welding one piece of metal on another piece of metal at a right angle, **Figure 14-26.** After the joint is tack welded together, the slag is chipped from the tack welds. If the slag is not removed, it will cause a slag inclusion in the final weld.

Holding thick plates tightly together on tee joints may cause underbead cracking or lamellar tearing, **Figure 14-27.** On thick plates the weld shrinkage can be great enough to pull the metal apart well below the bead or its heat-affected zone. In production welds cracking can be controlled by not assembling the plates tightly together. The space between the two plates can be set by placing a small wire spacer between them, **Figure 14-28.**

A fillet welded lap or tee joint can be strong if it is welded on both sides, even without having deep penetration, **Figure 14-29.** Some tee joints may be prepared for welding by cutting either a bevel or a J-groove in the vertical plate. This cut is not required for strength but may be necessary because of design limitations. Unless otherwise instructed, most fillet welds will be equal in size to the plates welded. A fillet weld will be as strong as the base plate if the size of the two welds equals the total thickness of the base plate. The weld bead should have a flat or slightly concave appearance to ensure the greatest strength and efficiency, **Figure 14-30.**

The root of fillet welds must be melted to ensure a completely fused joint. A notch along the root of the

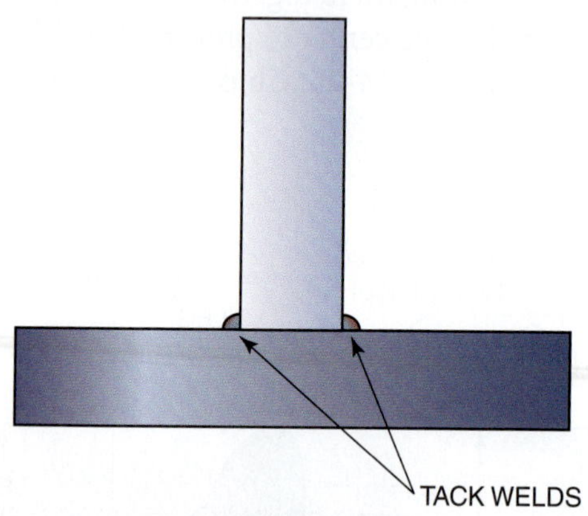

FIGURE 14-26 Tack welding both sides of a tee joint will help keep the tee square for welding.

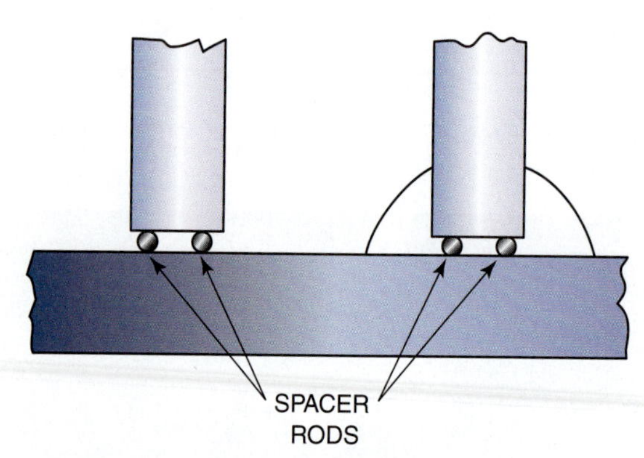

FIGURE 14-28 Base plate cracking can be controlled by placing spacers in the joint before welding.

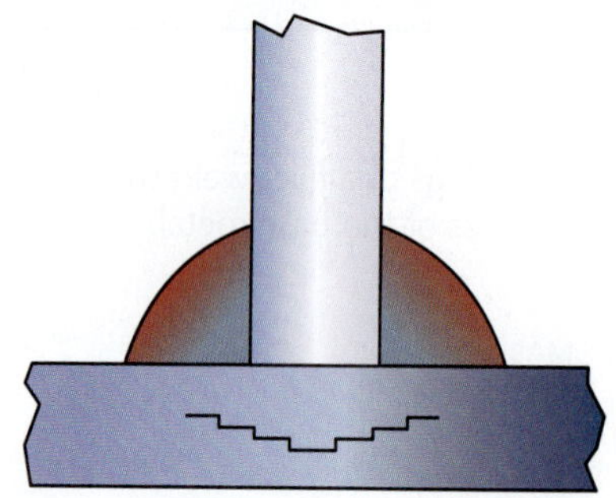

FIGURE 14-27 Underbead cracking or lamellar tearing of the base plate.

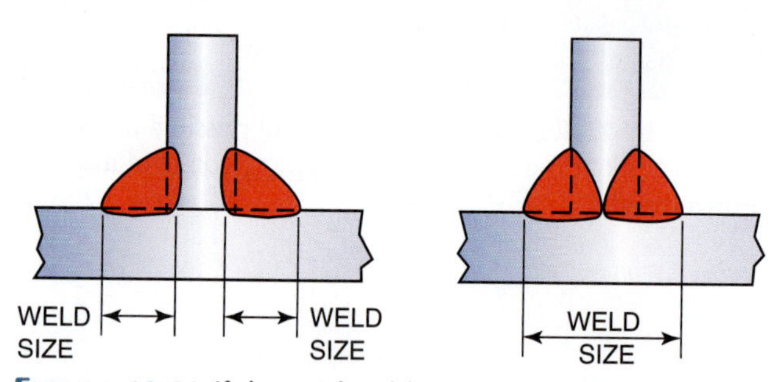

FIGURE 14-29 If the total weld sizes are equal, then both tee joints would have equal strength.

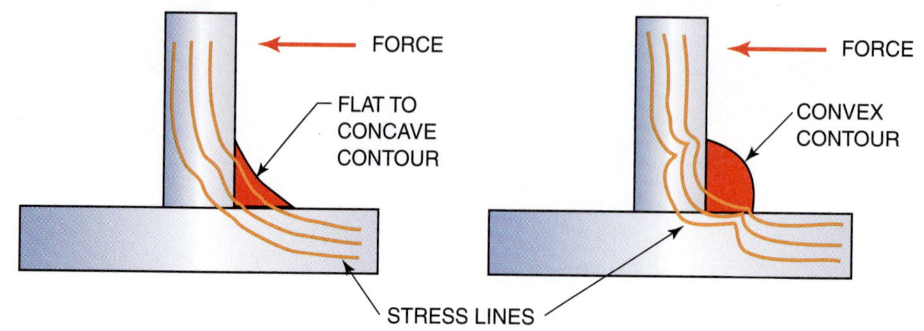

FIGURE 14-30 The stresses are distributed more uniformly through a flat or concave fillet weld.

weld pool is an indication that the root is not being fused together, **Figure 14-31**. To achieve complete root fusion, move the arc to a point as close as possible to the leading edge of the weld pool, **Figure 14-32**. If the arc strikes the unmelted plate ahead of the molten weld pool, it may become erratic, which will increase weld spatter.

PRACTICE 14-6

Lap Joint and Tee Joint 1F

Using a properly set up and adjusted FCA welding machine, proper safety protection, 0.035-in. and/or 0.045-in. (0.9-mm and/or 1.2-mm) -diameter E70T-1 and/or E70T-5 electrodes, and one or more pieces of mild steel plate, 12-in. (305-mm) -long and 3/8-in. (9.5-mm) -thick beveled plate, you will make a fillet weld in the flat position.

Tack weld the pieces of metal together and brace them in position. When making the lap or tee joints in the flat position, the plates must be at a 45° angle so that the surface of the weld will be flat, **Figure 14-33(A)** and **Figure 14-33(B)**. Starting at one end, make a weld along the entire length of the joint.

Repeat each type of joint with both classifications of electrodes as needed until consistently defect-free

FIGURE 14-31 Watch the root of the weld bead to be sure there is complete fusion. Courtesy of Larry Jeffus.

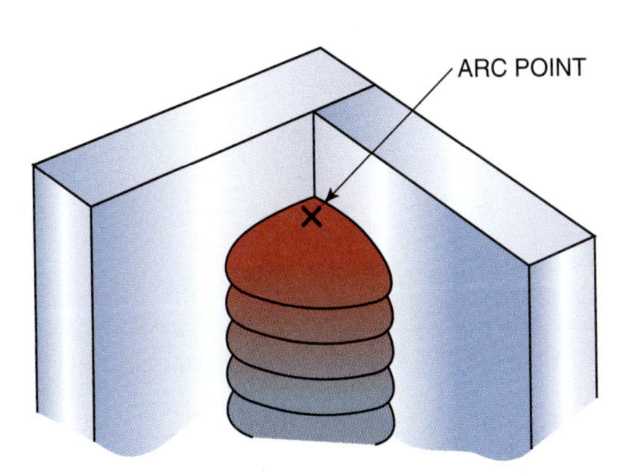

FIGURE 14-32 Moving the arc as close as possible to the leading edge of the weld will provide good root fusion.

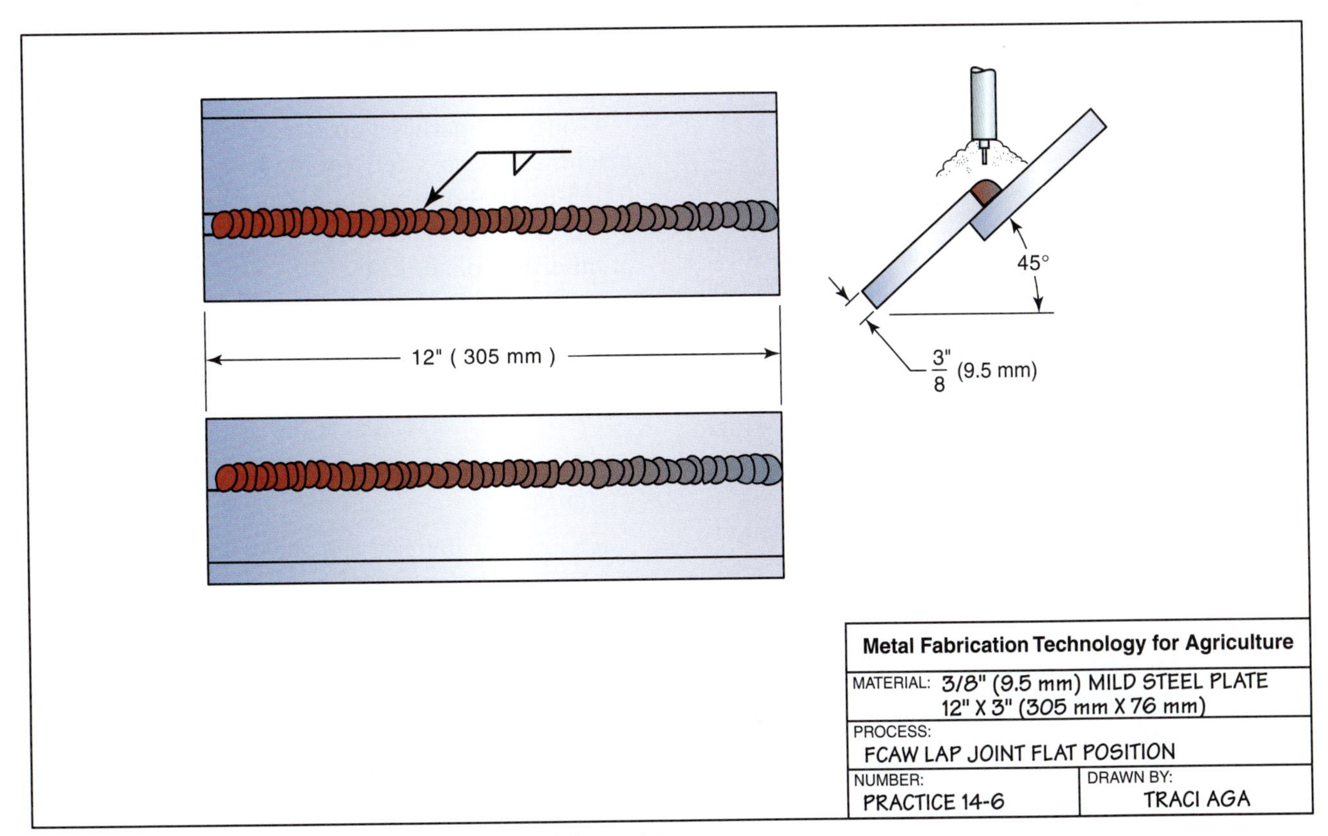

FIGURE 14-33(A) FCAW lap joint 3/8 in. mild steel flat position.

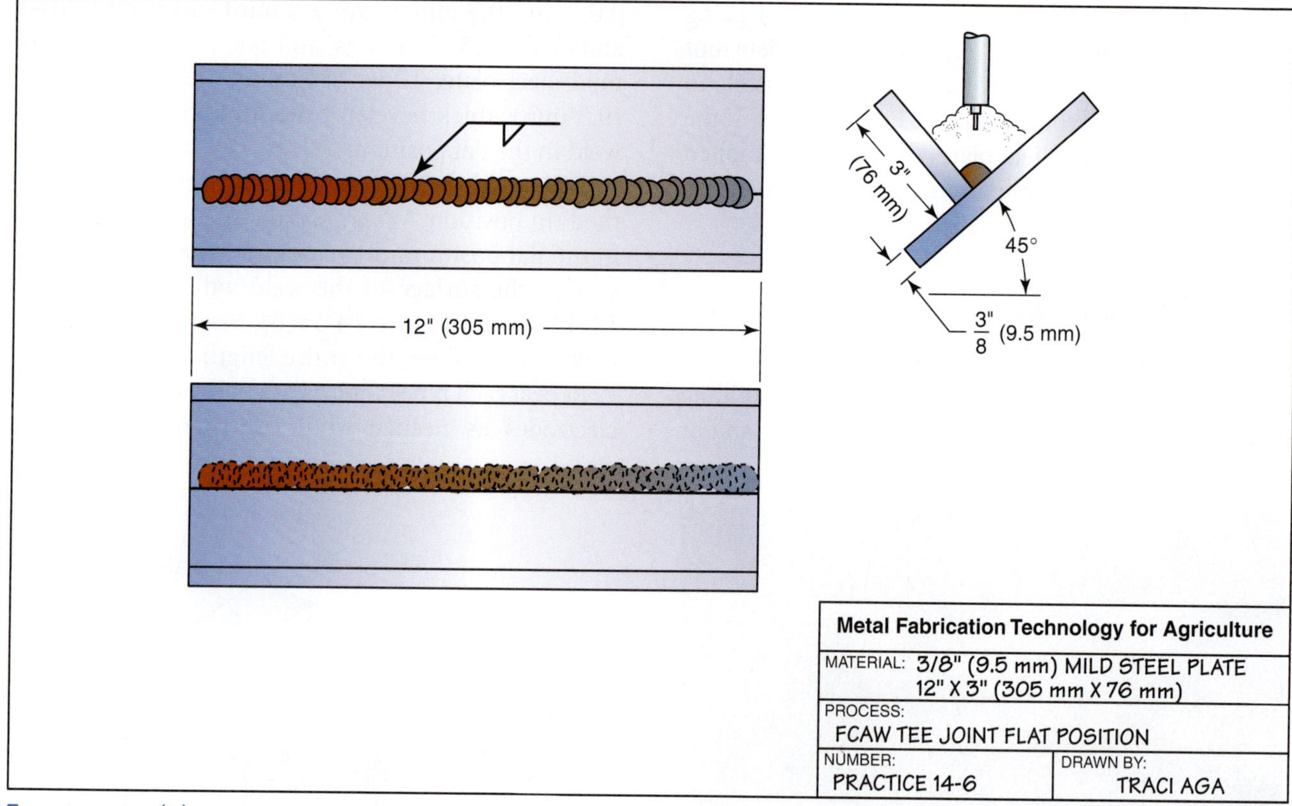

FIGURE 14-33(B) FCAW tee joint 3/8 in. mild steel flat position.

welds can be made. Turn off the welding machine and shielding gas and clean up your work area when you are finished welding.

Complete a copy of the "Student Welding Report" listed in Appendix I or provided by your instructor. ◆

PRACTICE 14-7

Tee Joint 1F

Using a properly set up and adjusted FCA welding machine, proper safety protection, 0.035-in. and/or through 1/16-in. (0.9-mm and/or through 1.6-mm) -diameter E70T-1 and/or E70T-5 electrodes, and one or more pieces of mild steel plate, 7-in. (178-mm) -long and 3/4-in. (19-mm) -thick or thicker beveled plate, you will make fillet weld in the flat position.

Following the same instructions for the assembly and welding procedure outlined in Practice 14-6, repeat each type of joint with both classifications of electrodes as needed until consistently defect-free welds can be made. Turn off the welding machine and shielding gas and clean up your work area when you are finished welding.

Complete a copy of the "Student Welding Report" listed in Appendix I or provided by your instructor. ◆

Vertical Welds

PRACTICE 14-8

Butt Joint at a 45° Vertical Up Angle

Using a properly set up and adjusted FCA welding machine, proper safety protection, 0.035-in. and/or 0.045-in. (0.9-mm and/or 1.2-mm) -diameter E70T-1 and/or E70T-5 electrodes, and one or more pieces of mild steel plate, 12 in. (305 mm) long and 1/4 in. (6 mm) thick or thinner, you will increase the plate angle gradually as you develop skill until you are making satisfactory welds in the vertical up position, **Figure 14-34**.

- Start practicing this weld with the plate at a 45° angle.
- Gradually increase the angle of the plate to vertical as skill is gained in welding this joint. A straight stringer bead or slight zigzag will work well on this joint.
- Establish a molten weld pool in the root of the joint.
- Cool, chip, and inspect the weld for uniformity and defects.

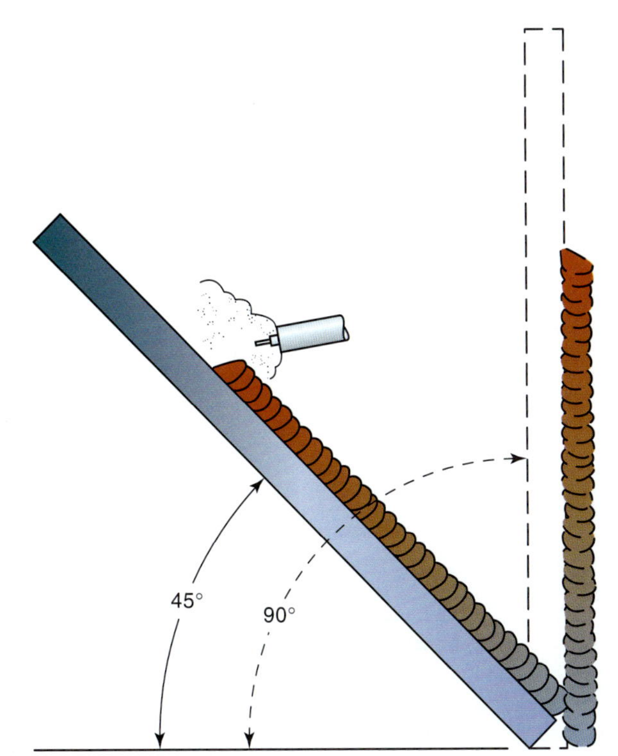

FIGURE 14-34 Start making welds with the plate at a 45° angle. As your skill develops, increase the angle until the plate is vertical.

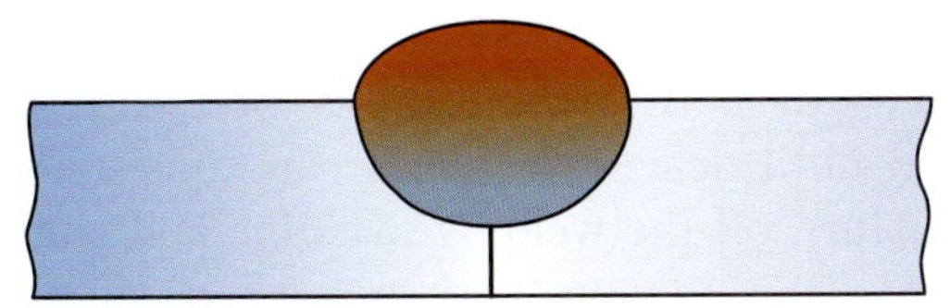

FIGURE 14-35 Low amperage causes too much buildup and not enough penetration.

It is easier to make a quality weld in the vertical up position if both the amperage and voltage are set at the lower end of their ranges. This will make the molten weld pool smaller, less fluid, and easier to control. A problem with lower power settings is that the weld bead often can be very convex, **Figure 14-35**. Faster travel speed and/or slightly wider weave patterns can be used to control the bead shape.

Start at the bottom of the plate and hold the welding gun at a slight upward angle to the plate, **Figure 14-36**. Brace yourself, lower your hood, and begin to weld. Depending on the machine settings and type of electrode used, you will make a weave pattern.

If the molten weld pool is large and fluid (hot), use a C or J weave pattern to allow a longer time for the molten weld pool to cool, **Figure 14-37**. Do not make the weave so long or fast that the electrode is allowed to strike the metal ahead of the molten weld pool. If

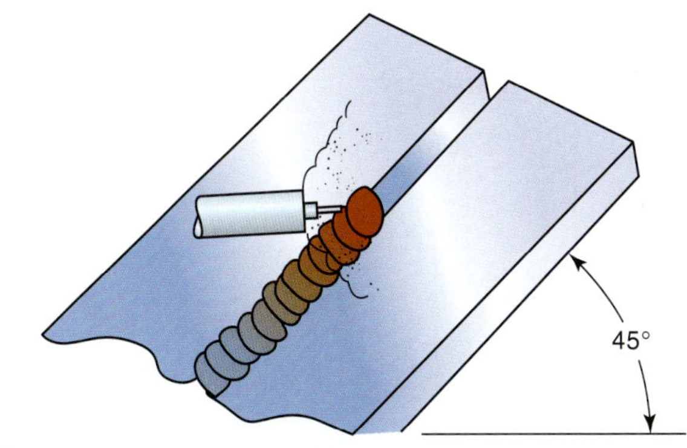

FIGURE 14-36 45° vertical up.

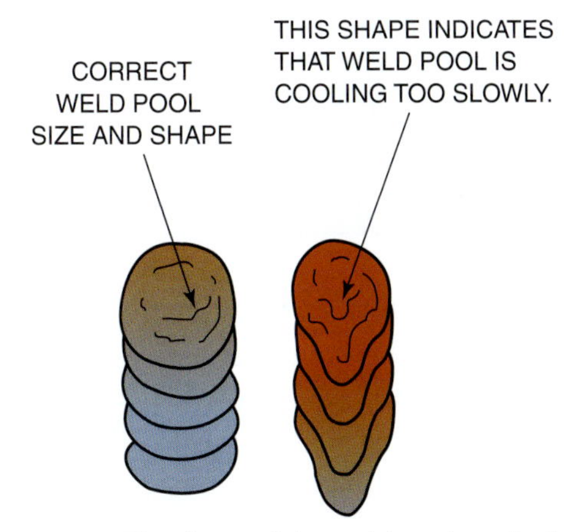

FIGURE 14-37 The shape of the weld pool can indicate the temperature of the surrounding base metal.

this happens, spatter increases and a spot or zone of incomplete fusion may occur.

A weld that is high and has little or no fusion is too "cold." Changing the welding technique will not correct this problem. The welder must stop welding and make the needed adjustments to the power supply or electrode feeder. Continue to weld along the entire 12-in. (305-mm) length of plate.

Repeat welds with both electrodes as needed until defect-free welds can be consistently made vertically in the 1/4-in. (6-mm) -thick plate. Turn off the welding machine and shielding gas and clean up your work area when you are finished welding.

Complete a copy of the "Student Welding Report" listed in Appendix I or provided by your instructor. ◆

PRACTICE 14-9

Butt Joint 3G

Using a properly set up and adjusted FCA welding machine, proper safety protection, 0.035-in. and/or 0.045-in. (0.9-mm and/or 1.2-mm) -diameter E70T-1 and/or E70T-5 electrodes, and one or more pieces

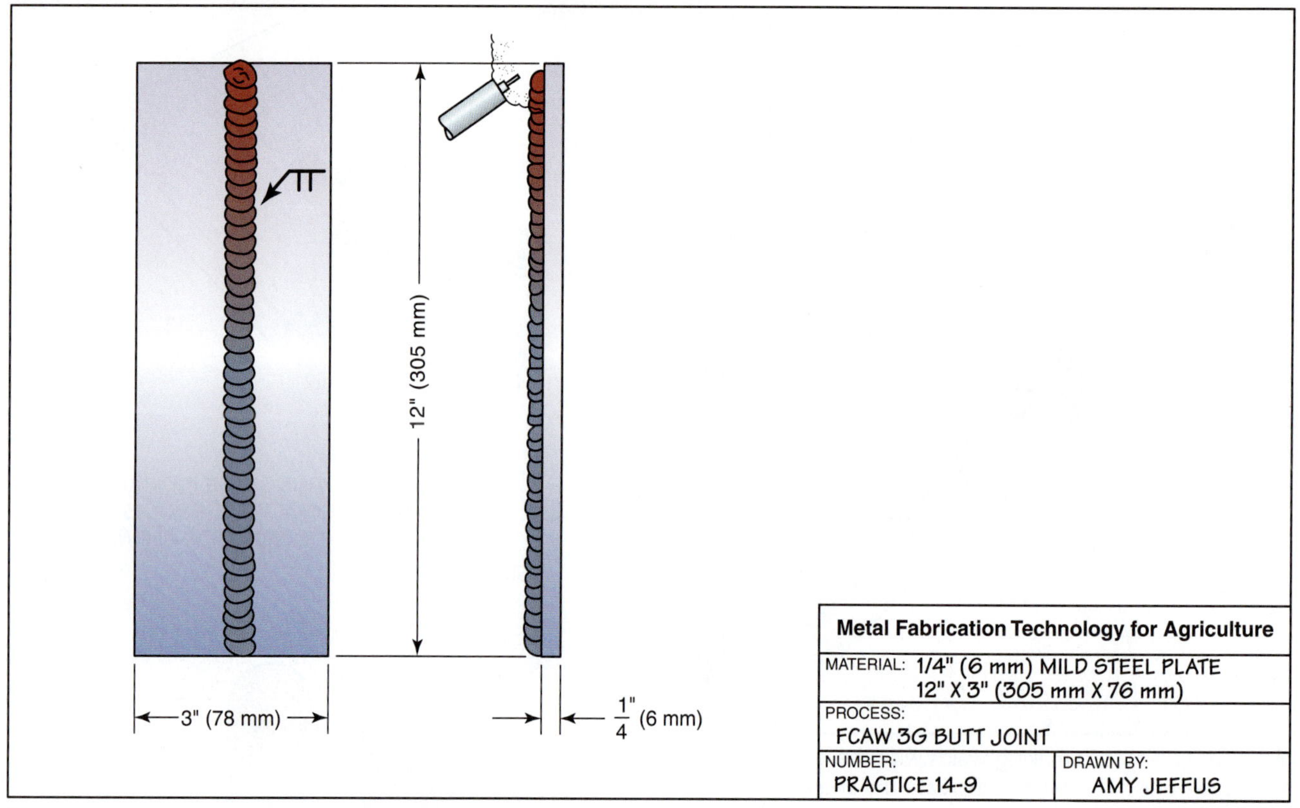

FIGURE 14-38 FCAW 3G butt joint 1/4 in. mild steel.

of mild steel plate, 12 in. (305 mm) long and 1/4 in. (6 mm) thick or thinner, you will make a groove weld in the vertical position, **Figure 14-38.**

Following the same instructions for the assembly and welding procedure outlined in Practice 14-8, repeat with both classifications of electrodes as needed until defect-free welds can be consistently made in the 1/4-in. (6-mm) -thick plate. Turn off the welding machine and shielding gas and clean up your work area when you are finished welding.

Complete a copy of the "Student Welding Report" listed in Appendix I or provided by your instructor. ◆

PRACTICE 14-10

Butt Joint 3G

Using a properly set up and adjusted FCA welding machine, proper safety protection, 0.035-in. and/or 0.045-in. (0.9-mm and/or 1.2-mm) -diameter E70T-1 and/or E70T-5 electrodes, and one or more pieces of mild steel plate, 12-in. (305-mm) -long and 3/8-in. (9.5-mm) -thick beveled plate, and a 14-in. (355-mm) -long, 1-in. (25-mm) -wide, and 1/4-in. (6-mm) -thick backing strip, you will make a groove weld in the vertical position.

Following the same instructions for assembly and welding procedure as outlined in Practice 14-8, repeat with both classifications of electrodes as needed until defect-free welds can consistently be made. Turn off the welding machine and shielding gas and clean up your work area when you are finished welding.

Complete a copy of the "Student Welding Report" listed in Appendix I or provided by your instructor. ◆

Horizontal Welds

PRACTICE 14-11

Lap Joint and Tee Joint 2F

Using a properly set up and adjusted FCA welding machine, proper safety protection, 0.035-in. and/or 0.045-in. (0.9-mm and/or 1.2-mm) -diameter E70T-1 and/or E70T-5 electrodes, and one or more pieces of mild steel plate, 12-in. (305-mm) -long and 3/8-in. (9.5-mm) -thick beveled plate, you will make a fillet weld in the horizontal position, **Figure 14-39(A)** and **Figure 14-39(B).**

The root weld must be kept small so that its contour can be controlled. Too large a root pass can trap slag under overlap along the lower edge of the weld, **Figure 14-40.** Clean each pass thoroughly before the weld bead is started. Follow the weld bead sequence shown

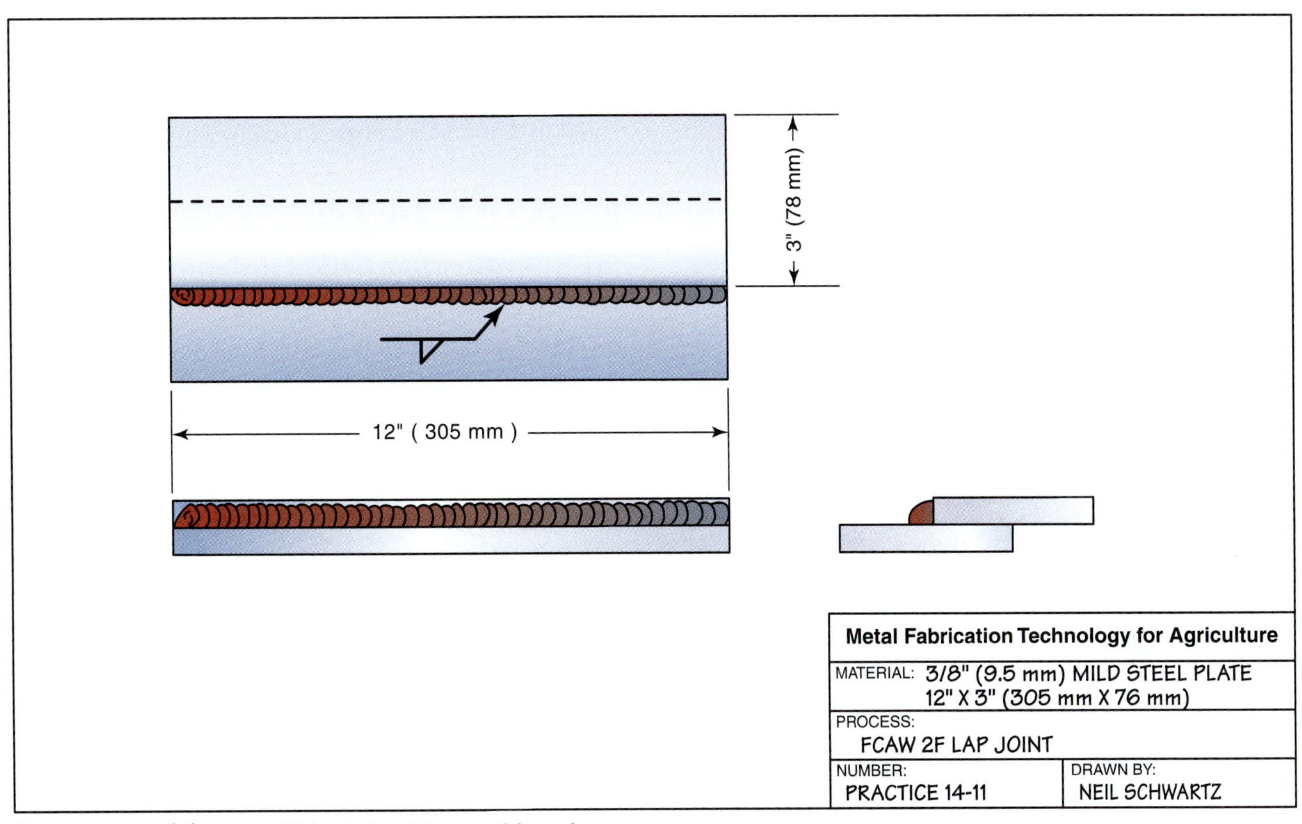

FIGURE 14-39(A) FCAW 2F lap joint 3/8 in. mild steel.

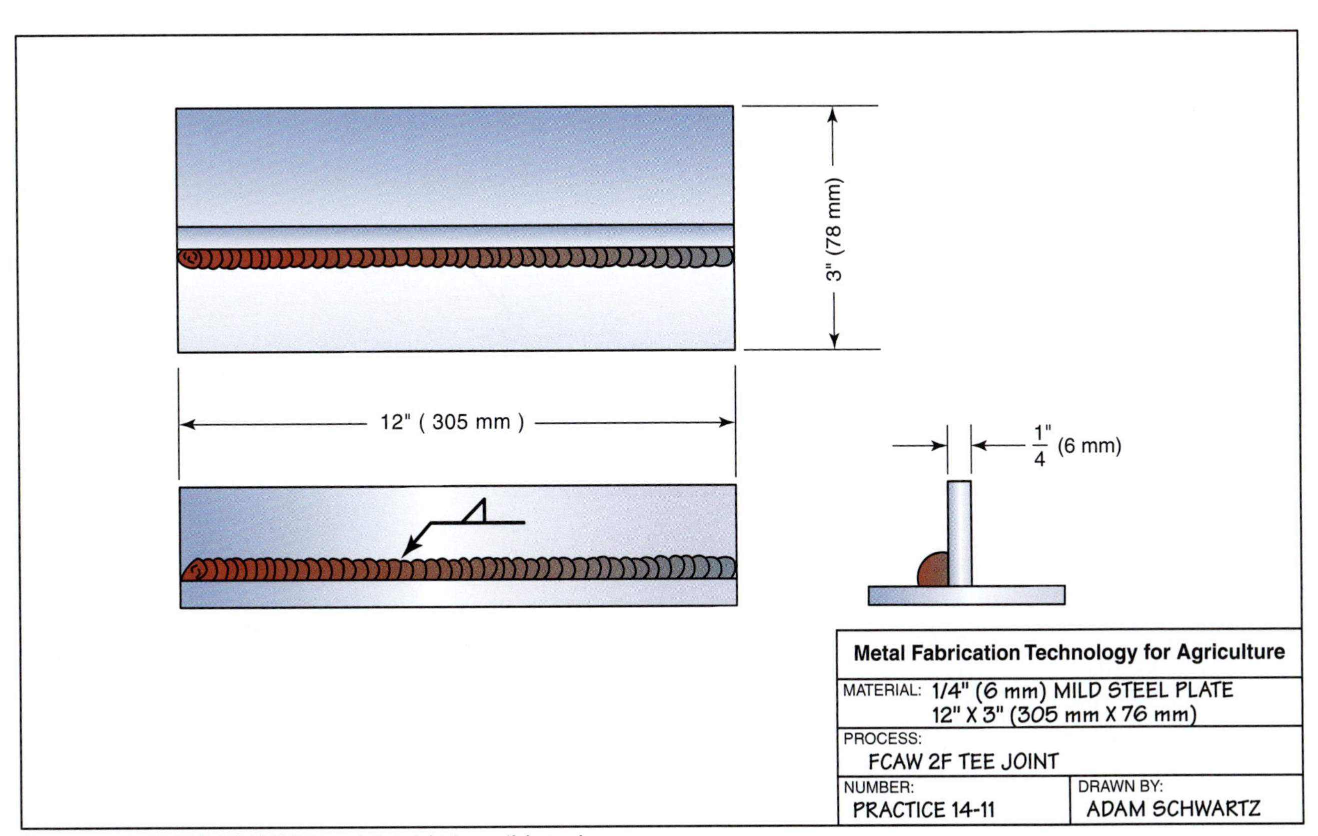

FIGURE 14-39(B) FCAW 2F tee joint 1/4 in. mild steel.

in **Figure 14-41**. Keeping all of the weld beads small will help control their contour.

Repeat each type of joint with both classifications of electrodes as needed until defect-free welds can consistently be made. Turn off the welding machine and shielding gas and clean up your work area when you are finished welding.

Complete a copy of the "Student Welding Report" listed in Appendix I or provided by your instructor. ◆

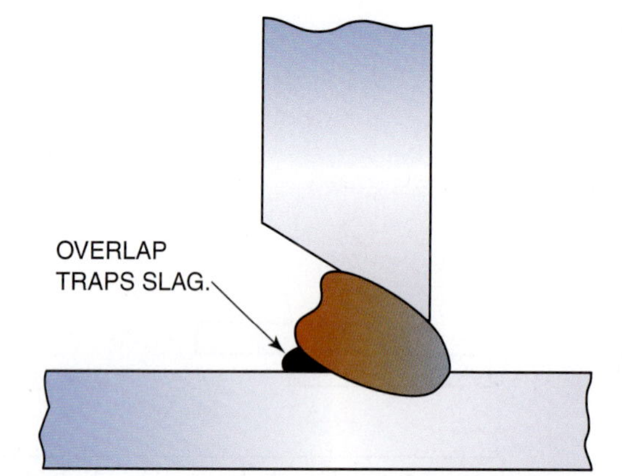

FIGURE 14-40 Slag can be trapped along the side of the root pass.

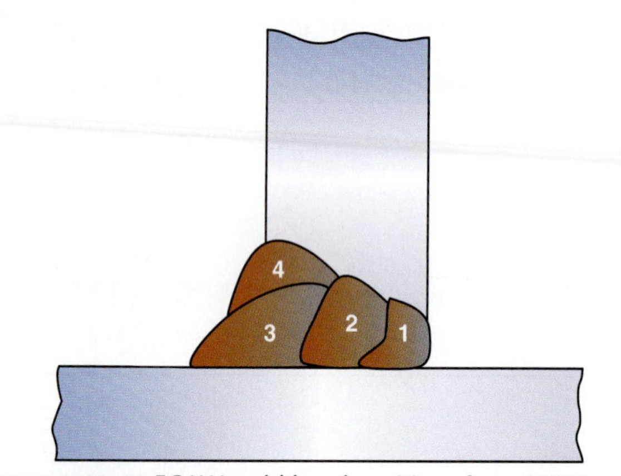

FIGURE 14-41 FCAW weld bead positions for a 100% penetration grooved tee joint.

PRACTICE 14-12

Stringer Bead at a 45° Horizontal Angle

Using a properly set up and adjusted FCA welding machine, proper safety protection, 0.035-in. and/or 0.045-in. (0.9-mm and/or 1.2-mm) -diameter E70T-1 and/or E70T-5 electrodes, and one or more pieces of mild steel plate, 12 in. (305 mm) long and 1/4 in. (6 mm) thick or thinner, you will increase the plate angle gradually as you develop skill until you are making satisfactory horizontal welds across the vertical face of the plate, **Figure 14-42.**

Repeat the weld using each electrode classification until welds using both electrodes can be made horizontally with uniform bead contours when the plates are vertical. Turn off the welding machine and shielding gas and clean up your work area when you are finished welding.

Complete a copy of the "Student Welding Report" listed in Appendix I or provided by your instructor. ◆

PRACTICE 14-13

Butt Joint 2G

Using a properly set up and adjusted FCA welding machine, proper safety protection, 0.035-in. and/or 0.045-in. (0.9-mm and/or 1.2-mm) -diameter E70T-1 and/or E70T-5 electrodes, and one or more pieces of mild steel plate, 12 in. (305 mm) long and 1/4 in.

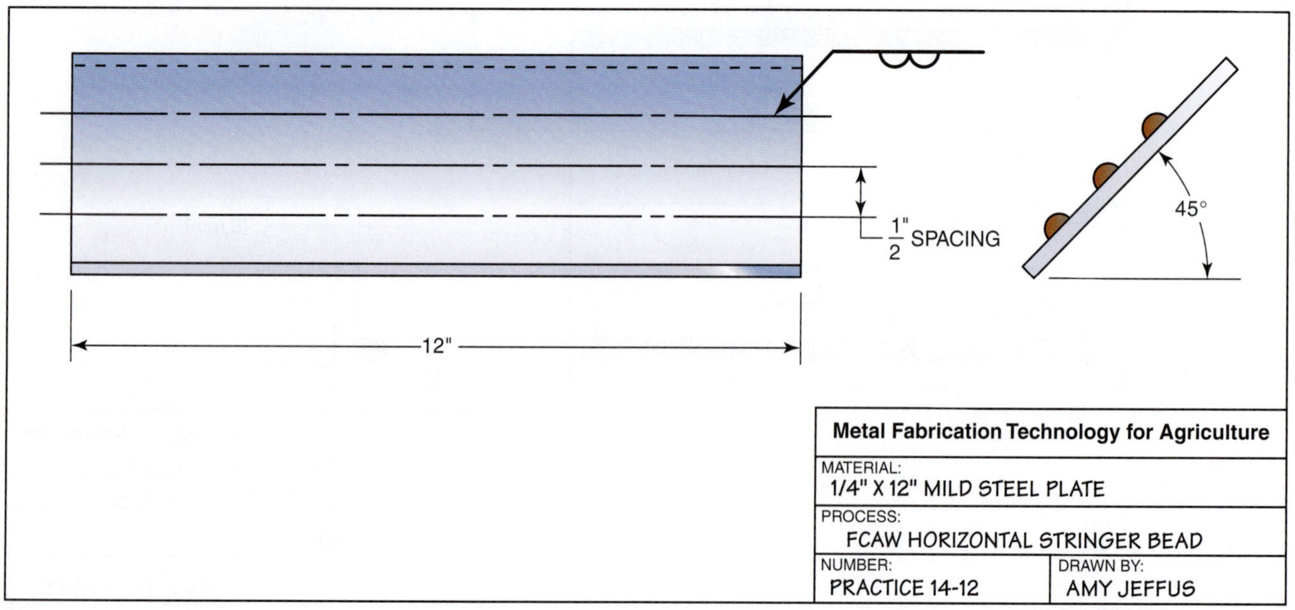

FIGURE 14-42 FCAW horizontal stringer bead.

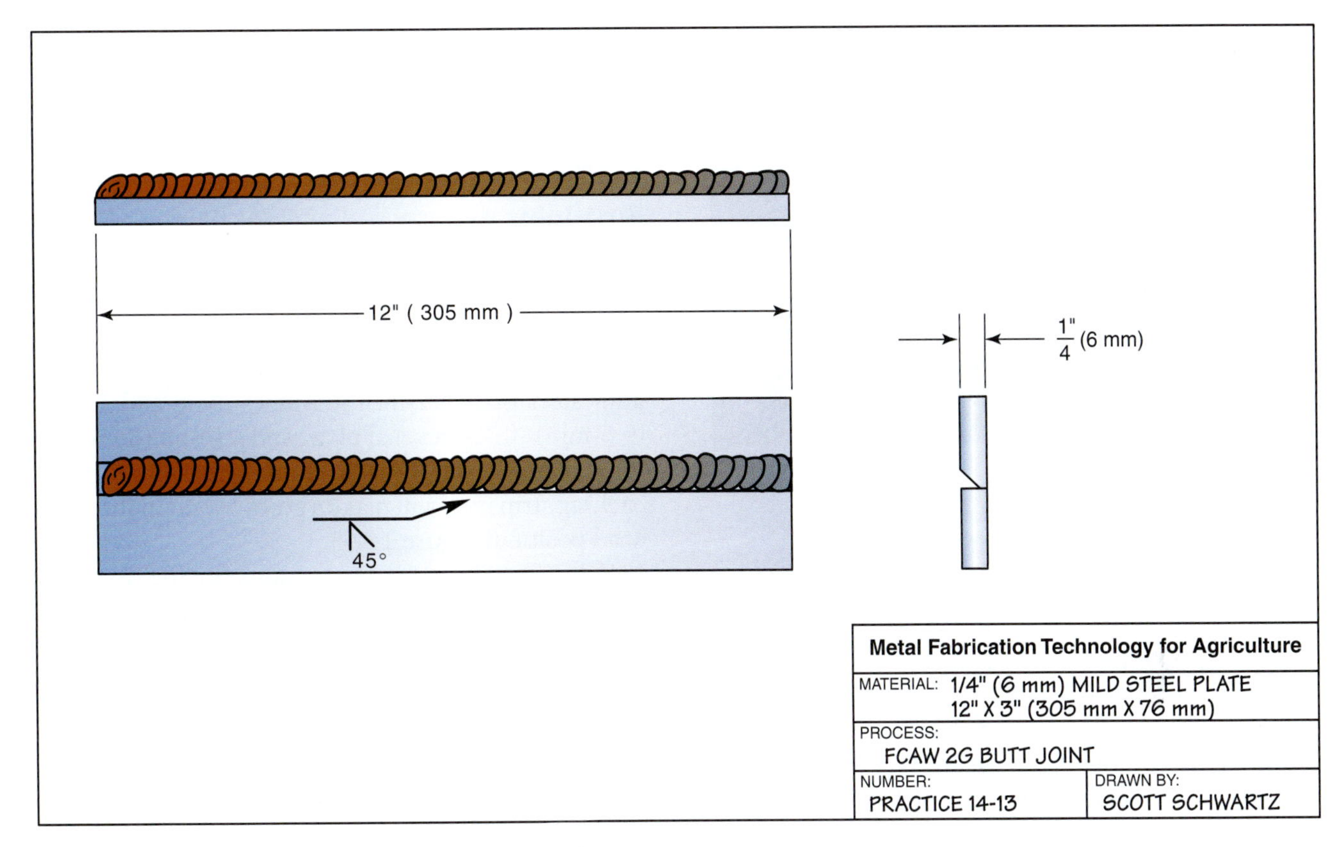

FIGURE 14-43 FCAW 2G butt joint 1/4 in. mild steel.

(6 mm) thick or thinner, you will make a groove weld in the horizontal position, **Figure 14-43.**

Following the same instructions for the assembly and welding procedure outlined in Practice 14-12, repeat with both classifications of electrodes as needed until defect-free welds can consistently be made in the 1/4-in. (6-mm) -thick plate. Turn off the welding machine and shielding gas and clean up your work area when you are finished welding.

Complete a copy of the "Student Welding Report" listed in Appendix I or provided by your instructor. ◆

PRACTICE 14-14

Butt Joint 2G

Using a properly set up and adjusted FCA welding machine, proper safety protection, 0.035-in. and/or 0.045-in. (0.9-mm and/or 1.2-mm) -diameter E70T-1 and/or E70T-5 electrodes, and one or more pieces of mild steel plate, 12-in. (305-mm) -long and 3/8-in. (9.5-mm) -thick beveled plate, and a 14-in. (355-mm) -long, 1-in. (25-mm) -wide, and 1/4-in. (6-mm) -thick backing strip, you will make a groove weld in the horizontal position.

Following the same instructions for the assembly and welding procedure outlined in Practice 14-12, repeat with both classifications of electrodes as needed until defect-free welds can consistently be made. Turn off the welding machine and shielding gas and clean up your work area when you are finished welding.

Complete a copy of the "Student Welding Report" listed in Appendix I or provided by your instructor. ◆

Overhead-position Welds

PRACTICE 14-15

Butt Joint 4G

Using a properly set up and adjusted FCA welding machine, proper safety protection, 0.035-in. and/or 0.045-in. (0.9-mm and/or 1.2-mm) -diameter E70T-1 and/or E70T-5 electrodes, and one or more pieces of mild steel plate, 12 in. (305 mm) long and 1/4 in. (6 mm) thick or thinner, you will make a groove weld in the overhead position.

The molten weld pool should be kept as small as possible for easier control. A small molten weld pool can be achieved by using lower current, faster traveling settings.

Lower current settings require closer control of gun manipulation to ensure that the electrode is fed into the molten weld pool just behind the leading

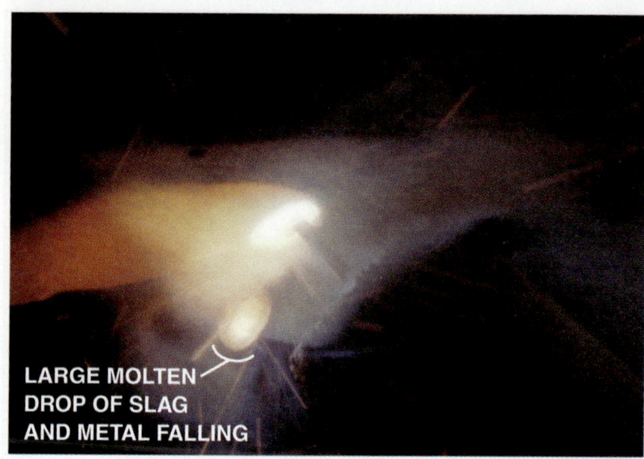

FIGURE 14-44 Hold the gun so that weld spatter will not fall onto the gun. Courtesy of Larry Jeffus.

edge. The low power will cause overlap and more spatter if this electrode-to-molten weld pool contact position is not closely maintained.

Faster travel speeds allow the welder to maintain a high production rate even if multiple passes are required to complete the weld. Weld penetration into the base metal at the start of the bead can be obtained by using a slow start or quickly reversing the weld direction. Both the slow start and reversal of weld direction put more heat into the weld start to increase penetration. The higher speed also reduces the amount of weld distortion by reducing the amount of time that heat is applied to a joint.

When welding overhead, extra personal protection is required to reduce the danger of burns. Leather sleeves or leather jackets should be worn.

Much of the spatter created during overhead welding falls into or on the nozzle and contact tube. The contact tube may short out to the gas nozzle. The shorted gas nozzle may arc to the work, causing damage both to the nozzle and to the plate. To control the amount of spatter, a longer stickout and/or a sharper gun-to-plate angle is required to allow most of the spatter to fall clear of the gun or nozzle, **Figure 14-44.**

Make several short weld beads using various techniques to establish the method that is most successful and most comfortable for you. After each weld, stop and evaluate it before making a change. When you have decided on the technique to be used, make a welded stringer bead that is 12 in. (305 mm) long.

Repeat with both classifications of electrodes as needed until defect-free welds can consistently be made in the 1/4-in. (6-mm) -thick plate. Turn off the welding machine and shielding gas and clean up your work area when you are finished welding.

Complete a copy of the "Student Welding Report" listed in Appendix I or provided by your instructor. ◆

PRACTICE 14-16

Butt Joint 4G

Using a properly set up and adjusted FCA welding machine, proper safety protection, 0.035-in. and/or 0.045-in. (0.9-mm and/or 1.2-mm) -diameter E70T-1 and/or E70T-5 electrodes, and one or more pieces of mild steel plate, 12-in. (305-mm) -long and 3/8-in. (9.5-mm) -thick beveled plate, and a 14-in. (355-mm) -long, 1-in. (25-mm) -wide, and 1/4-in. (6-mm) -thick backing strip, you will make a groove weld in the overhead position, **Figure 14-45.**

Following the same instructions for the assembly and welding procedure outlined in Practice 14-15, repeat with both classifications of electrodes as needed until defect-free welds can consistently be made. Turn off the welding machine and shielding gas and clean up your work area when you are finished welding.

Complete a copy of the "Student Welding Report" listed in Appendix I or provided by your instructor. ◆

PRACTICE 14-17

Lap Joint and Tee Joint 4F

Using a properly set up and adjusted FCA welding machine, proper safety protection, 0.035-in. and/or 0.045-in. (0.9-mm and/or 1.2-mm) -diameter E70T-1 and/or E70T-5 electrodes, and one or more pieces of mild steel plate, 12-in. (305-mm) -long and 3/8-in. (9.5-mm) -thick beveled plate, you will make a fillet weld in the overhead position, **Figure 14-46(A)** and **Figure 14-46(B).**

Following the same instructions for the assembly and welding procedure outlined in Practice 14-15, repeat each type of joint with both classifications of electrodes as needed until defect-free welds can consistently be made. Turn off the welding machine and shielding gas and clean up your work area when you are finished welding.

Complete a copy of the "Student Welding Report" listed in Appendix I or provided by your instructor. ◆

Thin-gauge Sheet Metal Welding

The introduction of small electrode diameters has allowed FCA welding to be used on thin sheet metal. Usually these welds will be a fillet type. Fillet welds are the easiest weld to make on thin stock. An effort

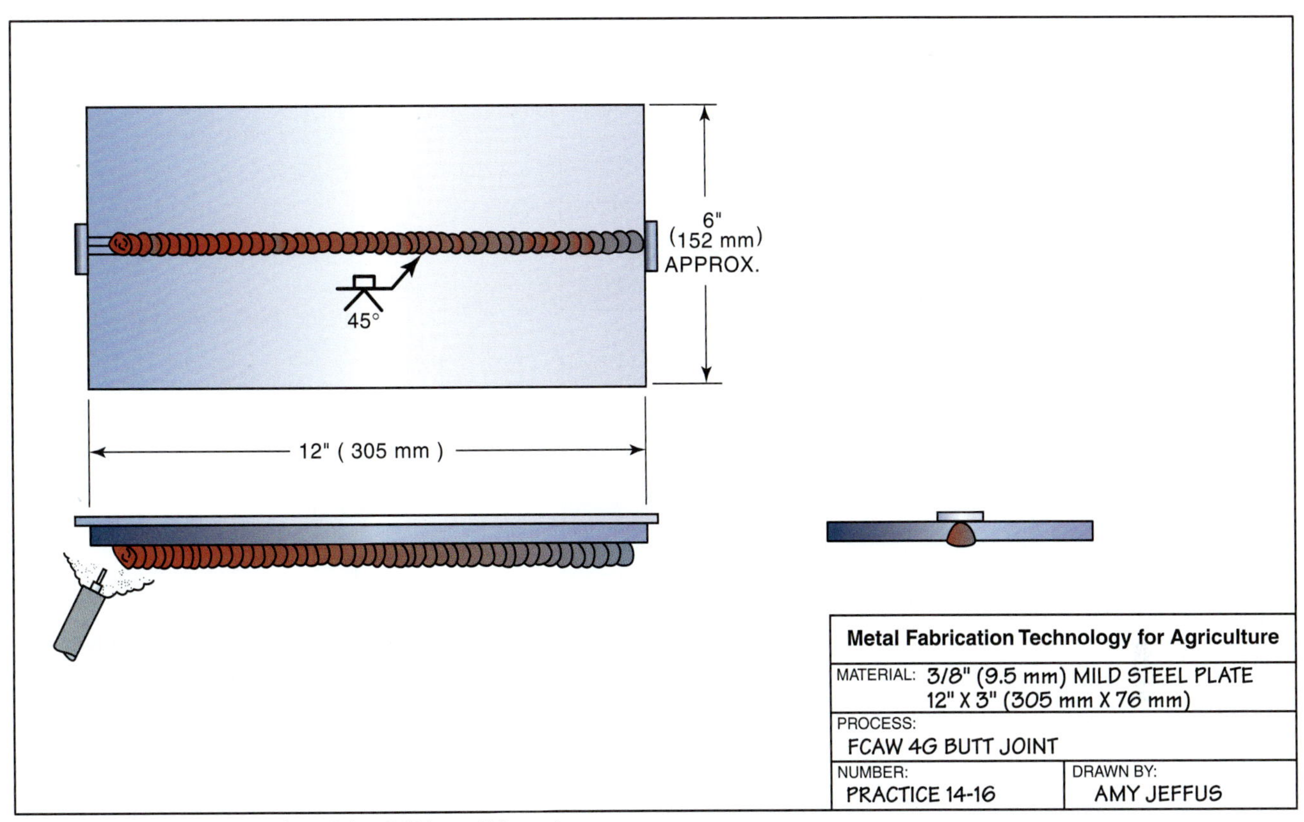

FIGURE 14-45 FCAW 4G butt joint 3/8 in. mild steel.

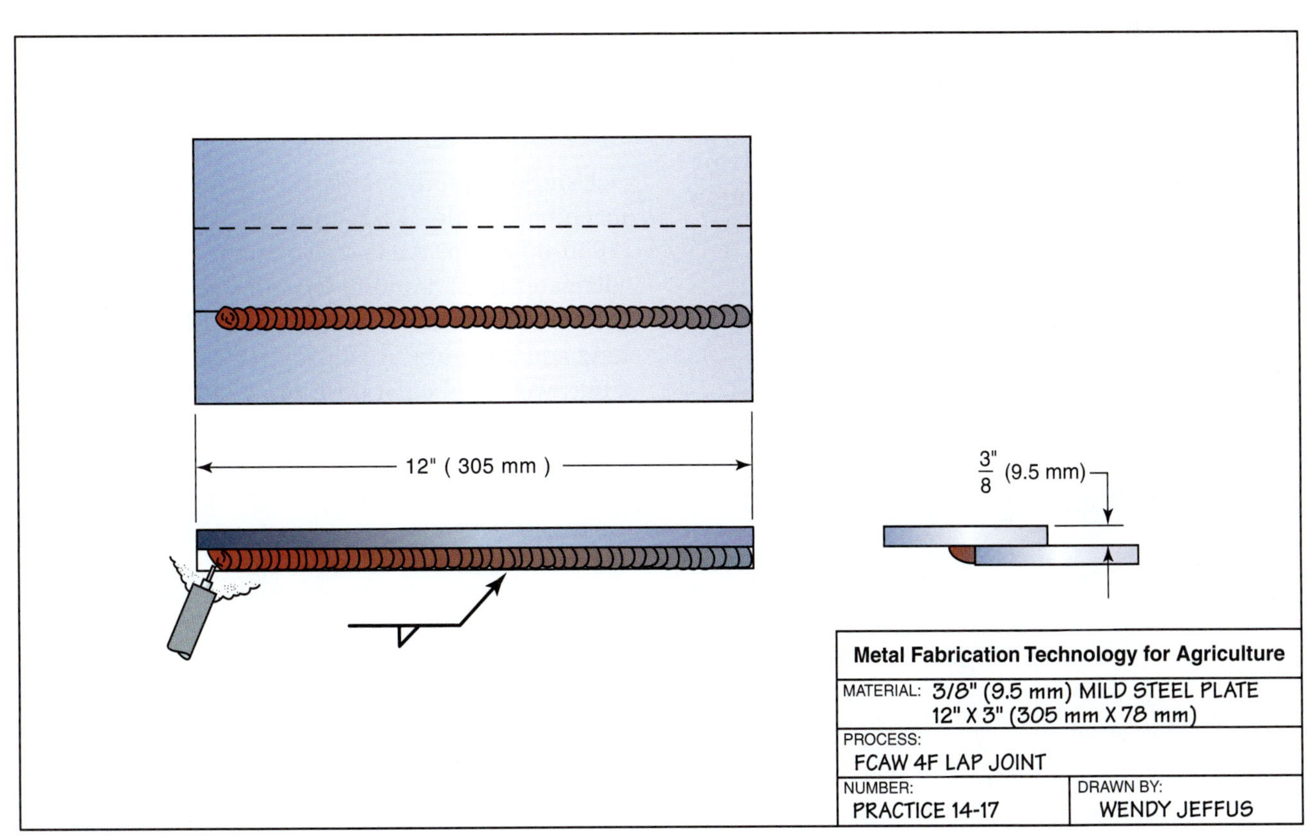

FIGURE 14-46(A) FCAW 4F lap joint 3/8 in. mild steel.

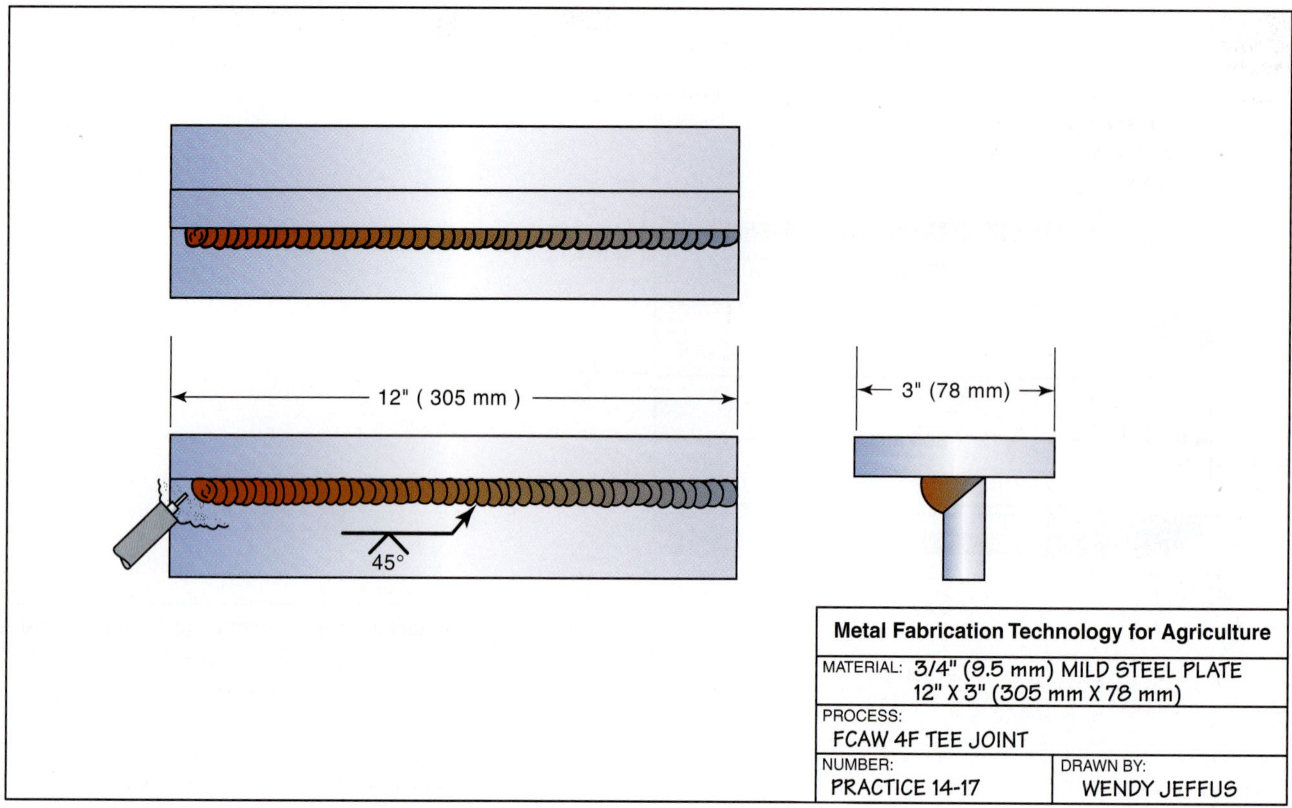

FIGURE 14-46(B) FCAW 4F tee joint 3/4 in. mild steel.

should be taken when possible to design the weld so it is not a butt-type joint. A common use for FCA welding on thin stock is to join it to a thicker member, **Figure 14-47**. This type of weld is used to put panels in frames.

The following practices include some butt-type joints. You will find that the vertical down welds are the easiest ones to make. If it is possible to position the weldment, production speeds can be increased if butt joints are required.

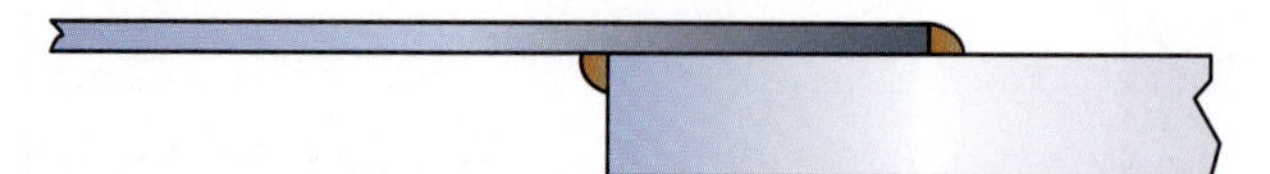

FIGURE 14-47 FCA welding thin to thick metal.

PRACTICE 14-18

Butt Joint 1G

Using a properly set up and adjusted FCA welding machine, **Table 14-2**, proper safety protection, 0.030-in. and/or 0.035-in. (0.8-mm and/or 0.9-mm) -diameter E70T-1 and/or E70T-5 electrodes, and one or more pieces of mild steel sheet, 12 in. (305 mm) long and 16-gauge to 18-gauge thick, you will make a butt weld in the flat position, **Figure 14-48**.

Electrode		Welding Power			Shielding Gas		Base Metal	
Type	Size	Amps	Wire Feed Speed IPM (cm/min)	Volts	Type	Flow	Type	Thick
E70T-1	0.030 in. (0.8 mm)	40 to 145	90 to 340 (228 to 864)	20 to 27	None	n/a	Low-carbon steel	16 gauge to 18 gauge
E70T-1	0.035 in. (0.9 mm)	130 to 200	288 to 576 (732 to 1,463)	20 to 28	None	n/a	Low-carbon steel	16 gauge to 18 gauge
E70T-5	0.035 in. (0.9 mm)	90 to 200	190 to 576 (483 to 1,463)	16 to 29	57% argon 25% CO_2	35 cfh	Low-carbon steel	16 gauge to 18 gauge

TABLE 14-2 FCA Welding Parameters for Use if Specific Settings Are Unavailable from Electrode Manufacturer.

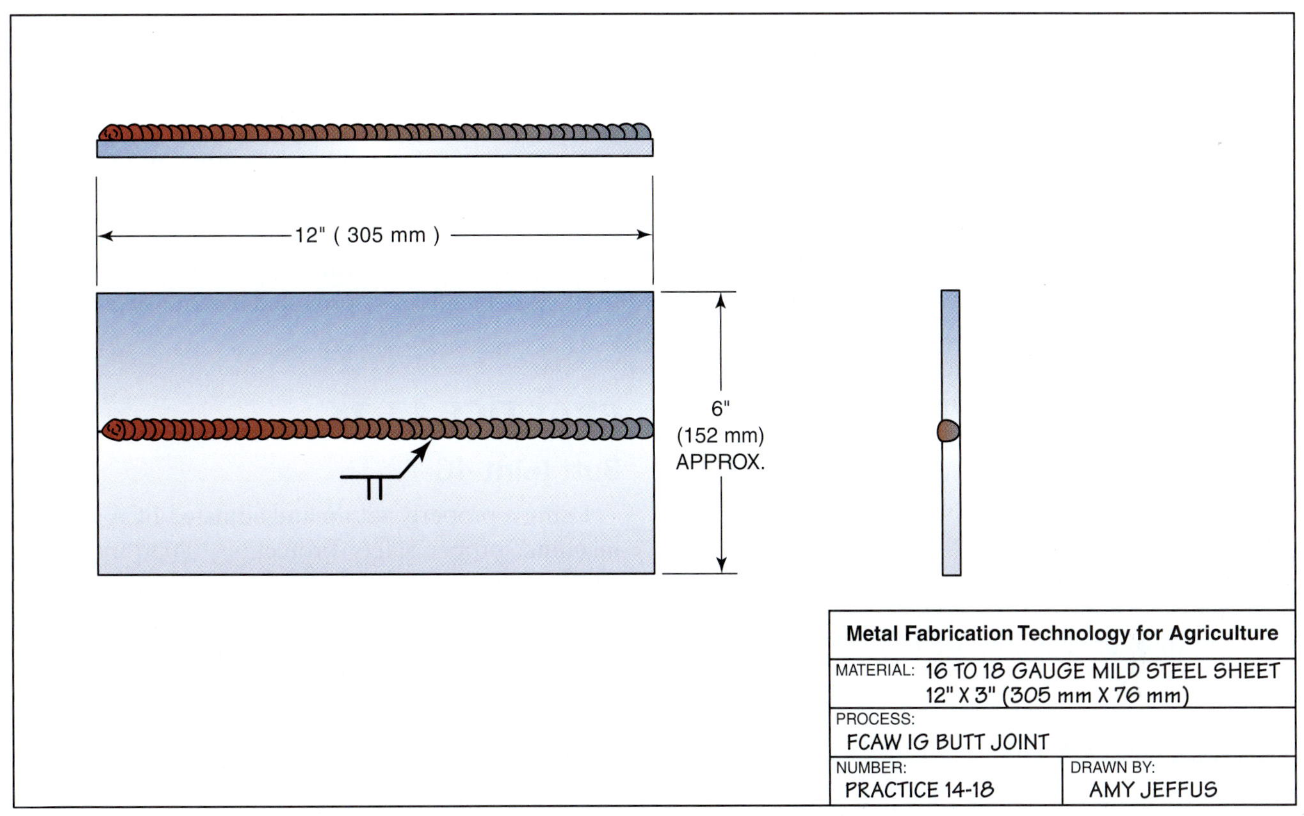

FIGURE 14-48 FCAW 1G butt joint 16- to 18-gauge mild steel.

Do not leave a root opening for these welds. Even the slightest opening will result in a burn-through. If a burn-through occurs, the welder can be pulsed off and on so that the hole can be filled. This process will leave a larger than usual buildup. Excessive buildup could be ground off if necessary as part of the postweld cleanup.

Repeat with both classifications of electrodes as needed until defect-free welds can consistently be made. Turn off the welding machine and shielding gas and clean up your work area when you are finished welding.

Complete a copy of the "Student Welding Report" listed in Appendix I provided by your instructor. ◆

PRACTICE 14-19

Lap Joint and Tee Joint 1F

Using a properly set up and adjusted FCA welding machine, proper safety protection, 0.030-in. and/or 0.035-in. (0.8-mm and/or 0.9-mm) -diameter E70T-1 and/or E70T-5 electrodes, and one or more pieces of mild steel sheet, 12 in. (305 mm) long and 16-gauge to 18-gauge thick, you will make a fillet weld in the flat position.

Following the same instructions for the assembly and welding procedure outlined in Practice 14-18, repeat each type of joint with both classifications of electrodes as needed until defect-free welds can consistently be made. Turn off the welding machine and shielding gas and clean up your work area when you are finished welding.

Complete a copy of the "Student Welding Report" listed in Appendix I or provided by your instructor. ◆

PRACTICE 14-20

Butt Joint 3G

Using a properly set up and adjusted FCA welding machine, proper safety protection, 0.030-in. and/or 0.035-in. (0.8-mm and/or 0.9-mm) -diameter E70T-1 and/or E70T-5 electrodes, and one or more pieces of mild steel sheet, 12 in. (305 mm) long and 16-gauge to 18-gauge thick, you will make a butt weld in the vertical up or down position.

Following the same instructions for the assembly and welding procedure outlined in Practice 14-18, repeat with both classifications of electrodes as needed until defect-free welds can consistently be made. Turn off the welding machine and shielding gas and clean up your work area when you are finished welding.

Complete a copy of the "Student Welding Report" listed in Appendix I or provided by your instructor. ◆

PRACTICE 14-21

Lap Joint and Tee Joint 3F

Using a properly set up and adjusted FCA welding machine, proper safety protection, 0.030-in. and/or 0.035-in. (0.8-mm and/or 0.9-mm) -diameter E70T-1 and/or E70T-5 electrodes, and one or more pieces of mild steel sheet, 12 in. (305 mm) long and 16 gauge to 18 gauge thick, you will make a fillet weld in the vertical up or down position.

Following the same instructions for the assembly and welding procedure outlined in Practice 14-18, repeat each type of joint with both classifications of electrodes as needed until defect-free welds can consistently be made. Turn off the welding machine and shielding gas and clean up your work area when you are finished welding.

Complete a copy of the "Student Welding Report" listed in Appendix I or provided by your instructor. ◆

PRACTICE 14-22

Lap Joint and Tee Joint 2F

Using a properly set up and adjusted FCA welding machine, proper safety protection, 0.030-in. and/or 0.035-in. (0.8-mm and/or 0.9-mm) -diameter E70T-1 and/or E70T-5 electrodes, and one or more pieces of mild steel sheet, 12 in. (305 mm) long and 16-gauge to 18-gauge thick, you will make a fillet weld in the horizontal position.

Following the same instructions for the assembly and welding procedure outlined in Practice 14-15, repeat each type of joint with both classifications of electrodes as needed until defect-free welds can consistently be made. Turn off the welding machine and shielding gas and clean up your work area when you are finished welding.

Complete a copy of the "Student Welding Report" listed in Appendix I or provided by your instructor. ◆

PRACTICE 14-23

Butt Joint 2G

Using a properly set up and adjusted FCA welding machine, proper safety protection, 0.030-in. and/or 0.035-in. (0.8-mm and/or 0.9-mm) -diameter E70T-1 and/or E70T-5 electrodes, and one or more pieces of mild steel sheet, 12 in. (305 mm) long and 16-gauge to 18-gauge thick, you will make a butt weld in the horizontal position.

Following the same instructions for the assembly and welding procedure outlined in Practice 14-15, repeat with both classifications of electrodes as needed until defect-free welds can consistently be made. Turn off the welding machine and shielding gas and clean up your work area when you are finished welding.

Complete a copy of the "Student Welding Report" listed in Appendix I or provided by your instructor. ◆

PRACTICE 14-24

Butt Joint 4G

Using a properly set up and adjusted FCA welding machine, proper safety protection, 0.030-in. and/or 0.035-in. (0.8-mm and/or 0.9-mm) -diameter E70T-1 and/or E70T-5 electrodes, and one or more pieces of mild steel sheet, 12 in. (305 mm) long and 16-gauge to 18-gauge thick, you will make a butt weld in the overhead position.

Following the same instructions for the assembly and welding procedure outlined in Practice 14-15, repeat with both classifications of electrodes as needed until defect-free welds can consistently be made. Turn off the welding machine and shielding gas and clean up your work area when you are finished welding.

Complete a copy of the "Student Welding Report" listed in Appendix I or provided by your instructor. ◆

PRACTICE 14-25

Lap Joint and Tee Joint 4F

Using a properly set up and adjusted FCA welding machine, proper safety protection, 0.030-in. and/or 0.035-in. (0.8-mm and/or 0.9-mm) -diameter E70T-1 and/or E70T-5 electrodes, and one or more pieces of mild steel sheet, 12 in. (305 mm) long and 16-gauge to 18-gauge thick, you will make a fillet weld in the overhead position.

Following the same instructions for the assembly and welding procedure outlined in Practice 14-15, repeat each type of joint with both classifications of electrodes as needed until defect-free welds can consistently be made. Turn off the welding machine and shielding gas and clean up your work area when you are finished welding.

Complete a copy of the "Student Welding Report" listed in Appendix I or provided by your instructor. ◆

Summary

In semiautomatic welding processes, the weld travel rate along the joint is controlled more by the process than by your welding technique. You must therefore learn how to travel at the proper rate to maintain the weld size. Flux cored arc welding is a relatively fast process. Therefore, your travel rates are much higher than for most other welding processes. This often causes new welders problems in that they have difficulty maintaining joint tracking as they are rapidly traveling along the groove. Practicing movement along the joint before you start is a good way of aiding in your development of these skills. The practice coupons in this section are 12 in. (305 mm) in length. However, if given the opportunity, you may want to weld longer joints after you have mastered the basic skills to further increase your joint tracking abilities.

Flux cored arc welding produces a large quantity of welding fumes. It is important that you position yourself so that you are not directly in line with the rising fumes. Make sure that you are welding so that your face is well out of this rising plume of welding fumes. In the field, welders sometimes use fans to gently blow the fumes away from them. However, if the fan is too close to the welding zone excessive air velocity will blow the shielding away from the weld, which may result in weld porosity. Take precautions to protect yourself from any potential health hazards.

Review

1. The major skill required for making consistently acceptable FCA welds is the ability to set up the welding system.
   A. true
   B. false
2. Because FCA welding is often a continuous process it produces more _________ than stick welding.
   A. scrap
   B. welding fumes
   C. welding electrode stubs waste
   D. noise
3. Why is it important that the feed rollers and conduit be aligned?
   A. So the shielding gas can flow down the welding cable assembly correctly.
   B. To prevent arcing of the rollers to the conduit.
   C. So the wire cast and helix will be set correctly.
   D. To prevent bird-nesting.
4. On most FCA welding gun assemblies, what parts attach to the gas diffuser?
   A. gun trigger and nozzle
   B. conduit and contact tube
   C. gas hose and gas nozzle
   D. electrode wire and feed rollers
5. What can occur if the wire feed tensioner is too loose?
   A. The wire feed will not start instantly when the gun trigger is pressed.
   B. The wire will slip out from under the rollers.
   C. The wire will bird-nest.
   D. The wire can be deformed out of round.
6. Full penetration FCA welds can be made in metal as thick as ______.
   A. 1/8 in. (3 mm)
   B. 1/4 in. (6 mm)
   C. 1/2 in. (13 mm)
   D. 16-gauge sheet metal

Review continued

7. For speed and economy try to avoid welds that require beveling the edges of plates.
   A. true
   B. false
8. Using Table 14-1 what wire feed speed range would be needed to produce an amperage range of 130 to 150 with 0.035 in. (0.9 mm) E70-T1 electrode wire?
   A. 200 to 400 IPM (508 to 1,016 cm/min)
   B. 288 to 567 IPM (732 to 1,463 cm/min)
   C. 288 to 380 IPM (732 to 975 cm/min)
   D. 200 to 300 IPM (508 to 762 cm/min)
9. What is the small flat spot at the tip of a beveled plate called?
   A. tow
   B. joint face
   C. flat tow
   D. root face
10. What is the name of the first weld made in a multi-pass groove weld?
    A. root pass
    B. hot pass
    C. filler pass
    D. cover pass
11. What is the name of the last weld made in a multi-pass groove weld?
    A. root pass
    B. hot pass
    C. filler pass
    D. cover pass
12. What type of inspection are you doing when you look at a weld for uniformity?
    A. appearance check
    B. shape and size inspection
    C. visual inspection
    D. check out inspection
13. Which of the following would not be checked in a visual inspection?
    A. arc strikes
    B. undercut
    C. filler pass fusion
    D. porosity
14. The leg of a fillet weld on a lap joint should be equal to __________.
    A. one-half the root depth
    B. the foot of the weld
    C. five times the electrode diameter
    D. the plate thickness
15. The higher the face of a fillet weld is built up on a tee joint the stronger the weld.
    A. true
    B. false
16. Which term is correct for the weld made on a tee joint in the flat position?
    A. 1F
    B. 2F
    C. 3F
    D. 4F
17. When the back of a vertical weld becomes pointed and not round the weld is __________.
    A. too cold
    B. too hot
    C. just right
    D. being underfilled
18. A vertical down weld that is too high and has little fusion is being made ___________.
    A. too cold
    B. too hot
    C. just right
    D. underfilled
19. When welding overhead extra personal protection is required to reduce the danger of burns.
    A. true
    B. false
20. What would the voltage range be for making a weld in 16-gauge sheet metal with 0.035 in (0.9 mm) E70T-1 electrodes?
    A. 15 to 20 volts
    B. 20 to 28 volts
    C. 90 to 200 volts
    D. 5 volts

Chapter 15

Filler Metal Selection

OBJECTIVES

After completing this chapter, the student should be able to

- ☑ explain how and when to use each type of filler metal.
- ☑ select the best filler metal to fit a specific welding job.
- ☑ list the forms filler metals come in.
- ☑ explain the significance of the filler metal prefixes.
- ☑ explain how to interpret the standard filler metal numbering systems.
- ☑ describe the effects alloys have on ferrous metals.

KEY TERMS

alloying elements
arc blow
core wire
fast freezing
filler metals
flux covering

INTRODUCTION

Manufacturers of filler metals may use any one of a variety of identification systems. There is not a mandatory identification system for filler metals. Manufacturers may use their own numbering systems, trade names, color codes, or a combination of methods to identify filler metal. They may voluntarily choose to use any one of several standardized systems.

The most widely used numbering and lettering system is the one developed by the American Welding Society (AWS). Other numbering and lettering systems have been developed by the American Society for Testing and Materials (ASTM) and the American Iron and Steel Institute (AISI). A system of using colored dots has also been developed by the National Electrical Manufacturers Association (NEMA). Some manufacturers have produced systems that are similar to the AWS system. Most major manufacturers include both the AWS identification and their own identification on the box, on the package, or directly on the filler metal.

Information that pertains directly to specific filler metal is readily available from most electrode manufacturers. The information given in charts, pamphlets, and pocket electrode guides is specific to their products and may or may not include standard AWS tests, terms, or classifications within their identification systems.

The AWS publishes a variety of books, pamphlets, and charts showing the minimum specifications for filler metal groups. It also publishes comparison charts that include all of the information manufacturers provide to the AWS regarding their filler metals. Both the literature on filler metal specifications and filler metal comparisons may be obtained directly from the AWS.

The AWS classification system is for minimum requirements within a grouping. Filler metals manufactured within a grouping may vary but still be classified under that grouping's classification.

A manufacturer may add elements to the metal or flux, such as more arc stabilizers. When one characteristic is improved, another characteristic may also change. The added arc stabilizer may make a smoother weld with less penetration. Other changes may affect the strength and ductility or other welding characteristics.

Because of the variables within a classification, some manufacturers make more than one type of filler metal that is included in a single classification. This and other information may be included in the data supplied by manufacturers.

Manufacturers' Electrode Information

The type of information given by different manufacturers ranges from general information to technical, chemical, and physical information. A mixture of different types of information may be given.

General information given by manufacturers may include some or all of the following: welding electrode manipulation techniques, joint design, prewelding preparation, postwelding procedures, types of equipment that can be welded, welding currents, and welding positions.

Understanding the Electrode Data

Technical procedures, physical properties, and chemical analysis information given by manufacturers include the following:

- Number of welding electrodes per pound
- Number of inches of weld per welding electrode
- Welding amperage range setting for each size of welding electrode
- Welding codes for which the electrode can be used
- Types of metal that can be welded
- Ability to weld on rust, oil, or paint
- Weld joint penetration characteristics
- Preheating and postheating temperatures
- Weld deposit physical strengths: ultimate tensile strength, yield point, yield strength, elongation, and impact strength
- Percentages of such alloys as carbon, sulfur, phosphorus, manganese, silicon, chromium, nickel, molybdenum, and other alloys

The information supplied by the manufacturer can be used for a variety of purposes, including the following:

- Estimates of the pounds of electrodes needed for a job
- Welding conditions under which the electrode can be used—for example, on clean or dirty metal
- Welding procedure qualification information regarding amperage, joint preparation, penetration, and welding codes
- Physical and chemical characteristics affecting the weld's strengths and metallurgical properties

Data Resulting from Mechanical Tests

Most of the technical information supplied is self-explanatory and easily understood. The mechanical properties of the weld are given as the results of standard tests. The following are some of the standard tests and the meaning of each test:

- Minimum tensile strength, psi—the load in pounds that would be required to break a section of soundweld that has a cross-sectional area of one square inch
- Yield Point, psi—the point in low and medium carbon steels at which the metal begins to stretch when force (stress) is applied after which it will not return to its original length
- Elongation, percent in 2 inches—the percentage that a 2-inch piece of weld will stretch before it breaks
- Charpy V notch, ft-lb—the impact load required to break a test piece of weld metal. This test may be performed on metal below room temperature at which point it is more brittle.

Data Resulting from Chemical Analysis

Chemical analysis of the weld deposit may also be included in the information given by manufacturers.

It is not so important to know what the different percentages of the alloys do, but it is important to know how changes in the percentages of the alloys affect the weld. Chemical composition can easily be compared from one electrode to another. The following are the major elements and the effects of their changes on the iron in carbon steel:

- Carbon (C)—As the percentage of carbon increases, the tensile strength increases, the hardness increases, and ductility is reduced. Carbon also causes austenite to form.
- Sulfur (S)—It is usually a contaminant, and the percentage should be kept as low as possible below 0.04%. As the percentage of carbon increases, sulfur can cause hot shortness and porosity.
- Phosphorus (P)—It is usually a contaminant, and the percentage should be kept as low as possible. As the percentage of phosphorus increases, it can cause weld brittleness, reduced shock resistance, and increased cracking.
- Manganese (Mn)—As the percentage of manganese increases, the tensile strength, hardness, resistance to abrasion, and porosity all increase; hot shortness is reduced. It is also a strong austenite former.
- Silicon (Si)—As the percentage of silicon increases, tensile strength increases, and cracking may increase. It is used as a deoxidizer and ferrite former.
- Chromium (Cr)—As the percentage of chromium increases, tensile strength, hardness, and corrosion resistance increase with some decrease in ductility. It is also a good ferrite and carbide former.
- Nickel (Ni)—As the percentage of nickel increases, tensile strength, toughness, and corrosion resistance increase. It is also an austenite former.
- Molybdenum (Mo)—As the percentage of molybdenum increases, tensile strengthens at elevated temperatures; creep resistance and corrosion resistance increase. It is also a ferrite and carbide former.
- Copper (Cu)—As the percentage of copper increases, the corrosion resistance and cracking tendency increases.
- Columbium (Cb)—As the percentage of columbium (niobium) increases, the tendency to form chrome-carbides is reduced in stainless steels. It is also a strong ferrite former.
- Aluminum (Al)—As the percentage of aluminum increases, the high temperature scaling resistance improves. It is also a good oxidizer and ferrite former.

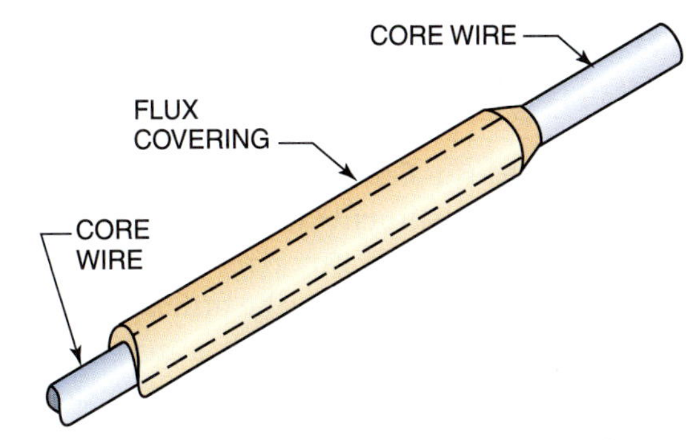

FIGURE 15-1 The two parts of a welding electrode.

SMAW Operating Information

Shielded metal arc welding electrodes, sometimes referred to as welding rods, arc welding rods, stick electrodes, or simply electrodes, have two parts. These two parts are the inner core wire and a flux covering, **Figure 15-1**.

The functions of the core wire include the following:

- To carry the welding current
- To serve as most of the filler metal in the finished weld

The functions of the flux covering include the following:

- To provide some of the alloying elements
- To provide an arc stabilizer (optional)
- To serve as an insulator
- To provide a slag cover to protect the weld bead and slow cooling rate
- To provide a protective gaseous shield during welding

Core Wire

A core wire is the primary metal source for a weld. For fabricating structural and low alloy steels, the core wires of the electrode use inexpensive rimmed or low carbon steel. For more highly alloyed materials, such as stainless steel, high nickel alloys, or nonferrous alloys, the core wires are of the approximate composition of the material to be welded. The core wire also supports the coating that carries the fluxing and alloying materials to the arc and weld pool.

Functions of the Flux Covering

Provides Shielding Gases Heat generated by the arc causes some constituents in the flux covering to decompose and others to vaporize, forming shielding gases. These gases prevent the atmosphere from contaminating the weld metal as it transfers across the arc gap. They also protect the molten weld pool as it cools to form solid metal. In addition, shielding gases and vapors greatly affect both the drops that form at the electrode tip and their transfer across the arc gap, **Figure 15-2.** They also cause the spatter from the arc and greatly determine arc stiffness and penetration. For example, the E6010 electrode contains cellulose. Cellulose decomposes into the hydrogen responsible for the deep electrode penetration so desirable in pipeline welding.

Alloying Elements Elements in the flux are mixed with the filler metal. Some of these elements stay in the weld metal as alloys. Other elements pick up contaminants in the molten weld pool and float them to the surface. At the surface, these contaminants form part of the slag, **Figure 15-3.**

Effect on Weld Welding fluxes can affect the penetration and contour of the weld bead. Penetration may be pushed deeper if the core wire is made to melt off faster than the flux melts. This forms a small chamber or crucible at the end of the electrode that acts as the combustion chamber of a rocket. As a result, the molten metal and hot gases are forced out very rapidly. The effect of this can be seen on the surface of the molten weld pool as it is blown back away from the end of the electrode. Some electrodes do not use this jetting action, and the resulting molten weld pool is much calmer (less turbulent) and may be rounded in appearance. In addition, the resulting bead may have less penetration, **Figure 15-4.**

Weld bead contour can also be affected by the slag formed by the flux. Some high-temperature slags, called refractory, solidify before the weld metal solidifies, forming a mold that holds the molten metal in place. These electrodes are sometimes referred to as **fast freezing** and are excellent for vertical, horizontal, and overhead welding positions.

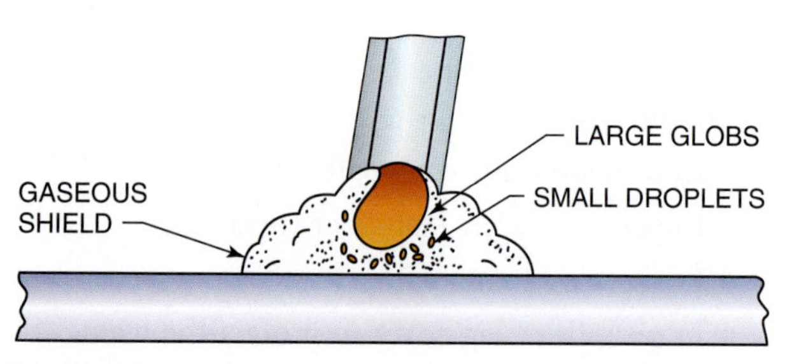

FIGURE 15-2 Methods of metal transfer during an arc.

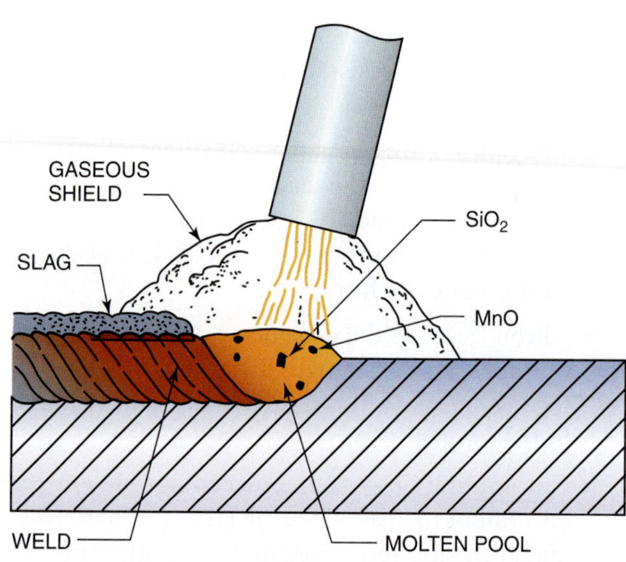

FIGURE 15-3 Silicon and manganese act as scavengers that combine with contaminants and float to the slag on top of the weld.

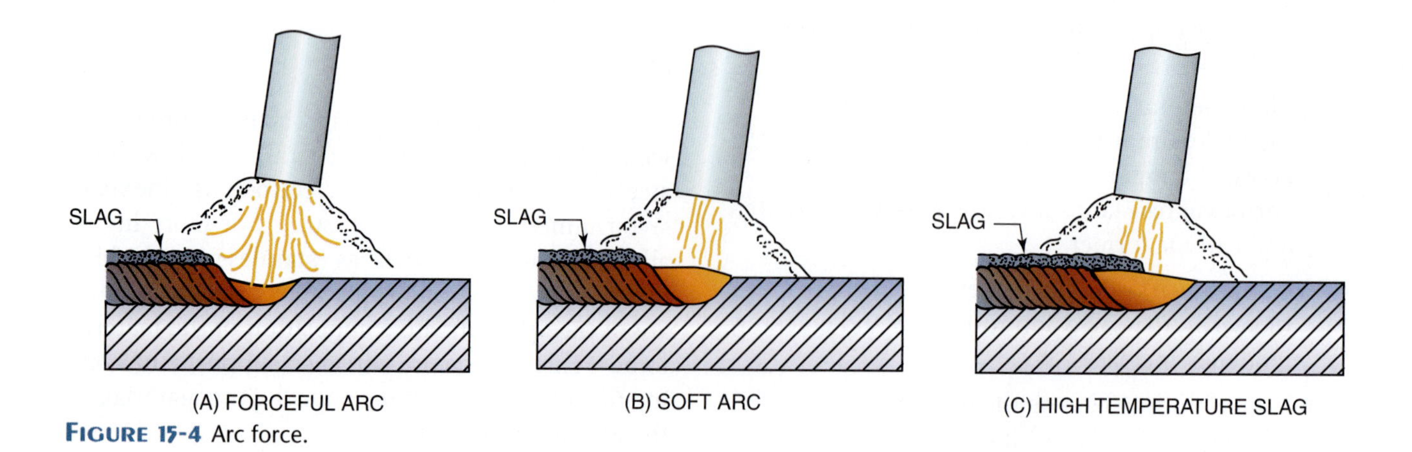

FIGURE 15-4 Arc force.

Filler Metal Selection

Selecting the best filler metal for a job is seldom delegated to the welder in large shops. The selection of the correct process and filler metal is a complex process. If the choice is given to the welder, it is one of the most important decisions the welder will make.

Covering all of the variables for selecting a filler metal would be well beyond the scope of this text. A sample of the types of things that must be considered for the selection of a SAW electrode follow. To further complicate things, welding electrodes have more than one application, and many welding electrodes may be used for the same type of work.

The following conditions that the welder should consider when choosing a welding electrode are not in order of importance. They are also not all of the factors that must be considered.

Shielded Metal Arc Welding Electrode Selection

- Type of electrode—What electrode has been specified in the blueprints or in the contract for this job?
- Type of current—Can the welding power source supply AC only, DC only, or both AC and DC?
- Power range—What is the amperage range on the welder and its duty cycle? Different types of electrodes require different amperage settings even for the same size welding electrode. For example, the amperage range for a 1/8-in. (3-mm) -diameter E6010 electrode is 75 A to 125 A, and the amperage range for a 1/8-in. (3-mm) -diameter E7018 electrode is 90 A to 165 A.
- Type of metal—Some welding electrodes may be used to join more than one similar type of metal. Other electrodes may be used to join together two different types of metal. For example, an E309-15 electrode can be used to join 305 stainless steel to 1020 mild steel.
- Thickness of metal—The penetration characteristics of each welding electrode may differ. Selecting one electrode that will weld on a specific thickness of material is important. For example, E6013 has very little penetration and is therefore good for welding on sheet metal.
- Weld position—Some welding electrodes can be used to make welds in all positions. Other electrodes may be restricted to making flat, horizontal, and/or vertical position welds; a few electrodes may be used to make flat position welds only.
- Joint design—The type of joint and whether it is grooved or not may affect the performance of the welding electrode. For example, the E7018 electrode does not produce a large, gaseous cloud to protect the molten metal. For this reason, the electrode movement is restricted so that the molten weld pool is not left unprotected by the gaseous cloud.
- Surface condition—It is not always possible to work on new, clean metal. Some welding electrodes will weld on metal that is rusty, oily, painted, dirty, or galvanized.
- Number of passes—The amount of reinforcement needed may require more than one welding pass. Some welding electrodes will build up faster, and others will penetrate deeper. The slag may be removed more easily from some welds than from others. For example, E6013 will build up a weld faster than E6010, and the slag is also more easily removed between weld passes.
- Distortion—Welding electrodes that will operate on low-amperage settings will have less heat input and cause less distortion. Welding electrodes that have a high rate of deposition (fills the joint rapidly) and can travel faster will also cause less distortion. For example, the flux on an E7024 has 50% iron powder, which gives it a faster fill rate and allows it to travel faster, resulting in less distortion of the metal being welded.
- Preheat or postheat—On low carbon steel plate 1 in. (25 mm) thick or more, thick preheating is required with most welding electrodes. Postheating may be required to keep a weld zone from hardening or cracking when using some welding electrodes. However, no postheating may be required when welding low alloy steel using E310-15.
- Temperature service—Weld metals react differently to temperature extremes. Some welds become very brittle and crack easily in cryogenic (low-temperature) service. A few weld metals resist creep and oxidation at high temperatures. For example, E310Mo-15 can weld on most stainless and mild steels without any high-temperature problems.
- Mechanical properties—Mechanical properties such as tensile strength, yield strength, hardness, toughness, ductility, and impact strength can be modified by the selection of specific welding electrodes.

- Postwelding cleanup—The hardness or softness of the weld greatly affects any grinding, drilling, or machining. The ease with which the slag can be removed and the quantity of spatter will affect the time and amount of cleanup required.
- Shop or field weld—The general working conditions such as wind, dirt, cleanliness, dryness, and accessibility of the weld will affect the choice of welding electrode. For example, the E7018 electrode must be kept very dry, but the E6010 electrode is not greatly affected by moisture.
- Quantity of welds—If a few welds are needed, a more expensive welding electrode requiring less skill may be selected. For a large production job requiring a higher skill level, a less expensive welding electrode may be best.

After deciding the specific conditions that may affect the welding, the welder has most likely identified more than one condition that needs to be satisfied. Some of the conditions will not interfere with others. For example, the type of current and whether a welder makes one or more weld passes have little or no effect on each other. However, if a welder needs to machine the finished weld, hardness is a consideration. When two or more conditions conflict, the welder is seldom the person who will make the decision. It may be necessary to choose more than one welding electrode. When welding pipe, E6010 and E6011 are often used for the root pass because of their penetration characteristics, and E7018 is used for the cover pass because of its greater strength and resistance to cracking.

Each AWS electrode classification has its own welding characteristics. Some manufacturers have more than one welding electrode in some classifications. In these cases, the minimum specifications for the classification have been exceeded. An example of more than one welding electrode in a single classification is Lincoln's Fleetweld 35®, Fleetweld 35LS®, and Fleetweld 180R®. These electrodes are all in AWS classification E6011. For the manufacturer's complete description of these electrodes, consult **Table 15-1**.

The characteristics of each manufacturer's filler metals can be compared to one another by using data sheets supplied by the manufacturer. General comparisons can be made easily using an electrode comparison chart.

When making an electrode selection, many variables must be kept in mind, and the performance characteristics must be compared before making a final choice.

AWS Filler Metal Classifications

The AWS classification system uses a series of letters and numbers in a code that gives the important information about the filler metal. The prefix letter is used to indicate the filler's form, a type of process the filler is to be used with, or both. The prefix letters and their meanings are as follows:

- E—Indicates an arc welding electrode. The filler carries the welding current in the process. We most often think of the E standing for an SMA "stick" welding electrode. It also is used to indicate wire electrodes used in GMAW, FCAW, SAW, ESW, EGW, etc.
- R—Indicates a rod that is heated by some source other than electric current flowing directly through it. Welding rods are sometimes referred to as being "cut length" or "welding wire." It is often used with OFW and GTAW.
- ER—Indicates a filler metal that is supplied for use as either a current-carrying electrode or in rod forms. The same alloys are used to produce the electrodes and the rods. This filler metal may be supplied as a wire on a spool for GMAW or as a rod for OFW or GTAW.
- EC—Indicates a composite electrode. These electrodes are used for SAW. Do not confuse an ECu, copper arc welding electrode, for an ECNi2, which is a composite nickel submerged arc welding wire.
- B—Indicates a brazing filler metal. This filler metal is usually supplied as a rod, but it can come in a number of other forms. Some of the forms it comes in are powder, sheets, washers, and rings.
- RB—Indicates a filler metal that is used as a current-carrying electrode, as a rod, or both. The form the filler is supplied in for each of the applications may be different. The composition of the alloy in the filler metal will be the same for all of the forms supplied. This filler can be used for processes like arc braze welding or oxyfuel brazing.
- RG—Indicates a welding rod used primarily with oxyfuel welding. This filler can be used with all of the oxyfuels, and some of the fillers are used with the GTAW process.
- IN—Indicates a consumable insert. These are most often used for welding on pipe. They are preplaced in the root of the groove to supply

Electrode Identification and Operating Data

Coating Color	AWS Number on Coating	(L)Lincoln	Electrode	Electrode Polarity	Sizes and Current Ranges (Amps) (Electrodes Are Manufactured in These Sizes for Which Current Ranges Are Given) 3/32" Size	1/8" Size	5/32" Size	3/16" Size	7/32" Size	1/4" Size
Brick red	6010		Fleetweld 5P	DC+	40–75	75–130	90–175	140–225	200–275	220–325[1]
Gray	6011		Fleetweld 35	AC	50–85	75–120	90–160	120–200	150–260	190–300
				DC+	40–75	70–110	80–145	110–180	135–235	170–270
Red brown	6011	Green	Fleetweld 35LS	AC		80–130	120–160			
				DC±		70–120	110–150			
Brown	6011		Fleetweld 180	AC	40–90	60–120	115–150			
				DC±	40–80	56–110	105–135			
Pink	7010-A1		Shield-arc 85	DC+	50–90	75–130	90–175	140–225		
Pink	7010-A1	Green	Shield-arc 85P	DC+				140–225		
Tan	7010-G		Shield-arc Hyp	DC+		75–130	90–185	140–225	160–250	
Gray	8010-G		Shield-arc 70+	DC+		75–130	90–185	140–225		

[1]Range for 5/16" size is 240–400 amps. DC+ is Electrode Positive. DC– is Electrode Negative.

All tests were performed in conformance with specifications AWS A5.5 and ASME SFA.5.5 in the aged condition for the E7010-G and E8010-G electrodes, and in the stress-relieved condition for Shield-Arc 85 & 85P. Tests for the other products were performed in conformance with specifications AWS A5.1 and ASME SF A.5.1 for the as-welded condition

Typical Mechanical Properties

Low figures in the stress-relieved tensile and yield strength ranges below for Shield-Arc 85 and 85P are AWS minimum requirements.
Low figures in the as-welded tensile and yield strength ranges below for the other products are AWS minimum requirements.

	Fleetweld 5P	Fleetweld 35	Fleetweld 35LS	Fleetweld 180	Shield-arc 85	Shield-arc 85P	Shield-arc Hyp	Shield-arc 70+
As welded tensile strength—psi	62–69,000	62–68,000	62–67,000	62–71,000	70–78,000	70–78,000	70–84,000	80–92,000
Yield point—psi	52–62,000	50–62,000	50–60,000	50–64,000	60–71,000	57–63,000	60–77,000	67,000–83,000
ductility—% elong. in 2"	22–32	22–30	22–31	22–31	22–26	22–27	22–23	19–24
Charpy V-notch toughness —ft. lb	20–60 @–20°F	20–90 @–20°F	20–57 @–20°F	20–54 @–20°F	68 @ 70°F	68 @ 70°F	30 @ 20°F	40 @ 50°F
Hardness, Rockwell B (avg) [5]	76–82	76–85	73–88	75–85			83–92	88–93
Stress-relieved @ 1150°F tensile strength—psi	60–69,000	60–66,000	60-65,000		70–83,000	70–74,000	80–82,000	80–84,000
Yield point—psi	46–56,000	46–56,000	46–51,000		57–69,000	57–65,000	72–76,000	71–76,000
ductility—% elong. in 2"	28–36	28–36	28–33		22–28	22–27	24–27	22–26
Charpy V-notch toughness —ft. lb	71 @ 70°F		120 @ 70°F		64 @ 70°F	68 @ 70°F	30 @ –20°F	30 @ –50°F
Hardness, Rockwell B (avg) [5]					80–89	80–87		

Conformances and Approvals

See *Lincoln Price Book* for certificate numbers, size, and position limitations, and other data.

Conforms to test requirements of AWS—A5.1 and ASME—SFA5.1 AWS—A5.5 and ASME—SFA5.5		*FW-5P* E6010	*FW-35* E6011	*FW-35LS* E6011	*FW-180* E6011	*SA-85* E7010-A1[4]	*SA-85P* E7010-A1	*SA-HYP* E7010-G	*SA-70+* E8010-G
ASME boiler code	Group	F3	F3	F3	F3	F3	F3	F3	F3
	Analysis	A1		A1	A1	A2	A2	A2	
American Bureau of Shipping and U.S. Coast Guard		Approved	Approved	Approved		Approved			
Conformance certificate available[4]		Yes	Yes	Yes	Yes	Yes	Yes	Yes	Yes
Lloyds		Approved	Approved						
Military specifications		MIL-QQE-450	MIL-QQE-450			MIL-E-22200/7			

[3] Also meets the requirements for E7010-G and E6010 in 3/32" size.

[4] "Certificate of Conformance" to AWS classification test requirements is available. These are needed for Federal Highway Administration projects.

[5] Hardness values obtained from welds made in accordance with AWS A5.1.

TABLE 15-1 Fleetweld 35®, Fleetweld 35LS®, and Fleetweld 180® Lincoln Electrodes. Source: The Lincoln Electric Company.

both filler metal and support for the root pass. The inserts may provide for some joint alignment and spacing.

The next two classifications are not filler metal. They are classified under the same system because they are welding consumables. The GTA welding tungsten is not a filler, but it is consumed, very slowly, during the welding process.

- EW—Indicates a nonconsumable tungsten electrode. The GTAW electrode is obviously not a filler metal, but it falls under the same classification system.
- F—Indicates a flux used for SAW. The composition of the weld metal is influenced by the flux. There are alloys and agents in the flux used for SAW that are dissolved into the weld metal. For this reason, the filler metal and flux are specified together with the filler metal identification first and the flux second.

In addition to the prefix, there are some suffix identifiers. The suffix may be used to indicate a change in the alloy in a covered electrode or the type of welding current to be used with stainless steel–covered electrodes.

Carbon Steel

Carbon and Low Alloy Steel-Covered Electrodes

The AWS specification for carbon steel-covered arc electrodes is A5.1, and for low alloy steel-covered arc electrodes it is A5.5. Filler metal classified within these specifications are identified by a system that uses the letter *E* followed by a series of numbers to indicate the minimum tensile strength of a good weld, the position(s) in which the electrode can be used, the type of flux coating, and the type(s) of welding current, **Figure 15-5.**

The tensile strength is given in pounds per square inch (psi). The actual strength is obtained by adding three zeroes to the right of the number given. For example, E60XX is 60,000 psi, E110XX is 110,000 psi, etc.

The next number located to the right of the tensile strength—1, 2, or 4—designates the welding position capable—for example,

- 1—In an E601X means all positions flat, horizontal, vertical, and overhead.
- 2—In an E602X means horizontal fillets and flat.
- 3—Is an old term no longer used; it meant flat only.
- 4—In an E704X means flat, horizontal, overhead, and vertical-down.

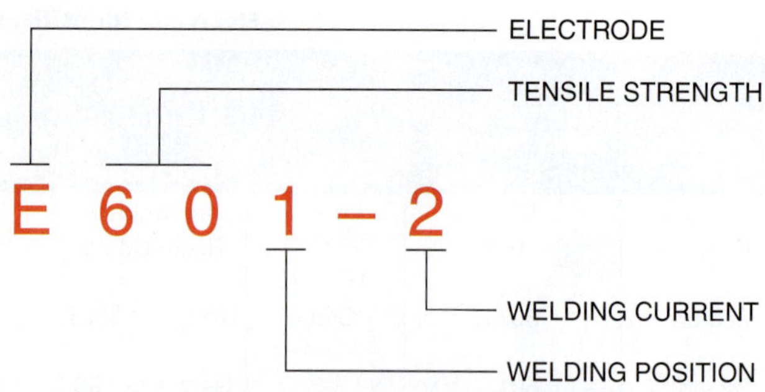

FIGURE 15-5 AWS numbering system for A5.1 and A5.5 carbon and low alloy steel covered electrodes.

The last two numbers together indicate the major type of covering and the type of welding current. For example, EXX10 has an organic covering and uses DCEP polarity. The AWS classification system for A5.1 and A5.5 covered arc welding electrodes is shown in **Table 15-2.** The type of welding current for any electrode may be expanded to include currents not listed if a manufacturer adds additional arc stabilizers to the electrode covering, **Table 15-3.**

On some covered arc electrodes, a suffix may be added to indicate the approximate alloy in the deposit as welded. For example, the letter *A* indicates a 1/2% molybdenum addition to the weld metal deposited. **Table 15-4** is a complete list of the major alloying elements in electrodes.

Some of the more popular arc welding electrodes and their uses in these specifications are as follows:

E6010

The E6010 electrodes are designed to be used with DCEP polarity and have an organic-based flux (cellulose, $C_6H_{10}O_5$). They have a forceful arc that results in deep penetration and good metal transfer in the vertical and overhead positions, **Figure 15-6.** The electrode is usually used with a whipping or stepping motion. This motion helps remove unwanted surface materials such as paint, oil, dirt, and galvanizing. Both the burning of the organic compound in the flux to form CO_2, which protects the molten metal, and the rapid expansion of the hot gases force the atmosphere away from the weld. A small amount of slag remains on the finished weld, but it is difficult to remove, especially along the weld edges. E6010 electrodes are commonly used for welding on pipe and in construction jobs, and for doing repair work.

E6011

The E6011 electrodes are designed to be used with AC or DCEP reverse polarity and have an organic-based flux. These electrodes have many of the welding

AWS Classification	Type of Covering	Capable of Producing Satisfactory Welds in Positions Shown[a]	Type of Current[b]
	E60 Series Electrodes		
E6010	High cellulose sodium	F, V, OH, H	DCEP
E6011	High cellulose potassium	F, V, OH, H	AC or DCEP
E6012	High titania sodium	F, V, OH, H	AC or DCEN
E6013	High titania potassium	F, V, OH, H	AC or DC, either polarity
E6020		H-fillets	AC or DCEN
E6022[c]	High iron oxide	F	AC or DC, either polarity
E6027	High iron oxide, iron powder	H-fillets, F	AC or DCEN
	E70 Series Electrodes		
E7014	Iron powder, titania	F, V, OH, H	AC or DC, either polarity
E7015	Low hydrogen sodium	F, V, OH, H	DCEP
E7016	Low hydrogen potassium	F, V, OH, H	AC or DCEP
E7018	Low hydrogen potassium, iron powder	F, V, OH, H	AC or DCEP
E7024	Iron powder, titania	H-fillets, F	AC or DC, either polarity
E7027	High iron oxide, iron powder	H-fillets, F	AC or DCEN
E7028	Low hydrogen potassium, iron powder	H-fillets, F	AC or DCEP
E7048	Low hydrogen potassium, iron powder	F, OH, H, V-down	AC or DCEP

[a]The abbreviations, F, V, V-down, OH, H, and H-fillets indicate the welding positions as follows:

F = Flat
H = Horizontal
H-fillets = Horizontal fillets
V-down = Vertical down
V = Vertical
OH = Overhead

[b]Reverse polarity means the electrode is positive; straight polarity means the electrode is negative.

[c]Electrodes of the E6022 classification are for single-pass welds.

For electrodes 3/16 in (4.8 mm) and under, except 5/32 in (4.0 mm) and under for classifications E7014, E7015, E7016, and E7018.

TABLE 15-2 Electrode Classification. Courtesy of the American Welding Society.

EXXX0—DCRP only
EXXX1—AC and DCEP
EXXX2—AC and DCEN
EXXX3—AC and DC
EXXX4—AC and DC
EXXX5—DCEP only
EXXX6—AC and DCEP
EXXX8—AC and DCEP

TABLE 15-3 Welding Currents.

Suffix Symbol	Molybdenum (Mo) %	Nickel (Ni) %	Chromium (Cr) %	Manganese (Mn) %	Vanadium (Va) %
A 1	0.5				
B 1	0.5		0.50		
B 2	0.5		1.25		
B 3	1.0		2.25		
B 4	0.5		2.00		
C 1		2.5			
C 2		3.5			
C 3		1.0			
D 1	0.3			1.5	
D 2	0.3			1.75	
G	0.2	0.5	0.30	1.00	0.1*
M					

*Only one of these alloys may be used.

TABLE 15-4 Major Alloying Elements in Electrodes.

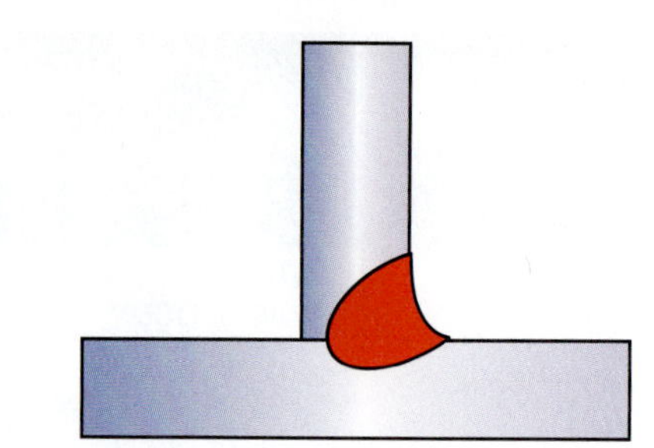

FIGURE 15-6 E6010.

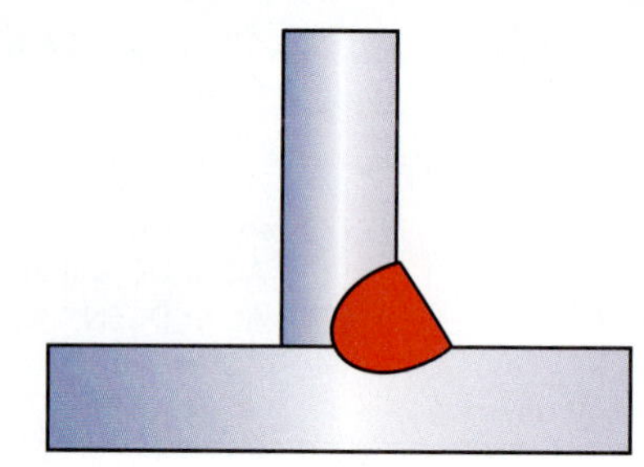

FIGURE 15-7 E6011.

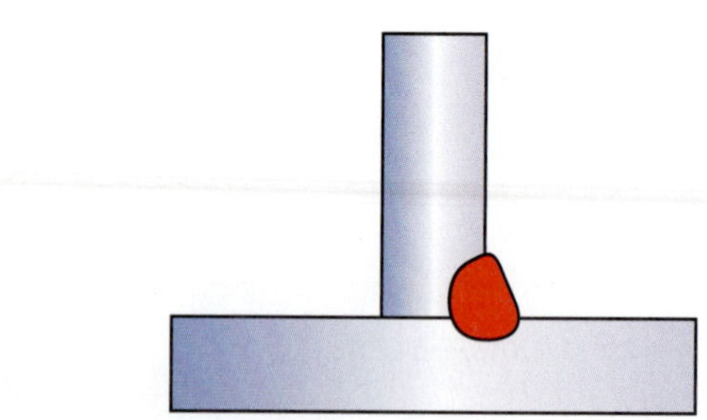

FIGURE 15-8 E6012.

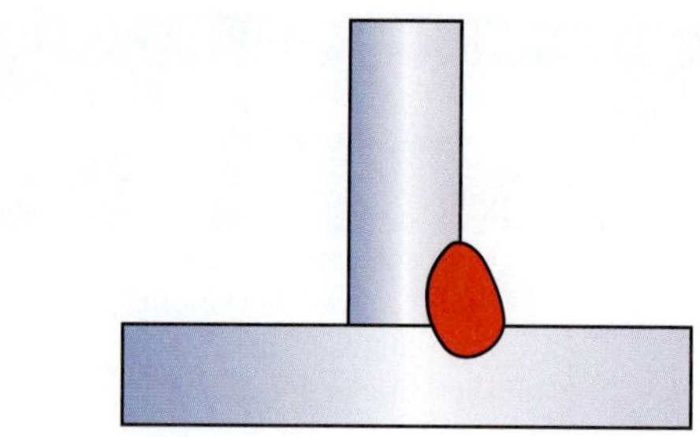

FIGURE 15-9 E6013.

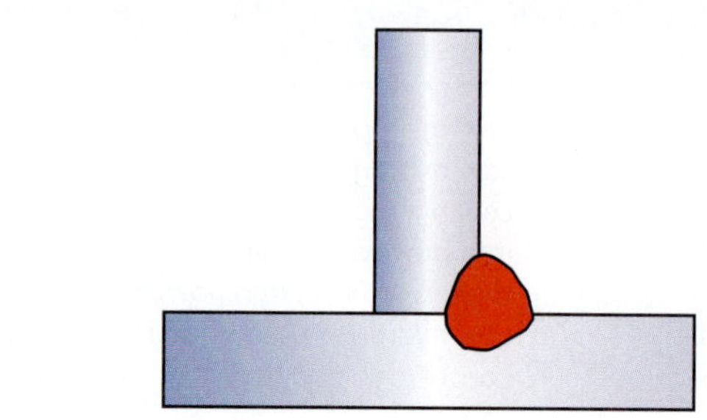

FIGURE 15-10 E7014.

characteristics of E6010 electrodes, **Figure 15-7**. In most applications, the E6011 is preferred. The E6011 has added arc stabilizers, which allow it to be used with AC. Using this welding electrode on AC only slightly reduces its penetration but will help control any arc blow problem. Arc blow, or arc wander, is the magnetic deflection of the arc from its normal path. When welding with either E6010 or E6011, the weld pool may be slightly concave from the forceful action of the rapidly expanding gas. This forceful action also results in more spatter and sparks during welding. E6011 is the most commonly used electrode for agricultural welding.

E6012

The E6012 electrodes are designed to be used with AC or DCEN polarity and have rutile-based flux (titanium dioxide TiO_2). This electrode has a very stable arc that is not very forceful, resulting in a shallow penetration characteristic, **Figure 15-8**. This limited penetration characteristic helps with poor-fitting joints or thin materials. Thick sections can be welded, but the joint must be grooved. Less smoke is generated with this welding electrode than with E6010 or E6011, but a thicker slag layer is deposited on the weld. If the weld is properly made, the slag can be removed easily and may even free itself after cooling. Spatter can be held to a minimum when using both AC and DC. E6012 electrodes are commonly used for all new work, including storage tanks, machinery fabrication, ornamental iron, and general repair work.

E6013

The E6013 electrodes are designed to be used with AC or DC, either polarity. They have a rutile-based flux. The E6013 electrode has many of the same characteristics of the E6012 electrode, **Figure 15-9**. The slag layer is usually thicker on the E6013 and is easily removed. The arc of the E6013 is as stable, but there is less penetration, which makes it easier to weld very thin sections. The weld bead will also be built up slightly higher than the E6012. E6013 electrodes are commonly used for sheet metal fabrication, metal buildings, surface buildup, truck and tractor work, and other farm equipment.

E7014

The E7014 electrodes are designed to be used with AC or DC, either polarity. They have a rutile-based flux with iron powder added. The E7014 electrode has many arc and weld characteristics that are similar to those of the E6013 electrode, **Figure 15-10**. Approximately 30% iron powder is added to the flux to allow it to build up a weld faster or have a higher travel speed. The penetration characteristic is light.

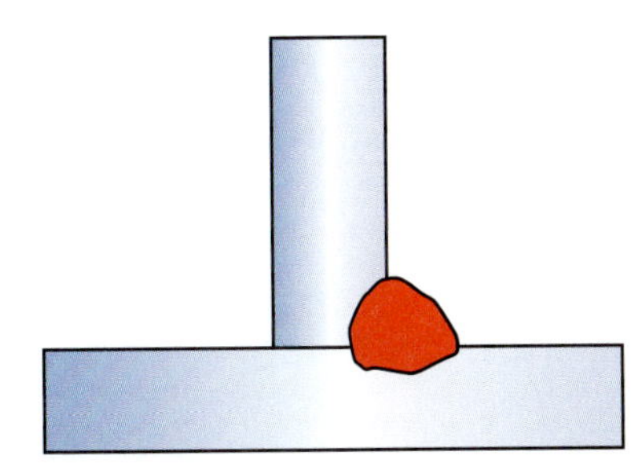

FIGURE 15-11 E7024.

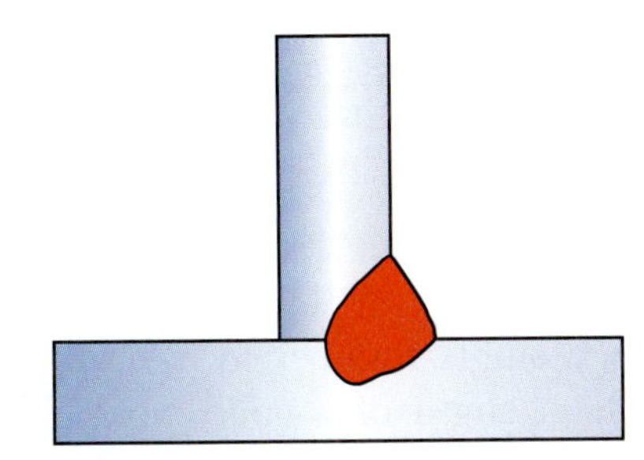

FIGURE 15-12 E7016.

This welding electrode can be used on metal with a light coating of rust, dirt, oil, or paint. The slag layer is thick and hard but can be completely removed with chipping. E7014 electrodes are commonly used for welding on heavy sheet metal, ornamental iron, machinery, frames, and general repair work.

E7024

The E7024 electrodes are designed to be used with AC or DC, either polarity. They have a rutile-based flux with iron powder added. This welding electrode has a light penetration and fast fill characteristic, **Figure 15-11.** The flux contains about 50% iron powder, which gives the flux its high rate of deposition. The heavy flux coating helps control the arc and can support the electrode so that a drag technique can be used. The drag technique allows this electrode to be used by welders with less skill. The slag layer is heavy and hard but can easily be removed. If the weld is performed correctly, the slag may remove itself. Because of the large, fluid molten weld pool, this electrode is equally used in the flat and horizontal position only, although it can be used on work that is slightly vertical. E7024 electrodes are commonly used in welding new equipment.

E7016

The E7016 electrodes are designed to be used with AC or DCEP polarity. They have a low-hydrogen-based (mineral) flux. This electrode has moderate penetration and little buildup, **Figure 15-12.** There is no iron powder in the flux, which helps when welding in the vertical or overhead positions. Welds

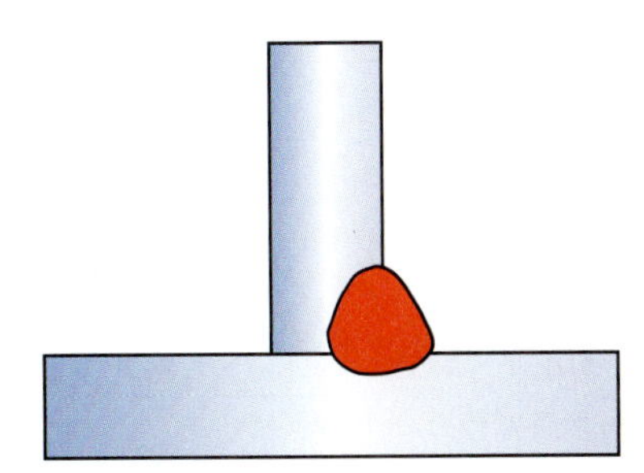

FIGURE 15-13 E7018.

on high sulfur and cold-rolled metals can be made with little porosity. Low alloy and mild steel heavy plates can be welded with minimum preheating. E7016 electrodes are commonly used for building construction and equipment fabrication.

E7018

The E7018 electrodes are designed to be used with AC or DCEP polarity. They have a low-hydrogen-based flux with iron powder added. The E7018 electrodes have moderate penetration and buildup, **Figure 15-13.** The slag layer is heavy and hard but can be removed easily by chipping. The weld metal is protected from the atmosphere primarily by the molten slag layer and not by rapidly expanding gases. For this reason, these electrodes should not be used for open root welds. The atmosphere may attack the root, causing a porosity problem. The E7018 welding electrodes are very susceptible to moisture, which may lead to weld porosity. These electrodes are commonly used for pipe, plate, trailer axles, and low-temperature equipment. E7018 electrodes are sometimes referred to as Lo-Hi rods, because they allow very little hydrogen into the weld pool.

Wire-type Carbon Steel Filler Metals

Solid Wire

The AWS specifications for carbon steel filler metals for gas shielded welding wire is A5.18. Filler metal classified within these specifications can be used for GMAW, GTAW, and PAW processes. Because in GTAW and PAW the wire does not carry the welding current, the letters *ER* are used as a prefix. The *ER* is followed by two numbers to indicate the minimum tensile strength of a good weld. The actual strength is obtained by adding three zeroes to the right of the number given. For example, ER70S-x is 70,000 psi.

The *S* located to the right of the tensile strength indicates that this is a solid wire. The last number—2, 3, 4, 5, 6, or 7—or the letter *G* is used to indicate the filler metal composition and the weld's mechanical properties, **Figure 15-14.**

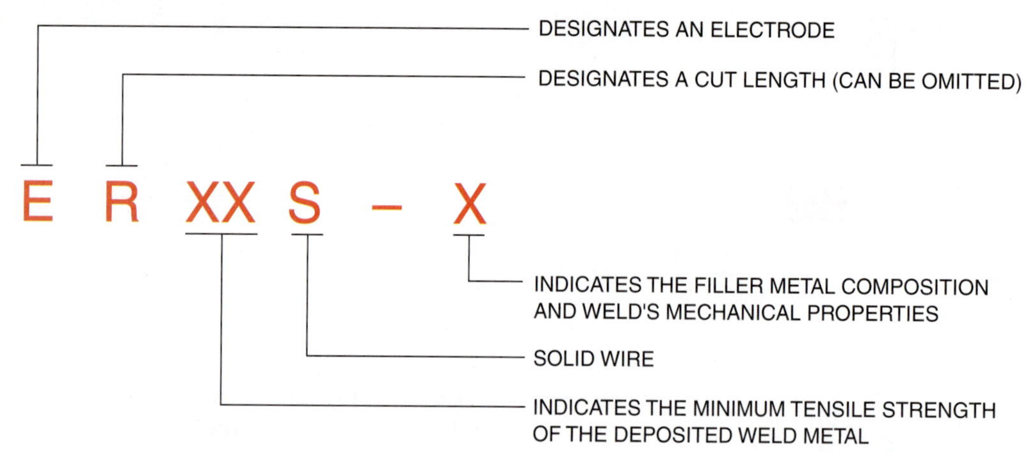

FIGURE 15-14 AWS numbering system for carbon steel filler metal for GMAW. Courtesy of the American Welding Society.

ER70S-2

This is a deoxidized mild steel filler wire. The deoxidizers allow this wire to be used on metal that has light coverings of rust or oxides. There may be a slight reduction in the weld's physical properties if the weld is made on rust or oxides, but this reduction is only slight, and the weld will usually still pass the classification test standards. This is a general-purpose filler that can be used on killed, semikilled, and rimmed steels. Argon-oxygen, argon-CO_2, and CO_2 can be used as shielding gases. Welds can be made in all positions.

ER70S-3

This is a popular agricultural filler wire. It can be used in single or multiple-pass welds in all positions. ER70S-3 does not have the deoxidizers required to weld over rust, over oxides, or on rimmed steels. It produces high-quality welds on killed and semi-killed steels. Argon-oxygen, argon-CO_2, and CO_2 can be used as shielding gases.

ER70S-6

This is a good general-purpose filler wire. It has the highest levels of manganese and silicon. The wire can be used to make smooth welds on sheet metal or thicker sections. Welds over rust, oxides, and other surface impurities will lower the mechanical properties, but not normally below the specifications of this classification. Argon-oxygen, argon-CO_2, and CO_2 can be used as shielding gases. Welds can be made in all positions.

Tubular Wire

The AWS specification for carbon steel filler metals for flux cored arc welding wire is A5.20. Filler metal classified within this specification can be used for the FCAW process. The letter *E*, for *electrode*, is followed by a single number to indicate the minimum tensile strength of a good weld. The actual strength is obtained by adding four zeroes to the right of the number given. For example, E6xT-x is 60,000 psi, and E7xT-x is 70,000 psi.

The next number, 0 or 1, indicates the welding position. Ex0T is to be used in a horizontal or flat position only. Ex1T is an all-position filler metal.

The *T* located to the right of the tensile strength and weld position numbers indicates that this is a tubular, flux cored wire. The last number—2, 3, 4, 5, 6, 7, 8, 10, or 11—or the letter *G* or *GS* is used to indicate if the filler metal can be used for single- or multiple-pass welds. The electrodes with the numbers ExxT-2, ExxT-3, ExxT-10, and ExxT-GS are intended for single-pass welds only, **Figure 15-15**.

E70T-1, E71T-1

E70T-1 and E71T-1 have a high-level deoxidizer in the flux core. It has high levels of silicon and manganese, which allow it to weld over some surface contaminations such as oxides or rust. This filler metal can be used for single- or multiple-pass welds. Argon 75% with 25% CO_2 or 100% CO_2 can be used as the shielding gas. It can be used on ASME A36, A106, A242, A252, A285, A441, and A572 or similar metals. Applications include railcars, heavy equipment, earth-moving equipment, shipbuilding, and general fabrication. The weld metal deposited has a chemical and physical composition similar to that of E7018 low hydrogen electrodes.

E70T-2, E71T-2

E70T-2 and E71T-2 are highly deoxidized flux cored filler metal that can be used for single-pass welds only. The high levels of deoxidizers allow this electrode to be used over mill scale and light layers of rust and still produce sound welds. Because of the high level of

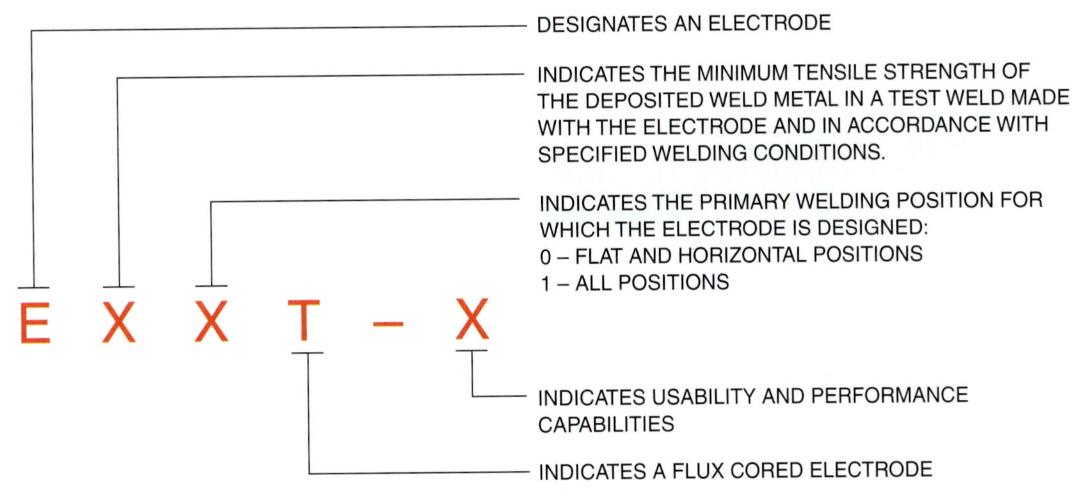

FIGURE 15-15 Identification system for mild steel FCAW electrodes. Courtesy of the American Welding Society.

manganese, if the filler is used for multiple-pass welds, there might be manganese-caused centerline cracking of the weld; 100% CO_2 can be used as the shielding gas. E70T-2 can be used on ASME A36, A106, A242, A252, A285, A441, and A572 or similar metals. Applications include repair and maintenance work and general fabrication.

E70T-4, E71T-4

E70T-4 and E71T-4 are self-shielding, flux cored filler metal. The fluxing agents produce a slag, which allows a larger than usual molten weld pool. The large weld pool permits high deposition rates. Weld deposits are ductile and have a high resistance to cracking. E70T-4 can be used to weld joints that have larger than usual root openings. Applications include large weldments and earth-moving equipment.

E71T-7

E71T-7 is a self-shielding, all position, flux cored filler metal. The fluxing system allows the control of the molten weld pool required for out-of-position welds. The high level of deoxidizers reduces the tendency for cracking in the weld. It can be used for single- or multiple-pass welds.

Stainless Steel Electrodes

The AWS specification for stainless steel covered arc electrodes is A5.4 and for stainless steel bare, cored, and stranded electrodes and welding rods is A5.9. Filler metal classified within the A5.4 uses the letter *E* as its prefix, and the filler metal within the A5.9 uses the letters *ER* as its prefix, **Table 15-5**.

Following the prefix, the American Iron and Steel Institute's (AISI), three-digit stainless steel number is used. This number indicates the type of stainless steel in the filler metal.

To the right of the AISI number, the AWS adds a dash followed by a suffix number. The number 15 is used to indicate there is a lime base coating, and the DCEP polarity welding current should be used. The number 16 is used to indicate there is a titania-type coating, and AC or DCEP polarity welding currents can be used. Examples of this classification system are E308-15 and E308-16 electrodes.

The letter *L* may be added to the right of the AISI number before the dash and suffix number to indicate a low carbon stainless welding electrode. E308L-15 and E308L-16 arc welding electrodes and ER308L and ER316L are examples of the use of the letter *L*, **Table 15-6**.

Stainless steel may be stabilized by adding columbium (Cb) as a carbide former. The designation Cb is added after the AISI number for these electrodes, such as E309Cb-16. Stainless steel filler metals are stabilized to prevent chromium-carbide precipitation.

E308-15, E308-16, E308L-15, E308-16, ER308, and ER308L

All are filler metals for 308 stainless steels. 308 stainless steels are used for food or chemical equipment, tanks, pumps, hoods, and evaporators. All E308 and ER308 filler metals can be used to weld on all 18-8-type stainless steels such as 301, 302, 302B, 303, 304, 305, 308, 201, and 202.

E309-15, E309-16, E309Cb-15, E309Cb-16, ER309, and ER309L

All are filler metals for 309 stainless steels. 309 stainless steels are used for high-temperature service, such as furnace parts and mufflers. All E309 filler metals can be used to weld on 309 stainless or to join mild steel to any 18-8-type stainless steel.

AISI TYPE NUMBER	442 446	430F 430FSE	430 431	501 502	403 405 416 416SE	321 410 420 414	348 347	317	316L	316	314	310 310S	309 309S	303 304 L	201 202 301 303S E 305 308	302 302B 304	Mild Steel
201–202–301 302–3028–304 305–308	**310** 312 309	**310** 312 309	**310** 312 309	**310** 312 309	**309** 310 312	**309** 310 312	**308**	**308**	**308**	**308**	**308**	**308**	**308**	**308**	**308**	**308**	**312** 310 309
303 303SE	**310** 309 312	**310** 309 312	**310** 309 312	**310** 309 312	**309** 310 312	**309** 310 312	**308**	**308**	**308**	**308**	**308**	**308**	**308**	**308**	**308-15**	**308**	**312** 310 309
304L	**310** 309 312	**310** 309 312	**310** 309 312	**310** 309 312	**309** 310 312	**309** 310 312	**308**	**308**	**308-L**	**308**	**308**	**308**	**308**	**308-L**	**308**	**308**	**312** 310 309
309 309S	**310** 309 312	**310** 309 312	**310** 309 312	**310** 309 312	**309** 310 312	**309** 310 312	**308**	**317** 316 309	**316**	**316**	**309**	**309**	**309**	**308**	**308**	**308**	**309** 310 312
310 310S	**310** 309 312	**310** 309 312	**310** 309 312	**310** 309 312	**310** 309 312	**310** 309 312	**308**	**317** 316 309	**316**	**316**	**310**	**310**	**309** 310	**308**	**308**	**308**	**310** 309 312
314	**310** 312 309	**310** 312 309	**310** 312 309	**310** 312 309	**310** 312 309	**310** 309 312	**309** 310 308	**309** 310	**309** 310	**309** 310	**310-15**	**310**	**309** 310	**309** 310	**309** 310	**309** 310	**310** 309 312
316	**310** 309 312	**310** 309 312	**310** 309 312	**310** 309 312	**309** 310 312	**309** 310 312	**308**	**316**	**316**	**316**	**309** 310 316	**310** 309 316	**309** 310 316	**308** 316	**308** 316	**308** 316	**309** 310 312
316L	**310** 309 312	**310** 309 312	**310** 309 312	**310** 309 312	**309** 310 312	**309** 310 312	**308**	**316** 317 308	**316-L**	**316**	**309** 310 316	**310** 309 316	**316** 309	**308** 316	**308** 316	**308** 316	**309** 310 312
317	**310** 309 312	**310** 309 312	**310** 309 312	**310** 309 312	**309** 310 312	**309** 310 312	**308**	**317**	**316** 308	**316** 308	**309** 310 317	**317** 316 309	**317** 316 309	**308** 316 317	**308** 316 317	**308** 316 317	**309** 310 312
321 348 347	**310** 309 312	**310** 309 312	**310** 309 312	**310** 309 312	**309** 310 312	**309** 310 312	347	**308** 347	**347** 308	**347** 308	**309** 310 347	**347** 308	**347** 308	**347** 308-L	**347** 308	**347** 308	**309** 310 312
403–405 410–420 414	**310** 309 312	**310** 309 312	**310** 309 312	**310** 309 312	**309** 310	**410**† 309††	**309** 310	**309** 310	**309** 310	**309** 310	**310** 309	**310** 309	**309** 310	**309** 310	**309** 310	**309** 310	**309** 310 312
416 416SE	**310** 309	**310** 309	**310** 309	**310**	**410-15**†	**410-15**† 309†† 310††	**309** 310	**309** 310 312	**309** 310 312	**309** 310 312	**309** 310 312	**310** 309 312	**309** 310 312	**309** 310 312	**309** 310 312	**309** 310 312	**309** 310 312
501 502	**310**	**310**	**310**	**502**† 310††	**310**	**310**	**310** 309	**310** 309	**310** 309	**310** 309	**310** 309	**310** 309	**310** 309	**310** 309	**310** 309	**310** 309	**310** 312 309
430 431	**310** 309	**310** 309	**430-15**† 310†† 309††	**310**	**310**	**310** 309	**310** 309	**310** 309	**310** 309	**310** 309	**310** 309	**310** 309	**310** 309	**310** 309	**310** 309	**310** 309	**310** 309 312
430F 430FSE	**310** 309	**410-15**†	**310** 309	**310** 309	**310** 309 312	**310** 309 312	**309** 310 312	**309** 310 312	**310** 309 312	**310** 309 312	**310** 309 312	**310** 309 312	**310** 309 312	**310** 309 312	**310** 309 312	**310** 309 312	**310** 309 312
442 443	**309** 310	**309** 310 312	**310** 309 312	**310** 309 312	**310** 309 312	**310** 309 312	**310** 309 312	**310** 309 312	**310** 309 312	**310** 309 312	**310** 309 312	**310** 309 312	**310** 309 312	**310** 309 312	**310** 309 312	**310** 309 312	**310** 309 312

†Preheat.

††No preheat necessary.

Bold numbers indicate first choice; light numbers indicate second and third choices. This choice can vary with specific applications and individual job requirements.

TABLE 15-5 Filler Metal Selector Guide for Joining Different Types of Stainless to the Same Type or Another Type of Stainless. Courtesy of Thermacote Welco.

UTP Designation	AWS/SFA5.4 Covered	AWS/SFA5.9 TIG and MIG	Description and Applications
6820	E 308-16	ER 308	For welding conventional 308 type SS
68 Kb	E 308-15		Low hydrogen coating
6820 Lc	E 308 L-16	ER 308L	Low carbon grade, prevents carbide precipitation adjacent to weld
308L Fe Hp	E 308 L-16		Fast depositing for maintenance and production coating
68 LcHL	E 308 L-16		High-performance electrode with rutile-acid coating, core wire alloyed, for stainless and acid-resisting CrNi steels
68 LcKb	E 308 L-15		Low carbon electrode for stainless, acid-resisting CrNi-steels
6824	E 309-16	ER 309	For welding 309 type SS and carbon steel to SS
6824 Kb	E 309-15		Special lime-coated electrode for corrosion and heat-resistant 22/12 CrNi-steels
6824 Lc	E 309 L-16	ER 309L	Same as 309, but with low carbon content
6824 Nb	E 309 Cb-16		Corrosion and heat-resistant 22/12 CrNi-steels
6824 MoNb	E309MoCb-16		Corrosion and heat-resistant 22/12 CrNi-steels
309L Fe Hp	E 309 L-16		High deposition rate, easy to use
6824 Mo Lc	E 309 L-16		For welding similar and dissimilar SS
68H	E 310-16	ER 310	For high-temperature service and cladding steel
6820 Mo	E 316-16	ER 316	For welding acid-resistant stainless steels
6820 Mo Lc	E 316 L-16	ER 316L	Low carbon grade, prevents intergranular corrosion
68 TI Mo	E 316 L-16		Most efficient type, for maintenance and production, high performance
68 MoLcHL	E 316 L-16		High-performance electrode with rutile-acid coating, core wire alloyed, for stainless and acid-resisting CrNi-Mo-steels
68 MoLcKb	E 316 L-15		Low carbon electrode for stainless and acid-resisting CrNiMo-steels
317 Lc Titan	E 317 L-16	ER 317L	Deposit resist sulphuric acid corrosion
317LFe Hp	E 317 L16		Fast melt-off rate, excellent for overlays, easy to use, high performance
68 Mo	E 318-16		Versatile stainless all-position electrode
320 Cb	E 320-15		For welding similar acid-resistant SS
3320 Lc	E 320		A rutile-coated electrode for welding in all positions except vertical down
6820 Nb	E 347-16	ER 347	Stabilized grade, prevents carbide precipitation
347 FeHp	E 347-16		High-performance stainless steel electrode of class E 347-16 for welding stabilized Cr Ni alloys
66	E 410-15		Low hydrogen electrode for corrosion and heat-resistant 14% Cr-steels
1915 HST			Low hydrogen, fully austenitic electrode with 0% ferrite content
1925			Extremely corrosion-resistant to phosphoric and sulfuric acids
68 Hcb	E 310 Cb	ER 310 Cb	For high heat applications and joining steels to stainless steel
2535 NbSn			Electrode is a lime-type special electrode and is used for surfacing and joining heat-resistant base metals, especially cast steel
E 330-16	E 330-16		Excellent for welding furnace parts
6805			For welding of base material 17 -4 Ph
6808 Mo			Rutile lime-type austenitic-ferritic electrode with low carbon content suited for joining and surfacing on corrosion-resistant steels and cast steel type with austenitic-ferritic structure (Duplex-steels)
6809 Mo			Rutile-basic austenitic-ferritic electrode with low carbon content suited for joining and surfacing on corrosion-resistant steels and cast steel types with an austenitic-ferritic structure (Duplex-steels)

TABLE 15-6 Stainless Steel Electrodes, Filler Metals, and Wires. Courtesy of UTP Welding Materials, Inc.

E310-15, E310-16, E309Cb-15, E309Cb-16, E310Mo-15, E310Mo-16, and ER310

All are filler metals for 310 stainless steels. 310 stainless steels are used for high-temperature service where low creep is desired, such as for jet engine parts, valves, and furnace parts. All E310 filler metals can be used to weld 309 stainless or to join mild steel to stainless or to weld most hard-to-weld carbon and alloy steels. E310Mo-15 and 16 electrodes have molybdenum added to improve their strength at high temperatures and to resist corrosive pitting.

E316-15, E316-16, E3116L-15, E316L-16, ER316, ER316L, and ER316L-Si

All are filler metals for 316 stainless steels. 316 stainless steels are used for high-temperature service where high strength with low creep is desired. Molybdenum is added to improve these properties and to resist corrosive pitting. E316 filler metals are used for welding tubing, chemical pumps, filters, tanks, and furnace parts. All E316 filler metals can be used on 316 stainless steels or when weld resistance to pitting is required.

Nonferrous Electrode The AWS identification system for covered nonferrous electrodes is based on the atomic symbol or symbols of the major alloy(s) or the metal's identification number. The alloy having the largest percentage appears first in the identification. The atomic symbol is prefixed by the letter *E*. For example, ECu is a covered copper arc welding electrode, and ECuNiAl is a copper-nickel-aluminum alloy covered arc welding electrode. A letter, number, or letter-number combination may be added to the right of the atomic symbol to indicate some special alloys. For example, ECuAl-A2 is a copper-aluminum welding electrode that has 1.59% iron added.

Aluminum and Aluminum Alloys

The AWS specifications for aluminum and aluminum alloy filler metals are A5.3 for covered arc welding electrodes and A5.10 for bare welding rods and electrodes. Filler metal classified within the A5.3 uses the atomic symbol Al, and in the A5.10 the prefix ER is used with the Aluminum Association number for the alloy, Table 15-7.

Aluminum Covered Arc Welding Electrodes

Al-2 and Al-43

The aluminum electrodes do not use the letter *E* before the electrode number. Aluminum covered arc welding electrodes are designed to weld with DCEP polarity. These electrodes can be used on thin or thick sections, but thick sections must be preheated to between 300°F (150°C) and 600°F (315°C). The preheating of these thick sections allows the weld to penetrate immediately when the weld starts. Aluminum arc welding electrodes can be used on 2024, 3003, 5052, 5154, 5454, 6061, and 6063 aluminum. When welding on aluminum, a thin layer of surface oxide may not prevent welding. Thicker oxide layers must be removed mechanically or chemically. Excessive penetration can be supported by carbon plates or carbon paste. Most arc welding electrodes can also be used for oxyfuel gas welding of aluminum.

Aluminum Bare Welding Rods and Electrodes

ER1100

1100 aluminum has the lowest percentage of alloy agents of all of the aluminum alloys, and it melts at 1215°F. The filler wire is also relatively pure. ER1100 produces welds that have good corrosion resistance and high ductility, with tensile strengths ranging from 11,000 to 17,000 psi. The weld deposit has a high resistance to cracking during welding. This wire can be used with OFW, GTAW, and GMAW. Preheating to 300°F to 350°F is required for GTA welding on plate or pipe 3/8 in. and thicker to insure good fusion. Flux is required for OFW. 1100 aluminum is commonly used for items such as food containers, food-processing equipment, storage tanks, and heat exchangers. ER1100 can be used to weld 1100 and 3003 grade aluminum.

ER4043

ER4043 is a general-purpose welding filler metal. It has 4.5% to 6.0% silicon added, which lowers its melting temperature to 1155°F. The lower melting temperature helps promote a free-flowing molten weld pool. The welds have high ductility and a high resistance to cracking during welding. This wire can be used with OFW, GTAW, and GMAW. Preheating to 300°F to 350°F is required for GTA welding on plate or pipe 3/8 in. and thicker to ensure good fusion. Flux is required for OFW. ER4043 can be used to weld on 2014, 3003, 3004, 4043, 5052, 6061, 6062, and 6063 and cast alloys 43, 355, 356, and 214.

ER5356

ER5356 has 4.5% to 5.5% magnesium added to improve the tensile strength. The weld has high ductility but only an average resistance to cracking during welding. This wire can be used for GTAW and GMAW. Preheating to 300°F to 350°F is required for GTA welding on plate or pipe 3/8 in. and thicker to insure good fusion. ER5356 can be used to weld on 5050, 5052, 5056, 5083, 5086, 5154, 5356, 5454, and 5456.

Base Metal	319 355	43 356	214	6061 6063 6151	5456	5454	5154 5254	5086	5083	5052 5652	5005 5050	3004	1100 3003	1060
1060	4145 4043 4047	4043 4047 4145	4043 5183 4047	4043 4047	5356 4043	4043 5183 4047	4043 5183 4047	5356 4043	5356 4043	4043 4047	1100 4043	4043	1100 4043	1260 4043 1100
1100 3003	4145 4043 4047	4043 4047 4145	4043 5183 4047	4043 4047	5356 4043	4043 5183 4047	4043 5183 4047	5356 4043	5356 4043	4043 5183 4047	4043 5183 5356	4043 5183 5356	1100 4043	
3004	4043 4047	4043 4047	5654 5183 5356	4043 5183 5356	5356 5183 5556	5654 5183 5356	5654 5183 5356	5356 5183 5556	5356 5183 5556	4043 5183 4047	4043 5183 5356	4043 5183 5356		
5005 5050	4043 4047	4043 4047	5654 5183 5356	4043 5183 5356	5356 5183 5556	5654 5183 5356	5654 5183 5356	5356 5183 5556	5356 5183 5556	4043 5183 4047	4043 5183 5356			
5052 5652	4043 4047	4043 5183 4047	5654 5183 5356	5356 5183 4043	5356 5183 5556	5654 5183 5356	5654 5183 5356	5356 5183 5556	5356 5183 5556	5654 5183 4043				
5083	NR	5356 4043 5183	5356 5183 5556	5356 5183 5556	5183 5356 5556	5356 5183 5556	5356 5183 5556	5356 5183 5556	5183 5356 5556					
5086	NR	5356 4043 5183	5356 5183 5556	5356 5183 5556	5356 5183 5556	5356 5183 5554	5356 5183 5554	5356 5183 5556						
5154 5254	NR	4043 5183 4047	5654 5183 5356	5356 5183 4043	5356 5183 5554	5654 5183 5356	5654 5183 5356							
5454	4043 4047	4043 5183 4047	5654 5183 5356	5356 5183 4043	5356 5183 5554	5554 4043 5183								
5456	NR	5356 4043 5183	5356 5183 5556	5356 5183 5556	5556 5183 5356									
6061 6063 6151	4145 4043 4047	4043 5183 4047	5356 5183 4043	4043 5183 4047										
214	NR	4043 5183 4047	5654 5183 5356											
43 356	4043 4047	4145 4043 4047												
319 355	4145 4043 4047													

Note: *First filler alloy listed in each group is the all-purpose choice. NR means that these combinations of base metals are not recommended for welding.*

TABLE 15-7 Recommended Filler Metals for Joining Different Types of Aluminum to the Same Type or a Different Type of Aluminum. Courtesy of Thermacote Welco.

ER5556

ER5556 has 4.7% to 5.5% magnesium and 0.5% to 1.0% manganese added to produce a weld with high strength. The weld has high ductility and only average resistance to cracking during welding. This wire can be used for GTAW and GMAW. Preheating to 300°F to 350°F is required for GTA welding on plate or pipe 3/8 in. and thicker to ensure good fusion. ER5556 can be used to weld on 5052, 5083, 5356, 5454, and 5456.

Special-purpose Filler Metals

ENi

The nickel arc welding electrodes are designed to be used with AC or DCEP polarity. These arc welding electrodes are used for cast iron repair. The carbon in cast iron will not migrate into the nickel weld metal, thus preventing cracking and embrittlement. The cast iron may or may not be preheated. A very short arc length and a fast travel rate should be used with these electrodes.

ECuAl

The aluminum bronzed welding electrodes are designed to be used with DCEP polarity. This welding electrode has copper as its major alloy. The aluminum content is at a much lower percentage. Iron is usually added but at a percentage that is very low. These electrodes are sometimes referred to as arc brazing electrodes, although this is not an accurate description. Stringer beads and a short arc length should be used with these electrodes. Aluminum bronze welding electrodes are used for overlaying bearing surfaces; welding on castings of manganese, bronze, brass, or aluminum bronze; or assembling dissimilar metals.

Surface and Buildup Electrode Classification Hardfacing or wear-resistant electrodes are the most popular special-purpose electrodes; however, there are also cutting and brazing electrodes. Specialty electrodes may be identified by manufacturers' trade names. Most manufacturers classify or group hardfacing or wear-resistant electrodes according to their resistance to impact, abrasion, or corrosion. Occasionally, electrode resistance to wear at an elevated temperature is listed. One electrode may have more than one characteristic or type of service listed.

EFeMn-A

The EFeMn-A electrodes are designed to be used with AC or DCEP polarity. This electrode is an impact-resistant welding electrode. It can be used on hammers, shovels, and spindles and in other similar applications.

ECoCr-C

The ECoCr-C electrodes are designed to be used with AC or DCEP polarity. This electrode is a corrosion- and abrasion-resistant welding electrode. It also maintains its resistance to elevated temperatures. ECoCr-C is commonly used for engine cams, seats and valves, chain saw bars, bearings, and dies.

Magnesium Alloys The joining of magnesium alloys by torch welding or brazing is possible without a fire hazard because the melting point of magnesium is 1202°F (651°C) to 858°F (459°C) below its boiling point, where magnesium may start to burn.

ER AZ61A

The ER AZ61A filler metal can be used to join most magnesium wrought alloys. This filler has the best weldability and weld strength for magnesium alloys AZ31B, HK31A, and HM21A.

ER AZ92A

The ER AZ92A filler metal can be used on cast alloys, Mg-Al-Zn and AM 100A. This filler metal has a somewhat higher resistance to cracking.

Summary

Proper filler metal selection is one of the most important factors affecting the successful welding of a joint. Many factors affect the selection of the most appropriate filler metal for a job. In some cases cost is the greatest factor and in others it is structural strength. For example, if you were building an ultralight aircraft, you would be more concerned with strength than cost. However, if you were building an iron fence, you might be more concerned with cost. Every application is different, so it may be a help for you to list the items you feel are most important for selecting a filler metal. This will help you select the most appropriate filler metal for your needs.

Manufacturers' literature on filler metals can be divided into two general sections. One section of the literature is technical and the other is advertisement. In the technical section you are provided with specific information on each filler metal's operation, performance, and uses. In the advertisement section you are provided with marketing information and claims regarding performance. Knowing the types of information in both sections will help you evaluate new material as you select filler metal.

If you are considering a large purchase of filler metals, it is advantageous for you to request samples of the various filler metals from manufacturers, so that you can test their performance in your applications. Pretesting of the products in your applications will give you an opportunity to determine which filler metal is going to give you the best value for your money. It may also be necessary to qualify the filler metal for your welding certification program before you make the purchase and begin using the product.

Review

1. Where might you find welding information about an electrode that would give the manufacturers' recommended preheating and postheating temperatures for welding?
   A. Mechanical Test Data
   B. Electrode Data
   C. Chemical Analysis Data
   D. MSDS Data
2. Which mechanical test data would give the minimum load that it would take to break a section of sound weld that has a cross-sectional area of one square inch?
   A. tensile strength test
   B. yield point test
   C. elongation percent test
   D. charpy V notch test
3. What chemical increases the tensile strength and hardness of the weld deposit?
   A. carbon (C)
   B. phosphorus (P)
   C. silicon (Si)
   D. copper (Cu)
4. As the percentage of nickel increases, tensile strength, toughness, and corrosion resistance increase.
   A. true
   B. false
5. Which of the following is not a name used to describe SMA welding electrodes?
   A. welding rods
   B. arc welding rods
   C. welding wire
   D. stick electrodes
6. What is the primary metal source for an SMA weld?
   A. the flux covering
   B. the flux core
   C. the powder added to the shielding
   D. the core wire
7. Which of the following is not a function of the flux covering of SMA welding rods?
   A. serve as an insulator
   B. provide a slag cover on the weld
   C. carry the welding current
   D. provide arc stabilizers
8. What are the elements in fluxes called that combine with contaminants and float them to the surface of the molten weld pool?
   A. arc stabilizers
   B. scavengers
   C. alloying elements
   D. filler metals
9. The type of joint and whether it is grooved or not may affect the performance of the welding electrode.
   A. true
   B. false
10. What are the properties of a weld such as tensile strength, yield strength, hardness, toughness, ductability, and impact strength collectively called?
    A. chemical
    B. mechanical
    C. electrical
    D. thermal
11. The prefix letter _____ indicates a rod that is heated by some source other than electric current flowing through it.
    A. R
    B. W
    C. B
    D. X
12. Using Table 15-1, what is the as-welded tensile strength in psi for an E6011 electrode?
    A. 52–62,000
    B. 62–68,000
    C. 70–78,000
    D. 50–64,000
13. Using Table 15-1 what is the DC voltage range for a 1/8 in. E6010 electrode?
    A. 56-110
    B. 140-225
    C. 40-90
    D. 75-130
14. The prefix letters _____ indicate a welding rod used primarily with oxyfuel welding.
    A. ER
    B. BCuZn
    C. IN
    D. RG

Review continued

15. According to Table 15-2, which electrode can only be used in the H-Fillet and F welding positions?
    A. E7024
    B. E6022
    C. E7016
    D. E7028
16. The ______ electrodes are designed to be used with AC or DCEN polarity and have rutile-based flux.
    A. E6010
    B. E6011
    C. E6012
    D. E7018
17. The ______ electrodes are sometimes referred to as Lo-Hi rods because they allow very little hydrogen into the weld pool.
    A. E6011
    B. E7018
    C. E6012
    D. E7024
18. What are the AWS specifications for carbon steel filler metals for flux cored arc welding wire?
    A. A5.3
    B. A5.9
    C. A5.15
    D. A5.20
19. The letter _______ added to the right of the AISI number before the dash and suffix number indicates a low-carbon stainless welding electrode.
    A. R
    B. L
    C. W
    D. T
20. Aluminum covered arc welding electrodes are designed to weld with ____ polarity.
    A. AC
    B. DCSP
    C. DCEN
    D. DCEP

Chapter 16

Weldability of Metals

OBJECTIVES

After completing this chapter, the student should be able to

- ☑ list the methods used to weld most ferrous metals.
- ☑ list the methods used to weld four nonferrous metals.
- ☑ explain the precautions that must be taken when welding various metals and alloys.
- ☑ describe the effects of preheating and postheating on welding.

KEY TERMS

alloy steels
aluminum
hardfacing
plain carbon steels
stainless steels
steel classification systems
weldability

INTRODUCTION

All metals can be welded, although some metals require far more care and skill to produce acceptably strong and ductile joints. The term **weldability** has been coined to describe the ease with which a metal can be welded properly. Good weldability means that almost any process can be used to produce acceptable welds and that little effort is needed to control the procedures. Poor weldability means that the processes used are limited and that the preparation of the joint and the procedure used to fabricate it must be controlled very carefully or the weldment will not function as intended.

Knowing why a part broke is as important to the repair process as knowing the type of metal it is made with. Parts break because they are worn out, damaged in an accident, underdesigned for the work, or defective. Welders have the greatest success fixing parts that are worn out or damaged in an accident. These parts were giving good service before they broke. Parts that were underdesigned or defective are more likely to break again if they are welded without fixing the design problem or defect first. For example, a bracket that vibrates too much and cracks will continue to vibrate and will crack again if you just weld the crack. Instead, see what can be done to stop the vibration to prevent it from cracking again. In other words, fix the problem before fixing the part—or you will be fixing the part again and again.

Welding processes produce a thermal cycle in which the metals are heated over a range of temperatures. Cooling of the metal

FIGURE 16-1 Improper welding technique for filler metal resulting in lack of fusion. Courtesy of Larry Jeffus.

to ambient temperatures then follows. The heating and cooling cycle can set up stresses and strains in the weld. Whatever the welding process used, certain metallurgical, physical, and chemical changes also take place in the metal. A wide range of welding conditions can exist for welding methods when joining metals with good weldability. However, if weldability is a problem, adjustments usually will be necessary in one or more of the following factors:

- Filler metal—If the wrong filler metal is selected, the weld can have major defects and not be fit for service. Common defects include porosity, cracks, and filler metal that just will not stick, **Figure 16-1.** The cracks can be in the filler metal or in the base metal alongside the weld. Chapter 15 lists various types of filler metals and their applications. If you are not sure which filler metal to use, a good general rule is that a little stronger or higher grade can be used successfully, but a lower strength or grade seldom works.
- Preheat and postheat—Cracking is a common problem when welding on brittle metals such as cast iron or some high-strength alloys. Preheating the part before starting the weld will reduce the stress caused by the weld and will help the filler metal flow. The most commonly used preheat temperature range is between 250°F and 400°F (120°C and 200°C) for most steel, **Table 16-1.** The preheat temperature can be as high as 1200°F (650°C) when welding cast iron. Preheating is required anytime the metal to be welded is below 70°F (20°C), because the cold metal quenches the weld. Postheating slows the cool-down rate following welding, which will prevent post-weld cracking of brittle metals. Postheating also reduces weld stresses that can result in cracks forming some time after the part is repaired.
- Welding procedure—The size of the weld bead, the number of welds, and the length of the welds all affect the weld. When large welds are needed, it is better to make more small welds than a few large welds. The small welds serve to postheat the weld. They reduce stresses and result in less distortion. Sometimes a series of short back stepping welds can be made, **Figure 16-2.** These short welds can be on very brittle metals like cast iron.

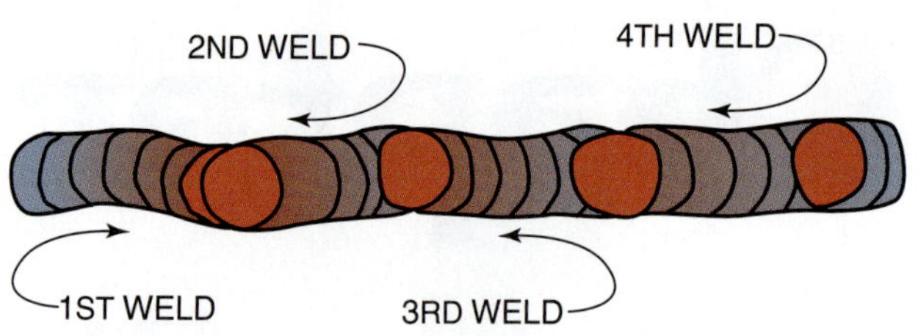

FIGURE 16-2 Welding sequence to produce the minimum weld stresses.

Steel Classification and Identification

SAE and AISI Classification Systems Two primary numbering systems have been developed to classify the standard construction grades of steel, including both carbon and alloy steels. These systems classify the types of steel according to their basic chemical composition. One classification sys-

Alloy Name	% Carbon*	Major Properties
Iron	0.0% to 0.03%	Soft, easily formed, not hardenable
Low carbon	0.03% to 0.30%	Strong, formable
Medium carbon	0.30% to 0.50%	High strength, tough
High carbon	0.50% to 0.90%	Hard, tough
Tool steel	0.80% to 1.50%	Hard, brittle
Cast iron	2% to 4%	Hard, brittle, most types resist oxidation

TABLE 16-1 Recommended Minimum Preheat Temperature for Carbon Steel.

tem was developed by the Society of Automotive Engineers (SAE). The other system is sponsored by the American Iron and Steel Institute (AISI).

The numbers used in both systems are now just about the same. However, the AISI system uses a letter before the number to indicate the method used in the manufacture of the steel.

Both numbering systems usually have a four-digit series of numbers. In some cases, a five-digit series is used for certain alloy steels. The entire number is a code to the approximate composition of the steel.

In both steel classification systems, the first number often, but not always, refers to the basic type of steel, as follows:

1XXX Carbon

2XXX Nickel

3XXX Nickel chrome

4XXX Molybdenum

5XXX Chromium

6XXX Chromium vanadium

7XXX Tungsten

8XXX Nickel chromium vanadium

9XXX Silicomanganese

The first two digits together indicate the series within the basic alloy group. There may be several series within a basic alloy group, depending upon the amount of the principle alloying elements. The last two or three digits refer to the approximate permissible range of carbon content. For example, the metal identified as 1020 would be 1XXX carbon steel with a XX20 0.20% range of carbon content, and 5130 would be 5XXX chromium steel with an XX30 0.030% range of carbon content.

The letters in the AISI system, if used, indicate the manufacturing process as follows:

- C—Basic open-hearth or electric furnace steel and basic oxygen furnace steel
- E—Electric furnace alloy steel

Table 16-2 shows the AISI and SAE numerical designations of alloy steels.

13XX	Manganese 1.75
23XX**	Nickel 3.50
25XX**	Nickel 5.00
31XX	Nickel 1.25; chromium 0.65
E33XX	Nickel 3.50; chromium 1.55; electric furnace
40XX	Molybdenum 0.25
41XX	Chromium 0.50 or 0.95; molybdenum 0.12 or 0.20
43XX	Nickel 1.80; chromium 0.50 or 0.80; molybdenum 0.25
E43XX	Same as above, produced in basic electric furnace
44XX	Manganese 0.80; molybdenum 0.40
45XX	Nickel 1.85; molybdenum 0.25
47XX	Nickel 1.05; chromium 0.45; molybdenum 0.20 or 0.35
50XX	Chromium 0.28 or 0.40
51XX	Chromium 0.80, 0.88, 0.93, 0.95, or 1.00
E5XXXX	High carbon; high chromium; electric furnace bearing steel
E50100	Carbon 1.00; chromium 0.50
E51100	Carbon 1.00; chromium 1.00
E52100	Carbon 1.00; chromium 1.45
61XX	Chromium 0.60, 0.80, or 0.95; vanadium 0.12, or 0.10, or 0.15 minimum
7140	Carbon 0.40; chromium 1.60; molybdenum 0.35; aluminum 1.15
81XX	Nickel 0.30; chromium 0.40; molybdenum 0.12
86XX	Nickel 0.55; chromium 0.50; molybdenum 0.20
87XX	Nickel 0.55; chromium 0.50; molybdenum 0.25
88XX	Nickel 0.55; chromium 0.50; molybdenum 0.35
92XX	Manganese 0.85; silicon 2.00; 9262-chromium 0.25 to 0.40
93XX	Nickel 3.25; chromium 1.20; molybdenum 0.12
98XX	Nickel 1.00; chromium 0.80; molybdenum 0.25
14BXX	Boron
50BXX	Chromium 0.50 or 0.28; boron
51BXX	Chromium 0.80; boron
81BXX	Nickel 0.33; chromium 0.45; molybdenum 0.12; boron
86BXX	Nickel 0.55; chromium 0.50; molybdenum 0.20; boron
94BXX	Nickel 0.45; chromium 0.40; molybdenum 0.12; boron

Note: The elements in this table are expressed in percent.

*Consult current AISI and SAE publications for the latest revisions.

**Nonstandard steel.

TABLE 16-2 AISI and SAE Numerical Designation of Alloy Steels.

Unified Numbering System (UNS) A unified numbering system is presently being promoted for all metals. This system eventually will replace the AISI and other systems.

Carbon and Alloy Steels

Steels alloyed with carbon and only a low concentration of silicon and manganese are known as plain carbon steels. These steels can be classified as low carbon, medium carbon, and high carbon steels. The division is based upon the percentage of carbon present in the material.

Plain carbon steel is basically an alloy of iron and carbon. Small amounts of silicon and manganese are added to improve their working quality. Sulfur and phosphorus are present as undesirable impurities. All steels contain some carbon, but steels that do not include alloying elements other than low levels of manganese or silicon are classified as plain carbon steels. Alloy steels contain specified larger proportions of alloying elements.

The AISI has adopted the following definition of carbon steel: "Steel is classified as carbon steel when no minimum content is specified or guaranteed for aluminum, chromium, columbium, molybdenum, nickel, titanium, tungsten, vanadium, or zirconium; and when the minimum content of copper which is specified or guaranteed does not exceed 0.40%; or when the maximum content which is specified or guaranteed for any of the following elements does not exceed the respective percentages hereinafter stated: manganese 1.65%, silicon 0.60%, copper 0.60%." Under this classification will be steels of different composition for various purposes.

Many special alloy steels have been developed and sold under various trade names. These alloy steels usually have special characteristics, such as high tensile strength, resistance to fatigue, corrosion resistance, or the ability to perform at high temperatures. Basically, the ability of carbon steel to be welded is a function of the carbon content, Table 16-3. (Other factors to be considered include thickness and the geometry of the joint.) All carbon steels can be welded by at least one method. However, the higher the carbon content of the metal, the more difficult it is to weld the steel. Special instructions must be followed in the welding process.

Low Carbon and Mild Steel Low carbon steel has a carbon content of 0.15% or less, and mild steel has a carbon content range of 0.15% to 0.30%. Both steels can be welded easily by all welding processes. The resulting welds are of extremely high quality. Oxyacetylene welding of these steels can be done by using a neutral flame. Joints welded by this process are of high quality, and the fusion zone is not hard or brittle.

Both low carbon and mild steels can be welded readily by the shielded metal arc (stick welding) method. The selection of the correct electrode for the particular welding application helps to ensure high strength and ductility in the weld.

The gas metal arc (MIG) and flux cored arc welding processes are used for welding both low and medium carbon steels due to the ease of welding and because they prevent contamination of the weld. The high productivity and lower cost make them increasingly popular welding processes.

Medium Carbon Steel The welding of medium carbon steels, having 0.30% to 0.50% carbon content, is best accomplished by the various fusion processes, depending upon the carbon content of the base metal. The welding technique and materials used are dictated by the metallurgical characteristics of the metal being welded. For steels containing more than 0.40% carbon, preheating and subsequent heat treatment generally are required to produce a satisfactory weld. Stick welding electrodes of the type used on low carbon steels can be used for welding this type of steel. The use of an electrode with a special low hydrogen coating may be necessary to reduce the tendency toward underbead cracking.

Common Name	Carbon Content	Typical Use	Weldability
Ingot iron	0.03% max.	Galvanizing and deep drawing sheet and strip	Excellent
Low carbon steel	0.15% max.	Welding electrodes, special plate and shapes, sheet and strip	Excellent
Mild steel	0.15% to 0.30%	Structural shapes, plate and bar	Good
Medium carbon steel	0.30% to 0.50%	Machinery parts, draw bars	Fair*
High carbon steel	0.50% to 1.00%	Springs, dies, and equipment tracks	Poor**

*Preheat and frequently postheat required.

**Difficult to weld without adequate preheat and postheat.

Table 16-3 Iron-Carbon Alloys, Uses, and Weldabilities.

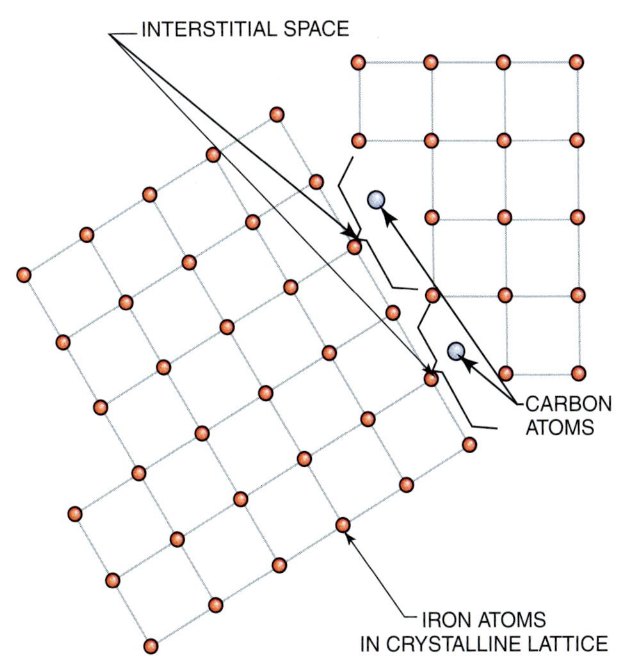

FIGURE 16-3 Preheating range for welding. Tempil Division, Big Three Industries, Inc.

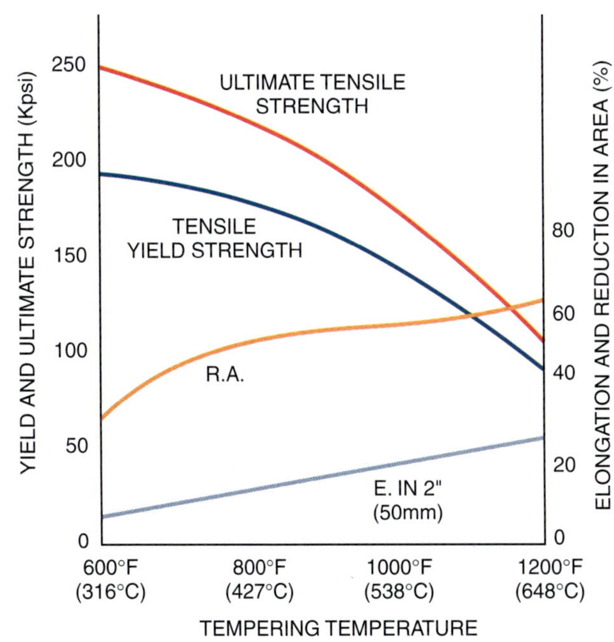

FIGURE 16-4 Stress-relieving range. Tempil Division, Big Three Industries, Inc.

Note: Electrodes for welding medium and high carbon steels, stainless steel, and cast iron are available. Some of these electrodes are very expensive, costing more than $100 a pound. Small quantities of these specialty stick electrodes are available from some agricultural supply houses and most welding suppliers. They are often prepackaged in small quantities, and some suppliers will even allow you to purchase one or two electrodes. When a single electrode can cost $20 or more, being able to purchase just the number of electrodes you need for a repair weld is important. MIG and flux cored electrode wires come on spools and are not available in very small quantities. Therefore, MIG and flux core welding may not be very cost effective for most agricultural repair welding.

High Carbon Steel High carbon steels usually have a carbon content of 0.50% to 0.90%. These steels are much more difficult to weld than either the low or medium carbon steels. Because of the high carbon content, the heat-affected zone next to the weld can become very hard and brittle. The welder can avoid this by using preheat and by selecting procedures that produce high-energy inputs to the weld. Refer to **Figure 16-3** for the preheat temperature for the specific carbon content. The martensite that does form is tempered by postweld heat treatments such as stress relief anneal. Refer to **Figure 16-4** for the temperature for stress-relief annealing between 1125°F and 1250°F (600°C and 675°C).

In arc welding high carbon steel, mild-steel shielded arc electrodes are generally used. However, the weld metal does not retain its normal ductility because it absorbs some of the carbon from the steel.

Welding on high carbon steels is often done to build up a worn surface to original dimensions or to develop a hard surface. In this type of welding, preheating or heat treatment may not be needed because of the way in which heat builds up in the part during continuous welding.

Stainless Steels

Stainless steels consist of four groups of alloys: austenitic, ferritic, martensitic, and precipitation hardening. The austenitic group is by far the most common. Its chromium content provides corrosion resistance, while its nickel content produces the tough austenitic microstructure. These steels are relatively easy to weld, and a large variety of electrode types are available.

The most widely used stainless steels are the chromium-nickel austenitic types. They are used for items such as dairy equipment, including milk tanks; hand sprayer tanks; and poultry medicine infusion system pumps, and are usually referred to by their chromium-nickel content as 18/8, 25/12, 25/20, and so on. For example, 18/8 contains 18% chromium and 8% nickel, with 0.08% to 0.20% carbon. To improve weldability, the carbon content should be as low as possible. Carbon should not be more than 0.03%, with the maximum being less than 0.10%.

Keeping the carbon content low in stainless steel will also help reduce carbide precipitation. Carbide

Nominal Composition of Stainless Steels						
AISI Type	Nominal Composition %					
	C	Mn Max	Si Max	Cr	Ni	Other
304	0.08 max.	2.0	1.0	18–20	8–12	
304L	0.03 max.	2.0	1.0	18–20	8–12	
316	0.08 max.	2.0	1.0	16–18	10–14	2.0–3.0 Mo
316L	0.03 max.	2.0	1.0	16–18	10–14	2.0–3.0 Mo

TABLE 16-4 Comparison of Standard-Grade and Low-Carbon Stainless Steels.

precipitation occurs when alloys containing both chromium and carbon are heated. The chromium and carbon combine to form chromium carbide (Cr_3C_2).

The combining of chromium and carbon lowers the chromium that is available to provide corrosion resistance in the metal. This results in a metal surrounding the weld that will oxidize, or rust. The amount of chromium carbide formed is dependent on the percentage of carbon, the time that the metal is in the critical range, and the presence of stabilizing elements.

If the carbon content of the metal is very low, little chromium carbide can form. Some stainless steel types have a special low carbon variation. These low carbon stainless steels are the same as the base type but with much lower carbon content. To identify the low carbon from the standard AISI number the *L* is added as a suffix. See examples 304 and 304L, **Table 16-4.**

Chromium carbides form when the metal is between 800°F and 1500°F (625°C and 815°C). The quicker the metal is heated and cooled through this range the less time that chromium carbides can form. Since austenitic stainless steels are not hardenable by quenching, the weld can be cooled using a chill plate. The chill plate can be water-cooled for larger welds.

Some filler metals have stabilizing elements added to prevent carbide precipitation. Columbium and titanium are both commonly found as chromium stabilizers. Examples of the filler metals are E310Cb and ER309Cb.

In fusion welding, stainless austenitic steels may be welded by all of the methods used for plain carbon steels.

Since *ferritic stainless* steels contain almost no nickel, they are cheaper than austenitic steels. They are used for ornamental or decorative applications such as architectural trim and at elevated temperatures such as used for heat exchanging. However, ferritic stainless steels also tend to be brittle unless specially deoxidized. Special high-purity, high-toughness ferritic stainless steels have been developed, but careful welding procedures must be used with them to prevent embrittlement. This means very carefully controlling nitrogen, carbon, and hydrogen.

Martensitic stainless steels are also low in nickel but contain more carbon than the ferritic. They are used in applications requiring both wear resistance and corrosion resistance. Items such as surgical knives and razor blades are made of them. Quality welding requires very careful control of both preheat and tempering immediately after welding.

Precipitation hardening stainless steels can be much stronger than the austenitic, without losing toughness. Their strength is the result of a special heat treatment used to develop the precipitate. They can be solution treated prior to welding and given the precipitation treatment after welding.

The closer the characteristics of the deposited metal match those of the material being welded, the better is the corrosion resistance of the welded joint. The following precautions should be noted:

- Any carburization or increase in carbon must be avoided, unless a harder material with improved wear resistance is actually desired. In this case, there will be a loss in corrosion resistance.
- It is important to prevent all inclusions of foreign matter, such as oxides, slag, or dissimilar metals.

In welding with the metal arc process, direct current is more widely used than alternating current. Generally, reverse polarity is preferred where the electrode is positive and the workpiece is negative.

The diameter of the electrode used to weld steel that is thinner than 3/16 in. (4.8 mm) should be equal to, or slightly less than, the thickness of the metal to be welded.

When setting up for welding, material .050 in. (1.27 mm) and less in thickness should be clamped firmly to prevent distortion or warpage. The edges should be butted tightly together. All seams should be accurately aligned at the start. It is advisable to tack weld the joint at short intervals as well as to use clamping devices.

The electrode should always point into the weld in a backhand or drag angle, **Figure 16-5.** Avoid using a figure-8 pattern or excessive side weaving motion such as that used in welding carbon steel. Best results are obtained with a stringer bead with little or very slight weaving motion and steady forward

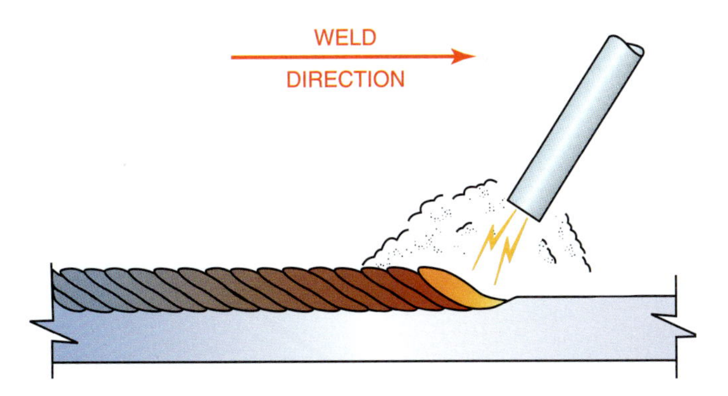

FIGURE 16-5 Backhand or drag angle.

travel, with the electrode holder leading the weld pool at about 60° in the direction of travel.

To weld stainless steels, the arc should be as short as possible. **Table 16-5** can be used as a guide.

Cast Iron

Cast iron is widely used in agriculture for engine components such as blocks, heads, and manifolds; for drive components such as transmission cases, gear boxes, transfer cases, and differential cases; and for equipment such as pumps, planters, drills, and pulleys. Cast iron is hard and ridged, which makes it ideal for any size casing or frame that must hold its shape even under heavy loads. For example, if a transmission case were to bend under a load, the gears and shafts inside would bind and stop turning.

All five types of cast iron have high carbon contents, usually ranging from 1.7% to 4%. The most common grades contain about 2.5% to 3.5% total carbon. The carbon in cast iron can be combined with iron or be in a free state. As more of the carbon atoms in the cast iron combine with iron atoms the cast iron becomes harder and more brittle. The five common types of cast iron are as follows:

- Gray cast iron—it is the most widely used type. It contains so much free carbon that a fracture surface has a uniform dark gray color. Gray cast iron is easily welded but because it is somewhat porous it can absorb oils into the surface which must be baked out before welding.
- White cast iron—is the hardest and most brittle of the cast irons because almost all of the carbon atoms are combined with the iron atoms. The surface of a fractured piece of this cast iron looks silvery white and may appear shiny. White cast iron is practically unweldable.
- Malleable cast iron—is white cast iron that has undergone a transformation as the result of a long heat treating process to reduce the brittleness. The fractured surface of malleable cast iron has a light, almost white, thin rim around the dark gray center. If malleable cast iron is heated above its critical temperature, about 1700°F (925°C), the carbon will recombine with the iron, transforming back into white cast iron. Malleable cast iron can easily be welded. To prevent it from reverting back to white cast iron do not preheat above 1200°F (650°C).
- Alloy cast iron—has alloying elements such as chromium, copper, manganese, molybdenum, or nickel added to obtain special properties. Various quantities and types of alloys are added to improve alloy cast iron's tensile strength and heat and corrosion resistance. Almost all grades of alloy cast iron can be easily welded if care is taken to slowly preheat and postcool the part to prevent changes in the carbon and iron structure.
- Nodular cast iron—sometimes called ductile cast iron, has its carbon formed into nodules or tiny round balls. These nodules are formed by adding an alloy. Nodular cast iron has greater tensile strength than gray cast iron and some of the corrosion resistance of alloy cast iron. Nodular cast iron is weldable, but proper preheating and postweld cooling temperatures and rates must be maintained or the nodular properties will be lost.

Not all cracks or breaks in cast iron present the same degree of difficulty to making welded repairs. Breaks across ears or tabs do not have nearly as much stresses as a crack in a surface, **Figure 16-6**. Cracks may increase in length when you try to weld them unless you drill a small hole, about 1/8 in. (3 mm), at both ends of the crack, **Figure 16-7**.

Metal Thickness		Electrode Diameter		Current	Voltage
in.	(mm)	in.	(mm)	(Amperes)	Open Circuit
.050	(1.27)	5/64	(1.98)	25–50	30–35
.050–.0625	(1.27–1.58)	3/32	(2.38)	30–90	35–40
.0625–.1406	(1.58–3.55)	3/32–1/8	(2.38–3.17)	50–100	40–45
.1406–.1875	(3.55–4.74)	1/8–5/32	(3.17–3.96)	80–125	45–50
.250 and up	(6.35 and up)	3/16	(4.76)	100–175	55–60

TABLE 16-5 Shielded Metal Arc Welding Electrode Setup for Stainless Steel.

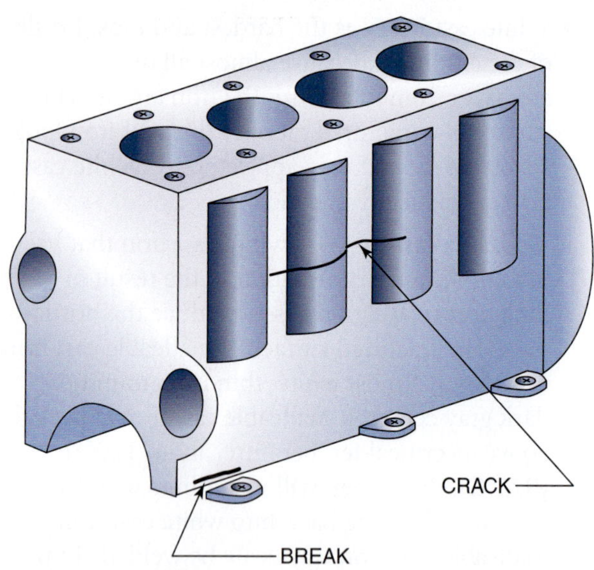

FIGURE 16-6 Cracks on engines can occur in the water jacket due to freezing, and ears can be broken off as a result of accidents or overlightening of misaligned parts.

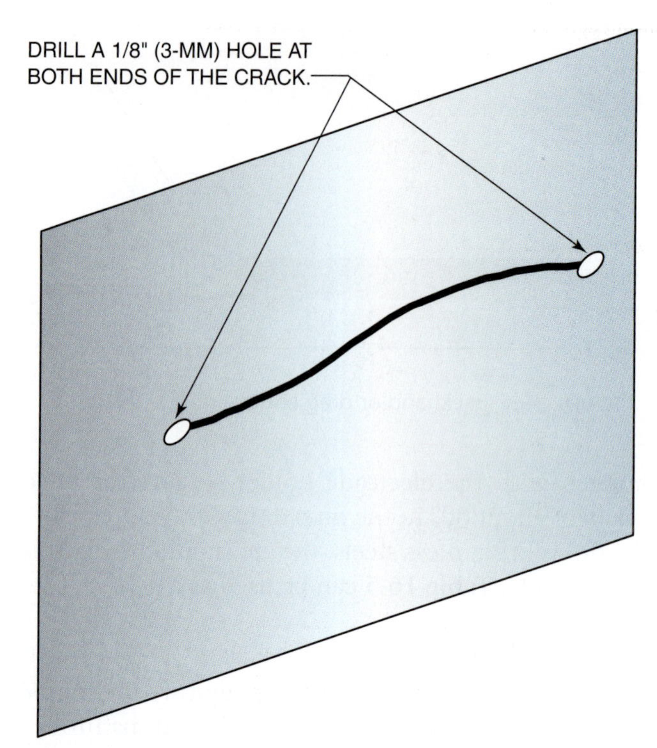

FIGURE 16-7 Stop drill the ends of a crack.

Preweld and Postweld Heating of Cast Iron

The major purpose of preheating and postheating of cast iron is to control the rate of temperature change. The level of temperature and the rate of change of temperature affects the hardness, brittleness, ductility, and strength of iron-carbon based metals such as steel and cast iron, **Table 16-6**.

Preheating the casting before welding reduces the internal stresses caused by the rapid or localized heating caused by welding, **Table 16-7**. Welding stresses occur because as metal is heated and cooled, it expands and contracts. Unless the heating and cooling cycles are slow and uniform, stresses within brittle materials will cause them to crack. In some aspects the brittleness of cast iron is much like that of glass. Both cast iron and glass will crack if they are heated or cooled unevenly or too quickly.

Property	Description
Hardness	The resistance to penetration or shaping by machining or drilling.
Brittleness	The ease with which a metal will crack or break apart without noticeable deformation.
Ductility	The ability of a metal to be permanently twisted, drawn out, bent, or changed in shape without cracking or breaking.
Strength	The ability of a metal to resist deformation or reshaping due to tensile, compression, shear, or torsional forces.

TABLE 16-6 Mechanical Properties of Metal.

	Preheat Temperatures for Weldable Cast Irons		
Joining Process	Temperature Range	Preferred Temperature	Minimum Temperature
Stick Welding	600°F to 1500°F (315°C to 815°C)	900°F to 1200°F (480°C to 650°C)	400°F (200°C)
Gas Welding	400°F to 1100°F (200°C to 600°C)	500°F (260°C)	400°F (200°C)
Braze Welding	500°F to 900°F (260°C to 480°C)	900°F (480°C)	500°F (260°C)

Note: Maintain the preheat temperature for 30 minutes after it is first reached to allow the core of the casting to reach this temperature.

TABLE 16-7 Welding Should be Performed at or Above the Preferred Temperature. (It can be performed at the minimum temperature, but some hardening and cracking may occur.)

Postweld heating changes the rate of cooling. Rapid cooling of a metal from a high temperature is called quenching. The faster an iron-carbon metal is quenched the harder, more brittle, less ductile, and higher in strength the metal will become. The slow cooling of an iron-carbon metal from a high temperature is called annealing. The slower an iron-carbon metal is cooled from a high temperature, the softer, less brittle, more ductile, and lower in strength the metal will become.

To reduce welding stresses, maintain the casting at the same temperature used for preheat or higher for 30 minutes following welding. The casting should cool slowly over the next 24 hours. Cover the casting to prevent the part from being cooled too rapidly by the surrounding air following welding. A firebrick or heavy metal box can be used to keep cool air away from the casting.

Practice Welding Cast Iron

Because there are a few differences between repairing a break and a crack, the following practices alternate between repairing breaks and cracks in cast iron. For example, other than not clamping the parts together, there would be little difference between using Practice 16-1 for welding a break and for welding a crack. In the field, you can use whichever welding procedure is most appropriate for repairing breaks or cracks.

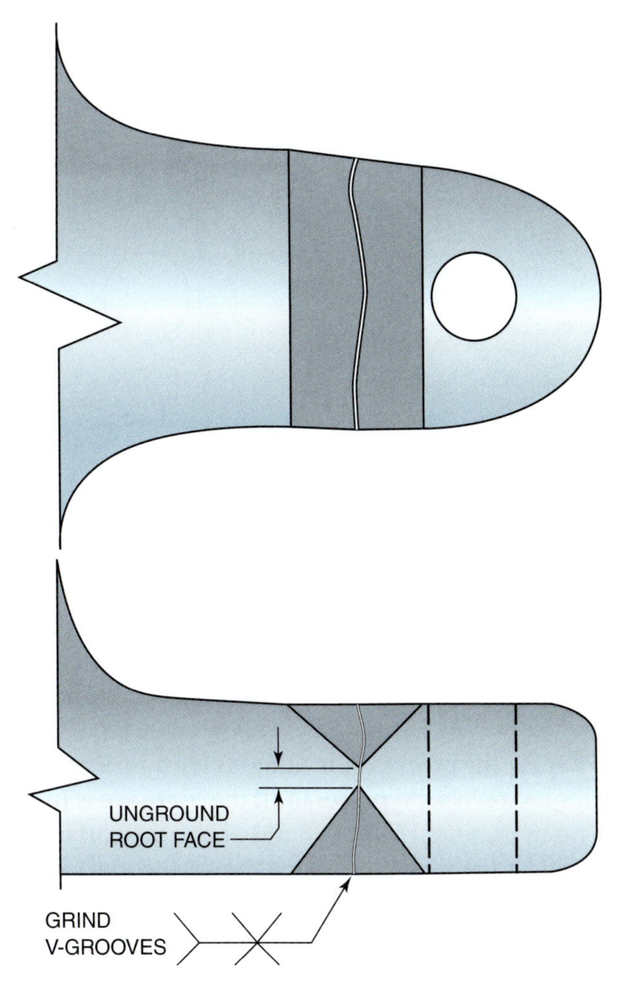

FIGURE 16-8 Grind the V groove, leaving a small root face so the part can be realigned.

PRACTICE 16-1

Arc Welding a Cast Iron Break with Preheating and Postheating

Using a properly set up stick welding station, a gas torch with a heating tip, firebricks, a right angle grinder, ENi electrodes, a C-clamp, a 900°F (482°C) temperature marking crayon, a chipping hammer, a wire brush, a broken piece of gray cast iron, a welding helmet, gas welding goggles, safety glasses, and all other required safety equipment, you are going to repair a cast iron break.

Grind the brake into a V-groove, leaving a 1/8-in. (3-mm) root face of the broken surface, **Figure 16-8.**

Note: Because cast iron is brittle it does not bend before it breaks, so broken parts can usually be fitted back together like the pieces of a puzzle.

Use a C-clamp to hold the broken piece in place. Mark the parts with a 900°F (482°C) temperature marking crayon.

Using a properly lit and adjusted oxyacetylene heating torch, begin preheating the part. Keep the flame moving all around the cast iron part so that it heats evenly. Do not point the flame directly onto the temperature indicator mark or it will turn color before the part is actually preheated. When the temperature mark turns black the part is properly preheated. With the flame off the part, remark the part to check it for proper preheat.

On thick castings, keep the flame on the part for 30 more minutes so the inside of the part is properly preheated.

Note: Starting the welds on the ends of the break or crack and welding to the center concentrates the weld stresses in the center of the weld. This reduces the chances of a crack forming at the end of the weld.

Strike an arc and make the first weld bead, starting at one edge. Weld to the center of the break. Break the arc, and chip and wire brush the flux off the weld. Strike another arc on the opposite end of the V-groove and make another weld back to the crack of the center, **Figure 16-9.** Turn the part over and make the same two welds on the back side of the break.

Note: A number of small welds are better than one or two large welds because the small welds do not

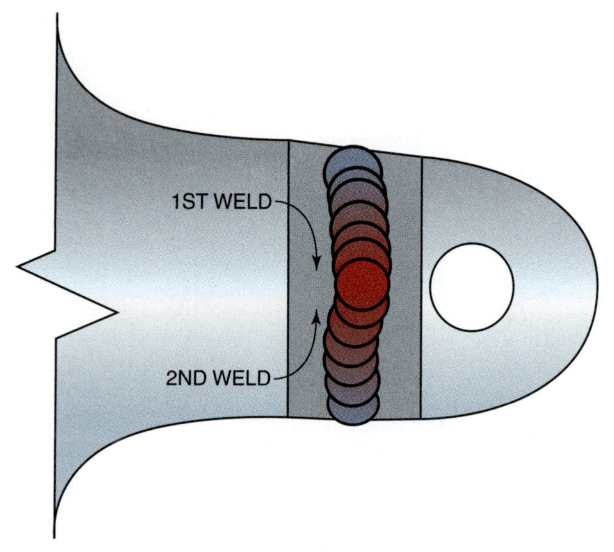

FIGURE 16-9 Start all welding beads on the edge of the break and end them in the middle.

have as much stress and are less likely to cause postweld cracking.

Repeat the process of making welds by alternating sides between weld passes until the V-groove is filled to no more than 1/8 in. (3 mm) above the surface.

Build a firebrick box around the part and place the torch so the flame will fill the box and the part can be postweld heated. Keep heating the part for 30 minutes after completing the weld. Close any gaps in the firebrick box and allow the part to cool slowly over the next 24 hours. Turn off the welder and torch set and clean up your work area when you are finished welding.

Complete a copy of the "Student Welding Report" listed in Appendix I or provided by your instructor. ◆

Welding without Preheating or Postheating

Cracks in large castings and castings that are to be repaired in place cannot be preheated and postheated to the desired temperatures but can still be repaired. One method of repairing these cracks is to drill and tap a series of overlapping holes along the crack, **Figure 16-10**. This is an excellent way to repair non–load-bearing cracks like those in the water jacket of an engine. Two-part epoxy patch material can also be used on non–load-bearing cracks. Read and follow the manufacturer instructions and safety rules when using epoxy repair kits.

Cracks in parts that cannot be preheated to the desired level can still be welded, but the welds will be very hard and are more likely to recrack. However, welding cracks in engine blocks, pump housings, and other large expensive castings, even if they might crack again, is more desirable than simply

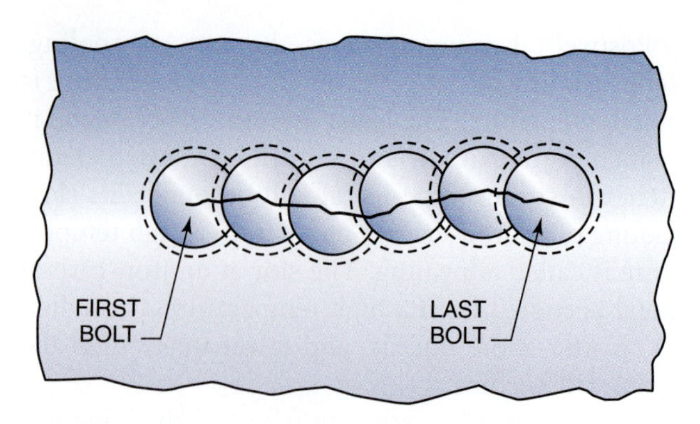

FIGURE 16-10 A crack in cast iron can be plugged by drilling and tapping overlapping holes. Each bolt overlaps the previous bolt. Only the last one must have a locking compound to prevent it from loosening.

buying a new part. The new cracks that might form may even be small enough to be patched with epoxy.

PRACTICE 16-2

Arc Welding a Cast Iron Crack without Preheating or Postheating

Using a properly set up stick welding station, a gas torch with a heating tip, firebricks, a portable drill with a 1/8-in. (3-mm) drill bit, a right angle grinder, ENi electrodes, a chipping hammer, a wire brush, a cracked piece of gray cast iron, a welding helmet, gas welding goggles, safety glasses, and all other required safety equipment, you are going to repair a cast iron crack.

Locate the ends of the crack and drill stop the crack by drilling 1/8-in. (3-mm) holes at both ends of the crack. Use the edge of the grinding disk to cut a V- or U-groove into the crack.

Note: Even though the part cannot be preheated to the desired level, it cannot be welded cold. It must be heated to at least 75°F (24°C) or higher before starting to weld. Engine blocks can be preheated by letting them run for a short time until they are hot.

Strike the arc on the casting just before the end of the crack and make a 1-in. (25-mm) long weld, **Figure 16-11**. Stop the weld and repeat the process starting at the other end of the crack.

Note: A series of short welds will create less stress than one large weld. The small welds are less likely to crack.

Chip and wire brush the welds.

Note: The next series of welds that are made to close the crack are done in a back stepping sequence. Back stepping welds are short welds that start ahead of the ending point of the first weld and go back to the end of

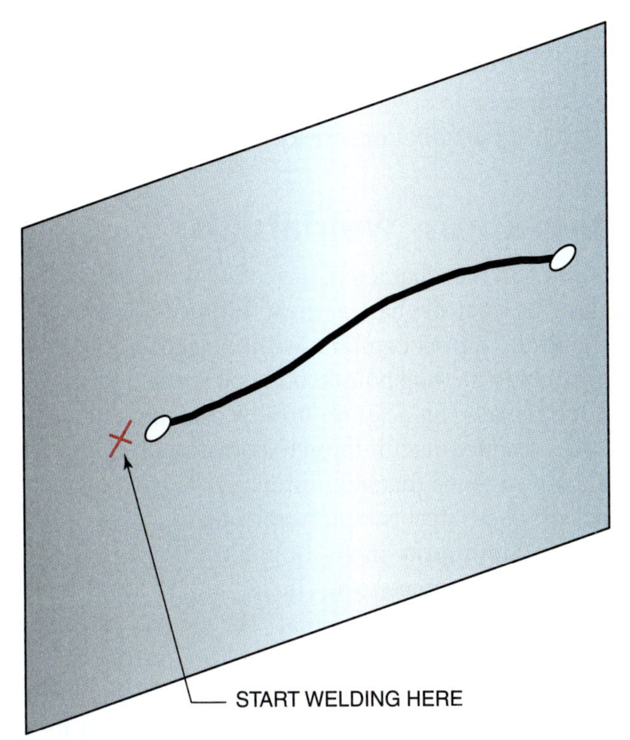

FIGURE 16-11 Start the weld on the casting's surface outside of the crack.

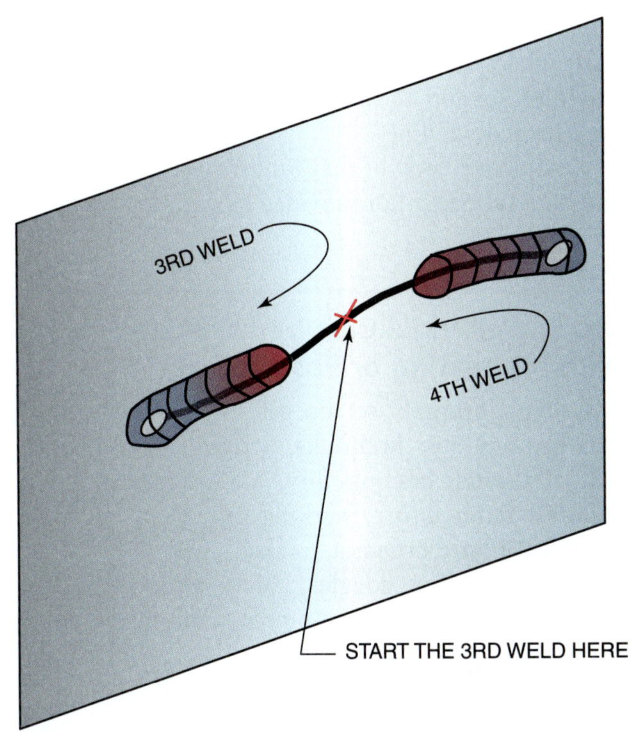

FIGURE 16-12 After both ends have been welded, use a back step welding process to complete the weld.

the first weld. Back stepping welds are used because they reduce weld stresses and are less likely to crack.

Start the third weld about 1 in. (25-mm) in front of the end of the first weld and move back to the end of the first weld, **Figure 16-12**. Vigorously chip the slag off the weld immediately after the arc stops. This both cleans the weld and mechanically works the weld surface, called peening, to reduce weld stresses. Repeat the backstep sequence of welding and peening until the crack is completely covered with welds. Turn off the welder and torch set and clean up your work area when you are finished welding.

Complete a copy of the "Student Welding Report" listed in Appendix I or provided by your instructor. ◆

PRACTICE 16-3

Gas Welding a Cast Iron Break with Preheating or Postheating

Using a properly set up gas torch with welding and heating tips, firebricks, a right angle grinder, cast iron gas welding rods, high temperature cast iron welding flux, a C-clamp, a 500°F (260°C) temperature marking crayon, a chipping hammer, a wire brush, a broken piece of gray cast iron, gas welding goggles, safety glasses, and all other required safety equipment, you are going to repair a cast iron break.

Grind the break into a V-groove, leaving a 1/8-in. (3-mm) root face on the broken surface. Use a C-clamp to hold the broken piece in place. Mark the parts with a 500°F (260°C) temperature marking crayon. Using a properly lit and adjusted oxyacetylene heating torch, begin preheating the part. Keep the flame moving all around the cast iron part so that it heats evenly. Do not point the flame directly onto the temperature indicator mark or it will turn color before the part is actually preheated. When the temperature mark turns black, the part is properly preheated. With the flame off the part, remark the part to check it for proper preheat.

On thick castings, keep the flame on the part for 30 more minutes so that the inside of the part is properly preheated. Flux the end of the cast iron filler rod by heating it with the torch and dipping it into the flux. As the weld progresses, occasionally re-dip the end of the filler rod back into the flux so new flux can be added to the weld. Start welding at one end of the crack. The molten weld pool formed by cast iron is not bright and shiny like that formed on mild steel. The cast iron molten weld pool looks dull and a little lumpy. As the weld progresses, move the tip of the filler rod around in the molten weld pool to keep it stirred up and to make it flatter.

Fill the crater at the end of the weld with a little extra filler metal. Turn the part over and make the same weld on the back side of the break. Build a firebrick box around the part and place the torch so the flame will fill the box and the part can be postweld heated. Keep heating the part for 30 minutes after completing the weld.

Close any gaps in the firebrick box and allow the part to cool slowly over the next 24 hours. Turn off the torch set and clean up your work area when you are finished welding.

Complete a copy of the "Student Welding Report" listed in Appendix I or provided by your instructor. ◆

PRACTICE 16-4

Braze Welding a Cast Iron Crack with Preheating or Postheating

Using a properly set up gas torch with welding and heating tips, firebricks, a right angle grinder, BRCuZn rods, brazing flux, a C-clamp, a 900°F (480°C) temperature marking crayon, a chipping hammer, a wire brush, a cracked piece of gray cast iron, gas welding goggles, safety glasses, and all other required safety equipment, you are going to repair a cast iron crack.

Locate the ends of the crack and drill stop the crack by drilling 1/8-in. (3-mm) holes at both ends of the crack. Use the edge of the grinding disk to cut a V- or U-groove into the crack. Mark the parts with a 900°F (480°C) temperature marking crayon. Using a properly lit and adjusted oxyacetylene heating torch, begin preheating the part. Keep the flame moving all around the cast iron part so that it heats evenly. Do not point the flame directly onto the temperature indicator mark or it will turn color before the part is actually preheated. When the temperature mark turns black, the part is properly preheated. With the flame off the part, remark the part to check it for proper preheat. On thick castings, keep the flame on the part for 30 more minutes so that the inside of the part is properly preheated.

Flux the end of the brazing rod by heating it with the torch and dipping it into the flux. As the weld progresses, occasionally re-dip the end of the filler rod back into the flux so new flux can be added to the weld. Start welding at one end of the crack. Heat the groove until it is dull red, to about 1800°F (980°C). Touch the tip of the brazing rod into the groove occasionally to test it for the proper brazing temperature. When the braze metal begins to flow, move the tip of the rod around in the molten braze pool to help it wet the surface of the groove. When the braze reaches the end of the groove, add a little extra fill to the crater at the end of the braze weld. Turn the part over and make the same weld on the back side of the break.

Build a firebrick box around the part and place the torch so the flame will fill the box and the part can be postweld heated. Keep heating the part for 30 minutes after completing the weld. Close any gaps in the firebrick box and allow the part to cool slowly over the next 24 hours. Turn off the torch set and clean up your work area when you are finished welding.

Complete a copy of the "Student Welding Report" listed in Appendix I or provided by your instructor. ◆

Aluminum Weldability

One of the characteristics of aluminum and its alloys is that it has a great affinity for oxygen. Aluminum atoms combine with oxygen in the air to form a high melting point oxide that covers the surface of the metal. This feature, however, is the key to the high resistance of aluminum to corrosion. It is because of this resistance that aluminum can be used in applications where steel is rapidly corroded.

Pure aluminum melts at 1200°F (650°C). The oxide that protects the metal melts at 3700°F (2037°C). This means that the oxide must be cleaned from the metal before welding can begin.

When the MIG welding process is used, the stream of inert gas covers the weld pool, excluding all air from the weld area. This prevents reoxidation of the molten aluminum. MIG welding does not require a flux.

Aluminum can be arc welded using aluminum welding rods. These rods must be kept in a dry place because the flux picks up moisture easily. Because aluminum melts so easily, use a piece of clean steel plate as a backing to weld on thin sections. The steel backing plate can support the root of the weld without the aluminum weld sticking to the steel plate. Thick aluminum casting must be preheated to about 400°F (200°C) before welding. The preheating helps the weld flow and reduces weld spatter.

Aluminum has high thermal conductivity. Aluminum and its alloys can rapidly conduct heat away from the weld area. For this reason, it is necessary to apply the heat much faster to the weld area to bring the aluminum to the welding temperature. Therefore, the intense heat of the electric arc makes this method best suited for welding aluminum.

When aluminum welds solidify from the molten state, they will shrink about 6% in volume. The stress that results from this shrinkage may create excessive joint distortion unless allowances are made before joining the metal. Cracking can occur because the thermal contraction is about two times that of steel. The heated parent metal expands when welding occurs. This expansion of the metal next to the weld area can reduce the root opening on butt joints during the process. The contraction that results upon cooling, plus the shrinkage of the weld metal, creates a tension and increases cracking.

The shape of the weld groove and the number of beads can affect the amount of distortion. Less distortion occurs with two-pass square butt welds. Other factors that have an influence on the weld are the speed of welding, the use of properly designed jigs and fixtures to support the aluminum while it is being welded, and tack welding to hold parts in alignment.

Repair Welding

Repair, or maintenance, welding is one of the most difficult types of welding. Some of the major problems include preparing the part for welding, identifying the material, and selecting the best repair method.

The part is often dirty, oily, and painted, and it must be cleaned before welding. There are many hazardous compounds that might be part of the material on the part. These compounds may or may not be hazardous on the part, but when they are heated or burned during welding they can become life threatening.

CAUTION

It is never safe to weld on any part that has not been cleaned before welding. All surface contamination must be removed before welding to prevent the possibility of injuring your health from exposure to materials released during welding. Some chemicals can be completely safe until they are exposed to the welding. The smoke or fumes they produce can be an irritant to the skin or eyes; it can be absorbed through the skin or lungs. If you are exposed to an unknown contamination, get professional help immediately.

Contamination can be removed by sand blasting, grinding, or using solvents. If a solvent is used be sure it does not leave a dangerous residue. Clean the entire part if possible or a large enough area so that any remaining material is not affected by the welding.

Before the joint can be prepared for welding you must try to identify the type of metal. There are several ways to determine metal type before welding. One method is to use a metal identification kit. These kits use a series of chemical analyses to identify the metal. Some kits can not only identify a type of metal but can also tell the specific alloy.

Another way to identify metal is to look at its color, test for magnetism, and do a spark test. The spark test should be done using a fine grinding stone. With experience, it is often possible to determine specific types of alloys with great accuracy. The sparks given off by each metal and its alloy are so consistent that the U.S. Bureau of Mines uses a camera connected to a computer to identify metals to aid in recycling. For the beginner it is best if you use samples of a known alloy and compare the sparks to your unknown. The test specimen and the unknown should be tested using the same grinding wheel and the same force against the wheel.

EXPERIMENT 16-1

Identifying Metal Using a Spark Test

In this experiment you will be working in a small group to identify various metals using a spark test, **Figure 16-13.** You will need to use proper eye safety equipment, a grinder, several different known and unknown samples of metal, and a pencil and paper to identify the unknown metal samples. Starting with the known samples, make several tests and draw the spark patterns as described in the following paragraphs. Next test the unknown samples and compare the drawings with the drawings from the known samples. See how many of the unknowns you can identify.

There are several areas of the spark test pattern that you must observe carefully, **Figure 16-14(1).** The first area is the grinding stone: are there sparks that are being carried around the wheel, or are all of the sparks leaving the wheel, **Figure 16-14(2)?**

The next area is immediately adjacent to the wheel where the spark stream leaves the wheel. Note the color of this area; it may vary from white to dull red.

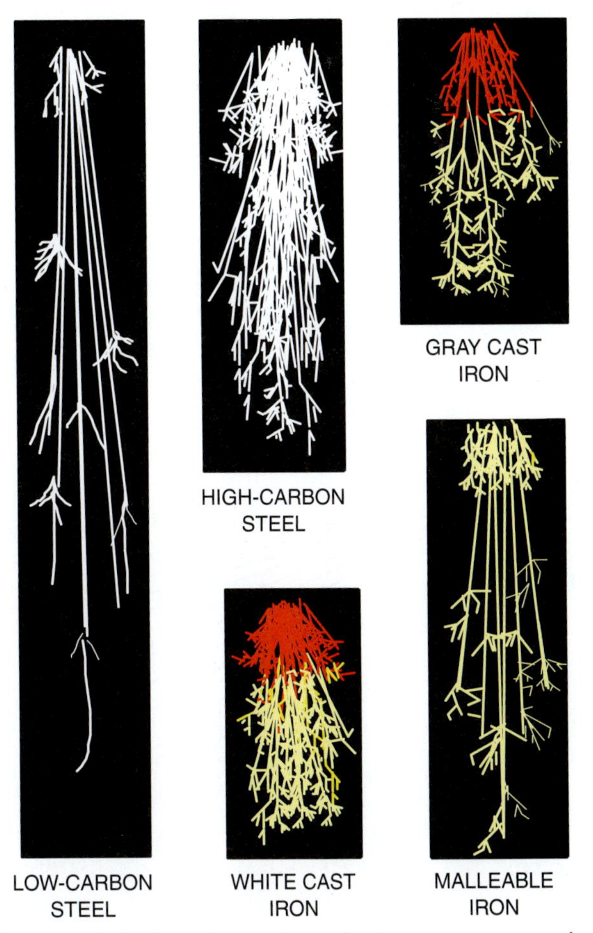

FIGURE 16-13 Spark test patterns for five common metals.

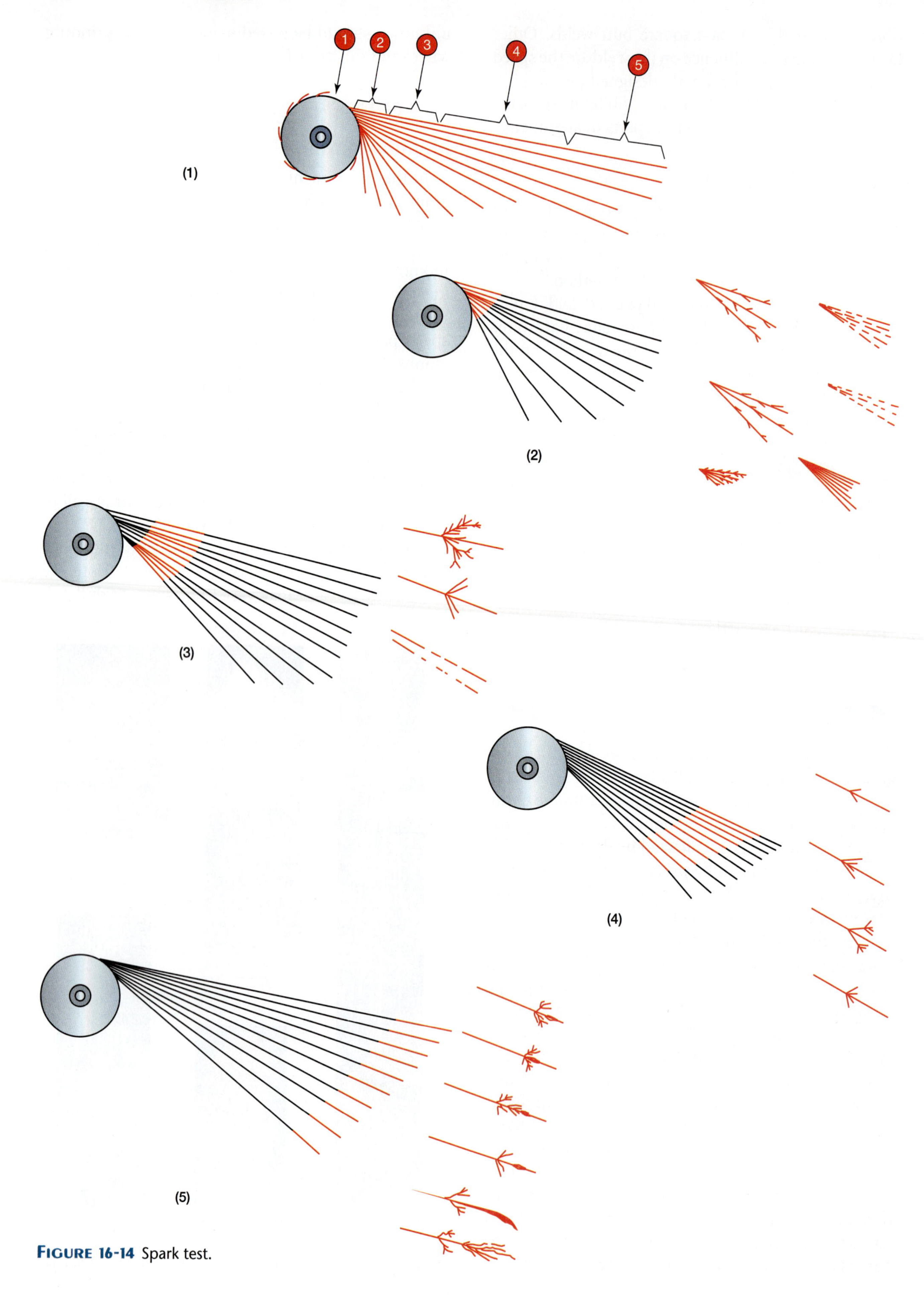

FIGURE 16-14 Spark test.

Metal	Thickness			
	1/8–1/4	1/4–1/2	1/2–1	Over 1
Mild steel	Square	V	V	V
Aluminum	Square	U	U	U
Magnesium	U	U	U	U
Stainless steel	Square	V	V	V
Cast iron	V	V	U	U

TABLE 16-8 Groove Shapes.

Also note the stream to see if the sparks are small, medium, or large and whether the column of sparks is tightly packed or spread out. Draw a sketch of what the spark stream looks like here, **Figure 16-14(3)**.

As the spark stream moves away from the wheel a few inches it will begin to change. This change may be in color, speed of the sparks, or size of the sparks; the sparks may divide into smaller, separate streams, explode in a burst of tiny fragments, or just stop glowing, **Figure 16-14(4)**. Sketch these changes you see in the sparks.

When the sparks end they may simply stop glowing, change color, change shape, explode, or divide into smaller parts, **Figure 16-14(5)**. Sketch these changes you see in the sparks.

Repeat this experiment with other types of metal as they become available.

Once the type of metal to be repaired is determined, to the best of your ability decide on the type of weld groove that is needed. Some breaks may need to be ground into a V- or U-groove and others may not need to be grooved at all, **Table 16-8**.

Often thin sections of most metals can be repaired without the need for the break to be grooved. Thicker sections of most metals will be easier to repair if the crack is ground into a groove. To help in the realignment of the part, it is a good idea to leave a small part of the crack unground so it can be used to align the parts, **Figure 16-15**.

After the part is ready to be welded, a test tab can be welded onto the part. This test tab should be welded in a spot that will not damage the part. Once the tab is welded on, it can be broken off to see if the weld will hold, **Figure 16-16**. If the test tab shows good strength, the repair should continue. If the test tab fails, a new welding procedure should be tried.

Complete a copy of the "Student Welding Report" listed in Appendix I or provided by your instructor. ◆

Hardfacing

Hardfacing is defined as the process of obtaining desired properties or dimensions by applying, using oxyfuel or arc welding, an integral layer of metal of one composition onto a surface, an edge, or the point of a base metal or another composition. The hardfacing operation makes the surface highly resistant to abrasion.

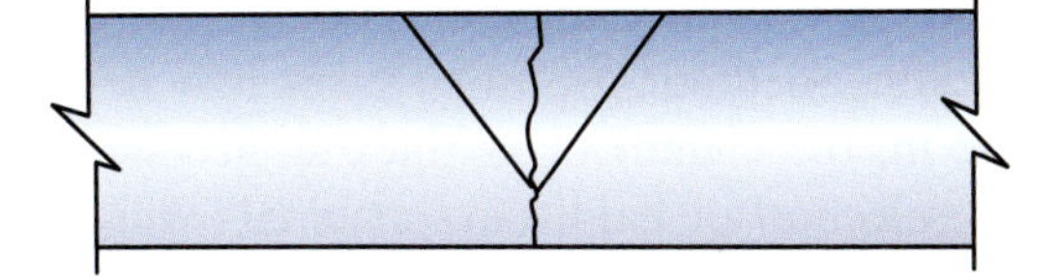

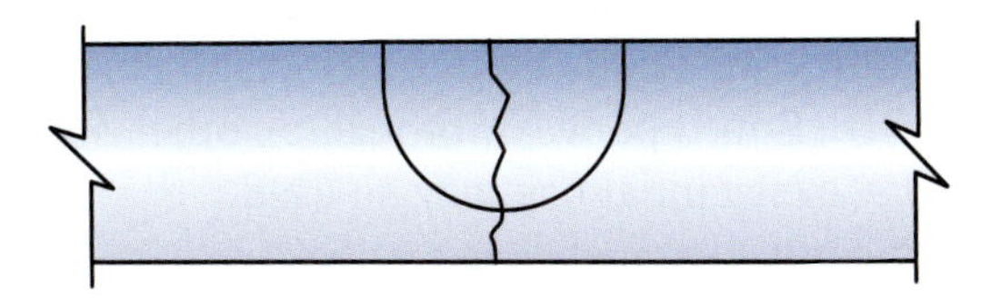

FIGURE 16-15 A small section of the bottom of the break is left so the parts can be realigned.

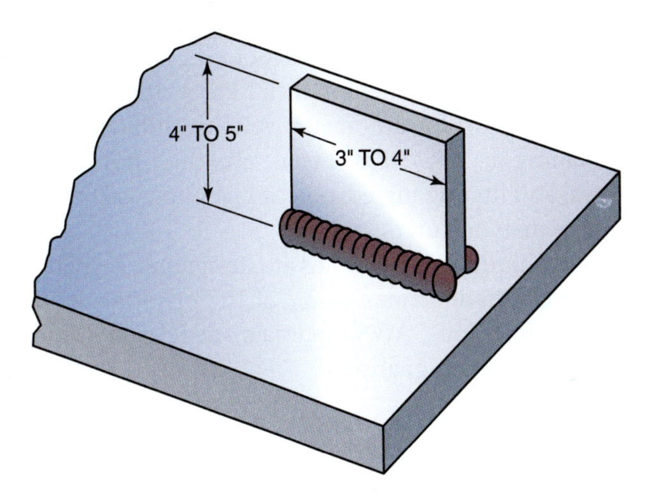

FIGURE 16-16 Tab test.

There are various techniques of hardfacing. Some apply a hard surface coating by fusion welding. In other techniques, no material is added but the surface metal is changed by heat treatment or by contact with other materials.

Several properties are required of surfaces that will be subjected to severe wearing conditions, including hardness, abrasion resistance, impact resistance, and corrosion resistance.

Hardfacing may involve building up surfaces that have become worn. Therefore, it is necessary to know how the part will be used and the kind of wear to expect. In this way, the proper type of wear-resistant material can be selected for the hardfacing operation.

When a part is subjected to rubbing or continuous grinding, it undergoes abrasion wear. When metal is deformed or lost by chipping, crushing, or cracking, impact wear results.

Selection of Hardfacing Metals Many different types of metals and alloys are available for hardfacing applications. Most of these materials can be deposited by any conventional manual or automatic arc or oxyfuel welding method. Deposited layers may be as thin as 1/32 in. (0.79 mm) or as thick as necessary. The proper selection of hardfacing materials will yield a wide range of characteristics.

Steel or special hardfacing alloys should be used where the surface must resist hard or abrasive wear. Where surfacing is intended to withstand corrosion-type or friction-type wear, bronze or other suitable corrosion-resistant alloys may be used.

Most hardfacing metals have a base of iron, nickel, copper, or cobalt. Other elements that can be added include carbon, chromium, manganese, nitrogen, silicon, titanium, and vanadium. The alloying elements have a tendency to form carbides. Hardfacing metals are provided in the form of rods for oxyacetylene welding, electrodes for shielded metal arc welding, or in hard wire form for automatic welding. Tubular rods containing a powdered metal mixture, powdered alloys, and fluxing ingredients can be purchased from various manufacturers.

Many hardfacing materials are designated by manufacturers' trade names. Some of the materials have AWS designations. AWS materials are classified into the following designations:

- High-speed steel
- Austenitic manganese steel
- Austenitic high chromium iron
- Cobalt-base metals
- Copper-base alloy metals
- Nickel-chromium boron metal
- Tungsten carbides

The coding system identifies the important elements of the hardfacing metal. The prefix *R* is used to designate a welding rod, and the prefix *E* indicates an electrode. Certain materials are further identified by the addition of digits after a suffix.

Hardfacing Welding Processes

Oxyfuel Welding

In hardfacing operations, oxyfuel welding permits the surfacing layer to be deposited by flowing molten filler metal into the underlying surface. This method of surfacing is called sweating or tinning, **Figure 16-17**.

With the oxyacetylene flame, small areas can be hardfaced by applying thin layers of material. In addition, the alloy can be easily flowed to the corners and edges of the workpiece without overheating or building up deposits that are too thick. Placement of the metal can be controlled accurately.

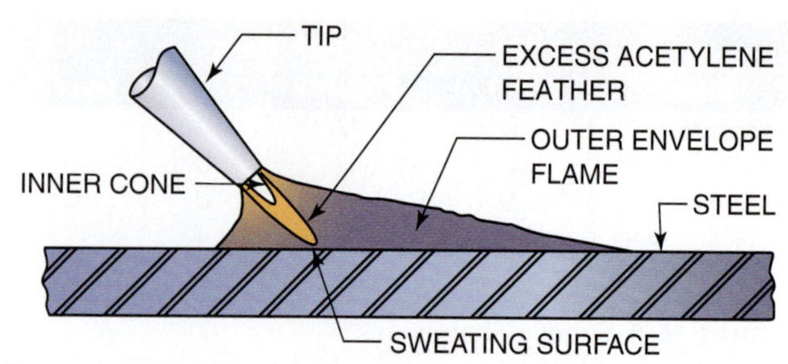

FIGURE 16-17 An example of how to produce sweating.

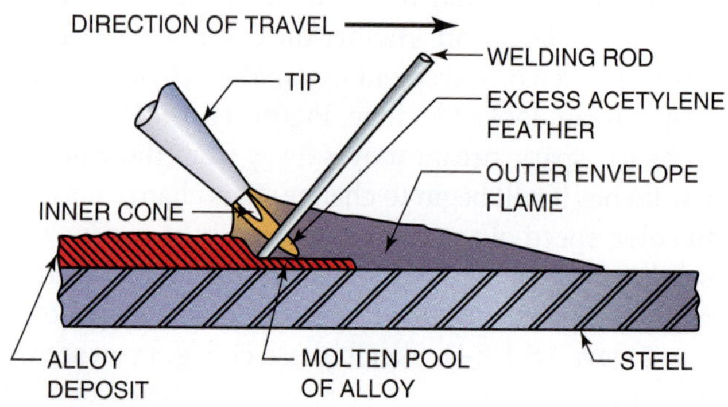

FIGURE 16-18 Approximate relationship of the tip, rod, and molten weld pool for forehand hardfacing.

The size of the weld is affected by many factors. These factors include the rate of travel, degree of preheat, type of metal being deposited, and thickness of the work.

Figure 16-18 shows the approximate relationship of the tip, rod, molten pool, and base metal during the hardfacing operation.

Iron, nickel, and cobalt-base alloys require an excess acetylene flame. Copper alloys and bronze call for a neutral or slightly oxidizing flame. Laps, blowholes, and poor adhesion of deposits can be prevented by a flame characteristic that is soft and quiet.

In all types of surfacing operations, the metal should be cleaned of all loose scale, rust, dirt, and other foreign substances before the alloy is applied. The best method of removing these impurities is by grinding or machining the surface. Fluxes may be used to maintain a clean surface. They also help to overcome oxidation that may develop during the operation.

Conventional methods may be used in holding the torch and rod. **Figure 16-19** shows the backhand method of hardfacing.

If the base metal is cast iron, it will not "sweat" like steel. Therefore, slightly less acetylene should be used. Alloys do not flow as readily on cast iron as they do on steel. Usually, it is necessary to break the surface crust on the metal with the end of the rod. A cast-iron welding flux is generally necessary. The best method is to apply a thin layer of the alloy and then build on top of it.

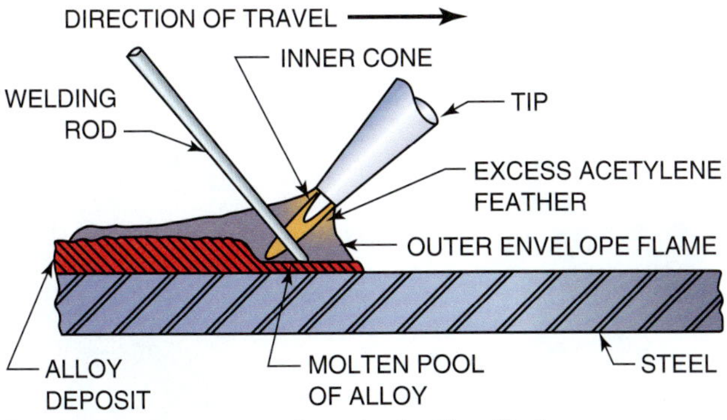

FIGURE 16-19 Backhand method of hardfacing. Courtesy of Praxair, Inc.

The oxyacetylene process is preferred for small parts. Cracking can be minimized by using adequate preheat, postheat, and slow cooling. Metal arc welding is preferred for large parts.

Arc Welding

Hardfacing by arc welding may be accomplished by shielded metal arc, gas metal arc, gas tungsten arc, submerged arc, plasma arc, or other processes.

The techniques employed for any one of these processes are similar to those used in welding for joining. The factor of dilution must be carefully considered because the composition of the added metal will differ from the base metal. The least amount of dilution of filler metal with base metal is an important goal, especially where the two metals differ greatly. Little dilution means that the deposited metal maintains its desired characteristics. When using high-melting-point alloys, dilution of the weld metal is usually kept well below 15%.

Hardfacing by the arc welding method has many advantages, including high rates of deposition, flexibility of operation, and ease of mechanization.

Hardfacing may be applied to many types of metals, including low and medium carbon steels, stainless steels, manganese steels, high-speed steels, nickel alloys, white cast iron, malleable cast iron, gray and alloy cast iron, brass, bronze, and copper.

Quality of Surfacing Deposit The type of service to which a part is to be exposed governs the degree of quality required of the surfacing deposit. Some applications require that the deposited metal contain no pinholes or cracks. For other applications, these requirements are of little importance. In most cases, the quality of the deposited metals can be very high. Steel-base alloys do not tend to crack, while other materials, such as high-alloy cast steels, are subject to cracking and porosity.

Hardfacing Electrodes The proper type of surfacing electrode must be selected, as one type of electrode will not meet all requirements. Most electrodes are sold under manufacturers' trade names.

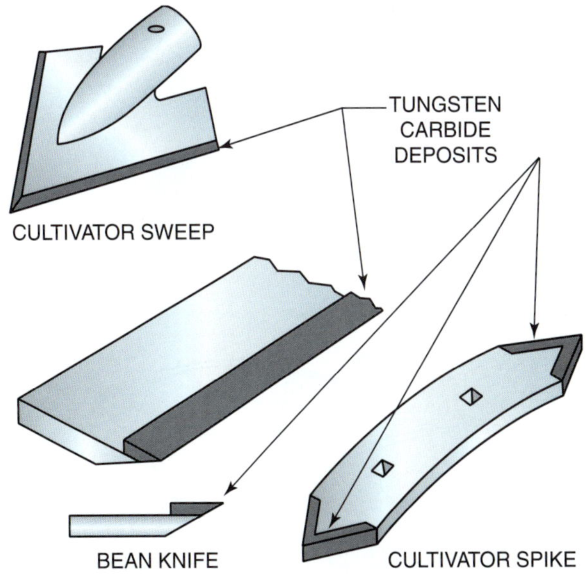

FIGURE 16-20 Farm and ranch tools that can be hardfaced with tungsten carbide electrodes to increase the life of the cutting or digging tool.

Electrodes may be classified into the following three general groups:

- Resistance to severe abrasion
- Resistance to both impact and moderate abrasion
- Resistance to severe impact and moderately severe abrasion

Tungsten carbide and chromium electrodes are included in the first group. The material deposited is very hard and abrasive-resistant. These electrodes can be one of two types, either coated tubular or regular coated cast alloy. The tubular types contain a mixture of powdered metal, powdered ferroalloys, and fluxing materials. The tubes are the coated type. These electrodes are used with the metallic arc.

Electrodes contain small tungsten carbide crystals embedded in the steel alloy. After this material is applied to a surface, the steel wears away with use, leaving the very hard tungsten carbide particles exposed. This wearing away of steel results in a self-sharpening ability of the surfacing material. Cultivator sweeps and scraping tools are among parts that are surfaced with this material, **Figure 16-20.**

Chromium carbide electrodes are tougher than tungsten carbide–type electrodes. However, chromium carbide electrodes are not as hard and are less abrasion-resistant. This material is too hard to be machined, but it has good corrosion-resistant qualities.

The electrodes in the second group are the high carbon type. When used for surfacing, these electrodes leave a tough and very hard deposit. Examples

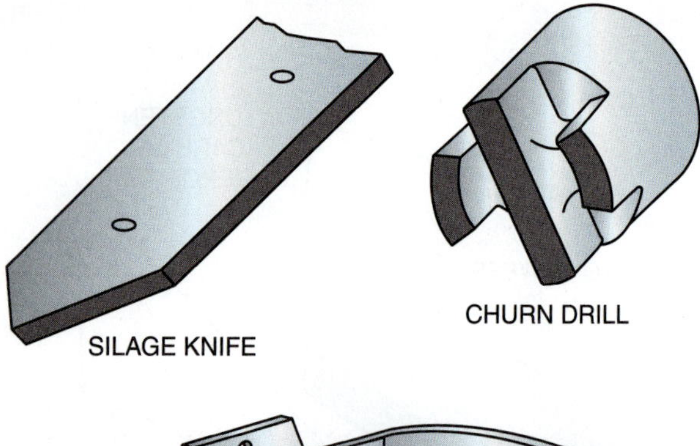

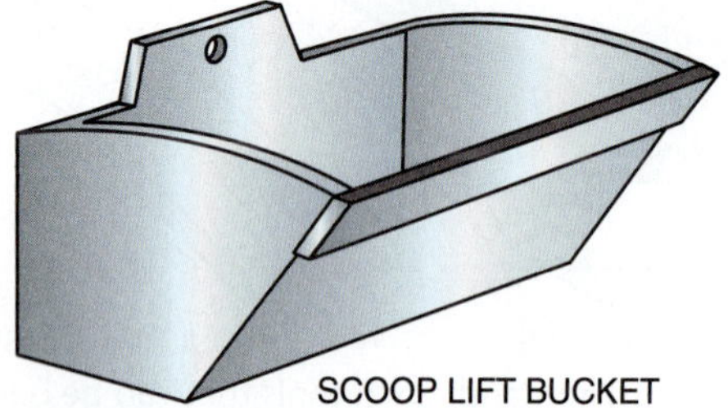

FIGURE 16-21 Products that are hardfaced to produce moderate impact resistance and severe abrasion resistance.

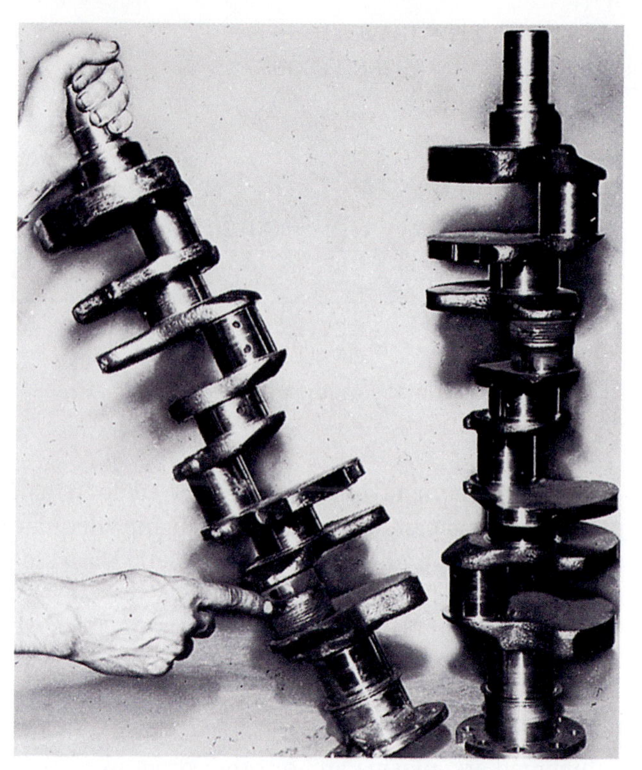

FIGURE 16-22 Bearing surfaces can be rebuilt on crankshafts and other rotating equipment surfaces. Courtesy of Hobart Brothers Company.

of hardfaced products in this group include gears, tractor lugs, and scraper blades.

The third group of electrodes is used for surfacing rock-crusher parts, links, pins, railroad track components, and parts where severe abrasion resistance is a requirement, **Figure 16-21** and **Figure 16-22**. Deposits from these electrodes are very tough but not hard. It is this quality that seems to work-harden the hardfacing material but leaves the material underneath in a softened condition. Therefore, cracking generally is not a problem.

Shielded Metal Arc Method

1. Start the process by cleaning the surface.
2. Since most hardfacing electrodes are too fluid for out-of-position welding, the work should be arranged in the flat position.
3. Set the amperage so that just enough heat is provided to maintain the arc. Too much heat will cause excessive dilution.
4. Hold a medium-long arc, using either a straight or welding pattern. When a thin bead is required, use the weave pattern and keep the weave to a width of 3/4 in. (19 mm).
5. If more than one layer is required, remove all slag before placing other layers.

Hardfacing with Gas Shielded Arc GTA, GMA, and FCA welding processes may be used in hardfacing operations. These three processes, in many instances, are better methods of hardfacing because of the ease with which the metal can be deposited. In addition, the hardfacing materials may be deposited to form a porosity-free, smooth, and uniform surface.

Where the job calls for cobalt-base alloys, the GTA method does an effective job. Very little preheating of the base is required. The GMA and FCA welding processes are somewhat faster than surfacing by GTA due to the fact that continuous wire is used.

Care must be exercised when using the GMA, FCA, and GTA welding processes for hardfacing in order to avoid dilution of the weld. Helium or a mixture of helium-argon normally produces a higher arc voltage than pure argon. For this reason, the dilution of the weld metal increases. An argon and oxygen mixture should be used for surfacing with the gas metal arc processes and argon used with gas tungsten arc processes. When using FCAW, shielding may be provided as either shielding gas or self-shielding. The self-shielding hardsurfacing process is used when working outdoors because of its ability to better resist the effect of light winds.

Carbon Arc Method

1. Clean the surface of all rust, scale, dirt, and other foreign particles. Place the workpiece in a flat position.
2. Using a commercial hardfacing paste, spread the paste evenly over the area to be surfaced. Allow the paste to dry somewhat.
3. Regulate the heat carefully so that just enough heat is provided to obtain a free-flowing molten pool. Too high a welding current tends

to create an excessive dilution of the base metal, thus lowering the hardness of the finished surface.

4. The carbon electrode should be moved in a circular motion. When sufficient heat is obtained, the paste will melt and fuse with the base metal.

Summary

All metals are weldable. The only limitation in the fabrication and repair of parts is the cost. It takes a skilled welder with an understanding of all the various characteristics of the metals and types of welding to fix worn or damaged parts. Being able to recognize the differences among the various classifications of metal will allow you to select the most appropriate welding repair procedure. Because of the complexity of this process, you may often be required to research through the original manufacturer of the equipment the types of processes and materials used in the weldment's construction.

Things change, so sometimes after a part is placed in service there is a need to change the welding procedures used in the part's original construction for the repair welding. In addition the original manufacturer may no longer have the welding procedure. In order to make a successful weld in such cases, the welder must be able to establish a new welding procedure. As part of your welding procedure you may perform tests to enhance the longevity of your repairs. Your welding experience will help you to be more efficient in producing the new welding procedure.

Repair welds do not have to look pretty; they only have to hold. Years ago when the author was welding in a small agricultural welding shop in Madisonville, Tennessee, a local farmer said, "Go on and try and weld it. It's broken anyway, and if you can't fix it it's no use to me anyway." He was right—if you have a broken part, and it is going to cost more to have someone else weld it than to buy a new part, you might as well try welding it yourself. You have nothing to lose and everything to gain; if you fix it, you are in luck and if you burn it up, you would have had to buy a replacement part anyway. Make sure you can get a replacement part before you start welding on the broken one. If the original part is damaged during the weld, a professional might not be able to repair it.

Last, a tendency of many welders is to overweld when making a repair. Keeping the weld sized correctly will make the part stronger. A too large weld can cause the part to be brittle and break.

Review

1. Why is it important to know why a part broke?
   A. If the part was underdesigned, simply welding it will not stop it from breaking again.
   B. It is the only way to know which filler metal to use for the repair.
   C. So you know who to blame.
   D. So you can tell the welding shop what happened, so they can help other farmers who may have the same equipment.
2. What can cause porosity in a weld?
   A. welding in the wrong direction
   B. using the wrong filler metal
   C. not preheating the part
   D. welding with a welding rod that is too small
3. In both steel classification systems, the first number often refers to the basic type of steel.
   A. true
   B. false
4. ASA 1020 steel would be classified as a ________________-type steel.
   A. nickel
   B. chromium
   C. carbon
   D. molybdenum

Review continued

5. What is the range of carbon in mild steel?
   A. 1.5% to 3.0%
   B. 0.01% to 0.15%
   C. 3% to 5%
   D. 0.15% to 0.30%
6. Why is stick welding the most practical way for ranchers and farmers to weld medium and high carbon steels?
   A. They do not have MIG welding equipment.
   B. They can buy only the number of rods needed for a job.
   C. It is the fastest way of welding any metal.
   D. Stick welders are easier to use than MIG welders, so they do not have to learn new skills to make the repair.
7. What flame setting should be used for oxy-acetylene welding of low-carbon and mild steel?
   A. a neutral flame
   B. an oxidizing flame
   C. a carbonizing flame
   D. a hot flame
8. According to Figure 16-3 what should the approximate preheat temperature be for a steel with 0.6% carbon?
   A. 600°F
   B. 900°F
   C. 100°F
   D. 350°F
9. According to Figure 16-4 what should the approximate stress-relieving temperature be for a steel with 0.1% carbon?
   A. 600°F
   B. 800°F
   C. 1000°F
   D. 1200°F
10. According to Table 16-3, which iron-carbon alloy would be used to make equipment track?
    A. high carbon steel
    B. ingot iron
    C. medium carbon steel
    D. mild steel
11. Which of the following is not a type of stainless steel?
    A. austenitic
    B. ferritic
    C. rust proof
    D. precipitation hardening
12. What does the term 18/8 refer to in stainless steel?
    A. The standard and SI (metric) weights of a specific sheet thickness.
    B. The percentage of chromium and nickel alloys.
    C. The type of filler metal required.
    D. The arc voltage and wire feed speed for MIG welding.
13. Which type of stainless can be hardened for use as corrosion-resistant knife blades?
    A. austenitic
    B. ferritic
    C. martensitic
    D. precipitation hardening
14. What is the range of carbon in cast iron?
    A. 0.5% to 0.6%
    B. 11% to 15%
    C. 0.001% to 0.2%
    D. 1.7% to 4%
15. What is the most widely used cast iron?
    A. malleable
    B. nodular
    C. white
    D. gray
16. Which type of cast iron is sometimes called ductile cast iron?
    A. malleable
    B. nodular
    C. white
    D. gray
17. Which of these cast iron parts would be the least difficult to make a welding repair on?
    A. A broken-off hold-down ear on a gear box.
    B. A crack in an engine water jacket.
    C. A break across the rear axle of a tractor.
    D. A crack in a transmission casting on a truck.
18. What must be done before welding a crack?
    A. preheat the casting to 1800°F (980°C)
    B. drill stop the crack
    C. mark the crack with soapstone so you can see it to weld
    D. clamp the crack with a C-clamp
19. About what temperature should a large aluminum casting be preheated to before arc welding a crack?
    A. 600°F (315°C)
    B. 1200°F (650°C)
    C. 400°F (200°C)
    D. 900°F (480°C)
20. Which of the following is not a way of helping to identify a type of metal?
    A. color
    B. magnetism
    C. sparks
    D. smell

Chapter 17

Welding Joint Design, Welding Symbols, and Fabrication

OBJECTIVES

After completing this chapter, the student should be able to

- ☑ understand the basics of joint design.
- ☑ identify the major parts of a welding symbol.
- ☑ explain the parts of a groove preparation.
- ☑ describe how nondestructive test symbols are used.
- ☑ list the five major types of joints.
- ☑ list seven types of weld grooves.
- ☑ lay out a welding project.

KEY TERMS

edge preparation
(G) and (F)
joint dimensions
joint type
weld joint
weld location
weld types
welding symbols

INTRODUCTION

The joint design affects the quality and cost of the completed weld. Selecting the most appropriate joint design for a welding job requires special attention and skill. The eventual design selection can be influenced by a number of factors.

Every weld joint selected for a job requires some compromises. For example, the compromises may be between strength and cost, equipment available and welder skill, or any two, three, or more variables. Because there are so many factors, a good design requires experience. Even with experience, trial welds are necessary before selecting the final joint configuration and welding parameters.

This chapter will familiarize welders with the most important factors and give them some appreciation of joint design and fabrication. Experience in the welding field will help a welder become a better joint designer and fabricator.

Welding symbols are the language used to let the welder know exactly what welding is needed. The welding symbol is used as

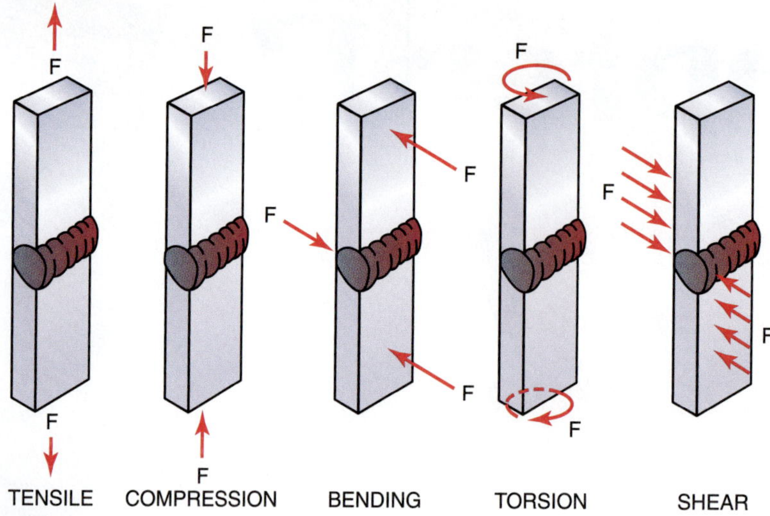

FIGURE 17-1 Forces on a weld.

a shorthand and can provide the welder with all of the required information to make the correct weld. The emphasis in this chapter is on using and interpreting welding symbols so the welder will develop a welder's "vocabulary."

Weld Joint Design

The selection of the best joint design for a specific weldment requires careful consideration of a variety of factors. Each factor, if considered alone, would result in a part that might not be able to be fabricated. For example, a narrower joint angle requires less filler metal, and that results in lower welding cost. But if the angle is too small for the welding process being used the weld cannot be made.

The purpose of a weld joint is to join parts together so that the stresses are distributed. The forces causing stresses in welded joints are tensile, compression, bending, torsion, and shear, **Figure 17-1**. The ability of a welded joint to withstand these forces depends upon both the joint design and the weld integrity. Some joints can withstand some types of forces better than others.

The basic parts of a weld joint design that can be changed include:

- Joint type—The type of joint is considered by the way the joint members come together, **Figure 17-2**.
- Edge preparation—The faying surfaces of the mating members that form the joint are shaped for a specific joint. This preparation may be the same on both members of the joint, or each side can be shaped differently, **Figure 17-3**.

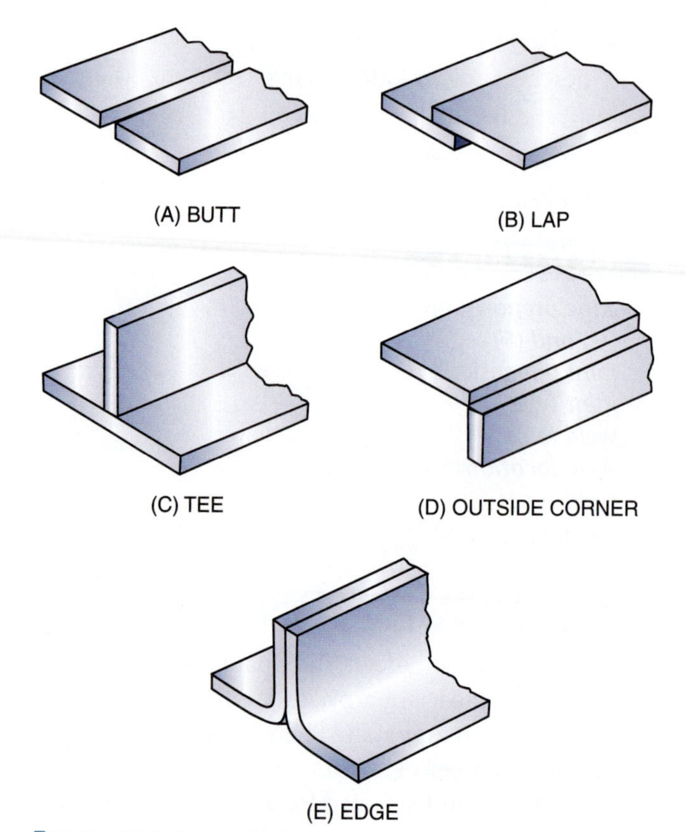

FIGURE 17-2 Types of joints.

- Joint dimensions—The depth and/or angle of the preparation and the joint spacing can be changed to make the weld, **Figure 17-4**.

Welding Process The welding process to be used has a major effect on the selection of the joint design. Each welding process has characteristics that affect its performance. Some processes are easily used in any position; others may be restricted to one or more positions. The rate of travel, penetration, deposition rate, and

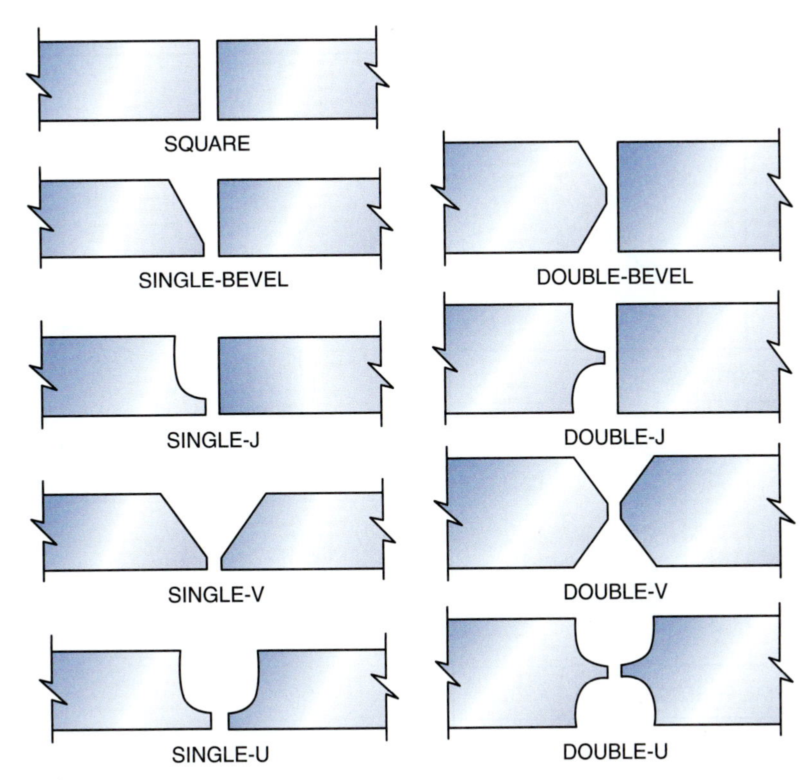

FIGURE 17-3 Edge preparation.

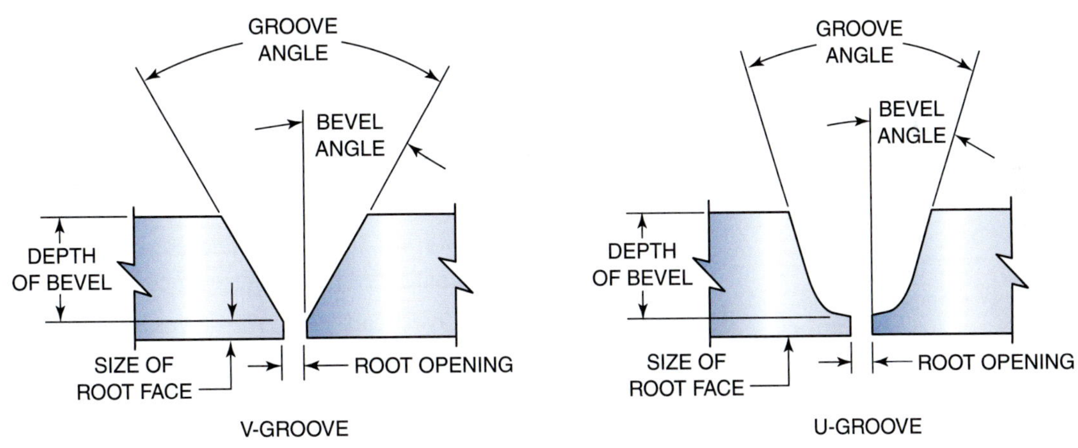

FIGURE 17-4 Groove terms.

heat input also affects the welds used on some joint designs. For example, a square butt joint can be made in very thick plates using either electroslag or electrogas welding, but not many other processes can be used on such a joint design.

Base Metal Because some metals have specific problems with things like thermal expansion, crack sensitivity, or distortion, the joint selected must control these problems. For example, magnesium is very susceptible to postweld stresses, and the U-groove works best for thick sections.

Plate Welding Positions The most ideal welding position for most joints is the flat position, because it allows for larger molten weld pools to be controlled. Usually, the larger the weld pool, the faster the joint can be completed. It is not always possible to position the part so that all the welds can be made in the flat position. Some joint design must be used for some out-of-position welding. For example, the bevel joint is often the best choice for horizontal welding, **Figure 17-5**.

The American Welding Society has divided plate welding into four basic positions for grooves (G) and fillet (F) welds as follows:

- Flat 1G or 1F—Welding is performed from the upper side of the joint, and the face of the weld

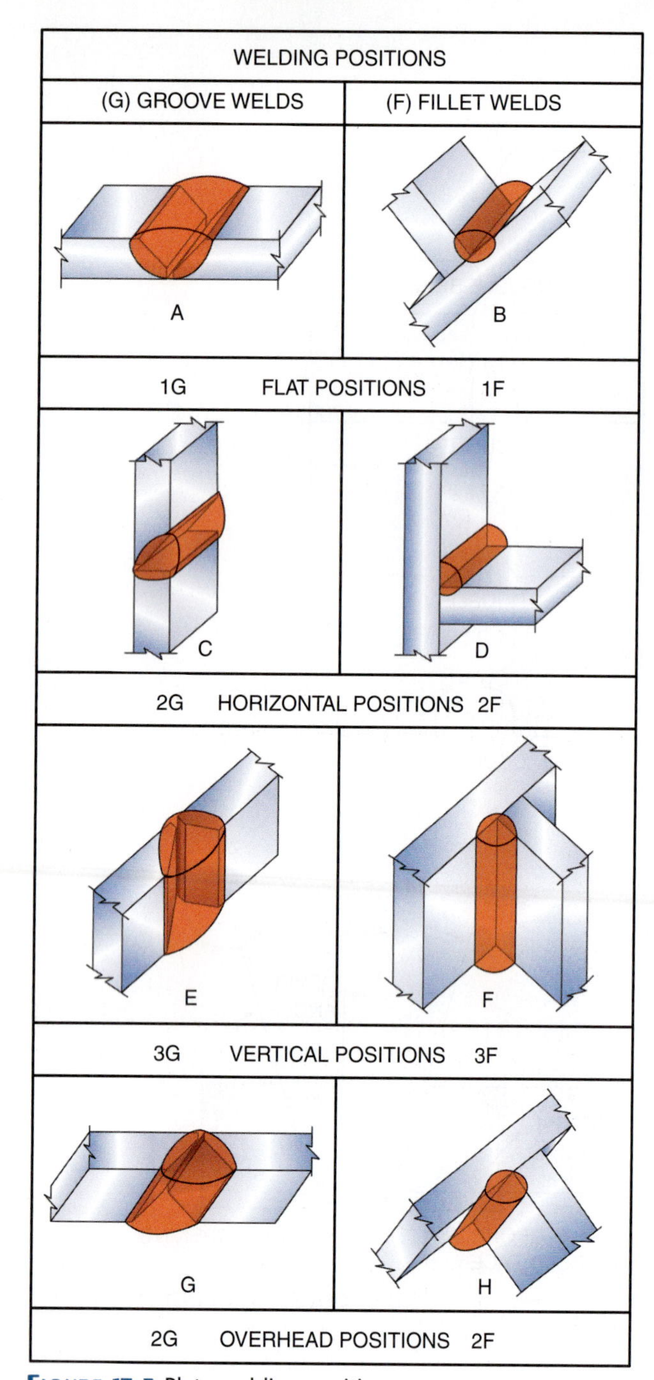

FIGURE 17-5 Plate welding positions.

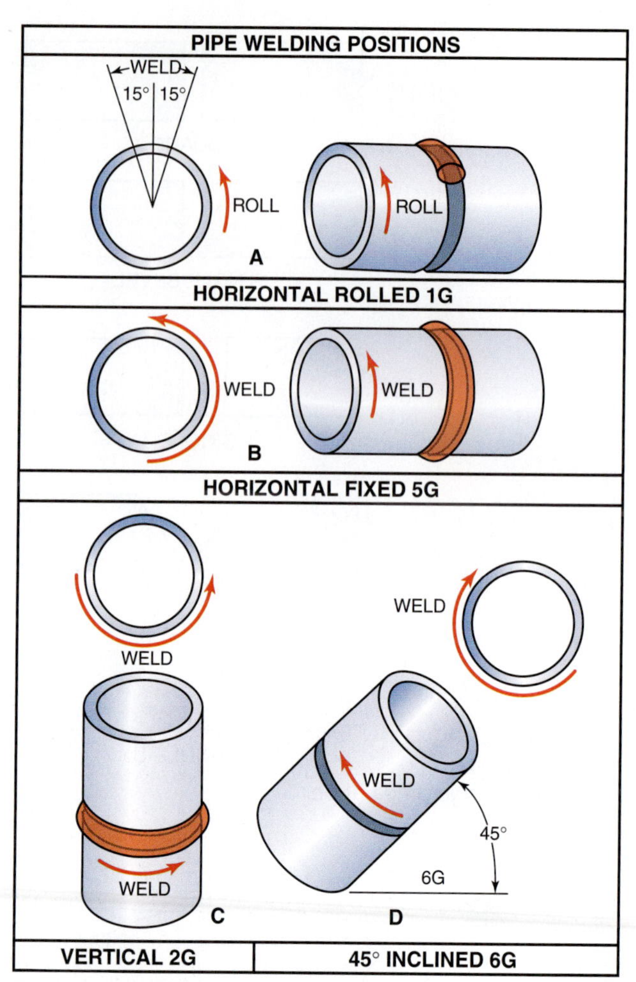

FIGURE 17-6 Pipe welding positions.

is approximately horizontal, **Figure 17-5A** and **Figure 17-5B**.

- Horizontal 2G or 2F—The axis of the weld is approximately horizontal, but the type of the weld dictates the complete definition. For a fillet weld, welding is performed on the upper side of an approximately vertical surface. For a groove weld, the face of the weld lies in an approximately vertical plane, **Figure 17-5C** and **Figure 17-5D**.
- Vertical 3G or 3F—The axis of the weld is approximately vertical, **Figure 17-5E** and **Figure 17-5F**.
- Overhead 4G or 4F—Welding is performed from the underside of the joint, **Figure 17-5G** and **Figure 17-5H**.

Pipe Welding Positions The American Welding Society Pipe has divided pipe welding into five basic positions:

- Horizontal rolled 1G—The pipe is rolled either continuously or intermittently so that the weld is performed within 0° to 15° of the top of the pipe, **Figure 17-6A**.
- Horizontal fixed 5G—The pipe is parallel to the horizon, and the weld is made vertically around the pipe, **Figure 17-6B**.
- Vertical 2G—The pipe is vertical to the horizon, and the weld is made horizontally around the pipe, **Figure 17-6C**.
- Inclined 6G—The pipe is fixed in a 45° inclined angle, and the weld is made around the pipe, **Figure 17-6D**.

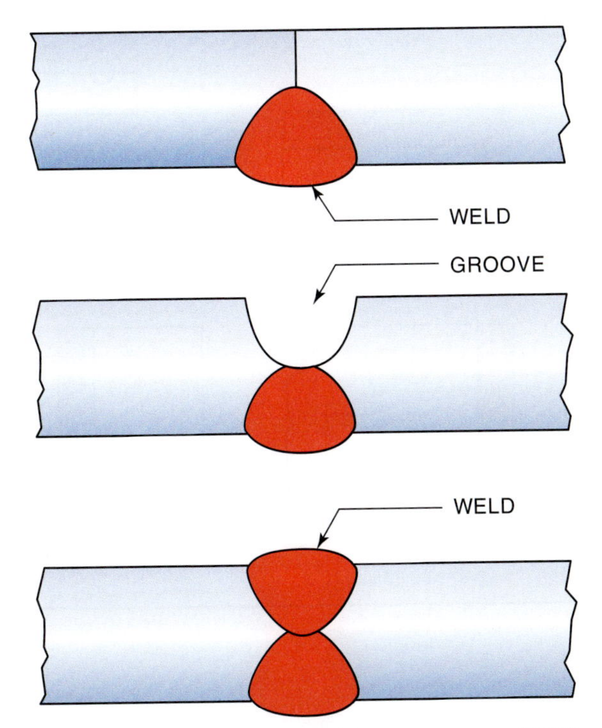

FIGURE 17-7 Back gouging a weld joint to ensure 100% joint penetration.

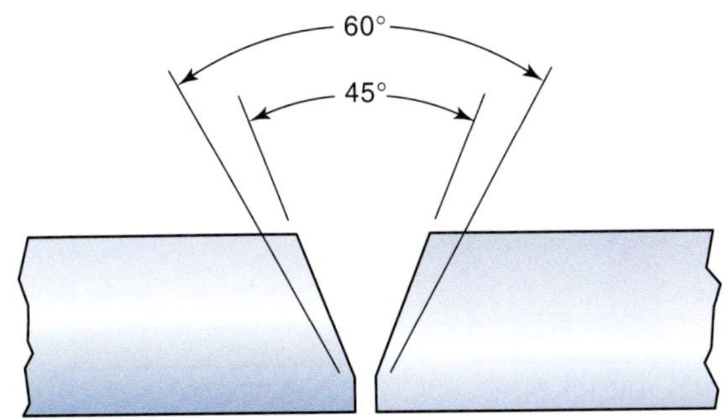

FIGURE 17-8 A smaller groove angle reduces both weld time and weld metal.

Metal Thickness As the metal becomes thicker the joint design must change. On thin sections it's often possible to make full penetration welds using a square butt joint. But with thicker plates or pipe the edge must be prepared with a groove on one or both sides. The edge may be shaped with a bevel, V-groove, J-groove, or U-groove. The choice of shape depends on the type of metal, its thickness, and whether it is made before or after assembly.

When welding on thick plate or pipe, it is often impossible for the welder to get 100% penetration without using some type of groove. The groove may be cut into either one of the plates or pipes or both. On some plates double-sided grooves can be cut on the joint, **Figure 17-3.** The groove may be ground, flame cut, gouged, sawed, or machined on the edge of the plate before or after the assembly. Bevels and V-grooves are best if they are cut before the parts are assembled. J-grooves and U-grooves can be cut either before or after assembly, **Figure 17-3.** The lap joint is seldom prepared with a groove, because little or no strength can be gained by grooving this joint.

For most welding processes, plates that are thicker than 3/8 in. (10 mm) may be grooved on both the inside and outside of the joint. Whether to groove one or both sides is most often determined by joint design, position, code, and application. Plates in the flat position are usually grooved on only one side unless they can be repositioned or are required to be welded on both sides. Tee joints in thick plates are easier to weld and have less distortion if they are grooved on both sides.

Sometimes plates are either grooved and welded or just welded on one side and then back gouged and welded, **Figure 17-7.** Back gouging is a process of cutting a groove in the back side of a joint that has been welded. Back gouging can ensure 100% joint fusion at the root and remove discontinuities of the root pass.

Code or Standards Requirements The type, depth, angle, and location of the groove are usually determined by a code or standard that has been qualified for the specific job. Organizations such as the American Welding Society, the American Society of Mechanical Engineers, and the American Bureau of Ships are among the agencies that issue such codes and specifications. The most common code or standards are the AWS D1.1 and the ASME Boiler and Pressure Vessel (BPV), Section IX.

The joint design for a specific set of specifications often must be what is known as prequalified. These joints have been tested and found to be reliable for the weldments for specific applications. The joint design can be modified, but the cost to have the new design accepted under the standard being used is often prohibitive.

Welder Skill Often the skills or abilities of the welder are a limiting factor in joint design. A joint must be designed in a manner so that the welders can reliably reproduce it. Some joints have been designed without adequate room for the welder to see the molten weld pool or room to get the electrode or torch into the joint.

Acceptable Cost Almost any weld can be made in any material in any position. A number of factors can affect the cost of producing a weld. Joint design is one major way to control welding cost. Changes in the design can reduce cost yet still meet the weldment's strength requirements. Reducing the groove angle can also help, **Figure 17-8.** It will decrease the welding

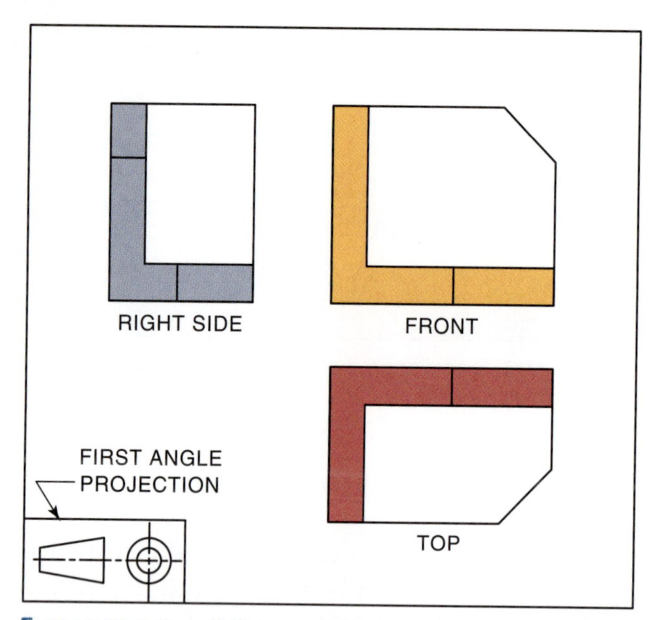

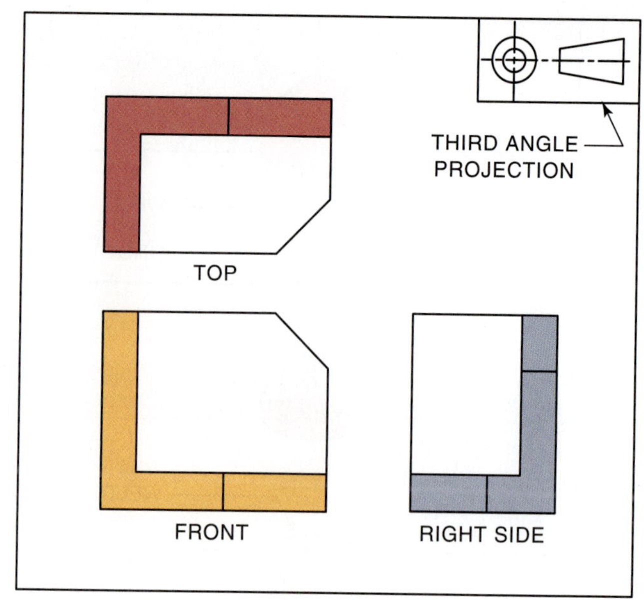

FIGURE 17-9 Two different methods used to rotate drawing views.

filler metal required to complete the weld as well as decrease the time required to fill the larger groove opening. Joint design must be a consideration in order for any project to be competitive and cost effective.

Mechanical Drawings

Mechanical drawings have been around for centuries. Leonardo da Vinci (1452–1519) used mechanical drawings extensively in his inventive works. Many of his drawings still exist today and are as easily understood now as when they were drawn. For that reason mechanical drawings have been called the universal language; they are produced in a similar format worldwide. Despite the few differences in how the views are laid out, **Figure 17-9**, the drawings are understandable. Notwithstanding different languages and measuring systems, the basic shape of an object and location of components can be determined from any good drawing.

A group of drawings, known as a *set of drawings,* should contain enough information to enable a welder to produce the weldment. The set of drawings may contain various pages showing different aspects of the project to aid in its fabrication. The pages may include the following: title page, pictorial, assembly drawing, detailed drawing, and exploded view, **Figure 17-10.**

In addition to the actual shape as described by the various lines, a set of drawings may contain additional information such as the title box and bill of materials. The *title box,* which appears in one corner of the drawing, should contain the name of the part, the company name, the scale of the drawing, the date of the drawing, who made the drawing, the drawing number, the number of drawings in this set, and tolerances.

A *bill of materials* can also be included in the set of drawings. This is a list of the various items that will be needed to build the project, **Table 17-1.**

Computers and Drawings Computers have made it much easier to draw plans for projects. The welded bird house project shown in **Figure 17-11** was drawn on AutoCAD LT®. A number of computer drawing programs, such as AutoCAD LT®, are available. Drafting programs use vector lines, unlike most drawing programs, which use raster art. Computers see vector lines as lines and raster lines as if they were part of a picture. Because vector drawings are seen as lines by the computer, lines stay crisp and sharp while zooming in and out, measuring, resizing, reshaping, and rotating the drawing. For example, as the lines on the roof of the vector-drawn bird house in **Figure 17-12** are magnified 300 and 500 times they stay the same.

Computers see raster images as a series of small squares called pixels. Raster drawing is commonly known as bit-map drawing because the computer maps the location of every little bit (pixel) of the drawing. When these pixels are very small the eye sees them as a line, but after zooming in they look like many colored squares. For example as the lines on the roof of the bit-map drawn bird house in **Figure 17-13** are magnified 300 and 500 times they look like a group of colored squares—not even recognizable as lines. Bit-map lines have a softer appearance and they work best in art and photographic programs. The sharp crisp lines of vector drawing work best for mechanical drawings.

Two-dimensional drafting programs, abbreviated 2D like AutoCAD LT® allow you to make mechanical drawings accurately for projects. Vector drafting programs allow you to draw trailers, barns, or other large projects more accurately than they can be built. Using the pull-down and dimensioning options, the precision for this birdhouse drawing could be set at 1/256″ (0.01 mm), **Figure 17-14.** Accuracy is important when parts must fit together and move without interfering with each other. It is also helpful when planning

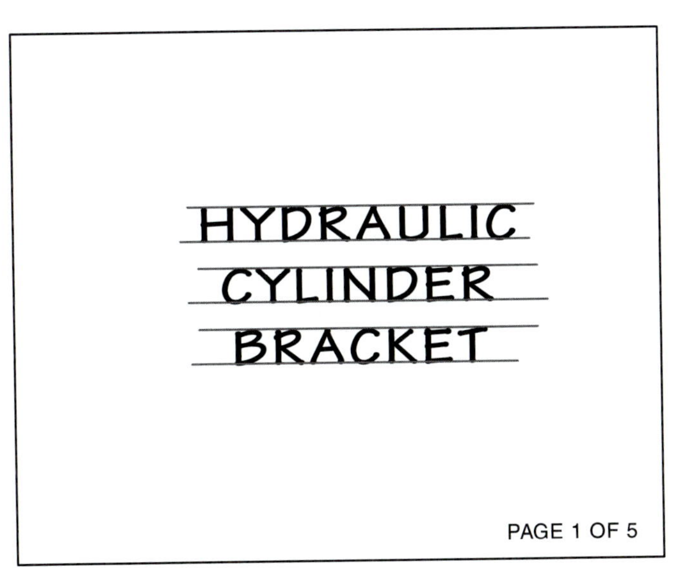

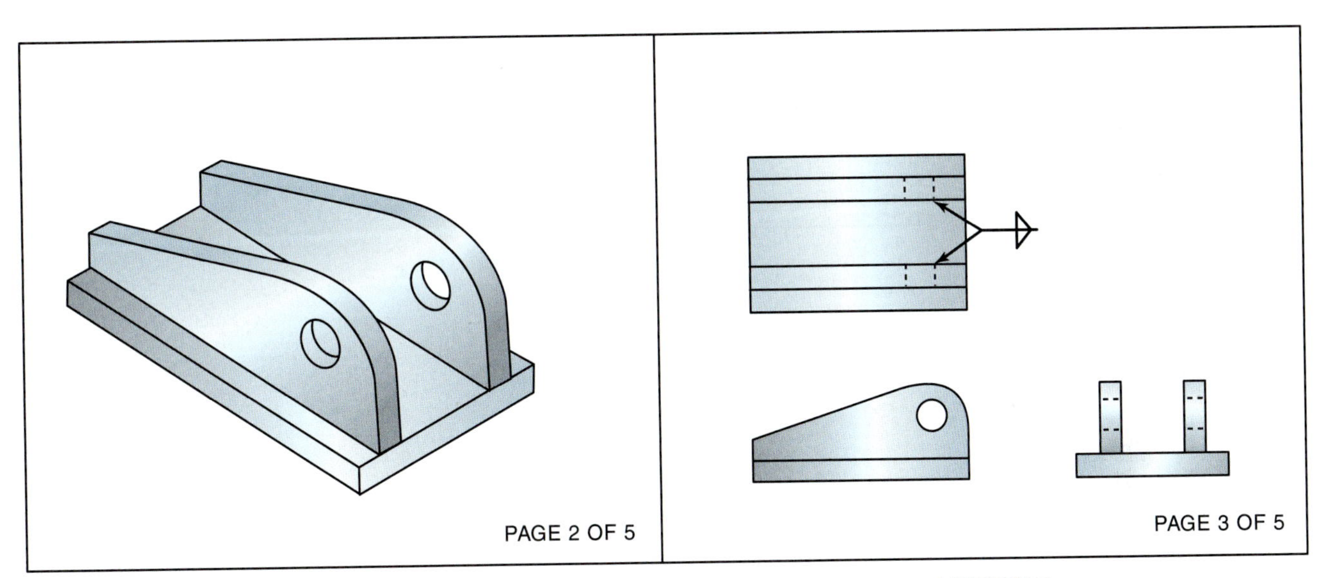

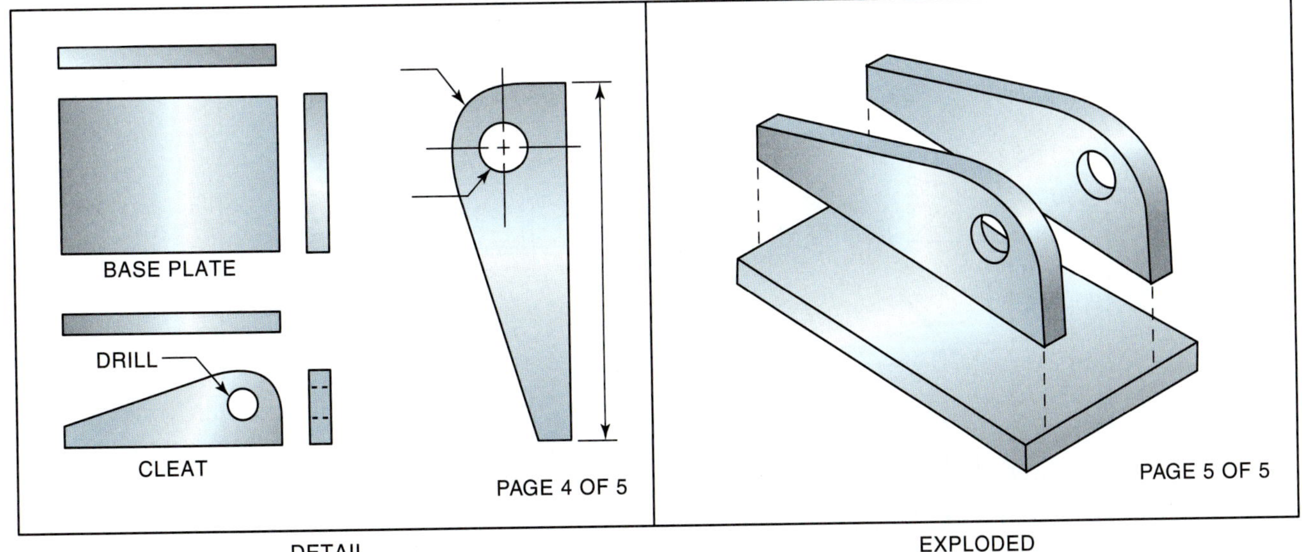

FIGURE 17-10 Drawings that can make up a set of drawings.

Part	Number Required	Type of Material	Size (Standard Units)	(SI Units)
Base	1	Hot roll steel	1/2″ × 5″ × 8″	12.7 mm × 127 mm × 203.2 mm
Cleat	2	Hot roll steel	1/2″ × 4″ × 8″	12.7 mm × 101.6 mm × 203.2 mm

TABLE 17-1 Bill of Materials.

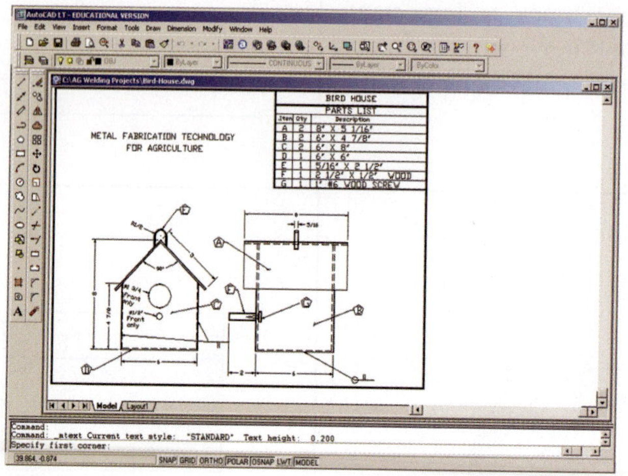

FIGURE 17-11 AutoCAD LT® drafting program used to design a welded project.

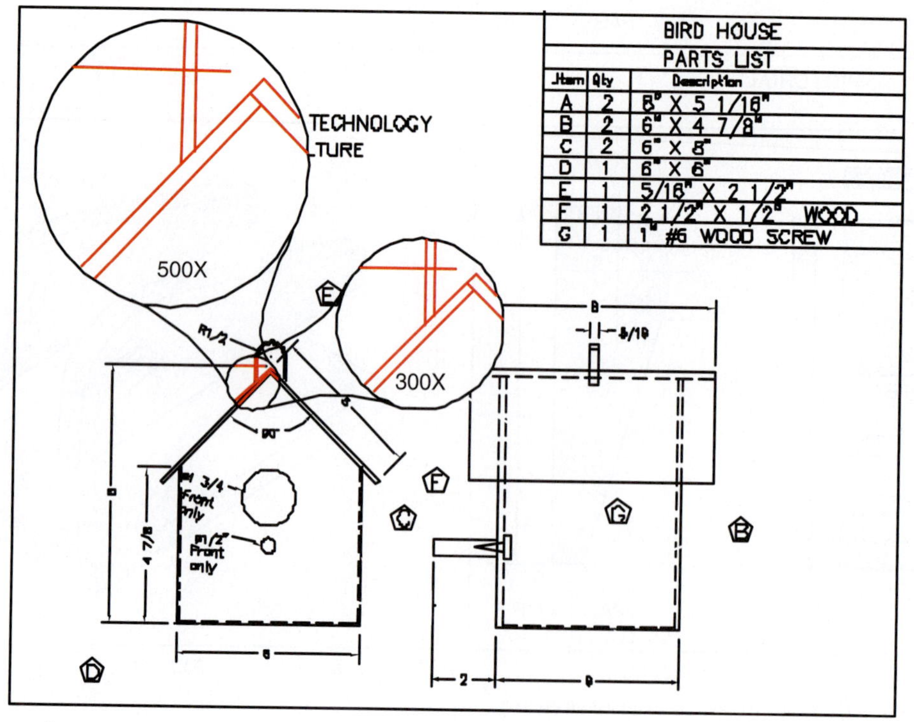

FIGURE 17-12 Vector line art drawing.

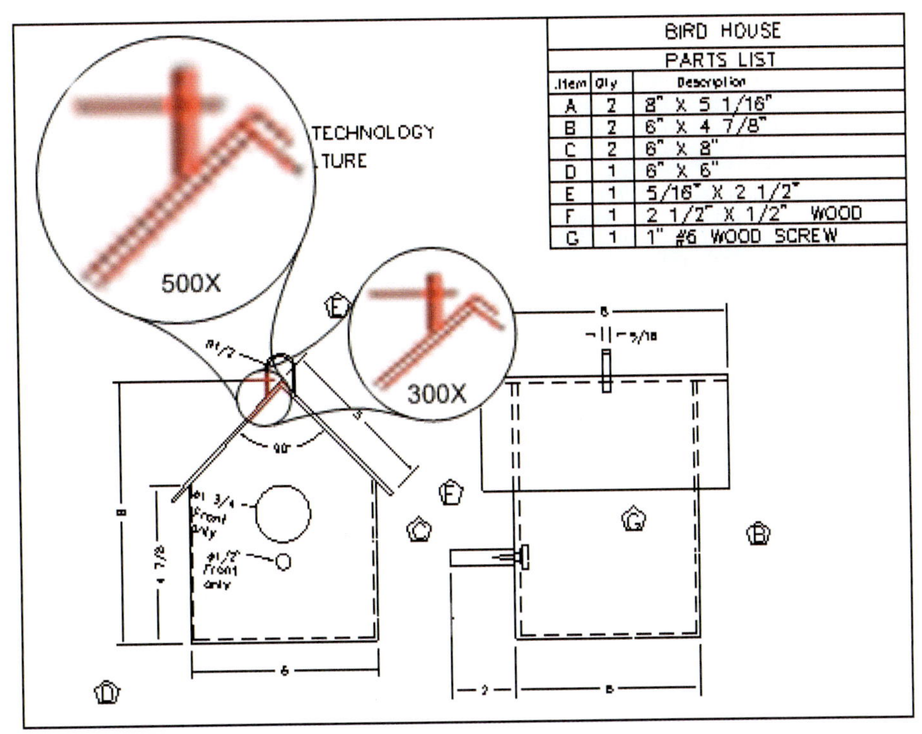

FIGURE 17-13 Bit map line drawing.

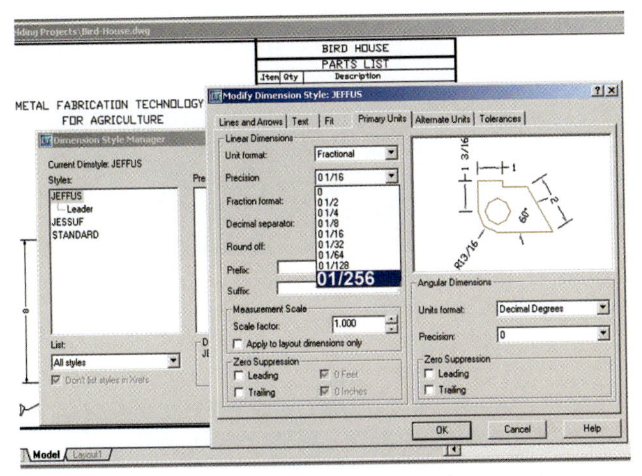

FIGURE 17-14 The drawing precision can be set on AutoCAD LT® using the dimension style manager.

FIGURE 17-15 Well designed agricultural trailer for both hauling hay and equipment. Courtesy of Larry Jeffus.

projects like the trailer in **Figure 17-15**. The back of the trailer is dropped with ramps built on that can be flipped down for easier loading of equipment. The ramps are hinged to the back of the trailer so that when they are in the up position the dropped back end of the bed is level. AutoCAD LT® makes designing the trailer easier, and leveling the bed makes it easier to haul hay, **Figure 17-16**.

Lines on a mechanical drawing connect to other lines on the drawing. It is that connection between lines that allows the object being drawn to take shape. Very rarely would a mechanical drawing line be drawn just by itself. Computerized drafting programs make it easy to make lines connect. As the cursor nears the end of a line in AutoCAD LT®, as with many vector drafting programs, the end of the line changes colors, **Figure 17-17**. Moving closer makes the line "jump" to join the end of the first line drawn. This feature of "joining lines" like other features can be adjusted or turned off as needed. This ability to customize the settings on a drafting program makes the program much easier and faster to use.

The readability and ease of understanding a drawing is dependent on the drawing's layout and location

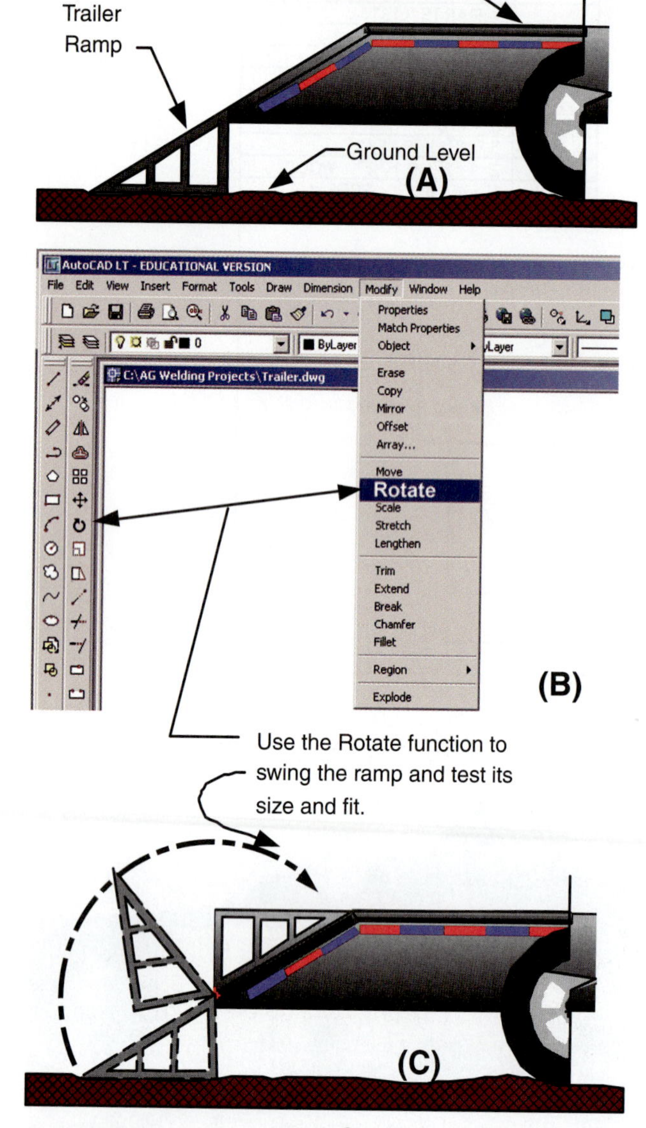

FIGURE 17-16 The AutoCAD LT® feature of rotation can be used used to check part alignment and fit such as the trailer ramp on this drawing.

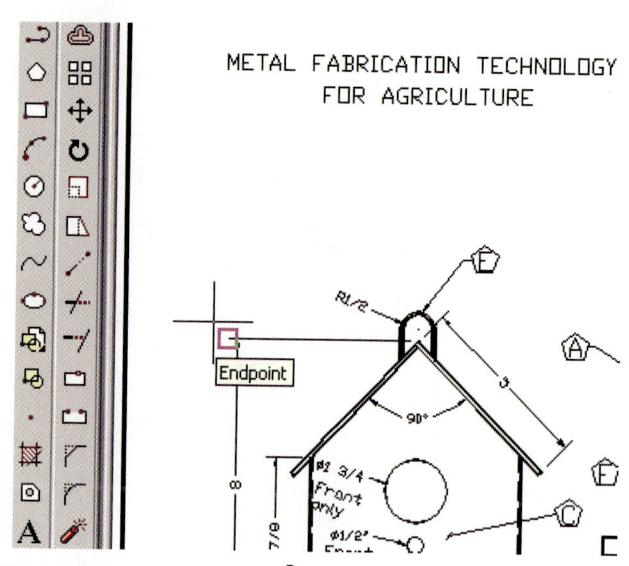

FIGURE 17-17 AutoCAD LT® makes it easy to connect lines by its end points connecting feature.

FIGURE 17-18 Hovering above an icon will provide a one or two word hint.

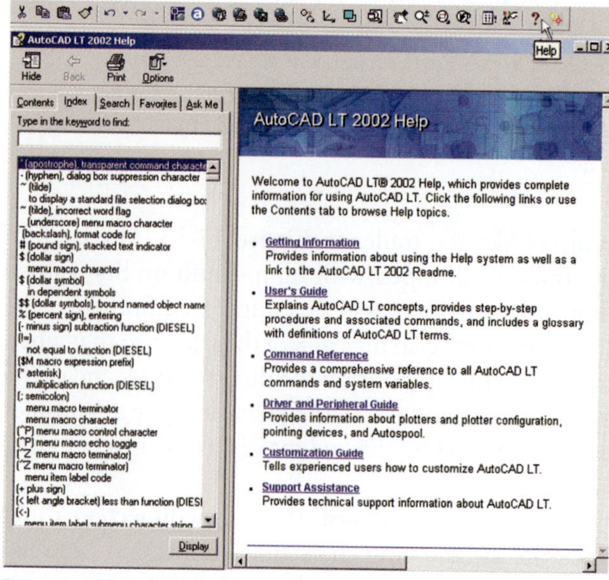

FIGURE 17-19 On-screen help will guide you through many of the computer program's features.

of diminutions and notes. Drafting programs allow you to move things around to make them clearer and easier to understand. This feature is important because pencil drawing, erasing, and redrawing can result in a messy drawing that is not clearer and easier to understand.

On-screen help with computer drafting is available in several general ways. One method of getting on-screen help is to hover the cursor over a function button for a moment. A one or two word description of the button will then appear, **Figure 17-18.** A more complete listing of help is available on screen by clicking on the "?" button on the tool bar at the top of the screen, **Figure 17-19.** Some programs like AutoCAD LT® provide an Active Assistant that pops up the first time a function button is clicked, **Figure 17-20**. The Active Assistant provides step-by-step instructions on how to use that function.

Lines To understand drawings, you need to know what the different types of lines represent. The language of drawing uses lines to represent its alphabet and the various parts of the object being illustrated. The various line types are collectively known as the *alphabet of lines*, **Table 17-2** and **Figure 17-21**.

Types of Drawings Drawings used for most welding projects can be divided into two categories, orthographic projections and pictorial. Projection drawings are made as though one were looking through the sides

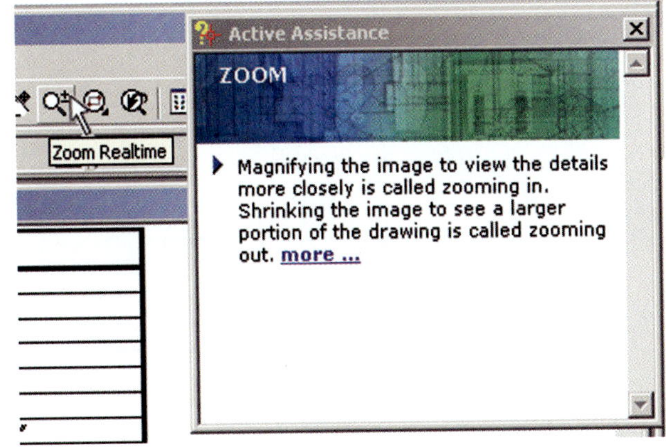

FIGURE 17-20 Active Assistant details the specific functions available.

Line Type	Description	Purpose
OBJECT LINE	Solid bold line	To show the intersection of surfaces or the extent of a curved surface
HIDDEN LINE	Broken medium line	To show the intersection of surfaces or the extent of a curved surface that occurs below the surface and hidden from view
CENTER LINE	Fine broken line made up of longer line sections on both sides of a short, dashed line	To show the center of a hole, curve, or symmetrical object
EXTENSION	Extension lines (fine line) extending from near the surface of the object	Extension lines extend from an object line or a hidden line to locate dimension points.
DIMENSION LINE	Dimension lines (medium line) extending between extension lines or object lines	Dimension lines touch the extension lines and/or object lines that represent the points being dimensioned.
CUTTING PLANE LINE	Bold broken lines with arrowheads pointing in the direction of the cut surface	These lines extend all the way across the surface that is being imaginarily cut. The arrowhead ends point in the direction in which the cut surface will be shown in the sectional drawing.
SECTION LINES (STEEL, CAST IRON)	Series of fine lines drawn at an angle to the object lines. The line angle usually changes from one part to another. The cast iron section lines are used universally for most sections.	Used to indicate a surface that has been imaginarily cut or broken. The spacing and pattern can be used to indicate the type of material that is being viewed.
LEADER OR ARROW LINE	Medium line with an arrowhead at one end	Leader and arrow lines are used to locate points on the drawing to which a specific note, dimension, or welding symbol refers.
LONG BREAK LINE	Bold straight line with intermittent zigzag	To indicate that a portion of the part has not been included in the drawing either to conserve space or because the omitted portion was not significant to this specific drawing
SHORT BREAK LINE	Bold freehand irregular line	Used for the same purposes as the long break above except on parts not wide enough to allow the long break lines with their zigzags to be used clearly

TABLE 17-2 Alphabet of Lines.

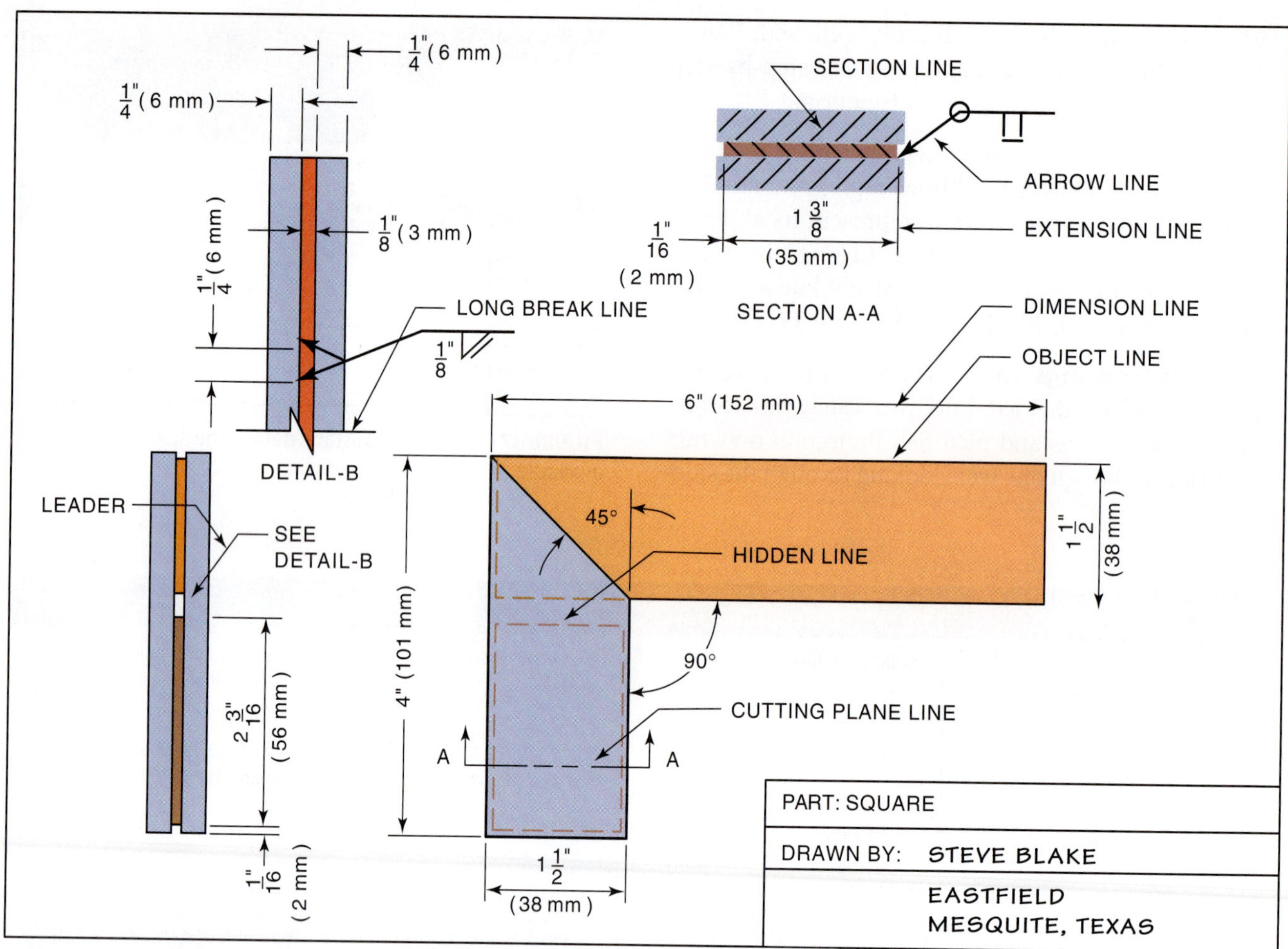

FIGURE 17-21 Drawing showing alphabet of lines.

of a glass box at the object and tracing its shape on the glass, **Figure 17-22**. If all the sides of the object were traced and the box unfolded and laid out flat, six basic views would be shown, **Figure 17-23**.

Pictorial drawings present the object in a more realistic or understandable form and usually appear as one of two types, isometric or cavalier, **Figure 17-24**. The more realistic, perspective drawing form is seldom used for welding projects.

Projection Drawings Usually not all of the six views are required to build the weldment. Only those needed are normally provided, usually only the front, right side, and top views. Sometimes only one or two of these views are needed.

The front view is not necessarily the front of the object. A view is selected as the front view because its overall shape is best described when the object is viewed from this direction. As an example, the front view of a car or truck would probably be the side of the vehicle because viewing the vehicle from its front may not show enough detail to let you know whether

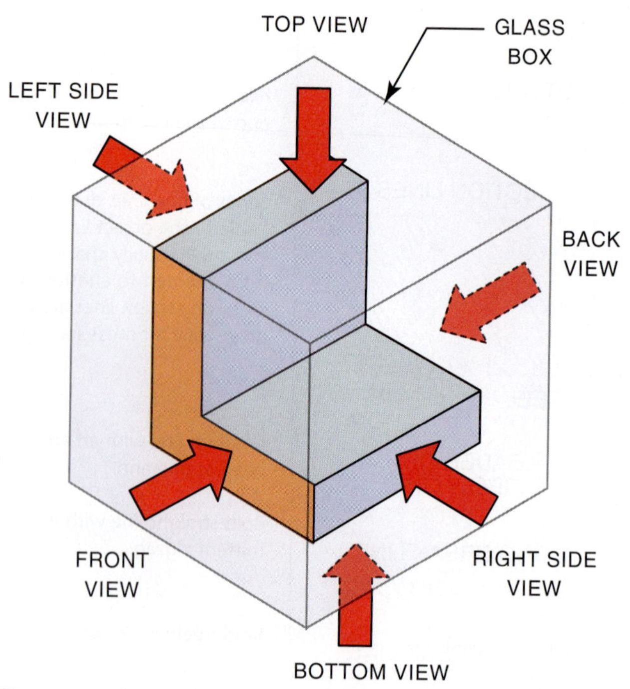

FIGURE 17-22 Viewing an object as if it were inside a glass box.

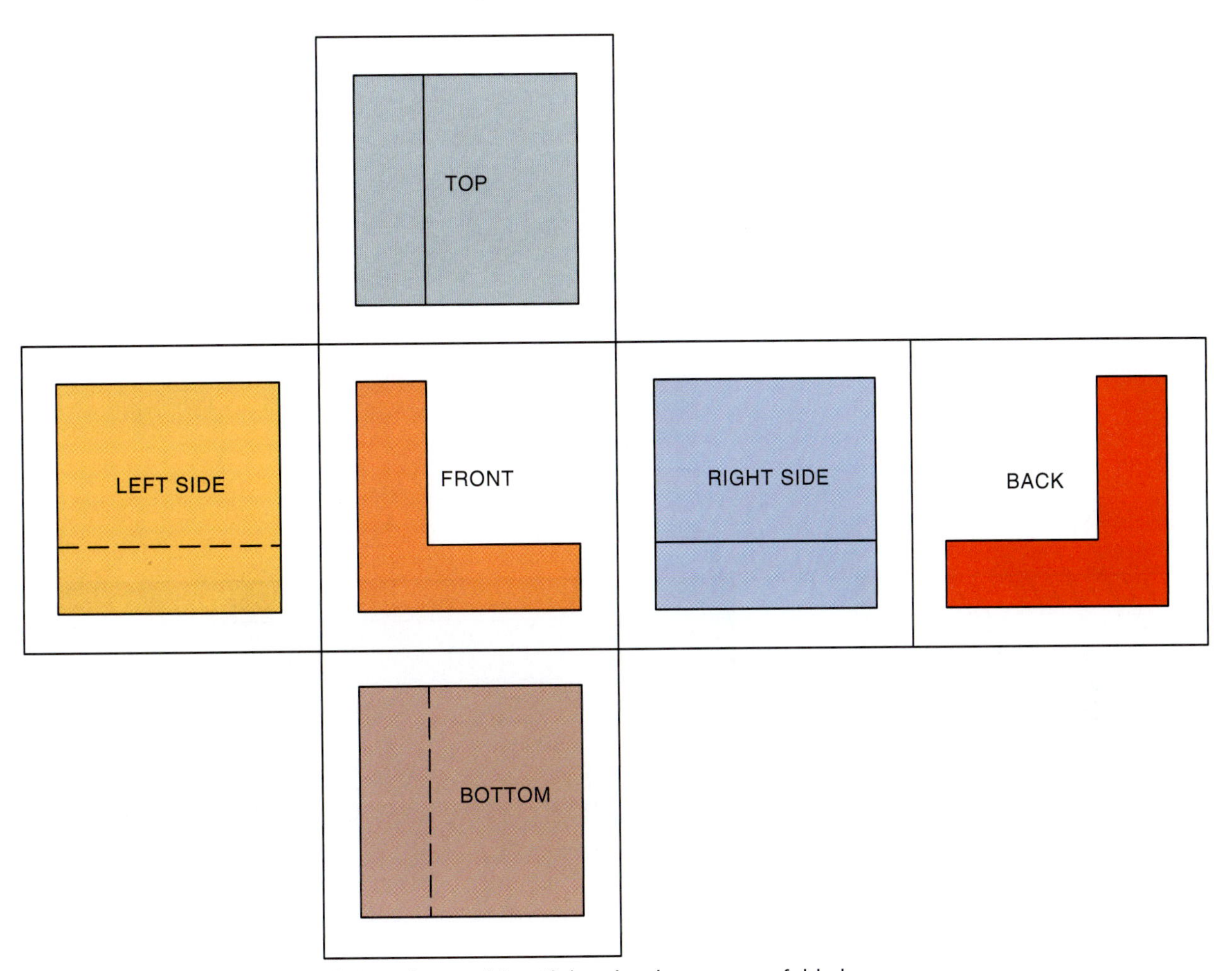

FIGURE 17-23 The arrangement of views for an object if the glass box were unfolded.

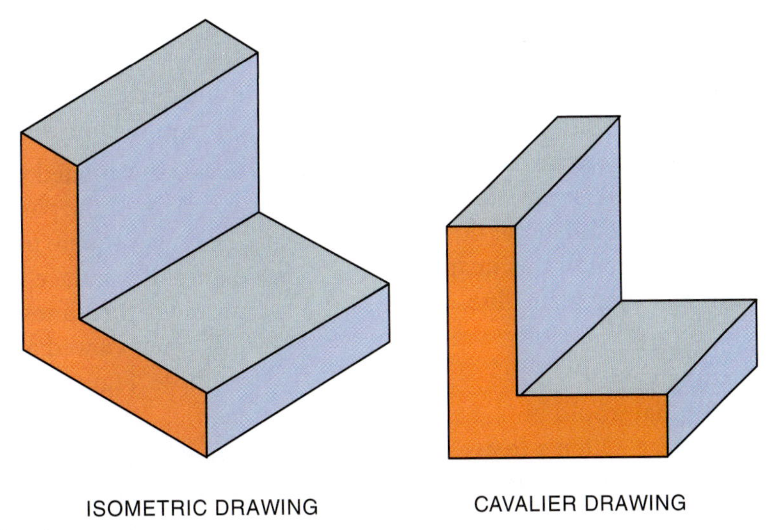

FIGURE 17-24 Pictorial drawing types.

it is a car, light truck, station wagon, or van. From the front, most vehicles look very similar.

Special Views Special views may be included on a drawing to help describe the object so it can be made accurately. Special views on some drawings may include:

- *Sections view:* The section view is drawn as if part of the object were sawn away to reveal internal details, **Figure 17-25A.** This view is useful when the internal details would not be as clear if they were shown as hidden lines. Sections can be either fully across the object or

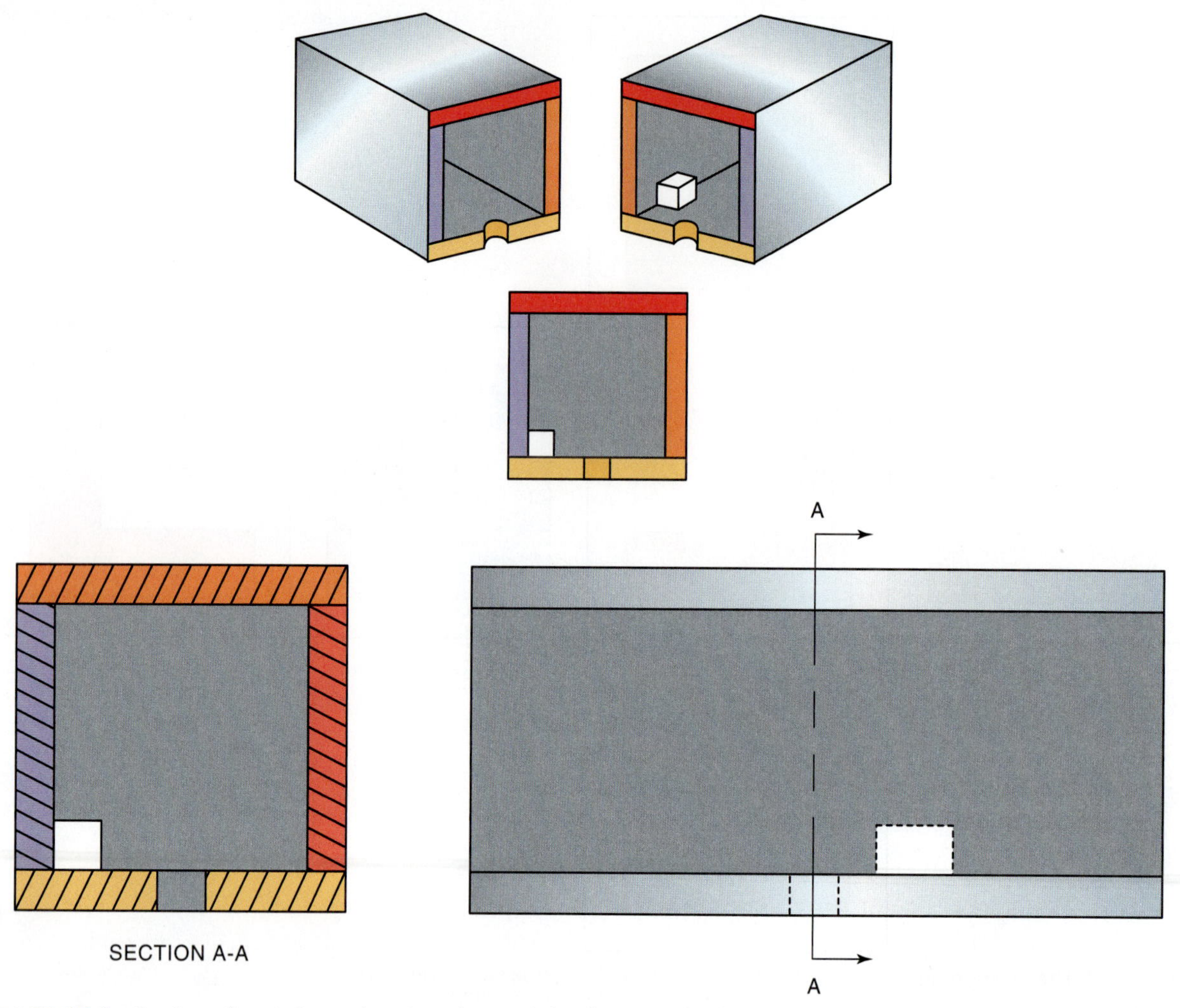

FIGURE 17-25 Sectioning of parts is used to show internal details more clearly.

just partially across it. The imaginary cut surface is set off from other noncut surfaces by section lines drawn at an angle on the cut surfaces. Some drawings use specific types of section lines to illustrate the type of material the part was made with. The location of this imaginary cut is shown using a cutting plane line, **Figure 17-25B.**

- *Cut-a-ways:* The cut-a-way view is used to show detail within a part that would be obscured by the part's surface. Often a freehand break line is used to outline the area that has been imaginarily removed to reveal the inner workings.
- *Detail views:* The detail view is usually an external view of a specific area of a part. Detail views show small details of a part's area and negate the need to draw an enlargement of the entire part. If only a small portion of a view has significance, this area can be shown in a detail view, either at the same scale or larger if needed. By showing only what is needed within the detail, the part drawn can be clearer and not require such a large page.
- *Rotated views:* A rotated view can be used to show a surface of the part that would not normally be drawn square to any of the six normal view planes. If a surface is not square to the viewing angle, then lines may be distorted. For example, when viewed at an angle, a circle looks like an ellipse, **Figure 17-26.**

Dimensioning Often it is necessary to look at other views to locate all of the dimensions required to build the object. By knowing how the views are arranged, it becomes easier to locate dimensions. Length dimensions can be found on the front and top views. Height dimensions can be found on the front and right side views. Width dimensions can be found on the top and right side views, **Figure 17-27.** The locations of dimensions on these views are consistent with both the first angle perspective or third angle perspective layouts.

If the needed dimensions cannot be found on the drawings, do not try to obtain them by measuring

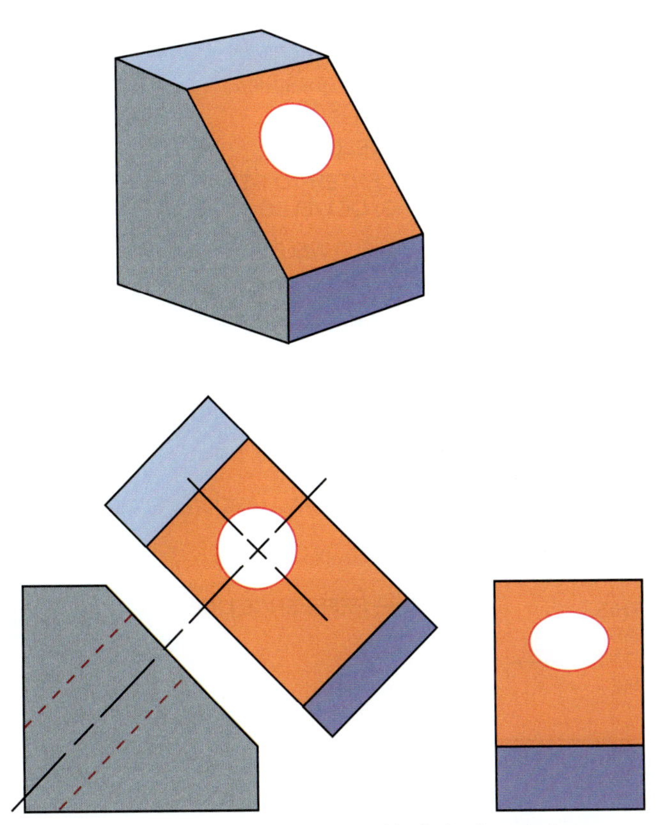

FIGURE 17-26 Notice that the round hole looks misshapen, or elliptical, in the right side view.

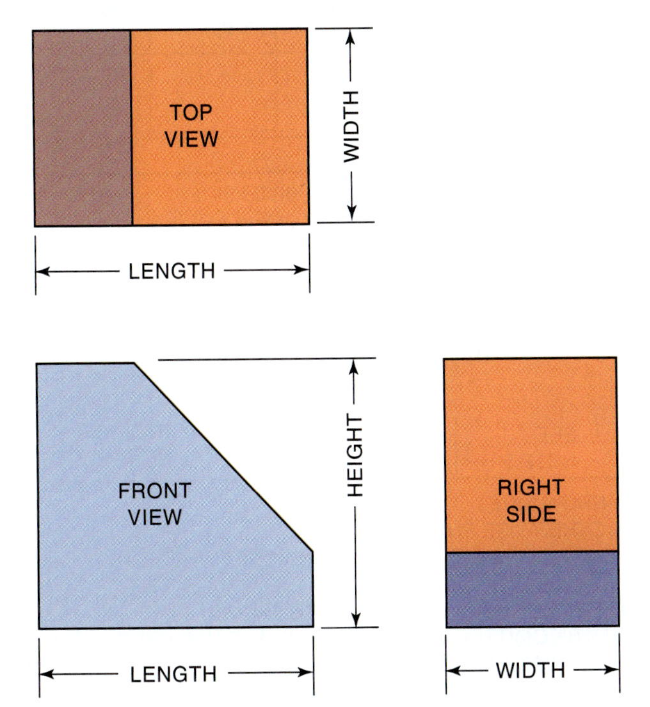

FIGURE 17-27 Drawing dimension locations.

the drawing itself. Even if the original drawing was made very accurately, the paper it is on changes with changes in humidity. Copies of the original drawing are never the exact same size. The most acceptable way of determining missing dimensions is to contact the person who made the drawing.

Keep the drawing clean and well away from any welding. Avoid writing or doing calculations on the drawing. Often a drawing will be filed following the project for use at a later date. The better care you take with the drawings, the easier it will be for someone else to use them.

Welding Symbols

The use of welding symbols enables a designer to indicate clearly to the welder important detailed information regarding the weld. The information in the welding symbol can include the following details for the weld: length, depth of penetration, height of reinforcement, groove type, groove dimensions, location, process, filler metal, strength, number of welds, weld shape, and surface finishing. All this information would normally be included on the welding assembly drawings.

Welding symbols are a shorthand language for the welder. They save time and money and serve to ensure understanding and accuracy. Welding symbols have been standardized by the American Welding Society. Some of the more common symbols for welding are reproduced in this chapter. If more information is desired about symbols or how they apply to all forms of manual and automatic machine welding, these symbols can be found in the complete material, *Standard Symbols for Welding, Brazing and Nondestructive Examination*, ANSI/AWS A2.4, published as an American National Standard by the American Welding Society.

Figure 17-28 shows the basic components of welding symbols, consisting of a reference line with an arrow on one end. Other information relating to various features of the weld are shown by symbols, abbreviations, and figures located around the reference line. A tail is added to the basic symbol as necessary for the placement of specific information.

Indicating Types of Welds

Weld types are classified as follows: fillets, grooves, flange, plug or slot, spot or protecting, seam, back or backing, and surfacing. Each type of weld has a specific symbol that is used on drawings to indicate the weld. A fillet weld, for example, is designated by a right triangle. A plug weld is indicated by a rectangle. All of the basic symbols are shown in **Figure 17-29**.

Weld Location

Welding symbols are applied to the joint as the basic reference. All joints have an arrow side (near side) and another side (far side). Accordingly, the

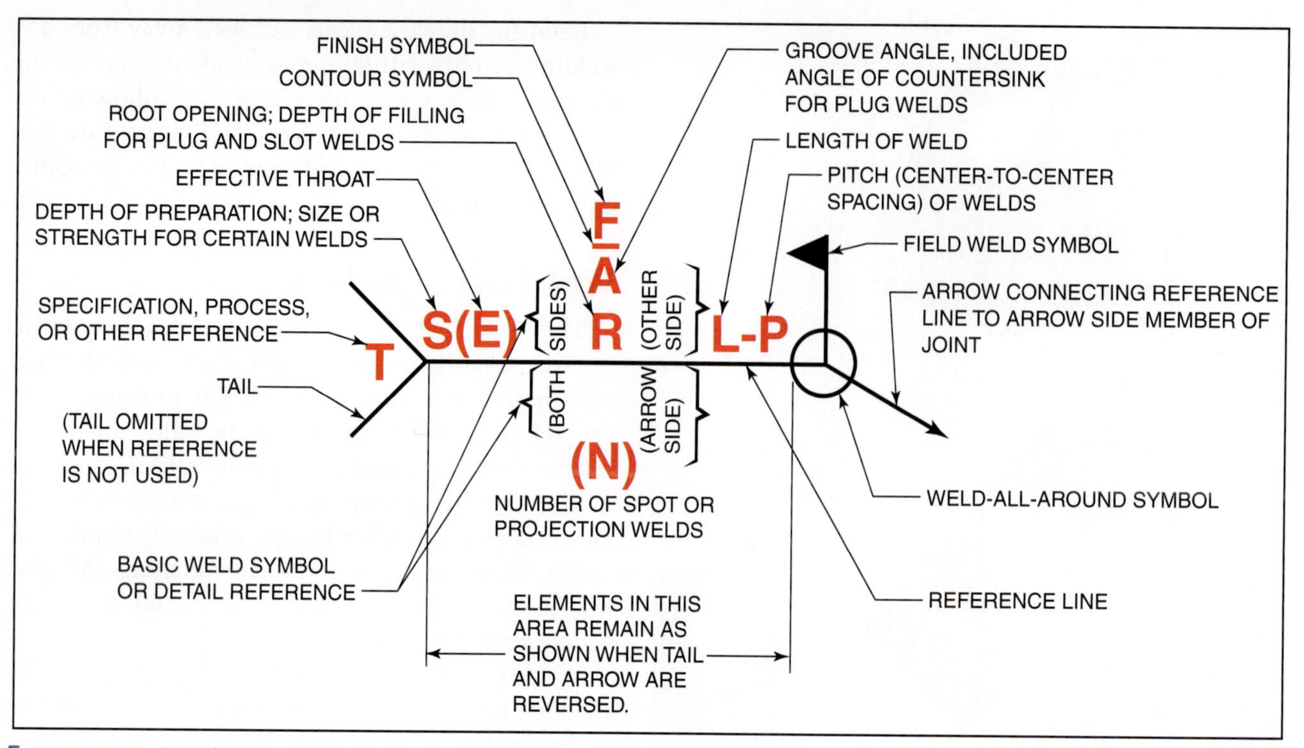

FIGURE 17-28 Standard location of elements of a welding symbol. Courtesy of the American Welding Society.

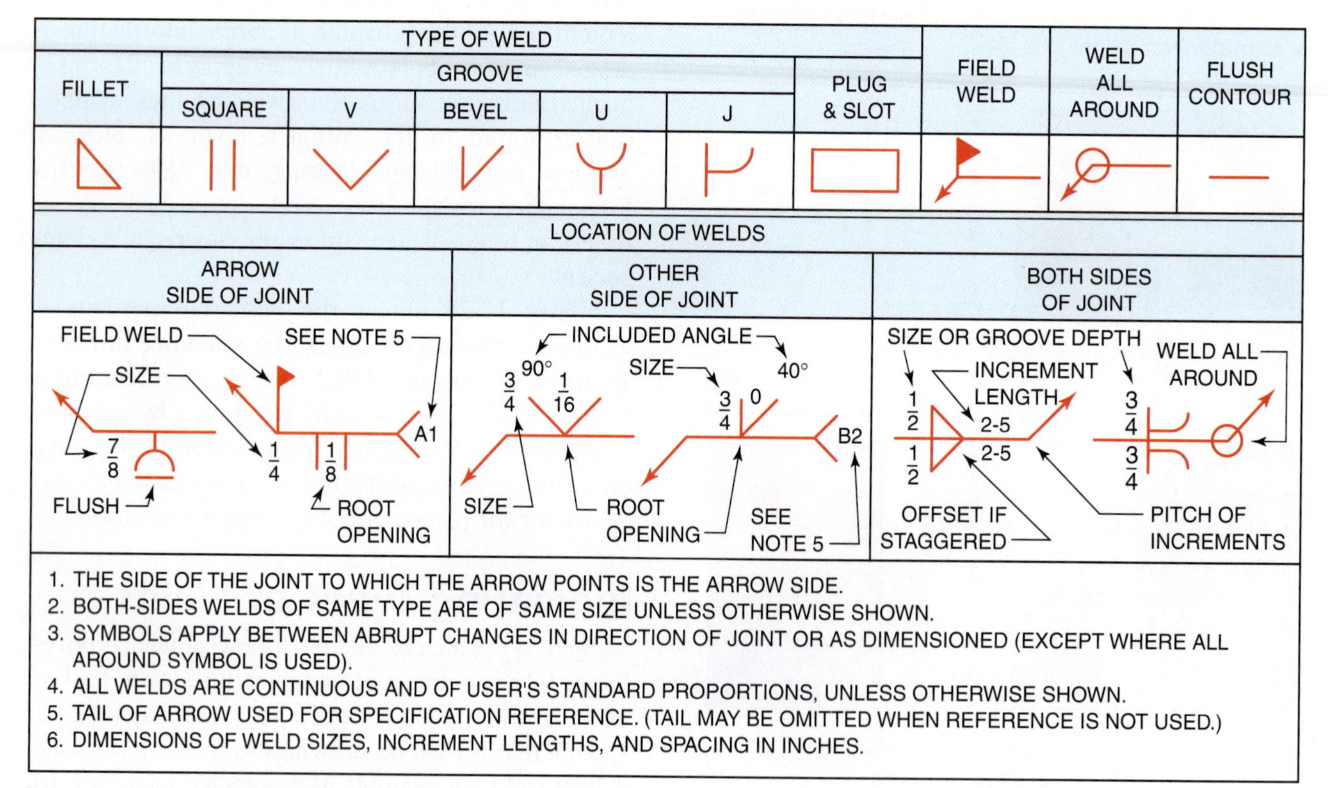

FIGURE 17-29 Welding symbols for different types of welds.

terms *arrow side, other side,* and *both sides* are used to indicate the weld location with respect to the joint. The reference line is always drawn horizontally. An arrow line is drawn from one end or both ends of a reference line to the location of the weld. The arrow line can point to either side of the joint and extend either upward or downward.

If the weld is to be deposited on the arrow side of the joint (near side), the proper weld symbol is placed below the reference line, **Figure 17-30A.**

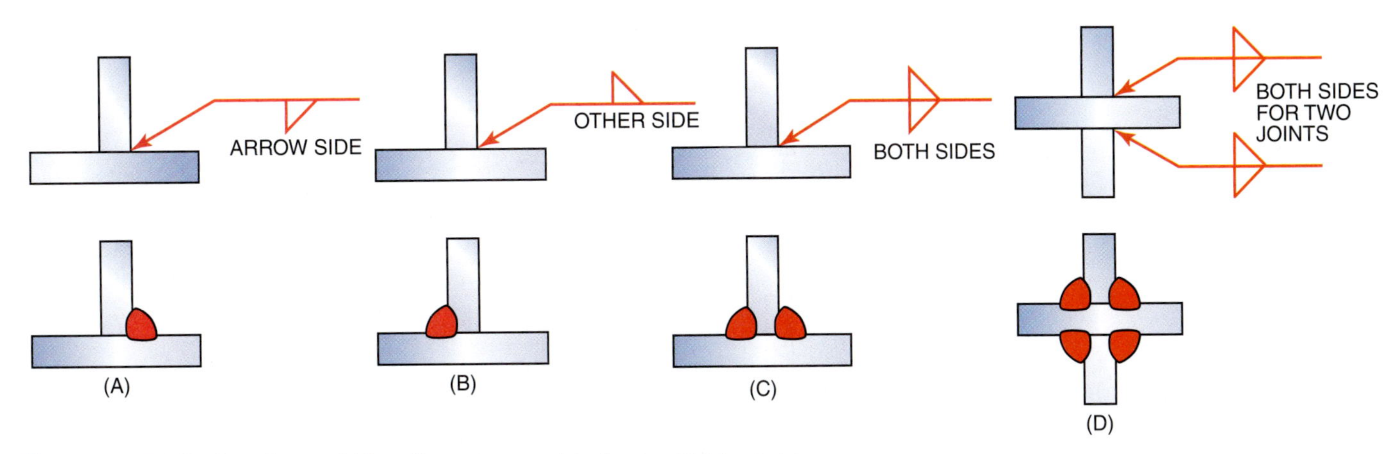

FIGURE 17-30 Designating weld locations. Courtesy of the American Welding Society.

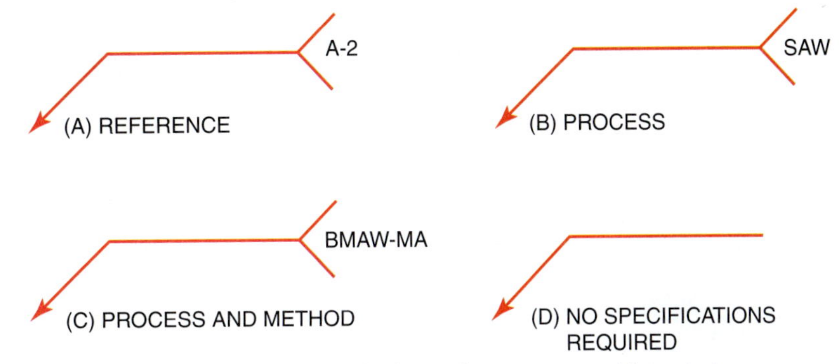

FIGURE 17-31 Locations of specifications, processes, and other references on weld symbols.

If the weld is to be deposited on the other side of the joint (far side), the weld symbol is placed above the reference line, **Figure 17-30B**. When welds are to be deposited on both sides of the same joint, the same weld symbol appears above and below the reference line, **Figure 17-30C** and **Figure 17-30D**, along with all detailed information.

The tail is added to the basic welding symbol when it is necessary to designate the welding specifications, procedures, or other supplementary information needed to make the weld, **Figure 17-31**. The notation placed in the tail of the symbol may indicate the welding process to be used, the type of filler metal needed, whether or not peening or root chipping is required, and other information pertaining to the weld. If notations are not used, the tail of the symbol is omitted.

For joints that are to have more than one weld, a symbol is shown for each weld.

Location Significance of Arrow

In the case of fillet and groove welding symbols, the arrow connects the welding symbol reference line to one side of the joint. The surface of the joint the arrow point actually touches is considered to be the arrow side of the joint. The side opposite the arrow side of the joint is considered to be the other (far) side of the joint.

On a drawing, when a joint is illustrated by a single line and the arrow of a welding symbol is directed to the line, the arrow side of the joint is considered to be the near side of the joint.

For welds designated by the plug, slot, spot, seam, resistance, flesh, upset, or projection welding symbols, the arrow connects the welding symbol reference line to the outer surface of one of the members of the joint at the center line of the desired weld. The member to which the arrow points is considered to be the arrow side member. The remaining member of the joint is considered to be the other side member.

Fillet Welds

Dimensions of fillet welds are shown on the same side of the reference line as the weld symbol and are shown to the left of the symbol, **Figure 17-32A**. When both sides of a joint have the same size fillet welds they are dimensioned as shown in **Figure 17-32B**. When both sides of a joint have different size fillet welds, both are dimensioned, **Figure 17-32C**. When the dimensions of one or both welds differ

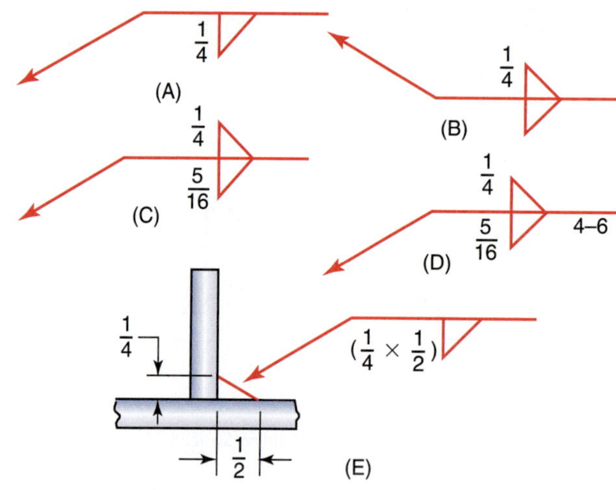

FIGURE 17-32 Dimensioning the fillet weld symbol.

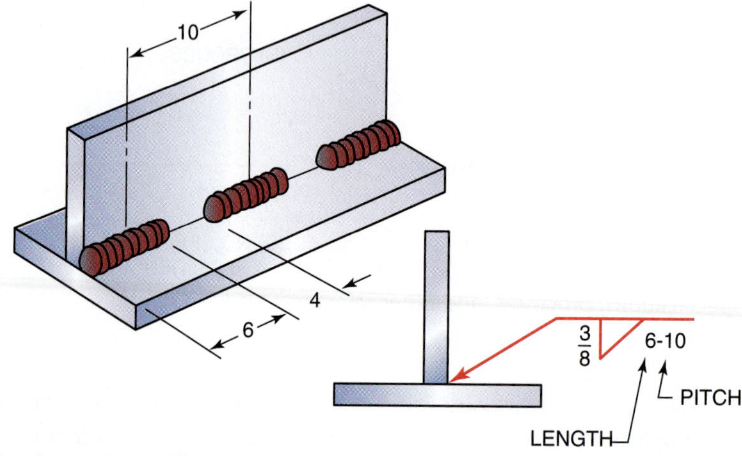

FIGURE 17-33 Dimensioning intermittent fillet welds.

from the dimensions given in the general notes, both welds are dimensioned. The size of a fillet weld with unequal legs is shown in parentheses to the left of the weld symbol, **Figure 17-32E**. The length of a fillet weld, when indicated on the welding symbol, is shown to the right of the weld symbol, **Figure 17-32D**. In intermittent fillet welds, the length and pitch increments are placed to the right of the weld symbol, **Figure 17-33**. The first number represents the length of the weld, and the second number represents the pitch or the distance between the centers of two welds.

Groove Welds

Joint strengths can be improved by making some type of groove preparation before the joint is welded. There are seven types of grooves. The groove can be made in one or both plates or on one or both sides. By cutting the groove in the plate, the weld can penetrate deeper into the joint. This helps to increase the joint strength without restricting flexibility.

The grooves can be cut in base metal in a number of different ways. The groove can be cut using an oxyfuel cutting torch, air carbon arc cutting, plasma arc cutting, machined, or saws.

The various types of groove welds are classified as follows:

- Single-groove and symmetrical double-groove welds that extend completely through the members being joined. No size is included on the weld symbol, **Figure 17-34A** and **Figure 17-34B**.
- Groove welds that extend only part way through the parts being joined. The size as measured from the top of the surface to the bottom (not including reinforcement) is included to the left of the welding symbol, **Figure 17-34C**.
- The size of groove welds with a specified effective throat is indicated by showing the depth of

groove preparation with the effective throat appearing in parentheses and placed to the left of the weld symbol, **Figure 17-34D**. The size of square groove welds is indicated by showing the root penetration. The depth of chamfering and the root penetration is read in that order from left to right along the reference line, **Figure 17-34E** and Figure **17-34F**.

- The root face's main purpose is to minimize the burn-through that can occur with a feather edge. The size of the root face is important to ensure good root fusion, **Figure 17-35**.
- The size of flare groove welds is considered to extend only to the tangent points of the members, **Figure 17-36**.
- The root opening of groove welds is the user's standard unless otherwise indicated. The root opening of groove welds, when not the user's standard, is shown inside the weld symbol, **Figure 17-34E** and **Figure 17-34F**.

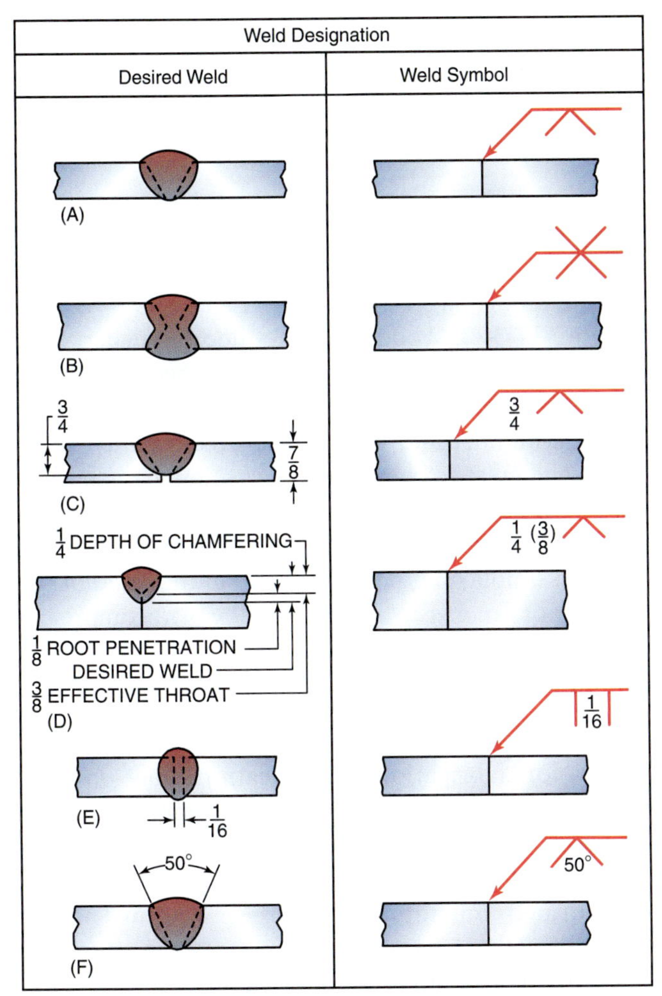

Figure 17-34 Described weld and welding symbols for grooved welds. Courtesy of the American Welding Society.

Backing

A backing (strip) is a piece of metal that is placed on the back side of a weld joint. The backing must be thick enough to withstand the heat of the root pass as it is burned-in. A backing strip may be used on butt joints, tee joints, and outside corner joints, **Figure 17-37**.

The backing may be either left on the finished weld or removed following welding. If the backing is to be removed, the letter *R* is placed in the backing symbol, **Figure 17-38**. The backing is often removed for a finished weld because it can be a source of stress concentration and a crevice to promote rusting.

Fabrication

The two ways that welding is used in agriculture are new fabrication and repair work. Fabrication is the process of cutting, shaping, and assembling material in order to produce a needed part or structure. Repair welding is the process of making a damaged or worn part useful through application of welds. Welding is a great fabrication tool because of its speed, flexibility and permanence. As a repair tool it can save both the time it would take to remove a part and the cost of replacement parts. Often farmers and ranchers have to fabricate parts for repair work.

Metal is a great construction material that can be used to build many useful additions to any farm or ranch. However, wood can be used with metal for many projects, and together they are far more practical

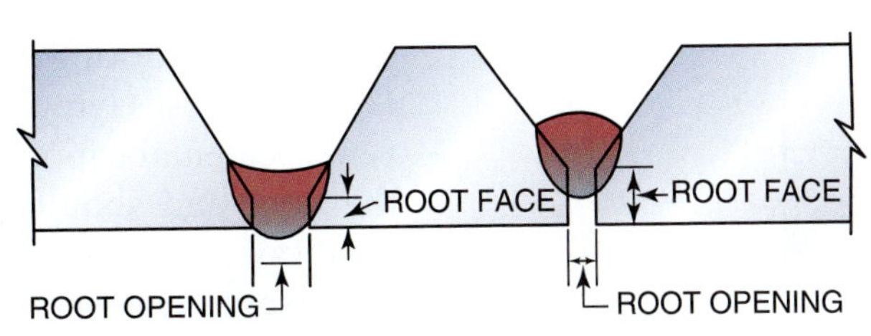

Figure 17-35 Effect root dimensioning can have on groove penetration.

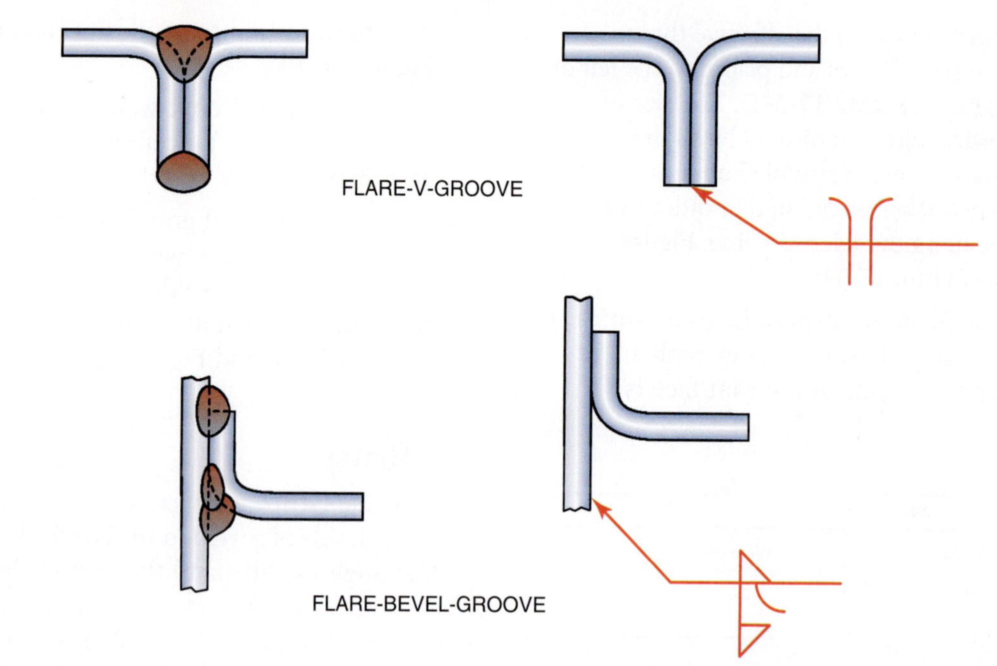

FIGURE 17-36 Designating flare-V- and flare-bevel-groove welds. Courtesy of the American Welding Society.

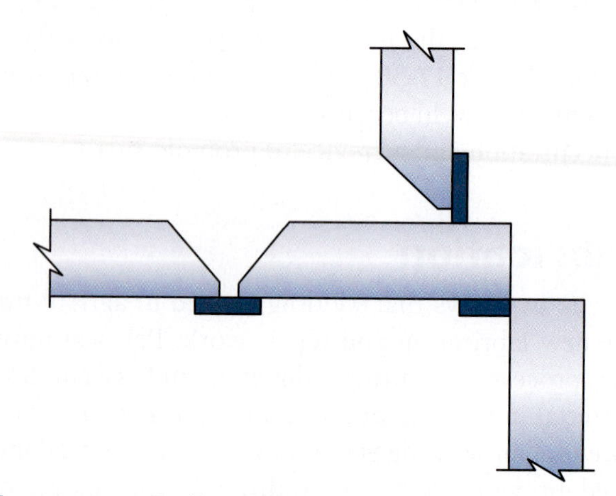

FIGURE 17-37 Backing strips.

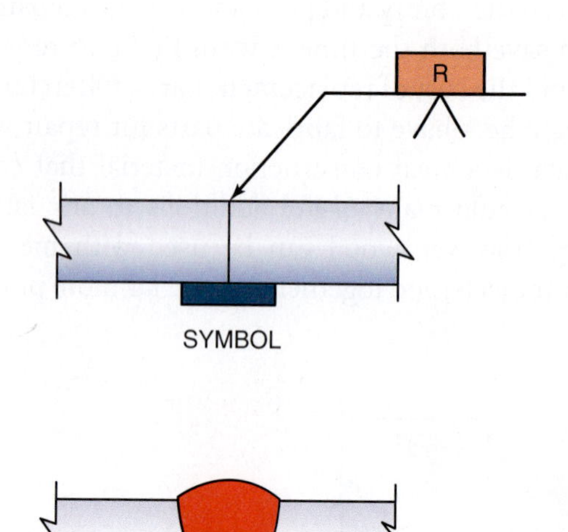

FIGURE 17-38 Butt weld with backing plate.

for building larger structures, **Figure 17-39A** and **Figure 17-39B**. Wood is a better flooring material for trailers and cattle chutes than metal because it provides the animals a better footing, **Figure 17-40**.

The number and type of steps required to take a plan and create a completed fabrication vary depending on the complexity and size of the finished project. All fabrications start with a plan. This plan can range from a simple one that exists only in the mind of the farmer or rancher to a very complex plan comprising a set of drawings. You must learn how to follow a set of drawings to produce a finished fabrication.

Safety As with any job, safety is of primary concern for fabrications. Fabrication may present some safety problems not normally encountered in straight welding. Unlike most practice welding, much of the larger fabrication work may need to be performed outside an enclosed welding booth. Additionally, several welders may be working simultaneously on the same structure, so extra care must be taken to prevent burns to you or the other welders from the arc or hot sparks. Ventilation is also important because the normal shop ventilation may not extend to the fabrication area. Often you will be working in an area with welding cables and torch hoses lying scattered on the floor. To prevent accidental tripping, these lines must be flat on the floor and should be covered if they are in a walkway.

These and other safety concerns are covered in Chapter 2, Welding Safety. You should also read any

(A)

(B)

FIGURE 17-39 (A) A cotton shed fabricated using wood and steel construction. (B) Inside steel girders with wooden rafters covered with sheet metal parts. Courtesy of Larry Jeffus.

FIGURE 17-40 Wooden flooring on a metal cattle chute. Courtesy of Larry Jeffus.

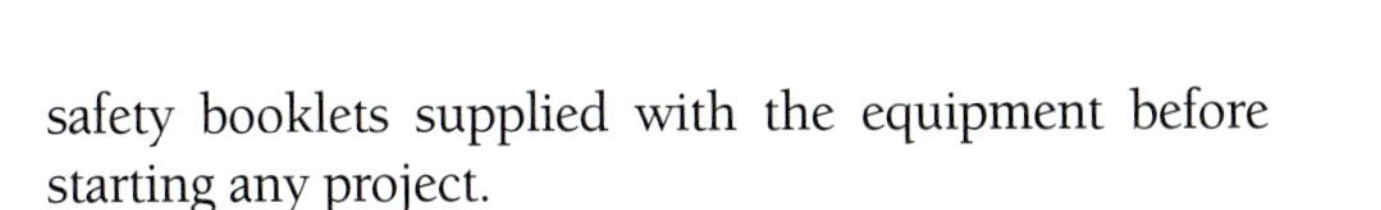

safety booklets supplied with the equipment before starting any project.

Shop Math

Measuring

Measuring for most welded fabrications does not require accuracies greater than what can be obtained with a steel rule or a steel tape, **Figure 17-41A** and **Figure 17-41B**. Both steel rules and steel tapes are available in standard and metric units. Standard unit rules and tapes are available in fractional and decimal units, **Figure 17-42**.

Tolerances

All measuring, whether on a part or on the drawing, is in essence an estimate, because no matter how accurate the measurement is there will always be a more accurate way of taking it. The more accurate the measurement, the more time it takes. To save time while still making an acceptable part, dimensioning tolerances have been established. Most drawings will state a dimensioning tolerance, the amount by which the part can be larger or smaller than the stated dimensions and still be acceptable. Tolerances are usually expressed as plus (+) and

FIGURE 17-41(A) Steel tape measures are available in lengths ranging from 6 feet to 100 feet. Courtesy of the Stanley Tool Div.

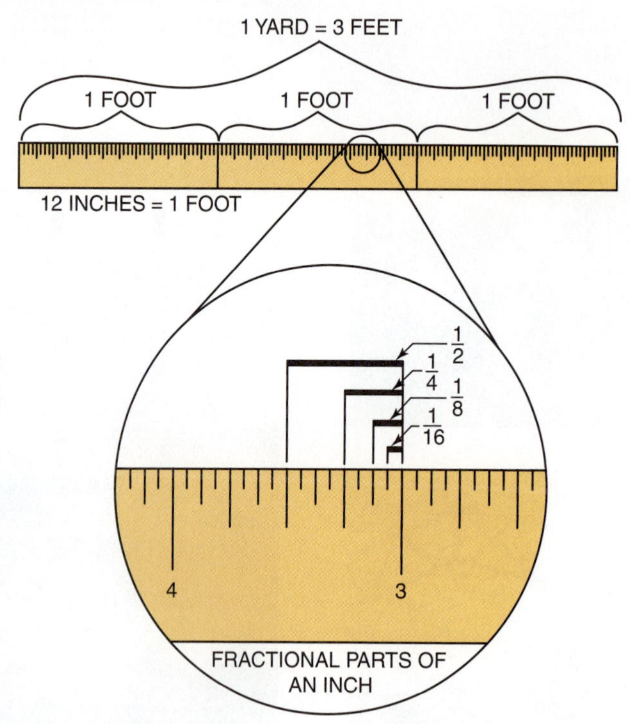

FIGURE 17-41(B) The standard system of linear measure is based on the yard. The yard is divided into three feet, each foot into twelve inches, and each inch into fractional parts. Courtesy of Mark Huth.

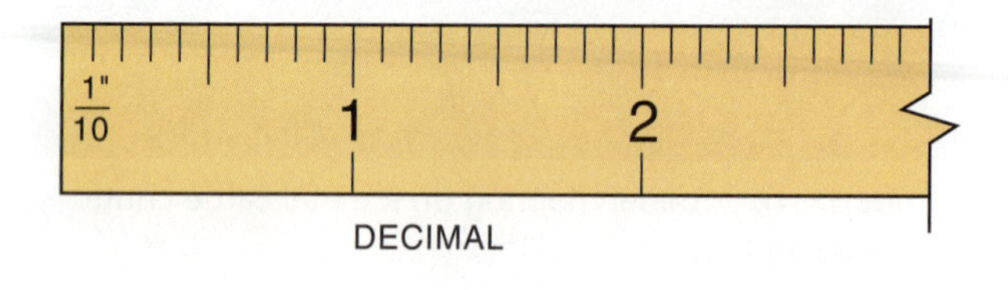

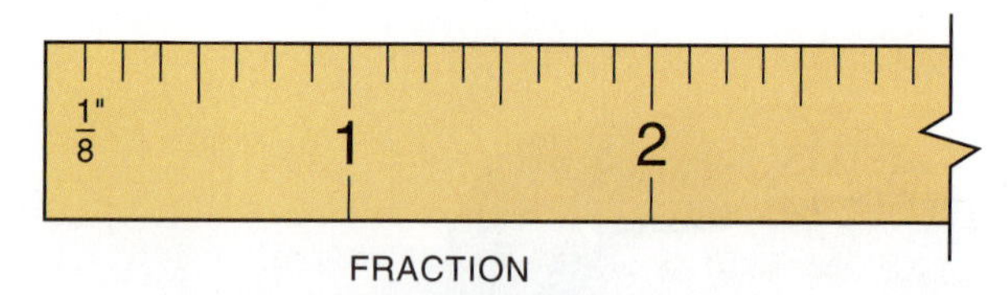

FIGURE 17-42 Two ways tapes can be dimensioned.

		Acceptable Dimensions	
Dimension	**Tolerance**	**Minimum**	**Maximum**
10″	±1/8″	9 7/8″	10 1/8″
2′ 8″	±1/4″	2′ 7 3/4″	2′ 8 1/4″
10′	±1/8″	9′ 11 7/8″	10′ 1/8″
11″	±0.125	10.875″	11.125″
6′	±0.25	5′ 11.75″	6′ 0.25″
250 mm	±5 mm	245 mm	255 mm
300 mm	+ 5 mm–0 mm	300 mm	305 mm
175 cm	±10 mm	174 cm	176 cm

TABLE 17-3 Dimension tolerances.

minus (−). If the tolerance is the same for both the plus and the minus, it can be written using the symbol ±, **Table 17-3**. In addition to the tolerance for a part, there may be an overall tolerance for the completed weldment. This dimension ensures that, if all the parts are either too large or too small, their cumulative effect will not make the completed weldment too large or too small.

Adding and Subtracting

Although most drawings are made with as many dimensions as possible, the welder may have to do some basic math to complete the project. Adding and subtracting fractions and mixed numbers can be accomplished quickly by following a simple rule.

RULE: Fractions that are to be added or subtracted must have the same denominator, or bottom number.

Add 1/2 + 1/4 + 3/8

1/2 = 4/8
1/4 = 2/8
+ 3/8 = 3/8
9/8 = 1 1/8

Add 6 1/2 + 3 3/4

6 1/2 = 6 2/4
+ 3 3/4 = 3 3/4
9 5/4 = 10 1/4

Subtract 5/8 from 3/4

3/4 = 6/8
− 5/8 = 5/8
1/8

Subtract 7 7/8 from 9 1/4

$$\begin{array}{r} 9\ 1/4 = 9\ 2/8 = 8\ 10/8 \\ -\ 7\ 7/8 \quad = \quad 7\ 7/8 \\ \hline \hline 1\ 3/8 \end{array}$$

Reducing Fractions

Some fractions can be reduced to a lower denominator. For example, 2/4 is the same as 1/2. When working with a drawing and making measurements, you can locate either one easily on the scale. Usually, such reductions are necessary only when you are working with several different dimensions or various fractional units. For reducing fractions in the shop, it is often easiest to divide both the numerator and denominator by 2. This method will simplify the reduction because all the fractional units found on shop rules and tapes are divisible by 2, for example, halves, fourths, eighths, sixteenths, and thirty-seconds. Using this method may require more than one reduction, but the simplicity of dividing by 2 offsets the time needed to repeat the reduction. Reduction of fractions will become easier with practice.

To reduce 4/8 inch:

$$\frac{4}{8} = \frac{4 \div 2}{8 \div 2} = \frac{2}{4}$$

The new fraction is 2/4 inch.
2/4 can be reduced again.

$$\frac{2}{4} = \frac{2 \div 2}{4 \div 2} = \frac{1}{2}$$

The new fraction is 1/2 inch, the lowest form.

Rounding Numbers

When multiplying or dividing numbers, we often get a whole number followed by a long decimal fraction. When we divide 10 by 3, for example, we get 3.3333333. For all practical purposes, we need not lay out weldments to an accuracy greater than the second decimal place. We would therefore round off this number to 3.33, a dimension that would be easier to work with in the welding shop.

RULE: When rounding off a number, look at the number to the right of the last significant place to be used. If this number is less than 5, drop it and leave the remaining number unchanged. If this number is 5 or greater, increase the last significant number by 1 and record the new number.

Round off 15.6549 to the second decimal place.

15.6549. Because the number in the third place is less than 5, the new number would be 15.65

Round off 8.2764 to the second decimal place.

8.2764. Because the number in the third place is 5 or more, the new number would be 8.28.

Round off 0.8539 to the third decimal place.

0.8539. Because the number in the fourth place is 5 or more, the new number would be 0.854.

Round off 156.8244 to the first decimal place.

156.8244. Because the number in the second place is less than 5, the new number would be 156.8.

Converting Fractions to Decimals

From time to time it may be necessary to convert fractional numbers to decimal numbers. A fraction-to-decimal conversion is needed before most calculators can be used to solve problems containing fractions. There are some calculators that will allow the inputting of fractions without converting them to decimals.

RULE: To convert a fraction to a decimal, divide the numerator (top number in the fraction) by the denominator (bottom number in the fraction).

To convert 3/4 to a decimal:

$3 \div 4 = 0.75$

To convert 7/8 to a decimal:

$7 \div 8 = 0.875$

Converting Decimals to Fractions

This process is less exact than the conversion of fractions to decimals. Except for specific decimals, the conversion will leave a remainder unless a small enough fraction is selected. For example, if you are converting 0.765 to the nearest 1/4 in., 3/4 in. would be acceptable and this conversion would leave a remainder of 0.015 in. ($0.765 - 0.75 = 0.015$). If you are working to a ±1/8-in. tolerance that has up to a 1/4-in. difference from the minimum to maximum dimensions, a measurement of 3/4 is acceptable. More accurately, 0.765 can be converted to 49/64 in., a dimension that would be hard to lay out and impossible to cut using a hand torch.

RULE: To convert a decimal to a fraction, multiply the decimal by the denominator of the fractional units desired; that is, for 8ths (1/8) use 8, for 4ths (1/4) use 4, and so on. Place the whole number (dropping or rounding off the decimal remainder) over the fractional denominator used.

To convert 0.75 to 4ths:

$0.75 \times 4 = 3.0$ or 3/4

To convert 0.75 to 8ths:

$0.75 \times 8 = 6.0$ or 6/8, which will reduce to 3/4

To convert 0.51 to 4ths:

$0.51 \times 4 = 2.04$ or 2/4, which will reduce to 1/2

To convert 0.47 to 8ths:

$0.47 \times 8 = 3.76$ or 3/8

(Note that the 0.76 of the 3.76 is more than 0.5, so it could be rounded up, giving 4 or 4/8, which will reduce to 1/2.)

Layout

Parts for fabrication may require that the welder lay out lines and locate points for cutting, bending, drilling, and assembling. Lines may be marked with a soapstone or a chalkline, scratched with a metal scribe, or punched with a center punch. If a piece of soapstone is used, it should be sharpened properly to increase accuracy, **Figure 17-43**. A chalk line will make a long, straight line on metal and is best used on large jobs, **Figure 17-44**. Both the scribe and punch can be used to lay out an accurate line, but the punched line is easier to see when cutting. A punch can be held as shown in **Figure 17-45**, with the tip just above the surface of the metal. When the punch is struck with a lightweight hammer, it will make a mark. If you move your hand along the line and rapidly strike the punch, it will leave a series of punch marks for the cut to follow.

Always start a layout as close to a corner of the material as possible. By starting in a corner or along the edge, you can take advantage of the preexisting cut as well as reduce wasted material.

It is easy to cut the wrong line. In welding shops one person may lay out the parts and another may make the cuts. Even when one person does both jobs, it is easy to cut the wrong line, either because of the restricted view through cutting goggles or because of the large number of lines on a part. To avoid making a cutting mistake, always identify whether lines are being used for cutting, for locating bends, as drill centers, or as assembly locations. The lines not to be cut may be marked with an *X*, or they may be identified by writing directly on the part. Mark the side of the line that is scrap so that when the kerf is removed from that side the

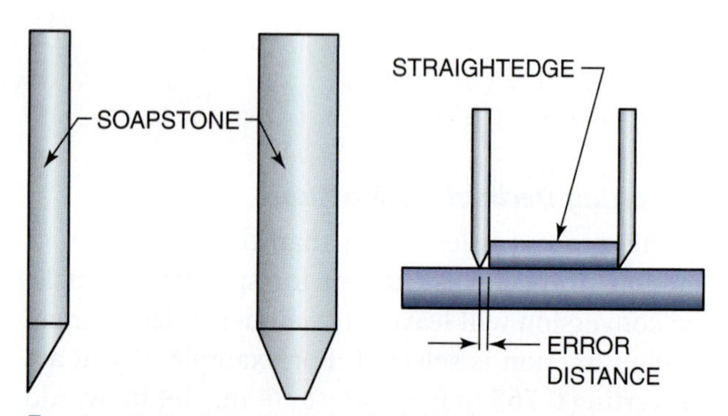

FIGURE 17-43 Proper method of sharpening a soapstone.

FIGURE 17-44(B) Check to see that the line is dark enough. Courtesy of Larry Jeffus.

FIGURE 17-44(A) Pull the chalk line tight and then snap it. Courtesy of Larry Jeffus.

FIGURE 17-44(C) Chalk line reel. Courtesy of Larry Jeffus.

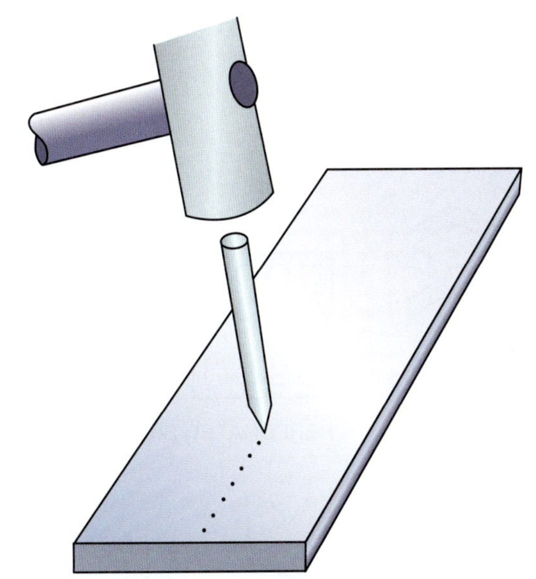

FIGURE 17-45 Holding the punch slightly above the surface allows the punch to be struck rapidly and moved along a line to mark it for cutting.

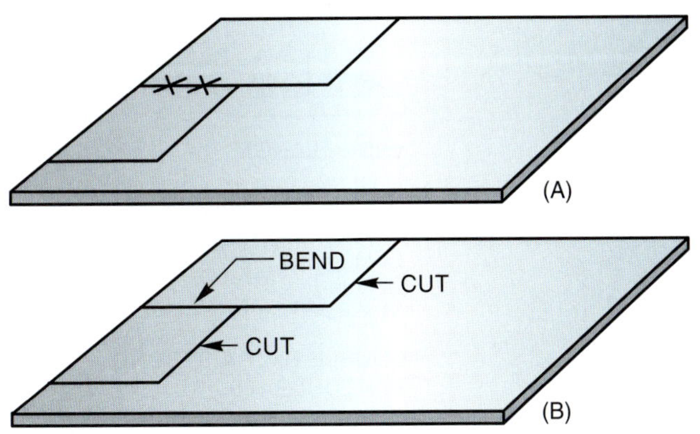

FIGURE 17-46 Identifying layout lines to avoid mistakes during cutting.

part will be the proper size, **Figure 17-46.** Any lines that have been used for constructing the actual layout line or to locate points for drilling or are made in error must be erased completely or clearly marked to avoid confusion during cutting and assembly.

Some shops have their own shorthand methods for identifying layout lines, or you can develop your own system. Failure to develop and use a system for identifying lines will ultimately result in a mistake. In a welding shop you will find only those who have made the wrong cut and those who will make the wrong cut. When it does happen, check with the welding shop supervisor to see what corrective steps can be taken. One advantage for most welding assemblies is that many cutting errors can be repaired by welding. Prequalified procedures are often established for just such an event, so check before deciding to scrap the part.

The process of laying out a part may be affected by the following factors:

- *Material shape:* **Figure 17-47** lists the most common metal shapes used for fabrication. Flat stock such as sheets and plates are easiest to lay out, and pipes and round tubing are the most difficult shapes to work with.
- *Part shape:* Parts with square and straight cuts are easier to lay out than are parts with angles, circles, curves, and irregular shapes.
- *Tolerance:* The smaller or tighter the tolerance that must be maintained, the more difficult the layout.
- *Nesting:* The placement of parts together in a manner that will minimize the waste created is called *nesting.*

Parts with square or straight edges are the easiest to lay out. Simply measure the distance and use a square or straight edge to lay out the line to be cut, **Figure 17-48.** Straight cuts that are to be made parallel to an edge can be drawn by using a combination square and a piece of soapstone. Set the combination square to the correct dimension and drag it along the edge of the plate while holding the soapstone at the end of the combination square's blade, **Figure 17-49.**

PRACTICE 17-1

Layout Square, Rectangular, and Triangular Parts

Using a piece of metal or paper, soapstone or pencil, tape measure, and square, you will lay out the parts shown in **Figure 17-50.** The parts must be laid out within ±1/16 in. of the dimensions. Convert the dimensions into S.I. metric units of measure.

Circles, arcs, and curves can be laid out by using either a compass or a circle template, **Figure 17-51.** The diameter is usually given for a hole or round part, and the radius is usually given for arcs and curves, **Figure 17-52.** The center of the circle, arc, or curve may be located using dimension lines and center lines. Curves and arcs that are to be made tangent to another line may be dimensioned with only their radiuses, **Figure 17-52.**

Complete a copy of the "Student Welding Report" listed in Appendix I or provided by your instructor. ◆

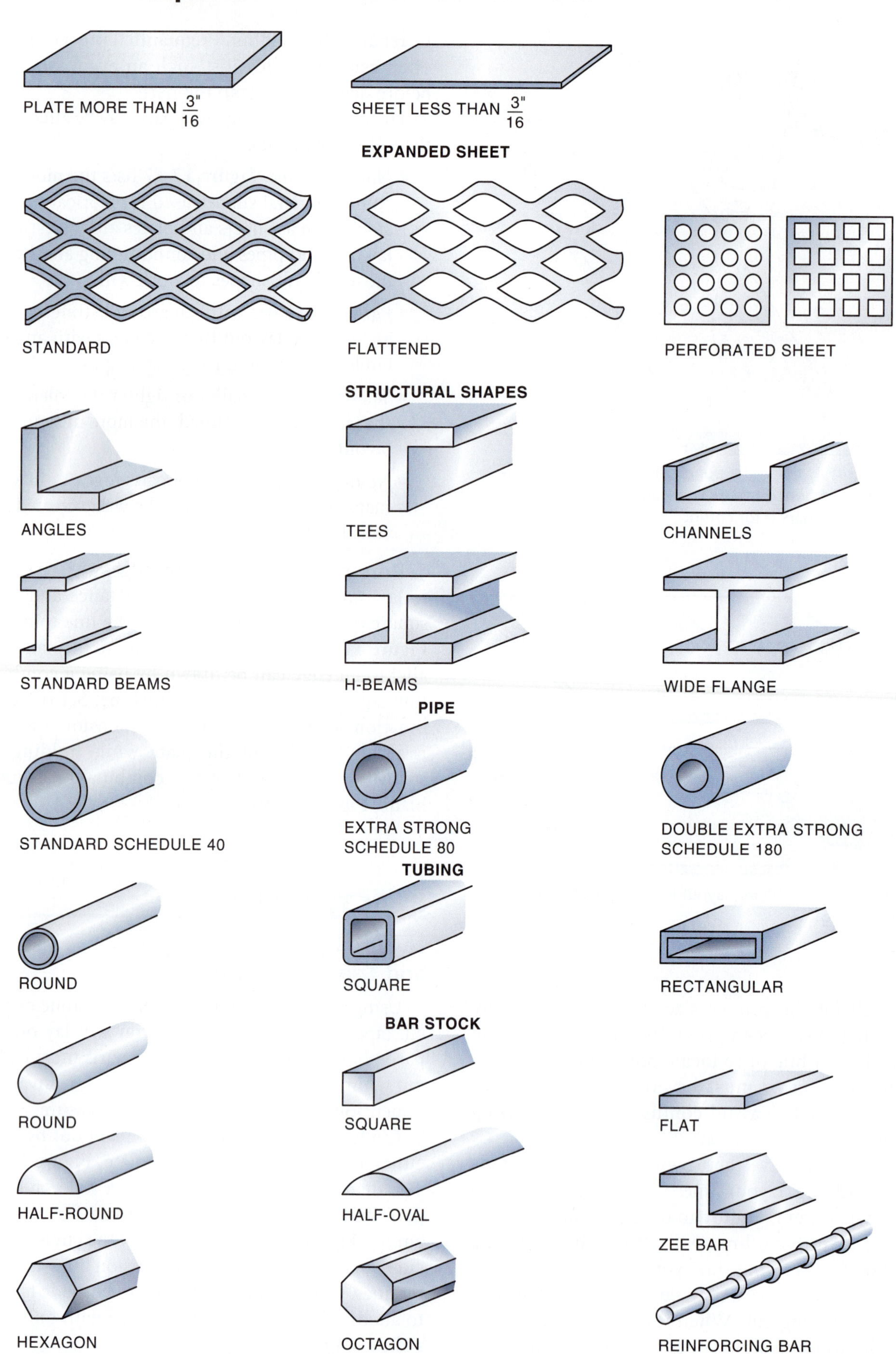

FIGURE 17-47 Standard metal shapes; most are available with different surface finishes, such as hot-rolled, cold-rolled, or galvanized.

FIGURE 17-48 Using a square to draw a straight line.
Courtesy of Larry Jeffus.

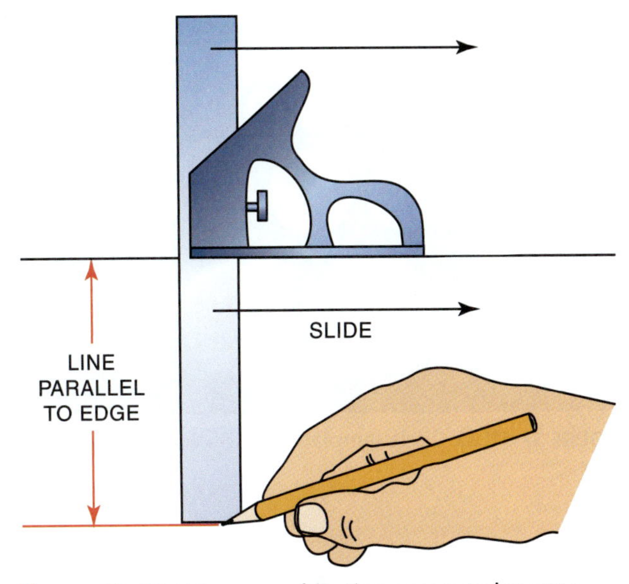

FIGURE 17-49 Using a combination square to lay out a strip of metal.

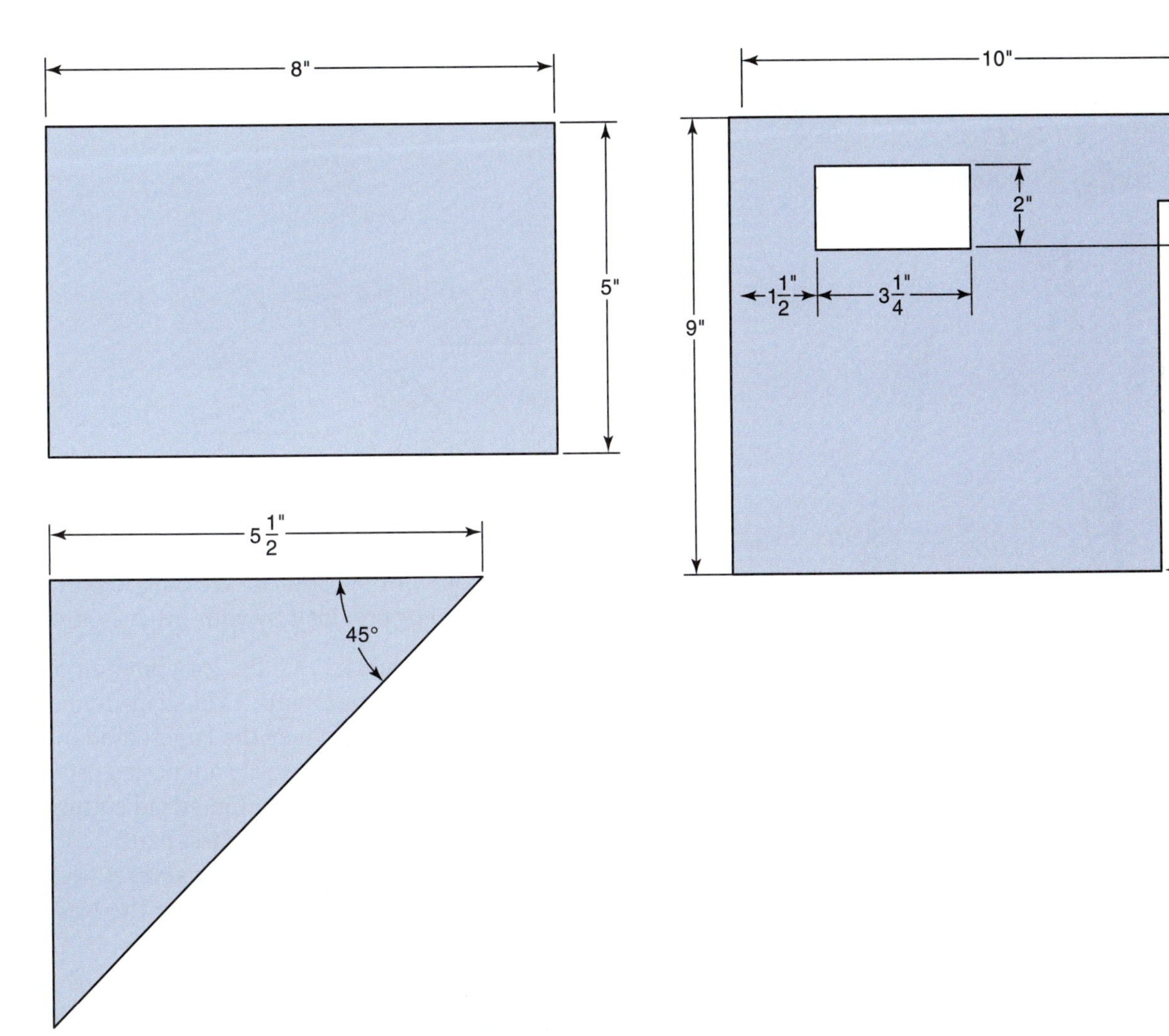

FIGURE 17-50 Layout parts for Practice 17-1.

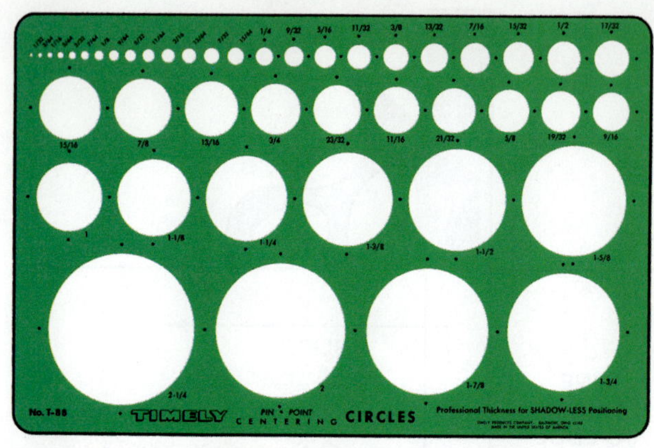

FIGURE 17-51(A) Circle template. Courtesy Timely Products Co., Inc.

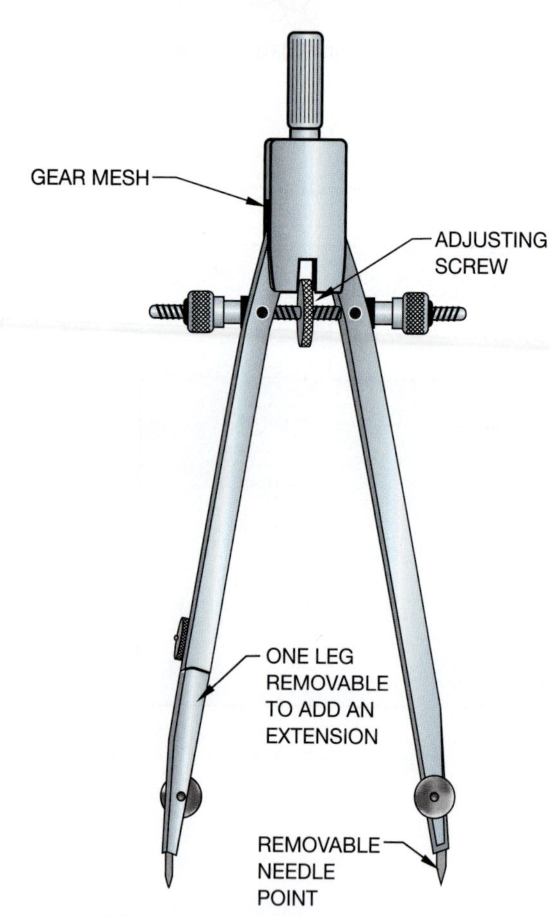

FIGURE 17-51(B) Compass. Courtesy of J. S. Staedtler, Inc.

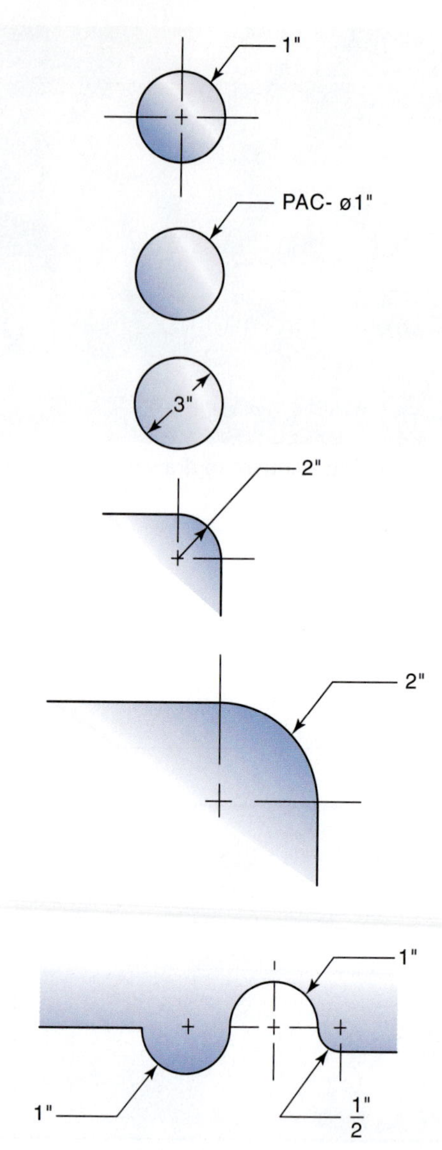

FIGURE 17-52 Dimensioning for arcs, curves, radii, and circles.

PRACTICE 17-2

Laying Out Circles, Arcs, and Curves

Using a piece of metal or paper, soapstone or pencil, tape measure, compass, or circle template and square, you will lay out the parts shown in **Figure 17-53**. The parts must be laid out within ±1/16 in. of the dimensions. Convert the dimensions into S.I. metric units of measure.

Complete a copy of the "Student Welding Report" listed in Appendix I or provided by your instructor. ◆

Nesting Laying out parts so that the least amount of scrap is produced is important. Odd-shaped and unusual-sized parts often produce the largest amount of scrap. Computers can be used to lay out nested parts with a minimum of scrap. Some computerized cutting machines can also be programmed to nest parts.

Manual nesting of parts may require several tries at laying out the parts in order to achieve the least possible scrap.

PRACTICE 17-3

Nesting Layout

Using metal or paper that is 8 1/2 in. × 11 in., soapstone or pencil, tape measure, and square, you will lay out the parts shown in **Figure 17-54** in a

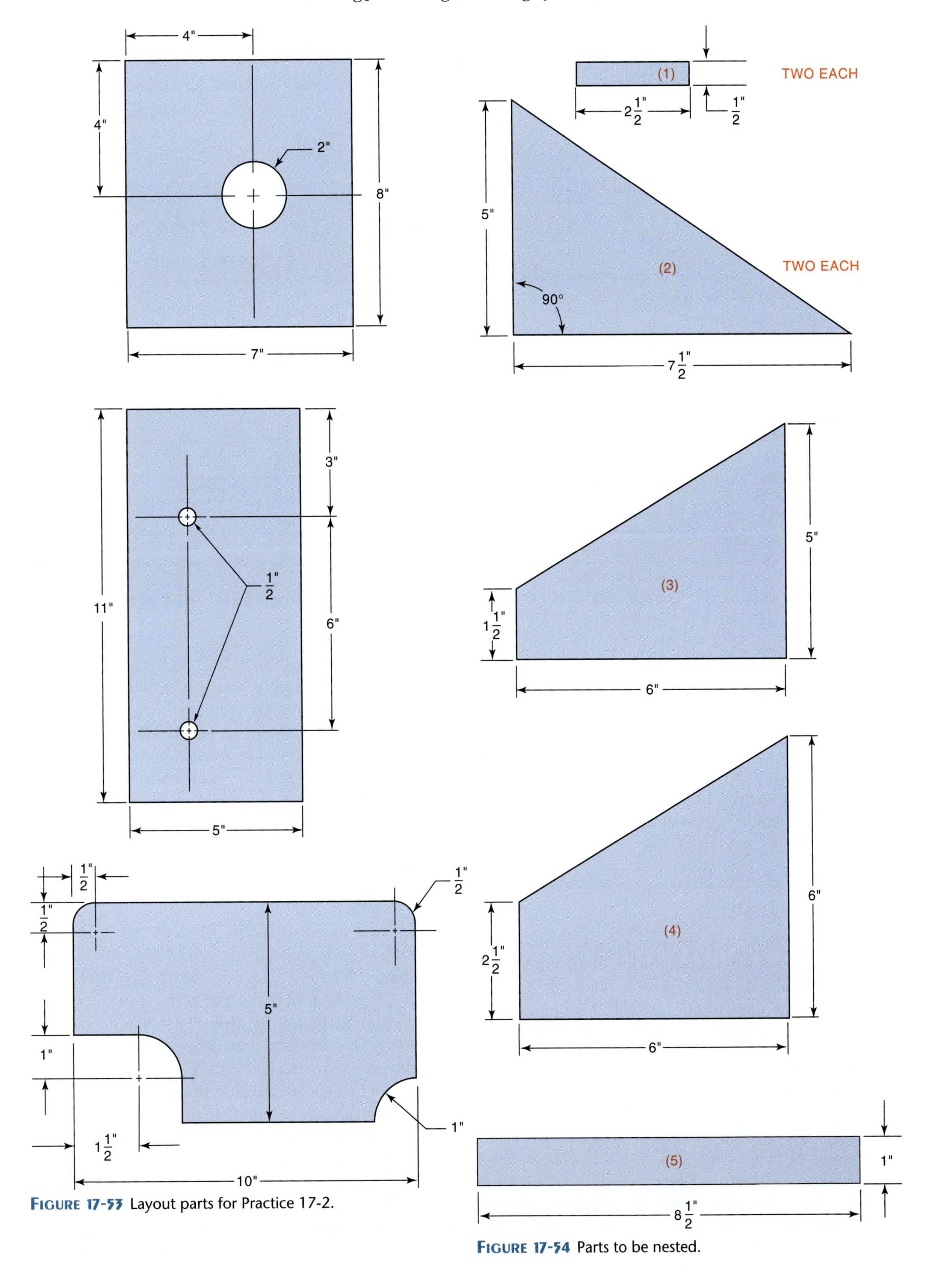

FIGURE 17-53 Layout parts for Practice 17-2.

FIGURE 17-54 Parts to be nested.

Part	Number Required	Type of Material	Size	
			Standard Units	S. I. Units

Table 17-4 Bill of Materials Form.

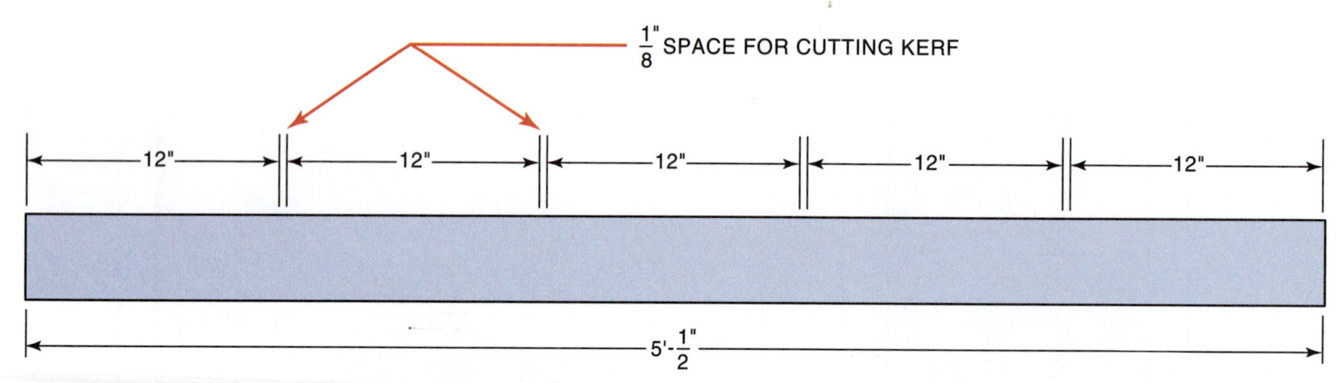

Figure 17-55 Because of the kerf, an additional 1/2 in. of stock would be required to make these five 1-ft pieces.

manner that will result in the least scrap. Assume a 0" kerf width. Use as many 8 1/2-in. × 11-in. pieces of stock as would be necessary to produce the parts using your layout. The parts must be laid out within ±1/16 in. of the dimensions. Convert the dimensions into S.I. metric units of measure.

Complete a copy of the "Student Welding Report" listed in Appendix I or provided by your instructor. ◆

PRACTICE 17-4

Bill of Materials

Using the parts laid out in Practice 17-3 and paper and pencil, you will fill out the bill of materials form shown in **Table 17-4.**

Complete a copy of the "Student Welding Report" listed in Appendix I or provided by your instructor. ◆

Kerf Space Because all cutting processes, except shearing, produce a kerf during the cut, this space must be included in the layout when parts are laid out side by side. The *kerf* is the space created as material is removed during a cut. The width of a kerf varies depending on the cutting process used. Of the cutting processes used in most shops, the metal saw will produce one of the smallest kerfs and the hand-held oxyfuel cutting torch can produce one of the widest.

When only one or two parts are being cut, the kerf width may not need to be added to the part dimension. This space may be taken up during assembly by the root gap required for a joint. If a large number of parts are being cut out of a single piece of stock, the kerf width can add up and increase the stock required for cutting out the parts, **Figure 17-55.**

PRACTICE 17-5

Allowing Space for the Kerf

Using a pencil, 8 1/2-in. × 11-in. paper, measuring tape or rule, and square, you will lay out four rectangles 2 1/2 in. × 5 1/4 in. down one side of the paper, leaving 3/32 in. for the kerf.

Two methods can be used to provide for the kerf spacing. One way is to draw a double line on the side of the part where the kerf is to be made, **Figure 17-56.** The other way is to lay out a single line and place an *X* on the side of the line that the cut is to be made, **Figure 17-57.** Note that no kerf space need be left along the sides made next to the edge of the paper or next to the scrap. What is the total length and width of material needed to lay out these four parts?

Parts can be laid out by tracing either an existing part or a template, **Figure 17-58.** When using either

FIGURE 17-56 Kerf is made between the lines.

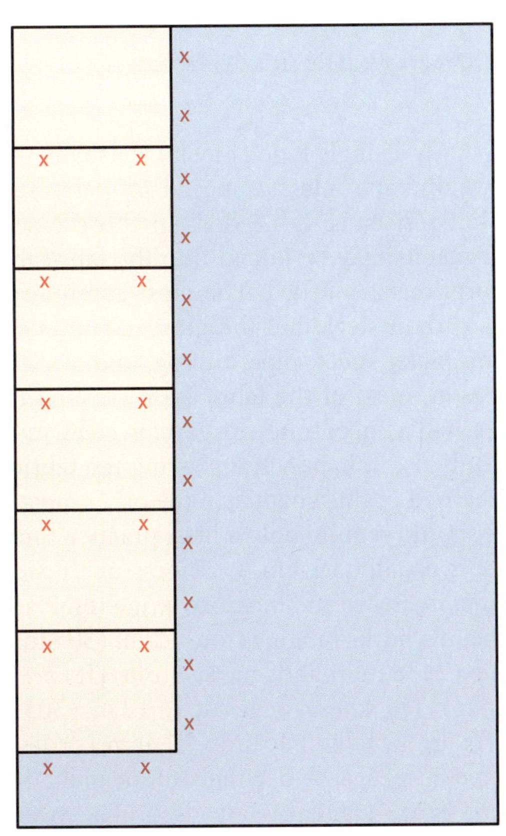

FIGURE 17-57 Xs mark the side of the line on which the kerf is to be made.

FIGURE 17-58(A) Tracing a part. Courtesy of Larry Jeffus.

FIGURE 17-58(B) Tracing a template. Courtesy of Larry Jeffus.

FIGURE 17-59 Be sure that the soapstone is held tightly into the part being traced. Courtesy of Larry Jeffus.

process, be sure the line you draw is made as tight as possible to the part's edge, **Figure 17-59**. The inside edge of the line is the exact size of the part. Make the cut on the line or to the outside so that the part will be the correct side once it is cleaned up, **Figure 17-60**. Sometimes a template is made of a part. Templates are useful if the part is complex and needs to fit into an existing weldment. They are also helpful when a large

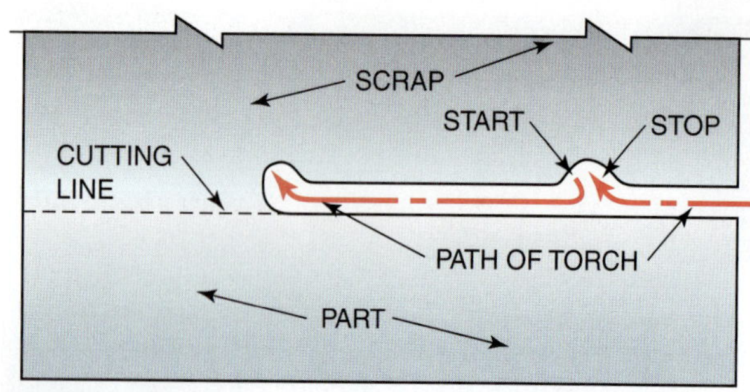

FIGURE 17-60 Turning out into scrap to make stopping and starting points smoother.

FIGURE 17-61 Pipe lateral being laid out with contour marker.
Courtesy of Larry Jeffus.

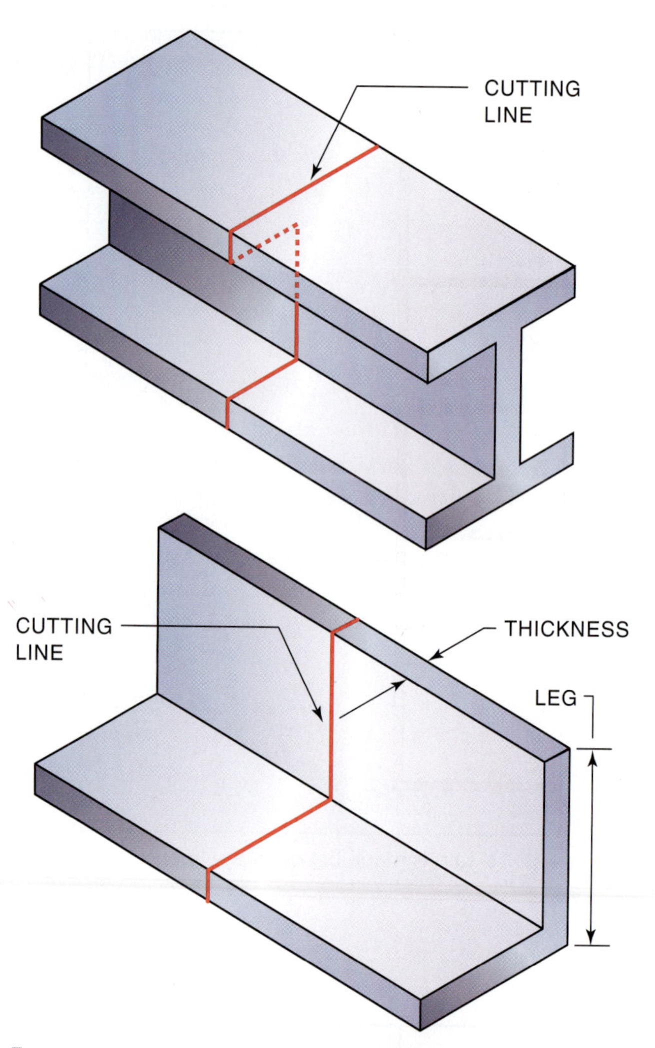

FIGURE 17-62 Layout for structural shapes.

number of the same part are to be made or when the part is only occasionally used. The advantage of using templates is that once the detailed layout work is completed exact replicates can be made anytime they are needed. Templates can be made out of heavy paper, cardboard, wood, sheet metal, or other appropriate material. The sturdier the material, the longer the template will last.

Special tools have been developed to aid in laying out parts; one such tool is the *contour marker,* **Figure 17-61.** These markers are highly accurate when properly used, but they do require a certain amount of practice. Once familiar with this tool, the user can lay out an almost infinite variety of joints within the limits of the tool being used. One advantage of tools like the contour marker is that all sides of a cut in structural shapes and pipe can be laid out from one side without relocating the tool, **Figure 17-62.**

Complete a copy of the "Student Welding Report" listed in Appendix I or provided by your instructor. ◆

Material Shapes

Metal stock can be purchased in a wide variety of shapes, sizes, and materials. Weldments may be constructed from combinations of sizes and/or shapes of metals. Only a single type of metal is usually used in most weldments unless a special property such as corrosion resistance is needed. In those cases dissimilar metals may be joined into the fabrication at such locations as needed. The most common metal used is carbon steel, and the most common shapes used are plate, sheet, pipe, tubing, and angles. For that reason, most of the fabrication covered in this chapter will concentrate on carbon steel in those commonly used shapes. Transferring the fabrication skills learned in this chapter to the other metals and shapes should require only a little practice time.

Plate is usually 3/16 in. (4.7 mm) or thicker and measured in inches and fractions of inches. Plates are available in widths ranging from 12 in. (305 mm) up to 96 in. (2438 mm) and lengths from 8 ft (2.4 m) to 20 ft (6 m). Thickness ranges up to 12 in. (305 mm).

Sheets are usually 3/16 in. (4.7 mm) or less and measured in gauge or decimals of an inch. Several different gauge standards are used. The two most common are the Manufacturer's Standard Gauge for Sheet Steel, used for carbon steels, and the American

Wire Gauge, used for most nonferrous metals such as aluminum and brass.

Pipe is dimensioned by its diameter and schedule or strength. Pipe that is smaller than 12 in. (305 mm) is dimensioned by its inside diameter, and the outside diameter is given for pipe that is 12 in. (305 mm) in diameter and larger, **Figure 17-63.** The strength of pipe is given as a schedule. Schedules 10 through 180 are available; schedule 40 is often considered a standard strength. The wall thickness for pipe is determined by its schedule (pressure range). The larger the diameter of the pipe, the greater its area. Pipe is available as welded (seamed) or extruded (seamless).

Tubing sizes are always given as the outside diameter. The desired shape of tubing, such as square, round, or rectangular, must also be listed with the ordering information.

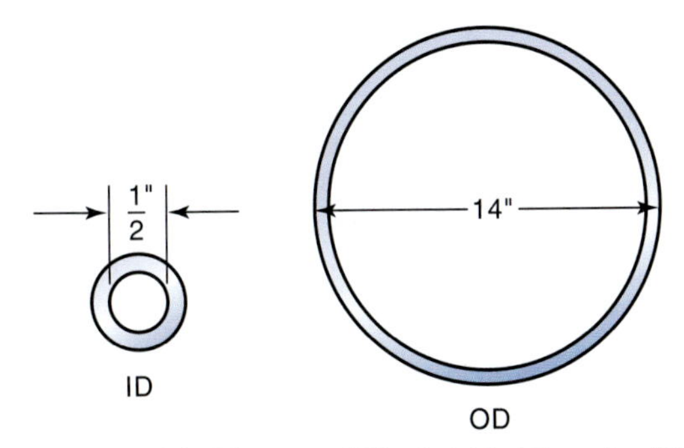

FIGURE 17-63 Inside Diameter (ID); Outside Diameter (OD).

The wall thickness of tubing is measured in inches (millimeters) or as Manufacturer's Standard Gauge for Sheet Metal. Tubing should also be specified as rigid or flexible. The strength of tubing may also be specified as the ability of tubing to withstand compression, bending, or twisting loads.

Angles, sometimes referred to as angle iron, are dimensioned by giving the length of the legs of the angle and their thickness, **Figure 17-62.** Stock lengths of angles are 20 ft, 30 ft, 40 ft, and 60 ft (6 m, 9.1 m, 12.2 m, and 18.3 m).

Assembly

The assembling process, bringing together all the parts of the weldment, requires a proficiency in several areas. You must be able to read the drawing and interpret the information provided there to properly locate each part. An assembly drawing has the necessary information, both graphically and dimensionally, to allow the various parts to be properly located as part of the weldment. If the assembly drawings include either pictorial or exploded views, this process is much easier for the beginning assembler; however, most assembly drawings are done as two, three, or more orthographic views, **Figure 17-64.** Orthographic views will be more difficult to interpret until you have developed an understanding of their various elements.

On very large projects such as buildings or ships, a corner or centerline is established as a baseline. This is the point where all measurements for all part

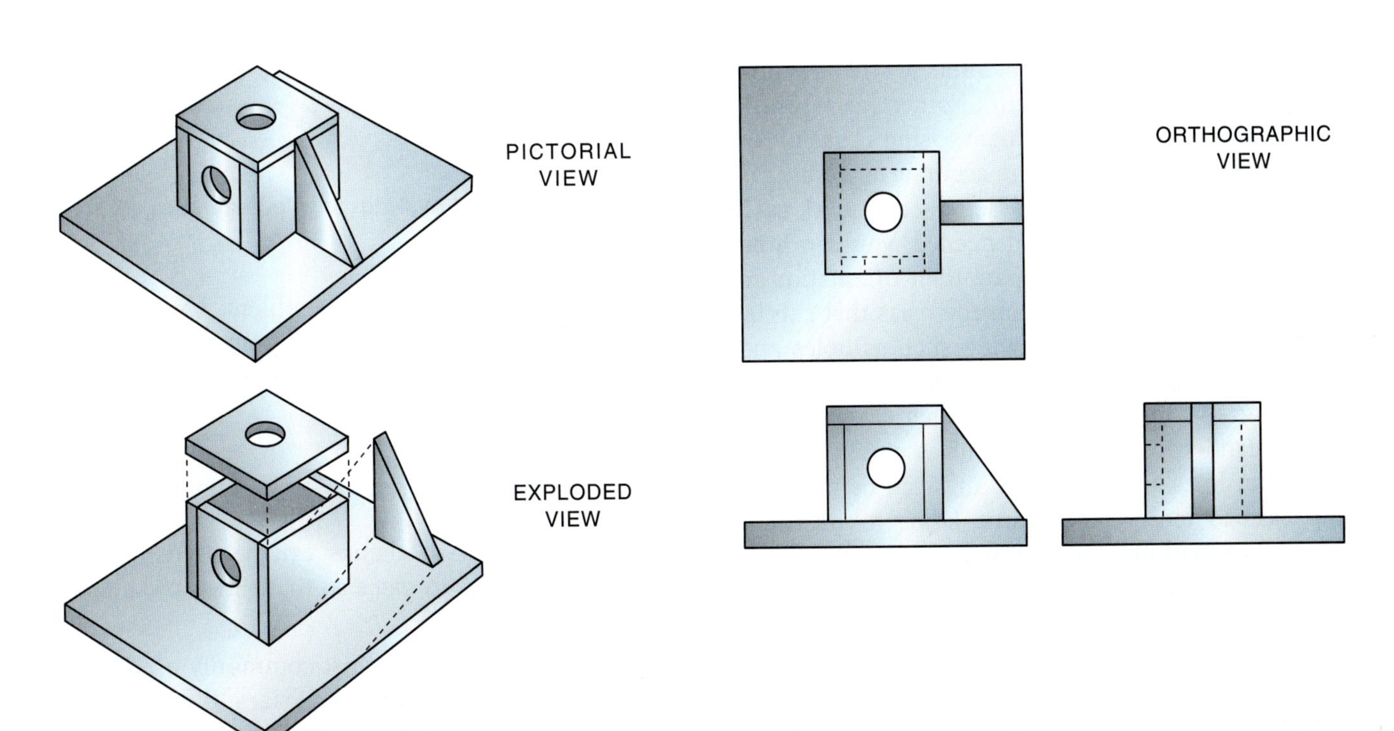

FIGURE 17-64 Types of drawings that can be used to show a weldment assembly.

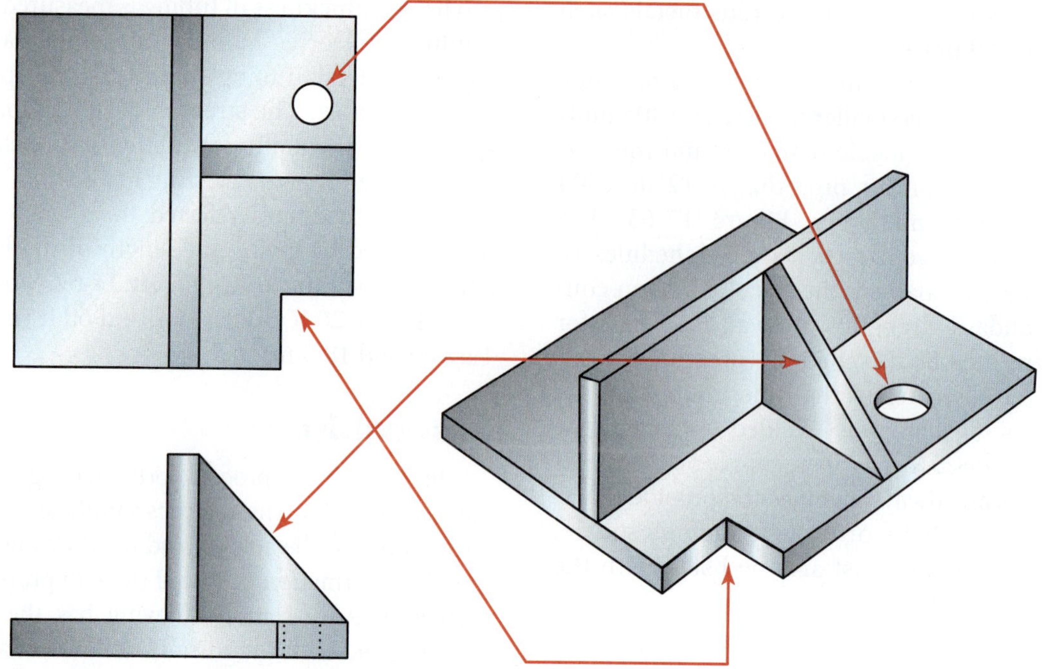

FIGURE 17-65 Identify unique points to aid in assembly.

location begins. When working with smaller weldments, a single part may be selected as such a starting point. Often, selecting the base part is automatic because all other parts are to be joined to this central part. On other weldments, however, the selection of a base part is strictly up to the assembler.

To start the assembly, select the largest or most central part to be the base for your assembly. All other parts will then be aligned to this one part. Using a base also helps to prevent location and dimension errors. Otherwise, a slight misalignment of one part, even within tolerances, will be compounded by the misalignment of other parts, resulting in an unacceptable weldment. Using a baseline or base part will result in a more accurate assembly.

Identify each part of the assembly and mark each piece for future reference. If needed, you can hold the parts together and compare their orientation to the drawing. Locate points on the parts that can be easily identified on the drawing such as holes and notches, **Figure 17-65**. Now mark the location of these parts—top, front, or other such orientation—so you can locate them during the assembly.

Layout lines and other markings can be made on the base to locate other parts. Using a consistent method of marking will help prevent mistakes. One method is to draw parallel lines on both parts where they meet, **Figure 17-66**.

After the parts have been identified and marked, they can be either held or clamped into place. Holding the parts in alignment by hand for tack welding is fast but often leads to errors and thus is not recommended for beginning assemblers. Experienced assemblers recognize that clamping the parts in place before tack welding is a much more accurate method, **Figure 17-67**.

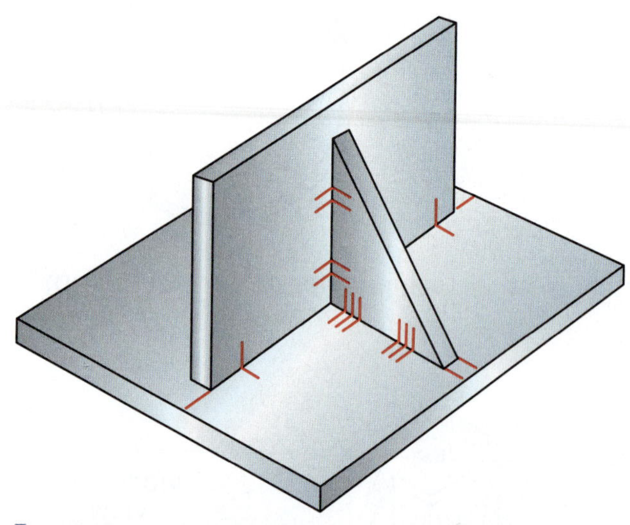

FIGURE 17-66 Layout markings to help locate the parts for tack welding.

Assembly Tools

Clamps A variety of clamps can be used to temporarily hold parts in place so that they can be tack welded.

- *C-clamps,* one of the most commonly used clamps, come in a variety of sizes, **Figure 17-68**. Some C-clamps have been specially designed for welding. Some of these clamps

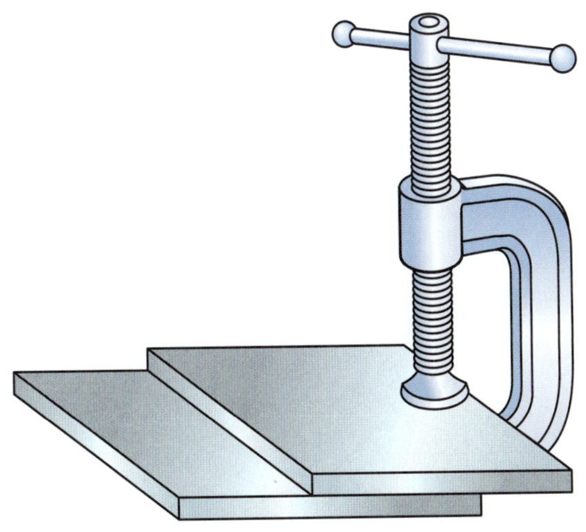

FIGURE 17-67 C-clamp being used to hold plates for tack welding.

FIGURE 17-68 C-clamps. Courtesy of Larry Jeffus.

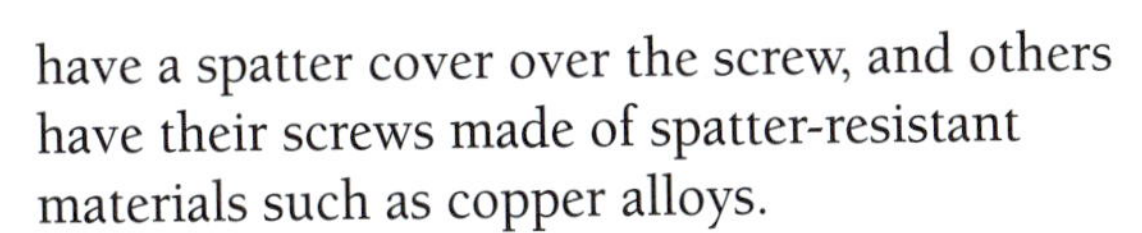

have a spatter cover over the screw, and others have their screws made of spatter-resistant materials such as copper alloys.

- *Bar clamps* are useful for clamping larger parts. Bar clamps have a sliding lower jaw that can be positioned against the part before tightening the screw clamping end, **Figure 17-69**. They are available in a variety of lengths.
- *Pipe clamps* are very similar to bar clamps. The advantage of pipe clamps is that the ends can be attached to a section of standard 1/2-in. pipe. This feature allows for greater flexibility in length, and the pipe can easily be changed if it becomes damaged.
- *Locking pliers* are available in a range of sizes with a number of various jaw designs, **Figure 17-70**. The versatility and gripping strength make locking pliers very useful. Some locking pliers have a self-adjusting feature that allows them to be moved between different thicknesses without the need to readjust them.

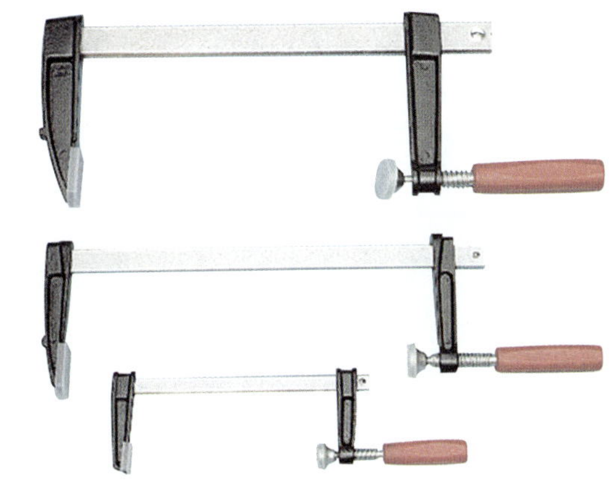

FIGURE 17-69 Bar clamps. Courtesy of Woodworker's Supply Inc.

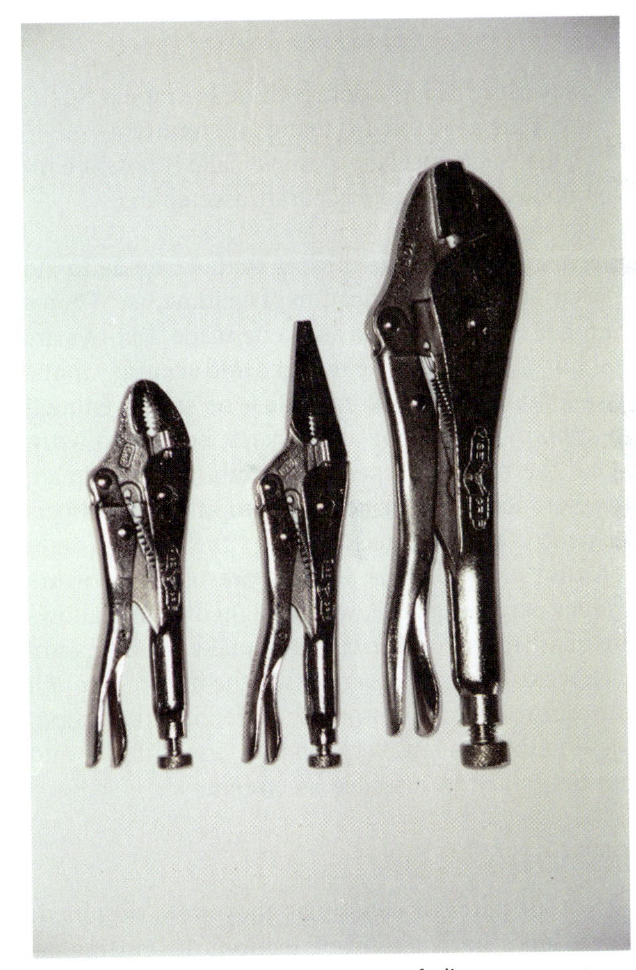

FIGURE 17-70 Three common types of pliers. Courtesy of Larry Jeffus.

- *Cam-lock clamps* are specialty clamps that are often used in conjunction with a jig or a fixture. They can be preset, allowing for faster work, **Figure 17-71**.

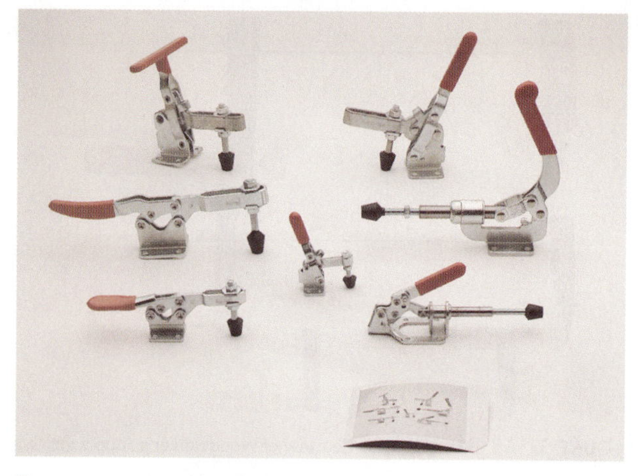

FIGURE 17-71 Toggle clamps. Courtesy of Woodworker's Supply Inc.

- *Specialty clamps* such as those for pipe welding, are available for many different types of jobs. Such specialty clamps make it possible to do faster and more accurate assembling.

Fixtures Fixtures are devices that are made to aid in assemblies and fabrication of weldments. When a number of similar parts are to be made, fixtures are helpful. They can increase speed and accuracy in the assembly of parts. Fixtures must be strong enough to support the weight of the parts, be able to withstand the rigors of repeated assemblies, and remain in tolerance. They may have clamping devices permanently attached to speed up their use. Often, locating pins or other devices are used to ensure proper part location. A well-designed fixture allows adequate room for the welder to make the necessary tack welds. Some parts are left in the fixture through the entire welding process to reduce distortion. Making fixtures for every job is cost prohibitive and not necessary for a skilled assembler.

Fitting

Not all parts fit exactly as they were designed. There may be slight imperfections in cutting or distortion of parts due to welding, heating, or mechanical damage. Some problems can be solved by grinding away the problem area. Hand grinders are most effective for this type of problem. Other situations may require that the parts be forced into alignment.

CAUTION

Never operate a hand grinder without the safety guard and eye protection.

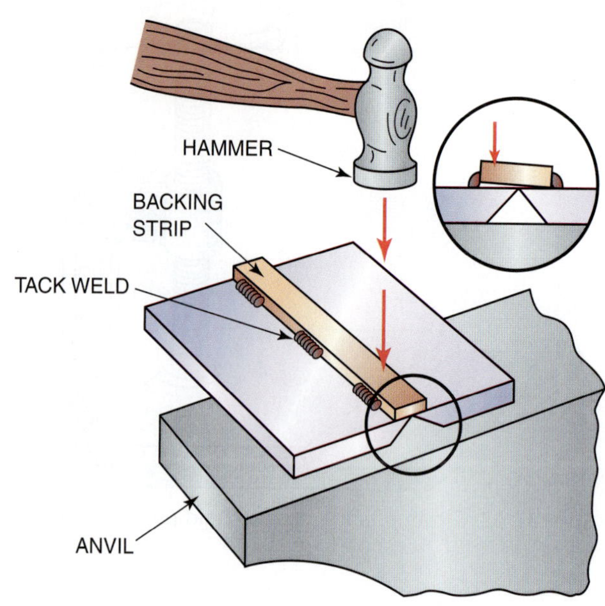

FIGURE 17-72 Using a hammer to align the backing strip and weld plates.

A simple way of correcting slight alignment problems is to make a small tack weld in the joint and then use a hammer and possibly an anvil to pound the part into place, **Figure 17-72.** Small tacks applied in this manner will become part of the finished weld. Be sure not to strike the part in a location that will damage the surface and render the finished part unsightly or unusable.

More aligning force can be applied using cleats or dogs with wedges or jacks. Cleats or dogs are pieces of metal that are temporarily attached to the weldment's parts to enable them to be forced into place. Jacks will do a better job if the parts must be moved more than about 1/2 in. (13 mm). Anytime cleats or dogs are used, they must be removed and the area ground smooth.

Some codes and standards will not allow cleats or dogs to be welded to the base metal. In these cases more expensive and time-consuming fixtures must be constructed to help align the parts if needed.

Tack Welding

Tack welding is a temporary method of holding the parts in place until they can be completely welded. Usually, all of the parts of a weldment should be assembled before any finishing welding is started. This will help reduce distortion. Tack welds must be strong enough to withstand any pounding or forcing during assembly and any forces caused by weld distortion during final welding. They must also be small enough to be incorporated into the final weld without causing a discontinuity in its size or shape, **Figure 17-73.**

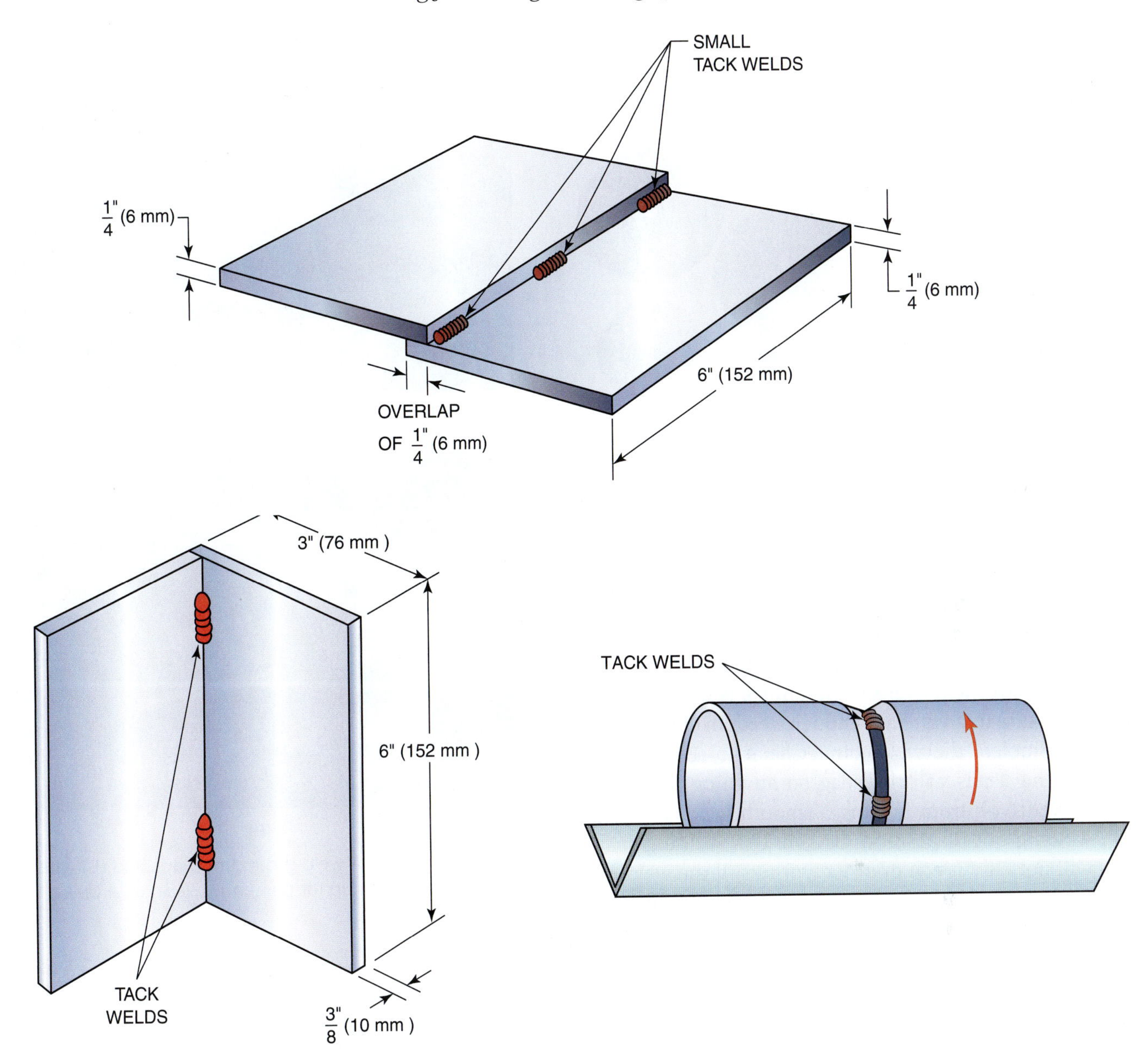

FIGURE 17-73 Make tack welds as small as possible.

Tack welds must be made with an appropriate filler metal, in accordance with any welding procedure. They must be located well within the joint so that they can be refused during the finish welding. Posttack welding cleanup is required to remove any slag or impurities that may cause flaws in the finished weld. Sometimes the ends of a tack weld must be ground down to a taper to improve their tie-in to the finished weld metal.

A good tack weld is one that does its job by holding parts in place yet is undetectable in the finished weld.

Welding

Good welding requires more than just filling up the joints with metal. The order and direction in which welds are made can significantly affect distortion in the weldment. Generally, welding on an assembly should be staggered from one part to another. This will allow both the welding heat and welding stresses to dissipate.

Keep the arc strikes in the welding joint so that they will be remelted as the weld is made. This will make the finished weldment look neater and reduce postweld cleanup. Some codes and standards do not allow arcs to be made outside of the welding joint.

Striking the arc in the correct location on an assembly is more difficult than working on a welding table. When working on an assembly, you will often be in an awkward position, which makes it harder to strike the arc correctly. Several techniques will help you improve your arc starting accuracy. You can use

FIGURE 17-74 Wire brushes and grinding stones used to clean up welds. Courtesy of Larry Jeffus.

your free hand to guide the electrode in to the correct spot. Resting your arm, shoulder, or hip against the weldment can also help. Practicing starting the weld with the power off sometimes is helpful.

Be sure that you have enough freedom of movement to complete the weld joint. Check to see that your welding leads will not snag on anything that would prevent you from making a smooth weld. If you are welding out of position, be sure that welding sparks will not fall on you or other workers. If the weldment is too large to fit into a welding booth, portable welding screens should be used to protect other workers in the area from sparks and welding light.

Follow all safety and setup procedures for the welding process. Practice the weld to be sure that the machine is set up properly before starting on the weldment.

Finishing

Depending on the size of the shop, the welder may be responsible for some or all of the finish work. Such work may vary from chipping, cleaning, or grinding the welds to applying paint or other protective surfaces.

Grinding of welds should be avoided if pieceable by properly sizing the weld as it is made. Grinding can be an expensive process, adding significant cost to the finished weldment. Sometimes it is necessary to grind for fitting purposes or for appearance, but even in these cases it should be minimized if possible.

Most grinding and wire brushing is done with a hand angle grinder, **Figure 17-74**. These grinders can be used with a flat or cupped grinding stone or sandpaper. As the grinder is used the stone will wear down and must be discarded once it has worn down to the paper backing. It is a good practice to hold the grinder at an angle so that, if anything is thrown off the stone or metal surface, it will not strike you or others in the area. Because of the speed of the grinding stone, any such object can cause serious injury.

(A)

(B)

FIGURE 17-75 (A) A weld ground to shape and (B) wire brushed to clean it up for painting. Courtesy of Larry Jeffus.

The grinder must be held securely so that there is a constant pressure on the work. If the pressure is too great, the grinder motor will overheat and may burn out. If the pressure is too light, the grinder may bounce, which could crack the grinding stone. Move the grinder in a smooth pattern along the weld, **Figure 17-75A**. Watch the weld surface as it begins to take the desired shape and change your pattern as needed. Final cleanup of the weld before painting can be done with a wire brush on a grinder, **Figure 17-75B**.

Painting and other finishes release fumes such as volatile organic compounds (VOC), which are often regulated by local, state, and national governments. Special ventilation is required for most paints. Such a ventilation system will remove harmful fumes from the air before it is released back into the environment. Check with your local, state, or national regu-

CAUTION

When using a portable grinder, be sure that it is properly grounded and that the sparks will not fall on others, cause damage, or start a fire. Always maintain control to prevent the stone from catching and gouging the part or yourself.

CAUTION

Be sure that any stone or sandpaper used is rated for a speed in revolutions per minute (RPM) that is equal to or greater than the speed of the grinder itself. Using a stone with a lower-rated RPM can result in its flying apart with an explosive force.

lating authority before using such products. Read and follow all manufacturer's instructions for the safe use of its product, **Figure 17-76**.

CAUTION

Most paints are flammable and must be stored well away from any welding.

PRACTICE 17-6

Utility Trailer Bill of Materials

Using a pencil, paper, calculator, and the information on the drawing in **Figure 17-77** and **Table 17-5**, you are going to make a bill of materials for a utility trailer.

To build the trailer, you will need to identify the size and length of each piece of material needed in the construction. There are two ways of listing the materials needed. One is called a parts list and the other is the bill of materials. The *parts list* is used to identify the number and size of each piece of material needed in the project. The *bill of materials* is a list of all of the material needed for the project's construction. The parts list is used to identify the parts needed for the assembly, and the bill of materials is used for ordering the material. In some cases the parts list and bill of materials can be combined into a single table.

FIGURE 17-76 Spray paint the finished weld to keep it from rusting. Courtesy of Larry Jeffus.

Parts List

Make a parts list of all the pieces of material needed to build this trailer, **Figure 17-78**. Identify and list on your sheet the type and dimensions of all the pieces of material shown in **Figure 17-77**.

Bill of Materials

Because the bill of materials will be used to purchase the material needed for the trailer, you will have to find the total lengths needed of each type of material and include that on the list, **Figure 17-79**.

The standard lengths of structural shapes like the angle and channel used in this trailer are 20 ft. Nesting the various lengths of each type of material will minimize the number of 20-ft sections of steel needed for the trailer. This will reduce the amount of waste of materials.

The 2-in. thick, 12-ft long lumber of the floor of the trailer comes in nominal widths of 6 in. and 8 in. Use **Table 17-5**, which lists the actual widths of each of these boards, to determine which width boards would make the best fit with the least waste.

Complete a copy of the "Student Welding Report" listed in Appendix I or provided by your instructor. ◆

PRACTICE 17-7

Laying Out the Utility Trailer Parts

Using a steel tape, square, soapstone, and the parts list from Practice 17-6, you are going to lay out the parts needed to construct the utility trailer.

When laying out parts on a section of structural steel, you can measure the pieces one at a time or make all of the measurements from one end. In most cases making the measurements from the end is the more accurate way to lay out parts, **Figure 17-80**.

Using the tape and soapstone, make a mark at the correct length for each of the parts. Remember to leave a space for the cut's kerf.

Use the square to draw a line all the way across all the surfaces that will be cut. In the case of the angle, you will draw a line on two sides, and you will draw on three sides of the channel, **Figure 17-81**.

Complete a copy of the "Student Welding Report" listed in Appendix I or provided by your instructor. ◆

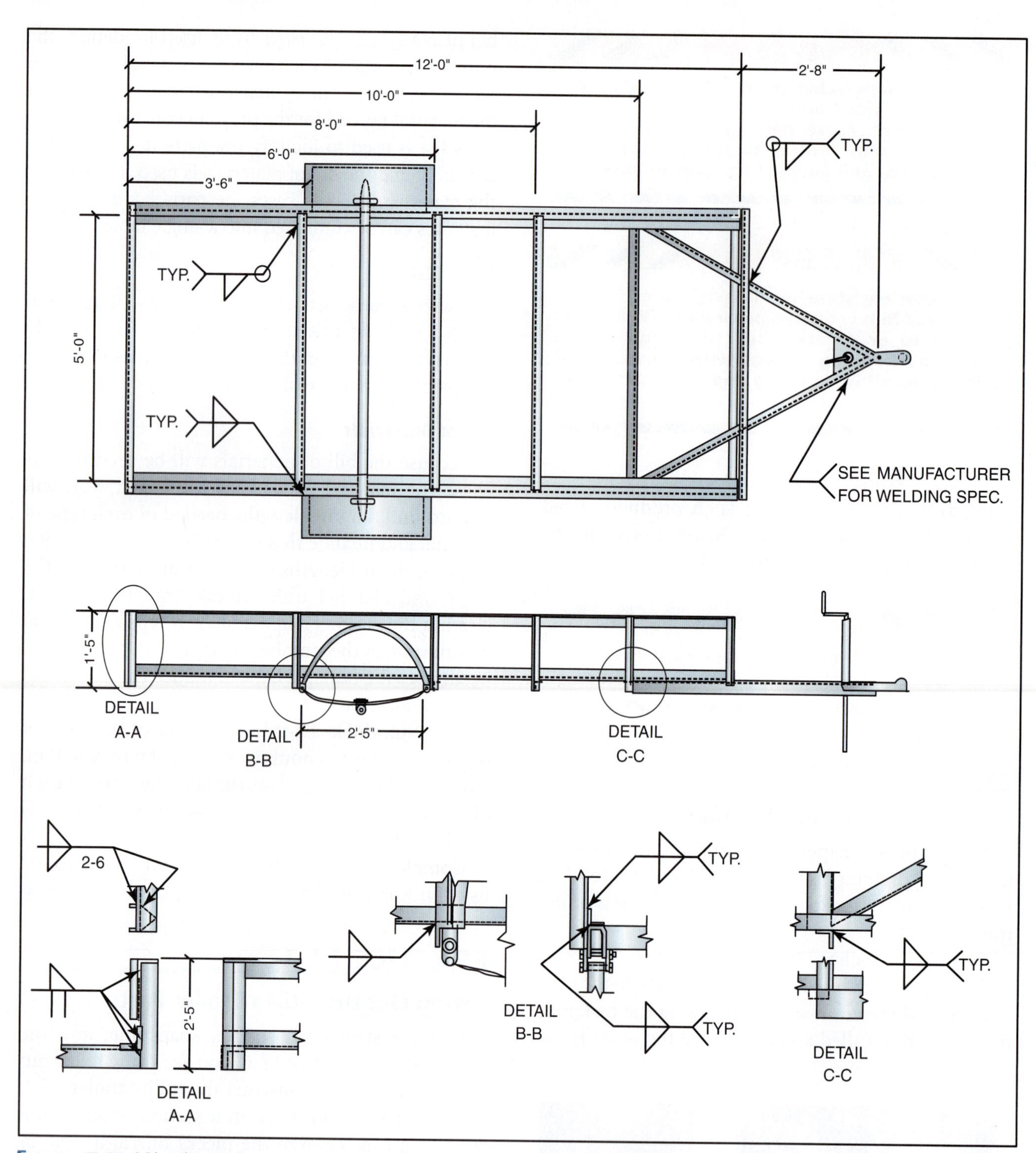

FIGURE 17-77 12′ utility trailer.

PRACTICE 17-8

Cutting Out the Utility Trailer Parts

Using a properly set up and adjusted oxyacetylene cutting set, gas welding goggles, chipping hammer, right angle grinder, safety glasses, gloves, and all other required personal protection equipment, you are going to cut out the parts for the utility trailer.

Place the metal to be cut on a cutting table at a comfortable height. Check to make certain there are not combustible materials in the area and that it is safe to cut in the area.

The parts can be cut so they remain on the cutting table or fall to the ground. If they are being allowed to fall, make sure they will not damage anything when they drop.

Start the cut on the edge of the metal and cut across one side. Move the torch to the other side and restart the cut. By cutting from the edges to the center the part being cut will drop off more easily, **Figure 17-82.**

Steel Sizes			
Shape	Size in. (mm)	Length ft (M)	Weight lb/ft (kg/M)
Angle	2 × 2 × 1/4 (50 × 50 × 5)	20 (5)	3.19 (3.77)
Angle	3 × 3 × 1/4 (75 × 75 × 5)	20 (5)	4.9 (7.29)
Channel	2 × 1 × 3/16 (50 × 25 × 4.5)	20 (5)	2.57 (3.82)

Common Dimensional Lumber Dimensions	
Nominal Size (metric*)	Actual Size (metric*)
2 × 6 (50 mm × 150 mm)	1 1/2 × 5 1/2 (37 mm × 137 mm)
2 × 8 (50 mm × 200 mm)	1 1/2 × 7 1/4 (37 mm × 181 mm)
2 × 10 (50 mm × 250 mm)	1 1/2 × 9 1/4 (37 mm × 231 mm)
2 × 12 (50 mm × 300 mm)	1 1/2 × 11 1/4 (37 mm × 281 mm)

TABLE 17-5 Standard Size of Construction Materials That Could Be Used for Building of the Utility Trailer.

Parts List		
Item	Description	Total

FIGURE 17-78 Parts list form.

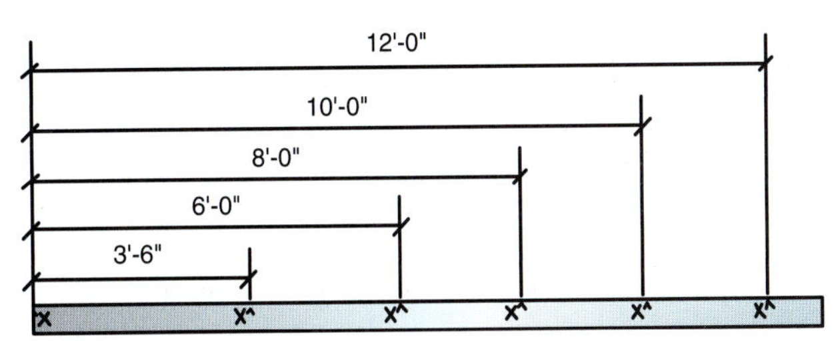

FIGURE 17-80 Mark all of the parts from the end and measure each part from the last section marked.

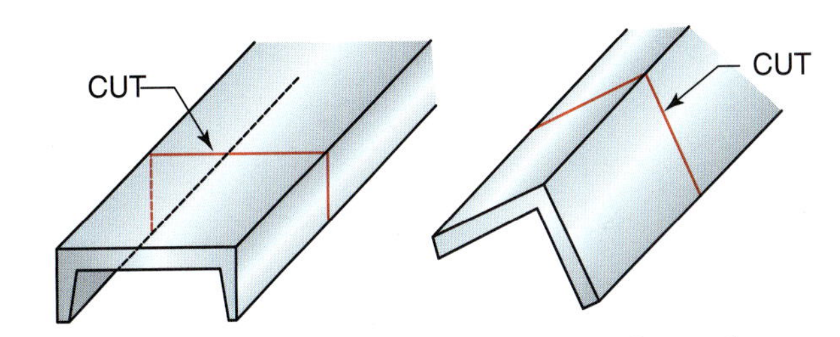

FIGURE 17-81 Marking the structural shapes for cutting.

Bill of Materials		
Item	Description	Total

FIGURE 17-79 Bill of materials form.

Several of the ends of the angles and one of the channels must be cut to fit together with other parts, **Figure 17-83**. Chip and grind the cut ends of the parts as needed to clean them up and get them ready for assembly.

Complete a copy of the "Student Welding Report" listed in Appendix I or provided by your instructor. ◆

PRACTICE 17-9

Assembling Trailer

Using the parts cut out in Practice 17-8, an oxyacetylene cutting set, a stick welding machine, a MIG welding machine or a flux cored welding machine, a steel tape measure, a chipping hammer, a square, a soapstone, C-clamps, a right-angle grinder, gas welding goggles, a welding hood, safety glasses, leather gloves, and all other required safety equipment, you are going to assemble the utility trailer.

To lay out a project, you must have sufficient area. Small projects can be laid out on layout tables,

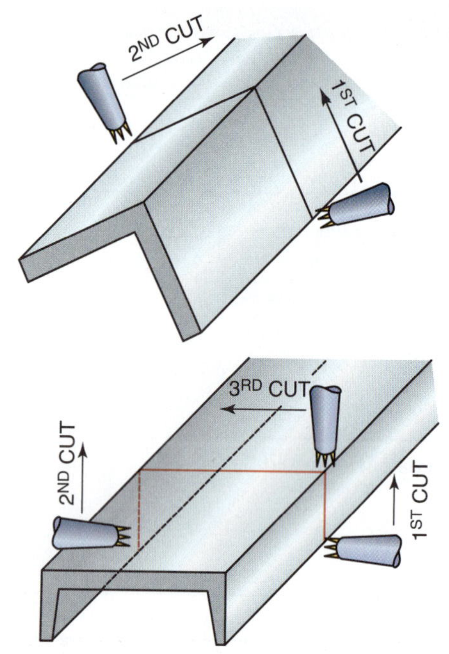

FIGURE 17-82 Cutting sequence for angle and channel structure shapes.

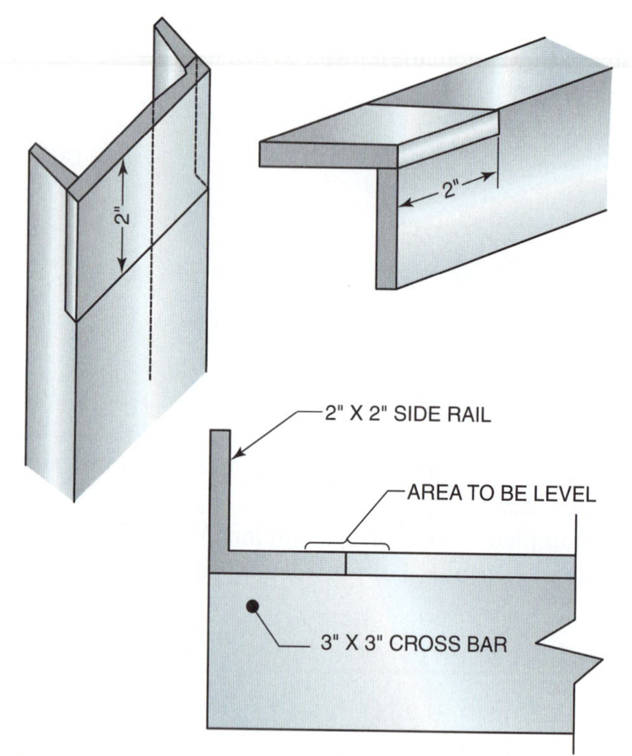

FIGURE 17-83 Coping the ends of the angles so they fit together properly.

Figure 17-84. Larger projects like this trailer must be laid out in an open area, preferably concreted so that leveling and squaring is easier. For a project this size you need at least 10 ft of additional width and 10 ft of additional length for adequate work area around

FIGURE 17-84 Layout table. Courtesy of Larry Jeffus.

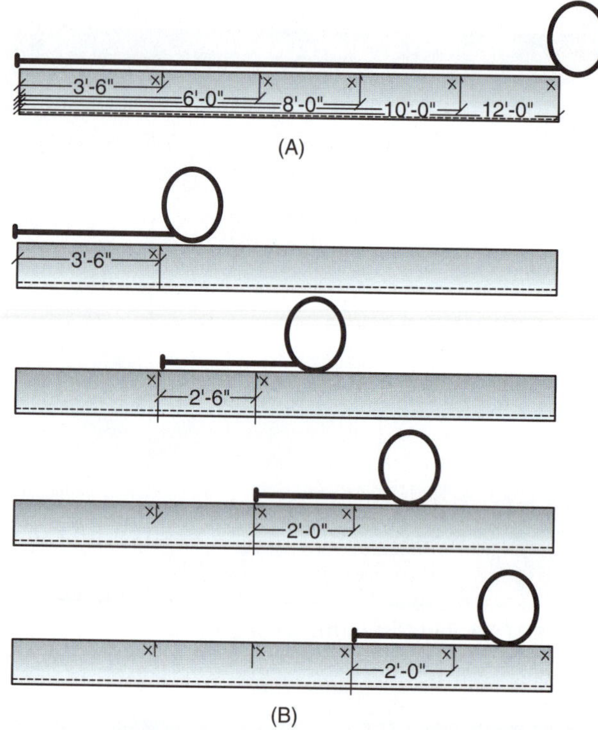

FIGURE 17-85 (A) is a more accurate way of measuring for a layout than (B).

the project. The total layout area then should be approximately 15 ft. × 24 ft.

Lay out the cross bar locations on the side angles. Make all of the measurements from one end, **Figure 17-85A**, because measuring them individually from each other can compound measuring errors, **Figure 17-85B.**

Place two 12-ft long pieces of 2 in. × 2 in. × 1/4 in. angles parallel approximately 5-ft apart. Use a framing square to square the 5-ft long piece of angle at one end of the side rail. Use C-clamps to secure the corners of the angles together.

Repeat this process on the opposite end with another 5-ft long piece of angle.

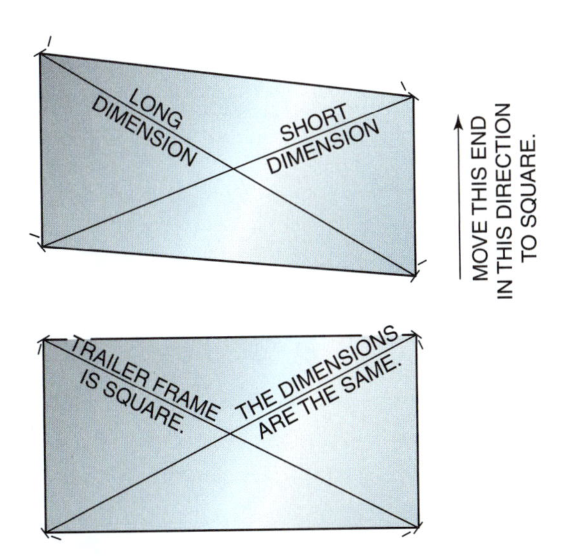

FIGURE 17-86 Cross measuring can be used to accurately square larger structures.

Move the second 12-ft long piece of angle into place and clamp it. Adjust the frame as needed so that both diagonal measurements are the same, **Figure 17-86.** Locate the remaining cross angle. Repeat this process until all 4 angle crossbars have been tack welded into place.

Recheck to see that the trailer is square by measuring diagonally from corner to corner. Tack weld all of the angles in place. Remove the C-clamps.

Measure and mark the location of the 17-in. long pieces of angle that support the top rail. Note that some of the angles are turned so that they are open toward the front, and some are turned so they are open toward the back. Four pieces are turned so that they are facing the trailer body, **Figure 17-77.**

Use a framing square to get each piece of angle in place. Use a C-clamp to hold them securely. Repeat this process with each of the remaining vertical angles.

Lay the 12-ft long top rail angle next to the frame. On this piece of angle, mark the location of each of the verticals. This will ensure that each vertical, when clamped in place, will be at the proper location.

Premarking the angles like this will speed up the assembly process. Clamp the angle in place on top of the vertical pickets.

With all of the vertical pickets clamped, check the squareness of the sides by cross measuring. Clamp the front top rail angle in place to the vertical pickets. Check the front rail for squareness. Tack weld the angle pickets in place and remove the C-clamps.

Installing the trailer's tongue will require some additional cutting on the angles. The trailer tongue is laid out and cut to shape so that it extends 2 ft 8 in. in front of the trailer. Cut and grind the angle as necessary to fit together.

Clamp the angle tongue into place, and check the alignment by cross measuring from the tip of the tongue to the corners of the trailer. Adjust the tongue as necessary so it is square. Tack weld the trailer tongue in place.

Place the preformed fender so that it is centered above the axle location. Clamp the fender in place. Use the MIG or flux-cored machine to make tack welds to hold the fender in place. These intermittent welds will be strong enough to hold the fender in place and will help prevent crack propagation to the part.

Complete a copy of the "Student Welding Report" listed in Appendix I or provided by your instructor. ◆

PRACTICE 17-10

Welding the Utility Trailer

Using the frame assembled in Practice 17-9, an oxyacetylene cutting set, a stick welding machine, a MIG welding machine or a flux cored welding machine, a steel tape measure, a chipping hammer, a square, a soapstone, C-clamps, a right angle grinder, gas welding goggles, a welding hood, safety glasses, leather gloves, and all other required safety equipment, you are going to weld the utility trailer.

CAUTION

When welding and cutting on concrete you must protect the concrete from direct exposure to the arc or cutting flame. Concrete contains moisture and the moisture can rapidly change to steam, causing the concrete surface to explode violently. To prevent this from happening, protect the concrete with a piece of scrap sheet metal when working in close proximity.

Weld around all the joints as specified in **Figure 17-77.** Do not over weld.

When all of the welds are complete on the top, either raise the frame so the welds on the bottom can be welded in the overhead position, or turn the frame over.

Make intermittent MIG or flux cored arc welds on the fender. Intermittent welds are strong enough to hold the fender in place and will help prevent crack propagation to the parts.

Complete a copy of the "Student Welding Report" listed in Appendix I or provided by your instructor. ◆

PRACTICE 17-11

Assembling the Springs and Axle

Using the frame welded in Practice 17-10, oxyacetylene cutting set, a stick welding machine, a MIG welding machine or a flux cored welding machine, a steel tape measure, a chipping hammer, a square, a soapstone, C-clamps, a right angle grinder, gas welding goggles, a welding hood, safety glasses, leather gloves, and all other required safety equipment, you are going to attach the springs and axle to the utility trailer.

Bolt the spring shackles to the spring and place the spring and axle assembly in place on the trailer frame. Use a steel tape to cross measure from the ends of the axle to the tongue of the trailer to square the axle to the frame.

Clamp the spring shackles and tack weld them in place. Recheck the squareness of the frame and axle before removing the springs and axle assembly from the framed shackles.

Removing the spring from the shackles will prevent excessive heat from damaging the rubber gaskets on the spring. Weld up the spring shackles. When the shackles have cooled, reinstall the spring and axle assembly and recheck the squareness.

Using a floor jack, if necessary, raise the frame and install the tires and rims. Lower the trailer onto the tires and tighten the lug nuts.

Complete a copy of the "Student Welding Report" listed in Appendix I or provided by your instructor. ◆

PRACTICE 17-12

Installing the Hitch

Using the frame welded in Practice 17-10, an oxyacetylene cutting set, a stick welding machine, a MIG welding machine or a flux cored welding machine, a steel tape measure, a chipping hammer, a square, a soapstone, C-clamps, a right angle grinder, an electric drill with an assortment of sizes of drill bits, gas welding goggles, a welding hood, safety glasses, leather gloves, and all other required safety equipment, you are going to attach the springs and axle to the utility trailer.

The trailer hitch should be designed for a 2-in. ball. Read the instructions that come with the trailer hitch. Some manufacturers make hitches that can be welded or bolted to the tongue.

Fasteners A variety of fasteners can be used to hold material together. The most common are nails, but screws, bolts with nuts, and adhesives are also widely used. The two most important considerations when selecting a type of fastener are how long it takes to install and whether it meets the strength requirements. Nails are the fastest fasteners to use, but a single nail has the least strength. A single bolt and washers with a nut are the strongest fasteners but take the longest to install. Nails are used for most agricultural framing construction because they have more than enough strength for the job.

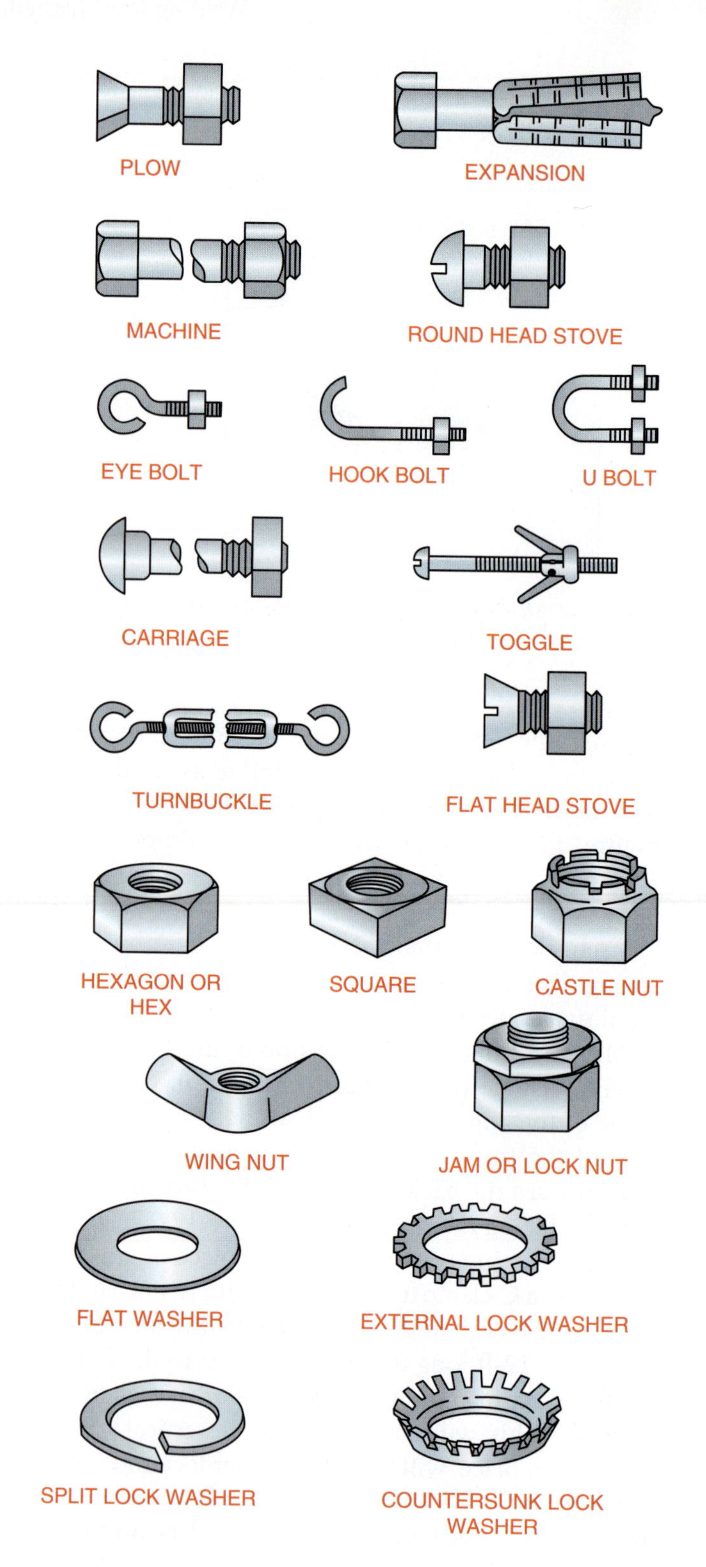

FIGURE 17-87 Common types of bolts, nuts, and washers.

Bolts, Nuts, and Washers

Bolts are threaded so they will fit into matching sized nuts or threaded holes. A number of types of bolts, nuts, and washers are available for agricultural fabrication, **Figure 17-87**. The most commonly used bolts in agricultural fabrication are machine and carriage bolts. Of the various types of nuts, the hex, or

Standard and Metric Thread Pitch						
Row	Standard Diameter	Course Threads*	Fine Threads*	Metric Diameter	Course Threads**	Fine Threads**
1	#10	24	32	4mm	0.7	—
2	#12	24	28	5 mm	0.8	—
3	1/4″	20	28	6 mm	1.0	—
4	3/8″	16	24	10 mm	1.5	1.25
5	1/2″	13	20	12 mm	1.75	1.25
6	5/8″	11	18	16 mm	2.0	1.5
7	3/4″	10	16	20 mm	2.5	1.5
8	1″	8	14	24 mm	3.0	2.0

TABLE 17-6 The Standard and Metric Bolts Listed on Each Row Are About the Same Size.

*Standard pitch is given in threads per inch.
**Metric pitch is given in threads per millimeter.

Identification Grade Mark	Specification	Description	Material	Max. Load Psi (KPa)	Tensile Strength Min psi (KPa)
	Un-graded	Utility Bolts	Low or Median Steel	33,000 (227,375)	60,000 (413,409)
	Grade 5	Bolts, Screws, Studs	Medium Carbon Steel	85,000 (585,663)	105,000 (723,466)
	Grade 8	Bolts, Screws, Studs	Medium Carbon Alloy Steel	120,000 (826,819)	150,000 (1,033,520)

TABLE 17-7 In 1990 Congress Passed a Law Standardizing These and Other Bolt Gradings and Bold Head Identification Markings.

hexagonal, and square nut types are most often used for fabrication. Flat and split lock washers are also in common use in agriculture.

The size of bolts and nuts is given as their diameter and pitch. The pitch is the number of threads per length. Standard-sized bolt and nut threads are expressed as the number of threads per inch. Metric-sized bolt and nut threads are expressed as the number of threads per millimeter. A 3-in. long carriage bolt that is 1/2 in. in diameter with 13 threads per in. would be expressed as a "3 1/2″-13 carriage bolt." A 75-mm long hex head 12-mm diameter bolt with 1.75 threads per mm would be expressed as a "24 mm 10 – 1.5 hex head bolt." Small diameter standard bolts and nuts, called machine screws, are sized by a numbering system from #000 to #20. Standard bolts and nuts starting at 1/8 in. and larger are measured in fractions of inches. Each size of bolt and nut can come with fine or coarse threads, **Table 17-6**. Coarse threads are sometimes referred to as standard threads because they are the ones most commonly used in the field.

The strength of bolts and nuts is expressed as its grade, **Table 17-7**. There are a number of grades of bolts and nuts available; however, the most commonly stocked types are ungraded, Grade 5, and Grade 8. Ungraded bolts are used for most agricultural fabrication. Graded bolts should always be used to connect heavy loads or attachments.

Washers are divided into several groups based on their application. The two most commonly used groups of washers are flat and lock washers. Flat washers spread out the load so the nut and bolt can be made tighter. On steel, they act as bearing surfaces so tightening is easier. In wood, flat washers keep the bolt head or nut from damaging the wood surface and/or prevent them from pulling through the wood. Fender washers are simply larger diameter flat washers for the same size bolt. Fender washers are particularly helpful if you are using a softer wood.

Locking washers help prevent the bolt and nut from becoming loose. The main type used in agriculture is the split lock washer. The sharp ends of the washer cut into the metal and bolt or nut when they are turned counterclockwise or loosened. Locking washers work best when used against metal and do not work well against wood.

Follow the specifications given in the literature. Install the tongue jack and bolt it to the hitch.

Complete a copy of the "Student Welding Report" listed in Appendix I or provided by your instructor. ◆

PRACTICE 17-13

Details and Finishing the Utility Trailer

Using an oxyacetylene cutting set; a stick welding machine; a MIG welding machine or a flux cored welding machine; a right angle grinder; an electric drill with assorted sizes of drill bits; a chipping hammer; a wire brush; a square; a soapstone; C-clamps; screw drivers; electrical pliers; an assortment of hand tools; general carpentry tools including hammers, a power circular saw, a hand saw, and measuring tape; chalk line; gas welding goggles; a welding hood; safety glasses; and all required personal protective equipment along with a drawing, you are going to finish the utility trailer by installing lights and safety chains, rounding the corners on the side rail, painting, and installing the wood decking.

Trailer Lights The trailer lights will be mounted on 10-in. long pieces of angle that are welded to the vertical picket immediately behind the wheel fender, **Figure 17-88.**

Using a C-clamp, secure the angle in place so that it is square. Tack weld the angle to the vertical picket. Make two welds as shown in **Figure 17-88.**

Trailer lights and wiring on agricultural trailers are subject to frequent damage because the trailers are often pulled off-road through fields where brush or debris can snag the wiring. To prevent damage, the wiring should be run through electrical mechanical tubing (EMT), **Figure 17-89A.** EMT comes in 10-ft lengths and can be hand bent to shape.

The EMT will be attached to the angle frame of the trailer by MIG or Flux cored welding. Bend the EMT so it will fit under the trailer frame and run up to a point just behind the tongue angle, **Figure 17-89B.**

Hold the EMT in place and make small tack welds at each angle. Because EMT is a thin-walled tubing, you will need to direct most of the arc's heat on the angle to keep from burning through tubing. EMT has

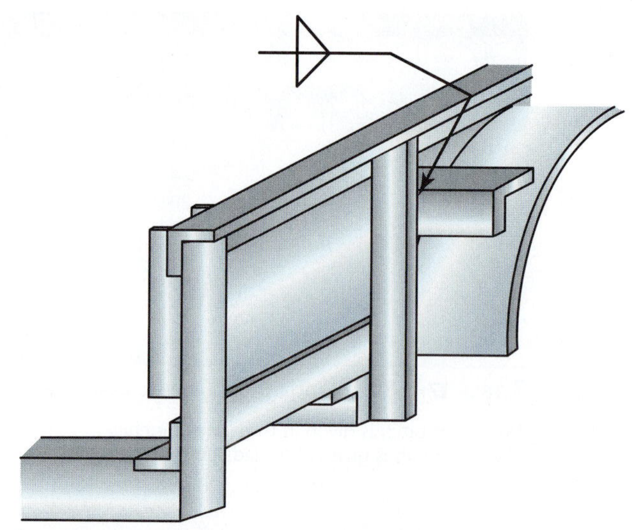

FIGURE 17-88 Make a fillet weld on both sides of the trailer light support angle.

FIGURE 17-89(A) EMT protects trailer wiring. Courtesy of Larry Jeffus.

a galvanized coating, so it is important that you not breathe the welding fumes.

With the tubing in place, an electrician's snake can be used to pull the light wiring. Mount the trailer lights according to manufacturer's specifications. An additional piece of angle or a short section of pipe can be used to protect the lights from damage.

The normal trailer four-wire color code is as follows:

- white is the ground wire
- brown is the tail lights
- green is the right-hand directional signal
- yellow is the left-hand directional signal

Safety Chain A safety chain must be secured to the tongue of the trailer. This short piece of chain is connected to the bumper of the tow vehicle in case the hitch comes off the tow ball. The chain must be located so it will not bind as the trailer is pulled around corners.

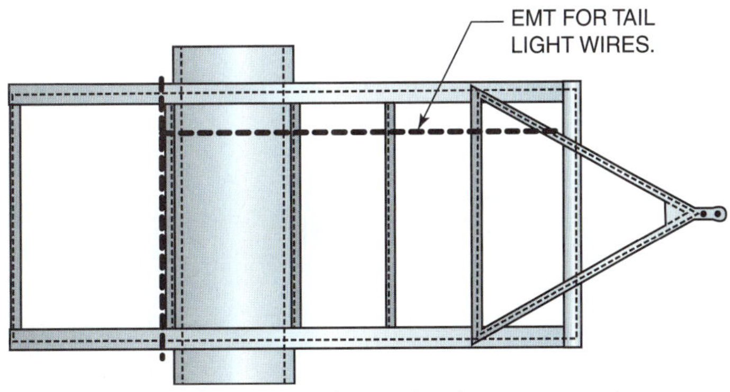

FIGURE 17-89(B) Run the EMT under the frame to a point just under the tongue angle.

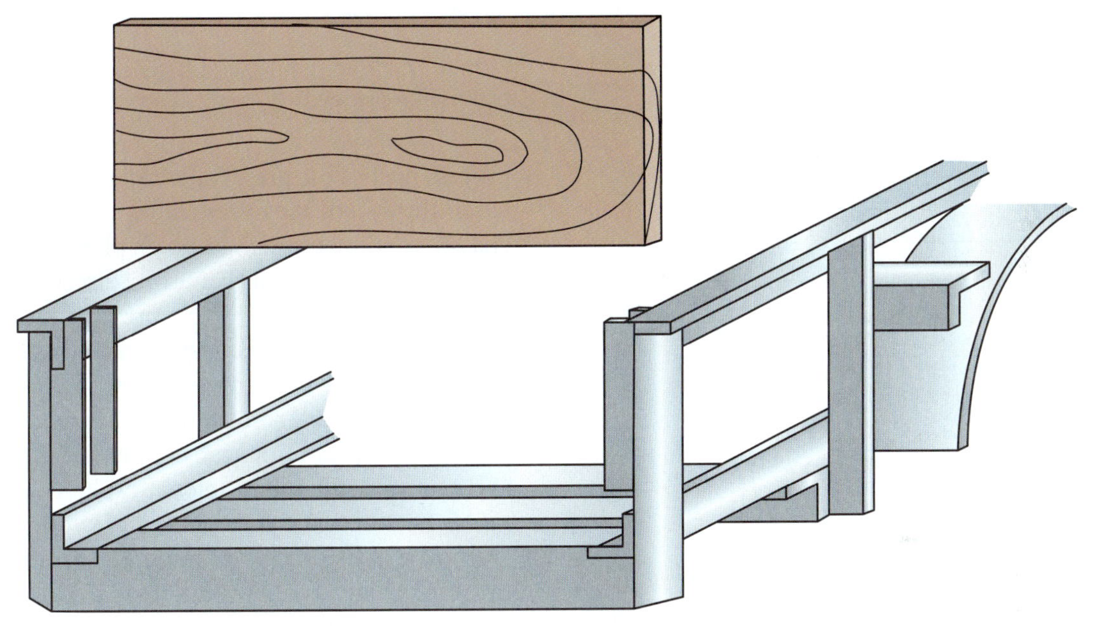

FIGURE 17-90 Trailer tailgate assembly.

Use an 8-ft length of 3/8 in. chain with hooks on both ends for the safety chain. Drill a 1/2-in. diameter hole in the back edge of the trailer hitch. Using a grade 5 or 8 bolt, nut, flat washer, and lock washer, bolt the center link of the chain to the trailer tongue.

Tailgate The trailer tailgate will slide in a 1 in. × 2 in. channel rail welded inside the back side rail as a guide for the tailgate. The rail will be made out of channel, and the tailgate will be made out of a 5-ft-long piece of 2 in. × 12 in. pressure-treated lumber, **Figure 17-90**.

Cut two 11-in. long pieces of 1 in. × 2 in. channel. Notch out a 2-in. long piece of the back of the channel so it will fit flush with the top rail angle. Chip the slag and grind the cut as necessary to get the channel to fit properly.

Clamp the channel squarely in place with C-clamps. Tack weld the channel to the vertical angle picket. Remove the C-clamps and weld the channel according to the weld specifications shown in **Figure 17-77**.

CAUTION

Read all of the circular saw manufacturer's safety material and saw operating instructions before beginning any work. You must follow all of these safety rules and operating instructions in order to work safely.

Use the steel measuring tape, square, and pencil to lay out the saw line on the two-by-twelve at 4 ft 11 1/2 in. Making the tailgate a little short will make it easier to install.

Because this is pressure-treated lumber, you will have to wear a mask and safety glasses or facemask when sawing. Place the board securely on sawhorses so the beams are at a comfortable work height. Make sure the board is supported so the saw blade will not bind at the end of the cut, **Figure 17-91**.

Put on your safety equipment before starting the cut. Align the saw blade with the line to be cut, but make sure the saw blade is not touching the wood.

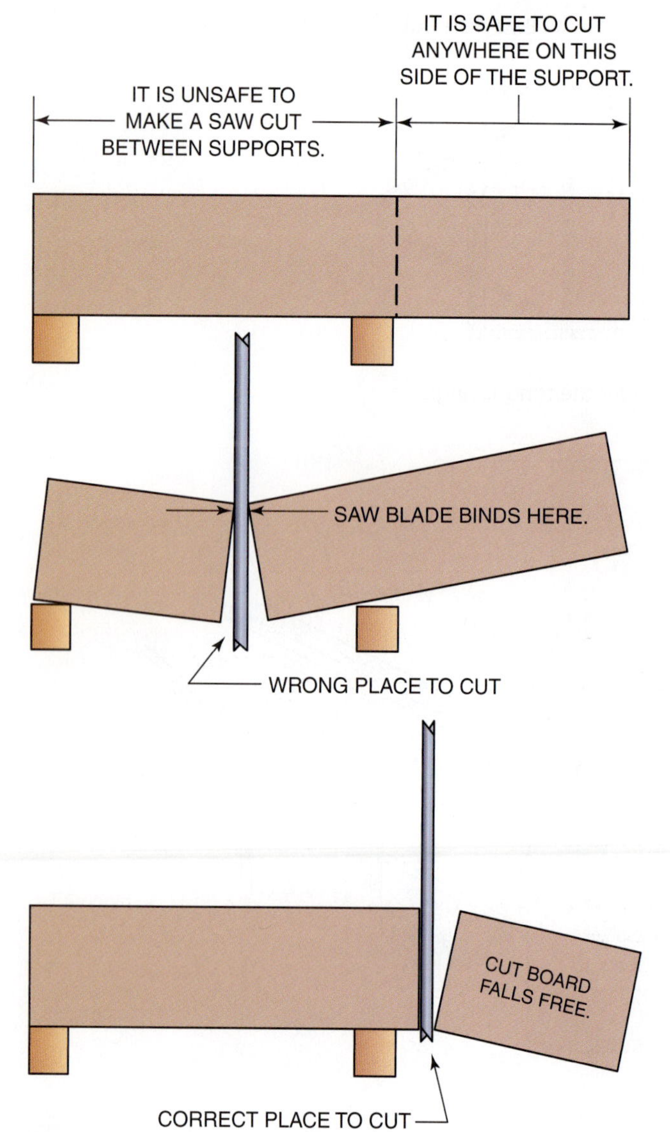

FIGURE 17-91 Properly support wood before sawing to prevent binding.

Pull the trigger to start the saw. Slowly begin the cut by sliding the saw across the board along the line. Once the cut is complete, put the saw down and unplug the power. Try the tailgate board in the channel and trim as necessary to get a smooth, easy fit.

Finishing Chip and wire brush all welds to remove slag. Following all local, state, and federal codes pertaining to spray painting, apply a rust proofing primer to the frame. Once the primer has dried, apply the finished paint. Be sure that the paint covers all of the small interior corners of the trailer so that it will not rust out prematurely.

Installing the Trailer Bed Pressure treated wood is treated under pressure with a preservative to protect it from rot and insect destruction. During the pressure-treating process, wood is placed inside a pressure vessel where the preservative is forced into the wood. This process pushes the preservative all the way to the center of the wood. Pressure-treated woods can be purchased as dimensional lumber, plywood, and trim boards. Any wood that comes in direct contact with the soil must be treated with a preservative to prevent it from rotting or being infested with termites.

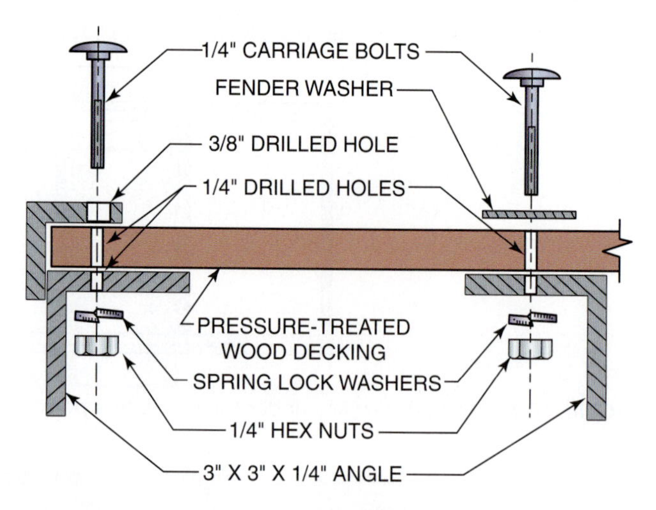

FIGURE 17-92 Bolts, nuts, and fasteners used to hold boards in trailer bed.

The five most commonly used wood preservatives are creosote, pentachlorophenol, acid copper chromate (ACC), ammoniacal copper arsenate (ACA), and chromated copper arsenate (CCA). Because of rising health concerns, check with your local agricultural extension agent about local restrictions on the use of any preservative-treated wood on your farm or ranch. There are some restrictions on the use of some quantities and types of treated wood around some crops and livestock.

CAUTION

Chemicals used as wood preservatives in pressure-treated woods are toxic and have been linked to health problems. Never handle or saw pressure-treated wood without proper personal protection equipment. Obtain the MSDS, which describes the correct procedures required when working with each specific type of wood preservative, from the wood manufacturer or supplier.

Because the trailer bed lumber has chemicals impregnated into it as a preservative, you must not weld adjacent to the wood, which would cause these chemicals to vaporize in the smoke.

The wood planks on the trailer are held in place with 1/4-in. carriage bolts. This is done so the wood can periodically be removed and the trailer bed replaced in the future. The back of the trailer's decking is held in place with a 5-ft long piece of 2 in. × 2 in. × 1/4 in. angle that is bolted in place, **Figure 17-92**.

Clamp the angle in place and drill 1/4-in. diameter holes through the angle, wood, and trailer frame. Because a 1/4-in. carriage bolt will be used to secure this in place, the hole in the angle will need to be enlarged to 3/8 in. so the head will fit flat. Put one carriage bolt at each end of the angle and one in the center of each wood plank.

Drill a 1/4 in. hole in the center of each trailer bed plank where they meet each of the 3 in. × 3 in. × 1/4 in. cross braces. Use a flat washer under the head of each bolt before it is installed.

Complete a copy of the "Student Welding Report" listed in Appendix I or provided by your instructor. ◆

Summary

Welding is an essential tool for the farmer or rancher because it can save time and money. With the ever-rising cost of equipment and replacement parts, being a skilled welder can sometimes make the difference between making a profit and having a loss for a season. Welding might avert the high cost of a breakdown or failure that occurs during a harvest or roundup. Such failures could jeopardize your delivery time, which could create additional expenses.

Many things can be made from either wood or metal. However, a rancher in Maine told me "because of our severe winters here, unless you wanted to build it eight times you had better build it out of metal." This may be a little bit of an exaggeration, but it does point out the permanency of welded structures.

Before beginning a project, it is a good idea to make a sketch or drawing as part of your planning. Making drawings will help identify problem areas in the construction. Drawings also can help you create a more accurate bill of materials. Even small projects can benefit from a sketch. Sketches do not have to be elaborate; they just need to provide enough material information to make construction easier.

Additional welding projects can be found in *101 Welding Projects* and *Welding Sculptures*, publishing through Delmar Learning in 2003.

Review

1. What do welding symbols show?
   A. when to make the weld
   B. who should make the weld
   C. the welder certification needed to make the weld
   D. exactly what welding is needed.
2. Which of the following is not a welded joint type?
   A. tee joint
   B. back joint
   C. butt joint
   D. lap joint
3. Some metals being welded have problems with ________.
   A. thermal expansion
   B. color match
   C. thickness
   D. arc slag
4. The welding process to be used has a major effect on the selection of the joint design.
   A. true
   B. false
5. The most ideal welding position for most joints is the ________ because it allows for larger molten weld pools to be controlled.
   A. vertical up position
   B. overhead position
   C. flat position
   D. horizontal position

Review Questions continued

6. __________ is when the pipe is parallel to the horizon, and the weld is made vertically around the pipe.
   A. 1G
   B. 2G
   C. 5G
   D. 6G
7. When welding on thick plate or pipe, it is often impossible for the welder to get 100% penetration without using some type of __________.
   A. preheat
   B. groove or bevel
   C. fillet
   D. base metal weld
8. What is the name of the joint in which one piece of metal is standing vertically on the other piece of metal?
   A. tee joint
   B. lap joint
   C. outside corner joint
   D. butt joint
9. A list of the various items that will be needed to build a weldment is known as __________.
   A. a shopping list
   B. a piece list
   C. an order form
   D. a bill of materials
10. Which view shows the most information about a part to be built?
    A. the top view
    B. the front view
    C. the right side view
    D. the bottom view
11. If the needed dimensions cannot be found on the drawings, just obtain them by measuring the drawing itself.
    A. true
    B. false
12. Why are some drawings made as sections?
    A. so they will fit on the page
    B. so they look more high tech
    C. to show the inside better
    D. so the color shows better
13. What type of weld does a V-shaped weld indicate?
    A. a fillet weld
    B. a V-groove weld
    C. a flat weld
    D. a beveled weld
14. What is the main purpose of the root face?
    A. to make the weld face look good
    B. to make starting the arc easier
    C. to keep the sharp root edge from cutting someone
    D. to minimize burn-through
15. What are the two areas of agriculture in which welding is used?
    A. fabrication and repair welds
    B. building things and making parts
    C. fixing broken parts and resurfacing worn parts
    D. as a hobby and as a job
16. What does 1/2 + 1/4 + 3/4 equal?
    A. 5/4
    B. 1 1/2
    C. 2
    D. 2 1/4
17. If a 12-in. long part can be made ± 1/2 inch, how small can it be and how large can it be and still be acceptable?
    A. 11 1/4 in. and 11 1/2 in.
    B. 11 in. and 13 in.
    C. 11 1/2 in. and 12 1/2 in.
    D. 11 in. and 12 in.
18. When converting a fraction to a decimal, divide the numerator (top number in the fraction) by the denominator (bottom number in the fraction).
    A. true
    B. false
19. According to Table 17-5 what is the actual size of a two-by-eight?
    A. 1 1/2 × 3 1/2 (37mm × 87mm)
    B. 1 1/2 × 5 1/2 (37mm × 137mm)
    C. 1 1/2 × 7 1/4 (37mm × 181mm)
20. The placement of parts in a manner that will minimize the waste created is called __________.
    A. stacking
    B. placement
    C. grouping
    D. nesting

Chapter 18

Plastic and Other Nonmetallic Fabrication Techniques

OBJECTIVES

After completing this chapter, the student should be able to

- ☑ describe solvent welding.
- ☑ describe the basics of hot-gas and airless welding.
- ☑ explain tack-welding, hand-welding, and speed-welding techniques.
- ☑ perform a single V-groove weld.
- ☑ perform a double V-groove weld.
- ☑ explain how to make a hot surface weld on pipe.

KEY TERMS

Acrylonitrile/butadiene/styrene (ABS)
chlorinated polyvinyl chloride (CPVC)
heat welding
Polybutylene (PB)
Polyethylene (PE)
Polypropylene (PP)
Polyvinyl chloride (PVC)
solvent welding
thermoplastics
thermoset plastics

INTRODUCTION

Metal is a great construction material; however, other materials like plastics are needed around the farm or ranch for many applications. Plastic has replaced steel and aluminum for most tanks, livestock feeders, and poultry waterers because it is less expensive and does not rust or corrode, **Figure 18-1**.

Projects consisting of metal and plastic often need to be constructed and repaired, **Figure 18-2**. Plastic can be assembled and repaired using adhesives, epoxies, and welding. In this chapter, you will learn to make decisions as to which material is best for a project and informed choices concerning assembly and repair methods.

FIGURE 18-1 Herbicide shipped in HDPE drums. Courtesy of Larry Jeffus.

FIGURE 18-2 Cattle feeder made from LDPE drum. Courtesy of Larry Jeffus.

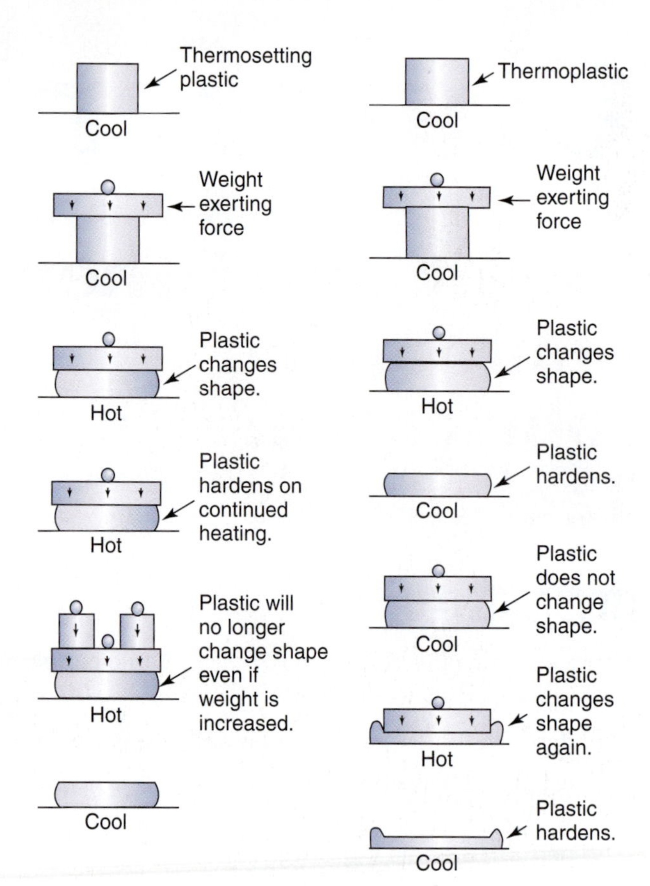

FIGURE 18-3 The effect of heat on plastics.

Types of Plastics

Two types of plastics are used today in agricultural equipment:

1. **Thermoplastics**. These plastics are capable of being repeatedly softened and hardened by heating and cooling, with no change in their appearance or chemical makeup. They soften or melt when heat is applied to them, and therefore are weldable with a plastic welder.
2. **Thermoset Plastics**. These plastics undergo a chemical change by the action of heating, a catalyst, or ultraviolet light, leading to an infusible state. They are set into a permanent shape that cannot be altered either by reapplying heat or with catalysts. The two types of plastic have similar sounding names; to keep them straight, remember that the set in thermoset means the plastic is set for its life. Thermoset plastics are not heat weldable, although they can be solvent welded.

Figure 18-3 explains the relationship of heat to plastics more fully. In general, the recommended repair techniques are chemical adhesive bonding for thermosetting plastics and welding for thermoplastics. Without a doubt, the ever-growing use of both of these materials has helped to transform plastic welding from a curious technique practiced by a select few into an important skill that has been accepted as an integral part of today's ag shop. Following are brief descriptions of the most common agricultural plastics.

FIGURE 18-4 Vacuum formed stock water trough. Courtesy of Larry Jeffus.

FIGURE 18-5 HDPE drums. Courtesy of Larry Jeffus.

Acrylonitrile/Butadiene/ Styrene (ABS)

Acrylonitrile/butadiene/styrene (ABS), is a thermoplastic that is semirigid with excellent toughness and forming properties. It can be injection molded, molded, extruded into pipe or tubing, and vacuum formed, **Figure 18-4**. ABS is available in normal and high-temperature types, either of which can be welded.

Polyethylene (PE)

Polyethylene (PE) is a very commonly used plastic. PE is available in four different types: polyethylene terephthalate (PET or PET-P), linear low-density polyethylene (LLDPE), low density polyethylene (LDPE), and high density polyethylene (HDPE). Each type has unique mechanical and thermal properties and is used for different applications. PEs can be heat welded.

Polyethylene terephthalate has excellent wear resistance and high strength, and is resistant to moderate acids. It can be injection molded into bottles or precision mechanical parts. PET machine parts are used as rollers or as rotors in pumps.

Linear low-density polyethylene is stretchy and tough. LLDPE is used in very thin sheets called film to make plastic grocery bags and some trash bags, and is injection molded to produce soft drink bottles and food jars, for example.

Low density polyethylene (LDPE) is tough, flexible, and fairly transparent. LDPE is used as a film to make trash bags and food bags, and can be injection molded into flexible items such as lids, squeeze bottles, and so forth.

High density polyethylene (HDPE) is stiff, resistant to chemicals, strong, and tough. Uncolored HDPE has a milky white appearance. It is easily colored to almost any color but is often seen as blue, red, or black. HDPE can be injection molded, molded, and extruded. It is injection molded to make bottles and containers such as those used for herbicides, pesticides, and mild acids and alkalines, **Figure 18-5**. Pump-up sprayer tanks, from 1 gal. to 4 gal. (4 L to 16 L) sizes, are often made from injection molded HDPE. HDPE can be molded into small and large tanks for water, chemicals, and fuel or gas "cans." It is extruded into small-diameter coils and large-diameter cut lengths of pipe for agricultural and irrigation piping. Some thicker HDPE pipes can withstand pressures up to 200 psi (1,300 kPa).

Polyvinyl Chloride (PVC) and Chlorinated Polyvinyl Chloride (CPVC)

Polyvinyl chloride (PVC) and chlorinated polyvinyl chloride (CPVC) are thermoplastics often used for plumbing pipes and electrical conduit. Both have a glossy surface. PVC is usually white, beige, or gray in color. CPVC is usually beige or gray in color. CPVC and PVC have the same characteristics, except that CPVC can withstand higher temperatures. Because they are semirigid, they can be used for some light-duty structures like framing for small animal hutches. PVC and CPVC can be molded into parts for equipment or containers such as those used for filtration or chemical injection systems. Both can be heat and solvent welded.

Polypropylene (PP)

Polypropylene (PP) is a low-gloss, semirigid plastic that is easily colored. It is a higher temperature thermoplastic that is quite similar to polyethylene;

polypropylene is susceptible to stress cracking and oxidation. Molded PP is used for parts around engines such as battery cases and electronic covers and is extruded into fibers for rope and woven sacks. PP can be heat welded.

Polybutylene (PB)

Polybutylene (**PB**) is a low-gloss, semirigid thermoplastic that has good toughness and high impact strength. Because of its resistance to creep at elevated temperatures, it is approved for hot and cold water. It is easily colored and can be seen as beige, black, yellow, or gray. PB can be heat welded.

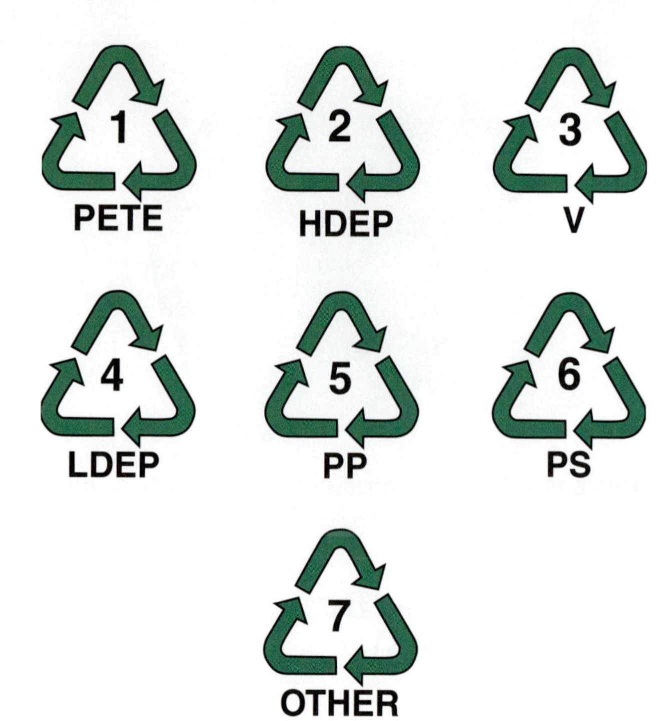

FIGURE 18-6 Recycle symbols for plastic.

Plastic Identification

Before a plastic weld can be made, you must know what type of material you are working with. With such a variety of plastics in existence—and new ones constantly being tested and refined—it is easy to be confused. This, in turn, can jeopardize the entire job because a repair that is done based on incorrect identification will likely yield unsatisfactory results. A repair that initially appears to be sound can quickly delaminate, crack, or discolor if the job was performed based on incorrect identification of the plastic.

The best way to identify plastic is to look on the part for the name of the type of plastic, look in the equipment literature, or call the manufacturer.

Other ways to identify an unknown plastic are to look for the recycling code, **Figure 18-6**; be familiar with plastic uses, **Table 18-1**; or use the burn test. The so-called burn test or melt test identifies different plastics by their different burn characteristics. Some also produce unique odors.

To do a burn test, scrape off a small sliver of the material from the backside or underside of the part, hold it with pliers or on the end of a piece of wire, and try to ignite it with a match or propane torch, **Figure 18-7**. The burn characteristics of common agricultural plastics are given in **Table 18-2**.

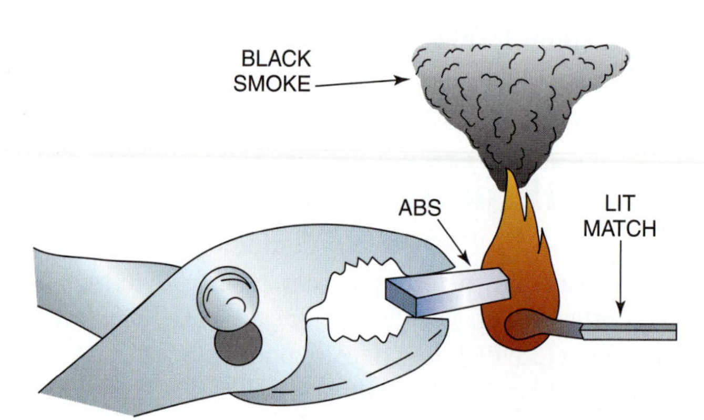

FIGURE 18-7 Conducting a burn test on ABS plastic.

The burn test has fallen out of favor as a technique for identifying plastics for several reasons. Having an open flame in a shop environment creates a potential fire hazard, and although the burn test does not generally pose a health threat it is a good idea to avoid inhaling the fumes. Another reason the burn test is not as popular as it once was is because it is not always reliable. For example, it is difficult to determine the difference between polypropylene and polyethylene because both burn with the same characteristics. Furthermore, some parts are now being manufactured from hybrids, which are blended plastics that use more than one ingredient. A burn test is of no help here.

The last possible way of identifying an unknown plastic (assuming it is probably a thermoplastic that is potentially weldable) is to make a trial-and-error weld, **Figure 18-8**. Try several different filler rods until you find one that sticks. Most suppliers have a sample pack of plastic welding rods that you can try to determine which type works best. The rods are color-coded, so once the rod that works is found the base material is identified. For more specifics on working with plastics, see **Table 18-3**.

Symbol	Chemical Name	Common Name	Design Applications	Thermosetting of Thermoplastic
ABS	Acrylonitrile-butadiene-styrene	ABS, Cycolac, Abson, Kralastic, Lustran, Absafil, Dylel	Body panels, dash panels, grilles, headlamp doors	Thermoplastic
ABS/MAT	Hard ABS reinforced with fiberglass	—	Body panels	Thermosetting
ABS/PVC	ABS/Polyvinyl chloride	ABS Vinyl	—	Thermoplastic
EP	Epoxy	Epon, EPO, Epotuf, Araldite	Fiberglass body panels	Thermosetting
EPDM	Ethylene-propylene-dienemonomer	EPDM, Nordel	Bumper impact strips, body panels	Thermosetting
PA	Polyamide	Nylon, Capron, Zytel, Rilsan, Minlon, Vydyne	Exterior finish trim panels	Thermosetting
PC	Polycarbonate	Lexan, Merlon	Grilles, instrument panels, lenses	Thermoplastic
PE	Polyethylene	Dylan, Fortiflex, Marlex Alathon,Hi-fax, Hosalen, Paxon	Inner fender panels, interior trim panels, valances, spoilers	Thermoplastic
PP	Polypropylene	Profax, Olefo, Marlex, Olemer, Aydel, Dypro	Interior mouldings, interior trim panels, inner fenders, radiator shrouds, dash panels, bumper covers	Thermoplastic
PRO	Polyphenylene oxide	Noryl, Olefo	Chromed plastic parts, grilles, headlamp doors, bezels, ornaments	Thermosetting
PS	Polystyrene	Lustrex, Dylene, Styron, Fostacryl, Duraton	—	Thermoplastic
PUR	Polyurethane	Castethane, Bayflex	Bumper covers, front and rear body panels, filler panels	Thermosetting
PVC	Polyvinyl chloride	Geon, Vinylete, Pilovic	Interior trim, soft filler panels	Thermoplastic
RIM	"Reaction injection molded" polyurethane	—	Bumper covers	Thermosetting
R RIM	Reinforced RIM-polyurethane	—	Exterior body panels	Thermosetting
SAN	Styrene-acrylonitrite	Lustran, Tyril, Fostacryl	Interior trim panels	Thermosetting
TPR	Thermoplastic rubber	—	Valance panels	Thermosetting
TPUR	Polyurethane	Pellethane, Estane, Roylar, Texin	Bumper covers, gravel deflectors, filler panels, soft bezels	Thermoplastic
UP	Polyester	SMC, Premi-glas, Selection Vibrin-mat	Fiberglass body panels	Thermosetting

TABLE 18-1 Standard Symbol, Chemical Name, Trade Name, and Design Applications of Most Commonly Used Plastics.

Plastic	Burn Characteristic
Polypropylene (PP)	Burns with no visible smoke and continues to burn once the flame source is removed. Produces a smell like burned wax. Bottom of flame is blue and top is yellow.
Polyethylene (PE)	Also smells like burned wax, makes no smoke, and continues to burn once the flame source is removed. Bottom of flame is blue and top is yellow.
ABS	Burns with a thick, black, sooty smoke and continues to burn when the flame source is removed. Produces a sweet odor when burned. Flame is yellowish orange.
PVC	Only chars and does not support a flame when you try to burn it. Gives off gray smoke and an acid-like smell. End of flame is yellowish green.
Thermoplastic Polyurethane (TPUR) and Thermosetting Polyurethane (PUR)	Burns with a yellow-orange sputtering flame and gives off black smoke. The thermoset version of polyurethane, however, will not support a flame.

TABLE 18-2 Burn Characteristics of Common Plastics.

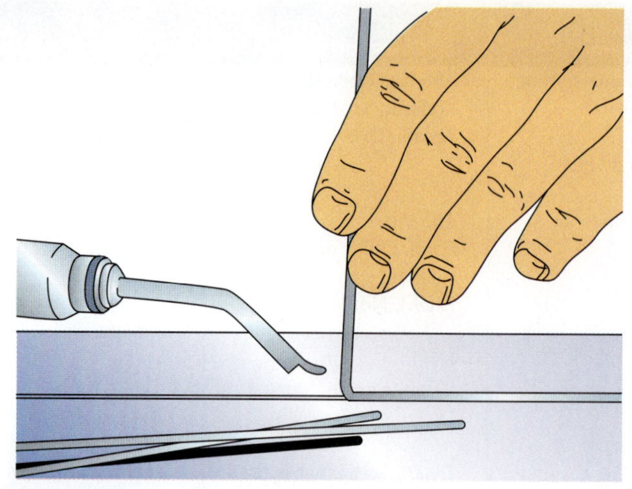

FIGURE 18-8 Making a trial-and-error weld.

Code	Material Name	Heat Resisting Temperature* °F	Resistance To Alcohol or Gasoline	Notes
AAS	Acrylonitrile Acrylic Rubber Styrene Resin	176	Alcohol is harmless if applied only for short time in small amounts (example, quick wiping to remove grease).	Avoid gasoline and organic or aromatic solvents.
ABS	Acrylonitrile Butadiene Styrene Resin	176	Alcohol is harmless if applied only for short time in small amounts (example, quick wiping to remove grease).	Avoid gasoline and organic or aromatic solvents.
AES	Acrylonitrile Ethylene Rubber Styrene Resin	176	Alcohol is harmless if applied only for short time in small amounts (example, quick wiping to remove grease).	Avoid gasoline and organic or aromatic solvents.
EPDN	Ethylene Propylene Rubber	212	Alcohol is harmless. Gasoline is harmless if applied only for short time in small amounts.	Most solvents are harmless, but avoid dipping in gasoline, solvents, etc.
PA	Polyamide (Nylon)	176	Alcohol and gasoline are harmless.	Avoid battery acid.
PC	Polycarbonate	248	Alcohol is harmless.	Avoid gasoline, brake fluid, wax, wax removers, and organic solvents.
PE	Polyethylene	176	Alcohol and gasoline are harmless.	Most solvents are harmless.
POM	Polyoxymethylene	212	Alcohol and gasoline are harmless.	Most solvents are harmless.
PP	Polypropylene	176	Alcohol and gasoline are harmless.	Most solvents are harmless.
PPO	Modified Polyphenylene Oxide	212	Alcohol is harmless.	Gasoline is harmless if applied only for quick wiping to remove grease.
PS	Polystyrene	140	Alcohol and gasoline are harmless if applied only for short time in small amounts.	Avoid dipping or immersing in alcohol, gasoline, solvents, etc.

*Above this temperature plastic will begin to soften.

TABLE 18-3 Handling Precautions for Plastics.

The Welding of Plastic

Plastic can be welded with heat or a solvent. In both methods, the surface of the plastic being joined is softened to a point where the two surfaces fuse together. Pressure to the joint or filler material is usually applied during welding. Heat welding can only be done on thermoplastic pipes, but solvent welding can be done on some of both types of plastic. Table 18-4 list several plastics commonly used in agriculture and their joining methods.

Heat Welding Heat welding is performed by heating or warming the surfaces to be joined to a temperature that will allow fusion to occur. The two methods of heat-welding plastic for agriculture are with hot gases and hot surfaces. Other methods of heat-welding include ultrasound and friction.

Plastic	Color	Properties	Method of Joining	Applications
ABS	Black	Rigid	Heat or solvent weldable	Sewer, and drain pipe
PE	Orange, blue, black, milky	Flexible	Heat weldable or mechanically coupled	Cold water only outdoor piping, buried
PCV	White, gray, beige, and many others	Rigid with high chemical resistance	Heat or solvent weldable	DWV, sewer, and drain pipe; cold-water buried pressure pipe; tubular goods
CPVC	Beige	Rigid, heat-resistant	Heat or solvent weldable	Hot and cold water supply tubes, indoors and buried
PP	Beige	Semi-rigid with high heat and chemical resistance	Heat weldable or mechanically coupled	Tubular drainage products for fixtures
PB	Beige, black, yellow, gray	Flexible, heat-resistant	Heat weldable or mechanical couplings	Hot and cold water supply tubes, indoors and buried; riser tubes
S or RS	Black, milky, or white	Rigid	Solvent weldable	Drain pipe outdoors and buried

TABLE 18-4 Types of Plastic Pipe and Their Applications.

Solvent Welding Solvent welding, sometimes called cement welding, uses a chemical that softens the plastic surfaces to allow them to fuse under light pressure. Solvent welding can be done as a one-step or two-step process. In the one-step method, a solvent/cleaner combination solution is used and, in the two-step method, separate cleaner and solvent cement are used. Many codes require the two-step method.

CAUTION

Keep in mind that although the techniques described in this chapter are proven repair methods, no technique is 100% reliable when used to repair a fuel tank or for other critical structural applications.

Heat Welding

The welding of plastics is not unlike the welding of metals. Both use a heat source, welding rod, and similar types of finished welds (butt joints, fillet welds, lap joints, and the like). Joints are prepared in much the same manner and are similarly evaluated for strength. Due to differences in the physical characteristics of each material, however, there are notable differences between welding metal and welding plastics.

When welding metal, the rod and base metal are made molten and puddled into a joint. Metals have a sharply defined melting point, but plastics have a wide melting range between the temperature at which they soften and the temperature at which they char or burn. Also unlike metals, plastics are poor conductors of heat and therefore difficult to heat uniformly. Because of this, the plastic filler rod and the surface of the plastic will char or burn before the material below the surface becomes fully softened. The decomposition time at welding temperature is shorter than the time required to completely soften many plastics for fusion welding. The result is that a plastic welder must work within a much smaller temperature range than the metal welder.

Because a plastic welding rod does not become completely molten and appears much the same before and after welding, a plastic weld might appear incomplete to the shop technician who is used to welding only metal. The explanation is simple. Since only the outer surface of the rod has become molten and the inner core has remained hard, the welder is able to exert pressure on the rod to force it into the joint and create a permanent bond. When heat is taken away, the rod reverts to its original form. Therefore, even though a strong and permanent bond has been obtained between the rod and base material, the appearance of the rod is much the same as before the weld was made, except for molten flow patterns on either side of the bead.

The proper combination of heat and pressure are required to fuse materials together when welding plastics. With the conventional hand-welding method, this combination is achieved by applying pressure on the welding rod with one hand, while at the same time applying heat and a constant fanning motion to the rod and base material with hot gas from the welding torch, **Figure 18-9.** Successful welds require that both pressure and heat be kept constant and in proper balance. Too much pressure on the rod tends to stretch the bead and produce unsatisfactory results; too much heat will char, melt, or distort the plastic. With practice, plastic welding can be mastered as completely as metal welding.

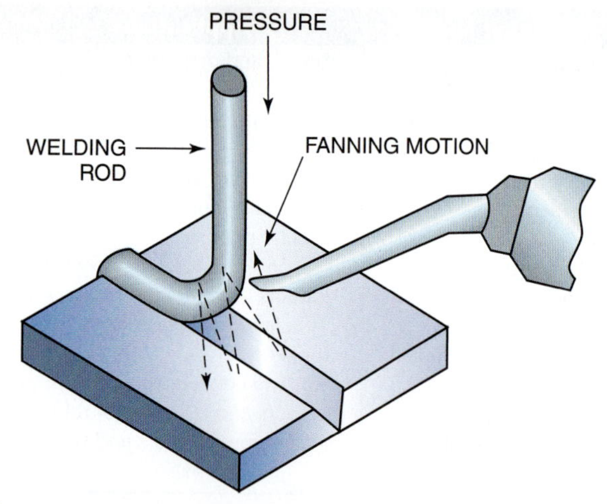

FIGURE 18-9 Successful plastic welding requires the proper combination of heat and pressure.

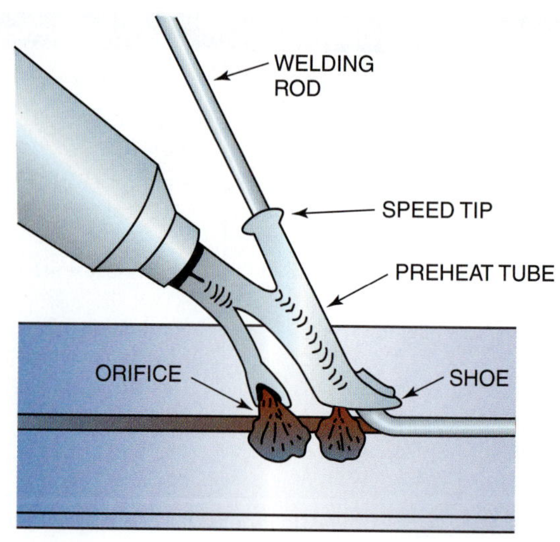

FIGURE 18-10 Plastic welder utilizing a high-speed tip.

High-speed Welding

High-speed welding incorporates the basic methods used in hand welding. Its primary difference lies in the use of a specially designed and patented high-speed tip, **Figure 18-10**, that enables the welder to produce more uniform welds and work at a much higher rate of speed. As with hand welding, constant heat and pressure must be maintained.

The increased efficiency of high-speed welding is made possible by preheating both the rod and base material before the point of fusion. The rod is preheated as it passes through a tube in the speed tip; the base material is preheated by a stream of hot gas passing through a vent in the tip ahead of the fusion point. A pointed shoe on the end of the tip applies pressure on the rod, thus eliminating the need for the operator to apply pressure. At the same time it smoothes out the rod, creating a more uniform appearance in the finished weld.

In high-speed welding, the conventional two-hand method becomes a faster and more uniform one-hand operation. Once started, the rod is fed automatically into the preheating tube as the welding torch is pulled along the joint. High-speed tips are designed to provide the constant balance of heat and pressure necessary for a satisfactory weld. The average welding speed is about 40 in. per min.

High-speed welding does have its disadvantages. Because increased speeds must be maintained to achieve the best possible weld, the high-speed welding torch is not suited for small, intricate work. Also, when the operator is new to this technique, the position in which the welder is held might seem clumsy and difficult. However, experience will enable the operator to successfully make all welds that can be made with a hand welder, including butt welds, V-welds, corner welds, and lap joint welds. Speed welds can be made on circular as well as flat work. In addition, inside welds on tanks can be speed welded, provided the working space is not too small to manipulate the torch.

Filler Material

Plastic filler rods come in a variety of chemical compositions and densities to meet specific plastics requirements. These rods can be divided into several general groups. The most common are ABS welding rods, PE welding rods, PVC welding rods, and PP welding rods.

ABS welding rods are available in cut lengths and as longer lengths on wound-on spools. These rods are available in a variety of colors including white, blue, and black.

PE welding rods are used on PET, HDPE, and LDPE plastics. They are available in both low and high densities. The low-density rods are used on LDPE plastics, are strong and flexible, and offer an excellent chemically resistant weld. The high-density PE welding rods have extra hardness and rigidity and a greater softening temperature. It is important to use the proper density PE filler rods to ensure a quality, serviceable weld.

PVC welding rods are the most commonly used plastic welding rods in agriculture. They are economical, corrosion-resistant, and self-extinguishing and are good thermal and electrical insulators.

PP welding rods are used on PP plastics. These rods produce welds that have outstanding stress and crack resistance. The welds have excellent chemical resistance and a temperature limit higher than that of PE.

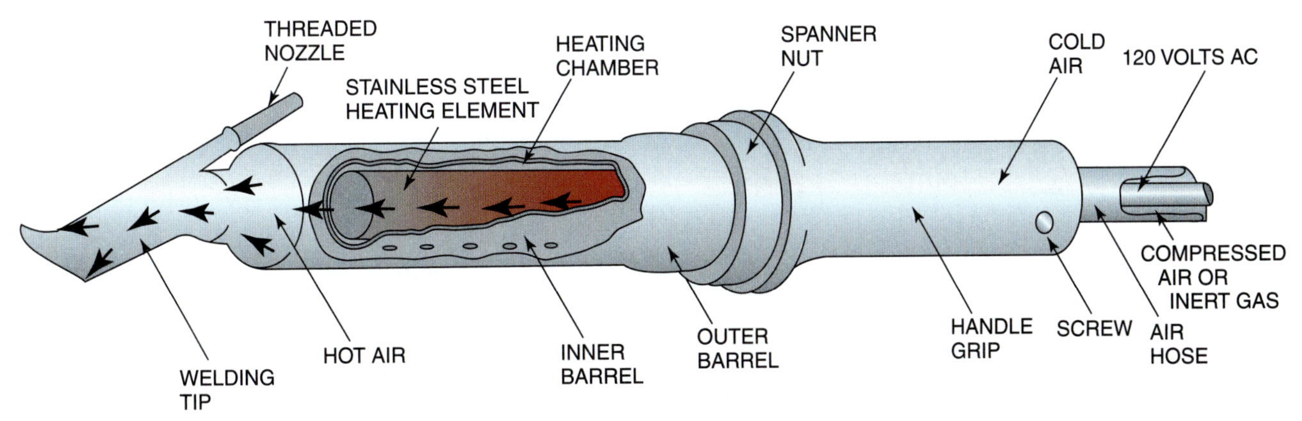

FIGURE 18-11 Typical hot-air welder.

Hot-gas Plastic Welding

The terms hot-gas and hot-air welders are often used interchangeably because the same welding gun can be used with both processes. Some plastics oxidize easily at welding temperatures and must be welded with an inert gas, usually nitrogen. Plastics that do not have oxide-forming problems are welded with hot gas. Because air is a gas, the term hot-gas can be used to refer to both processes.

A number of manufacturers make plastic hot-gas welding equipment for agricultural use, and all use the same basic technology. A ceramic or metal electric heating element is used to heat gas or air. The guns produce an adjustable gas or air stream between 450°F and 650°F (230°C to 340°C). The hot gas flow blows through a nozzle and onto the plastic. For plastics that oxidize quickly, nitrogen is supplied from a cylinder. If air is used, it can come from either the shop's air compressor or a self-contained portable compressor mounted in a carrying case that comes with the unit. Most hot-gas welders use a working pressure at the tip of around 3 psi. A pair of pressure regulators is required to reduce the air pressure first to around 50 psi, and then finally to the working pressure of 2 1/2 to 3 1/2 psi. A typical hot-gas welder is illustrated in **Figure 18-11.**

The barrel of the torch itself gets hot enough that skin contact could cause a burn if the hot gas were directed against the skin long enough. The torch is used in conjunction with the welding rod, which is normally 3/16 in. (4.5 mm) in diameter and made from the same material as the plastic being repaired. This will ensure that the strength, hardness, and flexibility of the repair is the same as that of the damaged part. Use of the proper welding rod is very important; an adequate weld is impossible if the wrong rod is used.

One of the problems with hot-gas welding is that the 3/16-in. (4.5-mm) diameter rod is often thicker than the panel to be welded. This can cause the panel to overheat before the rod has melted. Using a 1/8-in. (3-mm) rod with the hot-gas welder can often correct such warpage problems.

Three shapes of welding tips are available for use with most plastic welding torches. They are

- *Tacking welding tips.* These are used to tack broken sections together before welding. If necessary, tack welds can be easily broken apart for realigning.
- *Round welding tips.* These tips are used to fill small holes and make short welds, welds in hard-to-reach places, and welds on tight or particularly sharp corners.
- *Speed welding tips.* These are used for long, fairly straight welds. They hold the filler rod, automatically preheat it, and feed the rod into the weld, thus allowing for faster welding rates.

Setup, Shutdown, and Servicing

Naturally, no two hot-gas welders are exactly alike; their design can vary from one manufacturer to another. The setup, shutdown, and service procedures that follow, while typical of all hot-gas welders, should nonetheless be regarded as general guidelines only. For specific instruction, always refer to the owner's manual and other material provided by the welder manufacturer. Keep in mind that some manufacturers advise against using their welder on plastic that is any thinner than 1/8 in. (3 mm) because of the likelihood of distortion. In other cases, it is acceptable to weld plastics as thin as 1/16 in. (1.5 mm), provided they are supported from underneath

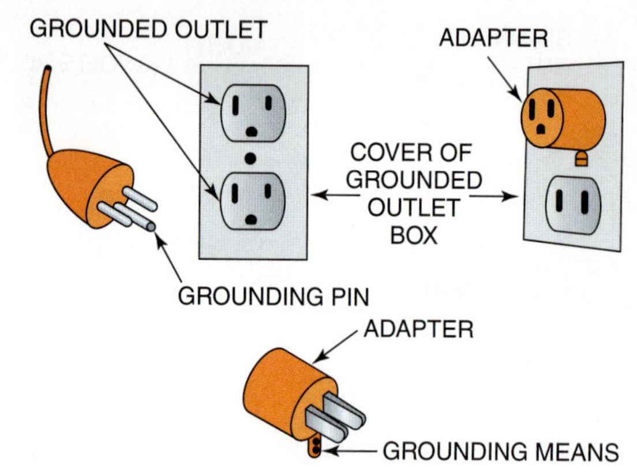

FIGURE 18-12 Methods of grounding a plastic welder.

during the operation. Again, it is very important to read and follow the specific directions for the welder being used.

PRACTICE 18-1

Set-up and Shut-down Hot-gas Welders

Using a hot-gas welder, air or gas supply, pressure regulator, thermometer, safety glasses, gloves, and other required personal protection equipment, you are going to set up a hot-gas welding system.

Close the air pressure regulator valve by turning the control handle counterclockwise until it is loose. This will prevent possible damage to the gauge from a sudden surge of excess air pressure.

Connect the regulator to a supply of either compressed air or inert gas. The standard rating for an air pressure regulator is 200 pounds of line pressure. If inert gas is used, a pressure-reducing valve is needed.

Turn on the air supply. The starting air pressure depends on the wattage of the heating element and the air pressure. The operating air pressure requires slightly less air.

Connect the welder to a common 120-volt AC outlet. A three-prong grounded plug or temporary adapter must be used with the welder at all times, as shown in **Figure 18-12.** Allow the welder to warm up at the recommended pressure. It is essential that either air or inert gas flows through the welder at all times, from warm up to cool down, to prevent burnout of the heating element and further damage to the gun.

Select the proper tip and insert it with pliers to avoid touching the barrel while hot. After the tip has been installed, the temperature will increase slightly due to backpressure. Allow 2 to 3 min for the tip to reach the required operating temperature.

Check the air temperature by holding a thermometer 1/4 in. (6 mm) from the hot-gas end of the torch. For most thermoplastics, the temperature should be in the 450°F to 650°F range. Information supplied with the welder usually includes a chart of welding temperatures.

If the temperature is too high to weld the material, increase the air pressure slightly until the temperature decreases. If the temperature is too low for the particular application, decrease the air pressure slightly until the temperature rises. When increasing and decreasing the air pressure, allow at least 1 to 3 min for the temperature to stabilize at the new setting.

Damage to the welder or heating element will not occur from too much air pressure; however, the element can become overheated by too little air pressure. When decreasing the air pressure, never allow the round nut that holds the barrel to the handle of the welder to become too hot to the touch. This is an indication of overheating.

A partial clogging of the dirt screen on the regulator or a fluctuation in the line voltage can also cause overheating or underheating. Watch for these symptoms.

If the threads at the end of the barrel become too tight, clean them with a good high-temperature grease to prevent seizing.

When the welding is finished, disconnect the electric supply and let the air flow through the welder for a few minutes until the barrel is cool to the touch. Then disconnect the air supply.

Complete a copy of the "Student Welding Report" listed in Appendix I or provided by your instructor. ◆

PRACTICE 18-2

Maintenance of Hot-gas Welders

Using a properly set up hot-gas welder, replacement heating element, spanner wrench, safety glasses, gloves, and other required personal protection equipment, you are going to perform routine cleanup and maintenance of a hot-gas welder.

While pushing the end of the barrel against a solid object, hold the handle tightly and push in. The pressure on the barrel will compress the element spring. Use a spanner wrench to loosen the spanner nut. Keep the pressure on the handle and back off the nut all the way by hand.

Hold the barrel and place the complete welder on a bench. Remove the barrel. Gently pull the element out of the handle. At the same time, unwind the cable (which has been spiraled into the handle) until it is completely out of the handle. Grasp the socket at the end of the wire tightly. Rock the element while pulling until the element is disconnected.

To install the new element, reverse the above procedure. Turn the element clockwise (about 1 1/2 turns) while pushing the wire gently back into the handle. This prevents kinking of the wire.

Other more involved servicing procedures are best left to qualified repair technicians. Many manufacturers make it clear that disassembling the welder automatically invalidates the warranty.

Complete a copy of the "Student Welding Report" listed in Appendix I or provided by your instructor. ◆

Airless Plastic Welding

Hot-gas and airless plastic welding use similar welding techniques. The major difference is that the hot-gas method uses 3/16-in.-diameter welding rods; 1/8-in. (3 mm) diameter rods are used with the airless method. This not only provides a quicker rod melt, it also helps eliminate two troublesome problems: panel warpage and excess rod buildup.

When setting up the typical airless welder, the first and most important step is to put the temperature dial at the appropriate setting, depending on the specific thermosetting plastic being worked on. It is crucial that the temperature setting is correct; otherwise, the entire welding operation will be jeopardized. It will normally take about three minutes for the welder to fully warm up. As for the selection of the welding rod, there is another factor to consider besides size—namely, compatibility. Make sure the rod is the same as the damaged plastic or the weld will more than likely not hold. To this end, many airless welder manufacturers provide rod application charts. When the correct rod has been chosen, it is good practice to run a small piece of it through the welder to clean out the tip before beginning.

Welding Temperatures

Each type of plastic has its own welding temperature. The welding temperature can differ somewhat among slight variations in composition of a type of plastic. Referring to **Table 18-5**, which lists some commonly welded plastics and their welding temperatures, is a good start, but you must make a small test weld to judge the correct temperature for the specific plastic you are currently welding.

ABS is easily welded using hot-gas or airless welding. ABS does not have a sharp melting point, but softens gradually after reaching its heat distortion point. Normal welding rod pressure is sufficient for holding an approximate 60° welding angle. Visual observation of the wavy flow pattern along the edges of the deposited rod is the best way to monitor ABS weld quality. The approximate temperatures for welding are 350°F to 400°F (175°C to 200°C), and high-temperature ABS at 500°F to 550°F (260°C to 290°C).

Thermoplastic Code	Welding Temperature °F (°C)
ABS	660 (350)
ABS/PC	660 (350)
PA	750 (400)
PBT	660 (350)
PC	660 (350)
PE hard (HDPE)	570 (300)
PE soft (LDPE)	520 (270)
PP	570 (300)
PP EPDM	570 (300)
PUR Thermoplastic	570/660 (300/350)
PVC hard	570 (300)
PVC soft	660 (350)

TABLE 18-5 Recommended Temperatures for Welding Plastic.

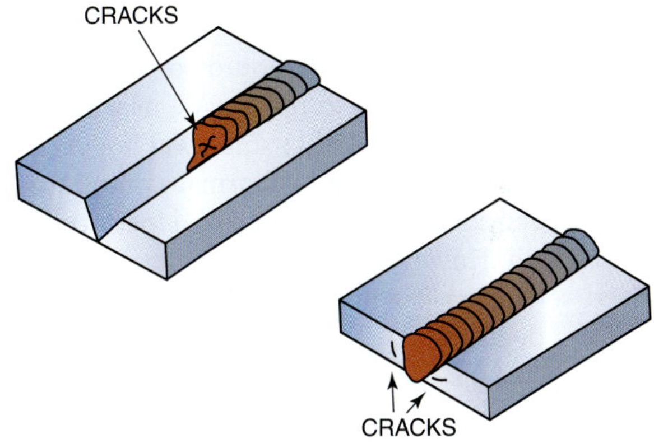

FIGURE 18-13 Examples of stress cracks.

PE plastics form oxide layers at elevated temperatures that are barriers to fusion. For this reason, nitrogen is the recommended gas for welding PE. It also helps achieve the maximum weld strength. Because a very thin coating of oxidized film can adversely affect the final welded product, best results are obtained by welding immediately after cutting the PE, then removing the oxidized film from the welding rod with fine sandpaper or by scraping with a knife.

The rod and base material should be of the same composition, since this affects the weld bond. Use large-sized rods for PE wherever possible, and inspect for poor tolerance and stress. The term *stress cracking*, often included in discussions about PE, refers to cracking or splitting of the material under certain conditions, including chemical reaction, heat, or stress. It results from welding materials of slightly different composition, welding at improper temperatures, and subjecting materials to undue stress or chemical attack, **Figure 18-13**. Correct weld speeds should always be used to avoid stress; maximum PE weld strength is achieved ten hours after the weld is completed.

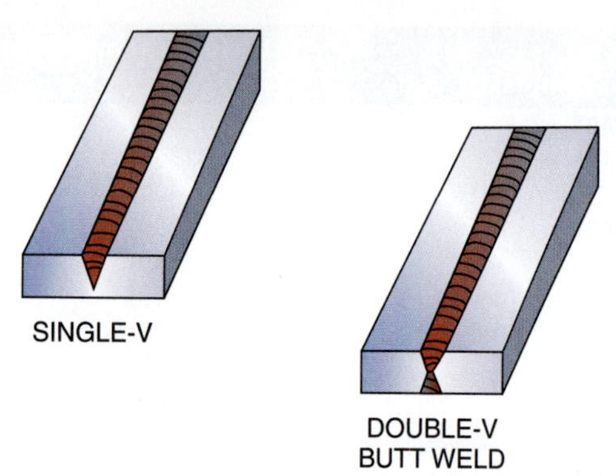

FIGURE 18-14 Single- and double-V butt welds produce strong joints.

Three important factors influence the welding of PVC: the type of PVC used (normal or high impact), the amount of plasticizer used, and the quality of the welding rod. Plasticizers are liquids or compounds added during extrusion to make PVC plastic more flexible. They are also used in some welding rods to improve weld quality. Under normal conditions, a 10% plasticized rod yields improved performance and strength. However, at high temperatures a plasticized rod decreases in strength, but an unplasticized rod does not.

PP requires the use of nitrogen to obtain maximum weld strength. Some splash of molten plastic will occur when welding PP, but this does not affect the weld and can be eliminated by throttling down the airflow.

Plastic Welding

When welding plastic, single- or double-V butt welds, **Figure 18-14**, produce the strongest joints; lap fillet welds are also good. When using a round- or V-shaped filler rod, the damaged area is prepared by slowly grinding, sanding, or shaving the adjoining surfaces with a sharp knife to produce a single or double V. For flat ribbon filler rods, V-grooving is not necessary. Wipe any dust or shavings from the joint with a clean, dry rag. The use of cleaning solvents is generally not recommended because they can soften the edges and cause poor welds.

Hot-gas and Airless Welding Techniques

Welding plastic is not difficult when it is done in a careful and thorough manner. Keep in mind the following general welding tips:

- The welding rod must be compatible with the base material for the strength, hardness, and flexibility of the repair to be the same as that of the part. To this end, the rods are color coded for easy identification.
- Always test a welding rod for compatibility with the base material. To do this, melt the rod onto an unexposed area of the cracked part, let the rod cool, and then try to pull it from the part. If the rod is compatible, it will adhere.
- A too-high temperature will burn or char the surface. Charred plastic will not hold.
- The travel speed along the joint must be correct so the filler plastic has time to bond to the base material.
- The rod must be kept at the correct angle so it fuses to the base plastic correctly.
- Pressure is required to make plastic welds but too much pressure stretches and distorts the weld.
- Pay close attention to the temperature setting of the welder; it must be correct for the type of plastic being welded.
- Never use oxygen or other flammable gases with a plastic welder.
- Never use a plastic welder, hot-gas welding gun, or similar tool in wet or damp areas. Remember, electric shock can kill.
- Become proficient at horizontal welds before attempting the more difficult vertical and overhead types.
- Make welds as large as they need to be. The greater the surface area of a weld, the stronger the bond.
- Before beginning an airless weld, run a small piece of the welding rod through the welder to clean out the torch tip.
- Consult a supplier for the brands of tools and materials that best fit your needs. Always read and follow the manufacturer's instructions carefully.

PRACTICE 18-3

Hot-gas Hand Welding Single V-Groove Joint

Using a properly set up hot-gas welder, two pieces of thermoplastic with matching filler rods, foil tape, safety glasses, gloves, and other required personal protection equipment, you are going to make a single V-groove weld joint using the hand welding technique.

Set the welder to the proper temperature, **Table 18-5**. Wash and clean the part with plastic cleaner. Make sure to remove all residue and dry the plastic. Place the pieces of plastic together and tack tape them, in alignment, with foil tape on the back side of

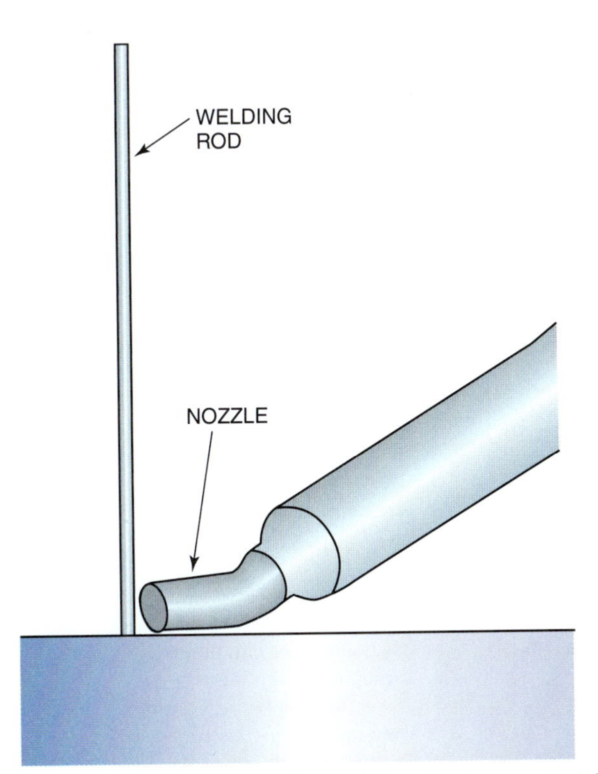

FIGURE 18-15 Keep the nozzle parallel to the base material and the rod at a right angle to the surface.

the joint. Using the welder tip, melt a V-groove in the joint. Select the welding tip best suited to the type and thickness of plastic. Select the proper welding rod for the type of plastic.

Prepare the rod for welding by cutting the end at an angle of approximately 60°. When starting a weld, the tip of the welder should be held about 1/4 in. to 1/2 in. above and parallel to the base material. The filler rod is held at a right angle to the work as shown in **Figure 18-15**, with the cut end of the rod positioned at the beginning of the weld.

Direct the hot gas from the tip alternately at the rod and the base, but concentrating more on the rod. Always keep the filler rod in line with the V while pressing it into the seam. Light pressure (about 3 psi) is sufficient for achieving a good bond. Once the rod begins to stick to the plastic, start to move the torch and use the heat to control the flow. Be careful not to melt or char the base plastic or to overheat the rod. As the welding continues, a small bead should form ahead of the rod along the entire weld joint. A good start is essential, because this is where most weld failures begin. For this reason, starting points on multiple-bead welds should be staggered whenever possible.

Once the weld has been started, the torch should continue to fan from rod to base material. Because the rod now has less bulk, a greater amount of heat must be directed at the base material. Experience will develop the proper technique.

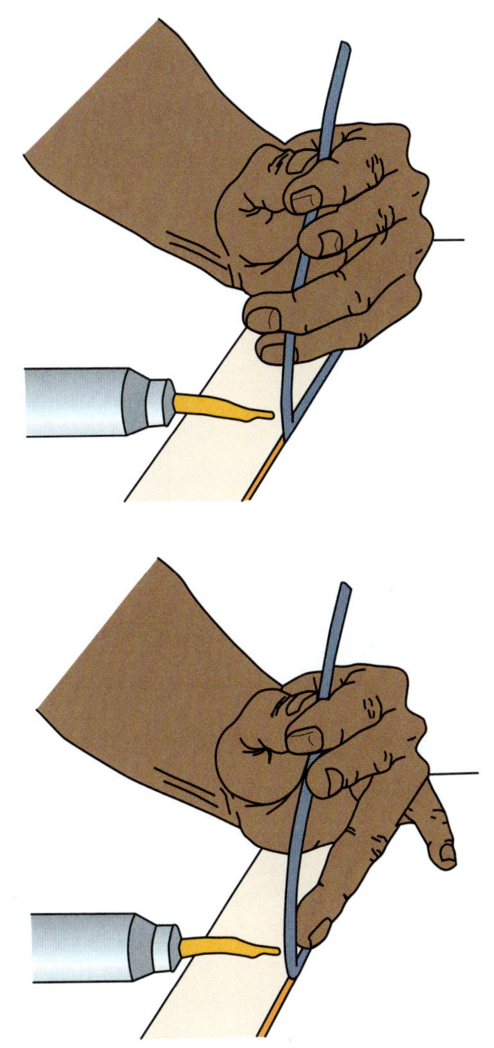

FIGURE 18-16 Methods of respositioning a grip on the rod.

Feed the rod in the weld as the weld progresses. The rod will gradually be used up, making it necessary for you to renew your grip on the rod. Unless this is done carefully, the momentary release of pressure might cause the rod to lift away from the weld and allow air to become trapped under the weld. As a result, the weld would be weakened. To prevent this from occurring, the welder must develop the ability to continuously apply pressure on the rod while repositioning the fingers. This can be done by applying pressure with the third and fourth fingers while moving the thumb and first finger up the rod, **Figure 18-16.** Another way is to hold the rod down into the weld with the third and fourth finger while repositioning the thumb and first finger. The rod is cool enough to do this because only the bottom of it should be heated. However, care should be observed in touching new welds or aiming the torch near the fingers. Allow the weld to cool and cure for about thirty minutes. Inspect the weld for uniformity in size and appearance. **Table 18-6** lists common problems and remedies for plastic

Problem	Cause	Remedy
Porous Weld	1. Porous weld rod 2. Balance of heat on rod 3. Welding too fast 4. Rod too large 5. Improper starts or stops 6. Improper crossing of beads 7. Stretching rod	1. Inspect rod. 2. Use proper fanning motion. 3. Check welding temperature. 4. Weld beads in proper sequence. 5. Cut rod at angle, but cool before releasing. 6. Stagger starts and overlap splices 1/2″.
Poor Penetration	1. Faulty preparation 2. Rod too large 3. Welding too fast 4. Not enough root gap	1. Use a 60-degree bevel. 2. Use small rod at root. 3. Check for flow liners while welding. 4. Use tacking tip or leave 1/32″ root or gap and clamp pieces.
Scorching	1. Temperature too high 2. Welding too slowly 3. Uneven heating 4. Material too cold	1. Increase air flow. 2. Hold constant speed. 3. Use correct fanning motion. 4. Preheat material in cold weather.
Distortion	1. Overheating at joint 2. Welding too slowly 3. Rod too small 4. Improper sequence	1. Allow each bead to cool. 2. Weld at constant speed; use speed tip. 3. Use larger-sized or triangular-shaped rod. 4. Offset pieces before welding. 5. Use double V or backup weld. 6. Backup weld with metal.
Warping	1. Shrinkage of material 2. Overheating 3. Faulty preparation 4. Faulty clamping of parts	1. Preheat material to relieve stress. 2. Weld rapidly—use backup weld. 3. Too much root gap. 4. Clamp parts properly; back up to cool. 5. For multilayer welds, allow time for each bead to cool.
Poor Appearance	1. Uneven pressure 2. Excessive stretching 3. Uneven heating	1. Practice starting, stopping, and finger manipulation on rod. 2. Hold rod at proper angle. 3. Use slow, uniform fanning motion, heating both rod and material (for speed welding: use only moderate pressure, constant speed, keep shoe free of residue).
Stress Cracking	1. Improper welding temperature 2. Undue stress on weld 3. Chemical attack 4. Rod and base material not same composition 5. Oxidation or degradation of weld	1. Use recommended welding temperature. 2. Allow for expansion and contraction. 3. Stay within known chemical resistance and working temperatures of material. 4. Use similar materials and inert gas for welding. 5. Refer to recommended application.

TABLE 18-6 Plastic Welding Troubleshooting Guide.

welding defects. Repeat this process until you can consistently make welds free of defects.

Turn off your hot-surface welder and clean up your work area when you are finished welding.

Complete a copy of the "Student Welding Report" listed in Appendix I or provided by your instructor. ◆

PRACTICE 18-4

Airless Hand Welding Single V-Groove Joint

Using a properly set up airless welder, two pieces of thermoplastic with matching filler rods, foil tape, safety glasses, gloves, and other required personal protection equipment, you are going to make a single V-groove weld joint using the hand welding technique.

Set the welder to the proper temperature, **Table 18-5**. Wash and clean the part with plastic cleaner. Make sure to remove all residue and dry the plastic. Place the pieces of plastic together and tack tape them, in alignment, with foil tape on the back side of the joint.

Using the welder tip melt a V-groove in the joint. Select the welding tip best suited to the type and thickness of plastic. Select the proper welding rod for the type of plastic. Use the same welding techniques as in Practice 18-3, steps 7 through 10.

Allow the weld to cool and cure for about 30 min. Inspect the weld for uniformity in size and appearance. **Table 18-6** lists common problems and remedies for plastic welding defects. Repeat this process until you can consistently make welds free of defects.

Turn off your hot surface welder and clean up your work area when you are finished welding.

Complete a copy of the "Student Welding Report" listed in Appendix I or provided by your instructor. ◆

PRACTICE 18-5

Hot-gas Speed Welding Single V-Groove

Using a properly set up hot-gas welder, two pieces of thermoplastic with matching filler rods, foil tape, safety glasses, gloves, and other required personal protection equipment, you are going to make a single V-groove weld joint using the speed welding torch.

Set the welder to the proper temperature, **Table 18-5**. Wash and clean the part with plastic cleaner. Make sure to remove all residue and dry the plastic. Place the pieces of plastic together and tack tape them, in alignment, with foil tape on the backside of the joint.

Using the welder tip, melt a V-groove in the joint. Select the welding tip best suited to the type and thickness of plastic. Select the proper welding rod for the type of plastic.

Start the weld holding the high-speed torch like a dagger with the hose on the outside of the wrist. Bring the tip over the starting point about 3 in. from the material so the hot gas will not affect the material, **Figure 18-17**. Cut the welding rod at a 60° angle, insert it into the preheating tube, and immediately place the pointed shoe of the tip on the material at the starting point. Hold the welder perpendicular to the material and push the rod through until it stops against the material at the starting point. If necessary, lift the torch slightly to allow the rod to pass under the shoe. Keeping a slight pressure on the rod with the left hand and only the weight of the torch on the shoe, pull the torch slowly toward you. The weld is now started.

CAUTION

Once the weld is started, do not stop. If the forward movement must stop for any reason, pull the tip off the rod immediately; if this is not done, the rod will melt into the tip's feed tube. Clean the feeder foot with a soft wire brush as soon as the welding is completed.

In the first inch or two of travel, the rod should be helped along by pushing it into the tube with slight pressure. Once the weld has been properly started, the torch is brought to a 45° angle; the rod will now feed automatically without further help. As the torch moves along, visual inspection will indicate the quality of the weld being produced.

The angle between the welder and the base material determines the welding rate. Since the preheater hole in the speed tip precedes the shoe, the angle of the welder to the material being welded determines how close the hole is to the base material and how much preheating is being done. For this reason, the torch is held at a 90° angle when starting the weld and at 45° thereafter, **Figure 18-18**. When a visual inspection of the weld indicates a welding rate that is too fast, the torch should be brought back to the 90° angle temporarily to slow down the welding rate, then gradually moved back to the desired angle for proper welding speed. It is important that the welder be held in such a way that the preheater hole and the shoe are always in line with the direction of the weld, so only the material in front of the shoe is preheated. A heat pattern on the base material will indicate the area being preheated. The rod should always be welded in the center of that pattern.

It is important to remember that, once started, speed welding must be maintained at a fairly constant rate of speed. The torch cannot be held still. To stop welding before the rod is used up, bring the torch back past the 90° angle and cut off the rod with

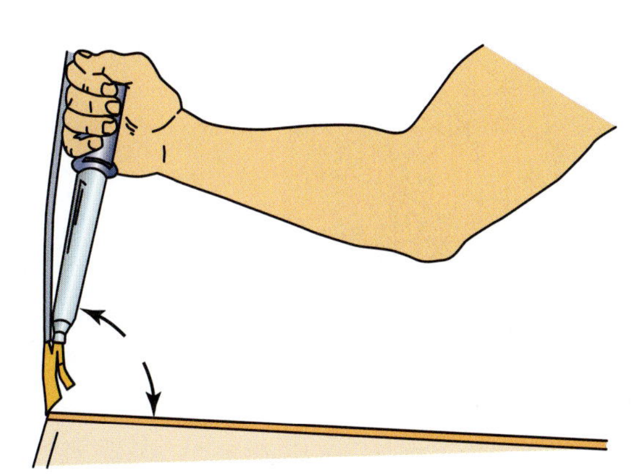

FIGURE 18-17 Starting a speed weld.

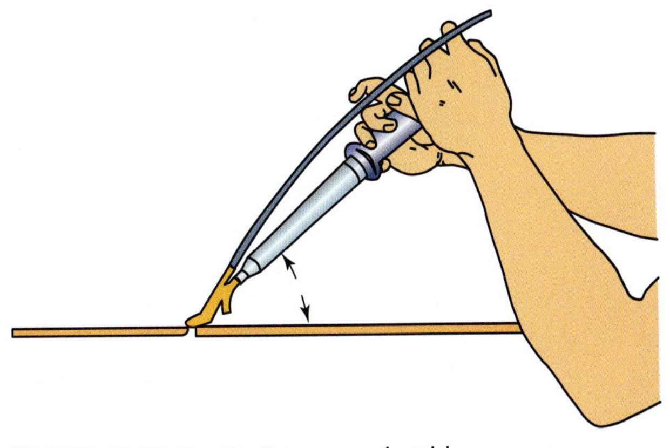

FIGURE 18-18 Continuing a speed weld.

the end of the shoe. This can also be accomplished by pulling the speed tip off the remaining rod. When cutting the rod with the shoe, the remaining rod must be removed promptly from the preheater tube. A rod not removed promptly from the preheater tube will char or melt, clogging the tube and making it necessary for the tube to be cleaned out by inserting a new rod in the tube.

A good speed weld in a V-joint will have a slightly higher crown and more uniformity than the normal hand weld. It should appear smooth and shiny, with a slight bead on each side. For best results and faster welding speed, the shoe on the speed tip should be cleaned occasionally with a wire brush to remove any residue that might cling to it and create drag on the rod.

Inspect the weld for uniformity in size and appearance. **Table 18-6** lists common problems and remedies for plastic welding defects. Repeat this process until you can consistently make welds free of defects.

Turn off your hot surface welder and clean up your work area when you are finished welding.

Complete a copy of the "Student Welding Report" listed in Appendix I or provided by your instructor. ◆

PRACTICE 18-6

Airless Speed Welding Single V-Groove

Using a properly set up airless welder, two pieces of thermoplastic with matching filler rods, foil tape, safety glasses, gloves, and other required personal protection equipment, you are going to make a single V-groove weld joint using the speed welding torch.

Set the welder to the proper temperature, **Table 18-5**. Wash and clean the part with plastic cleaner. Make sure to remove all residue and dry the plastic. Place the pieces of plastic together and tack tape them, in alignment, with foil tape on the back side of the joint.

Using the welder tip, melt a V-groove in the joint. Select the welding tip best suited to the type and thickness of plastic. Select the proper welding rod for the type of plastic.

Use the same welding techniques as in Practice 18-3, steps 7 through 11. Inspect the weld for uniformity in size and appearance. **Table 18-6** lists common problems and remedies for plastic welding defects. Repeat this process until you can consistently make welds free of defects.

Turn off your hot surface welder and clean up your work area when you are finished welding.

Complete a copy of the "Student Welding Report" listed in Appendix I or provided by your instructor. ◆

Double V-Groove Welds

On thick or hard plastics, double V-groove or two-sided welds are recommended to restore the total strength of the part. It is especially important that high-stress parts, such as large spray tanks, **Figure 18-19**, be double side welded.

FIGURE 18-19 450 gallon (1700 l) polyethylene tank air blast sprayer.

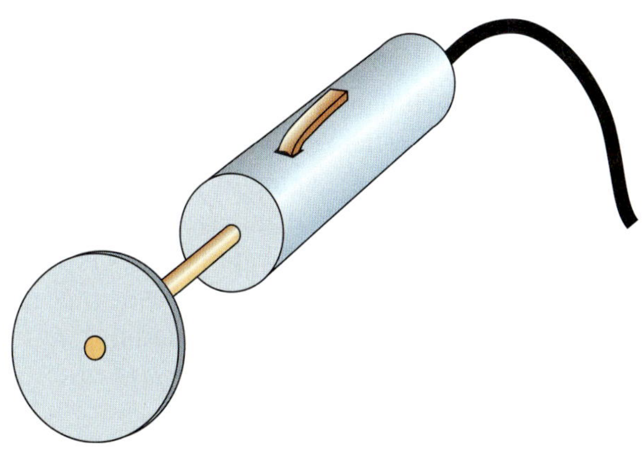

FIGURE 18-20 Die grinder with pad.

PRACTICE 18-7

Hot-Gas or Airless Speed Welding Double V-Groove

Using a properly set up hot-gas or airless welder, two pieces of thermoplastic with matching filler rods, foil tape, safety glasses, gloves, and other required personal protection equipment, you are going to make a double V-groove weld joint using the speed welding torch.

Set the welder to the proper temperature, **Table 18-5**. Wash and clean the part with plastic cleaner. Make sure to remove all residue and dry the plastic. Place the pieces of plastic together and tack tape them, in alignment, with foil tape on the back side of the joint. V-groove 50% of the way through the plastic, using a die grinder, a 1/4-in. (6 mm) drill with a butter bit or grinding pad, a small grinder, or a rotary file, **Figure 18-20**.

The first side weld should be done using the melt-flow method. With the rod inserted in the melt tube, place the flat shoe part of the tip directly over the V-groove and hold it in place until the melted rod begins to flow out around the shoe, **Figure 18-21**.

Let the rod melt out on its own; do not force it. You should feel the rod begin to collapse as it melts. Move the shoe very slowly and crisscross the groove as it fills with melted plastic. Do not move too fast or the welder will not have sufficient time to properly heat the base material and melt the rod.

Quickly cool the weld with a damp sponge or cloth. Remove the tape in preparation for the other side to be repaired. V-groove 50% of the way through the front of the panel and into the first weld. This will enable the two welds to be tied together for a strong repair. Inspect the weld for uniformity in size and appearance. **Table 18-6** lists common problems and remedies for plastic welding defects. Repeat this process until you can consistently make welds free of defects.

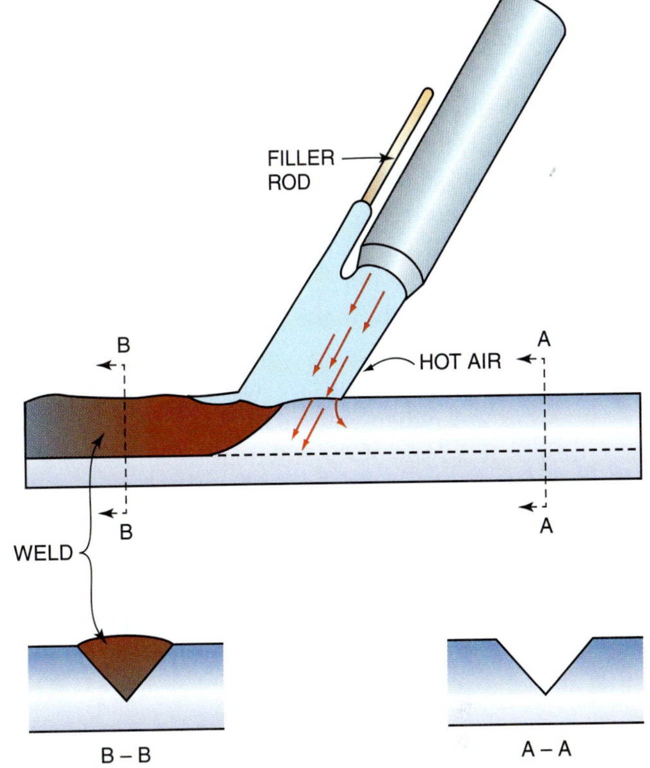

FIGURE 18-21 Using the melt-flow method of polypropylene.

Turn off your hot surface welder and clean up your work area when you are finished welding.

Complete a copy of the "Student Welding Report" listed in Appendix I or provided by your instructor. ◆

Tack Welding

Tack welds can be used for long tears where backup taping is difficult. These small tack welds can be made along the length of the tear to hold the two sides in place while doing the weld. For larger areas, a patch can be made from a piece of plastic and tacked in place.

PRACTICE 18-8

Hot-Gas or Airless Tack Welding

Using a properly set up hot-gas or airless welder, two pieces of thermoplastic with matching filler rods, foil tape, safety glasses, gloves and other required personal protection equipment, you are going to make a double V-groove weld joint using the speed welding torch.

Hold the damaged area in its correct position with clamps. Using a tacking welding tip, fuse the two sides to form a thin hinge weld along the root of the crack. This is especially useful for long cracks because it allows easy adjustment and alignment of the edges.

Start the tacking by drawing the point of the welding tip along the joint. Press the tip in firmly, making sure to contact both sides of the crack; draw the tip smoothly and evenly along the line of the crack.

The point of the tip will fuse both sides in a thin line at the root of the crack. The fused parts will hold the sides in alignment, though they can be separated and re-tacked if adjustment is necessary. Fuse the entire length of the crack. Repeat this process until you can consistently make tack welds to hold the parts in alignment.

Turn off your hot surface welder and clean up your work area when you are finished welding.

Complete a copy of the "Student Welding Report" listed in Appendix I or provided by your instructor ◆

Pipe Welding

Plastic pipe has replaced metal pipe in almost every area of the farm or ranch. Plastic pipe offers significant advantages over metal pipe in a number of areas. The problem of rust or corrosion, which is a constant problem with metal pipe, does not exist with plastic pipes. Plastic pipe is much lighter and easier to handle than metal pipe. Some plastics can withstand being buried or being exposed to direct sunlight without breaking down. Different types of plastic pipes are approved for use with fuels, pesticides, herbicides, acids, or food.

It is important to know the type of plastic pipe and its application before selecting a method of joining. **Table 18-4** lists the common types of plastic and their uses.

Heat Welding of Pipe

In addition to being welded using the basic hot-gas welding techniques, pipe can also be welded using hot surfaces. There are a number of different types of hot surface welding tools for pipe welding; two common types are used in agriculture. One uses a welding head with removable collar and sleeve dies that fit into and around the ends of the pipe and fittings, **Figure 18-22A.** This welding tool produces slip joint-type

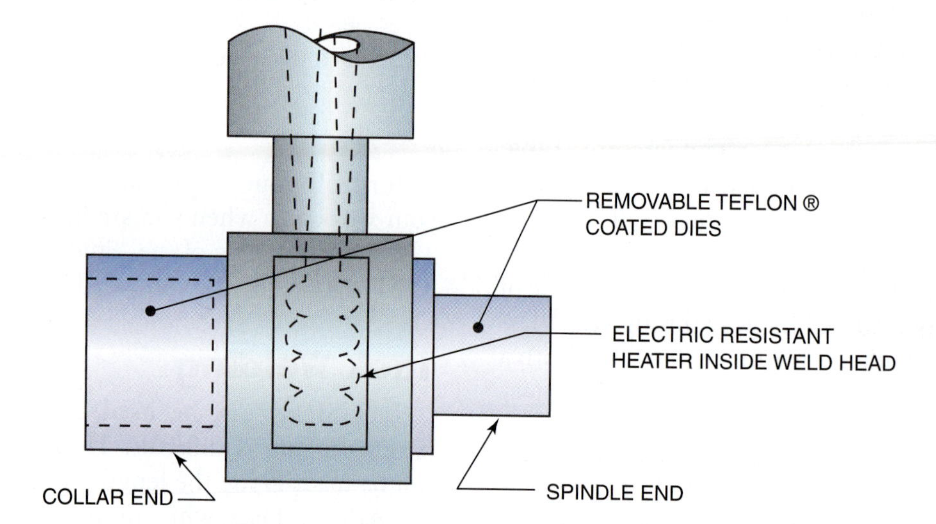

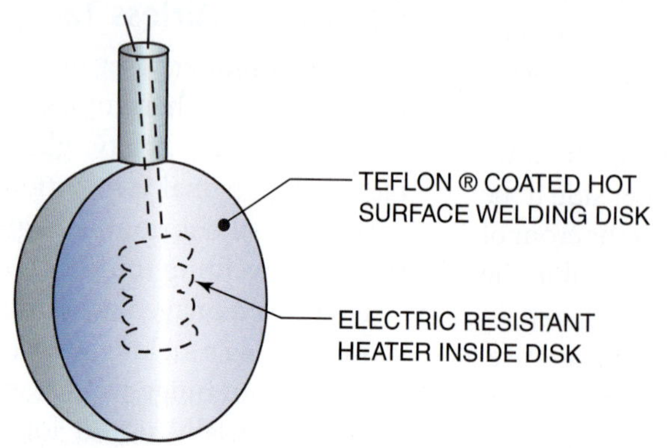

FIGURE 18-22 Hot surface welders.

FIGURE 18-23 Measure PVC 6 in. long; make a mark to be cut. Courtesy of Larry Jeffus.

FIGURE 18-24 Cut the PVC with ratchet scissors plastic tube cutter. Courtesy of Larry Jeffus.

welds and is effective on small diameter pipes. The other welding system uses a flat hot surface, **Figure 18-22B**. This welding tool makes butt joint-type welds. Plastic pipe up to several feet in diameter can be welded with hot surface welding tools.

PRACTICE 18-9

Slip Joint Hot Surface Welding

Using collar and spindle hot surface welding tool setup for welding a pipe sized anywhere between 1/2 in. and 1 in. (13 mm and 25 mm), a 6-in. (150 mm) long piece of thermoplastic pipe correctly sized for the welding tool, a coupling sized for the pipe, a stepping scissors plastic-pipe cutting tool, a knife, a hacksaw, gloves, and other personal protection equipment, you are going to make a slip-type weld in plastic pipe, **Figure 18-23** and **Figure 18-24**.

Cut the pipe to a length of 6 in. (150 mm). The best tool to use is a stepping scissors plastic-pipe cutting tool because, unlike a hacksaw cut, the cut is clean. If a hacksaw is used, the loose debris must be cleaned off with a knife, **Figure 18-25**.

Use a rag with soap and water, if needed, to clean the end of the pipe. Do not sand the pipe because this may remove too much of the surface resulting in a poor fitup. Dry the pipe.

Check the fit of the pipe and couple. The pipe should slide in most of the way freely before it begins to bind. This is called an interference fit, which means that the pipe is slightly larger than the back of the hole in the coupling.

Measure the depth of the coupling and make a pencil mark on the pipe at the correct depth. This will let you know that you have the fitting bottomed in the coupling without driving it in too far.

NOTE: The actual sizes of plastic pipe and fittings may vary among manufacturers for the same size pipe. Check to see if the pipe and fitting are from the same manufacturer or that they fit correctly.

Check the plastic pipe's manufacturer for the temperature and time, refer to the welder's manufacturer's recommended temperature setting and time settings, or refer to **Table 18-5**. Set the weld temperature and allow the required preheat time for the tool to reach welding temperature.

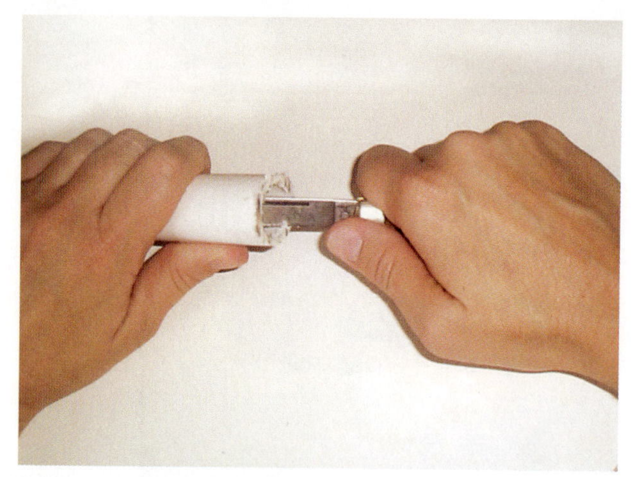

FIGURE 18-25 Clean off debris from hacksaw cut with a pocket knife. Courtesy of Larry Jeffus.

Slide the pipe into the collar end and slide the coupling over the spindle end. Do not place any bending or twisting force on the pipe, tool, or coupling, because as the pipe is warmed it will cause the plastic to deform. If your heater has a timer, set the timer; if not, watch the time so you do not underheat or overheat the plastic.

When it is time to make the weld, remove the pipe and coupling from the tool and quickly drive them together. Do not twist the pipe in the coupling, and do not push the pipe into the coupling farther than the depth of the fitting. Both underinsertion and overinsertion of the warm pipe are imperfections and may be considered defects.

Cool the pipe and use a hacksaw to cut the joint lengthwise down the center of the pipe and coupling. Inspect the pipe to see if the weld is acceptable. Look for the following problems:

a. Underinsertion leaves a gap in the base of the pipe and a weak joint, **Figure 18-26D.**

b. Overinsertion can cause plastic to be ejected from the base of the coupling, thereby restricting flow, **Figure 18-26E.**

c. Underheating will not allow the parts to be completely inserted, resulting in a lack of fusion and underinsertion, **Figure 18-26F.**

d. A correctly assembled weld will have complete joint fusion with a small ring of plastic showing at the joint end, **Figure 18-26G.**

Repeat this process until you can consistently make welds free of defects.

Turn off your hot surface welder and clean up your work area when you are finished welding.

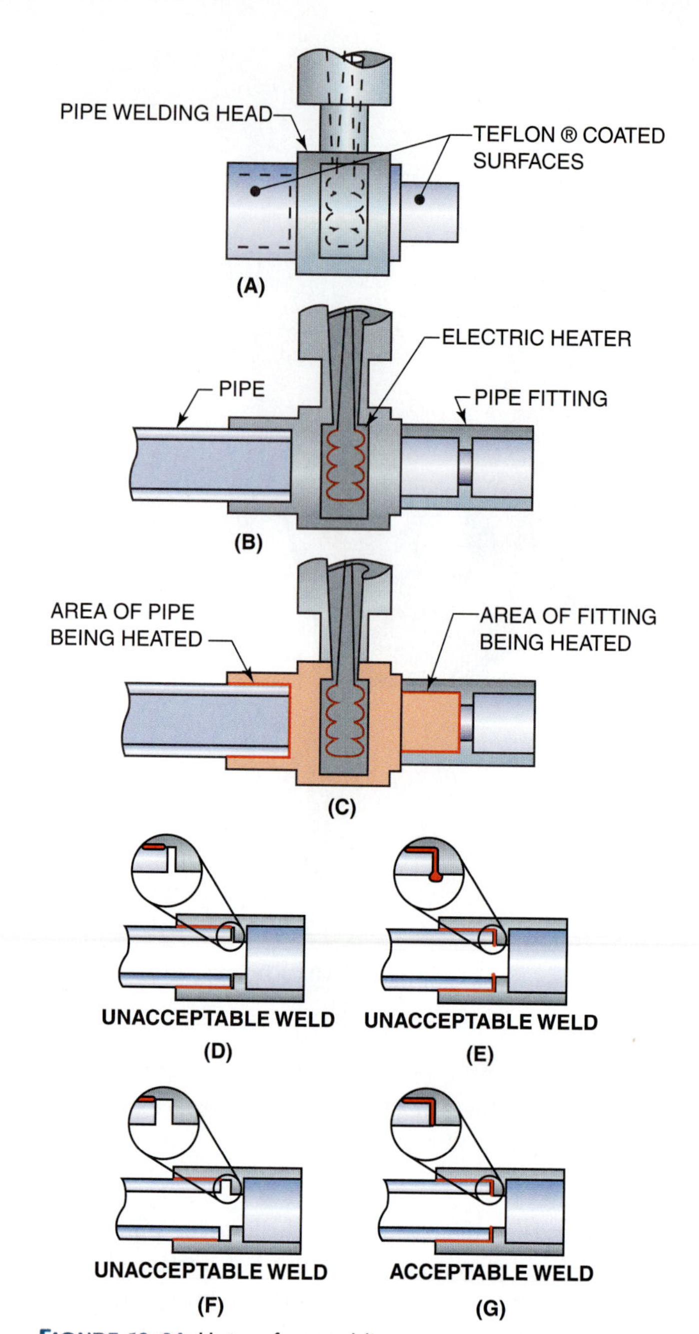

FIGURE 18-26 Hot surface welding.

Complete a copy of the "Student Welding Report" listed in Appendix I or provided by your instructor. ◆

PRACTICE 18-10

Butt Joint Hot Surface Welding

Using an approximately 3-in. (75 mm) diameter flat hot surface welding tool, two pieces of 6-in. (150-mm) long 1/2-in. to 1 in. (13-mm to 25-mm) diameter thermoplastic pipe, a stepping scissors plastic-pipe cutting tool, a hacksaw, gloves, and other personal protection equipment, you are going to make a butt weld in plastic pipe, **Figure 18-27.**

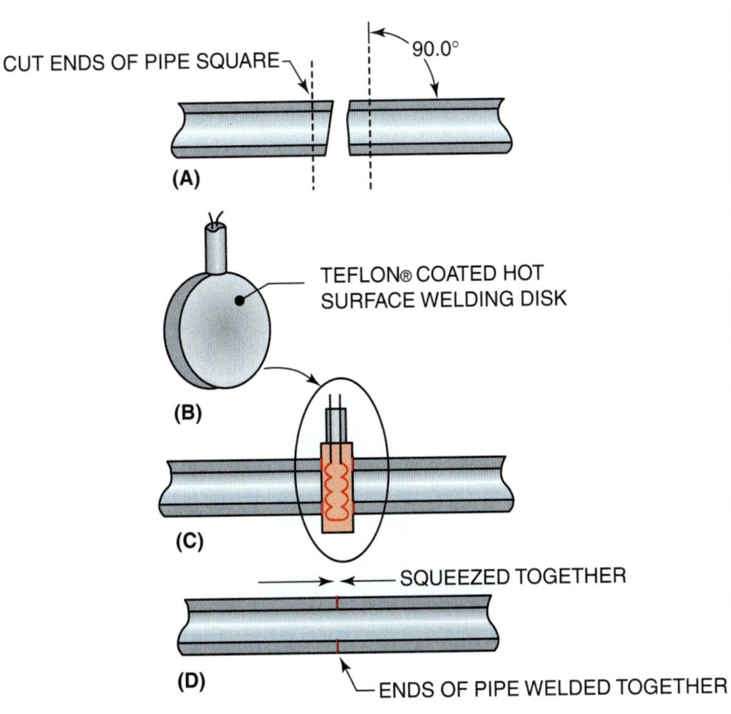

FIGURE 18-27 Disk type hot surface welder.

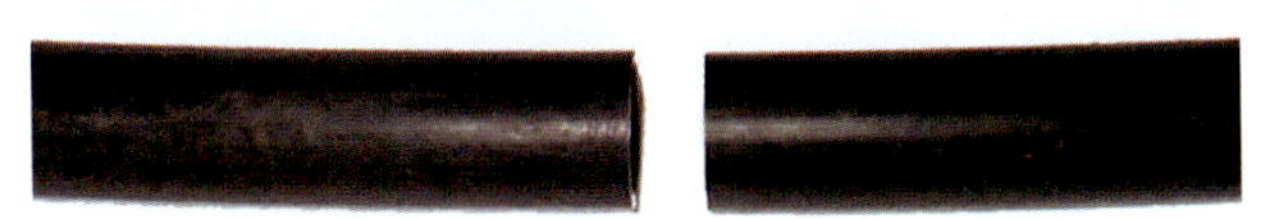

FIGURE 18-28 Cut pipe ends square before welding. Courtesy of Larry Jeffus.

FIGURE 18-29 ABS butt weld. Courtesy of Larry Jeffus.

FIGURE 18-30 PVC butt weld. Courtesy of Larry Jeffus.

Cut two pieces of pipe to a length of 6-in. (150-mm). The end of the pipe must be cut square. On small diameter pipe, 1-in. (25-mm) and less, the best tool to use is a stepping scissors plastic-pipe cutting tool. On pipe larger than 1-in. (25-mm) a special square cutting/dressing tool should be used to cut the pipe. These tools have rotating blades that cut the pipe squarely and dress the end smoothly at the same time, **Figure 18-28.**

Use a rag with soap and water, if needed, to clean the end of the pipe. Do not sand the pipe because this may remove too much of the surface, resulting in a poor fitup. Dry the pipe, both inside and out.

Check the squareness of the cut by placing the pipes end to end to see that they fit together smoothly and straight. Check the plastic pipe's manufacturer for the temperature and time, refer to the welder's manufacturer's recommended temperature setting and time settings, or refer to **Table 18-5.** Set the weld temperature and allow the required time for the tool to reach welding temperature.

Clamp the pipe in the hot surface pipe-welding tool. Make sure that the ends of the pipe can be brought together when the heating tool is removed. Small-diameter pipe ends are brought together with a manual lever once they are warmed. Large-diameter hot surface welding tools use hydraulic rams to bring the ends of the pipe together.

Set the weld timer or watch the time so you do not underheat or overheat the plastic. When it is time to make the weld, remove the hot surface tool and quickly drive the ends together. Hold squeezing pressure on the pipes for approximately 30 seconds or until the pipe weld has cooled. Either too little or too much squeezing force can result in a defective weld.

Cool the pipe and cut out a 1-in. (25-mm) long piece of pipe at the weld joint, **Figure 18-29** and **Figure 18-30.** Inspect the fusion line at the edge of the cut pipe, **Figure 18-31** and **Figure 18-32.** Repeat this process until you can consistently make welds that are free of defects.

Turn off your hot surface welder and clean up your work area when you are finished welding.

Complete a copy of the "Student Welding Report" listed in Appendix I or provided by your instructor. ◆

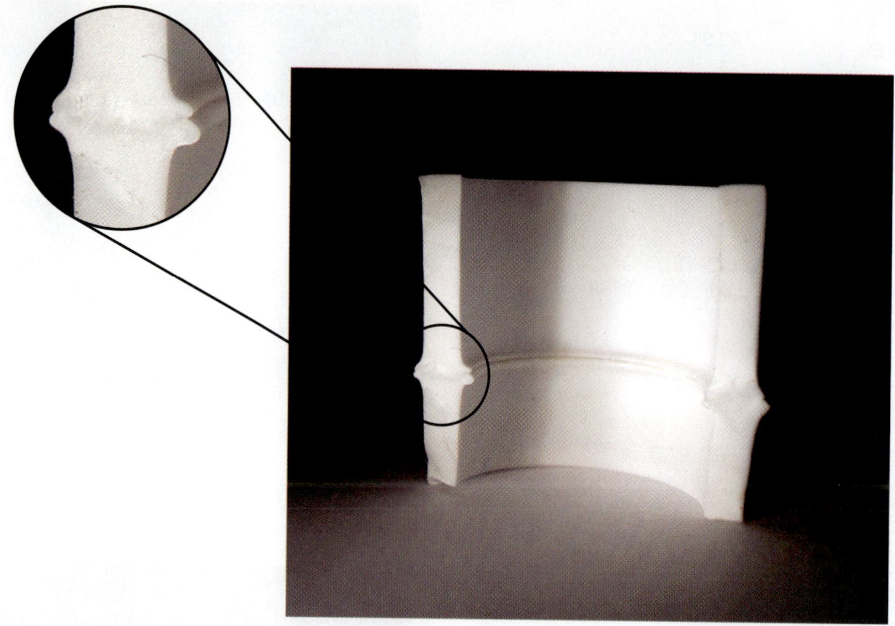

FIGURE 18-31 Section of PVC pipe at weld. Courtesy of Larry Jeffus.

FIGURE 18-32 Section of ABS pipe at weld. Courtesy of Larry Jeffus.

Type of Cement	Type of Pipe		
	ABS	PVC	CPV
ABS	R	R	R
PVC	NR	R	R
CPVC	NR	NR	R

TABLE 18-7 Common Types of Plastic Pipe and the Rcommended (R) and Non-Recommended (NR) Cement for Each Type of Pipe Material.

Solvent Welding

The surface of some plastics can be welded together using a solvent, **Table 18-7**. The solvent is used to soften the surface so that with light pressure the interfacing surfaces fuse together. Solvent welding can be done as a one-step or two-step process. In the one-step method, a solvent/cleaner combination solution is used; in the two-step method, separate cleaner and solvent cement are used. Many codes require the two-step method.

PRACTICE 18-11

Solvent Welding

Using PVC cleaner, PVC solvent, a 6-in. (150-mm) long piece of PVC pipe between 1/2-in. and 1-in. (13-mm and 25-mm) in diameter, a coupling sized for the pipe, PVC solvent cement, a stepping scissors plastic-pipe cutting tool, a hacksaw, a file, a knife, gloves, and other personal protection equipment, you are going to make a slip type weld in plastic pipe, **Figure 18-33**.

Cut the pipe to a length of 6 in. (150 mm). The best tool to use is a stepping scissors plastic-pipe cutting tool because, unlike a hacksaw cut, the cut is clean. If a hacksaw is used, the loose debris must be cleaned off with a knife. Use a file or pocket knife to cut a small chamfer on the outside of the end of the pipe. This chamfer helps the solvent flow behind the pipe as it is inserted, **Figure 18-34**.

Use a rag with soap and water, if needed, to clean the end of the pipe. Do not sand the pipe because this may remove too much of the surface, resulting in a poor fitup. Dry the pipe.

Check the pipe and couple for any damage such as nicks, gouges, or grooves and for fit. Recut the end of the pipe if necessary. The pipe should slide

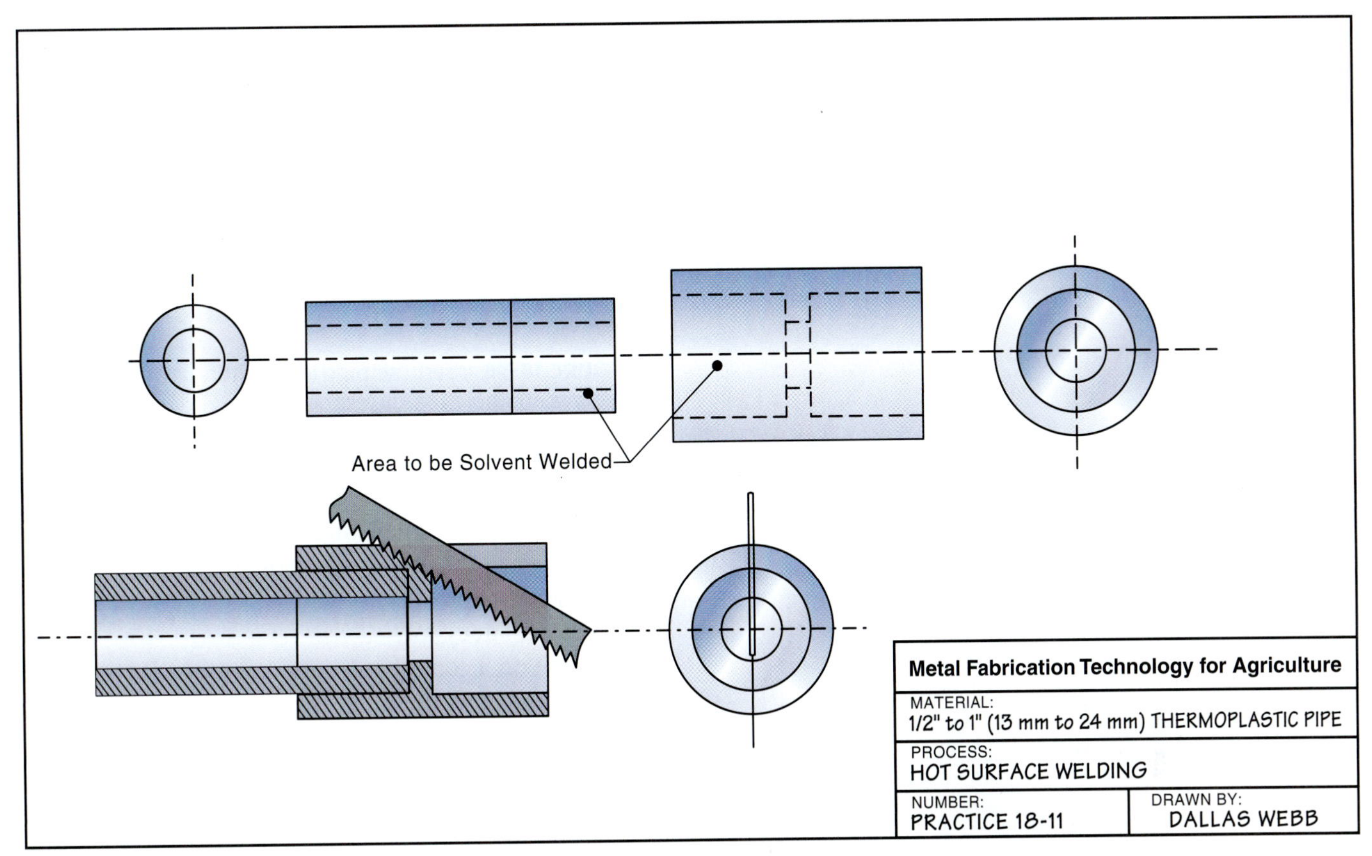

FIGURE 18-33 Practice 18-11.

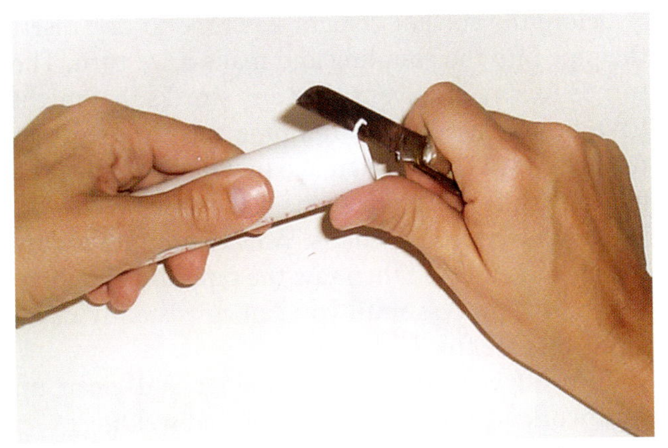

FIGURE 18-34 Chamfering the end of the PCV with a pocket knife. Courtesy of Larry Jeffus.

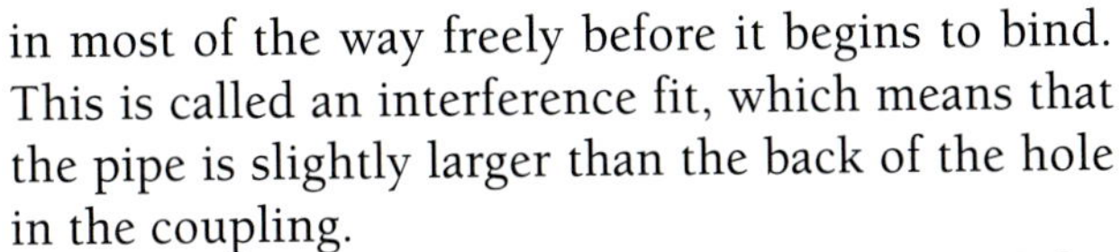

in most of the way freely before it begins to bind. This is called an interference fit, which means that the pipe is slightly larger than the back of the hole in the coupling.

NOTE: The actual sizes of plastic pipe and fittings may vary among manufacturers for the same size pipe. Check to see the pipe and fitting are from the same manufacturer or that they fit correctly.

Using the dauber brush attached to the inside of the cleaner can lid, wipe the cleaner around the outside of the pipe and inside of the coupling. Make several complete wipes with the cleaner brush around the pipe and coupling to completely remove any oil, dirt, and surface oxides. Be sure to clean an area on the pipe just a little further back than the pipe will slide into the coupling. If the pipe is not cleaned or not cleaned properly, the joint may hold now but will fail or leak at some time in the future, **Figure 18-35**.

FIGURE 18-35 Apply PVC cleaner. Courtesy of Larry Jeffus.

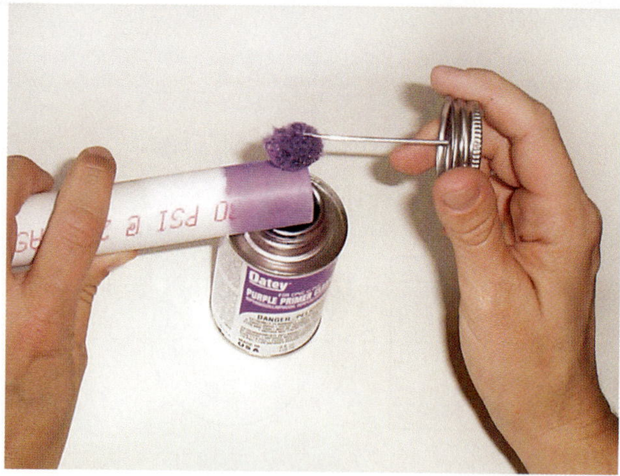

FIGURE 18-36 When the lettering has been removed, the pipe end is clean. Courtesy of Larry Jeffus.

FIGURE 18-37 Apply PVC solvent to the inside of the coupling. Courtesy of Larry Jeffus.

NOTE: Often PVC pipe has a manufacturer's ink stamping along one side of the pipe to identify the pipe type, size, schedule, and manufacturer. Wiping the cleaner around the end of the pipe enough times to remove this ink will ensure that you have totally cleaned the pipe, **Figure 18-36**.

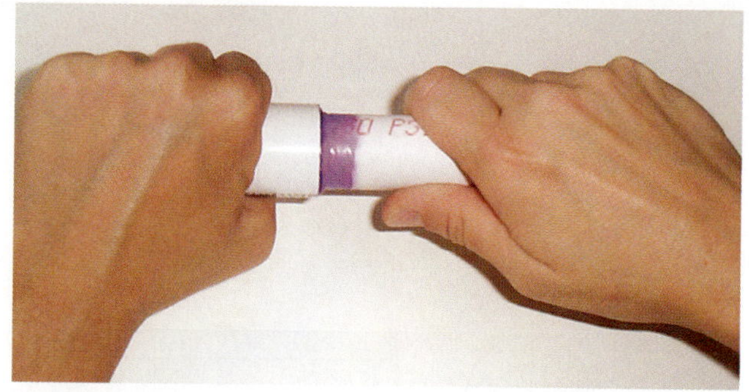

FIGURE 18-38 Insert the pipe in the fitting; rotate 90 degrees and hold for 15 seconds. Courtesy of Larry Jeffus.

Once the pipe is cleaned, do not touch the clean surface or lay it down where it might get dirty. Oil from your hand or dirt will cause the joint to fail sometime later.

Brush on a coat of solvent cement that is matched to the type of pipe and fitting you are using. It is important to use the right type of solvent cement, **Figure 18-37. Table 18-7** shows the various cements for solvent welding plastic pipe. Apply cement to the pipe end first and then apply a thin coat to the inside of the coupling, **Figure 18-38.** Be sure no bare spots are left that are not covered with solvent.

Before the solvent cement has time to dry, insert the pipe into the coupling and make a 1/4 turn. The turn is to spread solvent over any streaks that might have been caused when the pipe was inserted into the coupling. The interference fit will try to slide the joint apart, so hold pressure on the assembled pipe and coupling for 15 seconds. Wait 15 minutes for the solvents to cure and then saw the pipe and coupling. Repeat this process until you can consistently make welds that are free of defects.

Turn off your hot surface welder and clean up your work area when you are finished welding.

Complete a copy of the "Student Welding Report" listed in Appendix I or provided by your instructor. ◆

Review

1. Thermoplastic plastics can ______________.
   A. be cold formed easily
   B. not be melted
   C. be heat welded
   D. dissolve in water
2. Thermoset plastics cannot be remelted.
   A. true
   B. false
3. What is the abbreviation for acrylonitrile/butadiene/styrene plastic?
   A. ABS
   B. PVC
   C. LDPE
   D. PP
4. Which of the following is not a polyethylene plastic?
   A. PET
   B. LLDEP
   C. PE
   D. CPVC
5. Which plastic might be used for a pump-up spray tank?
   A. HDPE
   B. PVC
   C. PP
   D. PB
6. The best way to identify a plastic for welding is to _________.
   A. try to weld on it with a known filler rod
   B. know how it is used
   C. compare its color with a color chart
   D. look at the manufacturer's literature
7. According to the table on recycled plastic identification, which number would be used to identify a low-density polyethylene plastic?
   A. 1
   B. 2
   C. 3
   D. 4
8. Which two plastics have very similar properties?
   A. PB and HDPE
   B. PVC and CPVC
   C. ABS and PP
   D. LLDPE and PB
9. How can plastic be heated or warmed for welding?
   A. with a hot surface
   B. in hot water
   C. with an oxyacetylene torch
   D. with a propane torch
10. Solvent welding can only be done on thermoset plastics.
    A. true
    B. false
11. Never use a hot gas welding gun in a damp area.
    A. true
    B. false
12. What is put on the back of a plastic joint to hold the plastic in place for welding?
    A. electrical tape
    B. C clamps
    C. screws
    D. foil tape

Review continued

13. To start a hand weld the end of the filler rod should be cut at approximately ________.
    A. 60°
    B. 90°
    C. 45°
    D. a square
14. According to **Table 18-5**, what is the welding temperature of PA plastics?
    A. 520°F (270°C)
    B. 570°F (300°C)
    C. 660°F (350°C)
    D. 750°F (400°C)
15. According to **Table 18-6**, how can warping be remedied?
    A. make bigger welds
    B. preheat the material to relieve stress
    C. increase the air flow
    D. use an inert gas
16. What might happen if the weld is stopped and the rod is not removed immediately?
    A. The rod can melt into the tip's feed tube.
    B. The rod will melt off and fall on the floor.
    C. The weld will change colors and crack.
    D. There will be too much smoke to see the weld.
17. When should double V-groove welds be made?
    A. on all HDPE welds
    B. on all LLPE welds
    C. when the plastic is dyed a dark color
    D. on thick plastic parts
18. Tack welds are used on long tears when backup taping is difficult.
    A. true
    B. false
19. What would cause a large ring of plastic inside the coupling?
    A. underheating
    B. overheating
    C. underinsertion
    D. the wrong type of plastic pipe being used with the coupling
20. The PVC cleaner should be wiped around the pipe so that the pipe manufacturer's ink stamp is removed.
    A. true
    B. false

Appendix

- I. Student Welding Report
- II. Conversion of Decimal Inches to Millimeters and Fractional Inches to Decimal Inches and Millimeters
- III. Conversion Factors: U.S. Customary (Standard) Units and Metric Units (SI)
- IV. Abbreviations and Symbols
- V. Metric Conversions Approximations
- VI. Pressure Conversion
- VII. Welding Associations and Organizations
- VIII. Careers

Appendix I

Student Welding Report

I. STUDENT WELDING REPORT

Student Name:______________________________ Date:______________

Instructor:______________________________ Class:______________

Experiment or Practice #:______________________________ Process:______________

Briefly describe task: ____

____

____

____

INSPECTION REPORT

Inspection	Pass/Fail	Inspector's Name	Date
Safety:			
Equip. Setup:			
Equip. Operation:			
Welding	**Pass/Fail**	**Inspector's Name**	**Date**
Accuracy:			
Appearance:			
Overall Rating:			

Comments:

Student Grade:______________ Instructor Initials:______________ Date:______________

Appendix II

Conversion of Decimal Inches to Millimeters and Fractional Inches to Decimal Inches and Millimeters

Inches dec	mm	Inches dec	mm
0.01	0.2540	0.51	12.9540
0.02	0.5080	0.52	13.2080
0.03	0.7620	0.53	13.4620
0.04	1.0160	0.54	13.7160
0.05	1.2700	0.55	13.9700
0.06	1.5240	0.56	14.2240
0.07	1.7780	0.57	14.4780
0.08	2.0320	0.58	14.7320
0.09	2.2860	0.59	14.9860
0.10	2.5400	0.60	15.2400
0.11	2.7940	0.61	15.4940
0.12	3.0480	0.62	15.7480
0.13	3.3020	0.63	16.0020
0.14	3.5560	0.64	16.2560
0.15	3.8100	0.65	16.5100
0.16	4.0640	0.66	16.7640
0.17	4.3180	0.67	17.0180
0.18	4.5720	0.68	17.2720
0.19	4.8260	0.69	17.5260
0.20	5.0800	0.70	17.7800
0.21	5.3340	0.71	18.0340
0.22	5.5880	0.72	18.2880
0.23	5.8420	0.73	18.5420
0.24	6.0960	0.74	18.7960
0.25	6.3500	0.75	19.0500
0.26	6.6040	0.76	19.3040
0.27	6.8580	0.77	19.5580
0.28	7.1120	0.78	19.8120
0.29	7.3660	0.79	20.0660
0.30	7.6200	0.80	20.3200
0.31	7.8740	0.81	20.5740
0.32	8.1280	0.82	20.8280
0.33	8.3820	0.83	21.0820
0.34	8.6360	0.84	21.3360
0.35	8.8900	0.85	21.5900
0.36	9.1440	0.86	21.8440
0.37	9.3980	0.87	22.0980
0.38	9.6520	0.88	22.3520
0.39	9.9060	0.89	22.6060
0.40	10.1600	0.90	22.8600
0.41	10.4140	0.91	23.1140
0.42	10.6680	0.92	23.3680
0.43	10.9220	0.93	23.6220
0.44	11.1760	0.94	23.8760
0.45	11.4300	0.95	24.1300
0.46	11.6840	0.96	24.3840
0.47	11.9380	0.97	24.6380
0.48	12.1920	0.98	24.8920
0.49	12.4460	0.99	25.1460
0.50	12.7000	1.00	25.4000

Inches		mm	Inches		mm
frac	dec		frac	dec	
1/64	0.015625	0.3969	33/64	0.515625	13.0969
1/32	0.031250	0.7938	17/32	0.531250	13.4938
3/64	0.046875	1.1906	35/64	0.546875	13.8906
1/16	0.062500	1.5875	9/16	0.562500	14.2875
5/64	0.078125	1.9844	37/64	0.578125	14.6844
3/32	0.093750	2.3812	19/32	0.593750	15.0812
7/64	0.109375	2.7781	39/64	0.609375	15.4781
1/8	0.125000	3.1750	5/8	0.625000	15.8750
9/64	0.140625	3.5719	41/64	0.640625	16.2719
5/32	0.156250	3.9688	21/32	0.656250	16.6688
11/64	0.171875	4.3656	43/64	0.671875	17.0656
3/16	0.187500	4.7625	11/16	0.687500	17.4625
13/64	0.203125	5.1594	45/64	0.703125	17.8594
7/32	0.218750	5.5562	23/32	0.718750	18.2562
15/64	0.234375	5.9531	47/64	0.734375	18.6531
1/4	0.250000	6.3500	3/4	0.750000	19.0500
17/64	0.265625	6.7469	49/64	0.765625	19.4469
9/32	0.281250	7.1438	25/32	0.781250	19.8437
19/64	0.296875	7.5406	51/64	0.796875	20.2406
5/16	0.312500	7.9375	13/16	0.812500	20.6375
21/64	0.328125	8.3344	53/64	0.828125	21.0344
11/32	0.343750	8.7312	27/32	0.843750	21.4312
23/64	0.359375	9.1281	55/64	0.859375	21.8281
3/8	0.375000	9.5250	7/8	0.875000	22.2250
25/64	0.390625	9.9219	57/64	0.890625	22.6219
13/32	0.406250	10.3188	29/32	0.906250	23.0188
27/64	0.421875	10.7156	59/64	0.921875	23.4156
7/16	0.437500	11.1125	15/16	0.937500	23.8125
29/64	0.453125	11.5094	61/64	0.953125	24.2094
15/32	0.468750	11.9062	31/32	0.968750	24.6062
31/64	0.484375	12.3031	62/64	0.984375	25.0031
1/2	0.500000	12.7000	1	1.000000	25.4000

For converting decimal-inches to "thousandths," move decimal point in both columns to left.

Appendix III

Conversion Factors: U.S. Customary (Standard) Units and Metric Units (SI)

TEMPERATURE

Units

°F (each 1° change)	= 0.555°C (change)
°C (each 1° change)	= 1.8°F (change)
32°F (ice freezing)	= 0°Celsius
212°F (boiling water)	= 100°Celsius
–460°F (absolute zero)	= 0° Rankine
–273°C (absolute zero)	= 0° Kelvin

Conversions

°F to °C	_______ °F – 32	=	_______ × .555	=	_______ °C	
°C to °F	_______ °C × 1.8	=	_______ + 32	=	_______ °F	

LINEAR MEASUREMENT

Units

1 inch	= 25.4 millimeters
1 inch	= 2.54 centimeters
1 millimeter	= 0.0394 inch
1 centimeter	= 0.3937 inch
12 inches	= 1 foot
3 feet	= 1 yard
5280 feet	= 1 mile
10 millimeters	= 1 centimeter
10 centimeters	= 1 decimeter
10 decimeters	= 1 meter
1,000 meters	= 1 kilometer

Conversions

in. to mm	_______	in.	× 25.4	=	_______	mm
in. to cm	_______	in.	× 2.54	=	_______	cm
ft to mm	_______	ft	× 304.8	=	_______	mm
ft to m	_______	ft	× 0.3048	=	_______	m
mm to in.	_______	mm	× 0.0394	=	_______	in.
cm to in.	_______	cm	× 0.3937	=	_______	in.
mm to ft	_______	mm	× 0.00328	=	_______	ft
m to ft	_______	m	× .328	=	_______	ft

AREA MEASUREMENT

Units

1 sq in.	= 0.0069 sq ft
1 sq ft	= 144 sq in.
1 sq ft	= 0.111 sq yd
1 sq yd	= 9 sq ft
1 sq in.	= 645.16 sq mm
1 sq mm	= 0.00155 sq in.
1 sq cm	= 100 sq mm
1 sq m	= 1,000 sq cm

Conversions

sq in. to sq mm	_______	sq in. × 645.16	=	_______	sq mm
sq mm to sq in.	_______	sq mm × 0.00155	=	_______	sq in.

VOLUME MEASUREMENT

Units

1 cu in.	= 0.000578 cu ft
1 cu ft	= 1728 cu in.
1 cu ft	= 0.03704 cu yd
1 cu ft	= 28.32 L
1 cu ft	= 7.48 gal (U.S.)
1 gal (U.S.)	= 3.737 L
1 cu yd	= 27 cu ft
1 gal	= 0.1336 cu ft
1 cu in.	= 16.39 cu cm
1 L	= 1,000 cu cm

Appendix III

Conversion Factors: U.S. Customary (Standard) Units and Metric Units (SI) (continued)

1 L	= 61.02 cu in.
1 L	= 0.03531 cu ft
1 L	= 0.2642 gal (U.S.)
1 cu yd	= 0.769 cu m
1 cu m	= 1.3 cu yd

Conversions

cu in. to L	______ cu in.	× 0.01638	=	______ L
L to cu in.	______ L	× 61.02	=	______ cu in.
cu ft to L	______ cu ft	× 28.32	=	______ L
L to cu ft	______ L	× 0.03531	=	______ cu ft
L to gal	______ L	× 0.2642	=	______ gal
gal to L	______ gal	× 3.737	=	______ L

WEIGHT (MASS) MEASUREMENT

Units

1 oz	= 0.0625 lb
1 lb	= 16 oz
1 oz	= 28.35 g
1 g	= 0.03527 oz
1 lb	= 0.0005 ton
1 ton	= 2,000 lb
1 oz	= 0.283 kg
1 lb	= 0.4535 kg
1 kg	= 35.27 oz
1 kg	= 2.205 lb
1 kg	= 1,000 g

Conversions

lb to kg	______ lb	× 0.4535	=	______ kg
kg to lb	______ kg	× 2.205	=	______ lb
oz to g	______ oz	× 0.03527	=	______ g
g to oz	______ g	× 28.35	=	______ oz

PRESSURE and FORCE MEASUREMENTS

Units

1 psig	= 6.8948 kPa
1 kPa	= 0.145 psig
1 psig	= 0.000703 kg/sq mm
1 kg/sq mm	= 6894 psig
1 lb (force)	= 4.448 N
1 N (force)	= 0.2248 lb

Conversions

psig to kPa	______ psig	× 6.8948	=	______ kPa
kPa to psig	______ kPa	× 0.145	=	______ psig
lb to N	______ lb	× 4.448	=	______ N
N to lb	______ N	× 0.2248	=	______ psig

VELOCITY MEASUREMENTS

Units

1 in./sec	= 0.0833 ft/sec
1 ft/sec	= 12 in/sec
1 ft/min	= 720 in./sec
1 in./sec	= 0.4233 mm/sec
1 mm/sec	= 2.362 in./sec
1 cfm	= 0.4719 L/min
1 L/min	= 2.119 cfm

Conversions

ft/min to in./sec	______ ft/min	× 720	=	______ in./sec
in./min to mm/sec	______ in./min	× 0.4233	=	______ mm/sec
mm/sec to in./min	______ mm/sec	× 2.362	=	______ in./min
cfm to L/min	______ cfm	× 0.4719	=	______ L/min
L/min to cfm	______ L/min	× 2.119	=	______ cfm

Appendix IV

Abbreviations and Symbols

U.S. Customary (Standard) Units

°F	= degrees Fahrenheit
°R	= degrees Rankine = degrees absolute F
lb	= pound
psi	= pounds per square inch = lb per sq in.
psia	= pounds per square inch absolute = psi + atmospheric pressure
in.	= inches = i = "
ft	= foot or feet = f = '
sq in.	= square inch = $in.^2$
sq ft	= square foot = ft^2
cu in.	= cubic inch = $in.^3$
cu ft	= cubic foot = ft^3
ft-lb	= foot-pound
ton	= ton of refrigeration effect
qt	= quart

Metric Units (SI)

°C	= degrees Celsius
°K	= Kelvin
mm	= millimeter
cm	= centimeter
cm^2	= centimeter squared
cm^3	= centimeter cubed
dm^2	= decimeter
dm^3	= decimeter squared
dm	= decimeter cubed
m	= meter
m^2	= meter squared
m^3	= meter cubed
L	= liter
g	= gram
kg	= kilogram
J	= joule
kJ	= kilojoule
N	= newton
Pa	= pascal
kPa	= kilopascal
W	= watt
kW	= kilowatt
MW	= megawatt

Miscellaneous Abbreviations

P	= pressure
h	= hours
sec	= seconds
D	= diameter
r	= radius of circle
A	= area
π	= 3.1416 (a constant used in determining the area of a circle)
V	= volume
∞	= infinity

Appendix V

Metric Conversions Approximations

Metric Conversions Approximations
1/4 inch = 6 mm
1/2 inch = 13 mm
3/4 inch = 18 mm
1 inch = 25 mm
2 inches = 50 mm
1/2 gal = 2 L
1 gal = 4 L
1 lb = 1/2 K
2 lb = 1 K
1 psig = 7 kPa
1°F = 2°C

Appendix VI

Pressure Conversion

psi	kPa	psi	kPa
1	7 (6.9)	100	690 (689)
2	14 (13.7)	110	760 (758)
3	20 (20.6)	120	820 (827)
4	30 (27.5)	130	900 (896)
5	35 (34.4)	140	970 (965)
6	40 (41.3)	150	1030 (1034)
7	50 (48.2)	160	1100 (1103)
8	55 (55.1)	170	1170 (1172)
9	60 (62.0)	180	1240 (1241)
10	70 (69.9)	190	1310 (1310)
15	100 (103)	200	1380 (1379)
20	140 (137)	225	1550 (1551)
25	170 (172)	250	1720 (1723)
30	200 (206)	275	1900 (1896)
35	240 (241)	300	2070 (2068)
40	280 (275)	325	2240 (2240)
45	310 (310)	350	2410 (2413)
50	340 (344)	375	2590 (2585)
55	380 (379)	400	2760 (2757)
60	410 (413)	450	3100 (3102)
65	450 (448)	500	3450 (3447)
70	480 (482)	550	3790 (3792)
75	520 (517)	600	4140 (4136)
80	550 (551)	650	4480 (4481)
85	590 (586)	700	4830 (4826)
90	620 (620)	750	5170 (5171)
95	660 (655)	800	5520 (5515)
		850	5860 (5860)
		900	6210 (6205)
		950	6550 (6550)
		1000	6890 (6894)

Pounds per square inch (psi) converted to kilopascals (kPa). One psi equals 6.8948 kPa. In most applications the conversion from standard units of pressure to SI units can be rounded to an even number. The number in parentheses is the value before it is rounded off.

Appendix VII

Welding Associations and Organizations

AGA
American Gas Association
400 N. Capitol Street NW
Washington, DC 20001-1511

AISC
American Institute of Steel Construction
61 East Wacker Drive
Chicago, IL 60601-2604

AISI
American Iron and Steel Institute
1101 17th Street, NW
Washington, DC 20036-4708

ANSI
American National Standards Institute
25 West 42nd Street
New York, NY 10036-8007

ASME
American Society of Mechanical Engineers
3 Park Avenue
New York, NY 10016-5990

AWI
American Welding Institute
10628 Dutchtown Road
Knoxville, TN 37932-3205

AWS
American Welding Society
550 NW 42nd Avenue
Miami, FL 33126-5699

CWB
Canadian Welding Bureau
7250 West Credit Avenue
Mississauga, Ont. L5N 5N1
Canada

EWI
Edison Welding Institute
1250 Arthur E. Adams Drive
Columbus, OH 43221-3585

FFA
Future Farmers of America
P.O. Box 68960
6060 FFA Drive
Indianapolis, IN 46268-0960

IIW
International Institute of Welding
550 NW LeJeune Road
Miami, FL 33126-5699

National 4-H Council
7100 Connecticut Avenue
Chevy Chase, MD 20815

NSC
National Safety Council
1121 Spring Lake Business Park
Itasca, IL 60143-3201

TWI
The Welding Institute
Granta Park
Great Abington, Cambridge CB1 6AL
United Kingdom

USDA
United States Department of Agriculture
Room 9-E Jamie Whitten Bldg.
1400 Independence Avenue SW
Washington, DC 20250

Appendix VIII

Careers

Getting and Keeping a Good Welding Job

First impressions are important. It's easier than you think to make a positive one. Before going to a job interview, learn as much as you can about the company, its history, and its products to demonstrate your eagerness to become part of its employee team. A good place to start is by visiting the company's website and reading any literature, like catalogs, that it may have available.

The day of the interview, arrive about 10 minutes early. Dress appropriately. This means a little better than you would for the job. Remember, you're not dressing for work; you're dressing for the interview. Be prepared to demonstrate your welding skills. Be prepared to fill out all forms by knowing the pertinent information that may appear on a job application, such as your Social Security number.

Once you've secured employment, your attendance and punctuality are important, because your employer schedules work based on his or her expectations that you will be present, ready, and willing to work. If you know you are going to be absent or late to work, call your employer and ask permission. Don't just show up late or not show up. Every job should be done in a timely manner and performed with quality.

Most employers have additional expectations of their employees beyond their ability to competently perform welding skills. Appropriate work habits will make you a more valuable employee and increase your chances for advancement. You can show your employer that you are an exemplary employee by your willingness to learn new things and take advantage of any educational opportunities. Show initiative by staying busy and by being productive without your boss having to constantly tell you what to do next. Show enthusiasm when approaching new job tasks, and be eager to learn new skills.

Maintaining your work area, including tools, equipment, and supplies, in a neat, orderly, and clean manner will both improve safety and demonstrate a sense of respect for the value of your employer's property. This includes cleaning and returning borrowed tools to their proper place when the job is finished. Accidents may occur from time to time in the workplace, and equipment, tools, or supplies may be damaged or wasted. If this happens, taking responsibility for your involvement will demonstrate your integrity and honesty.

Teamwork is equally important, and your ability to get along with others will enhance your productivity and improve your overall working environment by reducing job stress and interpersonal conflicts. Working well with others and respecting their property is an essential part of teamwork. Horseplay and practical jokes, as well as profanity, have no place in the workplace. It is important to show good citizenship by having an open and receptive attitude to your coworkers' differences or special needs and by showing them proper respect.

All employees hope that their jobs will go on uninterrupted for many years; however, that may not be the case. When it comes time for you to leave, for whatever reason, such a new job or a layoff, do so professionally. Someday you may need to use your previous employer or former team members as professional references, or you might find it necessary to come back and ask for another job. Leaving a job with hard feelings or bitter words can create problems for any future employment. Instead, thank your employer for the opportunities you've had working with the company and move on to your next challenge. This may even be an occasion to assess your skills and start your own welding business.

Nichole Spearman

Agricultural Welding Instructor, Yamhill-Carlton High School, Yamhill, OR

Although I am an agricultural teacher, I did not grow up on a farm. I first discovered my interest in agriculture when a friend "roped" me into helping her at a sheep show. I learned how to weld while taking agriculture classes during all four years in high school. I improved my basic welding techniques at Oregon State University in Corvallis, OR, where I earned a Bachelor of Science degree in General Agriculture and a Masters degree in Teaching with an emphasis in Agricultural Education. I received distinction by being the first female assistant for small engines.

My favorite welding process is arc welding, but I am starting to enjoy MIG and TIG. Now I'm looking forward to improving my welding fabrication skills. I have been working mostly on small projects and repairs, such as welding tables back together and repairing trailers, using basic welding equipment. My students have shown an interest in building their own equipment, such as wood splitters, livestock panels, crib feeders, and small trailers.

People seem surprised when they find out that I am an agricultural teacher at Yamhill-Carlton High school, but I have noticed a growing number of female Ag teachers. Most of my students have never welded before, and I am looking forward to the opportunity to help them learn agricultural fabrication welding.

Randy Glover

Ranch Manager and Entrepreneur, Triple M Cattle Co., Grenville, NM

I learned how to weld in the vocational agricultural program in high school in Fredericksburg, TX in 1982. I use welding for ranch fabrication and repair, such as building pens for cattle, trailers, barn construction, repairing equipment–anything a large ranch requires. Having welding skills on my resume was a real plus in looking for ranch work. I manage a ranch 40 miles from the closest town, and it is invaluable to me and the ranch owner not to have to contract out welding work, which can be very costly and time consuming.

I also have a sideline business that I developed over a number of years using my welding skills to make custom silver-plated belt buckles, tack, and saddle hardware. This has now grown into a substantial source of extra income for my family, and I find time in the evenings and on snowy days to devote to this business. I market my work through a website, a saddle shop, and at occasional Western shows. The business is nice because it has no overhead, provides a little extra income, and allows me to be creative. Whether at the ranch or working on my evening business, I find welding really rewarding.

—Compiled by Marilyn Burris

Bilingual Glossary

The terms and definitions in this glossary are extracted from the American Welding Society publication AWS A3.0-80 Welding Terms and Definitions. The terms with an asterisk (*) are from a source other than the American Welding Society. Note: The English term and definition are given first, followed by the same term and definition in Spanish.

A

acceptable criteria. Agreed-upon standards that must be satisfactorily met.
criterios aceptables. Las normas sobre las que se ha llegado a un acuerdo y que deben cumplirse en forma satisfactoria.

acceptable weld. A weld that meets all the requirements and the acceptance criteria prescribed by welding specifications.
soldadura aceptable. Una soldadura que satisface los requisitos y el criterio aceptable prescribida por las especificaciónes de la soldadura.

***acetone.** A fragrant liquid chemical used in acetylene cylinders. The cylinder is filled with a porous material and acetone is then added to fill. Acetylene is then added and absorbed by the acetone, which can absorb up to 28 times its own volume of the gas.
acetona. Un liquido fragante químico que se usa en los cilindros del acetileno. El cilindro se llena de un material poroso y luego se le agrega la acetona hasta que se llene. El acetileno es absorbido por la acetona, la cual puede absorber 28 veces el propio volumen del gas.

***acetylene.** A fuel gas used for welding and cutting. It is produced as a result of the chemical reaction between calcium carbide and water. The chemical formula for acetylene is C_2H_2. It is colorless, is lighter than air, and has a strong garlic-like smell. Acetylene is unstable above pressures of 15 psig (1.05 kg/cm^2 g). When burned in the presence of oxygen, acetylene produces one of the highest flame temperatures available.
acetileno. Un gas combustible que se usa para soldar y cortar. Es producido a consecuencia de una reacción química de agua y calcio y carburo. La fórmula química para el acetileno es C_2H_2. No tiene color, es más ligero que el aire, y tiene un olor fuerte como a ajo. El acetileno es inestable en presiones más altas de 15 psig (1.05 kg/cm^2 g). Cuando se quema en presencia del oxígeno, el acetileno produce una de las llamas con una temperatura más alta que la que se utiliza.

acid copper chromate. A wood preservative.
cromato de cobre acido. Preservativo de la madera.

actual throat. See **throat of a fillet weld.**
garganta actual. Vea **garganta de soldadura filete.**

***adaptable.** Capable of making self-directed corrections; in a robot, this is often accomplished with visual, force, or tactile sensors.
adaptable. Capaz de hacer correcciones por instrucción propia de un robot, esto se lleva a cabo con sensores tangibles visuales, o de fuerza.

agriculture. Enterprises involving the production of plants and animals, along with supplies, services, mechanics, products, processing, and marketing related to those enterprises.
agricultura. Empresas que compreden la producción de plantas y animales, juntos con los articulos, servicios, mecanismos, productos, el elaborar, y la venta relativos a esas empresas.

air acetylene welding (AAW). An oxyfuel gas welding process that uses an air-acetylene flame. The process is used without the application of pressure. This is an obsolete or seldom-used process.
soldadura de aire acetileno. Un proceso de soldar con gas (oxi/combustible) que usa aire-acetileno sin aplicarse presión. Un proceso anticuado que es una rareza.

air carbon arc cutting (CAC-A). A carbon arc cutting process variation that removes molten metal with a jet of air.
arco de carbón con aire. Un proceso de cortar con arco de carbón variante que quita el metal derretido con un chorro de aire.

air compressor. A pump that increases pressure on air.
compresor de aire. Bomba que aumenta la presion sobre el aire.

air-dried lumber. Sawed lumber separated with wooden strips and protected from rain and snow for six months or more.
madera de construccion secado al aire libre. Madera aserrada separada con listones y protegida de la lluvia y la nieve por seis meses or mas.

Allen screw. A screw with a six-sided hole in the head.
tornillo de cabesa allen. Un tornillo cuyo cabeza consiste en agujero hexagonal.

***alloy.** A metal with one or more elements added to it, resulting in a significant change in the metal's properties.
aleación. Un metal en que se le agrega uno o más elementos resultando en un cambio significante en las propiedades del metal.

***alloying elements.** Elements in the flux that mix with the filler metal and become part of the weld metal. Major alloying elements are molydenum, nickel, chromium, manganese, and vanadium.
elementos de mezcla. Elementos en el flujo que se mezclan con el metal para rellenar y formar parte del metal soldado. Los elementos principales de mezcla son molibdeno, niquel, cromo, manganeso y vanadio.

ammoniacal copper arsenate. A wood preservative.
arsenato de cobre amoniaco. Preservativo de la madera.

***amperage.** A measurement of the rate of flow of electrons; amperage controls the size of the arc.
amperaje. Una medida de la proporción de la corriente de electrones; el amperaje controla el tamaño del arco.

amperage range. The lower and upper limits of welding power, in amperage, that can be produced by a welding machine or used with an electrode or by a process.
rango de amperaje. Los límites máximos y mínimos de poder de soldadura (en amperaje) que puede tener una máquina para soldar o que pueden usarse con un electrodo o a través de un proceso.

angle of bevel. See preferred term **bevel angle.**
ángulo del bisel. Es preferible que vea el término **ángulo del bisel.**

***anode.** Material with a lack of electrons; thus, it has a positive charge.
ánodo. Un material que carece electrones; por eso tiene una carga positiva.

anvil. A heavy steel object used to help bend, cut, and shape metal.
yunke. Objeto pesado de acero empleado para ayundar en remachar, embutir, y cortar el metal.

arc blow. The deflection of arc from its normal path because of magnetic forces.
soplo del arco. Desviación de un arco eléctrico de su senda normal a causa de fuerzas magnéticas.

arc cutting (AC). A group of thermal cutting processes that severs or removes metal by melting with the heat of an arc between an electrode and the workpiece.
corte con arco. Un grupo de procesos termales para cortar que desúne o quita el metal derretido con el calor del arco en medio del electrodo y la pieza de trabajo.

arc force. The axial force developed by a plasma.
fuerza del arco. La fuerza axial desarrollada por la plasma.

arc gouging. Thermal gouging that uses an arc cutting process variation to form a bevel or groove.
gubiadura con arco. Gubiadura termal que usa un proceso variante de corte con arco para formar un bisel o ranura.

arc length. The length from the tip of the welding electrode to the adjacent surface of the weld pool.
largura del arco. La distancia de la punta del electrodo a la superficie que colinda con el charco de la soldadura.

arc plasma. A state of matter found in the region of an electrical discharge (arc). See also **plasma.**
arco de plasma. Un estado de la materia encontrado en la región de una descarga eléctrica (arco). Vea también **plasma.**

arc spot weld. A spot weld made by an arc welding process.
soldadura de puntos por arco. Una soldadura de punto hecha por un proceso de soldadura de arco.

arc strike. A discontinuity consisting of any localized remelted metal, heat-affected metal, or change in the surface profile of any part of a weld or base metal resulting from an arc.
golpe del arco. Una discontinuidad que consiste de cualquier rederretimiento del metal localizado, metal afectado por el calor, o cambio en el perfil de la superficie de cualquier parte de la soldadura o metal base resultante de un arco.

arc voltage. The voltage across the welding arc.
voltaje del arco. El voltaje a través del arco de soldar.

arc welding (AW). A group of welding processes that produces coalescence of workpieces by heating them with an arc. The processes are used with or without the application of pressure and with or without filler metal.
soldadura de arco. Un grupo de procesos de soldadura que producen una unión de piezas de trabajo calentándolas con un arco. Los procesos se usan con o sin la aplicación de presión y con o sin metal para rellenar.

arc welding electrode. A component of the welding circuit through which current is conducted between the electrode holder and the arc. See also **arc welding (AW).**
electrodo para soldadura de arco. Un componente del circuito de soldadura que conduce la corriente a través del porta-electrodo y el arco. Vea también **soldadura de arco.**

arc welding gun. A device used to transfer current to a continuously fed consumable electrode, guide the electrode, and direct the shielding gas.
pistola de soldadura de arco. Aparato que se usa para transferir corriente eléctrica continuamente a un alimentador de electrodo consumible. También se usa para guiar al electrodo y dirigir el gas de protección.

***arm.** An interconnected set of links and powered joints comprising a manipulator, which supports or moves a wrist and hand or end effector.
brazo. Una entreconexión que une un juego de eslabones y coyunturas de potencia conteniendo un manipulador, que apolla o mueve una muñeca o mano o el que efectúa al final.

as-welded. Pertaining to the weld metal, welded joints, and weldments after welding but prior to any subsequent thermal, mechanical, or chemical treatments.
como-soldado. Pertenece a metal soldado, juntas soldadas, o soldaduras ya soldadas pero antes de que se les hagan tratamientos termales, mecánicos, o químicos.

***atmospheric pressure.** The pressure at sea level resulting from the weight of a column of air on a specified area; expressed for an area of 1 square inch or square centimeter; normally given as 14.7 psi (1.05 kg/cm^2).
presion atmosferica. La presión al nivel del mar que resulta del peso de una columna de aire en una area especificada; expresada para una área de una pulgada cuadrada o un centímetro cuadrado. Normalmente dado como 14.7 psi (1.05 kg/cm^2).

austenitic manganese steel. A steel alloy with a high carbon content containing 10% or more manganese that is very tough and that will harden when cold-worked.
acero al manganeso austenítico. Aleación de acero con un alto contenido de carbono, que contiene un 10% o más de manganeso, y que es muy tenaz y endurece cuando se lo trabaja en frío.

autogenous weld. A fusion weld made without filler metal.
soldadura autógena. Una soldadura fundida sin metal de rellenar.

axis of a weld. A line through the length of a weld, perpendicular to and at the geometric center of its cross section.
eje de la soldadura. Una linea a lo largo de la soldadura perpendicula a y al centro geométrico de su corte transversal.

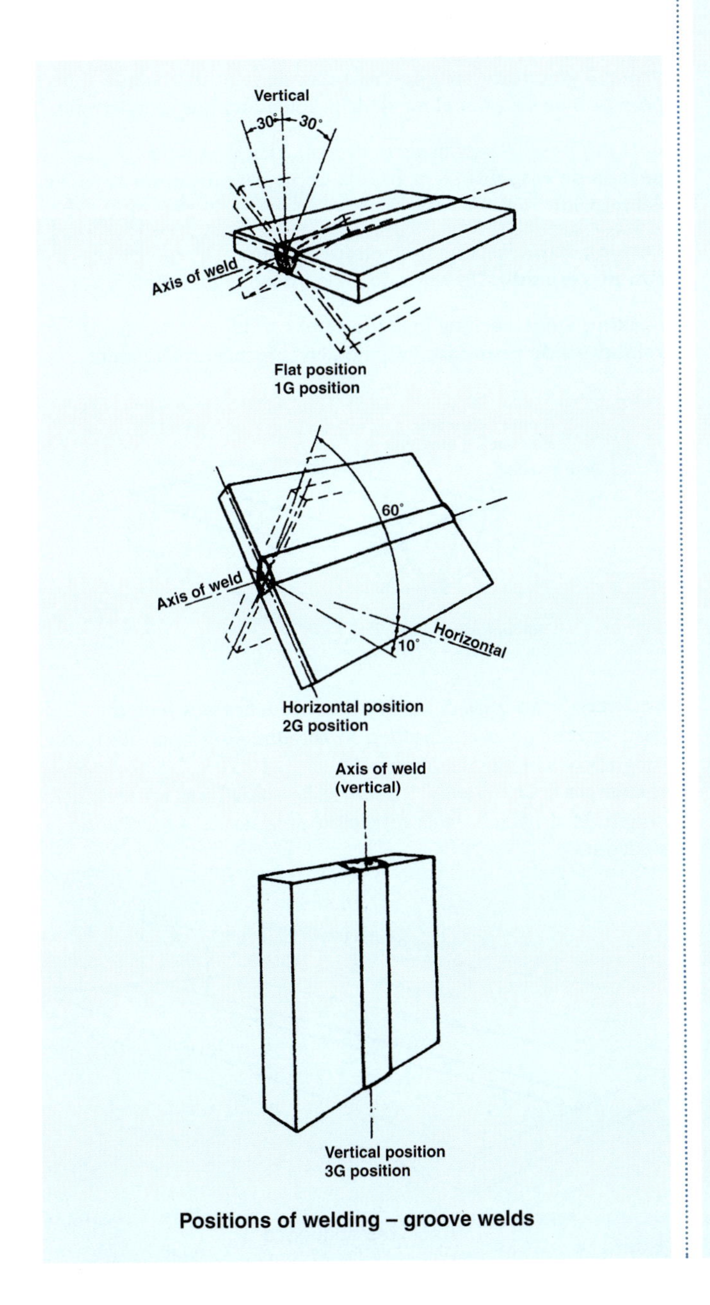

Positions of welding – groove welds

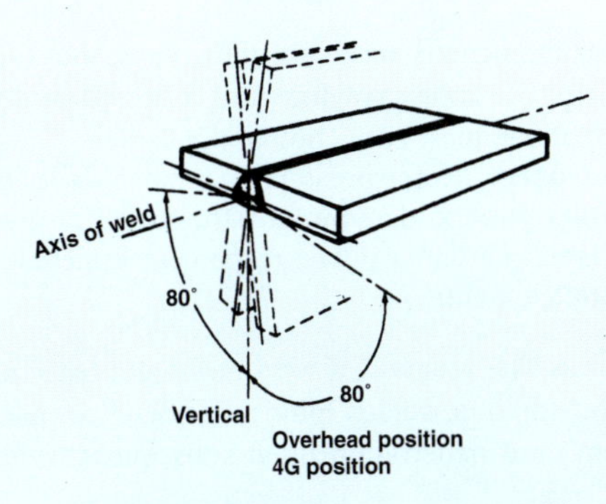

Positions of welding – groove welds

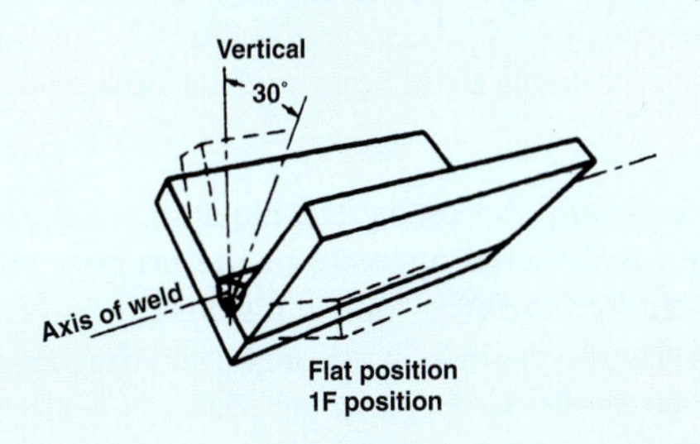

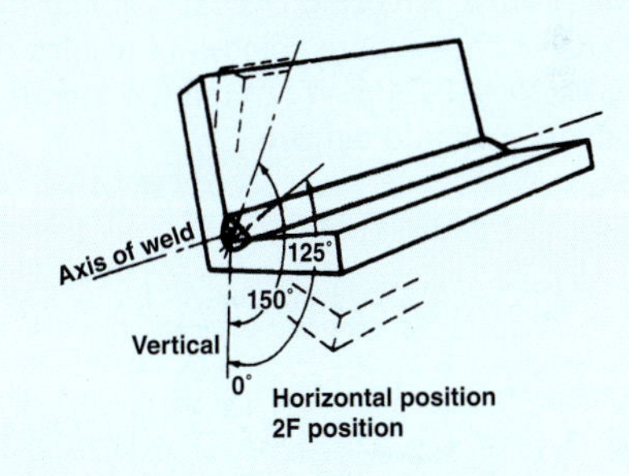

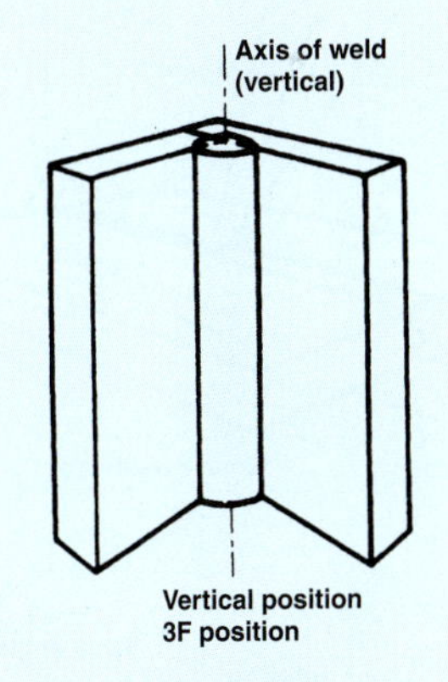

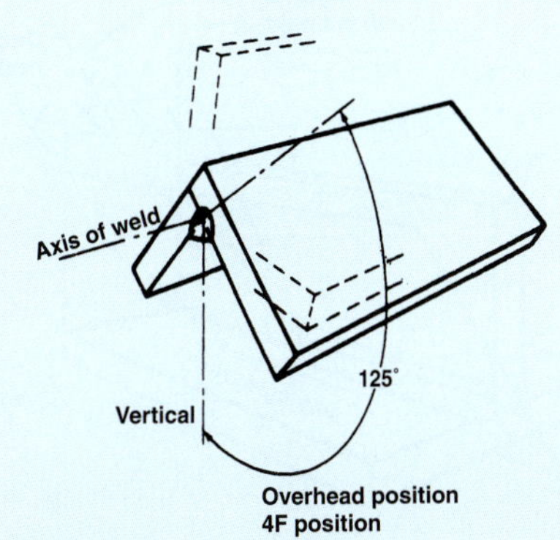

Positions of welding – fillet welds

B

backfire. The momentary recession of the flame into the welding tip, cutting tip, or flame spraying gun, followed by immediate reappearance or complete extinction of the flame.
llama de retroceso. El retroceso momentaneo de la llama dentro de la punta para soldar, punta para cortar, o pistola para rociar con llama. La llama puede reaparecer inmediatamente o apagarse completamente.

back gouging. The removal of weld metal and base metal from the weld root side of a welded joint to facilitate complete fusion and complete joint penetration upon subsequent welding from that side.
gubia trasera. Quitar el metal soldado y el metal base del lado de la raíz de una junta soldada para facilitar una fusión completa y penetración completa de la junta soldada subsecuente a soldar de ese lado.

backhand welding. A welding technique in which the welding torch or gun is directed opposite to the progress of welding. Sometimes referred to as the "pull gun technique" in GMAW and FCAW. See also **travel angle**, **work angle**, and **drag angle**.
soldadura en revés. Una técnica de soldar la cual el soplete o pistola es guiada en la dirección contraria al adelantamiento de la soldadura. A veces se refiere como una tecnica de "estirar la pistola" en GMAW y FCAW. Vea también **ángulo de avance**, **ángulo de trabajo**, y **ángulo del tiro**.

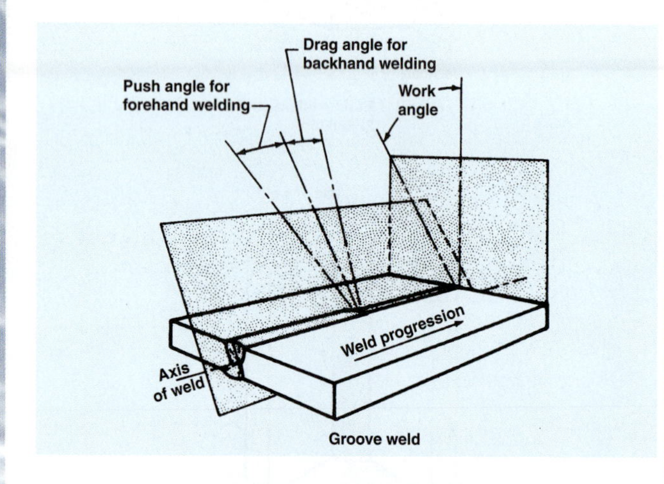

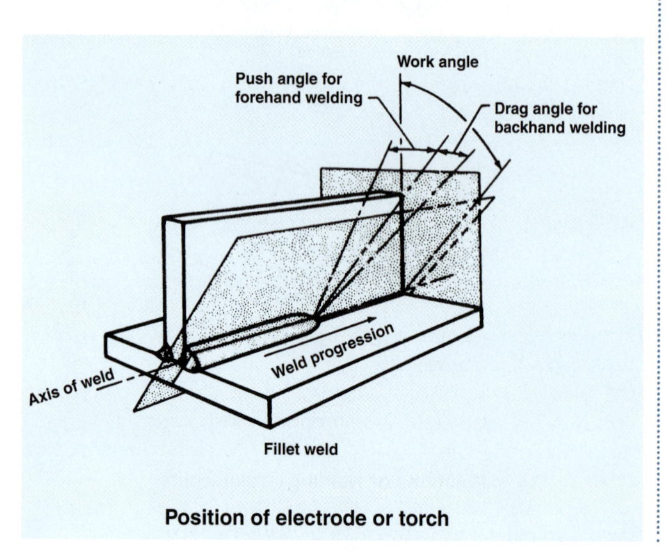

Position of electrode or torch

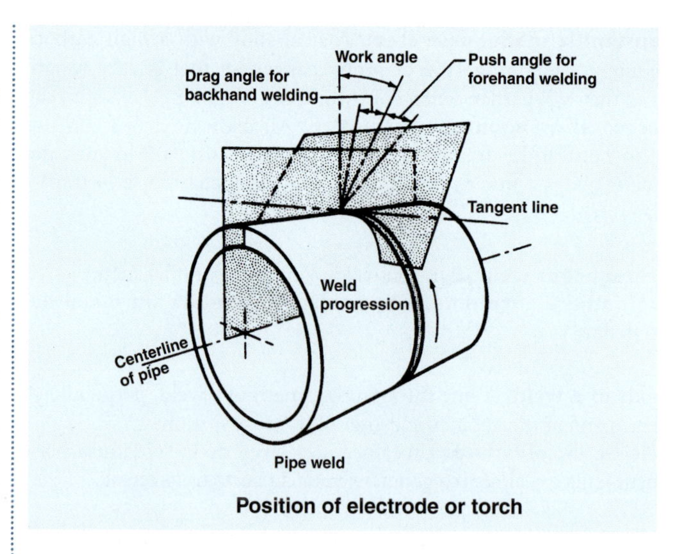

Position of electrode or torch

backing. A material (base metal, weld metal, carbon, or granular material) placed at the root of a weld joint for the purpose of supporting molten weld metal.
respaldo. Un material (metal base, metal de soldadura, carbón o material granulado) puesto en la raíz de la junta soldada con el proposito de sostener el metal de la soldadura que está derretido.

backing pass. A pass made to deposit a backing weld.
pasada de respaldo. Una pasada hecha para depositar la pasada del respaldo.

backing strip. Backing in the form of a strip.
tira de respaldo. Un respaldo en la forma de una tira.

backing weld. Backing in the form of a weld.
soldadura de respaldo. Respaldo en la forma de soldadura.

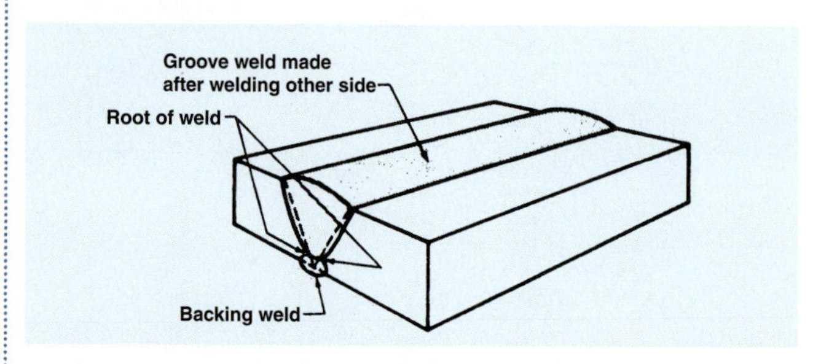

backstep sequence. A longitudinal sequence in which the weld bead increments are deposited in the direction opposite to the progress of welding the joint.
secuencia a la inversa. Una serie de soldaduras en secuencia longitudinal hechas en la dirección opuesta del progreso de la soldadura.

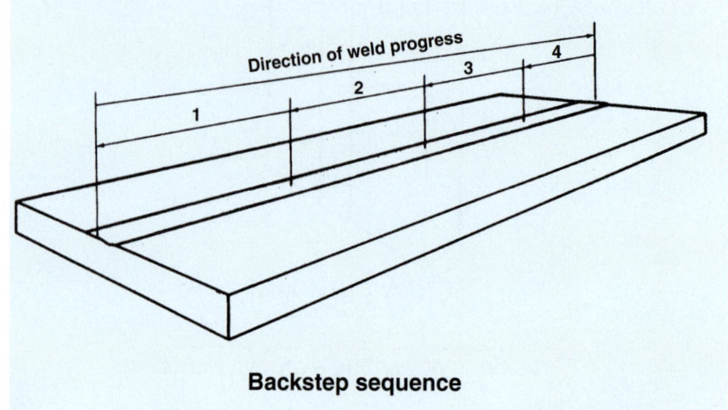

Backstep sequence

back weld. A weld deposited at the back of a single groove weld.
soldadura de atrás. Una soldadura que se deposita en la parte de atrás de una soldadura de ranura sencilla.

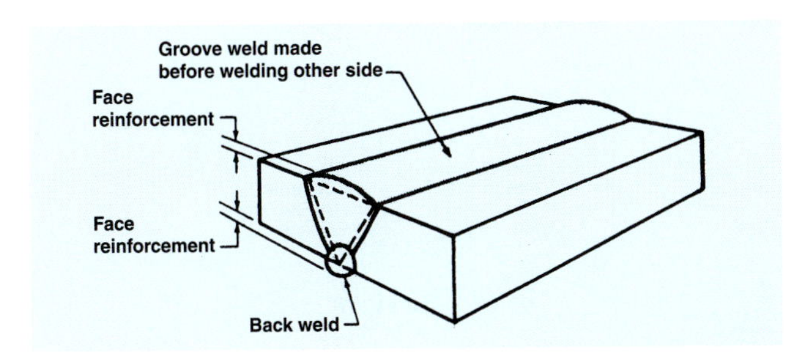

bare electrode. A filler metal electrode that has been produced as a wire, strip, or bar with no coating or covering other than that which is incidental to its manufacture or preservation.
electrodo descubierto. Un electrodo de metal para rellenar que se ha producido como alambre, tira, o barra sin revestimiento o cubierto con solo lo necesario para su fabricación y conservación.

base material. The material that is welded, brazed, soldered, or cut. See also **base metal** and **substrate.**
material base. El material que está soldado, soldado con soldadura fuerte, soldado con soldadura blanda, o cortado. Vea también **metal base** y **substrato.**

base metal. The metal or alloy that is welded, brazed, soldered, or cut. See also **base material** and **substrate.**
metal base. El metal que está soldado con soldadura fuerte, soldado con soldadura blanda, o cortado. Vea también **material base** y **substrato.**

bend-test. A test in which a specimen is bent to a specified bend radius. See also **face bend-test, root bend-test,** and **side bend-test.**
prueba de dobléz. Una prueba donde la probeta se dobla a una vuelta con un radio specificado. Vea también **prueba de dobléz de cara, prueba de dobléz de raíz,** y **prueba de dobléz de lado.**

bevel. An angular type of edge preparation.
bisel. Una preparación de tipo angular con filo.

bevel angle. The angle formed between the prepared edge of a member and a plane perpendicular to the surface of the member. Refer to drawings for **bevel.**
ángulo del bisel. El ángulo formado entre el corte preparado de un miembro y la plana perpendicular a la superficie del miembro. Refiera a los dibujos del **bisel.**

bill of materials. A listing of materials with specifications that are needed in a project.
lista de materiales. Una lista de los materiales y sus especificaciones que se necesitan para un proyecto.

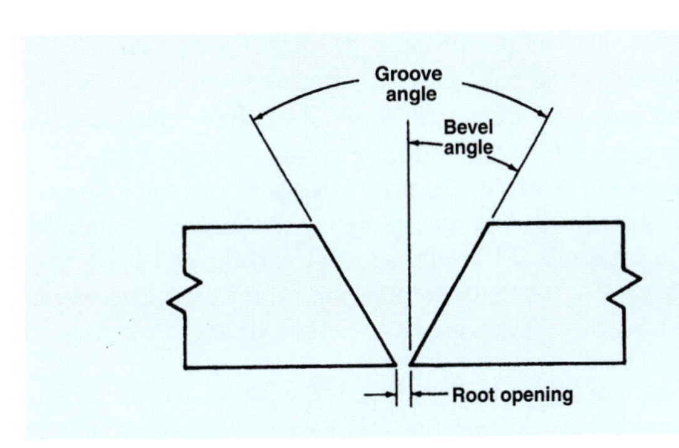

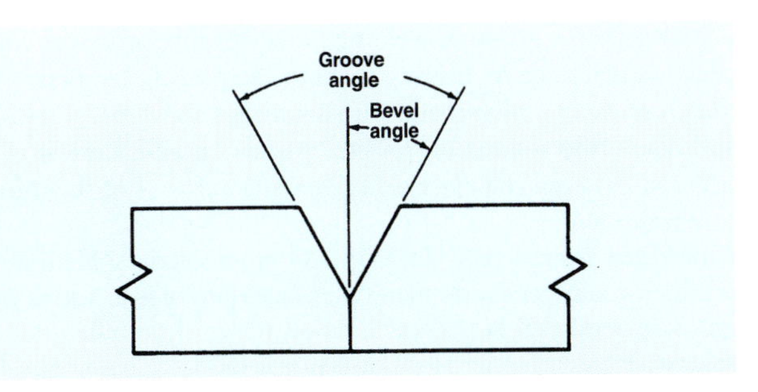

blind joint. A joint, no portion of which is visible.
junta ciega. Una junta en que no hay porción visible.

block sequence. A combined longitudinal and cross-sectional sequence for a continuous multiple-pass weld in which separated increments are completely or partially welded before intervening increments are welded.
secuencia de bloques. Una soldadura continua de pasadas multiples en sucesión combinadas longitudinal y sección transversa donde los incrementos separados son completamente o parcialmente soldados antes que los incrementos sean soldados.

braze. A weld produced by heating an assembly to suitable temperatures and by using a filler metal having a liquidus above 450°C (840°F) and below the solidus of the base metal. The filler metal is distributed between the closely fitted faying surfaces of the joint by capillary action.
soldadura de latón. Una soldadura producida cuando se calienta un montaje a una temperatura conveniente usando un metal de relleno que se liquida arriba de 840°F (450°C) y abajo del estado sólido del metal base. El metal de relleno es distribuido por acción capilar en una junta entre las superficies empalmadas montadas muy cerca.

brazeability. The capacity of a metal to be braced under the fabrication conditions imposed into a specific suitably designed structure and to perform satisfactorily in the intended service.
soldabilidad fuerte. La capacidad de un metal de refuerzo bajo las condiciones impuestas en la fabricación de una estructura diseñada especificamente para funcionar satisfactoriamente en los servicios intentados.

***braze buildup.** Braze metal added to the surface of a part to repair wear.
formación con bronce. Reparación de partes gastadas donde se agrega bronce.

brazement. An assembly whose component parts are joined by brazing.
montaje de soldadura fuerte. Un montaje donde las partes son unidas por soldadura fuerte.

braze metal. That portion of a braze that has been melted during brazing.
latón. La porción del bronce que se derrite cuando se solda.

braze welding. A welding process variation that uses a filler metal with a liquidus above 840°F (450°C) and below the solidus of the base metal. Unlike brazing, in braze welding the filler metal is not distributed in the joint by capillary action.
soldadura con bronce. Es un proceso de soldar variante que usa un metal de relleno con un liquido arriba de 840°F (450°C) y abajo del estado del metal base. Distinto a la soldadura fuerte, el metal de relleno no es distribuido por acción capilar.

brazing (B). A group of welding processes that produces coalescence of materials by heating them to the brazing temperature in the presence of a filler metal with a liquidus above 840°F (450°C) and below the solidus of the base metal. The filler metal is distributed between the closely fitted faying surfaces of the joint by capillary action.
soldadura fuerte (B). Un grupo de procesos de soldadura que produce coalescencia de materiales calentándolos a una temperatura de soldar en la presencia de un material de relleno el cual se derrite a una temperatura de 840°F (450°C) y bajo del estado sólido del metal base. El metal de relleno se distribuye por acción capilar de una junta entre las superficies empalmadas montadas muy cerca.

brazing filler metal. The metal that fills the capillary joint clearance and has a liquidus above 840°F (450°C) but below the solidus of the base metal.
metal de relleno para soldadura fuerte. El metal que rellena el espacio libre en la junta capilar y se derrite a una temperatura de 840°F (450°C) y bajo del estado sólido del metal base.

brazing rod. Filler metal used in the brazing process is supplied in the form of rods; the filler metal is usually an alloy of two or more metals; the percentages and types of metals used in the alloy impart different characteristics to the braze being made. Several classification systems are in use by manufacturers of filler metals. See also **brazing filler metal.**
varilla de latón. Metal de relleno usado en procesos de soldadura fuerte es surtida en forma de varillas; el metal para rellenar es usualmente un aleación de dos or más metales; los porcentajes y tipos de metales usados en la aleación imparten diferentes características a la soldadura que se está haciendo. Varios sistemás de clasificación se están usando por los fabricantes de metales de relleno. Vea también **metal de relleno para soldadura fuerte.**

brazing temperature. The temperature to which the base metal is heated to enable the filler metal to wet the base metal and form a brazed joint.
temperatura de soldadura fuerte. La temperatura a la cual se calienta el metal base para permitir que el metal de relleno moje al metal base y forme la soldadura fuerte.

buildup. A surfacing variation in which surfacing material is deposited to achieve the required dimensions. See also **hardfacing.**
recubrimiento. Una variación en la superficie donde el metal es depositado para que pueda obtener las dimensiones requeridas. Vea **endurecimiento de caras.**

bull float. A smooth board attached to a long handle used to smooth out newly poured concrete.
llana para hormigón. Madero liso ligado a un mango largo, que se usa para extender el hormigón que se acaba de derramar.

***burn-through.** Burning out of molten metal on the back side of the plate.
metal quemado que pasa al otro lado. Metal derretido que se quema en el lado de atrás del plato.

butt joint. A joint between two members aligned approximately in the same plane.
junta a tope. Una junta entre dos miembros alineados aproximadamente en el mismo plano.

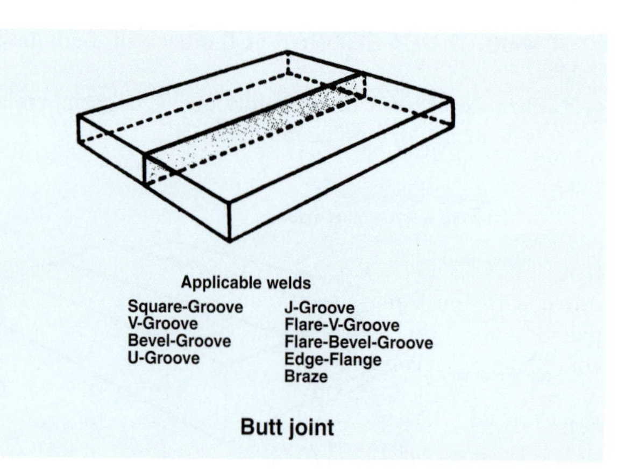

Butt joint

C

capillary action. The force by which liquid, in contact with a solid, is distributed between closely fitted faying surfaces of the joint to be brazed or soldered.
acción capilar. La fuerza por la que el liquido, en contacto con un sólido, es distribuido entre el empalme de las juntas del superficie para soldadura fuerte o blanda.

carbon arc cutting (CAC). An arc cutting process that uses a carbon electrode. See also **air carbon arc cutting.**
corte con arco y carbón. Un proceso de corte con arco que usa un electrodo de carbón. Vea también **arco de carbón con aire.**

carbon arc welding (CAW). An arc welding process that uses an arc between a carbon electrode and the weld pool. The process is used with or without shielding and without the application of pressure.
soldadura con arco de carbón. Un proceso de soldar de arco en que se usa un arco entre el electrodo de carbón y el metal derretido. El proceso se usa con o sin protección y sin aplicación de presión.

carbon electrode. A nonfiller metal electrode used in arc welding and cutting, consisting of a carbon or graphite rod, which may be coated with copper or other materials.
electrodo de carbón. Un electrodo de metal que no se rellena usado en soldaduras de arco y para cortes consistiendo de varillas de carbón o grafito que pueden ser cubiertas de cobre u otros materiales.

carbonizing (carburizing). A reducing oxyfuel gas flame in which there is an excess of fuel gas, resulting in a carbon-rich zone extending around and beyond the cone.
llama carburante. Una llama minorada de gas combustible a oxígeno donde hay un exceso de gas combustible, resultando en una zona rica de carboón extendiendose alrededor y al otro lado del cono.

***carbon steel.** Steel whose physical properties are primarily the result of the percentage of carbon contained within the alloy. Carbon content ranges from 0.04% to 1.4%, often referred to as plain carbon steel, low carbon steel, or straight carbon steel.
acero al carbono. Acero cuyas propiedades físicas son primariamente el resultado del porcentaje de carbón que es contenido dentro de la aleación. El contenido del carbón es clasificado entre 0.04% a 1.4%, frecuentemente es referido como el carbono de acero liso, acero de bajo carbón, o carbon de acero recto.

***cast.** The natural curve in the electrode wire for gas metal arc welding as it is removed from the spool; cast is measured by the

diameter of the circle that the wire makes when it is placed on a flat surface without any restraint.
distancia. La curva natural en el alambre electrodo para soldadura de arco metálico para gas cuando se aparta del carrete; la distancia es medida en el círculo que hace el alambre cuando es puesto en una superficie plana sin restricción.

***cast iron.** A combination of iron and carbon. The carbon may range from 2% to 4%. Approximately 0.8% of the carbon is combined with the iron. The remaining free carbon is found as graphite mixed throughout the metal. Gray cast iron is the most common form of cast iron.
acero vaciado. Una combinación de acero y carbono. El carbono puede ser clasificado de 2% a 4%. Aproximadamente 0.8% del carbono es combinado con el hierro. El resto de carbono libre es encontrado como grafito mezclado por todo el metal. El acero fundido gris es la forma más común.

cathode. A natural curve material with an excess of electrons, thus having a negative charge.
cátodo. Un material de curva natural con un exceso de electrones, por eso tiene una carga negativa.

CCA. Chromated copper arsenate.
ACC. Arsenate de cobre cromado.

cellulose-based electrode fluxes. Fluxes that use an organic-based cellulose ($C_6H_{10}O_5$) (a material commonly used to make paper) held together with a lime binder. When this flux is exposed to the heat of the arc, it burns and forms a rapidly expanding gaseous cloud of CO_2 that protects the molten weld pool from oxidation. Most of the fluxing material is burned, and little slag is deposited on the weld. E6010 is an example of an electrode that uses this type of flux.
fundentes para electrodos celulósicos. Fundentes que usan celulosa de base orgánica ($C_6H_{10}O_5$) (un material normalmente utilizado para fabricar papel), y que se mantienen unidos con un aglomerante de cal. Cuando a este fundente se lo expone al calor del arco, se consume y forma una nube gaseosa de CO_2 que se expande rápidamente y protege de la oxidación al charco de soldadura derretido. La mayor parte del material del fundente se consume, y se deposita poca escoria en la soldadura. El E6010 es un ejemplo de un electrodo que utiliza este tipo de fundente.

***center.** A manufacturing unit consisting of two or more cells and the materials transport and storage buffers that interconnect them.
centro. Una unidad manufacturera la cual consiste de dos o más celdas y el traslado de materiales y los amortiguadores del almacén que los entreconecta.

certified welders. Individuals who have demonstrated their welding skills for a process by passing a specific welding test.
soldador certificado. Personas que han demostrado, mediante una prueba específica de soldadura, su habilidad para soldar en un proceso.

CFM. Cubic feet per minute.
PCM. Pie cubico por minuto.

chain intermittent welds. Intermittent welds on both sides of a joint in which the weld increments on one side are approximately opposite those on the other side.
soldadura intermitente de cadena. Soldadura intermitente en los dos lados de una junta en cual los incrementos de soldadura están aproximadamente opuestos a los del otro lado.

chalk line. A cotton cord that, with chalk applied, is used to create long, straight lines.
cordel entizado. Una cuerda de alogodon entizado, que se usa para hacer lineas largas y derechas.

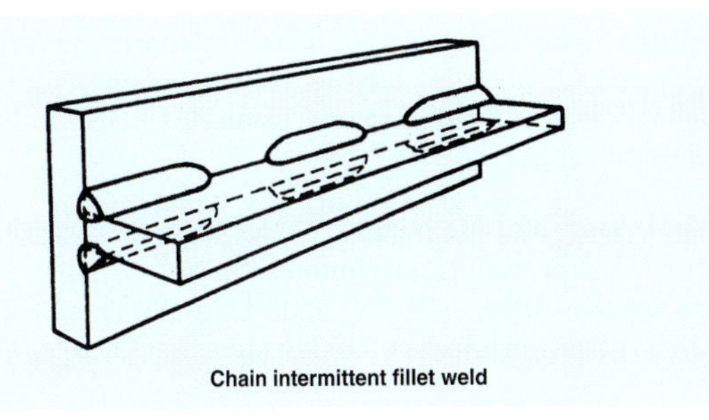
Chain intermittent fillet weld

***chill plate.** A large piece of metal used in welding to correct overheating.
plato desalentador. Una pieza de metal grande que se usa para corregir el sobrecalentamiento.

coalescence. The growing together or growth into one body of the materials being welded.
coelescencia. El crecimiento o desarrollo de un cuerpo de los materiales los cuales se están soldando.

coating. A relatively thin layer (< 1 mm [0.04 in.]) of material applied by surfacing for the purpose of corrosion prevention, resistance to high-temperature scaling, wear resistance, lubrication, or other purposes. See also **surfacing** and **hardfacing**.
revestimiento. Una capa de material relativamente delgado (< 1 mm [0.04 pulgadas]) aplicada por la superficie con el propósito de prevenir corrosión, resistencia a las altas temperaturas, resistencia a la deterioración, lubricación, o para otros propósitos. Vea también **capa de revestimiento, recubrimiento superficial**, y **endurecimiento de caras.**

cold soldered joint. A joint with incomplete coalescence caused by insufficient application of heat to the base metal during soldering.
junta soldada fría. Una junta con coalescencia incompleta causada por no haber aplicado suficiente calor al metal base durante la soldadura.

***combustion rate.** Also known as rate of propagation of a flame, this is the speed at which the fuel gas burns, in ft/sec (m/sec). The ratio of fuel gas to oxygen affects the rate of burning: a higher percentage of oxygen increases the burn rate.
velocidad de combustión. También es conocida como velocidad de propagación de una llama, ésta es la velocidad en la cual se quema el gas combustible, en pies/sec (m/sec). La proporción del gas combustible al oxígeno afecta la proporción de quemadura: un porcentaje más alto del oxígeno aumenta la proporción de quemarse.

complete fusion. Fusion over the entire fusion faces and between all adjoining weld beads.
fusión completa. Fusión sobre todas las caras de fusión y en medio de todos los cordónes de soldadura inmediatos.

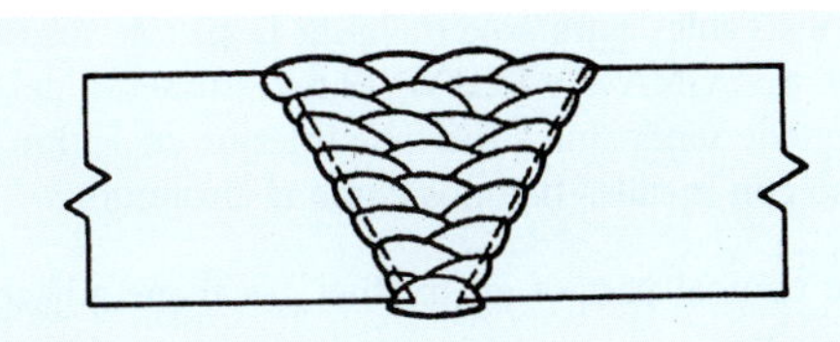

composite electrode. A generic term for multicomponent filler metal electrodes in various physical forms, such as stranded wires, tubes, and covered wire. See also **covered electrode**, **flux cored electrode**, and **stranded electrode**.
electrodo compuesto. Un término genérico para componentes múltiples para electrodos de metal de aporte en varias formas físicas, como cable de alambre, tubos, y alambre cubierto. Vea también **electrodo cubierto**, **electrodo de núcleo de fundente**, y **electrodo cable**.

compressed air. Air pumped under high pressure that may be carried by pipe, tubing, or reinforced hose.
aire comprimido. Aire inyectado bajo alta presion que puede ser llevado por medio de tubo o manguera reforzada.

concave fillet weld. A fillet weld with a concave face.
soldadura de filete cóncava. Soldadura de filete con cara cóncava.

concave root surface. A root surface that is concave.
superficie raíz cóncava. La superficie del cordón raíz con cara cóncava.

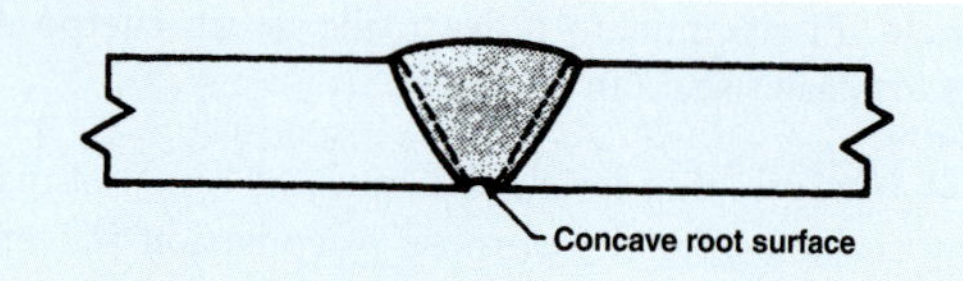

concavity. The maximum distance from the face of a concave fillet weld perpendicular to a line joining the toes.
concavidad. La distancia máxima de la cara de una soldadura de filete cóncava perpendicular a una linea que une con los pies.

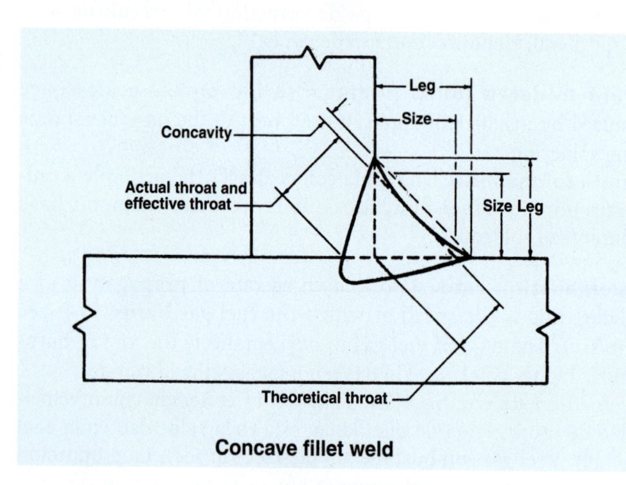

Concave fillet weld

conduit liner. A flexible steel tube that guides the welding wire from the feed rollers through the welding lead to the gun used for GMAW and FCAW welding. The steel conduit liner may have a nylon or Teflon® inner surface for use with soft metals such as aluminum.
revestimiento de conducto. Un tubo flexible de acero que guía el alambre para soldar desde los rodillos de alimentación, a través de los cables para soldar, hasta la pistola, usado en soldaduras de tipo GMAW y FCAW. El revestimiento del conducto de acero puede tener una superficie interior de Teflon® o nylon para su uso con metales blandos como el aluminio.

cone. The conical part of an oxyfuel gas flame adjacent to the orifice of the tip.
cono. La parte cónica de la llama del gas de oxígeno combustible que colinda con la abertura de la punta.

constricted arc. A plasma arc column that is shaped by the constricting orifice in the nozzle of the plasma arc torch or plasma spraying gun.
arco constreñido. Una columna de arco plasma que está formada por el constreñimiento del orificio en la lanza de la antorcha del arco plasma o pistola de rociado plasma.

constricting nozzle. A device at the exit end of a plasma arc torch or plasma spraying gun containing the constricting orifice.
boquilla de constreñimiento. Un aparato a la salida de la antorcha de un arco plasma o la pistola de rociado plasma que contiene la boquilla de constreñimiento.

constricting orifice. The hole in the constricting nozzle of the plasma arc torch or plasma spraying gun through which the arc plasma passes.
orificio de constreñimiento. El agujero en la boquilla del constreñimiento en la antorcha de arco plasma o de la pistola de rociado plasma por donde pasa el arco de plasma.

consumable electrode. An electrode that provides filler metal.
electrodo consumible. Un electrodo que surte el metal de relleno.

contact tube. A device that transfers current to a continuous electrode.
tubo de contacto. Un aparato que traslada corriente continua a un electrodo.

continuous weld. A weld that extends continuously from one end of a joint to the other. Where the joint is essentially circular, it extends completely around the joint.
soldadura continua. Una soldadura que se extiende continuamente de una punta de la junta a la otra. Donde la junta es esencialmente circular, se extiende completamente alrededor de la junta.

convex fillet weld. A fillet weld with a convex face.
soldadura de filete convexa. Una soldadura de filete con una cara convexa.

convexity. The maximum distance from the face of a convex fillet weld perpendicular to a line joining the toes.
convexidad. La distancia máxima de la cara de la soldadura convexa filete perpendicula a la linea que une los pies.

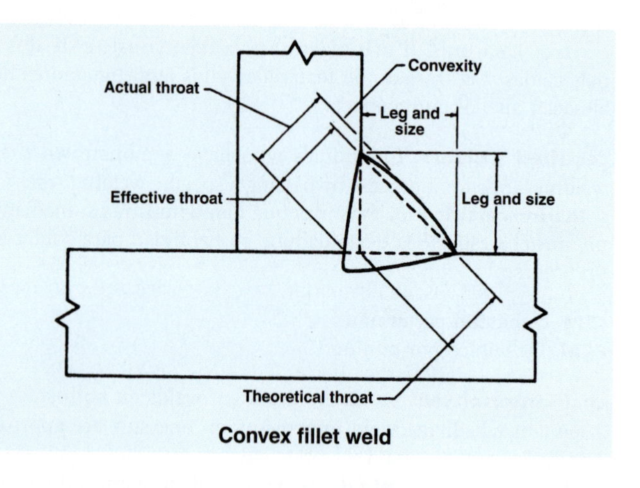

Convex fillet weld

convex root surface. A root surface that is convex.
raíz superficie convexa. La raíz que es convexa.

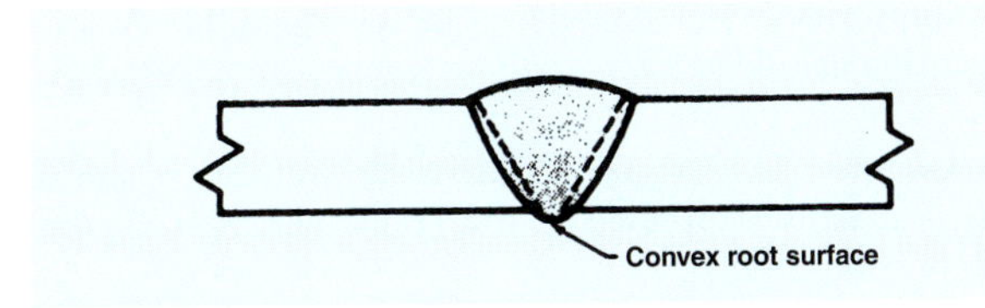

cordless. A tool containing a rechargeable battery pack to drive the unit when not plugged into an electrical outlet.
sin cordones. Herramienta que incluye una bateria recargable para impulsar el grupo cuando no esta enchufado a un enchufe electrico.

cored solder. A solder wire or bar containing flux as a core.
soldadura de núcleo. Un alambre o barra para soldar que contiene fundente en el núcleo.

***core wire.** The wire portion of the coated electrode for shielded metal arc welding. The wire carries the welding current and adds most of the filler metal required in the finished weld. The composition of the core wire depends upon the metals to be welded.
alambre del centro. La porción del alambre del electrodo forrado para proteger el metal de la soldadura de arco. El alambre lleva la corriente de la soldadura y añade casi todo el metal para rellenar que es requerido para terminar la soldadura. La composición del alambre del centro depende de los metales que se van a usar para soldar.

corner joint. A joint between two members located approximately at right angles to each other.
junta de esquina. Una junta dentro de dos miembros localizados aproximadamente a ángulos rectos de unos a otros.

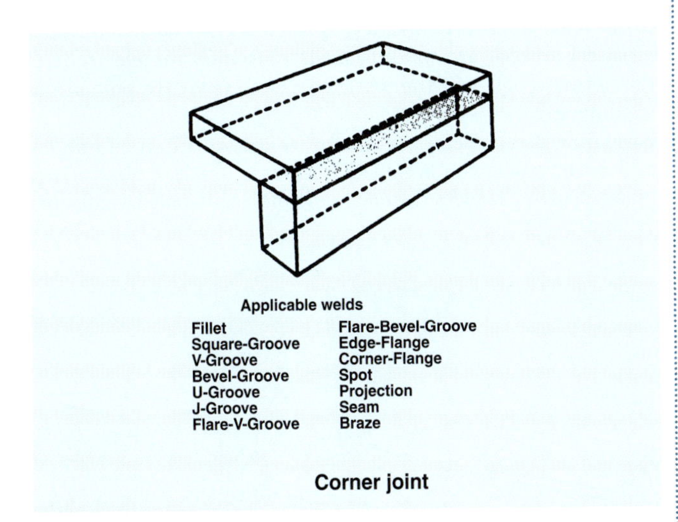

Corner joint

***corrosion resistance.** The ability of the joint to withstand chemical attack; determined by the compatibility of the base materials to the filler metal.
resistencia a la corrosión. La abilidad de una junta de resistir ataques químicos; determinado por la compatibilidad de los materiales bases al metal de relleno.

corrosive flux. A flux with a residue that chemically attacks the base metal. It may be composed of inorganic salts and acids, organic salts and acids, or activated rosins or resins.
fundente corrosivo. Un fundente con un residuo que ataca químicamente al metal base. Puede estar compuesto de sales y ácidos inorgánicos, sales y ácidos orgánicos, o abelinotes o resinas activados.

cosmetic pass. A weld pass made primarily to enhance appearance.
pasada cosmética. Una pasada que se le hace a la soldadura para mejorar la apariencia.

***coupling distance.** The distance to be maintained between the inner cones of the cutting flame and the surface of the metal being cut, in the range of 1/8 in. (3 mm) to 3/8 in. (10 mm).
distancia de acoplamiento. La distancia que debe de mantenerse entre los conos internos de la llama y la superficie del metal que se está cortando, varía de 1/8 pulgadas (3 mm) a 3/8 pulgadas (10 mm).

covered electrode. A composite filler metal electrode consisting of a core of a bare electrode or metal cored electrode to which a covering sufficient to provide a slag layer on the weld metal has been applied. The covering may contain materials providing such functions as shielding from the atmosphere, deoxidation, and arc stabilization and can serve as a source of metallic additions to the weld.
electrodo cubierto. Un electrodo compuesto de metal para rellenar que consiste de un núcleo de un electrodo liso o electrodo con núcleo de metal el cual se le agrega cubrimiento suficiente para proveer una capa de escoria sobre el metal de la soldadura que se le aplicó. El cubierto puede contener materiales que pueden proveer funciones como protección de la atmósfera, deoxidación, y estabilización del arco, y también puede servir como fuente para añadir metales adicionales a la soldadura.

cover lens. A round cover plate.
lente para cubrir. Un plato redondo de vidrio para cubrir el lente obscuro.

***cover pass.** The last layer of weld beads on a multipass weld. The final bead should be uniform in width and reinforcement, not excessively wide, and free of any visual defects.
pasada para cubrir. La última capa de cordónes soldadura de pasadas múltiples. La pasada final debe ser uniforme en anchura y refuerzo, no excesivamente ancha, y libre de defectos visuales.

cover plate. A removable pane of colorless glass, plastic-coated glass, or plastic that covers the filter plate and protects it from weld spatter, pitting, and scratching.
plato para cubrir. Una hoja removible de vidrio claro, vidrio cubierto con plástico o plástico que cubre el plato filtrado y lo protege de salpicadura, picaduras y de que se rayen.

crack. A fracture-type discontinuity characterized by a sharp tip and high ratio of length and width to opening displacement.
grieta. Una desunión discontinuidada de tipo fractura caracterizada por una punta filoza y proporción alta de lo largo y de lo ancho al desplazamiento de la abertura.

crater. A depression in the weld face at the termination of a weld bead.
crater. Una depresión en la superficie de la soldadura a donde se termina el cordón de soldadura.

crater crack. A crack in the crater of a weld bead.
grieta de crater. Una grieta en el crater del cordón de soldar.

critical weld. A weld so important to the soundness of the weldment that its failure could result in the loss or destruction of the weldment and injury or death.
soldadura crítica. Una soldadura tan importante para la calidad del conjunto de partes soldadas, que su fracaso podría ocasionar la pérdida o destrucción de dicho conjunto, así como también lesiones o muerte.

cup. A nonstandard term for gas nozzle.
tazón. Un término que no es la norma para de boquilla de gas.

cutting attachment. A device for converting an oxyfuel gas welding torch into an oxygen cutting torch.
equipo para cortar. Un aparato para convertir una antorcha para soldar en una antorcha para cortar con oxígeno.

cutting head. The part of a cutting machine in which a cutting torch or tip is incorporated.
cabeza de la antorcha para cortar. La parte de una maquina para cortar en donde una antorcha para cortar o una punta es incorporada.

cutting tip. The part of an oxygen cutting torch from which the gases issue.
punta para cortar. Esa parte de la antorcha para cortar con oxígeno por donde salen los gases.

cutting tip, high-speed. Designed to provide higher oxygen pressure, thus allowing the torch to travel faster.
punta para cortar a alta velocidad. Diseñada para proveer presión más alta de oxígeno, asi puede caminar la antorcha más rápidamente.

cutting torch. A device used for plasma arc cutting to control the position of the electrode, to transfer current to the arc, and to direct the flow of plasma and shielding gas.
antorcha para cortar. Un aparato que se usa para cortes de arco de plasma para el control de la posición del electrodo.

cylinder. A portable container used for transportation and storage of a compressed gas.
cilindro. Un recipiente portátil que se usa para transportar y guardar un gas comprimido.

cylinder manifold. A multiple header for interconnection of gas sources with distribution points.
conexión de cilindros múltiple. Una tuberia con conexiones múltiples que sirve como fuente de gas con puntos de distribución.

***cylinder pressure.** The pressure at which a gas is stored in approximately 2200 pounds per square inch (psi), and acetylene is stored at approximately 225 psi.
presión del cilindro. La presión del cilindro en el cual un gas se guarda en aproximadamente 2200 libras por pulgada cuadrada (psi), y el acetilino se guarda a aproximadamente 225 psi.

D

defect. A discontinuity or discontinuities that by nature or accumulated effect (for example, total crack length) render a part or product unable to meet minimum applicable acceptance standards or specifications. This term designated rejectability and flaw. See also **discontinuity** and **flaw.**
defecto. Una desunión o desuniónes que por la naturaleza o efectos acumulados (por ejemplo, distancia total de una grieta) hace que una parte o producto no esté de acuerdo con las normas o especificaciones mínimas para aceptarse. Este término designado no bueno y falta. Vea también **discontinuidad** y **falta.**

defective weld. A weld containing one or more defects.
soldadura defectuosa. Una soldadura que contiene uno o más defectos.

deposited metal. Filler metal that has been added during a welding operation.
metal depositado. Metal de relleno que se ha agregado durante una operación de soldadura.

deposition rate. The weight of material deposited in a unit of time. It is usually expressed as kilograms per hour (kg/hr) (pounds per hour [lb/h]).
relación de deposición. El peso del material depositado en una unidad de tiempo. Es regularmente expresado en kilogramos por hora (kg/hora) (libras por hora [lb/hora]).

depth of fusion. The distance that fusion extends into the base metal or previous bead from the surface melted during welding.
grueso de fusión. La distancia en que la fusión se extiende dentro del metal base o del cordón anterior de la superficie que se derretió durante la soldadura.

***destructive testing.** Mechanical testing of weld specimens to measure strength and other properties. The tests are made on specimens that duplicate the material and weld procedures required for the job.
prueba destructiva. Pruebas mecánicas de probetas de soldadura para medir la fuerza y otras propiedades. Las pruebas se hacen en probetas que duplican el material y los procedimientos de la soldadura requeridos para el trabajo.

dimensioning. The measurements of an object, such as its length, width, and height, or the measurements for locating such things as parts, holes, and surfaces.
acotación. Las medidas de un objeto, tal como su longitud, ancho, y altura, o las medidas para ubicar cosas como piezas, agujeros o superficies.

direct-current electrode negative (DCEN). The arrangement of direct-current arc welding leads on which the electrode is the negative pole and the workpiece is the positive pole of the welding arc.
corriente directa con electrodo negativo. El arreglo de los cables para soldar con la soldadura de arco donde el electrodo es polo negativo y la pieza de trabajo es polo positivo de la soldadura de arco.

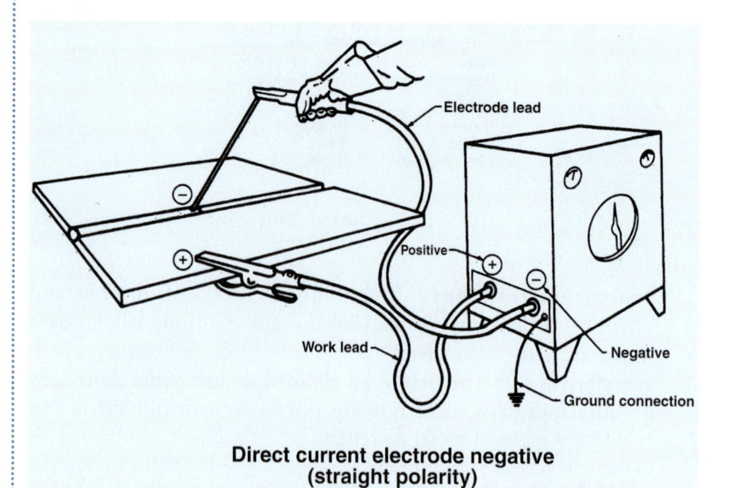

Direct current electrode negative (straight polarity)

direct-current electrode positive (DCEP). The arrangement of direct-current arc welding leads on which the electrode is the positive pole and the workpiece is the negative pole of the welding arc.

corriente directa con el electrodo positivo. El arreglo de los cables para soldar con la soldadura de arco con el electrodo es el polo positivo y la pieza de trabajo es el polo negativo de la soldadura de arco.

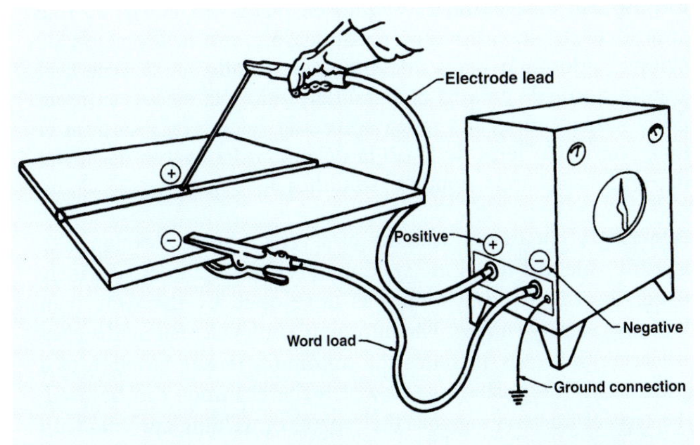

Direct current electrode positive (reverse polarity)

discontinuity. An interruption of the typical structure of a material, such as a lack of homogeneity in its mechanical, metallurgical, or physical characteristics. A discontinuity is not necessarily a defect. See also **defect** and **flaw**.
discontinuidad. Una interrupción de la estructura típica de un material, el que falta de homogenidad en sus caracteristicas mecánicas, metalúrgicas, o fisica. Vea también **defecto** y **falta**.

***distortion.** Movement or warping of parts being welded, from the prewelding position and condition compared to the postwelding condition and position.
deformacion. Movimiento o torcimiento de las partes que se están soldando, comparando la posición antes de soldar a la posicion despues de soldar.

double bevel-groove weld. A type of groove weld. Refer to drawing for **groove weld.**
soldadura de ranura con doble bisel. Es un tipo de soldadura de ranura. Refiérase al dibujo para **soldadura de ranura.**

double J-groove weld. A type of groove weld. Refer to drawing for **groove weld.**
soldadura de ranura de doble-J-. Un tipo de soldadura de ranura. Refiérase al dibujo para **soldadura de ranura.**

double U-groove weld. A type of groove weld. Refer to drawing for **groove weld.**
soldadura de ranura de doble-U-. Un tipo de soldadura de ranura. Refiérase al dibujo para **soldadura de ranura.**

double V-groove weld. A type of groove weld. Refer to drawing for **groove weld.**
soldadura de ranura de doble-V-. Un tipo de soldadura de ranura. Refiérase al dibujo para **soldadura de ranura.**

drag (thermal cutting). The offset distance between the actual and straight line exit points of the gas stream or cutting beam measured on the exit surface of the base metal.
tiro (corte termal). La distancia desalineada entre la actual y la linea recta del punto de salida del chorro de gas o el rayo de cortar medido a la salida de la superficie del metal base.

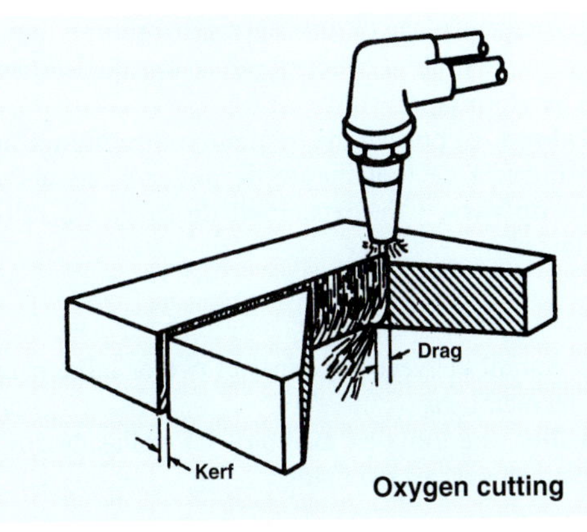

Oxygen cutting

drag angle. The travel angle when the electrode is pointing in a direction opposite to the progression of welding. This angle can also be used to partially define the position of guns, torches, rods, and beams.
ángulo del tiro. El ángulo de avance cuando el electrodo está apuntando en una dirección opuesta del progreso de la soldadura. Este ángulo también se puede usar para parcialmente definir la posición de pistolas, antorchas, varillas, y rayos.

***drag lines.** High-pressure oxygen flow during cutting forms lines on the cut faces. A correctly made cut has up and down drag lines (zero drag); any deviation from the pattern indicates a change in one of the variables affecting the cutting process; with experience the welder can interpret the drag lines to determine how to correct the cut by adjusting one or more variables.
lineas del tiro. La salida del oxígeno a presión elevada durante el corte forma lineas en las caras del corte. Un corte hecho correctamente tiene lineas hacia arriba y hacia abajo (zero tiro); cualquier desviación de la norma indica un cambio en uno de los variables que afectan el proceso de cortar; con experiencia el soldador puede interpretar las lineas de tiro y determinar como corregir el corte ajustando uno o más variables.

***drift.** The tendency of a system's response to gradually move away from the desired response.
deriva. La tendencia de la respuesta del sistema de retirarse gradualmente de la respuesta deseada.

***drooping output.** Volt-ampere characteristic of the shielded metal arc process power supply where the voltage output decreases as increasing current is required of the power supply. This characteristic provides a reasonably high voltage at a constant current.
reducción de potencia de salida. Característica de voltioamperios de la alimentación de poder de un proceso de soldadura de arco protegido donde la salida del voltaje disminuye mientras un aumento de la corriente es requerida de la alimentación de poder. Está característica proporciona un voltaje razonable alto con corriente constante.

***ductility.** As applied to a soldered or brazed joint, it is the ability of the joint to bend without failing.
ductilidad. Como aplicada a junta de soldadura fuerte o soldadura blanda, es la abilidad de la junta de doblarse sin fallar.

duplex receptacle. A double receptacle wired so that both outlets are on the same circuit.
receptáculo duplex. Receptáculo doble contectado para que las dos tomas corran en el mismo circuito.

duty cycle. The percentage of time during an arbitrary test period that a power source or its accessories can be operated at rated output without overheating.
ciclo de trabajo. El porcentaje de tiempo durante un período a prueba arbitraria de una fuente de poder y sus accesorios que pueden operarse a la capacidad de carga de salida sin sobrecalentarse.

E

edge joint. A joint between the edges of two or more parallel or nearly parallel members.
junta de orilla. Una junta en medio de las orillas de dos o más miembros paralelos o casi paralelos.

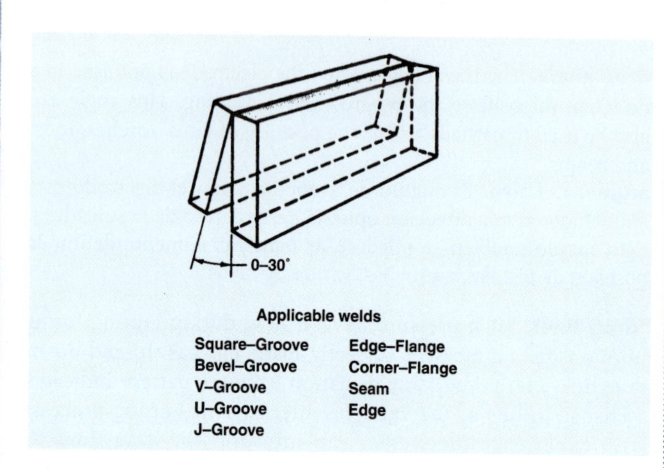

edge preparation. The surface prepared on the edge of a member for welding.
preparación de orilla. La superficie preparada en la orilla de un miembro que se va a soldar.

effective length of weld. The length of weld throughout which the correctly proportioned cross section exists. In a curved weld, it shall be measured along the axis of the weld.
distancia efectiva de soldadura. La distancia de una sección transversa correctamente proporcionada que existe por toda la soldadura. En una soldadura en curva, debe medirse por el axis de la soldadura.

effective throat. The minimum distance from the root of a weld to its face, less any reinforcement. See also **joint penetration.** Refer to drawing for **convexity.**
garganta efectiva. La distancia mínima de la raíz a la cara de una soldadura, menos el refuerzo. Vea también **penetración de junta.** Refiérase al dibujo para **convexidad.**

***elastic limit.** The maximum force that can be applied to a material or joint without causing permanent deformation or failure.
limite elástico. La fuerza máxima que se le puede aplicar a un material o junta sin causar deformación o falta permanente.

electrode. A component of the electrical circuit that terminates at the arc, molten conductive slag, or base metal.
electrodo. Un componente del circuito eléctrico que termina al arco, escoria derretida conductiva, o metal base.

***electrode angle.** The angle between the electrode and the surface of the metal; also known as the direction of travel (leading angle or trailing angle); leading angle pushes molten metal and slag ahead of the weld; trailing angle pushes the molten metal away from the leading edge of the molten weld pool toward the back, where it solidifies.
ángulo del electrodo. El ángulo en medio del electrodo y la superficie del metal; también conocido como la dirección de avance (apuntado hacia adelante o apuntado hacia atras); el ángulo apuntado empuja el metal derretido y la escoria enfrente de la soldadura; y el ángulo apuntado hacia atrás empuja el metal derretido lejos de la orilla delantera del charco del metal derretido hacia atrás, donde se solidifica.

***electrode classification.** Any of several systems developed to identify shielded metal arc welding electrodes. The most widely used identification system was developed by the American Welding Society (AWS). The information represented by the classification generally includes the minimum tensile strength of a good weld, the position(s) in which the electrode can be used, the type of flux coating, and the type(s) of welding currents with which the electrode can be used.
clasificación de electrodo. Cualquiera de los varios sistemas desarollados para identificar electrodos protegidos para soldadura de arco. El sistema de identificación que se usa mucho más fue desarrollado por la Sociedad de Soldadura Americana (AWS). La información representada por la clasificación generalmente incluye la fuerza tensible mínima de una soldadura, la posición(es) donde se puede usar el electrodo, el tipo de recubrimiento de fundente y los tipo(s) de corrientes para soldar con la cual se puede usar el electrodo.

electrode extension (GMAW, FCAW, SAW). The length of unmelted electrode extending beyond the end of the contact tube during welding.
extensión del electrodo (GMAW, FCAW, SAW). La distancia de extensión del electrodo que no está derretido más allá de la punta del tubo de contacto durante la soldadura.

electrode holder. A device used for mechanically holding and conducting current to an electrode during welding.
porta electrodo. Un aparato usado para detener mecánicamente y conducir corriente a un electrodo durante la soldadura.

electrode lead. The electrical conductor between the source of arc welding current and the electrode holder. Refer to drawing for **direct-current electrode negative.**
cable de electrodo. Un conductor eléctrico en medio de la fuente para la corriente de soldar con arco y el portaelectrodo. Refiérase al dibujo de **corriente directa con electrodo negativo.**

electrode setback. The distance the electrode is recessed behind the constricting orifice of the plasma arc torch or thermal spraying gun, measured from the outer face of the nozzle.
retroceso del electrodo. La distancia del hueco del electrodo que está detrás del orificio constringente de la antorcha de arco plasma o pistola de rocio termal, se mide de la cara de afuera a la boquilla.

***exhaust pickup.** A component of a forced ventilation system that has sufficient suction to pick up fumes, ozone, and smoke from the welding area and carry the fumes, etc., outside of the area.
recogedor de extracción. Un componente de un sistema de ventilación forzada que tiene suficiente succión para recoger vaho, ozono, y humo de la área de soldadura y lleva al vaho, etc., a fuera de la area.

F

face bend-test. A test in which the weld face is on the convex surface of a specified bend radius.
prueba de dobléz de cara. Una prueba donde la cara de la soldadura está en la superficie convexa al radio de dobléz especificado.

face of weld. The exposed surface of a weld on the side from which welding was done.
cara de la soldadura. La superficie expuesta de una soldadura del lado de donde se hizo la soldadura.

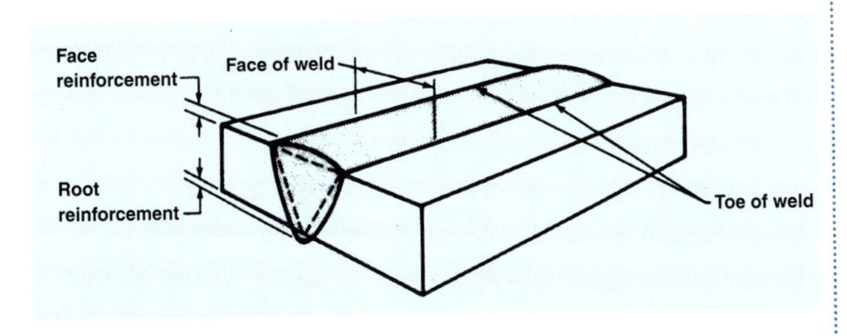

face reinforcement. Reinforcement of a weld at the side of the joint from which welding was done. See also **root reinforcement**. Refer to drawing for **face of weld**.
refuerzo de cara. Refuerzo de una soldadura en el lado de la junta de donde se hizo la soldadura. Vea también **refuerzo de raíz**. Refiérase al dibujo para **cara de la soldadura**.

face shield. A device positioned in front of the eyes and a portion of, or all of, the face, whose predominant function is protection of the eyes and face. See also **helmet**.
protector de cara sostenido a mano. Un aparato puesto en frente de los ojos y una porción, o en toda la cara, cuya función predominante es de proteger los ojos y la cara. Vea también **casco**.

***fast freezing electrode.** An electrode whose flux forms a high-temperature slag that solidifies before the weld metal solidifies, thus holding the molten metal in place. This is an advantage for vertical, horizontal, and overhead welding positions.
electrodo de congelación rápida. Un electrodo cuyo flujo forma una escoria a temperaturas altas que se puede solidificar antes de que el metal de soldadura se pueda solidificar, asi detiene el metal derretido en su lugar. Está es una ventaja en soldaduras de posiciones vertical, horizontal y sobrecabeza.

feed rollers. A set of two or four individual rollers which, when pressed tightly against the filler wire and powered up, feed the wire through the conduit liner to the gun for GMAW and FCAW welding.
rodillos de alimentación. Un conjunto de dos o cuatro rodillos individuales que al ser presionados fuertemente contra el alambre de relleno y ser accionados alimentan al alambre a través del revestimiento de canal hasta la pistola, en soldaduras tipo GMAW y FCAW.

filler metals. The metals or alloys to be added in making a welded, brazed, or soldered joint.
metales de aporte. Los metales o aleados que se agregan cuando se hace una soldadura blanda o soldadura fuerte.

***filler pass.** One or more weld beads used to fill the groove with weld metal. The bead must be cleaned after each pass to prevent slag inclusions.
pasada para rellenar. Uno o más cordones de soldadura usados para llenar la ranura con el metal de soldadura. El cordón debe ser limpiado después de cada pasada para prevenir inclusiones de escoria.

fillet weld. A weld of approximately triangular cross section joining two surfaces approximately at right angles to each other in a lap joint, tee joint, or corner joint. Refer to drawing for **convexity**.
soldadura de filete. Una soldadura de filete de sección transversa aproximadamente triangular que une dos superficies aproximademente en ángulos rectos de uno al otro en junta de traslape, junta en- T- o junta de esquina. Refiérase al dibujo para **convexidad**.

fillet weld break test. A test in which the specimen is loaded so that the weld root is in tension.
prueba de rotura en soldadura de filete. Una prueba en donde la probeta es cargada de manera en que la tensión esté sobre la soldadura.

fillet weld leg. The distance from the joint root to the toe of the fillet weld.
pierna de soldadura filete. La distancia de la raíz de la junta al pie de la soldadura filete.

fillet weld size. For equal leg fillet welds, the leg lengths of the largest isosceles right triangle that can be inscribed within the fillet weld cross section. For unequal leg fillet welds, the leg lengths of the largest right triangle that can be inscribed within the fillet weld cross section.
tamaño de soldadura filete. Para soldaduras de filete que tienen piernas iguales, lo largo de las piernas del isósceles más grande del triángulo recto que puede ser inscribido dentro de la sección. Para soldaduras de filete con piernas desiguales, lo largo de las piernas del triángulo recto más grande puede inscribirse dentro de la sección transversal.

filter plate. An optical material that protects the eyes against excessive ultraviolet, infrared, and visible radiation.
lente filtrante. Un material óptico que protege los ojos contra ultravioleta excesiva, infrarrojo, y radiación visible.

final current. The current after downslope but prior to current shut-off.
corriente final. La corriente después del pendiente en descenso pero antes de que la corriente sea cerrada.

fissure. A small, cracklike discontinuity with only slight separation (opening displacement) of the fracture surfaces. The prefixes *macro* and *micro* indicate relative size.
hendemiento. Una pequeña, discontinuidad como una grieta con solamente una separación (abertura desalojada) de las superficies fracturadas. El prefijo *marco* y *micro* indica el tamaño relativo.

***fixed inclined (6G) position.** For pipe welding, the pipe is fixed at a 45° angle to the work surface. The effective welding angle changes as the weld progresses around the pipe.
posición (6G) inclinado fijo. Para soldadura de tubo, el tubo se fija a un ángulo de 45° de la superficie del trabajo. El ángulo efectivo de la soldadura cambia cuando la soldadura progresa alrededor del tubo.

fixture. A device designed to hold parts to be joined in proper relation to each other.

fijación. Una devisa diseñada para detener partes que se van a unir en relación propia de una a la otra.

flame propagation rate. The speed at which flame travels through a mixture of gases.
cantidad de propagación de la llama. La rapidez en que la llama camina a través de una mezcla de gas.

flange weld. A weld made on the edges of two or more joint members, at least one of which is flanged.
soldadura de reborde. Una soldadura que se hace en las orillas de dos o más miembros de junta, donde por lo menos uno tiene reborde.

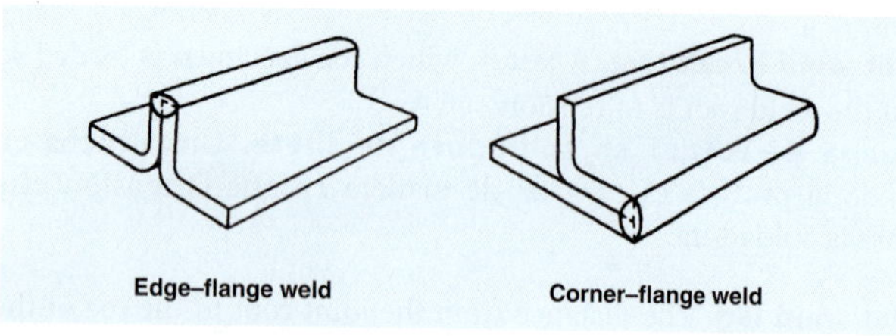

flash. The material that is expelled or squeezed out of a weld joint and that forms around the weld.
ráfaga. El material que es despedido o exprimido fuera de una junta de soldadura y se forma alrededor de la soldadura.

flashback. A recession of the flame into or back of the mixing chamber of the oxyfuel gas torch or flame spraying gun.
llamarada de retroceso. Una recesión de la llama adentro o atrás de la cámara mezcladora de una antorcha de gas oxicombustible o pistola de rociar a llama.

flashback arrester. A device to limit damage from a flashback by preventing propagation of the flame front beyond the point at which the arrester is installed.
válvula de retención. Un aparato para limitar el daño de una llamarada de retroceso para prevenir la propagación del frente de la llama más allá del punto donde se instala la válvula de retención.

flash welding (FW). A resistance welding process that produces a weld at the faying surfaces of a butt joint by a flashing action and by the application of pressure after heating is substantially completed. The flashing action, caused by the very high current densities at small contact points between the workpieces, forcibly expels the material from the joint as the workpieces are slowly moved together. The weld is completed by a rapid upsetting of the workpieces.
soldadura de relámpago. Un proceso de soldadura de resistencia que produce una soldadura en el empalme de la superficie de una junta tope por una acción de relampagueo y por la aplicación de presión después que el calentamiento este substancialmente acabado. La acción del relampagueo, causado por densidades de corrientes altas a unos puntos de contacto pequeños en medio de las piezas de trabajo, despiden fuertemente el material de la junta cuando las piezas de trabajo se mueven despacio. La soldadura es terminada por un acortamiento rápido de las piezas de trabajo.

flat position. The welding position used to weld from the upper side of the joint; the weld face is approximately horizontal.
posición plana. La posición de soldadura que se usa para soldar del lado de arriba de una junta; la cara de la soldadura está aproximadamente horizontal.

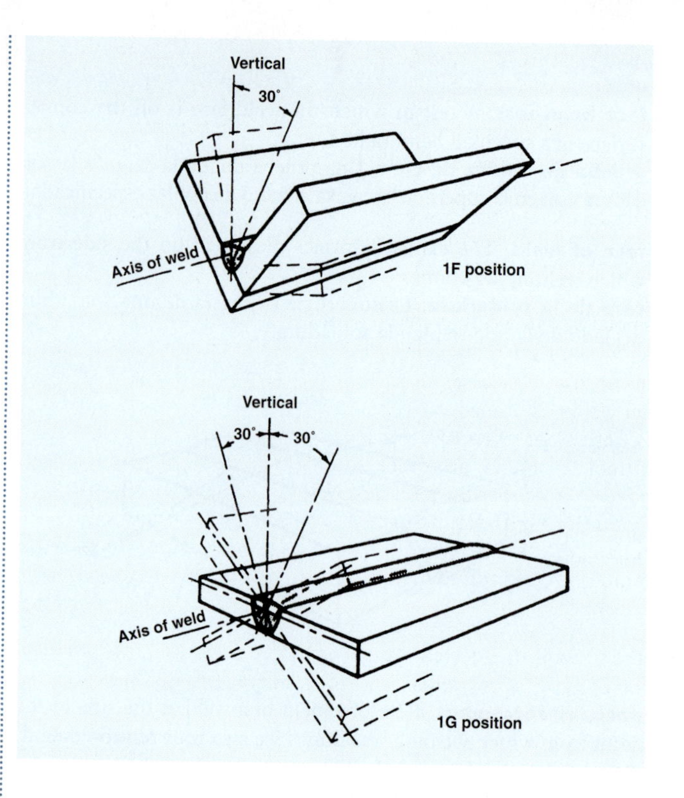

flaw. An undesirable discontinuity.
falta. Una discontinuidad indeseable.

flow rate. The rate at which a given volume of shielding gas is delivered to the weld zone. The units used for welding are cubic feet, inches, meters, and centimeters.
caudal. Velocidad a la cual llega un determinado volumen de gas protector a la zona de soldadura. Las unidades usadas para la soldadura son pies cúbicos, pulgadas, metros, y centímetros.

flux. A material used to hinder or prevent the formation of oxides and other undesirable substances in molten metal and on solid metal surfaces and to dissolve or otherwise facilitate the removal of such substances.
flujo. Un material que se usa para impedir o prevenir la formación de óxidos y otras substancias indeseables en el metal derretido y en las superficies del metal sólido, y para desolver o de otra manera facilitar el removimiento de dichas substancias.

flux cored arc welding (FCAW). An arc welding process that uses an arc between a continuous filler metal electrode and the weld pool. The process is used with shielding gas from a flux contained within the tubular electrode, with or without additional shielding from an externally supplied gas, and without the application of pressure.
soldadura de arco con núcleo de fundente. Un proceso de soldadura de arco que usa un arco entre medio de un electrodo de metal rellenado continuo y el charco de la soldadura. El proceso es usado con gas de protección del flujo contenido dentro del electrodo tubular, y sin usarse protección adicional de abastecimiento de gas externo, y sin aplicarse presión.

flux cored electrode. A composite tubular filler metal electrode consisting of a metal sheath and a core of various powdered materials, producing an extensive slag cover on the face of a weld bead. External shielding may be required.
electrodo de núcleo de fundente. Un electrodo tubular de metal para rellenar con una compostura que consiste de una envoltura de metal y un núcleo con varios materiales de polvo,

que producen un forro extensivo de escoria en la superficie del cordón de soldadura. Protección externa puede ser requerida.

flux cover. In metal bath dip brazing and dip soldering, a cover of flux over the molten filler metal bath.
tapa de fundente. En metal de baño soldadura fuerte y soldadura blanda por inmersión, una tapa de fundente sobre el baño del metal de relleno derretido.

***forced ventilation.** To remove excessive fumes, ozone, or smoke from a welding area, a ventilation system may be required to supplement natural ventilation. Where forced ventilation of the welding area is required, the rate of 200 cu ft (56 m^3) or more per welder is needed.
ventilación forzada. Para quitar excesivo vaho, ozono y humo de la área donde se solda, un sistema de ventilación puede ser requerido para suplementar la ventilación natural. Donde la ventilación forzada de la área de la soldadura es requerida, la rázon de 200 pies cúbicos (56 m^3) o más es requerido por cada soldador.

forehand welding. A welding technique in which the welding torch or gun is directed toward the progress of welding. See also **travel angle**, **work angle**, and **push angle**. Refer to drawing for **backhand welding**.
soldadura directa. Una técnica de soldar en cual la pistola o la antorcha para soldar es dirigida hacia al progreso de la soldadura. Vea también **ángulo de avance**, **ángulo de trabajo**, y **ángulo de empuje**. Refiérase al dibujo **soldadura en revés**.

forge welding (FOW). A solid state welding process that produces a weld by heating the workpieces to welding temperature and applying blows sufficient to cause permanent deformation at the faying surfaces.
soldadura por forjado. Un proceso de soldadura de estado sólido que produce una soldadura calentando las piezas de trabajo a una temperatura de soldadura y aplicando golpes suficientes para causar una deformación permanente en las superficies del empalme.

form. A metal or wooden structure that contains and shapes concrete until it hardens.
molde. Una estructura de metal o de madera que contiene y moldea el hormigón hasta que se frague.

framing square (carpenter's square). A flat square with a body and tongue.
escuadra. Un instrumento plano con cuerpo y legueta, usado para segurar ensambladuras.

frost line. The maximum depth that the soil freezes in a given locality.
deposito de escarcha. La parte mas profundo del suelo que se hiela en un sitio determinado.

fuel gases. Gases such as acetylene, natural gas, hydrogen, propane, stabilized methylacetylene propadiene, and other fuels normally used with oxygen in one of the oxyfuel processes and for heating.
gases combustibles. Gases como acetileno, gas natural, hidrógeno, propano, metilacetileno propodieno estabilizado, y otros combustibles normalmente usados con oxígeno en uno de los procesos de oxicombustible y para calentar.

full fillet weld. A fillet weld whose size is equal to the thickness of the thinner member joined.
soldadura de filete llena. Una soldadura de filete cuyo tamaño es igual de grueso como el miembro más delgado de la junta.

full penetration. A nonstandard term for complete joint penetration.
penetración llena. Un término fuera de la norma en vez de la penetración de junta.

fusion. The melting together of filler metal and base metal, or of base metal only, to produce a weld. See also **depth of fusion.**
fusión. El derretir el metal de relleno y el metal base juntos o el metal base solamente, para producir una soldadura. Vea también **grueso de fusión.**

fusion welding. Any welding process or method that uses fusion to complete the weld.
soldadura de fusión. Cualquier proceso de soldadura o método que usa fusión para completar la soldadura.

fusion zone. The area of base metal melted as determined on the cross section of a weld. Refer to drawing for **depth of fusion.**
zona de fusión. La área del metal base que se derritió como determinada en la sección transversa de la soldadura. Refiérase al dibujo **grueso de fusion.**

galvanized. Coated with zinc for rust resistance.
galvanizado. Cubierto con una capa de cinc para proteccion contra la corrosion.

gap. A nonstandard term when used for arc length, joint clearance, and root opening.
abertura. Un término fuera de norma cuando se usa en lugar del arco, despejo de junta, y abertura de raíz.

gas cup. A nonstandard term for gas nozzle.
tazón de gas. Un término fuera de norma en vez de boquilla de gas.

gas cylinder. A portable container used for transportation and storage of compressed gas.
cilindro de gas. Un recipiente portátil que se usa para transportación y deposito de gas comprimido.

gas metal arc welding (GMAW). An arc welding process that uses an arc between a continuous filler metal electrode and the weld pool. The process is used with shielding from an externally supplied gas and without the application of pressure.
soldadura de arco metálico con gas. Un proceso de soldar con arco que usa un arco en medio de un electrodo de metal para rellenar continuo y el charco de soldadura. El proceso usa protección de un abastecedor externo de gas y sin la aplicación de presión.

gas nozzle. A device at the exit end of the torch or gun that directs shielding gas.
boquilla de gas. Un aparato a la salida de la punta de la antorcha o pistola que dirige el gas protector.

gas regulator. A device for controlling the delivery of gas at some substantially constant pressure.
regulador de gas. Un aparato para controlar la salida de gas a una presión substancialmente constante.

gas tungsten arc welding (GTAW). An arc welding process that uses an arc between a tungsten electrode (nonconsumable) and the weld pool. The process is used with shielding gas and without the application of pressure.
soldadura de arco de tungsteno con gas. Un proceso de soldadura de arco que usa un arco en medio del electrodo tungsteno (no consumible) y el charco de la soldadura. El proceso es usado con gas de protección y sin aplicación de presión.

***gauge** (regulator). A device mounted on a regulator to indicate the pressure of the gas passing into the gauge. A regulator is provided with two gauges—one (high-pressure gauge) indicates the pressure of the gas in the cylinder; the second gauge (low-pressure gauge) shows the pressure of the gas at the torch.
manómetro (regulador). Un aparato montado en un regulador para indicar la presión del gas que está pasando por el manómetro. El regulador tiene dos manómetros—uno (manómetro de alta presión) indica la presión del gas en el cilindro; el segundo manómetro (manómetro de presión baja) enseña la presión del gas en la antorcha.

***gauge pressure.** The actual pressure shown on the gauge; does not take into account atmospheric pressure.
manómetro para presión. La presión actual que se enseña en el manómetro; no toma en cuenta la presión atmosférica.

globular transfer. The transfer of molten metal in large drops from a consumable electrode across the arc.
traslado globular. El traslado del metal derretido en gotas grandes de un electrodo consumible a través del arco.

gouging. The forming of a bevel or groove by material removal. See also **back gouging** and **arc gouging.**
escopleando con gubia. Formando un bisel o ranura removiendo el material. Vea también **gubia trasera y gubia dura con arco.**

groove. An opening or a channel in the surface of a part or between two components that provides space to contain a weld.
ranura. Una abertura o un canal en la superficie de una parte o en medio de dos componentes, la cual provee espacio para contener una soldadura.

groove angle. The total included angle of the groove between parts to be joined by a groove weld.
ángulo de ranura. El ángulo total incluido de la ranura entre partes para unirse por una soldadura de ranura.

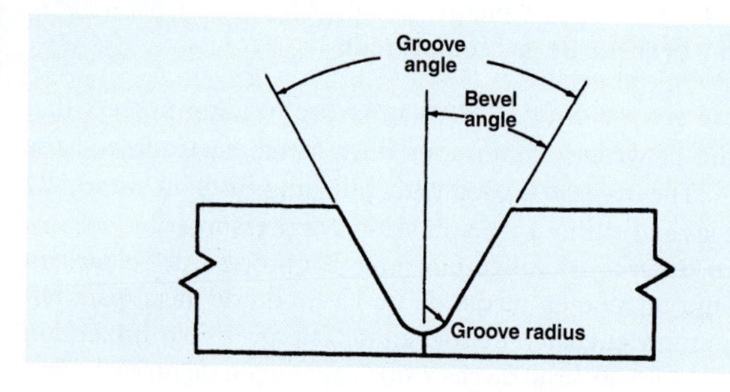

groove face. The surface of a joint member included in the groove.
cara de ranura. La superficie de un miembro de una junta incluido en la ranura.

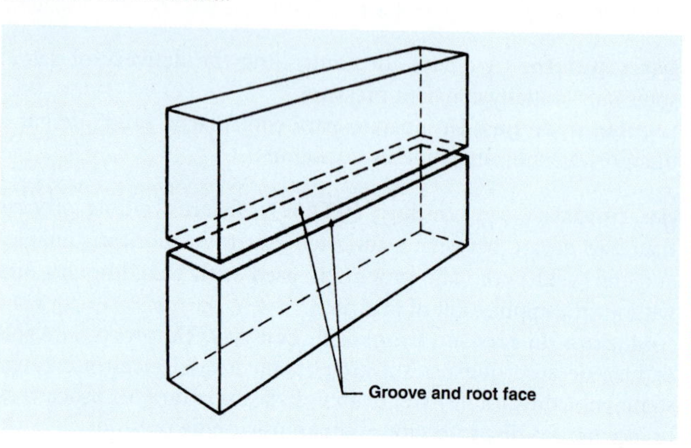

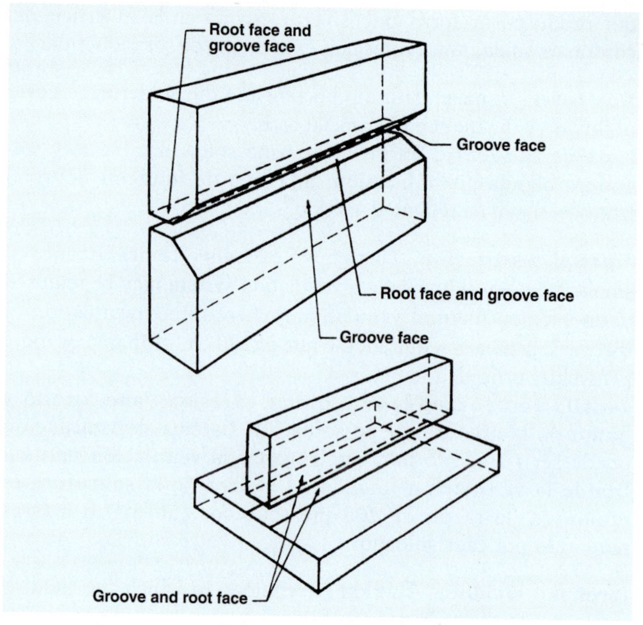

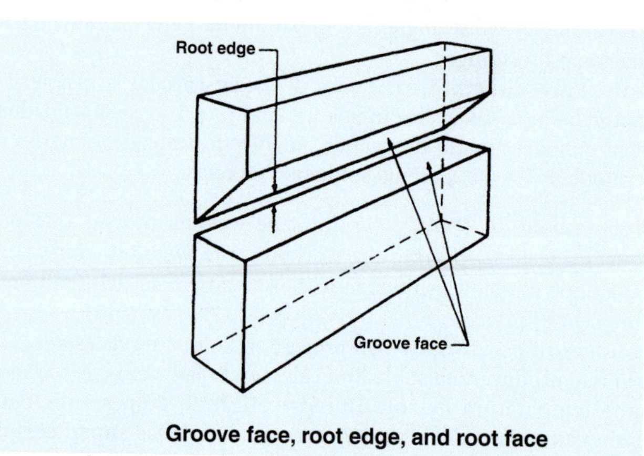

Groove face, root edge, and root face

groove radius. The radius used to form the shape of a J- or U-groove weld joint. Refer to drawings for **bevel.**
radio de ranura. La radio que se usa para formar una junta de una soldadura con una ranura de forma U o J. Refiérase a los dibujos para **bisel.**

groove weld. A weld made in the groove between two members to be joined. The standard types of groove welds are shown in the drawings.
soldadura de ranura. Una soldadura hecha en la ranura dentro de dos miembros que se unen. Los tipos normales de soldadura de ranura se ven en los dibujos.

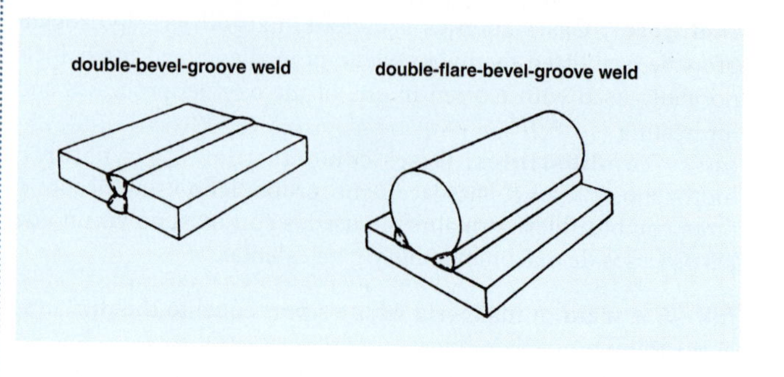

(figure continued on page 509)

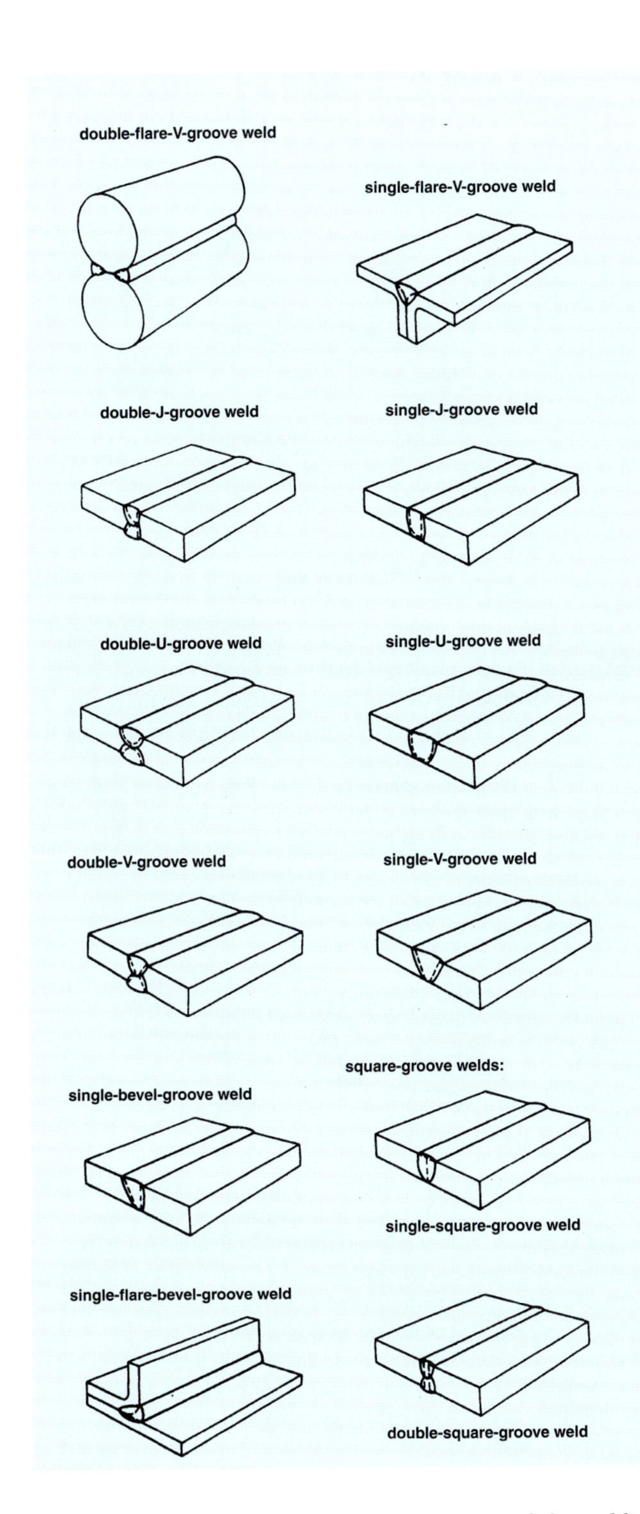

ground connection. An electrical connection of the welding machine frame to the earth for safety.
conexión a tierra. Una conexión eléctrica del marco de la máquina de soldar a la tierra para seguridad.

ground lead. A nonstandard and incorrect term for workpiece lead.
cable de tierra. Un término fuera de norma e incorrecto que se usa en vez de cable de pieza de trabajo.

***guided bend specimen.** Any bend specimen that will be bend-tested in a fixture that controls the bend radii, such as the AWS bend-test fixture.
probeta de dobléz guiada. Cualquier probeta de dobléz en la cual se va a hacer un dobléz guiado en una máquina que controla el radio del dobléz, como la máquina de dobléz guiado del AWS.

H

hand saw. A saw used to cut across boards or to rip boards and panels.
sierra de mano. Una sierra usada para cortar a traves de maderos o hender maderos y tableros.

hardfacing. A surfacing variation in which surfacing material is deposited to reduce wear.
endurecimiento de caras. Una variación superficial donde el material superficial es depositado para reducir el desgastamiento.

heat-affected zone. The portion of the base metal whose mechanical properties or microstructure has been altered by the heat of welding, brazing, soldering, or thermal cutting.
zona afectada por el calor. La porción del metal base cuya propiedad mecánica o microestructura ha sido alterada por el calor de soldadura, soldadura fuerte, soldadura blanda, o corte termal.

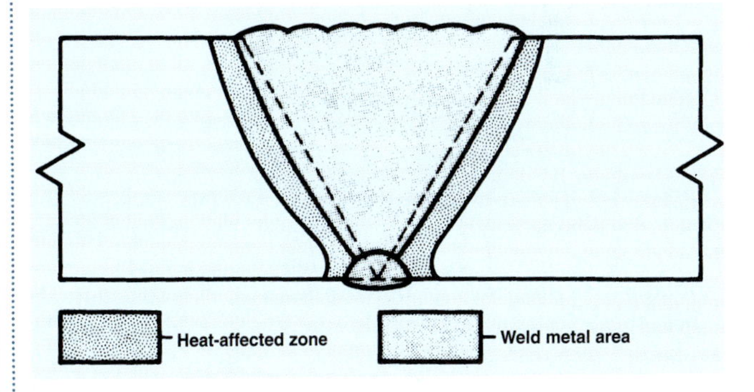

heating torch. A device for directing the heating flame produced by the controlled combustion of fuel gases.
antorcha de calentamiento. Un aparato para dirigir la llama de calentamiento que es producida por una combustión controlada de gases de combustión.

helmet. A device designed to be worn on the head to protect eyes, face, and neck from arc radiation, radiated heat, spatter, or other harmful matter expelled during arc welding, arc cutting, and thermal spraying.
casco. Un aparato diseñado para usarse sobre la cabeza para proteger ojos, cara y cuello de radiación del arco, calor radiado, salpicadura, u otra materia dañosa despedida durante la soldadura de arco, corte por arco, y rociado termal.

hinge. An object that pivots and permits a door or other object to swing back and forth or up and down.
bisagra. Un objeto que da vueltas sobre su eje y permite que una puerta u otro objeto gire de aca para alla o de arriba abajo.

horizontal fixed position (pipe welding). The position of a pipe joint in which the axis of the pipe is approximately horizontal and the pipe is not rotated during welding.
posición fija horizontal (soldadura de tubos). La posición de una junta de tubo la cual el axis del tubo es aproximadamente horizontal, y el tubo no da vueltas durante la soldadura.

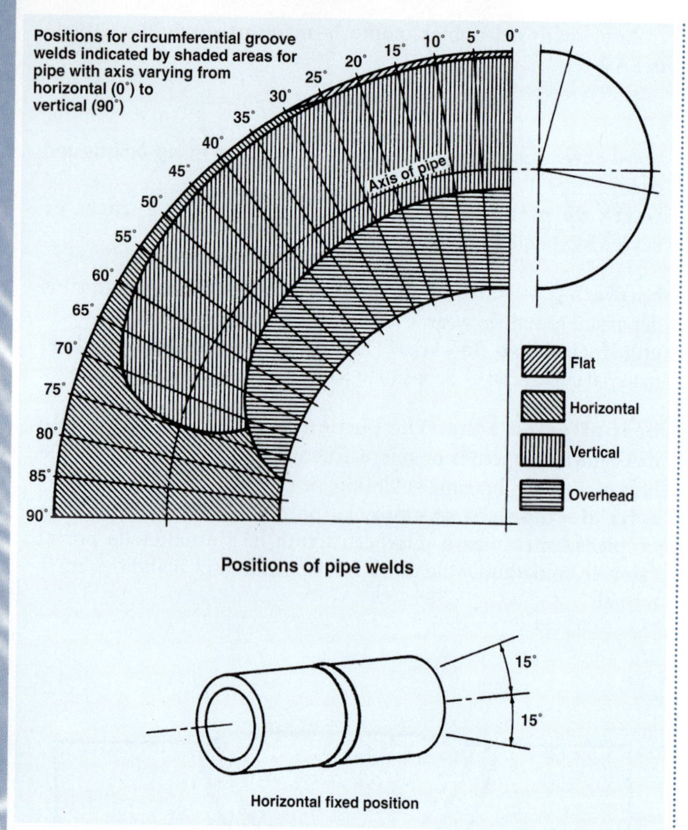

Positions of pipe welds

Horizontal fixed position

***horizontal fixed (5G) position weld.** For pipe welding, the pipe is fixed horizontally (cannot be rolled). The weld progresses from overhead, to vertical, to flat position around the pipe.
soldadura de posición fija horizontal (5G). Para soldadura de tubos, el tubo está fijo horizontalmente (no se pueder rodar). La soldadura progresa de sobre cabeza, a vertical, a la posición plana alrededor del tubo.

horizontal position (fillet weld). The position in which welding is performed on the upper side of an approximately horizontal surface and against an approximately vertical surface.
posición horizontal (soldadura de filete). La posición de la soldadura la cual es hecha en el lado de arriba de una superficie horizontal aproximadamente y junto a una superficie vertical aproximadamente.

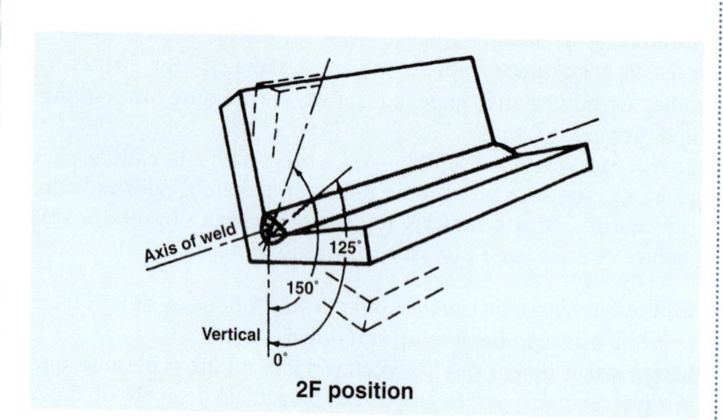

2F position

horizontal position (groove weld). The position of welding in which the weld axis lies in an approximately horizontal plane and the weld face lies in an approximately vertical plane.
posición horizontal (de ranura). La posición para soldar en la cual el axis de la soldadura está en una plana horizontal aproximadamente, y la cara de la soldadura está en una plana vertical aproximadamente.

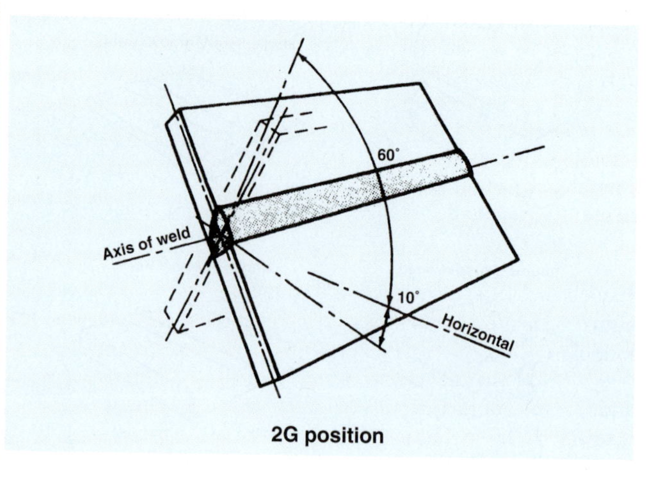

2G position

horizontal rolled position (pipe welding). The position of a pipe joint in which the axis of the pipe is approximately horizontal and welding is performed in the flat position by rotating the pipe.
posición horizontal rodada (soldadura de tubo). La posición de una junta de tubo en la cual el axis del tubo es horizontal aproximadamente, y la soldadura se hace en la posición plana con rotación del tubo.

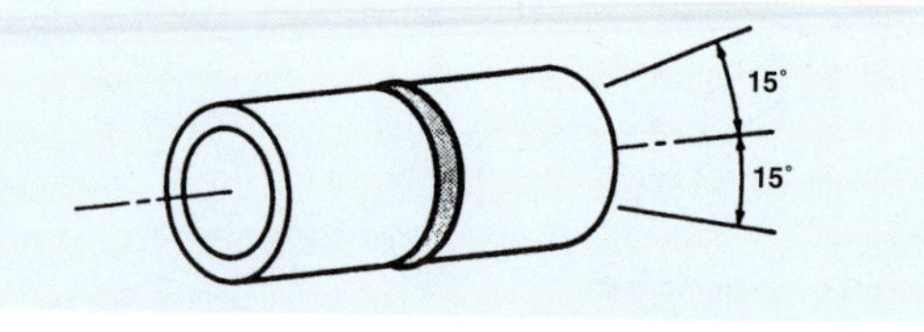

***horizontal rolled (1G) position.** For pipe welding, this position yields high-quality and high-quantity welds. Pipe to be welded is placed horizontally on the welding table in a fixture to hold it steady and permit each rolling. The weld proceeds in steps, with the pipe being rolled between each step, until the weld is complete. For plate, see **axis of a weld**.
posición (1G) horizontal rodada. Para soldadura de tubo, está posición produce soldaduras de alta calidad y alta cantidad. El tubo que se va a soldar se pone horizontalmente sobre la mesa de soldadura en una instalación fija que lo detiene seguro y permite cada rodadura. La soldadura procede en pasos, con el tubo siendo rodado entre cada paso, hasta que la soldadura esté completa. Para plato, vea **eje de la soldadura.**

hose. A flexible line that carries gases or liquids.
manga. Tubo largo flexible que lleva gases o liquidos.

hot crack. A crack formed at temperatures near the completion of solidification.
grieta caliente. Una grieta formada a temperaturas cerca de la terminación de la solidificación.

***hot pass.** The welding electrode is passed over the root pass at a higher than normal amperage setting and travel rate to reshape an irregular bead and turn out trapped slag. A small amount of metal is deposited during the hot pass so the weld bead is convex, promoting easier cleaning.

pasada caliente. El electrodo de soldadura se pasa sobre la pasada de raíz poniendo el amperaje más alto que lo normal y proporción de avance para reformar un cordón irregular y sacar la escoria atrapada. Una cantidad pequeña de metal es depositada durante la pasada caliente para que el cordón soldado sea convexo, promoviendo más fácil la limpieza.

hot start current. A very brief current pulse at arc initiation to stabilize the arc quickly.
corriente caliente para empezar. Un pulso muy breve de corriente a iniciación de arco para estabilizar el arco aprisa.

hydrogen embrittlement. The delayed cracking in steel that may occur hours, days, or weeks following welding. It is a result of hydrogen atoms that dissolved in the molten weld pool during welding.
fragilidad causada por el hidrógeno. El fisuramiento retardado en el acero que puede ocurrir horas, días o semanas después de la soldadura. Es el resultado de la disolución de átomos de hidrógeno en el charco de soldadura derretido durante la soldadura.

I

inclined position. The position of a pipe joint in which the axis of the pipe is at an angle of approximately 45° to the horizontal, and the pipe is not rotated during welding.
posición inclinada. Una posición de junta de tubo en la cual el axis del tubo está a un ángulo de aproximadamente 45° a la horizontal, y no se le da vueltas al tubo durante la soldadura.

included angle. A nonstandard term for groove angle.
ángulo incluido. Un término fuera de norma para ángulo de ranura.

inclusion. Entrapped foreign solid material, such as slag, flux, tungsten, or oxide.
inclusión. Material extraño atrapado sólido, como escoria, flujo, tungsteno, u óxido.

incomplete fusion. A weld discontinuity in which fusion did not occur between weld metal and fusion faces or adjoining weld beads.
fusión incompleta. Una discontinuidad en la soldadura en la cual no ocurrió fusión entre el metal soldado y caras de fusión o cordones soldados inmediatos.

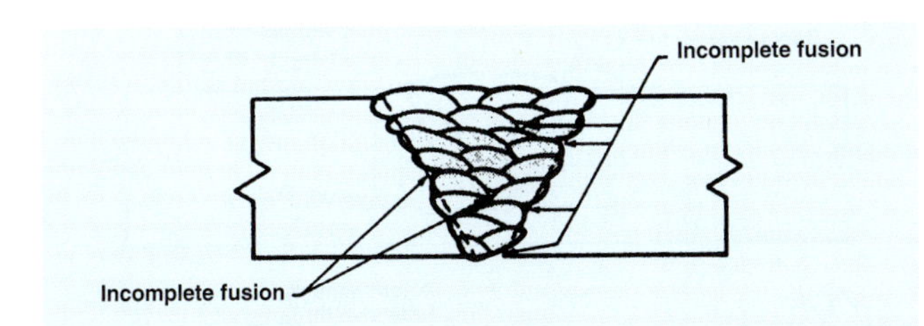

incomplete joint penetration. Joint penetration that is unintentionally less than the thickness of the weld joint.
penetración de junta incompleta. Penetración de la junta que no es intencionalmente menos de lo grueso de la junta de soldar.

inert gas. A gas that normally does not combine chemically with materials.
gas inerte. Un gas que normalmente no se combina químicamente con materiales.

***injector chamber.** One method of completely mixing the fuel gas and oxygen to form a flame. High-pressure oxygen is passed through a narrowed opening (venturi) to the mixing chamber. This action creates a vacuum, which pulls the fuel gas into the chamber and ensures thorough mixing. Used for equal gas pressures and is particularly useful for low-pressure fuel gases.
cámara de inyector. Un método de mezclar completamente el gas de combustión y el oxígeno para formar una llama. Oxígeno a alta presión es pasado por una abertura angosta (venturi) a la cámara de mezcla. Está acción hace un vacuo, la cual estira el gas combustible para dentro de la cámara y asegura una mezcla completa. Usada para presiones de gas que son iguales y es particularmente útil para gases combustibles de presión baja.

***inner cone.** The portion of the oxyacetylene flame closest to the welding tip. The primary combustion reaction occurs in the inner cone. The size and color of the cone serve as indicators of the type of flame (carburizing, oxidizing, neutral).
cono interno. La porción de la llama de oxiacetileno más cerca de la punta para soldar. La reacción de combustión principal ocurre en el cono interno. El tamaño y el calor del cono sirve como indicadores del tipo de la llama (carburante, oxidante, neutral).

interpass temperature. In a multipass weld, the temperature of the weld area between weld passes.
temperatura de pasada interna. En una soldadura de pasadas multiples, la temperatura en la área de la soldadura entre pasadas de soldaduras.

***iron.** An element. Very seldom used in its pure form. The most common element alloyed with iron is carbon.
hierro. Un elemento. Es muy raro que se use en forma pura. El elemento más común del aleado con hierro es carbón.

J

J-groove weld. A type of groove weld.
soldadura con ranura-J. Es un tipo de soldadura de ranura.

joint. The junction of members or the edges of members that are to be joined or have been joined.
junta. El punto en que se unen dos miembros o las orillas de los miembros que están para unirse o han sido unidos.

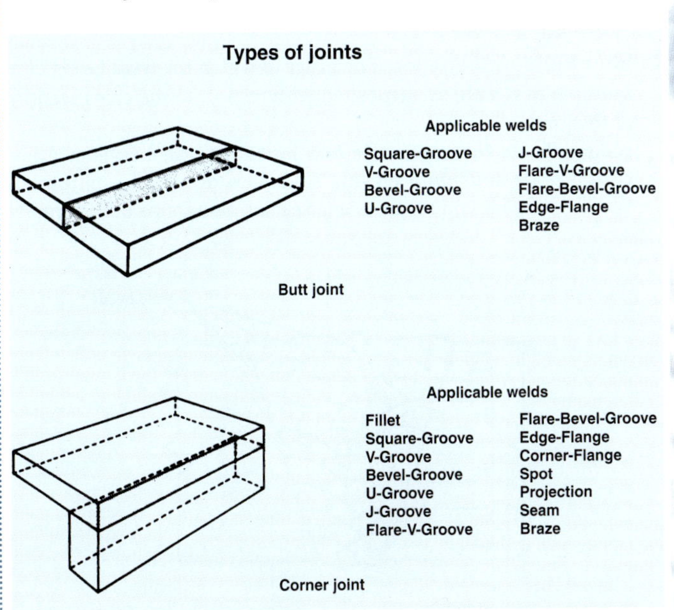

(figure continued on page 512)

(figure continued from page 511)

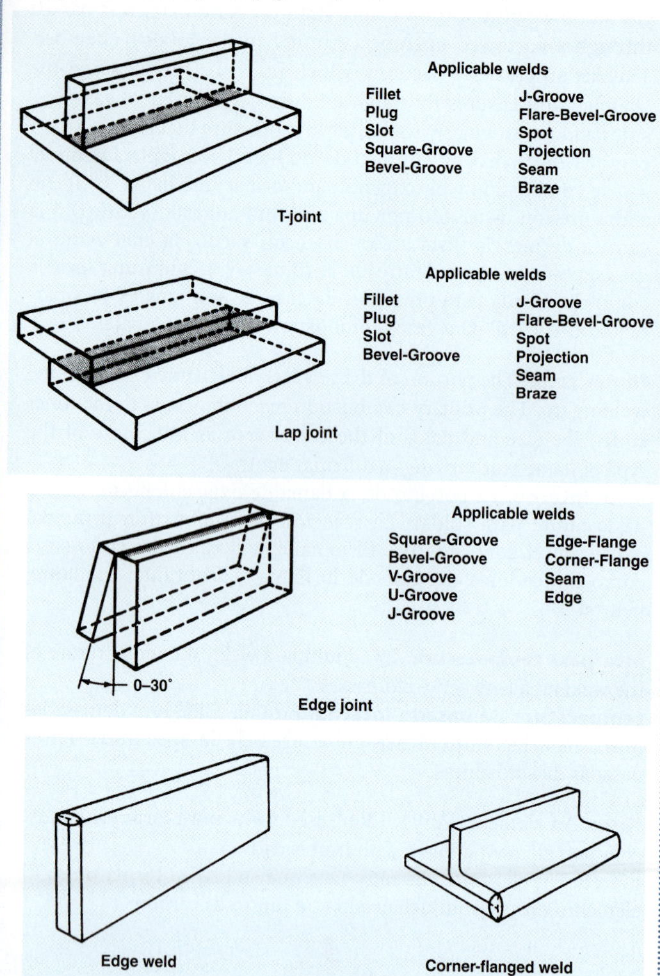

joint buildup sequence. The order in which the weld beads of a multiple-pass weld are deposited with respect to the cross section of the joint.
secuencia de formación de una junta. La orden en la cual los cordones de soldadura en una soldadura de pasadas múltiples son depositadas con respecto a la sección transversa de la junta.

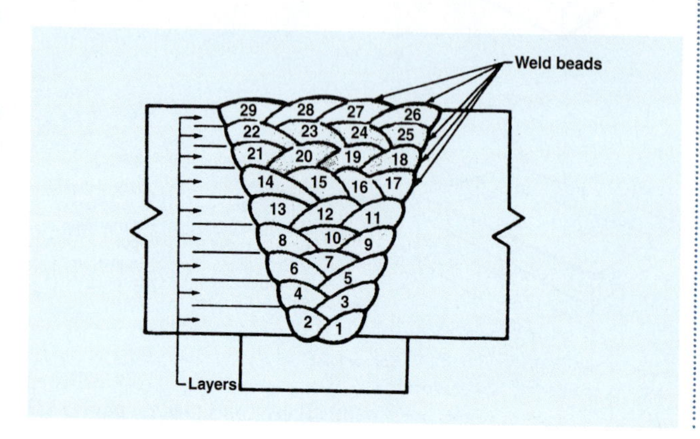

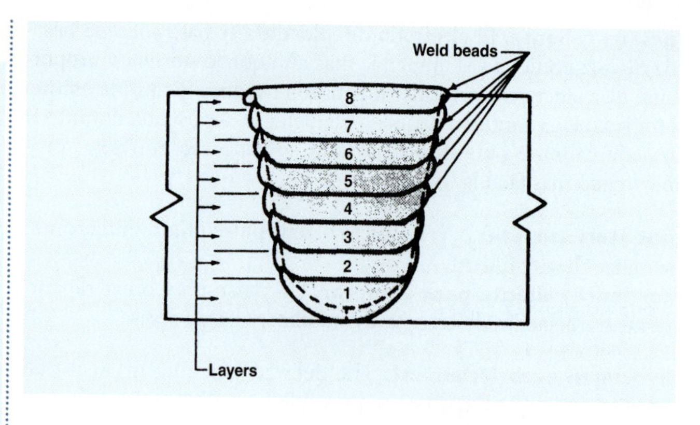

joint clearance. The distance between the faying surfaces of a joint.
despejo de junta. La distancia entre las superficies del empalme de una junta.

joint design. The joint geometry together with the required dimensions of the welded joint.
diseño de junta. La geometría de la junta junto con las dimensiones requeridas de la junta de la soldadura.

joint efficiency. The ratio of the strength of a joint to the strength of the base metal, expressed in percent.
eficiencia de junta. La razón de la fuerza de una junta a la fuerza del metal base, expresada en por ciento.

joint geometry. The shape and dimensions of a joint in cross section prior to welding.
geometría de junta. La figura y dimensión de una junta en sección transversa antes de soldarse.

joint penetration. The distance the weld metal extends from its face into a joint, exclusive of weld reinforcement.
penetración de junta. La distancia del metal soldado que se extiende de su cara hacia adentro de la junta, exclusiva de la soldadura de refuerzo.

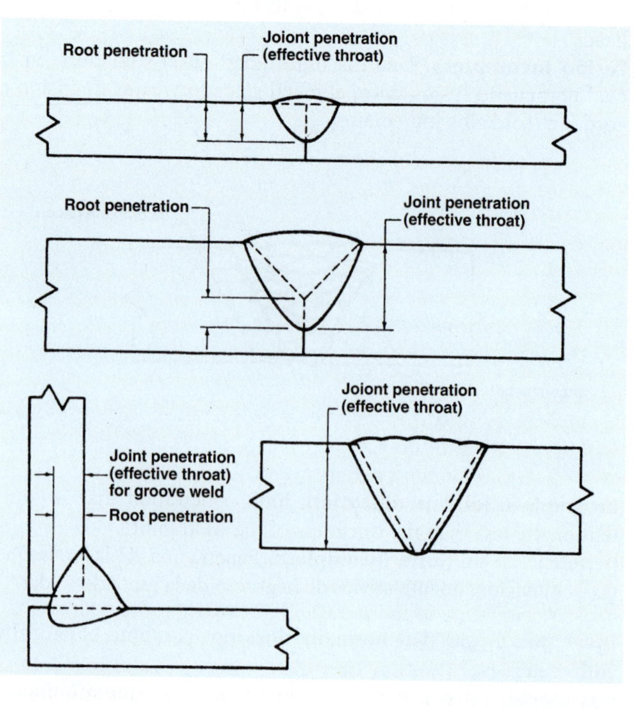

(figure continued on page 513)

(figure continued from page 512)

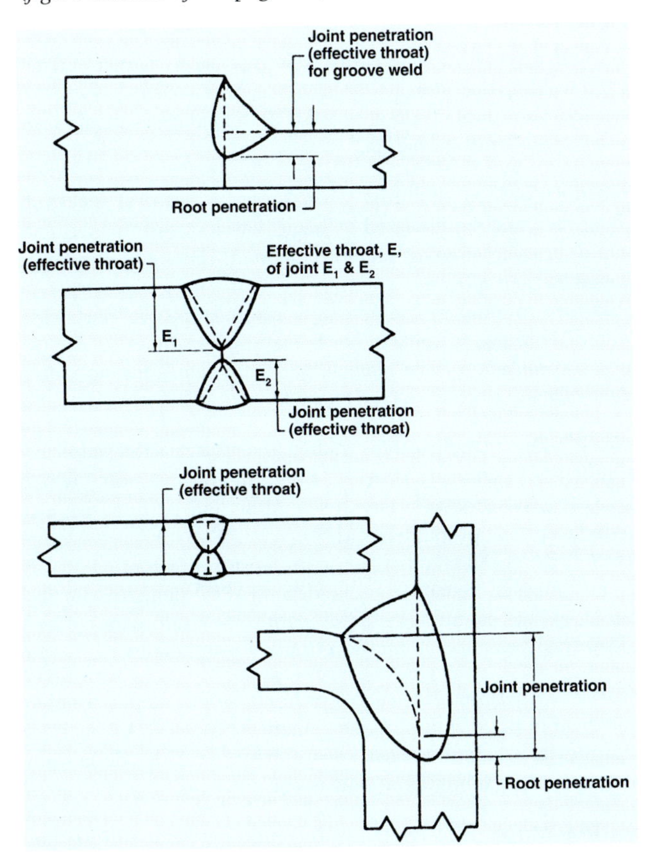

joint root. The portion of a joint to be welded where the members approach closest to each other. In cross section, the joint root may be a point, a line, or an area.
raíz de junta. Esa porción de una junta que está para soldarse donde los miembros están más cercanos uno del otro. En la sección transversa, la raíz de la junta puede ser una punta, una línea, o una área.

joint type. A weld joint classification based on the five basic arrangements of the component parts such as a butt joint, corner joint, edge joint, lap joint, and tee-joint.
tipo de junta. Una clasificación de una junta de soldadura basada en los cinco arreglos del componente de partes como junta a tope, junta en esquina, junta de orilla, junta de solape, y junta en T.

joint welding sequence. See preferred term **joint buildup sequence.**
secuencia para soldar una junta. Vea el término preferido **secuencia de formación de una junta.**

K

kerf. The width of the cut produced during a cutting process. Refer to drawing for **drag.**
cortadura. La anchura del corte producido durante un proceso de cortar. Refiérase al dibujo de **tiro.**

keyhole welding. A technique in which a concentrated heat source penetrates completely through a workpiece, forming a hole at the leading edge of the weld pool. As the heat source progresses, the molten metal fills in behind the hole to form the weld bead.
soldadura con pocillo. Una técnica en la cual una fuente de calor concentrado se penetra completamente a través de la pieza de trabajo, formando un agujero en la orilla del frente del charco de la soldadura. Asi como progresa la potencia de calor, el metal derretido rellena detrás del agujero para formar un cordón de soldadura.

kiln-dried. Lumber that has been dried by heat in a special oven called a kiln.
secado al horno. Dicese de los maderos que han sido secados por la calenture en un gran horno especial que se llama tostadero.

***kindling point.** The lowest temperature at which a material will burn.
punto de ignición. La temperatura más baja la cual un material se puede quemar.

L

lack of fusion. A nonstandard term for incomplete fusion.
falta de fusión. Un término fuera de norma para fusión incompleta.

lack of penetration. A nonstandard term for incomplete joint penetration.
falta de penetración. Un término fuera de norma para penetración de junta incompleta.

lag screws (lag bolts). Screws with coarse threads designed for use in structural timber or lead anchors.
tirafondo. Tornillos con roscas gruesas creados para usarse en maderos de construccion o ancoras de plomo.

land. See preferred term **root face.**
hombro. Vea el término preferido **cara de raíz.**

lap joint. A joint between two overlapping members.
junta de solape. Una junta entre dos miembros traslapadas.

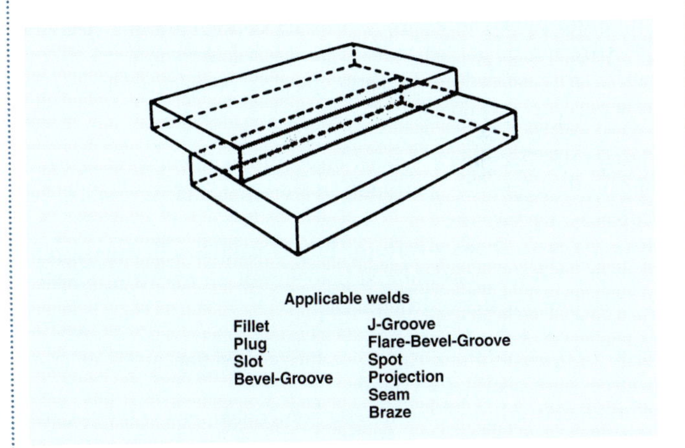

layer. A stratum of weld metal or surfacing material. The layer may consist of one or more weld beads laid side by side. Refer to drawing for **joint buildup sequence.**
capa. Un estrato de metal de soldadura o material de superficie. La capa puede consistir de uno o más cordones de soldadura depositados o puestos de lado. Refiérase al dibujo **secuencia de formación de una junta.**

***leak-detecting solution.** A solution, usually soapy water, that is brushed or sprayed on the hose fittings at the regulator and torch to detect gas leaks. If a small leak exists, soap bubbles form.
solución para descubrir escape. Una solución, por lo regular de agua enjabonada, que se acepilla o se rocía sobre las conexiones de las mangueras y los reguladores y antorcha para detectar escape de gas. Si existe un escape pequeño, se forman burbujas de jabón.

leg of a fillet weld. See fillet weld leg.
pierna de soldadura filete. Vea pierna de soldadura filete.

***liquid-solid phase bonding process.** Soldering or brazing where the filler metal is melted (liquid) and the base material does not melt (solid); the phase is the state at which bonding takes place between the solid base material and liquid filler metal. There is no alloying of the base metal.
proceso de ligación de fase líquido-sólido. Soldando con soldadura blanda o soldadura fuerte donde el metal de relleno se derrite (líquido) y el material base no se derrite (sólido); la fase es el estado la cual el ligamento se lleva a cabo entre el material base sólido y el metal de relleno (líquido). No se mezcla con el metal base.

local preheating. Preheating a specific portion of a structure.
precalentamiento local. El precalentamiento de una porción especificada de un estructura.

local stress relief heat treatment. Stress relief heat treatment of a specific portion of a structure.
tratamiento de calor para relevar la tensión local. Un tratamiento de calor el cual releva la tensión de una porción especificada de una estructura.

M

***machine operation.** Welding operations are performed automatically under the observation and correction of the operator.
operación de máquina. Operaciones de soldadura son ejecutadas automáticamente bajo la observación y corrección del operador.

machine welding. Welding with equipment that performs the welding operation under the constant observation and control of a welding operator. The equipment may or may not perform the loading and unloading of the work.
máquina para soldadura. Soldadura con equipo que ejecutan la operación de soldadura bajo la observación constante de un operador de soldadura. El equipo pueda o no ejecutar el cargar o descargar del trabajo.

***magnetic flux lines.** Parallel lines of force that always go from the north pole to the south pole in a magnet, and surround a DC current–carrying wire.
líneas magnéticas de flujo. Líneas paralelas de fuerza que siempre van del polo norte al polo sur en un magneto, y rodea un alambre que lleva corriente DC.

manifold. A multiple header for interconnection of gas or fluid sources with distribution points.
conexión múltiple. Una tuberia con conexiones múltiples que sirve como fuente de gas o flúido con puntos de distribución.

***manifold system.** Used when there are a number of work stations or a high volume of gas is required. A piping system that allows several oxygen and fuel-gas cylinders to be connected to several welding stations. Normally regulators are provided at the manifold and at the stations to provide control of the oxygen and fuel-gas pressures. Safety features such as reverse flow valves, flashback arrestors, and back pressure release must be provided at the manifold.
sistema de conexiones múltiples. Usado cuando hay un número de estaciones de trabajo o cuando se requiere un alto volumen de gas. Un sistema de tubos que permite que se conecten varios cilindros de oxígeno y gas combustible a varias estaciones de soldadura. Normalmente se usan reguladores en el tubo de conexiones múltiples y en las estaciones para mantener el control de la presión del oxígeno y el gas combustible. Normas de seguridad como válvulas de retención, protector de agua contra retroceso de llama, y escape de presión deben usarse en el tubo de conexiones múltiples.

***manual operation.** The entire welding process is manipulated by the welding operator.
operación manual. Todo el proceso de soldadura es manipulado por un operador de soldadura.

manual welding. Welding with the torch, gun, or electrode holder held and manipulated by hand. Variations of this term are *manual brazing, manual soldering, manual thermal cutting,* and *manual thermal spraying.*
soldadura manual. Soldando con la antorcha, pistola, porta electrodo detenido y manipulado por la mano. Variaciones de este término son *soldadura fuerte manual, soldadura blanda manual, cortes termal manual,* y *rociado termal manual.*

***MAPP®.** One manufacturer's trade name for a specific stabilized, liquefied MPS mixture. MAPP® has a distinctive odor, which makes it easy to detect; used for welding and cutting. See also **methylacetylene propadiene (MPS).**
MAPP®. Un nombre comercial de un fabricante para una específica estabilizada, licuada, mezcla MPS. MAPP® tiene un olor distintivo, el cual es muy fácil de descubrir; es usado para cortes y soldaduras. Vea también **metilacetileno y propadieno.**

masonry. Anything constructed of brick, stone, tile, or concrete units held in place with portland cement.
albañilería. Cualquiera cosa constuida de unidades de ladrillo, piedra, azulego, u homigon, unidas por pórtland.

melting range. The temperature range between solidus and liquidus.
variación de derretimiento. La variación de temperatura entre solidus y liquidus.

melting rate. The weight or length of electrode, wire, rod, or powder, melted in a unit of time.
cantidad de derretimiento. El peso o lo largo de un electrodo, alambre, varilla, o polvo derretido en una unidad de tiempo.

melt-through. Complete joint penetration for a joint welded from one side. Visible root reinforcement is produced.
derretir de un lado a otro. Una junta con penetración completa para una junta que está soldada de un lado. Refuerzo de raíz visible es producido.

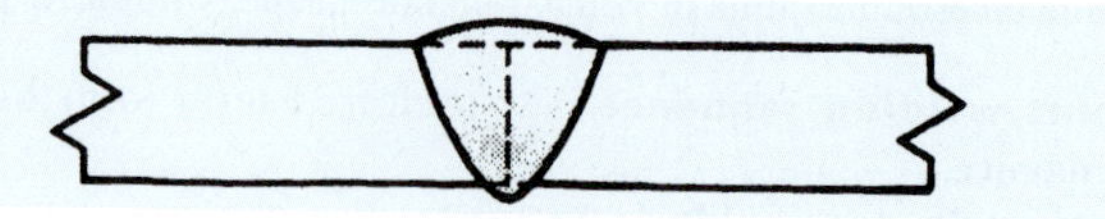

metal. An opaque, lustrous, elemental, chemical substance that is a good conductor of heat and electricity, usually malleable, ductile, and more dense than other elemental substances.
metal. Una opaca, brillante, elemental, substancia química que es una buena conductora de calor y electricidad, por lo regular es maleable, ductil, y es más densa que otras substancias elementales.

metal cored electrode. A composite tubular filler metal electrode consisting of a metal sheath and a core of various powdered materials, producing no more than slag islands on the face of a weld bead. External shielding may be required.
electrodo de metal de núcleo. Un electrodo de metal para rellenar tubular compuesto consistiendo de una envoltura de metal y núcleo de varios materiales en polvo, que producen nada más

que islas de escoria en la cara del cordón de soldadura. Protección externa puede ser requerida.

metal electrode. A filler or nonfiller metal electrode used in arc welding or cutting, which consists of a metal wire or rod that has been manufactured by any method and that is either bare or covered with a suitable covering or coating.
electrodo de metal. Un electrodo de metal que se usa para rellenar o para no rellenar la soldadura de arco o para cortar, que consiste de un alambre de metal o varilla que ha sido fabricada por cualquier método ya sea liso o cubierto con un cubierto o revestimiento propio.

***methylacetylene propadiene (MPS).** A family of fuel gases that are mixtures of two or more gases (propane, butane, butadiene, methylacetylene, and propadiene). The neutral flame temperature is approximately 5031°F (2927°C), depending upon the actual gas mixture. MPS is used for oxyfuel cutting, heating, brazing, and metallizing; rarely used for welding.
metilacetileno y propadieno. Una familia de gases de combustión que son mezclas de dos o más gases (propano, butano, butadiano, metilacetileno, propadieno). La temperatura de la llama natural es aproximadamente 5031°F (2927°C), dependiendo de la mezcla actual del gas. MPS es usado como gas de combustión para cortar, calentar, soldadura fuerte, y metalizar; es muy raro que se use para soldar.

mineral-based electrode fluxes. Fluxes that use inorganic compounds such as the rutile-based flux (titanium dioxide, TiO_2). These mineral compounds do not contain hydrogen, and electrodes that use these fluxes are often referred to as low hydrogen electrodes. Less smoke is generated with this welding electrode than with cellulose-based fluxes, but a thicker slag layer is deposited on the weld. E7018 is an example of an electrode that uses this type of flux.
fundentes para electrodos de base mineral. Fundentes que usan compuestos inorgánicos, como por ejemplo, el fundente a base de rutilo (bióxido de titanio, TiO_2). Estos compuestos minerales no contienen hidrógeno, y a los electrodos que usan estos fundentes se los llama con frecuencia electrodos de bajo hidrógeno. En la soldadura con electrodos se producen menos humos que en la que se realiza con fundentes celulósicos, pero se deposita una capa de escoria más gruesa en la soldadura. El F7018 es un ejemplo de un electrodo que usa este tipo de fundente.

mixing chamber. That part of a welding or cutting torch in which a fuel gas and oxygen are mixed.
cámara mezcladora. Esa parte de una antorcha para soldar y cortar por la cual el gas combustible y el oxígeno son mezclados.

mold. A high-temperature container into which liquid metal from the thermite welding process is poured and held until it cools and hardens into the container's interior shape.
molde. un contenedor de alta temperatura en el cual se vierte y se mantiene metal líquido del proceso de soldadura con termita hasta que éste se enfríe y se solidifique tomando la forma interior del contenedor.

molten weld pool. The liquid state of a weld prior to solidification as weld material.
charco de soldadura derretido. El estado líquido de una soldadura antes de solidificarse como material de soldadura.

***multipass weld.** A weld requiring more than one pass to ensure complete and satisfactory joining of the metal pieces.
soldadura de pasadas múltiples. Una soldadura que requiere más de una pasada para asegurar una completa y satisfactoria unión de las piezas de metal.

N

nail. A fastener that is driven into the material it holds.
clavo. Sujetador que se hinca en el material a que sujeta.

National Electrical Manufacturers Association (NEMA). The group that developed the system of color coding electrodes.
Asociación Nacional de Fabricantes Electrotecnicos (ANFE). El grupo que desarrollo el sistema de codificar electrodos por asignarios colores especificos.

neutral flame. An oxyfuel gas flame that has characteristics neither oxidizing nor reducing. Refer to drawing for **cone.**
llama neutral. Una llama de gas oxicombustible que no tiene características de oxidación ni de reducción. Refiérase al dibujo para **cono.**

nonconsumable electrode. An electrode that does not provide filler metal.
electrodo no consumible. Un electrodo que no provee metal de relleno.

noncorrosive flux. A soldering flux that in neither its original nor its residual form chemically attacks the base metal. It usually is composed of rosin or resin-base materials.
flujo no corrosivo. Un flujo para soldadura blanda que ni en su forma original ni en su forma restante químicamente ataca el metal base. Regularmente es compuesto de materiales de colofonia o resino de base.

***nondestructive testing.** Methods that do not alter or damage the weld being examined; used to locate both surface and internal defects. Methods include visual inspection, penetrant inspection, magnetic particle inspection, radiographic inspection, and ultrasonic inspection.
pruebas no destructivas. Métodos que no alteran ni dañan la soldadura que se está examinando. Se usa para encontrar ambos defectos internos y de superficie. Incluye métodos como inspección visual, inspección penetrante, inspección de partículas magnéticas, inspección de radiografía, inspección ultrasónica.

nozzle. A device that directs shielding media.
boquilla. Un aparato que dirige el medio de protección.

nugget. The weld metal joining the workpieces in spot, seam, and projection welds.
botón. El metal de soldadura que une a las piezas de trabajo en soldadura de puntos, costura, y proyección de soldaduras.

nugget size. The diameter or width of the nugget measured in the plane of the interface between the pieces joined.
tamaño del botón. El diámetro o lo ancho del botón medido en el plano del interfaze entre las piezas unidas.

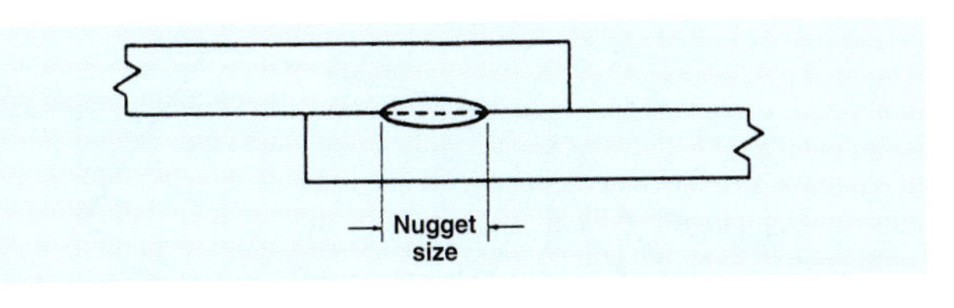

O

open circuit voltage. The voltage between the output terminals of the power source when no current is flowing to the torch or gun.
voltaje de circuito abierto. El voltaje entre los terminales de salida de una fuente de poder cuando la corriente no está corriendo a la antorcha o pistola.

open-root joint. An unwelded joint without backing or consumable insert.
junta de raíz abierta. Una junta que no está para soldarse sin respaldo o inserto consumible.

orifice. See constricting orifice.
orifice. Vea orifice de constreñimiento.

orifice gas. The gas that is directed into the plasma arc torch or thermal spraying gun to surround the electrode. It becomes ionized in the arc to form the arc plasma and issues from the constricting orifice of the nozzle as a plasma jet.
gas para orifice. El gas que es dirigido dentro de la antorcha de plasma o la pistola de rociado termal para rodear el electrodo. Se vuelve ionizado dentro del arco para formar el arco de plasma y sale de la orifice de constreñimiento a la boquilla como chorro de plasma.

orifice throat length. The length of the constricting orifice in the plasma arc torch or thermal spraying gun.
largo de garganta del orifice. Lo largo de la orifice constreñida en la antorcha de plasma o en la pistola de plasma para rociar.

***outer envelope.** The outer boundary of the oxyacetylene flame. The secondary combustion reaction occurs in the outer envelope.
envoltura externa. El límite de afuera de la llama de oxiacetileno. La reacción de la combustión secundaria ocurre en la envoltura externa.

***out-of-position welding.** Any welding position other than the flat position; includes vertical, horizontal, and overhead positions.
soldadura fuera de posición. Cualquier posición de soldadura menos la de la posición plana; incluye vertical, horizontal, y posiciones de sobrecabeza.

overhead position. The position in which welding is performed from the underside of the joint.
posición de sobrecabeza. La posición en la cual se hace la soldadura por el lado de abajo de la junta.

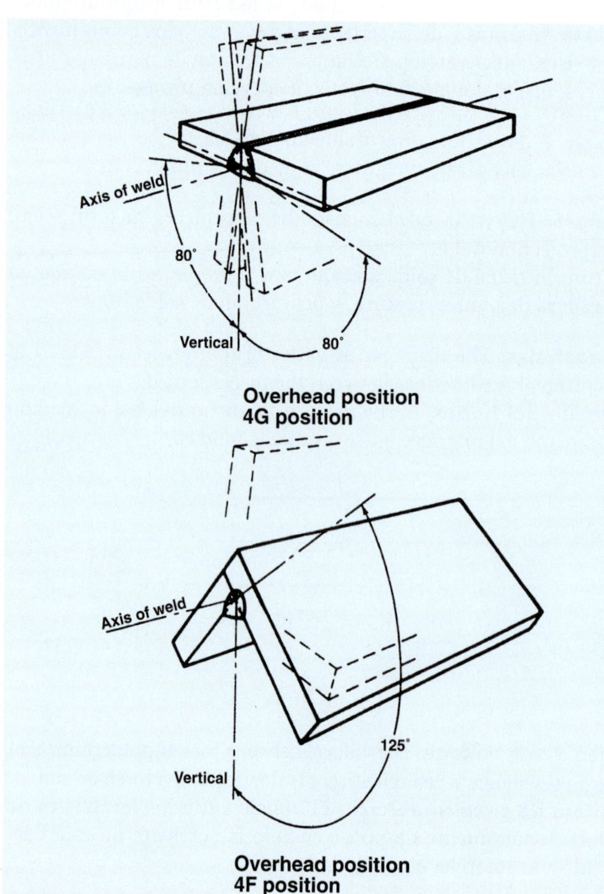

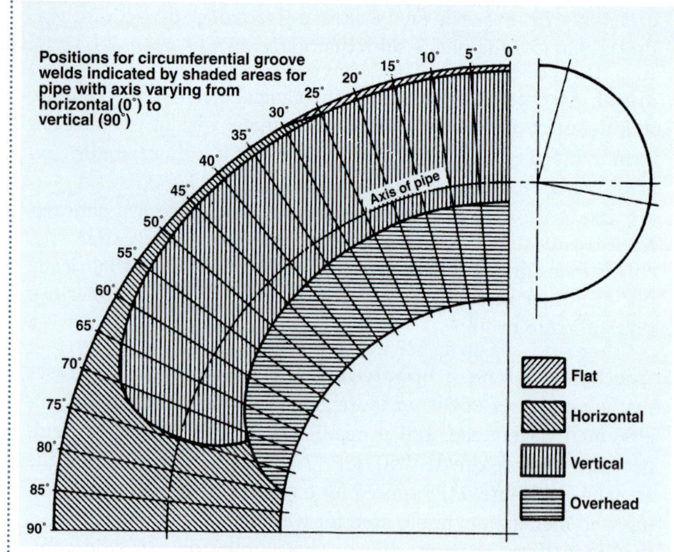

overlap. The protrusion of weld metal beyond the toe, face, or root of the weld; in resistance seam welding, the area in the preceding weld remelted by the succeeding weld.
traslapo. El metal de la soldadura que sobresale más allá del pie, cara, o de la raíz de una soldadura; en soldaduras de costuras por resistencia, la área de la soldadura anterior se rederrite por la soldadura subsiguiente.

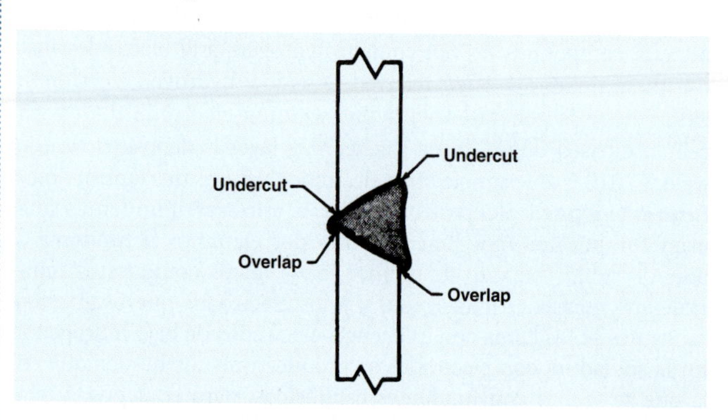

oxidizing flame. An oxyfuel gas flame in which there is an excess of oxygen, resulting in an oxygen-rich zone extending around and beyond the cone.
llama oxidante. Una llama de gas oxicombustible en la cual hay un exceso de oxígeno, resultando en una zona rica de oxígeno extendiéndose alrededor y más allá del cono.

oxyacetylene cutting (OFC-A). An oxyfuel cutting process variation that uses acetylene as the gas.
cortes con oxiacetileno. Un proceso de cortes con gas con variación que usa acetileno como gas de combustión.

***oxyacetylene hand torch.** Most commonly used oxyfuel gas cutting torch; may be a cutting torch only or a combination welding and cutting torch set. On the combination set different tips can be attached to the same torch body. The torch mixes the oxygen and fuel gas and directs the mixture to the tip. The torch can be an equal pressure-type (equal pressures of oxygen and fuel gas) or an injector-type (equal pressures of high-pressure oxygen and low-pressure fuel gas).
antorcha de mano oxiacetileno. La antorcha de gas oxicombustible es la que se usa más frecuentemente; puede ser una antorcha para hacer cortes solamente o una combinación de un juego de antorcha para soldar y hacer cortes. En el juego de

combinación diferentes puntas pueden ser conectadas al mismo mango de la antorcha. La antorcha mezcla el oxígeno y gas combustible y dirige la mezcla a la punta. La antorcha puede ser de tipo de presión igual (presiones iguales de oxígeno y gas combustible) o de tipo inyector (presiones iguales de oxígeno de alta presión y baja presión de gas combustible).

oxyacetylene welding (OAW). An oxyfuel gas welding process that uses acetylene as the fuel gas. The process is used without the application of pressure.
soldadura con oxiacetileno (OAW). Un proceso de soldadura de gas oxicombustible que usa acetileno con gas de combustión. El proceso se usa sin aplicación de presión.

***oxyfuel flame.** A flame resulting from the combustion of oxygen mixed with a fuel gas. This intense flame is applied to two pieces of metal to cause them to melt to form weld pools. When the edges of the weld pools run together and fuse, the two pieces of metal are joined.
llama oxicombustible. Una llama que resulta de una combustión de oxígeno mezclado con gas combustible. Está llama intensa es aplicada a dos piezas de metal para hacer que se derritan para formar charcos de soldadura. Cuando las orillas de los charcos de soldadura se juntan y se derriten, las dos piezas de metal se unen.

oxyfuel gas cutting (OFC). A group of oxygen cutting processes that uses heat from an oxyfuel gas flame. See also **oxygen cutting, oxyacetylene cutting,** and **oxypropane cutting.**
gas para cortar oxicombustible. Un grupo de procesos para cortar con oxígeno que usa calor de una llama de gas oxicombustible. Vea **cortes con oxigeno, cortes con oxiacetileno,** y **cortes con oxipropano.**

oxyfuel gas cutting torch. A device used for directing the preheating flame produced by the controlled combustion of fuel gases and to direct and control the cutting oxygen.
antorcha para cortes de gas oxicombustible. Un aparato que se usa para dirigir la llama precalentada producida por la combustión controlada de los gases de combustión y para dirigir y controlar el oxígeno para cortar.

oxyfuel gas welding (OFW). A group of welding processes that produces coalescence of workpieces by heating them with an oxyfuel gas flame. The processes are used with or without the application of pressure and with or without filler metal.
soldadura con gas oxicombustible. Un grupo de procesos de soldadura que produce coalescencia de las piezas de trabajo calentándolas con una llama de gas oxicombustible. Los procesos son usados sin la aplicación de presión y con o sin el metal para rellenar.

oxyfuel gas welding torch. A device used in oxyfuel gas welding, torch brazing, and torch soldering for directing the heating flame produced by the controlled combustion of fuel gases.
antorcha para soldar con gas oxicombustible. Un aparato que se usa para soldar con gas oxicombustible, soldadura fuerte con antorcha, soldadura blanda con antorcha y para dirigir la llama calentada producida por combustión controlada de gases de combustión.

oxygen arc cutting (AOC). An oxygen cutting process that uses an arc between the workpiece and a consumable tubular electrode, through which oxygen is directed to the workpiece.
cortes de oxígeno con arco. Es un proceso de cortar con oxígeno que usa un arco entre la pieza de trabajo y un electrodo tubular consumible, por el cual el oxígeno es dirigido a la pieza de trabajo.

oxygen cutting (OC). A group of thermal cutting processes that severs or removes metal by means of the chemical reaction between oxygen and the base metal at elevated temperature. The necessary temperature is maintained by the heat from an arc, an oxyfuel gas maintained by the heat from an arc, an oxyfuel gas flame, or other sources. See also **oxyfuel gas cutting.**
cortes con oxígeno. Un grupo de procesos termales que corta y quita el metal por medio de una reacción química entre el oxígeno y el metal base a una temperatura elevada. La temperatura necesaria es mantenida por el calor del arco, un gas oxicombustible mantenido por el calor del arco, una llama de gas oxicombustible, o de otras fuentes. Vea también **gas para cortar oxicombustible.**

oxypropane cutting (OFC-P). An oxyfuel gas cutting process variation that uses propane as the fuel gas.
cortes con oxipropano. Un proceso de cortar con gas combustible con variación que usa propano como gas combustible.

parent metal. See preferred term **base metal.**
metal de origen. Vea término preferido **metal base.**

***paste range.** The temperature range of soldering and brazing filler metal alloys in which the metal is partly solid and partly liquid as it is heated or cooled.
grados de la pasta. Los grados de la temperatura del metal para rellenar aleados para soldadura blanda o soldadura fuerte cuando se calienta o se enfria.

peel test. A destructive method of inspection that mechanically separates a lap joint by peeling.
prueba por pelar. Un método de inspección destructivo de pelar que separa mecánicamente una junta de solape.

peening. The mechanical working of metals using impact blows.
martillazos (con martillo de bola). Metales que se trabajan mecánicamente con golpes de impacto.

***penetration.** The depth into the base metal (from the surface) that the weld metal extends, excluding any reinforcement.
penetración. La profundidad de adentro del metal base (de la superficie) que el metal de soldadura se extiende, excluyendo cualquier refuerzo.

pilot arc. A low-current arc between the electrode and the constricting nozzle of the plasma arc torch to ionize the gas and facilitate the start of the welding arc.
piloto del arco. Un arco de corriente baja en medio del electrodo y la boquilla constreñida de la antorcha de arco de plasma para ionizar el gas y facilitar el arranque del arco para soldar.

pilot hole. A small hole drilled in material to guide the center point of larger drills; a hole drilled to receive the threaded part of a screw.
guia. Un agujero pequeño taladrado en el material par guiar el punto central de taladradoras mas grandes; agujero taladrado para recibir la parte roscada de un tornillo.

plasma. A gas that has been heated to an at least partially ionized condition, enabling it to conduct an electric current.
plasma. Un gas que ha sido calentado a lo menos parcialmente a una condicón ionizada permitiendo que conduzca una corriente eléctrica.

plasma arc cutting (PAC). An arc cutting process that uses a constricted arc and removes the molten metal with a high-velocity jet of ionized gas issuing from the constricting orifice.
cortes con arco de plasma. Un proceso de cortar con el arco que usa un arco constreñido y quita el metal derretido con un chorro de alta velocidad de gas ionizado que sale de la orifice constringente.

plasma arc welding (PAW). An arc welding process that uses a constricted arc between a nonconsumable electrode and the weld pool (transferred arc) or between the electrode and the constricting nozzle (nontransferred arc). Shielding is obtained from the ionized gas issuing from the torch, which may be supplemented by an auxiliary source of shielding gas. The process is used without the application of pressure.
soldadura con arco de plasma. Un proceso de soldadura de arco que usa un arco constreñido entre un electrodo que no se consume y el charco de la soldadura (arco transferido) o entre el electrodo y la lanza constreñida (arco no transferido). La protección es obtenida del gas ionizado que sale de la antorcha, el cual puede ser suplementado por una fuente auxiliar de gas para protección. El proceso es usado sin la aplicación de presión.

plug weld. A weld made in a circular hole in one member of a joint fusing that member to another member. A fillet-welded hole should not be construed as conforming to this definition.
soldadura de tapón. Una soldadura que se hace en un agujero circular en un miembro de una junta uniendo ese miembro con otro miembro. Un agujero de soldadura de filete no debe ser interpretado como confirmación de está definición.

plywood. Sheets of lumber made from veneer.
madera contrachapada. Tablas de maderos hechas de capas de madera terciada.

***point-to point control.** A control scheme whereby the inputs or commands specify only a limited number of points along a desired path of motion. The control system determines the intervening path segments.
control de punto a punto. Una esquema de control con que las entradas o las ordenes especifican solamente un número limitado de puntos a lo largo de la senda de moción deseada. El sistema de control determina el intervenio de los segmentos de la senda.

porosity. Cavity-type discontinuities formed by gas entrapment during solidification or in a thermal spray deposit.
porosidad. Un tipo de cavidad de desuniones formadas por gas atrapado durante la solidificación o en un deposito rociado termal.

portland cement. Dry powder made by burning limestone and clay followed by grinding and mixing.
pórtland. Polvo seco obtenido por calcinación de caliza y arcilla y en seguida pulverizandolo y mezclandolo.

postflow time. The time interval from current shut-off to shielding gas and/or cooling water shut-off.
tiempo de poscorriente. El intervalo de tiempo de cuando se cierra la corriente a cuando se cierra el gas de protección y o cuando se cierra el agua para enfriar.

postheating. The application of heat to an assembly after welding, brazing, soldering, thermal spraying, or thermal cutting. See also **postweld heat treatment.**
poscalentamiento. La aplicación de calor a una asamblea después de la soldadura, soldadura fuerte, soldadura blanda, rociado termal o corte termal. Vea también **tratamiento de calor postsoldadura.**

***postpurge.** Once welding current has stopped in gas tungsten arc welding, this is the time during which the gas continues to flow to protect the molten pool and the tungsten electrode as cooling takes place to a temperature at which they will not oxidize rapidly.
pospurgante. Cuando la corriente de soldar se ha dentenido en la soldadura de arco gas tungsteno, este es el tiempo durante en que el gas continua a salir para proteger el charco de soldadura derretido y el electrodo de tungsteno se enfrian a una temperatura donde no se oxidan rápidamente.

postweld heat treatment. Any heat treatment subsequent to welding.
tratamiento de calor postsoldadura. Cualquier tratamiento de calor subsiguiente a la soldadura.

powder flame spraying. A thermal spraying process variation in which the material to be sprayed is in powder form.
rociado de polvo con llama. Un proceso termal para rociar con variación el cual el material que está para rociar se está en forma de polvo.

power source. An apparatus for supplying current and voltage suitable for welding, thermal cutting, or thermal spraying.
fuente de poder. Un aparato para surtir corriente y voltaje conveniente para soldar, para hacer cortes termales, o rociado termal.

preheat. The heat applied to the base metal or substrate to attain and maintain preheat temperature.
precalentamiento. El calor aplicado al metal base o substrato para obtener y mantener temperatura de precalentamiento.

***preheat flame.** Brings the temperature of the metal to be cut above its kindling point, after which the high-pressure oxygen stream causes rapid oxidation of the metal to perform the cutting.
llama para precalentamiento. Sube la temperatura del metal que está para cortarse a una temperatura de encendimiento, después que la corriente del oxígeno de alta presión cause una oxidación rápida del metal para hacer el corte.

***preheat holes.** The cutting tip has a central hole through which the oxygen flows. Surrounding this central hole are a number of other holes called preheat holes. The differences in the type or number of preheat holes determine the type of fuel gas to be used in the tip.
agujeros para precalentamiento. La boquilla para cortar tiene un agujero central por donde corre el oxígeno. Rodeando este agujero central hay un numero de otros agujeros que se llaman agujeros para precalentar. Las diferencias en el tipo o número de agujeros percalentados determina el tipo de gas combustible que se usará en la boquilla.

preheating. The application of heat to the base metal immediately before welding, brazing, soldering, thermal spraying, or cutting.
precalentamiento. La aplicación de calor al metal base inmediatamente antes de la soldadura, soldadura fuerte, soldadura blanda, rociado termal o cortes.

preheat temperature. The temperature of the base metal or substrate in the welding, brazing, soldering, thermal spraying, or thermal cutting area immediately before these operations are performed. In a multipass operation, it is also the temperature in the area immediately before the second and subsequent passes are started.
temperatura de precalentamiento. La temperatura del metal base o substrato en la soldadura, soldadura fuerte, soldadura blanda, rociado termal, o en la área de los cortes termal inmedi-

atamente antes de que estas operaciones sean ejecutadas. En una operación multipasada, es también la temperatura en la área inmediatamente antes de empezar la segunda pasada y pasadas subsiguientes.

***primary combustion.** The first reaction in the chemical reaction resulting when a mixture of acetylene and oxygen is ignited. This reaction frees energy and forms carbon monoxide (CO) and free hydrogen.
combustión primaria. La primera reacción en una reacción química resulta cuando una mezcla de oxígeno y acetileno es encendida. Está reacción libra la energia y forma carbón monóxido (CO) e hidrógeno libre.

projection weld. A weld made by projection welding.
soldadura de proyección. Una soldadura hecha con soldadura de proyección.

protective atmosphere. A gas envelope surrounding the part to be brazed, welded, or thermal sprayed, with the gas composition controlled with respect to chemical composition, dew point, pressure, flow rate, etc. Examples are inert gases, combusted fuel gases, hydrogen, and vacuum.
atmósfera protectora. Una envoltura de gas que está alrededor de la parte que está para soldarse con soldadura fuerte, soldadura o rociada termal, con la composición del gas controlado con respecto a la química compuesta, punto de rocío, presión, cantidad de corriente, etc. Ejemplos son gas inerto, gases de combustión que ya están encendidos, hidrógeno, y vacuo.

puddle. See preferred term **weld pool.**
charco. Vea término preferido **charco de soldadura.**

***purged.** The process of opening first one cylinder valve and then the other to replace all air in the hoses with the appropriate gas prior to welding.
limpidor. El proceso de abrir primero una válvula de un cilindro y luego el otro para reemplazar todo el aire en las mangueras con un gas apropiado antes de empezar a soldar.

push angle. The travel angle when the electrode is pointing in the direction of weld progression. This angle can also be used to partially define the position of guns, torches, rods, and beams.
ángulo de empuje. El ángulo de avance cuando el electrodo apunta en la dirección en que la soldadura progresa. Este ángulo también puede ser usado para parcialmente definir la posición de pistolas, antorchas, varillas, y rayos.

push weld (resistance welding). A spot or projection weld made by push welding.
soldadura de empuje (soldadura de resistencia). Una soldadura de botón o proyección hecha por soldadura de empuje.

qualification. See preferred terms **welder performance qualification.**
calificación. Vea términos preferidos **calificación de ejecución del soldador.**

quick-coupling. Hose ends that snap together.
empalme automatico. Bocas de manga que cierran de golpe.

rafter. A single timber supporting a roof section.
par. Una viga unica que apoya una seccion del techo.

reciprocating saw. A saw with a stiff blade that moves back and forth.
sierra de movimiento alternativo. Sierra con una banda rigida que tiene movimiento alterno.

reducing flame. An oxyfuel gas flame with an excess of fuel gas.
llama de reducción. Una llama de gas oxicombustible con un exceso de gas combustible.

regulator. A device for controlling the delivery of gas at some substantially constant pressure.
regulador. Un aparato para controlar la expedición de gas a una presión substancialmente constante.

resistance spot welding (RSW). A resistance welding process that produces a weld at the faying surfaces of a joint by the heat obtained from resistance to the flow of welding current through the workpieces from electrodes that concentrate the welding current and pressure at the weld area.
soldadura de puntos por resistencia. Un proceso de soldar por resistencia que produce una soldadura en los empalmes de la superficie de una junta por el calor obtenido de la resistencia al correr la corriente a través de las piezas de trabajo de los electrodos que sirven para concentrar la corriente para soldar y la presión en la área de la soldadura.

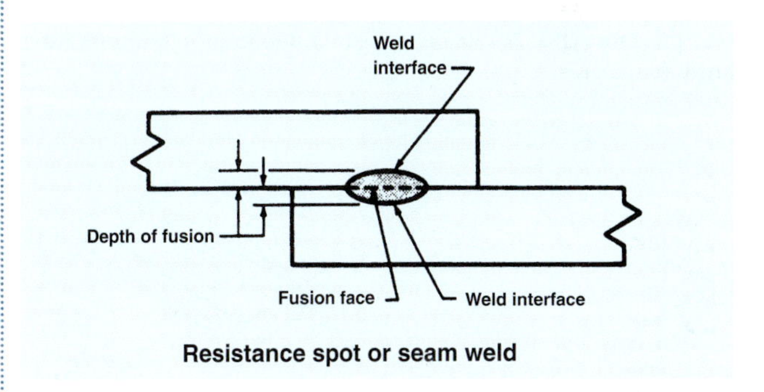

Resistance spot or seam weld

reverse polarity. The arrangement of direct-current arc welding leads with the work as the negative pole and the electrode as the positive pole of the welding arc. A synonym for direct-current electrode. Refer to drawing for **direct-current electrode positive.**
polaridad invertida. El arreglo de los cables para soldar con el arco con corriente directa con el cable de tierra como el polo negativo y el electrodo como polo positivo del arco para soldar. Un sinónimo para corriente directa electrodo. Refiérase al dibujo para **corriente directa con el electrodo positivo.**

rip saw. A saw with teeth filed to a knifelike edge and used to cut with the grain.
sierra de hender. Una sierra cuyos dientes son afilados a corte como cuchillo y que se emplea para cortar según la direccion de los hilos.

root. See preferred terms of **root of joint** and **root of weld.**
raíz. Vea las términos preferidos de **raíz de junta** y **raíz de soldadura.**

root bead. A weld bead that extends into, or includes part or all of, the joint root.
cordón de raíz. Un cordón de soldadura que se extiende adentro, o incluye parte o toda la junta de raíz.

root bend-test. A test in which the weld root is on the convex surface of a specified bend radius.

prueba de dobléz de raíz. Una prueba en la cual la raíz de la soldadura está en una superficie convexa de un radio especificado para el dobléz.

root crack. A crack in the weld or heat-affected zone occurring at the root of a weld.
grieta de raíz. Una grieta en la soldadura o en la zona afectada por el calor que ocurre en la raíz de la soldadura.

root edge. A root face of zero width. See also **root face.** Refer to drawing for **groove face.**
orilla de raíz. Una cara de raíz con una anchura de cero. Vea también **cara de raíz.** Refiérase al dibujo para **cara de ranura.**

root face. The portion of the groove face adjacent to the root of the joint. Refer to drawing for **groove face.**
cara de raíz. La porción de la cara de la ranura adyacente a la raíz de la junta. Refiérase al dibujo para **cara de ranura.**

root gap. See preferred term **root opening.**
rendija de raíz. Vea el término preferido **abertura de ráiz.**

root of joint. The portion of a joint to be welded where the members approach closest to each other. In cross section, the root of the joint may be a point, a line, or an area.
raíz de junta. saporción de una junta que está para soldarse donde los miembros se acercan muy cerca del uno al otro. En sección transversa, la raíz de una junta puede ser una punta, una línea, o una área.

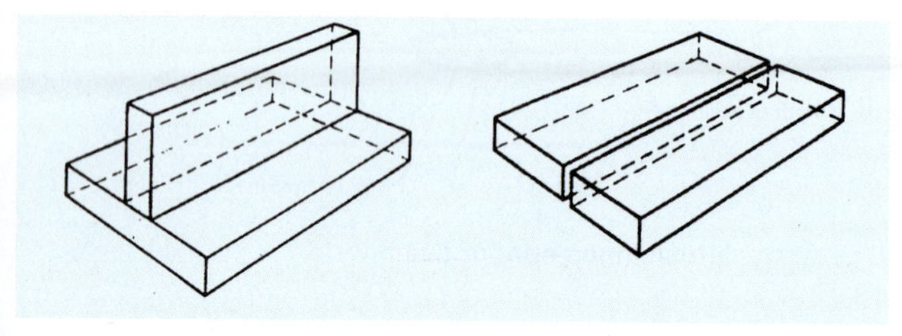

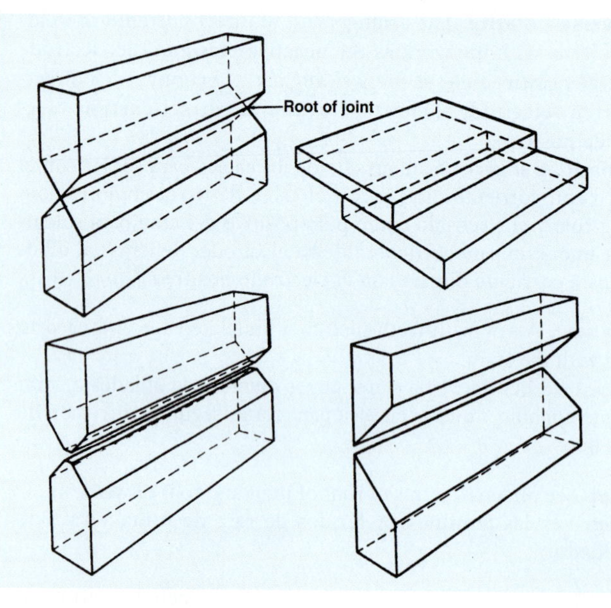

root of weld. The points, as shown in cross section, at which the back of the weld intersects the base metal surfaces.
raíz de soldadura. Las puntas, como ensena la sección transversa, donde la parte de atrás cruza con la superficie del metal base.

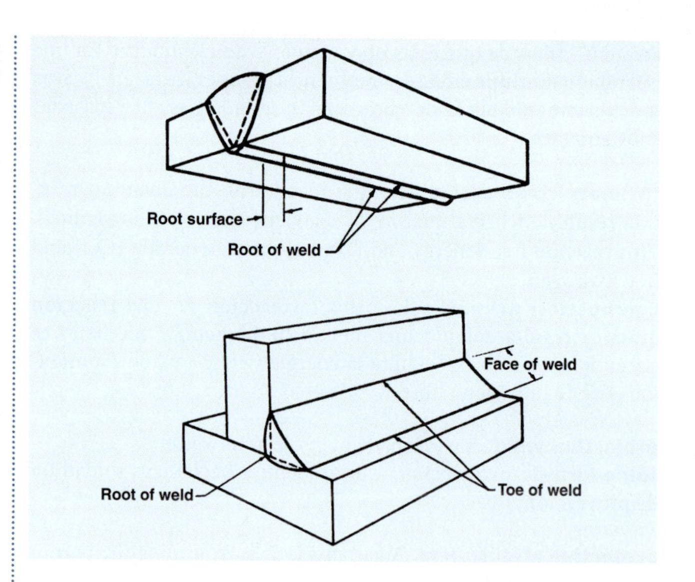

root opening. The separation between the members to be joined at the root of the joint. Refer to drawings for **bevel.**
abertura de raíz. La separación entre los miembros que están para unirse a la raíz de la junta. Refiérase al dibujo para **bisel.**

***root pass.** The first weld of a multipass weld. The root pass fuses the two pieces together and establishes the depth of weld metal penetration.
pasada de raíz. La primera soldadura de una soldadura de pasadas múltiples. La pasada de raíz funde las dos piezas juntas y establece la profundidad de la penetración del metal soldado.

root penetration. The distance the weld metal extends into the joint root.
penetración de raíz. La distancia que se extiende el metal de soldadura adentro de la junta de raíz.

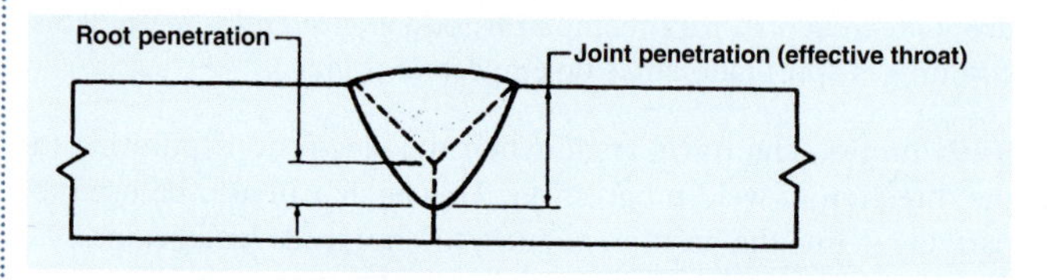

root radius. See preferred term **groove radius.**
radio de raíz. Vea el término preferido **radio de ranura.**

root reinforcement. Reinforcement of weld at the side other than that from which welding was done. Refer to drawing for **face of weld.**
refuerzo de raíz. Refuerzo de soldadura en el lado opuesto de donde se hizo la soldadura. Refiérase al dibujo para **cara de la soldadura.**

root surface. The exposed surface of a weld on the side other than that from which welding was done. Refer to drawings for **root of weld.**
superficie de raíz. La superficie expuesta de una soldadura en el lado opuesto de donde se hizo la soldadura. Refiérase a los dibujos para **raíz de soldadura.**

rough lumber. Lumber as it comes from the sawmill.
madero en bruto. Maderos asi como llegan del aserradero.

runoff weld tab. Additional material that extends beyond the end of the joint, on which the weld is terminated.

solera de carrera final de soldadura. Material adicional que se extiende más allá de donde se acaba la junta, en la cual la soldadura es terminada.

scarf joint. A form of butt joint.
junta de echarpe. Una forma de junta a tope.

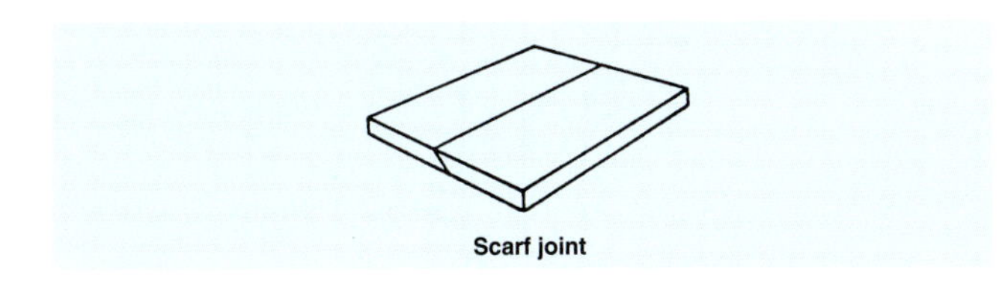
Scarf joint

***scavenger.** Elements in the flux that pick up contaminants in the molten weld pool and float them to the surface where they become part of the slag.
limpiadores o (expulsadores). Elementos en el flujo que levantan los contaminantes en el charco de soldadura derretida y los flotan a la superficie donde se forman parte de la escoria.

screwdriver. A turning tool with a straight tip, Phillips tip, or special tip.
destornillador. Una herramienta que da vueltas y que tiene punta llana, de phillips, o especial.

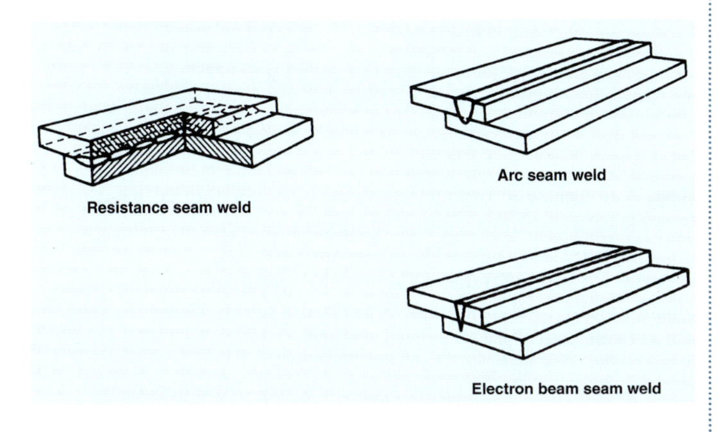
Resistance seam weld
Arc seam weld
Electron beam seam weld

***secondary combustion.** In the combustion of acetylene and oxygen, the secondary reaction unites oxygen and the free hydrogen to form water vapor (H_2O) and liberate more heat. The carbon monoxide unites with more oxygen to form carbon dioxide (CO_2).
combustión secundaria. En la combustión de acetileno y oxígeno, la reacción secundaria une el oxígeno y el hidrógeno libre para formar vapor de agua (H_2O) y liberar más calor. El carbón monóxido se une con más oxígeno para formar carbón bióxido (CO_2).

semiautomatic arc welding. Arc welding with equipment that controls only the filler metal feed. The advance of the welding is manually controlled.
soldadura de arco semiautomático. La soldadura de arco con equipo que controla solamente la alimentación del metal de relleno. El avance de la soldadura es controlado manualmente.

shielded metal arc welding (SMAW). An arc welding process with an arc between a covered electrode and the weld pool. The process is used with shielding from the decomposition of the electrode covering, without the application of pressure, and with filler metal from the electrode.
soldadura de arco metálico protegido. Un proceso de soldadura de arco con un arco en medio de un electrodo cubierto y el charco de soldadura. El proceso se usa con protección de descomposición del cubrimiento del electrodo sin la aplicación de presión, y con el metal de relleno del electrodo.

shielding gas. Protective gas used to prevent or reduce atmospheric contamination.
gas protector. El gas protector se usa para prevenir o reducir la contaminación atmosférica.

short arc. A nonstandard term for short-circuiting transfer arc welding.
arco corto. Un término fuera de la norma para transferir por corto circuito (soldadura de arco).

short-circuiting arc welding. A nonstandard term for short-circuiting transfer (arc welding).
soldadura de arco con corto circuito. Un término fuera de la norma para transferir por corto circuito (soldadura de arco).

short-circuiting transfer (arc welding). Metal transfer in which molten metal from a consumable electrode is deposited during repeated short circuits.
transferir por corto circuito (soldadura de arco). Transferir metal el cual el metal derretido del electrodo consumible es depositado durante repetidos cortos circuitos.

shoulder. See preferred term **root face.**
hombro. Vea término preferido **cara de raíz.**

side bend-test. A test in which the side of a transverse section of the weld is on the convex surface of a specified bend radius.
prueba de dobléz de lado. Una prueba en la cual el lado de una sección transversa de la soldadura está en la superficie convexa de un radio de dobléz especificado.

silver soldering, silver alloy brazing. Nonpreferred terms used to denote brazing with a silver-base filler metal.
soldadura blanda con plata, soldadura fuerte con aleación de plata. Términos no preferidos que se usan para denotar soldadura fuerte con metal para rellenar con base de plata.

single-bevel-groove weld. A type of groove weld. Refer to drawing for **groove weld.**
soldadura de ranura de un solo bisel. Tipo de soldadura de ranura. Refiérase al dibujo para **soldadura de ranura.**

single-flare V-groove weld. A type of groove weld. Refer to drawing for **groove weld.**
soldadura de ranura de una sola V acampanada. Un tipo de soldadura de ranura. Refiérase al dibujo para **soldadura de ranura.**

single J-groove weld. A type of groove weld. Refer to drawing for **groove weld.**
soldadura de ranura de una sola J. Un tipo de soldadura de ranura. Refiérase al dibujo para **soldadura de ranura.**

single square-groove weld. A type of groove weld. Refer to drawing for **groove weld.**
soldadura de ranura de una sola escuadra. Un tipo de soldadura de ranura. Refiérase al dibujo de **soldadura de ranura.**

single U-groove weld. A type of groove weld. Refer to drawing for **groove weld.**
soldadura de ranura de una sola U. Un tipo de soldadura de ranura. Refiérase al dibujo para **soldadura de ranura.**

single V-groove weld. A type of groove weld. Refer to drawing for **groove weld.**
soldadura de ranura de una sola V. Un tipo de soldadura de ranura. Refiérase al dibujo de **soldadura de ranura.**

size of weld.
groove weld. The joint penetration (depth of bevel plus the root penetration when specified). The size of a groove weld and its effective throat are one and the same.
fillet weld. For equal leg fillet welds, the leg lengths of the largest isosceles right triangle that can be inscribed within the fillet weld cross section. Refer to drawings for concavity and convexity. For unequal leg fillet welds, the leg lengths of the largest right triangle that can be inscribed within the fillet weld cross section.
Note: When one member makes an angle with the other member greater than 105°, the leg length (size) is of less significance than the effective throat, which is the controlling factor for the strength of a weld.
flange weld. The weld metal thickness measured at the root of the weld.
tamaño de la soldadura.
soldadura de ranura. La penetración de la junta (profundidad del bisel más la penetración de la raíz cuando está especificada). El tamaño de la soldadura de ranura y la garganta efectiva son una y la misma.
soldadura filete. Para soldaduras con piernas iguales de filete, lo largo de las piernas del triángulo recto con el isosceles más grande que puede ser inscrito dentro de la sección transversa de la soldadura de filete. Refiérase al dibujo para concavidad y convexidad. Para piernas de soldadura de filete desiguales, lo largo de las piernas del triángulo recto más grande que puede ser inscrito dentro de la sección transversa de la soldadura de filete.
Nota: Cuando un miembro hace un ángulo con otro miembro más grande de 105 grados, lo largo de la pierna (tamaño) es de menor significado que la garganta efectiva, la cual es el factor de control para la fuerza de una soldadura.
soldadura de brida. Lo grueso del metal de soldadura se mide a la raíz de la soldadura.

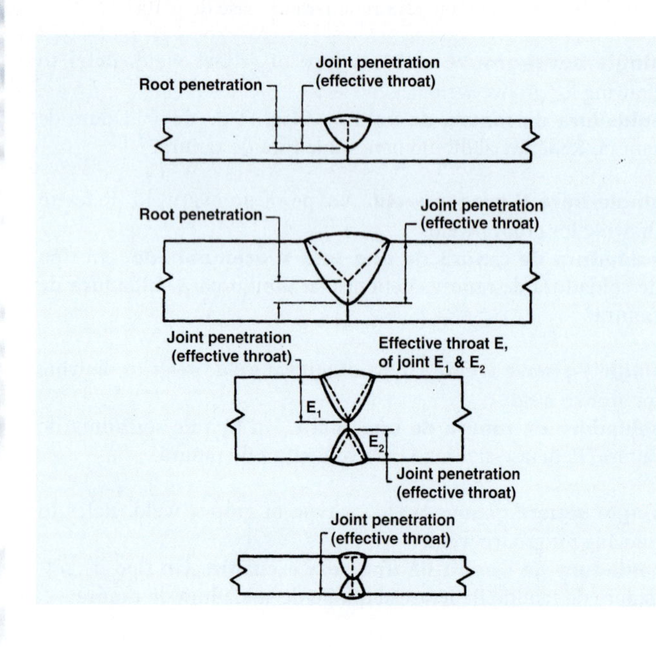

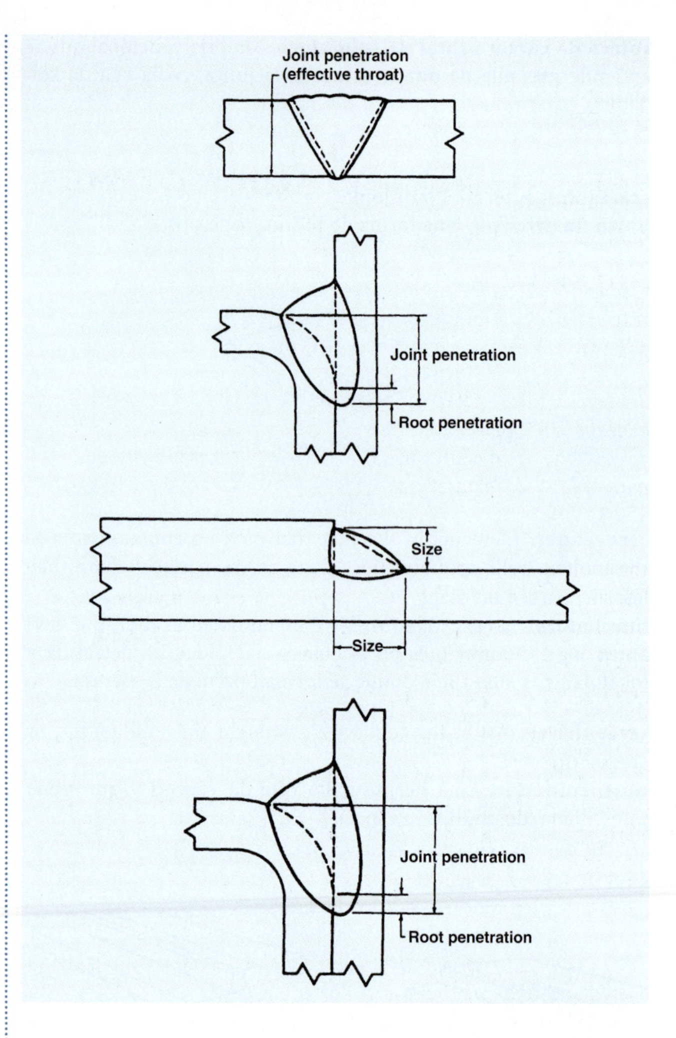

slag. A nonmetallic product resulting from the mutual dissolution of flux and nonmetallic impurities in some welding and brazing processes.
escoria. Un producto que no es metálico resultando de una disolución mutual del flujo y las impuridades no metálicas en unos procesos de soldadura y soldadura fuerte.

slag inclusion. Nonmetallic solid material entrapped in weld metal or between weld metal and base metal.
inclusion de escoria. Un material sólido que no es metálico atrapado en el metal de soldadura en medio del metal base.

***slope.** For gas metal arc welding, the volt-ampere curve of the power supply indicates that there is a slight decrease in voltage as the amperage increases; the rate of voltage decrease in the slope.
pendiente. Para soldadura de arco de metal con gas, la curva voltio-amperio de la fuente de poder indica que si hay un ligero decremento en voltaje cuando los amerios aumentan; la proporción del voltaje decrementa en el pendiente.

slot weld. A weld made in an elongated hole in one member of a lap or tee-joint joining that member to that portion of the surface of the other member that is exposed through the hole. The hole may be open at one end and may be partially or completely filled with weld metal. (A fillet-welded slot should not be construed as conforming to this definition.)
soldadura de ranura alargada. Una soldadura hecha en un agujero alargado en un miembro de una junta en solape o T uniendo ese miembro a esa porción de la superficie del otro miembro que está expuesto a través del agujero. El agujero

puede ser abierto en una punta y puede ser parcialmente o completamente rellenado con metal de soldadura. (Una ranura alargada con soldadura de filete no debe de interpretarse como conforme a está definición.)

slugging. The act of adding a separate piece or pieces of material in a joint before or during welding that results in a welded joint not complying with design, drawing, or specification requirements.
usar trozos de metal. El acto de agregar una pieza o piezas separadas de material en una junta antes o durante la soldadura que resulta en una junta soldada que no cumple con diseño, dibujo, o las especificaciones requeridas.

solder. A filler metal used in soldering that has a liquidus not exceeding 450°C (840°F).
soldadura (material para soldar). Un metal de relleno usado para soldadura blanda que tiene un liquidus que no excede de 450°C (840°F).

soldering (S). A group of welding processes that produces coalescence of materials by heating them to the soldering temperature and by using a filler metal with a liquidus not exceeding 450°C (840°F) and below the solidus of the base metals. The filler metal is distributed between closely fitted faying surfaces of the joint by capillary action.
soldadura blanda. Un grupo de procesos de soldadura que produce coalescencia de materiales calentándolos a una temperatura de soldar y usando un metal para rellenar con un liquidus que no exceda de 450°C (840°F) y más abajo del solidus de los metales base. El metal para rellenar es distribuido en medio de las superficies empalmadas acopladas muy cerca de la junta por acción capilar.

soldering gun. An electrical soldering iron with a pistol grip and a quick heating, relatively small bit.
pistola de soldar. Un fierro eléctrico para soldar con mango de pistola, rápido para calentarse, y tiene una punta relativamente pequeña.

spatter. The metal particles expelled during welding that do not form a part of the weld.
salpicadura. Las partículas de metal que se despidan cuando se está soldando y que no forman parte de la soldadura.

spatter loss. Metal lost due to spatter.
pérdida causa salpicadura. El metal perdido debido a la salpicadura.

spool. A type of filter metal package consisting of a continuous length of electrode wound on a cylinder (called the barrel), which is flanged at both ends. The flange extends below the inside diameter of the barrel and contains a spindle hole.
carrete. Un paquete de metal tipo filtro consistiendo de una extensión continua de un electrodo enrollado en un cilindro (llamado el barril), el cual tiene una brida en los dos extremos. La brida se extiende debajo del diámetro de adentro del barril y contiene un agujero huso.

spot weld. A weld made between or upon overlapping members in which coalescence may start and occur on the faying surfaces or may proceed from the surface of one member. The weld cross section (plan view) is approximately circular. See also **arc spot weld** and **resistance spot welding.**
soldadura de puntos. Una soldadura hecha en medio o sobre miembros traslapados en la cual la coalescencia puede empezar y ocurrir en las superficies empalmadas o puede continuar en la superficie de un miembro. La sección transversa (plan de vista) es aproximadamente circular. Vea también **soldadura de puntos por arco** y **soldadura de puntos por resistencia.**

spray arc. A nonstandard term for spray transfer.
arco para rociar. Un término fuera de norma para traslado rociado.

spray transfer (arc welding). Metal transfer in which molten metal from a consumable electrode is propelled axially across the arc in small droplets.
traslado rociado (soldadura de arco). Transferir el metal el cual el metal derretido de un electrodo consumible es propelado axialmente a traves del arco en gotitas pequeñas.

square. A device used to draw angles for cutting and to check the cuts for accuracy. Also the top of a wall.
escuadra. Instrumento que se emplea en trazar angulos para cortar y para revisar la percision de los cortes.

***square butt joint.** A joint made when two flat pieces of metal face each other with no edge preparation. See also **square groove weld.**
junta escuadra de tope. Una junta hecha cuando dos piezas planas de metal se enfrentan una a la otra sin preparación de orilla. Vea también **soldadura de ranura escuadra.**

square butt weld. See **butt joint.**
junta escuadra de tope. Vea **junta a tope.**

square-groove weld. A type of groove weld.
soldadura de ranura escuadra. Un tipo de soldadura de ranura.

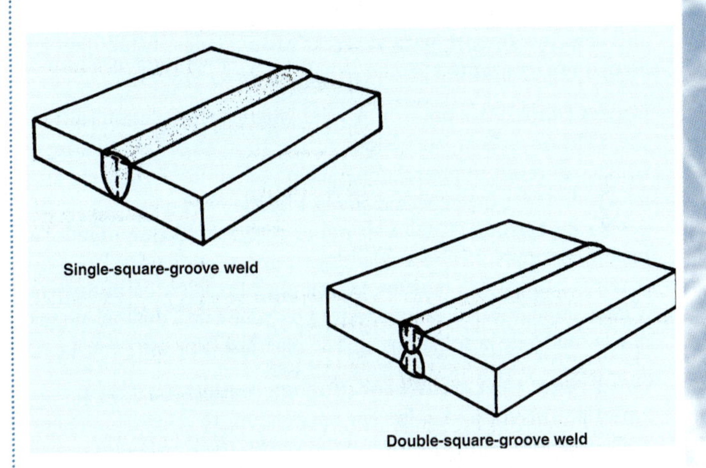

staggered intermittent welds. Intermittent welds on both sides of a joint in which the weld increments on one side are alternated with respect to those on the other side.
soldadura intermitente de cadena. Soldaduras intermitentes en los dos lados de una junta en cual los incrementos de soldadura son alternados de un lado con respecto a los del otro lado.

standoff distance. The distance between a nozzle and the workpiece.
distancia de alejamiento. La distancia entre la boquilla y la pieza de trabajo.

staple. A piece of wire with both ends sharpened and bent to form two legs of equal length.
grapa. Una pieza de alambre con ambos extremos afilados y combados para formar dos piernas de mismo longitud.

starting weld tab. Additional material that extends beyond the beginning of the joint, on which the weld is started.
solera para empezar a soldar. Material adicional que se extiende más allá del principio de la junta, en donde la soldadura es empezada.

***steel.** An alloy consisting primarily of iron and carbon. The carbon content may be as high as 2.2% but is usually less than 1.5%.
acero. Una aleación que consiste primeramente de hierro y carbón. El contenido del carbón puede ser tan alto como 2.2% pero es regularmente menos de 1.5%.

stick electrode. A nonstandard term for covered electrode.
electrodo de varilla. Un término fuera de norma por electrodo cubierto.

stickout. See preferred term **electrode extension.**
sobresalga. Vea término preferido **extensión del electrodo.**

straight polarity. The arrangement of direct-current arc welding leads in which the work is the positive pole and the electrode is the negative pole of the welding arc. A synonym for direct-current electrode negative. Refer to drawing for **direct-current electrode negative.**
polaridad directa. El arreglo de los cables de soldadura de arco con corriente directa donde el cable de la tierra es el polo positivo y el porta electrodo es el polo negativo del arco de soldadura. Un sinónimo para corriente directa con electrodo negativo. Refiérase al dibujo para **corriente directa con electrodo negativo.**

stranded electrode. A composite filler metal electrode consisting of stranded wires that may mechanically enclose materials to improve properties, stabilize the arc, or provide shielding.
electrodo cable. Electrodo de metal para rellenar compuesto que consiste de cable de alambres que pueden encerrar materiales mecánicamente para mejorar propiedades, estabilizar el arco, o proveer protección.

***stress point.** Any point in a weld where incomplete fusion of the weld on one or both sides of the root gives rise to stress, which can result in premature cracking or failure of the weld at a load well under the expected strength of the weld.
punto de tensión. Cualquier punto en una soldadura donde la fusión incompleta en la soldadura en uno o en los dos lados de la raíz le aumenta la tensión, la cual puede resultar en una grieta o falta prematura en la soldadura con una carga mucho menos que la fuerza de la soldadura que se esperaba.

stress relief heat treatment. Uniform heating of a structure or a portion thereof to a sufficient temperature to relieve the major portion of the residual stresses, followed by uniform cooling.
tratamiento de calor para relevar la tensión. Calentamiento uniforme de una estructura o una porción a una temperatura suficiente para relevar la mayor porción de las tensiones restantes, seguido por enfriamiento uniforme.

stringer bead. A type of weld bead made without appreciable weaving motion. See also **weave bead.**
cordón encordador. Un tipo de cordón de soldadura sin movimiento del tejido apreciable. Vea también **cordón tejido.**

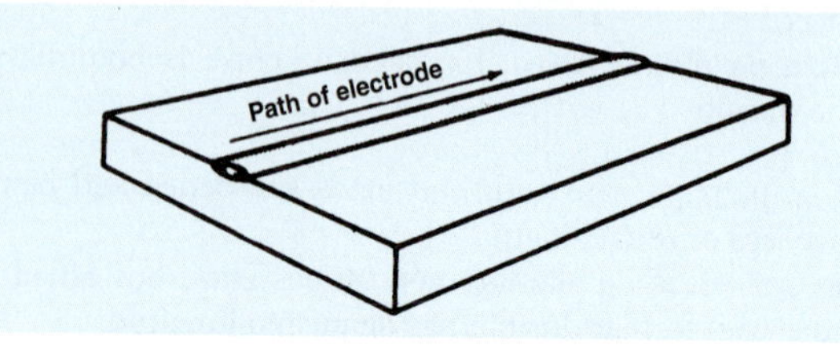

subfloor. The first layer of flooring.
subpiso. Primera capa de suelo.

substrate. Any base material to which a thermal sprayed coating or surfacing weld is applied.
substrato. Cualquier material base al cual se le aplica una capa termal o una soldadura de superficie.

suck back. See preferred term **concave root surface.**
succión del cordón de raíz. Vea el término preferido **superficie raíz concavo.**

surface preparation. The operations necessary to produce a desired or specified surface condition.
preparación de la superficie. Las operaciones necesarias para producir una deseada o una especificada condición de la superficie.

surfacing. The application by welding, brazing, or thermal spraying of a layer of material to a surface to obtain desired properties or dimensions, as opposed to making a joint. See also **coating** and **hardfacing.**
recubrimiento superficial. La aplicación a la soldadura, a la soldadura fuerte o rociado termal de una capa de material a la superficie para obtener las deseadas propiedades o dimensiones, contrario a la hechura de una junta. Vea también **revestimiento** y **endurecimiento de caras.**

T

tab. See **runoff weld tab, starting weld tab,** and **weld tab.**
solera. Vea **solera de carrera final de soldadura, solera para empezar a soldar,** y **solera para soldar.**

tack weld. A weld made to hold parts of a weldment in proper alignment until the final welds are made.
soldadura de puntos aislados. Una soldadura hecha para detener las partes en su propio alineamiento hasta que se hagan las soldaduras finales.

taps. Connections to a transformer winding that are used to vary the transformer turns ratio, thereby controlling welding voltage and current.
grifo. Conexiones al arrollamiento de un transformador que se usan para variar la proporción de vueltas del transformador, asi se puede controlar la corriente y el voltaje para soldar.

tee joint. A joint between two members located approximately at right angles to each other in the form of a *T*.
junta en T. Una junta en medio de dos miembros que están localizados aproximadamente a ángulos rectos de uno al otro en la forma de *T*.

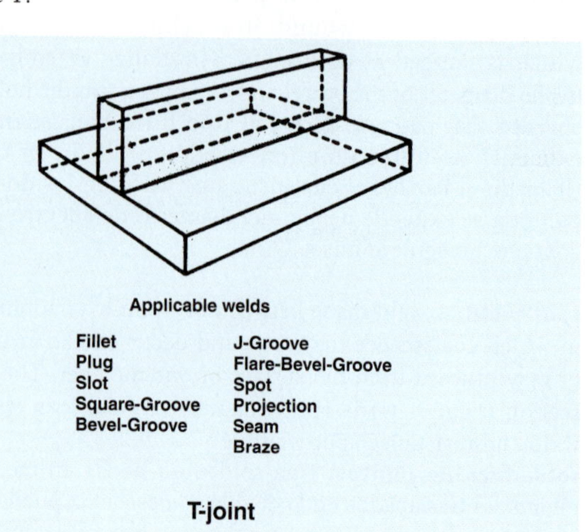

T-joint

***tempering.** Reheating hardened metal before it cools to room temperature to make it tough, not brittle.
templar. Recalentando un metal endurecido antes de que se enfrie a la temperatura del ambiente para hacerlo duro, no frágil.

***tensile strength.** As applied to a brazed or soldered joint, the ability of the joint to withstand being pulled apart.
resistencia a la tensión. Como es aplicada a una junta de soldadura fuerte o soldadura blanda, la capacidad de una junta que resista ser estirada hasta que se rompa en dos pedazos.

throat area. The area bounded by the physical parts of the secondary circuit in a resistance spot, seam, or projection welding machine. Used to determine the dimensions of a part that can be welded and to determine, in part, the secondary impedance of the equipment.
área de garganta. La área limitada por las partes físicas del circuito secundario en un punto de resistencia, costura, o una máquina de soldar de proyección. Se usa para determinar las dimensiones de una parte que puede ser soldada y determinar, en parte, la impedancia secundaria del equipo.

throat depth. In a resistance spot, seam, or projection welding machine, the distance from the centerline of the electrodes or platens to the nearest point of interference for flat sheets.
profundidad de garganta. En una punta de resistencia, costura, o máquina de soldar de proyección, la distancia de la línea del centro del electrodo o platinas al punto más cercano de interferencia para las hojas planas.

throat of a fillet weld.
actual throat. The shortest distance from the root of weld to its face. Refer to drawing for **convexity.**
effective throat. The minimum distance minus any reinforcement from the root of weld to its face. Refer to drawing for **convexity.**
theoretical throat. The distance from the beginning of the root of the joint perpendicular to the hypotenuse of the largest right triangle that can be inscribed within the fillet weld cross section. This dimension is based on the assumption that the root opening is equal to zero. Refer to drawing for **convexity.**
garganta de soldadura filete.
garganta actual. La distancia más corta de la raíz de una soldadura a su cara. Refiérase al dibujo para **convexidad.**
garganta efectiva. La distancia mínima menos cualquier refuerzo de la raíz de la soldadura a su cara. Refiérase al dibujo para **convexidad.**
garganta teórica. La distancia de donde empieza la raíz de la junta perpendicular a la hipotenusa del triángulo recto más grande que puede ser inscrito adentro de la sección transversa de una soldadura de filete. Esta dimensión está basada en la proposición que la abertura de la raíz es igual a cero. Refiérase al dibujo para **convexidad.**

TIG welding. A nonstandard term when used for gas tungsten arc welding.
soldadura TIG. Un término fuera de norma cuando es usado por soldadura de arco de tungsteno con gas.

toe crack. A crack in the base metal occurring at the toe of a weld.
grieta de pie. Una grieta en el metal base que ocurre al pie de la soldadura.

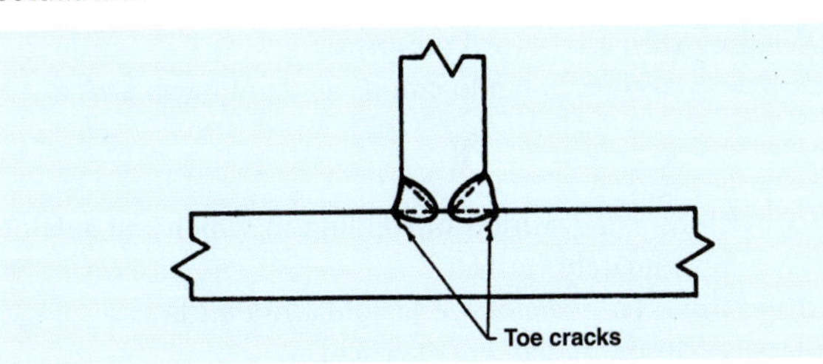

toe nail. To drive a nail at an angle near the end of one piece and into the face of another piece.
clavar en cruz. Clavar un clavo a un angulo cerca del borde de una pieza y por la cara de otra pieza.

toe of weld. The junction between the face of a weld and the base metal. Refer to drawing for **face of weld.**
pie de la soldadura. La unión entre la cara de la soldadura y el metal base. Refiérase al dibujo para **cara de la soldadura.**

tolerances. The allowable deviation in accuracy or precision between the measurement specified and the part as laid out or produced.
tolerancias. desviación permitida en la precisión entre la medida especificada y la pieza instalada o producida.

torch. See preferred terms **cutting torch** and **welding torch.**
antorcha. Vea el término preferido **antorcha para cortar** y **antorcha para soldar.**

***torch angle.** The angle between the centerline of the torch and the work surface; the ideal torch angle is 45°. The torch angle affects the percentage of heat input into the metal, thus affecting the speed of melting and the size of the molten weld pool.
ángulo de antorcha. El ángulo en medio de la línea del centro de la antorcha y la superficie del trabajo; el ángulo ideal de la antorcha es 45°. El ángulo de la antorcha afecta el por ciento de calor que entra dentro del metal, asi afectando la rapidez de derretimiento y el tamaño del charco del metal derretido de la soldadura.

torch brazing (TB). A brazing process that uses heat from a fuel-gas flame.
soldadura fuerte con antorcha. Un proceso de soldadura fuerte que usa calor de una llama de gas combustible.

***torch manipulation.** The movement of the torch by the operator to control the weld bead characteristics.
manipulacion de la antorcha. El movimiento de la antorcha por el operador para el control de las características del cordón de la soldadura.

torch soldering (TS). A soldering process that uses heat from a fuel-gas flame.
soldadura blanda con antorcha. Un proceso de soldadura blanda que usa calor de una llama de gas combustible.

transducer. A device that transforms one form of energy into another.
transducor. Un aparato que convierte una forma de energía a otra.

transferred arc. A plasma arc established between the electrode of the plasma arc torch and the workpiece.
arco transferido. Un arco de plasma establecido entre el electrodo de la antorcha de arco de plasma y la pieza de trabajo.

***transition current.** In gas metal arc welding, current above a critical level to permit spray transfer; the rate at which drops are transferred changes in relationship to the current. Transition current depends upon the alloy bearing welded and is proportional to the wire diameter.
corriente de transición. En soldadura de arco y metal con gas, corriente arriba de un nivel crítico para permitir el traslado del rociado; la proporción en la cual las gotas son transferidas cambia en relación a la corriente. La corriente de transición depende del aleado que se está soldando y es proporcional al diámetro del alambre.

transverse face-bend. See **face bend.**
doblez de cara transversal. Vea **doblez de cara.**

transverse root-bend. See root bend.
cordón de raíz transversal. Vea **cordón de raíz.**

travel angle. The angle less than 90° between the electrode axis and a line perpendicular to the weld axis, in a plane determined by the electrode axis and the weld axis. The angle can also be used to partially define the position of guns, torches, rods, and beams. See also **drag angle** and **push angle.** Refer to drawing for **backhand welding.**
ángulo de avance. El ángulo menos de 90° entre el eje del electrodo y una línea perpendicular al eje de la soldadura, en un plano determinado por el eje del electrodo y el eje de la soldadura. El ángulo también puede ser usado para parcialmente definir la posición de las pistolas, antorchas, varillas, y rayos. Vea también **ángulo del tiro** y **ángulo de empuje.** Refiérase al dibujo para **soldadura en revés.**

travel angle (pipe). The angle less than 90° between the electrode axis and a line perpendicular to the weld axis at its point of intersection with the extension of the electrode axis, in a plane determined by the electrode axis and a line tangent to the pipe surface of the same point. This angle can also be used to partially define the position of guns, torches, rods, and beams. Refer to drawing for **backhand welding.**
ángulo de avance (tubo). El ángulo menos de 90° entre el eje del electrodo y la línea perpendicular al eje a su punto de intersección con la extensión del eje del electrodo, en un plano determinado por el eje del electrodo y una línea tangente a la superficie del tubo del mismo punto. Este ángulo también puede ser usado para parcialmente definir la posición de las pistolas, varillas, y rayos. Refiérase al dibujo para **soldadura en revés.**

truss. Rigid framework.
cercha. Armazon rigida.

try square. A tool used to try or test the accuracy of cuts that have been made.
escuadra de comprobación. Un instrumento utilizado para la percision de los cortes de se han hecho.

tungsten electrode. A nonfiller metal electrode used in arc welding, arc cutting, and plasma spraying, made principally of tungsten.
electrodo de tungsteno. Un electrodo de metal que no se rellena que se usa para soldadura de arco, cortes por arco, rociado por plasma, y hecho principalmente de tungsteno.

***type A fire extinguisher.** An extinguisher used for combustible solids, such as paper, wood, and cloth. Identifying symbol is a green triangle enclosing the letter *A*.
extinguidor para incendios tipo A. Un extinguidor que se usa para combustibles sólidos como papel, madera, y tela. El símbolo de identificación es un triángulo verde con la letra *A* adentro.

***type B fire extinguisher.** An extinguisher used for combustible liquids, such as oil and gas. Identifying symbol is a red square enclosing the letter *B*.
extinguidor para incendios tipo B. Un extinguidor que se usa para liquidos combustibles, como aceite y gas. El símbolo de identificación es un cuadro rojo con la letra *B* adentro.

***type C fire extinguisher.** An extinguisher used for electrical fires. Identifying symbol is a blue circle enclosing the letter *C*.
extinguidor para incendios tipo C. Un extinguidor que se usa para incendios eléctricos. El símbolo de identificación es un círculo azul con la letra *C* adentro.

***type D fire extinguisher.** An extinguisher used on fires involving combustible metals, such as zinc, magnesium, and titanium. Identifying symbol is a yellow star enclosing the letter *D*.
extinguidor para incendios tipo D. Un extinguidor que se usa para incendios de metales combustibles, como zinc, magnesio, y titanio. El símbolo de identificación es una estrella amarilla con una letra *D* adentro.

U

U-groove weld. A type of groove weld.
soldadura de ranura en U. Un tipo de soldadura de ranura.

underbead crack. A crack in the heat-affected zone, generally not extending to the surface of the base metal.
grieta entre o bajo cordones. Una grieta en la zona afectada por el calor, generalmente no se extiende a la superficie del metal base.

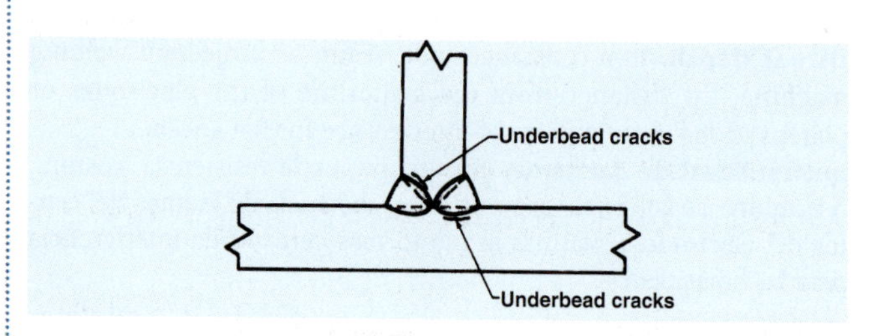

undercut. A groove melted into the base metal adjacent to the toe or root of a weld and left unfilled by weld metal. Refer to drawing for **overlap.**
socavación. Una ranura dentro del metal base adyacente al pie o raíz de la soldadura y se deja sin rellenar con el metal de soldadura. Refiérase al dibujo para **traslapo.**

underfill. A depression on the face of the weld or root surface extending below the surface of the adjacent base metal.
valle. Una depresión en la cara de la soldadura o la superficie de la raíz extendiéndose más abajo de la superficie del adyacente metal base.

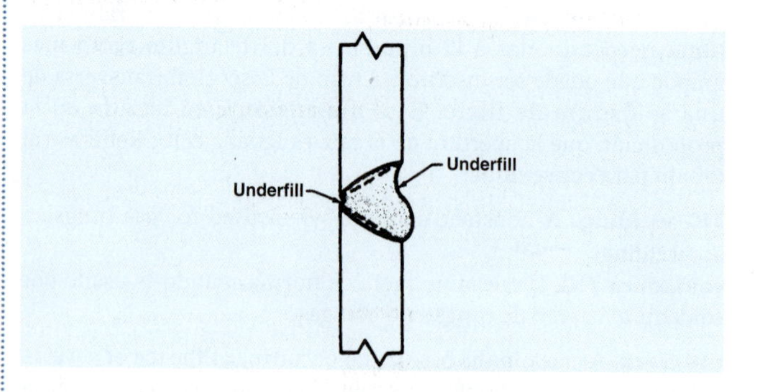

uphill. Welding with an upward progression.
soldando hacia arriba. Solando con progresión hacia arriba.

V

vertical position. The position of welding in which the axis of the weld is approximately vertical.
posición vertical. La posición de la soldadura en la cual el eje para soldarse es aproximadamente vertical.

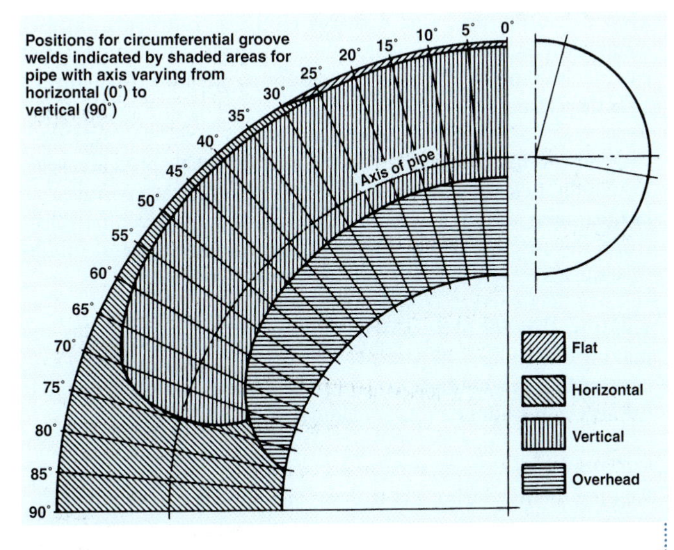

vertical position (pipe welding). The position of a pipe joint in which welding is performed in the horizontal position and the pipe may or may not be rotated.
posición vertical (soldadura de tubo). La posición de una junta de tubo en la cual la soldadura se hace en la posición horizontal y el tubo puede o no dar vueltas.

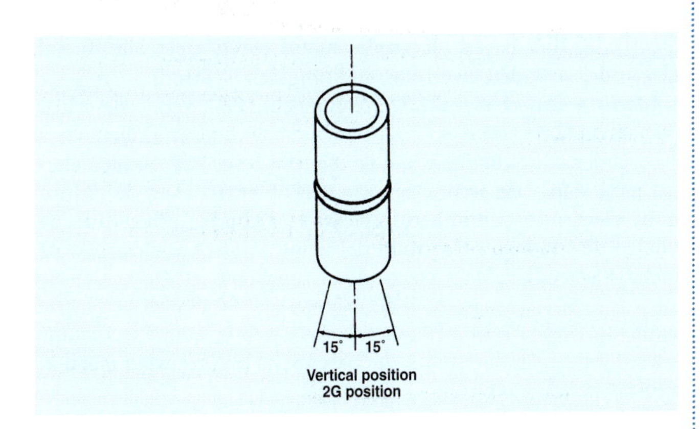

Vertical position
2G position

V-groove weld. A type of groove weld.
soldadura de ranura V. Un tipo de soldadura de ranura.

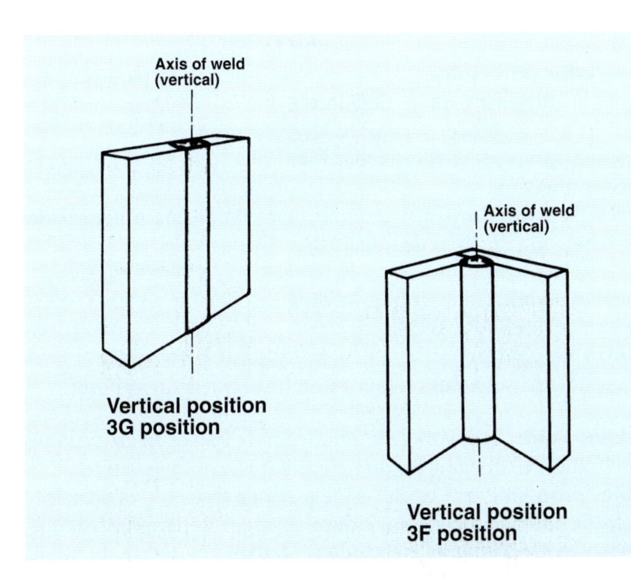

Vertical position
3G position

Vertical position
3F position

voltage range. The lower and upper limits of welding power, in volts, that can be produced by a welding machine or used with an electrode or by a process.
rango de voltaje. Los límites máximos y mínimos de poder de soldadura (en voltios) que puede tener una máquina para soldar o que pueden usarse con un electrodo o a través de un proceso.

voltage regulator. An automatic electrical control device for maintaining a constant voltage supply to the primary of a welding transformer.
regulador de voltaje. Un aparato de control eléctrico automático para mantener y proporcionar un voltaje constante a la primaria de un transformador de una soldadura.

***wagon tracks.** A pattern of trapped slag inclusions in the weld that show up as discontinuities in X rays of the weld.
huellas de carreta. Una muestra de inclusiones de escoria atrapadas en la soldadura que enseña que hay discontinuidades en los rayos-x de la soldadura.

***wattage.** A measurement of the amount of power in the arc; the wattage of the arc controls the width and depth of the weld bead.
número de vatios. Una medida de la cantidad de poder en el arco; el número de vatios del arco controla lo ancho y hondo del cordón de la soldadura.

weave bead. A type of weld bead made with transverse oscillation.
cordón tejido. Un tipo de cordón de soldadura hecha con oscilación transversa.

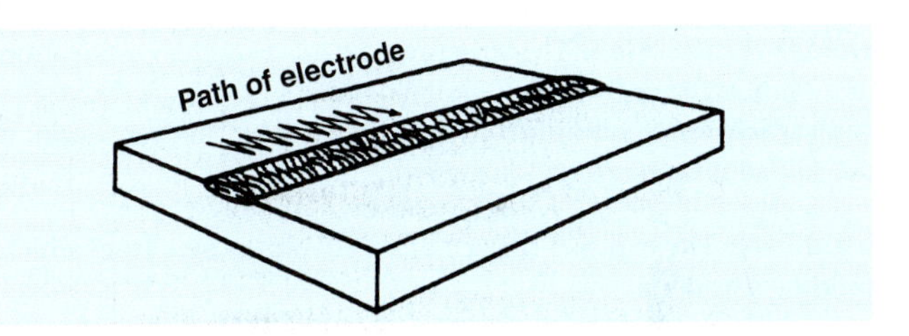

***weave pattern.** The movement of the welding electrode as the weld progresses; common weave patterns include circular, square, zigzag, stepped, C, J, T, and figure 8.
muestra de tejido. El movimiento del electrodo para soldar a como progresa la soldadura; las muestras de tejidos comunes incluyen circular, de cuadro, zigzag, de pasos, C, J, T y la figura 8.

weld. A localized coalescence of metals or nonmetals produced either by heating the materials to suitable temperatures, with or without the application of pressure, or by the application of pressure alone and with or without the use of the filler material.
soldar. Una coalescencia localizada de metales o metaloides producida al calentar los materiales a una temperatura adecuada, con o sin la aplicación de presión, o por la aplicación de presión solamente y con o sin el uso del material de relleno.

weldability. The capacity of a material to be welded under the fabrication conditions imposed into a specific, suitably designed structure and to perform satisfactorily in the intended service.
soldabilidad. La capacidad de un material para soldarse bajo las condiciones de fabricación impuestas en un específico, en un diseño de estructura adecuada y para ejecutar satisfactoriamente los servicios intentados.

weld axis. A line through the length of the weld, perpendicular to and at the geometric center of its cross section.
eje de la soldadura. Una línea a través de lo largo de la soldadura, perpendicular a y al centro geométrico de su sección transversa.

weld bead. A weld resulting from a pass. See also **stringer bead** and **weave bead.**
cordón de soldadura. Una soldadura que resulta de una pasada. Vea también **cordón encordador** y **cordón tejido.**

weld brazing. A joining method that combines resistance welding with brazing.
soldadura y soldadura fuerte. Un método de unir que combina soldadura de resistencia con soldadura fuerte.

weld crack. A crack in weld metal.
grieta en la soldadura. Una grieta en el metal de soldadura.

welder. One who performs manual or semiautomatic welding.
soldador. Uno que ejecuta soldadura manual o semiautomática.

welder certification. Written verification that a welder has produced welds meeting a prescribed standard of welder performance.
certificación del soldador. Verificación escrita de que un soldador ha producido soldaduras que cumplen con la norma prescrita de la ejecución del soldador.

welder performance qualification. The demonstration of a welder's ability to produce welds meeting prescribed standards.
calificación de ejecución del soldador. La demostración de la habilidad del soldador de producir soldaduras que cumplen con las normas prescritas.

welder registration. The act of registering a welder certification or a photostatic copy thereof.
registración del soldador. El acto de registrar una certificación del soldador o una copia fotostata de ello.

weld face. The exposed surface of a weld on the side from which welding was done.
cara de la soldadura. La superficie expuesta de una soldadura en el lado de donde se hizo la soldadura.

weld gauge. A device designed for checking the shape and size of welds.
instrumento para medir la soldadura. Un aparato diseñado para comprobar la forma y tamaño de las soldaduras.

weld groove. A channel in the surface of a workpiece or an opening between two joint members that provides space to contain a weld.
soldadura de ranura. Un canal en la superficie de una pieza de trabajo o una abertura entre dos miembros de junta que provee espacio para contener una soldadura.

welding. A joining process that produces coalescence of materials by heating them to the welding temperature, with or without the application of pressure or by the application of pressure alone, and with or without the use of filler metal.
soldadura. Un proceso de unión que produce coalescencia de materiales calentándolos a la temperatura de soldadura, con o sin la aplicación de presión o por la aplicación de presión solamente, y con o sin el uso del metal de relleno.

welding arc. A controlled electrical discharge between the electrode and the workpiece that is formed and sustained by the establishment of a gaseous conductive medium, called an arc plasma.
arco de soldadura. Una descarga eléctrica controlada entre el electrodo y la pieza de trabajo que es formada y sostenida por el establecimiento de un medio conductivo gaseoso, llamado un arco de plasma.

welding cables. The work cable and electrode cable of an arc welding circuit. Refer to drawing for direct current electrode positive.
cables para soldar. Los cables de pieza de trabajo y el portelectrodo de un circuito de soldadura de arco. Refierase al dibujo orriente directa con el electrodo positivo.

welding current. The current in the welding circuit during the making of a weld.
corriente para soldadura. La corriente en el circuito de soldar durante la hechura de una soldadura.

welding current (automatic arc welding). The current in the welding circuit during the making of a weld, but excluding upslope, downslope, start, and crater fill current.
corriente de soldadura (soldadura de arco automático). La corriente en el circuito de soldar durante la hechura de una soldadura, pero excluyendo el pendiente en ascenso, pendiente en descenso, empiezo, y corriente par llenar el crater.

welding electrode. A component of the welding circuit through which current is conducted and that terminates at the arc, molten conductive slag, or base metal. See also **arc welding electrode, bare electrode, carbon electrode, composite electrode, covered electrode, flux cored electrode, metal cored electrode, metal electrode, stranded electrode,** and **tungsten electrode.**
soldadura con electrodo. Un componente del circuito de soldar por donde la corriente es conducida y que termina en el arco, en la escoria derretida conductiva, o en el metal base. Vea también **electro para soldar de arco, electrodo descubierto, electrodo de carbón, electrodo compuesto, electrodo cubierto, electrodo de núcleo de fundente, electrodo de metal de núcleo, electrodo de metal, electrodo cable,** y **electrodo de tungsteno.**

welding filler metal. The metal or alloy to be added in making a weld joint that alloys with the base metal to form weld metal in a fusion welded joint.
metal de soldadura para rellenar. El metal o aleación que se va a agregar en la hechura de una junta de soldadura que se mezcla con el metal base para formar metal de soldadura en una junta de fusión de soldadura.

welding generator. A generator used for supplying current for welding.
generador para soldar. Un generador que se usa para proporcionar la corriente para la soldadura.

welding ground. A nonstandard and incorrect term for workpiece connection.
tierra de soldadura. Un término fuera de norma e incorrecto para conexión de pieza de trabajo.

welding head. The part of a welding machine in which a welding gun or torch is incorporated.
cabeza de soldar. La parte de una máquina para soldar la cual una pistola de soldadura o una antorcha se puede incorporar.

welding leads. The work lead and electrode lead of an arc welding circuit. Refer to drawing for **direct-current electrode positive.**
cables para soldar. Los cables de pieza de trabajo y el portelectrodo de un circuito de soldadura de arco. Refiérase al dibujo **corriente directa con el electrodo positivo.**

welding machine. Equipment used to perform the welding operation. For example, spot welding machine, arc welding machine, seam welding machine, etc.
máquina para soldar. El equipo que se usa para ejecutar la operación de soldadura. Por ejemplo, máquina de soldadura por puntos, máquina de soldadura de arco, máquina de soldadura de costura, etc.

welding operator. One who operates adaptive control, automatic, mechanized, or robotic welding equipment.
operador de soldadura. Uno que opera control adaptivo, automático, mecanizado, o equipo robótico para soldar.

welding position. See **flat position, horizontal position, horizontal fixed position, horizontal rolled position, overhead position, and vertical position.**
posición de soldadura. Vea **posición plana, posición horizontal, posición fija horizontal, posición horizontal rodada, posición de sobrecabeza, y posición vertical.**

POSITION OF WELDING

Flat. See flat position
Horizontal. See horizontal position, horizontal fixed position, and horizontal rolled position.
Vertical. See vertical position.
Overhead. See overhead position.

POSITION FOR QUALIFICATION

Plate welds

Groove welds

1G.	See flat position.
2G.	See horizontal position.
3G.	See vertical position.
4G.	See overhead position

Fillet welds

1F.	See flat position.
2F.	See horizontal position.
3F.	See vertical position.
4F.	See overhead position.

Pipe welds

Groove welds

1G.	See horizontal rolled position.
2G.	See vertical position.
3G.	See horizontal fixed position.
5G.	Inclined position.
5GR.	Inclined position.

Position

welding power source. An apparatus for supplying current and voltage suitable for welding. See also **welding generator, welding rectifier,** and **welding transformer.**
fuente de poder para soldar. Un aparato para surtir corriente y voltaje adecuado para soldar. Vea también **generador para soldar, rectificador para soldar,** y **transformador para soldar.**

welding procedure qualification record (WPQR). A record of welding variables used to produce an acceptable test weldment and the results of tests conducted on the weldment to qualify a welding procedure specification.
registro de calificación de procedimiento de la soldadura. Un registro de los variables usados para producir una probeta aceptable y los resultados de la prueba conducida en la probeta para calificar el procedimiento de especificación.

welding procedure specification (WPS). A document providing in detail the required variables for specific application to assure repeatability by properly trained welders and welding operators.
calificación de procedimiento de soldadura. Un documento que provee en detalle los variables requeridos para la aplicación específica para asegurar la habilidad de repetir el procedimiento por soldadores y operadores que estén propiamente preparados.

welding process. A materials joining process that produces coalescence of materials by heating them to suitable temperatures, with or without the application of pressure or by the application of pressure alone, and with or without the use of filler metal.
proceso para soldar. Un proceso para unir materiales que produce coalescencia calentándolos a una temperatura adecuada con o sin la aplicación de presión solamente y con o sin usarse material para rellenar.

welding rectifier. A device in a welding machine for converting alternating current to direct current.
rectificador para soldar. Un aparato en una máquina para soldar para convertir la corriente alterna a corriente directa.

welding rod. A form of welding filler metal, normally packaged in straight lengths, that does not conduct the welding current.
varilla para soldar. Una forma de metal de soldadura para rellenar, normalmente empaquetada en piezas derechas, que no conduce la corriente para soldar.

welding sequence. The order of making the welds in a weldment.
orden de sucesión (para soldar). La orden de hacer las pasadas de soldar en una soldadura.

welding symbol. A graphical representation of a weld.
símbolo de soldadura. Una representación gráfica de una soldadura.

welding technique. The details of a welding procedure that are controlled by the welder or welding operator.
ejecución de soldadura. Los detalles del procedimiento que son controlados por el soldador u operador de soldadura.

welding tip. A welding torch tip designed for welding.
boquilla (punta) para soldar. Una boquilla en la antorcha de soldadura que está diseñada para soldar.

welding torch (arc). A device used in the gas tungsten and plasma arc welding processes to control the position of the electrode, to transfer current to the arc, and to direct the flow of shielding and plasma gas.
antorcha para soldar (arco). Un aparato usado en los procesos de soldadura del gas tungsteno y arco plasma para controlar la posición del electrodo, para transferir corriente al arco, y para dirigir la corriente del gas protector y gas de la plasma.

welding torch (oxyfuel gas). A device used in oxyfuel gas welding, torch brazing, and torch soldering for directing the heating flame produced by the controlled combustion of fuel gases.
antorcha para soldar (gas oxicombustible). Un aparato usado en soldadura de gas oxicombustible, soldadura blanda con antorcha y soldadura fuerte con antorcha y para dirigir la llama para calentar producida por la combustión controlada de gases de combustión.

welding transformer. A transformer used to supplying current for welding.
transformador para soldar. Un transformador que se usa para dar corriente para la soldadura.

welding voltage. See **arc voltage.**
voltaje para soldar. Vea **voltaje del arco.**

welding wire. A form of welding filler metal, normally packaged as coils or spools, that may or may not conduct electrical current, depending upon the filler metal and base metal in a solid state weld with filler metal.
alambre para soldar. Una forma de metal para rellenar con soldadura, normalmente empaquetado en rollos o en carretes que pueda o pueda que no conducir corriente eléctrica, dependiendo en el metal de relleno y el metal base en una soldadura que está en estado sólido con metal de relleno.

weld length. See **effective length of weld.**
largura de la soldadura. Vea **distancia efectiva de soldadura.**

weldment. An assembly whose component parts are joined by welding.
conjunto de partes soldadas. Una asamblea cuyas partes componentes están unidas por la soldadura.

weld metal. The portion of a fusion weld that has been completely melted during welding.
metal de soldadura. La porción de una soldadura de fusión que se ha derretido completamente durante la soldadura.

weld metal area. The area of the weld metal as measured on the cross section of a weld. Refer to drawing for heat-affected zone.
área de metal de soldadura. La área del metal de la soldadura la cual fue medida en la sección transversa de la soldadura. Refiérase al dibujo para zona afectada por el calor.

weldor. See preferred term **welder.**
soldador. Vea el término preferido **soldador.**

weld pass. A single progression of welding along a joint. The result of a pass is a weld bead or layer.
pasada de soldadura. Una progresión singular de la soldadura a lo largo de una junta. El resultado de una pasada es un cordón o una capa.

weld pass sequence. The order in which the weld passes are made.
secuencia de pasadas de soldadura. La orden en que las pasadas de soldadura se hacen.

weld penetration. A nonstandard term for joint penetration and root penetration.
penetracion de soldadura. Un término fuera de norma para penetración de junta y penetración de raíz.

weld pool. The localized volume of molten metal in a weld prior to its solidification as weld metal.
charco de soldadura. El volumen localizado del metal derretido en una soldadura antes de su solidificación como metal de soldadura.

weld puddle. A nonstandard term for weld pool.
charco de soldadura. Un término fuera de norma para charco de soldadura.

weld reinforcement. Weld metal in excess of the quantity required to fill a joint. See also **face reinforcement** and **root reinforcement.**
refuerzo de soldadura. Metal de soldar en exceso de la cantidad requerida para llenar una junta. Vea también **refuerzo de cara** y **refuerzo de raíz.**

weld root. The points, shown in a cross section, at which the root surface intersects the base metal surfaces.
raíz de soldadura. Los puntos, enseñados en una sección transversa, la cual la superficie de la raíz se interseca con las superficies del metal base.

weld size. See preferred term **size of weld.**
tamaño de soldadura. Vea el término preferido **tamaño de la soldadura.**

***weld specimen.** A sample removed from a welded plate according to AWS specifications, which detail the preparation of the plate, the cutting of the plate, and the size of the specimen to be tested.
probeta de soldadura. Una prueba apartada del plato soldado de acuerdo con las especificaciones del AWS, las cuales detallan la preparación del plato, el corte del plato, y el tamaño de la probeta que se va a probar.

weld symbol. A graphical character connected to the welding symbol indicating the type of weld.
símbolo de soldadura. Un signo gráfico conectado al símbolo de soldadura indicando el tipo de soldadura.

weld tab. Additional material that extends beyond either end of the joint, on which the weld is started or terminated.
solera para soldar. Material adicional que se extiende más allá de cualquier punto de la junta en la cual la soldadura es empezada o terminada.

weld test. A welding performance test to a specific code or standard.
prueba de soldadura. Una prueba de ejecución de soldadura según una norma o código específico.

weld toe. The junction of the weld face and the base metal.
pie de la soldadura. La unión de la cara de la soldadura y el metal base.

wetting. The phenomenon whereby a liquid filler metal or flux spreads and adheres in a thin, continuous layer on a solid base metal.
exudación. El fenómeno de que un metal para rellenar líquido o un flujo se puede desparramar y adherirse en una capa delgada, capa continua en un sólido metal base.

wind brace. A construction member forming a triangle brace that strengthens the greenhouse structure.
conraviento. Pieza de construccion que forma una riostra triangular que fortalece el armazon del invernadero.

wire-feed speed. The rate at which wire is consumed in arc cutting, thermal spraying, or welding.
velocidad de alimentador de alambre. La velocidad que el alambre es consumido en cortes de arco, rociado termal o soldadura.

work angle. The angle less than 90° between a line perpendicular to the major workpiece surface and a plane determined by the electrode axis and the weld axis. In a tee joint or a corner joint, the line is perpendicular to the nonbutting member. This angle can also be used to partially define the position of guns, torches, rods, and beams.
ángulo de trabajo. El ángulo menos de 90° entre una línea perpendicular a la superficie de pieza de trabajo mayor y una plana determinada por el eje del electrodo y el eje de la soldadura. En una junta-T o en una junta de esquina, la línea es perpendicular a un miembro que no topa. Este ángulo puede ser usado también para parcialmente definir la posición de pistolas, antorchas, varillas y rayos.

work angle (pipe). The angle less than 90° between a line, which is perpendicular to the cylindrical pipe surface at the point of intersection of the weld axis and the extension of the electrode axis, and a plane determined by the electrode axis and a line tangent to the pipe at the same point. In a tee joint, the line is perpendicular to the nonbutting member. This angle can also be used to partially define the position of guns, torches, rods, and beams.
ángulo de trabajo (tubo). El ángulo menos de 90° entre una línea, la cual es perpendicular a la superficie de un tubo cilíndrico al punto de intersección del eje de la soldadura y la extensión del eje del electrodo, y un plano determinado por el eje del electrodo y una línea tangente al tubo al mismo punto. En una junta-T, la línea es perpendicular a un miembro que no topa. Este ángulo puede también usarse para definir parcialmente la posición de pistolas, antorchas, varillas, y rayos.

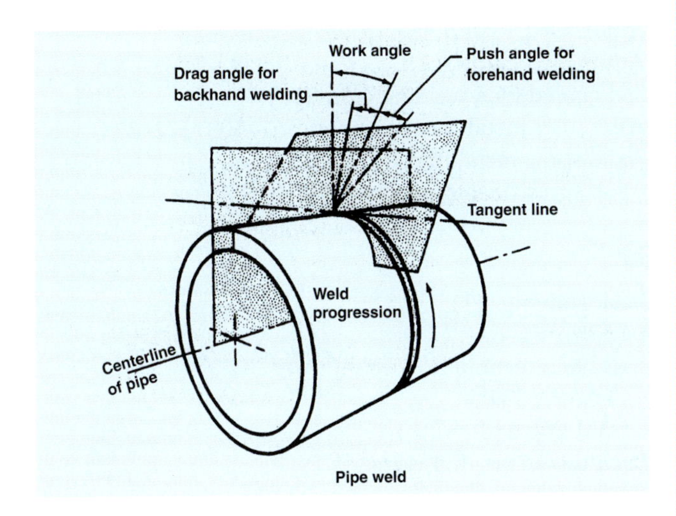

work connection. The connection of the work lead to the work. Refer to drawing for **direct-current electrode negative.**
pinza de tierra. La conexión del cable de trabajo (tierra) al trabajo. Refiérase al dibujo para **corriente directa con electrodo negativo.**

***working pressure.** The pressure at the low-pressure gauge, ranging from 0 to 45 psi (depending on the type of gas), used for welding and cutting.
presión de trabajo. La presión en el manómetro de baja presión, con escala de 0 a 45 psi (dependiendo en el tipo de gas), usado para cortar y soldar.

working range. All positions within the working envelope. The range of any variable within which the system normally operates.
extensión de trabajo. Todas las posiciones dentro del alcance del trabajo. El alcance de cualquier variable dentro del sistema que opera normalmente.

work lead. The electric conductor between the source of arc welding current and the work. Refer to drawing for **direct-current electrode negative.**
cable de tierra. Un conductor eléctrico entre la fuente de la corriente del arco y la pieza de trabajo. Refiérase al dibujo **corriente directa con electrodo negativo.**

workpiece. The part that is welded, brazed, soldered, thermal cut, or thermal sprayed.
pieza de trabajo. La parte que está soldada, con soldadura fuerte, soldadura blanda, corte termal, o rociado termal.

workpiece connection. The connection of the workpiece lead to the workpiece.
conexion de pieza de trabajo (pinzas). La conexión del cable de la pieza de trabajo a la pieza de trabajo.

workpiece lead. The electrical conductor between the arc welding current source and workpiece connection.
cable de pieza de trabajo. El conductor eléctrico entre la fuente de corriente de soldadura de arco y la conexión de la pieza de trabajo.

work station. A manufacturing unit consisting of one or more numerically controlled machine tools serviced by a robot.
estación de trabajo. Una unidad manufacturera de una o más herramienta numerada que es controlada por una máquina y abatecida por un robot.

yaw. The angular displacement of a moving body about an axis that is perpendicular to the line of motion and to the top side of the body.
guiñada. El desalojamiento de un cuerpo en movimiento alrededor de un eje que está perpendicular a la línea de movimiento y al lado más alto del cuerpo.

Index

C

D

E

F

G

H

I

J

K

L

M

R

S

T

U

V

W

Y

Z